Lecture Notes in Computer Science 4622

Commenced Publication in 1973
Founding and Former Series Editors:
Gerhard Goos, Juris Hartmanis, and Jan van Leeuwen

Editorial Board

T0189057

Alfred Menezes (Ed.)

Advances in Cryptology – CRYPTO 2007

27th Annual International Cryptology Conference
Santa Barbara, CA, USA, August 19-23, 2007
Proceedings

 Springer

Volume Editor

Alfred Menezes
University of Waterloo
Department of Combinatorics & Optimization
Waterloo, Ontario N2L 3G1, Canada
E-mail: ajmeneze@uwaterloo.ca

Library of Congress Control Number: 2007932207

CR Subject Classification (1998): E.3, G.2.1, F.2.1-2, D.4.6, K.6.5, C.2, J.1

LNCS Sublibrary: SL 4 – Security and Cryptology

ISSN   0302-9743
ISBN-10   3-540-74142-9 Springer Berlin Heidelberg New York
ISBN-13   978-3-540-74142-8 Springer Berlin Heidelberg New York

Springer is a part of Springer Science+Business Media

springer.com

© International Association for Cryptologic Research 2007
Printed in Germany

Typesetting: Camera-ready by author, data conversion by Scientific Publishing Services, Chennai, India
Printed on acid-free paper   SPIN: 12104802   06/3180   5 4 3 2 1 0

Preface

CRYPTO 2007, the 27th Annual International Cryptology Conference, was sponsored by the International Association for Cryptologic Research (IACR) in cooperation with the IEEE Computer Society Technical Committee on Security and Privacy, and the Computer Science Department of the University of California at Santa Barbara. The conference was held in Santa Barbara, California, August 19-23 2007. CRYPTO 2007 was chaired by Markus Jakobsson, and I had the privilege of serving as the Program Chair.

The conference received 186 submissions. Each paper was assigned at least three reviewers, while submissions co-authored by Program Committee members were reviewed by at least five people. After 11 weeks of discussion and deliberation, the Program Committee, aided by reports from over 148 external reviewers, selected 33 papers for presentation. The authors of accepted papers had four weeks to prepare final versions for these proceedings. These revised papers were not subject to editorial review and the authors bear full responsibility for their contents.

The Committee identified the following three papers as the best papers: "Cryptography with Constant Input Locality" by Benny Applebaum, Yuval Ishai and Eyal Kushilevitz; "Practical Cryptanalysis of SFLASH" by Vivien Dubois, Pierre-Alain Fouque, Adi Shamir and Jacques Stern; and "Finding Small Roots of Bivariate Integer Polynomial Equations: A Direct Approach" by Jean-Sébastien Coron. The authors of these papers received invitations to submit full versions to the *Journal of Cryptology*. After a close vote, the Committee selected Benny Applebaum, Yuval Ishai and Eyal Kushilevitz, the authors of the first paper, as recipients of the Best Paper Award.

The conference featured invited lectures by Ross Anderson and Paul Kocher. Ross Anderson's paper "Information Security Economics – And Beyond" has been included in these proceedings.

There are many people who contributed to the success of CRYPTO 2007. I would like the thank the many authors from around the world for submitting their papers. I am deeply grateful to the Program Committee for their hard work, enthusiasm, and conscientious efforts to ensure that each paper received a thorough and fair review. Thanks also to the external reviewers, listed on the following pages, for contributing their time and expertise. It was a pleasure working with Markus Jakobsson and the staff at Springer. I am grateful to Andy Clark, Cynthia Dwork, Arjen Lenstra and Bart Preneel for their advice. Finally, I would like to thank Dan Bernstein for organizing a lively Rump Session, and Shai Halevi for developing and maintaining his most useful Web Submission and Review Software.

June 2007   Alfred Menezes

CRYPTO 2007

August 19-23, 2007, Santa Barbara, California, USA

Sponsored by the
International Association for Cryptologic Research (IACR)

in cooperation with
IEEE Computer Society Technical Committee on Security and Privacy,
Computer Science Department, University of California, Santa Barbara

General Chair
Markus Jakobsson, Indiana University, USA

Program Chair
Alfred Menezes, University of Waterloo, Canada

Program Committee

Amos Beimel .... Ben-Gurion University, Israel
Alex Biryukov .... University of Luxembourg, Luxembourg
Xavier Boyen .... Voltage Security, USA
Yevgeniy Dodis .... New York University, USA
Orr Dunkelman .... Katholieke Universiteit Leuven, Belgium
Matt Franklin .... UC Davis, USA
Steven Galbraith .... Royal Holloway, University of London, UK
Rosario Gennaro .... IBM Research, USA
Martin Hirt .... ETH Zurich, Switzerland
Nick Howgrave-Graham .... NTRU, USA
Antoine Joux .... DGA and Université de Versailles, France
John Kelsey .... NIST, USA
Neal Koblitz .... University of Washington, USA
Kaoru Kurosawa .... Ibaraki University, Japan
Tanja Lange .... Technische Universiteit Eindhoven, Netherlands
Kristin Lauter .... Microsoft Research, USA
Kenny Paterson .... Royal Holloway, University of London, UK
David Pointcheval .... École Normale Supérieure, France
Bart Preneel .... Katholieke Universiteit Leuven, Belgium
Zulfikar Ramzan .... Symantec, USA
Omer Reingold .... Weizmann Institute of Science, Israel
Rei Safavi-Naini .... University of Calgary, Canada
Amit Sahai .... UCLA, USA
Palash Sarkar .... Indian Statistical Institute, India
Nigel Smart .... University of Bristol, UK
Adam Smith .... UCLA and Penn State University, USA
Rainer Steinwandt .... Florida Atlantic University, USA
Yiqun Lisa Yin .... Independent Consultant, USA

Advisory Members

Cynthia Dwork (CRYPTO 2006 Program Chair) ....Microsoft, USA
David Wagner (CRYPTO 2008 Program Chair) ....UC Berkeley, USA

External Reviewers

Michel Abdalla	Matthias Fitzi	Joseph Liu
Masayuki Abe	Georg Fuschbauer	Stefan Lucks
Joel Alwen	Nicolas Gama	Norbert Lütkenhaus
Elena Andreeva	Joachim von zur Gathen	Philip MacKenzie
Tomoyuki Asano	Willi Geiselmann	Tal Malkin
Nuttapong Attrapadung	Craig Gentry	Keith Martin
Georges Baatz	Marc Girault	Alexander Maximov
Lejla Batina	Mark Gondree	David Mireles
Aurélie Bauer	Jens Groth	Ilya Mironov
Zuzana Beerliová	Manabu Hagiwara	Anton Mityagin
Josh Benaloh	Iftach Haitner	Payman Mohassel
Waldyr Benits Jr.	Shai Halevi	David Molnar
Daniel J. Bernstein	Goichiro Hanaoka	Tal Moran
Jens-Matthias Bohli	Kristiyan Haralambiev	Moni Naor
Alexandra Boldyreva	Danny Harnik	Ashwin Nayak
Carl Bosley	Swee-Huay Heng	Adam O'Neill
Colin Boyd	Shoichi Hirose	Gregory Neven
Daniel R.L. Brown	Katrin Hoepper	Phong Nguyen
Ran Canetti	Susan Hohenberger	Jesper Buus Nielsen
David Cash	Thomas Holenstein	Kobbi Nissim
Dario Catalano	Emeline Hufschmitt	Wakaha Ogata
Denis Charles	Russell Impagliazzo	Rafail Ostrovsky
Lily Chen	Yuval Ishai	Elisabeth Oswald
Benoît Chevallier-Mames	Tetsu Iwata	Rafael Pass
Sherman Chow	Malika Izabachène	Maura Paterson
Carlos Cid	Shaoquan Jiang	Olivier Pereira
Henry Cohn	Charanjit Jutla	Giuseppe Persiano
Scott Contini	Jonathan Katz	Duong Hieu Phan
Jason Crampton	Aggelos Kiayias	Benny Pinkas
Joan Daemen	Eike Kiltz	Angela Piper
Quynh Dang	Darko Kirovski	Alf van der Poorten
Cécile Delerablée	Lars Knudsen	Manoj Prabhakaran
Alex Dent	Yuichi Komano	Bartosz Przydatek
Zeev Dvir	Hugo Krawczyk	Prashant Puniya
Morris Dworkin	Sébastien Kunz-Jacques	Tal Rabin
Phil Eagle	Brian LaMacchia	Dominik Raub
Pooya Farshim	Gaëtan Leurent	Oded Regev
Marc Fischlin	Yehuda Lindell	Jean-René Reinhard

Table of Contents

VI   Random Oracles

VII   Hash Functions

VIII   Theory II

IX   Quantum Cryptography

X   Cryptanalysis II

XI   Encryption

XII   Protocol Analysis

XIII   Public-Key Encryption

XIV   Multi-party Computation

Practical Cryptanalysis of SFLASH

Vivien Dubois[1], Pierre-Alain Fouque[1], Adi Shamir[1,2],
and Jacques Stern[1]

[1] École normale supérieure
Département d'Informatique 45, rue d'Ulm
75230 Paris cedex 05, France
Vivien.Dubois@ens.fr,
Pierre-Alain.Fouque@ens.fr, Jacques.Stern@ens.fr
[2] Weizmann Institute of Science
Adi.Shamir@weizmann.ac.il

Abstract. In this paper, we present a practical attack on the signature scheme SFLASH proposed by Patarin, Goubin and Courtois in 2001 following a design they had introduced in 1998. The attack only needs the public key and requires about one second to forge a signature for any message, after a one-time computation of several minutes. It can be applied to both SFLASH^{v2} which was accepted by NESSIE, as well as to SFLASH^{v3} which is a higher security version.

1 Introduction

In the last twenty years, multivariate cryptography has emerged as a potential alternative to RSA or DLOG [12,2] schemes. Many schemes have been proposed whose security appears somehow related to the problem of deciding whether or not a quadratic system of equations is solvable, which is known to be NP-complete [5]. An attractive feature of such schemes is that they have efficient implementations on smart cards, although the public and secret keys are rather large. Contrary to RSA or DLOG schemes, no polynomial quantum algorithm is known to solve this problem.

The SFLASH Scheme. SFLASH is based on the Matsumoto-Imai scheme (MI) [7], also called the C^* scheme. It uses the exponentiation $x \mapsto x^{q^\theta+1}$ in a finite field $\mathbb{F}_{q^n}$ of dimension n over a binary field $\mathbb{F}_q$, and two affine maps on the input and output variables. The MI scheme was broken by Patarin in 1995 [8]. However, based on an idea of Shamir [13], Patarin *et al.* proposed at CT-RSA 2001 [10] to remove some equations from the MI public key and called the resulting scheme C^{*-}. This completely avoids the previous attack and, although not appropriate for an encryption scheme, it is well-suited for a signature scheme. The scheme was selected in 2003 by the NESSIE European Consortium as one of the three recommended public key signature schemes, and as the best known solution for low cost smart cards.

A. Menezes (Ed.): CRYPTO 2007, LNCS 4622, pp. 1–12, 2007.

Previous Attacks on SFLASH. The first version of SFLASH, called SFLASH^{v1}, is a more efficient variant of C^{*-} using a small subfield. It has been attacked by Gilbert and Minier in [6]. However, the later versions (SFLASH^{v2} and SFLASH^{v3}) were immune to this attack.

Recently, Dubois, Fouque and Stern in [1] proposed an attack on a special class of SFLASH-like signatures. They show that when the kernel of the linear map $x \mapsto x + x^{q^\theta}$ is non-trivial, the C^{*-} scheme is not secure. The attack is very efficient in this case, but relies on some specific properties which are not met by the NESSIE proposals and which make the scheme look less secure.

Our Results. In this paper, we achieve a total break of the NESSIE standard with the actual parameters suggested by the designers: given only the public key, a signature for any message can be forged in about one second after a one time computation of several minutes. The asymptotic running time of the attack is $O(\log^2(q)n^6)$ since it only needs standard linear algebra algorithms on $O(n^2)$ variables, and n is typically very small. As in [1], the basic strategy of the attack is to recover additional independent equations in order to apply Patarin's attack [8]. To this end, both attacks use the differential of the public key. However, the attacks differ in the way the invariants related to the differential are found. The differential of the public key, also called its polar form, is very important since it transforms quadratic equations into linear ones. Hence, it can be used to find some linear relations that involve the secret keys. Its cryptanalytic significance had been demonstrated in [4].

Organization of the Paper. In section 2, we describe the SFLASH signature scheme and the practical parameters recommended by Patarin *et al.* and approved by NESSIE. Then, in section 3 we present the multiplicative property of the differential that we need. Next, in section 4 we describe how to recover linear maps related to multiplications in the finite field from the public key. In section 5, we show how to break the NESSIE proposal given only the public key. In section 6, we extend the attack to cover the case when up to half of the equations are removed, and finally in section 7, we compare our method with the technique of [1] before we conclude.

2   Description of SFLASH

In 1988, Matsumoto and Imai [7] proposed the C^* scheme for encryption and signature. The basic idea is to hide a quadratic easily invertible mapping F in some large finite field $\mathbb{F}_{q^n}$ by two secret invertible linear (or affine) maps U and T which mix together the n coordinates of F over the small field $\mathbb{F}_q$:

$$P = T \circ F \circ U$$

where $F(x) = x^{q^\theta + 1}$ in $\mathbb{F}_{q^n}$. This particular form was chosen since its representation as a multivariate mapping over the small field is quadratic, and thus the size of the public key is relatively small.

The secret key consists of the maps U and T; the public key P is formed by the n quadratic expressions, whose inputs and outputs are mixed by U and T, respectively. It can be seen that F and P are invertible whenever $\gcd(q^\theta + 1, q^n - 1) = 1$, which implies that q has to be a power of 2 since q is a prime power.

This scheme was successfully attacked by Patarin [8] in 1996. To avoid this attack and restore security Patarin *et al.* proposed in [11] to remove from the public key the last r quadratic expressions (out of the initial n), and called this variant of C^* schemes, C^{*-}. Furthermore, if the value of r is chosen such that $q^r \geq 2^{80}$, then the variant is termed C^{*--}. If we denote by Π the projection of n variables over $\mathbb{F}_q$ onto the first $n - r$ coordinates, we can represent the public key by the composition :

$$P_\Pi = \Pi \circ T \circ F \circ U = T_\Pi \circ F \circ U.$$

In the sequel, P denotes the public key of a C^* scheme whereas P_Π denotes a C^{*-} or C^{*--} public key. In both cases the secret key consists of the two linear maps T and U.

To sign a message m, the last r coordinates are chosen at random, and the signer recovers s such that $P_\Pi(s) = m$ by inverting T, U and F. A signature (m, s) can be checked by computing $P_\Pi(s)$ with the public key, which is extremely fast since it only involves the evaluation of a small number of quadratic expressions over the small finite field $\mathbb{F}_q$.

For the NESSIE project and in [10], Patarin *et al.* proposed two particular recommended choices for the parameters of C^{*--} :

- for SFLASH^{v2} : $q = 2^7$, $n = 37$, $\theta = 11$ and $r = 11$
- for SFLASH^{v3} : $q = 2^7$, $n = 67$, $\theta = 33$ and $r = 11$

SFLASH^{v3} was actually proposed to provide an even more conservative level of security than SFLASH^{v2} [10]. However, the designers made clear that they viewed SFLASH^{v2} as providing adequate security, and no attack on these two choices of parameters had been reported so far.

The important fact to notice here is that in both cases $\gcd(n, \theta) = 1$ and thus the attack described in [1] on a modified version of SFLASH in which $\gcd(n, \theta) > 1$ cannot be applied. The attack described in this paper shares with [1] the basic observation about the multiplicative property of C^{*-} schemes which is described in section 3, but proceeds in a completely different way. More discussion about the relationships between the two attacks can be found in section 7.

3   The Multiplicative Property of the Differential

The attack uses a specific multiplicative property of the differential of the public key of a C^{*-} scheme.

The differential of the internal quadratic system $F(x) = x^{q^\theta + 1}$ is a symmetric bilinear function in $\mathbb{F}_{q^n}$, called DF, and it is defined for all $a, x \in \mathbb{F}_{q^n}$ by the linear operator :

$$DF(a, x) = F(a + x) - F(a) - F(x) + F(0).$$

When $F(x) = x^{q^\theta + 1}$, we get for all $a, x \in \mathbb{F}_{q^n}$

$$DF(a, x) = ax^{q^\theta} + a^{q^\theta} x.$$

Note that this expression is bilinear since exponentiation by q^θ is a linear operation. This map has a very specific multiplicative property: for all $\xi \in \mathbb{F}_{q^n}$

$$DF(\xi \cdot a, x) + DF(a, \xi \cdot x) = (\xi + \xi^{q^\theta}) \cdot DF(a, x) \tag{1}$$

We now explain how this identity on the internal polynomial induces a similar one on the differential of the public keys in C^* and C^{*-}. Due to the linearity of the DP operator, we can combine it with the linear maps T and U to get that the differential of any C^* public key P is $DP(a, x) = T \circ DF(U(a), U(x))$. Then, equation (1) becomes for any $\xi \in \mathbb{F}_{q^n}$:

$$T \circ DF(\xi \cdot U(a), U(x)) + T \circ DF(U(a), \xi \cdot U(x))$$
$$= T \circ (\xi + \xi^{q^\theta}) \cdot DF(U(a), U(x))$$
$$= T \circ (\xi + \xi^{q^\theta}) \cdot T^{-1}(DP(a, x)).$$

We denote by M_ξ and $M_{L(\xi)}$ respectively the multiplications by ξ and by $L(\xi) = \xi + \xi^{q^\theta}$. Also, we let N_ξ denote the linear map $U^{-1} \circ M_\xi \circ U$ which depends on the secret key. We still use the word "multiplication" for N_ξ, even though this wording is not actually accurate since this is not the standard multiplication in $\mathbb{F}_{q^n}$, due to the action of the input transformation U. With these notations :

$$DP(N_\xi(a), x) + DP(a, N_\xi(x)) = T \circ M_{L(\xi)} \circ T^{-1}(DP(a, x)).$$

Finally, if DP_Π is the differential of a C^{*-} public key P_Π, then :

$$DP_\Pi(N_\xi(a), x) + DP_\Pi(a, N_\xi(x)) = T_\Pi \circ M_{L(\xi)} \circ T^{-1}(DP(a, x)).$$

Let $\Lambda(L(\xi))$ denote the linear map $T_\Pi \circ M_{L(\xi)} \circ T^{-1}$, then

$$DP_\Pi(N_\xi(a), x) + DP_\Pi(a, N_\xi(x)) = \Lambda(L(\xi)) (DP(a, x)). \tag{2}$$

This last equation is interesting since each coordinate of the left hand side is linear in the unknown coefficients of N_ξ and each coordinate of the right hand side is a linear combination by the unknown coefficients of $\Lambda(L(\xi))$ of the symmetric bilinear coordinate forms of the original DP, which are partially known since their first $(n - r)$ coordinates are public.

The heart of the attack consists in identifying some N_ξ, given the public key and equation (2), and then using its mixing effect on the n coordinates to recover the r missing quadratic forms from the $(n - r)$ known quadratic forms of the public key. In the next section, we will see how to recover some non-trivial multiplication N_ξ, in which ξ can be any value in $\mathbb{F}_{q^n} \setminus \mathbb{F}_q$.

4   Recovering Multiplications from the Public Key

Any linear mapping can be represented by an $n \times n$ matrix with n^2 entries from $\mathbb{F}_q$. Note that the multiplications N_ξ form a tiny subspace of dimension n within the space of all linear maps whose dimension is n^2.

The coordinates of DP_Π are known symmetric bilinear forms that can be seen as $n(n-1)/2$-dimensional vectors. They generate a $(n-r)$-dimensional subspace V_Π which is contained in the n-dimensional space V, generated by the full set of coordinates of DP in the original C^* public key.

Consider now the expression :

$$S_M(a, x) = DP_\Pi(M(a), x) + DP_\Pi(a, M(x))$$

where S_M is defined for any linear mapping M as a $(n - r)$-tuple of symmetric bilinear forms. Most choices of M do not correspond to any multiplication by a large field element ξ, and thus we do not expect them to satisfy the multiplicative property described in section 3. Due to relation (2), when M is a multiplication N_ξ, the $(n - r)$ coordinates of S_{N_ξ} are in V. It is unlikely that they are all in the subspace V_Π. However, there is a huge number of possible values for ξ, and it can be expected that for some choices of $\xi \in \mathbb{F}_{q^n} \setminus \mathbb{F}_q$, some of the bilinear forms in $S_M(a, x)$ will be contained in the known subspace V_Π. Our goal now is to detect such special multiplications.

Dimension of the Overall Linear Maps Space. Let us consider k of the published expressions, for instance the first k, and let us study the vector space $E(1, \ldots, k)$ of linear maps M such that the first k coordinates of $S_M(a, x)$ are all contained in V_Π. Since membership in V_Π is expressed by the vanishing of $n(n-1)/2 - (n-r)$ linear forms, the elements of this subspace satisfy a system of $k \cdot (n(n-1)/2 - (n-r))$ linear equations in the n^2 unknown coefficients of M. If all these equations were independent, the dimension of $E(1, \ldots, k)$ would be $n^2 - k \cdot (n(n-1)/2 - (n-r))$ which is clearly impossible as soon as $k \geq 3$. Otherwise, we can only claim that it is lower-bounded by this number. On the other hand, it can be seen that the space $E(1, \ldots, k)$ contains a subspace of multiplications, whose dimension is now to be computed.

Dimension of the Multiplications Space. For a multiplication N_ξ, thanks to equation (2), the coordinates of S_{N_ξ} are guaranteed to be linear combinations of the coordinates of DP, whose coefficients $\Lambda(L(\xi))$ are linear in $\xi + \xi^{q^\theta}$. Setting $\zeta = \xi + \xi^{q^\theta}$, the first k linear combinations are given by the k linear forms

$$\Lambda_i(\zeta) = \Pi_i \circ T \circ M_\zeta \circ T^{-1}$$

for $i = 1, \ldots, k$ where Π_i is the projection on the ith coordinate. Note that $\Lambda_i : \zeta \mapsto \Lambda_i(\zeta)$ are linear bijections from $\mathbb{F}_{q^n}$ to $(\mathbb{F}_q^n)^*$, the vector space of linear forms over $\mathbb{F}_q$. Indeed, the kernel of Λ_i consists of the elements ζ such that the ith row of $T \circ M_\zeta \circ T^{-1}$ is zero. Since $T \circ M_\zeta \circ T^{-1}$ is invertible for $\zeta \neq 0$, the kernel of Λ_i must be trivial. This implies that Λ_i is a linear bijection, and we will use this property. Note that this is the converse of the assumption underlying the attack in [1], and in this sense, our new attack and the old attack can be seen as complementary.

Let us consider the subspace L' of $(\mathbb{F}_q^n)^*$ generated by the first $(n-r)$ coordinate projections. In this case, the k conditions $\Pi_i \circ S_{N_\xi} \in V_\Pi$ become

$$\Lambda_i(L(\xi)) \in L', \; \forall i = 1, \ldots, k \tag{3}$$

which means that $\Lambda_i(L(\xi))$ only depends on the $(n-r)$ first rows of DP, *i.e.* only on the known DP_Π.

Consequently, when searching for a multiplication by ξ for which equation (3) holds, we get the following set of conditions on $\zeta = L(\xi) = \xi + \xi^{q^\theta}$:

(*i*) $\zeta \in \text{Im}(L)$
(*ii*) $\Lambda_i(\zeta) \in L'$ for $i = 1, \ldots, k$

Since $\zeta = \xi + \xi^{q^\theta}$ and $\gcd(n, \theta) = 1$, ζ is non-zero unless $\xi = 0$ or 1. This means that the kernel of L has dimension 1, hence ζ ranges over a space of dimension $n-1$. Condition (*i*) corresponds to a single linear relation over the coordinates of $L(\xi)$, since $\dim \text{Im}(L) = n-1$. Also, since Λ_i is a linear bijection and L' is of codimension r, each of the conditions in (*ii*) corresponds to r additional linear relations. Altogether, this means that we have $kr+1$ linear equations. Furthermore, since we are interested in the space of N_ξ's and not in the space of M_ζ's, the dimension is $n - kr - 1 + 1 = n - kr$ since the kernel of L is of dimension 1. This implies that whenever we add a condition (*i.e.* increase k by 1), we add about $n^2/2$ linear equations on the full space of linear maps, but their effect on the subspace of multiplications is to reduce its dimension only by r. Finally, the space of multiplications in $E(1, \ldots, k)$ includes at least one non-trivial multiplication, *i.e.* a multiplication by an element outside $\mathbb{F}_q$ whenever

$$n \geq kr + 2. \tag{4}$$

Consequently, the dimension of $E(1, \ldots, k)$ is

$$\max \left\{ n^2 - k \left(\frac{n(n-1)}{2} - (n-r) \right), n - kr, 1 \right\}.$$

Figure 1 describes the expected evolution of the dimension of the space of all linear maps and of the dimension of the subspace of multiplications for two different choices of r. The intuition behind our attack is that initially there are many "useless maps" and few multiplications. However, the number of useless maps drops rapidly as we add more equations, whereas the number of multiplications drops slowly (since many of the equations are linearly related on the subspace of multiplications). This leads to an elimination race, and we hope to get rid of all the "bad maps" before we inadvertently kill off all the "good maps" by imposing too many conditions.

Taking $k = 3$, it can be seen that the first expression of the max is not positive. This seems to indicate that $E(1, \ldots, k)$ consists entirely of multiplications.

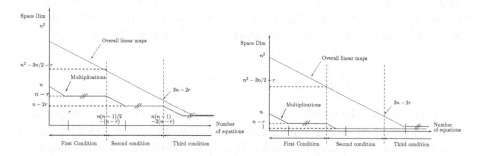

Fig. 1. Evolution of the dimensions of the overall linear maps and their subspace of multiplications when $r < n/3$ (left figure) and when $r \geq n/3$ (right figure), as we add more linear equations

This is demonstrated in the left figure. This subspace contains non-trivial multiplications, whenever $n - 3r > 1$. Therefore, the attack is expected to work for values of r up to $(n-2)/3$. The right figure shows a case in which r is too large, and thus the "good maps" are eliminated before the "bad maps". We will see in section 6 how to improve the attack and deal with values of r up to about $n/2$. Note that even without this improvement, our technique is already sufficient to recover non-trivial multiplications for the recommended parameters of SFLASH^{v2} and SFLASH^{v3}, since $r = 11$ is smaller than both $35/3$ and $65/3$. Of course, the argument that was offered is only heuristic. However, it was confirmed by a large number of experiments, in which the attack always behaved as expected by our heuristic analysis, and signatures were successfully forged.

5   Recovering a Full C^* Public Key

The final part of the attack is to recover a set P'_{Π} of additional equations which are independent of the first system P_{Π}. If the rank of the concatenation of the original P_{Π} and the newly computed r equations of P'_{Π} is full, then Patarin's attack on MI [8] can be mounted, although we do not necessarily reconstruct the r original equations of the full public key. This idea is the same as in [1].

Recovering a Full Rank System. To reconstruct a full rank system, we note that the action of the final linear map T is to compute different linear combinations of the full (*i.e.* non-truncated) internal quadratic polynomials $F \circ U$. Consequently, if we were able to mix by some linear mapping the internal quadratic coordinates $F \circ U$ before the action of T_{Π}, then we will be able to create new quadratic polynomials which could replace the r missing ones.

   When we compose the multiplication $N_{\xi} = U^{-1} \circ M_{\xi} \circ U$ (which was found in the previous part of the attack) with the truncated public key P_{Π}, the inputs of the internal quadratic mapping $F(x) = x^{q^{\theta}+1}$ are multiplied by ξ. Indeed,

$$P_{\Pi} \circ N_{\xi} = T_{\Pi} \circ F \circ M_{\xi} \circ U$$

since $P_\Pi \circ N_\xi(x) = T_\Pi \circ F \circ U \circ U^{-1} \circ M_\xi(U(x)) = T_\Pi(F(M_\xi(U(x))))$. Let us denote this new system by P'_Π. We can show that the outputs of the internal quadratic equations $F \circ U$ are multiplied by $\xi^{q^\theta+1}$. Indeed, $T_\Pi \circ F(\xi \cdot U(x)) = T_\Pi((\xi \cdot U(x))^{q^\theta+1}) = T_\Pi(\xi^{q^\theta+1} \cdot F(U(x)))$, and so :

$$P'_\Pi = P_\Pi \circ N_\xi = T_\Pi \circ M_{\xi^{q^\theta+1}} \circ F \circ U$$

Let us consider the special case $\xi \in \mathbb{F}_{q^n} \setminus \mathbb{F}_q$. In this situation, we say that N_ξ is non-trivial. Since F is a permutation and thus $F(\mathbb{F}_q) = \mathbb{F}_q$, $\xi^{q^\theta+1}$ is not in $\mathbb{F}_q$ either. Thus, the multiplication by $M_{\xi^{q^\theta}+1}$ is non-trivial, *i.e.* corresponds in particular to a non-diagonal matrix.

Therefore, in the sets P_Π and P'_Π the internal quadratic coordinates of $F \circ U$ are mixed with two different linear combinations, T_Π and $T_\Pi \circ M_{\xi^{q^\theta}+1}$. We hope that for some value $\xi \in \mathbb{F}_{q^n} \setminus \mathbb{F}_q$, r equations in the set P'_Π together with P_Π will form a full rank system. This special case is not necessary since we could use different values of ξ to add r different quadratic forms to the $(n - r)$ public ones. However, in our experiments it was always sufficient to use one ξ, and then Patarin's attack could be applied to forge actual signatures.

In practice, to determine if the new system of n equations is of full rank, we simply tested whether Patarin's attack succeeded. If not, another set of r equations was chosen amongst the $(n - r)$ equations of P'_Π. For each choice of r equations, the success probability was approximately $1 - 1/q$, which is close to 1 for $q = 2^7$

If $\xi \in \mathbb{F}_q$ (*i.e.* the multiplication is trivial), P'_Π is simply P_Π where each coordinate has been multiplied by the same element of $\mathbb{F}_q$, since $F(\mathbb{F}_q) = \mathbb{F}_q$ and multiplication by an element of $\mathbb{F}_q$ is a diagonal matrix. Thus, such trivial ξ are not interesting for our attack and this is the reason why they were discarded from our search for appropriate N_ξ in the previous section.

Practical Results. We carried our experiments on a 2GHz AMD Opteron PC using different parameters. The following table provides the time to recover a non-trivial multiplication and the time to recover an independent set of equations which form a full rank system. This computation has to be done only once per

n	37	**37**	67	**67**	131
θ	11	**11**	33	**33**	33
q	2	**128**	2	**128**	2
r	11	**11**	11	**11**	11
N_ξ Recovery	4s	**70s**	1m	**50m**	35m
C^* Recovery	7.5s	**22s**	2m	**10m**	7m
Forgery	0.01s	**0.5s**	0.02s	**2s**	0.1s

public key. Then Patarin's attack requires about one second to forge an actual signature for any given message. All these operations can be carried out by solving various systems of linear equations with a relatively small number of variables ($O(n^2)$ or $O(n)$, depending on the operation).

The two columns in bold font represent the time to attack SFLASH^{v2} and SFLASH^{v3}. The notation 's' is for seconds and 'm' is for minutes.

6   Breaking SFLASH When the Number of Deleted Quadratic Equations r Is Up to $n/2$

In this section, we deal with this problem by a technique which we call *distillation*, since it allows to gradually filter additional linear maps which are not multiplications. When $r \leq (n-2)/3$, we can use three conditions to eliminate all the useless linear maps, while retaining at least a two dimensional subspace of multiplications (since we reduce the initial n coordinates three times by r). When $r > (n-2)/3$, this will usually kill all the multiplications along with the useless linear maps.

Distillation is performed by relaxing the constraints, *i.e.* by forcing only two coordinates of S_M to be in V_Π. This will cancel a large fraction of useless linear maps, but not all of them. To clarify the situation, we use in the rest of this section angular brackets to demonstrate the stated number of dimensions for the SFLASH^{v3} parameters of $n = 67$ and $r = 11$.

After forcing the two conditions, the dimension of the space of linear maps is reduced to

$$n^2 - 2(n(n-1)/2 - (n-r)) = 3n - 2r \; \langle 179 \rangle$$

of the n^2 $\langle 4489 \rangle$ at the beginning, while the dimension of the good subspace (*i.e.* the subspace of multipications) is $n - 2r$ $\langle 45 \rangle$. Now, to find at least one non-trivial multiplication, we need to eliminate all the remaining useless linear maps. The new idea is that we can perform this process twice with different pairs of coordinates, *i.e.* coordinates 1 and 2 for the first time and coordinates 3 and 4 for the second, and get two different sets of linear maps, say VS_1 and VS_2, which contain both good and bad linear maps. Two random linear subspaces of dimension m in a linear space of dimension t are likely to have a nonzero intersection if and only if $m > t/2$, and then the dimension of the intersection is expected to be $2m - t$. We can apply this criterion separately to the space of all linear maps (in which $t = n^2$) and to the subspace of multiplications (in which $t = n$). In our example $VS_1 \cap VS_2$ is likely to contain non-trivial multiplications since $\langle 45 \rangle > \langle 67 \rangle /2$, but is not likely to contain other maps since $\langle 179 \rangle < \langle 4489 \rangle /2$. More generally, we may have to replace each one of SV_1 and SV_2 by the sum of several such linear subspaces in order to build up the dimension of the multiplications to more than $n/2$. For example, if each VS_i has only a $\langle 10 \rangle$-dimensional subspace of multiplications, we can replace it by the sum of four such linear subspaces to get the expected dimension up to $\langle 40 \rangle$, and the intersection of two such sums will have an expected dimension of $\langle 13 \rangle$, and thus many non-trivial multiplications.

Asymptotic Analysis. We now show how to deal with any $r < (1-\varepsilon)n/2$ for a fixed ε and large enough n. Note that our goal here is to simplify the description, rather than to provide the most efficient construction or tightest analysis. Since $n - 2r > \varepsilon n$, we can impose pairs of conditions and create linear subspaces VS_i of total dimension $O(n)$ which contain a subspace of multiplications of dimension $\varepsilon n \geq 2$. If we add $1/\varepsilon$ such subspaces, the dimension of the subspace of multiplications will increase to almost n, while the total dimension will remain n/ε, which is much smaller than n^2. Consequently, the intersection of two such sums is likely to consist entirely of multiplications.

Experimentations. We get the following timing results when r is close to $n/2$ and 's', 'm' and 'h' respectively denotes seconds, minutes and hours.

n	37	37	67	67
θ	11	11	33	33
q	2	128	2	128
r	17	16	32	31
N_ξ Recovery	$8s$	$4m$	$3.5m$	$10h$
C^* Recovery	$7.5s$	$22s$	$3m$	$10m$
Forgery	$0.01s$	$0.4s$	$0.02s$	$2s$

7   Comparison with the Method of Dubois *et al.* [1]

In both attacks, the basic strategy is to recover additional independent equations in order to apply Patarin's attack [8]. They both use the differential of the public key, but differ in the way the invariants of the differential are found. The method of [1] can only deal with schemes where $\gcd(n, \theta) > 1$, which implies that the kernel of $L(\xi) = \xi + \xi^{q^\theta}$ is of dimension strictly larger than 1.

To recover non-trivial multiplication in [1], skew-symmetric mappings with respect to a bilinear form B are considered, *i.e.* linear maps M such that $B(M(a), x) = -B(a, M(x))$. In fact, the authors show that skew-symmetric mappings related to the symmetric bilinear forms of a C^* public key are specific multiplications in the extension $\mathbb{F}_{q^n}$ by means of a suitable transformation depending on the secret key, namely $U^{-1} \circ M_\xi \circ U$ where $\xi \in \mathrm{Ker}\, L$. For such maps, we get $DP(M(a), x) + DP(a, M(x)) = 0$. Since DP can be computed from the public key, this equation defines linear equations in the unknowns of M. However, in the case considered in this paper, *i.e.* when $\dim \mathrm{Ker}\, L = 1$ or equivalently when $\gcd(n, \theta) = 1$, the only skew-symmetric maps are the trivial multiplications which are useless to recover new independent quadratic equations.

To recover non-trivial multiplications, we introduce here different and more elaborate conditions related to the vector space generated by the various images of the differential in public key coordinates. In this case, we are also able to detect images of multiplications. However, the multiplications to be found are not known in advance but are only shown to exist by counting arguments, and the way we find them is by setting up an elimination race between the multiplications and other linear maps.

8   Conclusion

Multivariate cryptographic schemes are very efficient but have a lot of exploitable mathematical structure. Their security is not fully understood, and new attacks against them are found on a regular basis. It would thus be prudent not to use them in any security-critical applications.

One of the most interesting open problems is whether the new techniques described in this paper can be applied to the HFE cryptosystem [9]. The main attacks discovered so far against HFE are based on Gröbner bases [3], and are very slow. So far, we could not find a way how to detect non-trivial multiplications in HFE, since it lacks the multiplicative property described in section 3, but this is a very promising line of attack which should be pursued further.

Acknowledgements

Part of this work is supported by the Commission of the European Communities through the IST program under contract IST-2002-507932 ECRYPT.

References

1. Dubois, V., Fouque, P.A., Stern, J.: Cryptanalysis of SFLASH with Slightly Modified Parameters. In: Naor, M. (ed.) EUROCRYPT 2007. LNCS, vol. 4515, pp. 264–275. Springer, Heidelberg (2007)
2. El Gamal, T.: A Public Key Cryptosystem and a Signature Scheme Based on Discrete Logarithms. IEEE Transactions on Information Theory IT–31(4), 469–472 (1985)
3. Faugère, J.-C., Joux, A.: Algebraic Cryptanalysis of Hidden Field Equation (HFE) Cryptosystems Using Gröbner Bases. In: Boneh, D. (ed.) CRYPTO 2003. LNCS, vol. 2729, pp. 44–60. Springer, Heidelberg (2003)
4. Fouque, P.A., Granboulan, L., Stern, J.: Differential Cryptanalysis for Multivariate Schemes. In: Cramer, R.J.F. (ed.) EUROCRYPT 2005. LNCS, vol. 3494, pp. 341–353. Springer, Heidelberg (2005)
5. Garey, M.R., Johnson, D.S.: Computers and Intractability, A Guide to the Theory of NP-Completeness. Freeman, New-York (1979)
6. Gilbert, H., Minier, M.: Cryptanalysis of SFLASH. In: Knudsen, L.R. (ed.) EUROCRYPT 2002. LNCS, vol. 2332, pp. 288–298. Springer, Heidelberg (2002)
7. Matsumoto, T., Imai, H.: Public Quadratic Polynomial-tuples for Efficient Signature-Verification and Message-Encryption. In: Günther, C.G. (ed.) EUROCRYPT 1988. LNCS, vol. 330, pp. 419–453. Springer, Heidelberg (1988)
8. Patarin, J.: Cryptanalysis of the Matsumoto and Imai Public Key Scheme of Eurocrypt'88. In: Coppersmith, D. (ed.) CRYPTO 1995. LNCS, vol. 963, pp. 248–261. Springer, Heidelberg (1995)

9. Patarin, J.: Hidden field equations (HFE) and isomorphisms of polynomials (IP): two new families of asymmetric algorithms. In: Maurer, U.M. (ed.) EUROCRYPT 1996. LNCS, vol. 1070, pp. 33–48. Springer, Heidelberg (1996)
10. Patarin, J., Courtois, N., Goubin, L.: FLASH, a Fast Multivariate Signature Algorithm. In: Naccache, D. (ed.) CT-RSA 2001. LNCS, vol. 2020, pp. 297–307. Springer, Heidelberg (2001)
11. Patarin, J., Goubin, L., Courtois, N.: C^*_{-+} and HM : Variations Around Two Sechemes of T. Matsumoto and H. Imai. In: Ohta, K., Pei, D. (eds.) ASIACRYPT 1998. LNCS, vol. 1514, pp. 35–49. Springer, Heidelberg (1998)
12. Rivest, R.L., Shamir, A., Adleman, L.M.: A method for obtaining digital signatures and public-key cryptosystem. Communications of the ACM 21(2), 120–126 (1978)
13. Shamir, A.: Efficient Signature Schemes Based on Birational Permutations. In: Stinson, D.R. (ed.) CRYPTO 1993. LNCS, vol. 773, pp. 1–12. Springer, Heidelberg (1994)

Full Key-Recovery Attacks on HMAC/NMAC-MD4 and NMAC-MD5

Pierre-Alain Fouque, Gaëtan Leurent, and Phong Q. Nguyen

École Normale Supérieure – Département d'Informatique,
45 rue d'Ulm, 75230 Paris Cedex 05, France
{Pierre-Alain.Fouque,Gaetan.Leurent,Phong.Nguyen}@ens.fr

Abstract. At Crypto '06, Bellare presented new security proofs for
HMAC and NMAC, under the assumption that the underlying compression function is a pseudo-random function family. Conversely, at Asiacrypt '06, Contini and Yin used collision techniques to obtain forgery
and partial key-recovery attacks on HMAC and NMAC instantiated with
MD4, MD5, SHA-0 and reduced SHA-1. In this paper, we present the
first full key-recovery attacks on NMAC and HMAC instantiated with
a real-life hash function, namely MD4. Our main result is an attack
on HMAC/NMAC-MD4 which recovers the full MAC secret key after
roughly 2^{88} MAC queries and 2^{95} MD4 computations. We also extend the
partial key-recovery Contini-Yin attack on NMAC-MD5 (in the related-
key setting) to a full key-recovery attack. The attacks are based on generalizations of collision attacks to recover a secret IV, using new differential
paths for MD4.

Keywords: NMAC, HMAC, key-recovery, MD4, MD5, collisions, differential path.

1   Introduction

Hash functions are fundamental primitives used in many cryptographic schemes
and protocols. In a breakthrough work, Wang et al. discovered devastating collision attacks [17,19,20,18] on the main hash functions from the MD4 family,
namely MD4 [17], RIPE-MD [17], MD5 [19], SHA-0 [20] and SHA-1 [18]. Such
attacks can find collisions in much less time than the birthday paradox. However,
their impact on the security of existing hash-based cryptographic schemes is unclear, for at least two reasons: the applications of hash functions rely on various
security properties which may be much weaker than collision resistance (such
as pseudorandomness); Wang et al.'s attacks are arguably still not completely
understood.

This paper deals with key-recovery attacks on HMAC and NMAC using collision attacks. HMAC and NMAC are hash-based message authentication codes
proposed by Bellare, Canetti and Krawczyk [3], which are very interesting to
study for at least three reasons: HMAC is standardized (by ANSI, IETF, ISO
and NIST) and widely deployed (e.g. SSL, TLS, SSH, Ipsec); both HMAC and

A. Menezes (Ed.): CRYPTO 2007, LNCS 4622, pp. 13–30, 2007.

NMAC have security proofs [2,3]; and both are rather simple constructions. Let H be an iterated Merkle-Damgård hash function. Its HMAC is defined by

$$\text{HMAC}_k(M) = H(\bar{k} \oplus \text{opad} \,||\, H(\bar{k} \oplus \text{ipad} \,||\, M)),$$

where M is the message, k is the secret key, $\bar{k}$ its completion to a single block of the hash function, opad and ipad are two fixed one-block values. The security of HMAC is based on that of NMAC. Since H is assumed to be based on the Merkle-Damgård paradigm, denote by H_k the modification of H where the public IV is replaced by the secret key k. Then NMAC with secret key (k_1, k_2) is defined by:

$$\text{NMAC}_{k_1,k_2}(M) = H_{k_1}(H_{k_2}(M)).$$

Thus, HMAC_k is essentially equivalent to $\text{NMAC}_{H(k\oplus\text{opad}), H(k\oplus\text{ipad})}$[1]. Attacks on NMAC can usually be adapted to HMAC (pending few modifications), except in the related-key setting[2].

HMAC/NMAC Security. The security of a MAC algorithm is usually measured by the difficulty for an attacker having access to a MAC oracle to forge new valid MAC-message pairs. More precisely, we will consider two types of attack: the existential forgery where the adversary must produce a valid MAC for *a message of its choice*, and the universal forgery where the attacker must be able to compute the MAC of *any message*.

The security of HMAC and NMAC was carefully analyzed by its designers. It was first shown in [3] that NMAC is a pseudorandom function family (PRF) under the two assumptions that (A1) the keyed compression function f_k of the hash function is a PRF, and (A2) the keyed hash function H_k is *weakly collision resistant*. The proof for NMAC was then lifted to HMAC by further assuming that (A3) the key derivation function in HMAC is a PRF. However, it was noticed that recent collision attacks [17,19,20,18] invalidate (A2) in the case of usual hash function like MD4 or MD5, because one can produce collisions for any public IV. This led Bellare [2] to present new security proofs for NMAC under (A1) only. As a result, the security of HMAC solely depends on (A1) and (A3). The security of NMAC as a PRF holds only if the adversary makes less than $2^{n/2}$ NMAC queries (where n is the MAC size), since there is a generic forgery attack using the birthday paradox with $2^{n/2}$ queries.

Since recent collision attacks cast a doubt on the validity of (A1), one may wonder if it is possible to exploit collision search breakthroughs to attack HMAC and NMAC instantiated with real-life hash functions. In particular, MD4 is a very tempting target since it is by far the weakest real-life hash function with respect to collision resistance. It is not too difficult to apply collision attacks on MD4 [17,21], to obtain distinguishing and existential forgery attacks on

---

[1] There is small difference in the padding: when we use $H(k||\cdot)$ instead of H_k, the length of the input of the hash function (which is included in the padding) is different.

[2] If we need an oracle $\text{NMAC}_{k_1,k_2+\Delta}$, we can not emulate it with an related-key HMAC oracle.

HMAC/NMAC-MD4: for instance, this was done independently by Kim *et al.* [9] and Contini and Yin [4]. The situation is more complex with MD5, because the differential path found in the celebrated MD5 collision attack [19] is not well-suited to HMAC/NMAC since it uses two blocks: Contini and Yin [4] turned instead to the much older MD5 pseudo-collisions of de Boer and Bosselaers [7] to obtain distinguishing and existential forgery attacks on NMAC-MD5 in the related-key setting. It is the use of pseudo-collisions (rather than full collisions) which weakens the attacks to the related-key setting.

Interestingly, universal forgery attacks on HMAC and NMAC seem much more difficult to find. So far, there are only two works in that direction. In [4], Contini and Yin extended the previous attacks to *partial* key-recovery attacks on HMAC/NMAC instantiated with MD4, SHA-0, and a step-reduced SHA-1, and related-key *partial* key-recovery attacks on NMAC-MD5. In [14], Rechberger and Rijmen improved the data complexity of Kim *et al.* [9] attacks, and extended them to a partial key recovery against NMAC-SHA-1. These attacks are only partial in the sense that the NMAC attacks only recover the second key k_2, which is not sufficient to compute new MACs of arbitrary messages; and the HMAC attacks only recover $H(k \oplus \text{ipad})$ where k is the HMAC secret key, which again is not sufficient to compute new MACs of arbitrary messages, since it does not give the value of k nor $H(k \oplus \text{opad})$. Note that recovering a single key of NMAC does not significantly improve the generic full key-recovery attack which recovers the keys one by one.

Very recently, Rechberger and Rijmen have proposed full key-recovery attacks against NMAC in the related-key setting in [15]. They extended the attack of [4] to a full key-recovery attack against NMAC-MD5, and introduced a full key-recovery attack against NMAC when used with SHA-1 reduced to 34 rounds.

Our Results. We present what seems to be the first universal forgery attack, *without related keys*, on HMAC and NMAC instantiated with a real-life hash function, namely MD4. Our main result is an attack on HMAC/NMAC-MD4 which recovers the full NMAC-MD4 secret key after 2^{88} MAC queries; for HMAC, we do not recover the HMAC-MD4 secret key k, instead we recover both $H(k \oplus \text{ipad})$ and $H(k \oplus \text{opad})$, which is sufficient to compute any MAC. We also obtain a full key-recovery attack on NMAC-MD5 in the related-key setting, by extending the attack of Contini and Yin [4]. This improvement was independently proposed in [14].

Our attacks have a complexity greater than the birthday paradox, so they are not covered by Bellare's proofs. Some MAC constructions have security proof against PRF-attacks, but are vulnerable to key-recovery attacks. For instance, the envelope method with a single key was proved to be secure, but a key-recovery attack using 2^{67} known text-MAC pairs, and 2^{13} chosen texts was found by Preneel and van Oorschot [12]. However, in the case of NMAC, we can prove that the security against universal forgery cannot be less than the security of the compression function against a PRF distinguisher (see the full version of this paper for the proof). This shows that NMAC offers good resistance beyond the birthday paradox: a universal forgery attack will have a time complexity of

2^n if there is no weakness in the compression function. Conversely, there is a generic attack against any iterated stateless MAC using a collision in the inner hash function to guess the two subkeys k_1 and k_2 independently, in the case of NMAC it requires $2^{n/2}$ queries and 2^{n+1} hash computations [11,13].

Our attacks on MD4 and MD5 are rather different from each other, although both are based on IV-recovery attacks, which allow to recover the IV when one is given access to a hash function whose IV remains secret. Such IV-recovery attacks can be exploited to attack HMAC and NMAC because the oracle can in fact be very weak: we do not need the full output of the hash function; essentially, we only need to detect if two related messages collide under the hash function. The MD5 related-key attack closely follows the Contini-Yin attack [4]: the IV-recovery attack is more or less based on message modification techniques. The MD4 IV-recovery attack is based on a new technique: we use differential paths which depend on a condition in the IV. The advantage of this technique is that it can be used to recover the outer key quite efficiently, since we only need to control the *difference* of the inputs and not the *values* themselves. This part of the attack shares some similar ideas with [15]. To make this possible *without related keys*, we need a differential path with a message difference only active in the first input words. We found such IV-dependant paths using an automated tool described in [8]. To make this attack more efficient, we also introduce a method to construct cheaply lots of message pairs with a specific hash difference.

Our results are summarized in the following table, together with previous attacks where "Data" means online queries and "Time" is offline computations:

Attacks		Data	Time	Mem	Remark
Generic	E-Forgery	$2^{n/2}$	–	–	[13] Collision based
	U-Forgery	$2^{n/2}$	2^{n+1}	–	[13] Collision based
		1	$2^{2n/3}$	$2^{2n/3}$	[1] TM tradeoff, 2^n precomputation
NMAC-MD4 HMAC-MD4	E-Forgery	2^{58}	–	–	[4] Complexity is actually lower [8]
	Partial-KR	2^{63}	2^{40}	–	[4] Only for NMAC
	U-Forgery	2^{88}	2^{95}	–	**New result**
NMAC-MD5 *Related keys*	E-Forgery	2^{47}	–	–	[4]
	Partial-KR	2^{47}	2^{45}	–	[4]
	U-Forgery	2^{51}	2^{100}	–	**New result** – Same as [15]

Like [4], we stress that our results on HMAC and NMAC do not contradict any security proof; on the contrary they show that when the hypotheses over the hash function are not met, an attack can be built.

Road Map. This paper is divided in five sections. In Section 2, we give background and notations on MD4, MD5 and collision attacks based on differential cryptanalysis. In Section 3, we explain the framework of our key-recovery attacks on HMAC and NMAC, by introducing IV-recovery attacks. In Section 4, we present key-recovery attacks on HMAC/NMAC-MD4. Finally, in Section 5, we present related-key key-recovery attacks on NMAC-MD5.

2 Background and Notation

Unfortunately, there does not seem to be any standard notation in the hash function literature. Here, we will use a notation similar to that of Daum [6].

2.1 MD4 and MD5

MD4 and MD5 follow the Merkle-Damgård construction. Their compression function are designed to be very efficient using 32-bit words and operations implemented in hardware in most processors:

- rotation $\lll$;
- addition mod 2^{32} $\boxplus$;
- bitwise boolean operations Φ_i. For MD4 and MD5, they are:
  - IF(x, y, z) $= (x \wedge y) \vee (\neg x \wedge z)$
  - MAJ$(x, y, z) = (x \wedge y) \vee (x \wedge z) \vee (y \wedge z)$
  - XOR$(x, y, z) = x \oplus y \oplus z$
  - ONX$(x, y, z) = (x \vee \neg y) \oplus z$.

MD4 uses IF(x, y, z), MAJ(x, y, z) and XOR(x, y, z), while MD5 uses IF (x, y, z), IF(z, x, y), XOR(x, y, z) and ONX(x, y, z).

The compression function cMD4 (resp. cMD5) of MD4 (resp. MD5) uses an internal state of four words, and updates them one by one in 48 (resp. 64) steps. Their input is 128 bits × 512 bits, and their output is 128 bits. Here, we will assign a name to every different value of these registers, following [6]: the value changed on step i is called Q_i. Then the cMD4 compression function is defined by:

$$\text{Step update: } Q_i = (Q_{i-4} \boxplus \Phi_i(Q_{i-1}, Q_{i-2}, Q_{i-3}) \boxplus m_i \boxplus k_i) \lll s_i$$
$$\text{Input: } Q_{-4}||Q_{-1}||Q_{-2}||Q_{-3}$$
$$\text{Output: } Q_{-4} \boxplus Q_{44}||Q_{-1} \boxplus Q_{47}||Q_{-2} \boxplus Q_{46}||Q_{-3} \boxplus Q_{45}$$

And the cMD5 compression function is given by:

$$\text{Step update: } Q_i = Q_{i-1} \boxplus (Q_{i-4} \boxplus \Phi_i(Q_{i-1}, Q_{i-2}, Q_{i-3}) \boxplus m_i \boxplus k_i) \lll s_i$$
$$\text{Input: } Q_{-4}||Q_{-1}||Q_{-2}||Q_{-3}$$
$$\text{Output: } Q_{-4} \boxplus Q_{60}||Q_{-1} \boxplus Q_{63}||Q_{-2} \boxplus Q_{62}||Q_{-3} \boxplus Q_{61}$$

The security of the compression function was based on the fact that such operations are not "compatible" and mix the properties of the input.

We will also use $x^{[k]}$ to represent the $k + 1$-th bit of x, that is $x^{[k]} = (x \ggg k) \bmod 2$ (note that we count bits and steps starting from 0).

2.2 Collision Attacks Based on Differential Cryptanalysis

It is natural to apply differential cryptanalysis to find collisions on hash functions based on block ciphers, like MD4 and MD5 (which both follow the Davies-Meyer construction). The main idea is to follow the differences in the internal state Q_i

of the compression function, when the inputs (the IVs or the messages) have a special difference.

Our attacks on NMAC-MD5 are based on the MD5 pseudo-collision of de Boer and Bosselaers [7], like [4] (this is the only known attack against MD5 compression function's pseudo-randomness). Let IV be a 128-bit value satisfying the so-called dBB condition: the most significant bit of the last three 32-bit words of IV are all equal. Clearly, a randomly chosen IV satisfies the dBB condition with probability $1/4$. It is shown in [7] that in such a case, a randomly chosen 512-bit message M satisfies with heuristic probability 2^{-46}:

$$\mathrm{cMD5}(IV, M) = \mathrm{cMD5}(IV', M),$$

where IV' is the 128-bit value derived from IV by flipping the most significant bit of each of the four 32-bit words of IV. The probability 2^{-46} is obtained by studying the most likely differences for the internal state Q_i, as is usual in differential cryptanalysis.

Our attacks on HMAC/NMAC-MD4 are based on recent collision search techniques for MD4 [17,21], which are organized as follows:

1. A precomputation phase:
   - choose a message difference Δ
   - find a differential path
   - compute a set of sufficient conditions
2. Search for a message M satisfying all the conditions for a given IV; then $\mathrm{cMD4}(\mathrm{IV}, M) = \mathrm{cMD4}(\mathrm{IV}, M \boxplus \Delta)$.

The differential path specifies how the computations of $\mathrm{cMD4}(\mathrm{IV}, M \boxplus \Delta)$ and $\mathrm{cMD4}(\mathrm{IV}, M)$ are related: it describes how the differences introduced in the message will evolve in the internal state Q_i. By choosing a special Δ with a low Hamming weight and extra properties, we can find differences in the Q_i which are very likely. Then we look at each step of the compression function, and we can express a set of sufficient conditions that will make the Q_i's follow the path. The conditions are on the Q_i's, and their values depends of the IV and the message M. For a given message M, we will have $\mathrm{cMD4}(\mathrm{IV}, M) = \mathrm{cMD4}(\mathrm{IV}, M \boxplus \Delta)$ if the conditions are satisfied; we expect this to happen with probability 2^{-c} for a random message if there are c conditions. Wang introduced some further ideas to make the search for such message more efficient, but they can't be used in the context of NMAC because the IV is unknown.

3   Key-Recovery Attacks on HMAC and NMAC

In this section, we give a high-level overview of our key-recovery attacks on HMAC and NMAC instantiated with MD4 and MD5. Detailed attacks will be given in the next two sections: Section 4 for MD4 and Section 5 for MD5. We will assume that the attacker can request the MAC of messages of its choice, for a fixed secret key, and the goal is to recover that secret key. In the related-key setting, we will assume like in [4] that the attacker can request the MAC of messages of its choice, for the fixed secret key as well as for other related secret keys (with a chosen relation). In fact, we will not even need the full output of MAC requests: we will only need to know if the two MACs of messages of our choice collide or not.

To simplify our exposition, we will concentrate on the NMAC case:

$$\text{NMAC}_{k_1,k_2}(M) = H_{k_1}(H_{k_2}(M)).$$

The NMAC-MD4 attack can easily be extended to HMAC-MD4, pending minor modifications. Our NMAC attack will first recover k_2, then k_1. We will collect NMAC collisions of a special shape, in order to disclose hash collisions with first k_2 then k_1.

3.1   Extracting Hash Collisions from NMAC Collisions

We will extract hash collisions from NMAC collisions, that is, pairs (M_1, M_2) of messages such that:

(C) $M_1 \neq M_2$ and $\text{NMAC}_{k_1,k_2}(M_1) = \text{NMAC}_{k_1,k_2}(M_2)$.

Our attacks are based on the elementary observation that H-collisions can leak through NMAC. More precisely, two messages M_1 and M_2 satisfy (C) if and only if they satisfy either (C1) or (C2):

(C2) $M_1 \neq M_2$ and $H_{k_2}(M_1) = H_{k_2}(M_2)$: we have a collision in the inner hash function;

(C1) $H_{k_2}(M_1) = N_1 \neq N_2 = H_{k_2}(M_2)$ and $H_{k_1}(N_1) = H_{k_1}(N_2)$: we have a collision in the outer hash function .

If we select M_1 and M_2 uniformly at random, then (C) holds with probability 2^{-128} if NMAC is a random function. However, if we select many pairs (M_1, M_2) in such a way that (C2) holds with a probability significantly higher than 2^{-128}, then whenever $\text{NMAC}_{k_1,k_2}(M_1) = \text{NMAC}_{k_1,k_2}(M_2)$, it will be likely that we also have (C2). More precisely, we have (since $Ci \cap C = Ci$):

$$\frac{\Pr(C2|C)}{\Pr(C1|C)} = \frac{\Pr(C2)}{\Pr(C1)}$$

and we expect that $\Pr(C2) \gg \Pr(C1) \approx 2^{-128}$.

Note that the Merkle-Damgård construction used in H leads to a simple heuristic way to distinguish both cases (without knowing the secret keys k_1 and k_2) if M_1 and M_2 have the same length, and therefore the same padding block P: if $H_{k_2}(M_1) = H_{k_2}(M_2)$, then for any M, we have $H_{k_2}(M_1||P||M) = H_{k_2}(M_2||P||M)$. In other words, the condition (C2) is preserved if we append $P||M$ to both M_1 and M_2 for a randomly chosen M, but that is unlikely for the condition (C1).

To illustrate our point, assume that we know a non-zero Δ such that for all keys k_2, a randomly chosen one-block message M_1 satisfies with probability 2^{-64} the condition $H_{k_2}(M_1) = H_{k_2}(M_2)$ where $M_2 = M_1 \boxplus \Delta$. If we select 2^{64} one-block messages M_1 uniformly at random and call the NMAC oracle on each M_1 and $M_1 \boxplus \Delta$, we are likely to find a pair $(M_1, M_2 = M_1 \boxplus \Delta)$ satisfying (C). By the previous reasoning, we expect that such a pair actually satisfies (C2).

Thus, the NMAC oracle allows us to detect collisions on H_{k_2}, if we are able to select messages which have a non-negligible probability of satisfying (C2). To detect collisions in H_{k_1}, we will use the values of k_2 (recovered using collisions in H_{k_2}): then, we can compute H_{k_2} and directly check whether the NMAC collision come from (C1). We now explain how to use such collision detections to recover the secret keys k_2 and k_1.

3.2   IV-Recovery Attacks

The previous subsection suggests the following scenario. Assume that a fixed key k is secret, but that one is given access to an oracle which on input M_1 and M_2, answers whether $H_k(M_1) = H_k(M_2)$ holds or not. Can one use such an oracle to recover the secret key k? If so, we have what we call an IV-recovery attack.

An IV-recovery attack would clearly reveal the second key k_2 of NMAC, because of (C2). But it is not clear why this would be relevant to recover the outer key k_1. To recover k_1 thanks to (C1), we would need the following variant of the problem. Namely, one would like to retrieve a secret key k_1 when given access to an oracle which on input M_1 and M_2, answers whether $H_{k_1}(H_{k_2}(M_1)) = H_{k_1}(H_{k_2}(M_2))$ holds or not, where k_2 is known. Since the messages are first processed through a hash function, the attacker no longer chooses the input messages of the keyed hash function, and this oracle is much harder to exploit than the previous one. We call such attacks composite IV-recovery attacks. In the attack on HMAC/NMAC-MD4, we will exploit the Merkle-Damgård structure of H_{k_2} to efficiently extend the basic IV-recovery attacks into composite IV-recovery attacks.

We will present two types of IV-recovery attacks. The first type is due to Contini and Yin [4] and uses related messages, while the second type is novel, based on IV-dependent differential paths.

Using related messages. We present the first type of IV-recovery attacks. Assume that we know a specific differential path corresponding to a message difference Δ and with total probability p much larger than 2^{-128}. In other words, a randomly chosen message M will satisfy with probability p:

$$H_k(M) = H_k(M \boxplus \Delta).$$

By making approximately $2/p$ queries to the H_k-oracle, we will obtain a message M such that $H_k(M) = H_k(M \boxplus \Delta)$. Contini and Yin [4] then make the heuristic assumption that the pair $(M, M \boxplus \Delta)$ must follow the whole differential path, and not just the first and last steps. Since they do not justify that assumption, let us say a few words about it. The assumption requires a strong property on our specific differential path: that there are no other differential paths with better (or comparable) probability. In some sense, differential cryptanalysis on block ciphers use similar assumptions, and to see how realistic that is, one makes experiments on reduced-round versions of the block cipher. However, one might argue that there are intuitively more paths in hash functions than in block ciphers because of the following facts:

- the message length of a compression function is much bigger than the length of the round key in a block cipher.
- a step of the compression function is usually much simpler than a block-cipher step.

Also, because the paths for hash functions have a different shape from those of block ciphers, experiments on reduced-round versions may not be as conclusive.

The paper [4] shows that for usual differential paths (like those of MD4), if $(M, M \boxplus \Delta)$ satisfies the whole path, then one can build plenty of messages M^* closely related to M such that:

- If a specific internal register Q_i (during the computation of $H_k(M)$) satisfies certain conditions, then the pair $(M^*, M^* \boxplus \Delta)$ follows the whole path with probability p or larger, in which case $H_k(M^*) = H_k(M^* \boxplus \Delta)$.
- Otherwise, the pair $(M^*, M^* \boxplus \Delta)$ will drift away from the path at some position, and the probability of $H_k(M^*) = H_k(M^* \boxplus \Delta)$ is heuristically 2^{-128}.

Thus, by sending to the oracle many well-chosen pairs $(M', M' \boxplus \Delta)$, one can learn many bits of several internal register Q_i's during the computation of $H_k(M)$. Applying exhaustive search on the remaining bits of such Q_i's, one can guess the whole contents of four consecutive Q_i's. By definition of cMD4 and cMD5, it is then possible to reverse the computation of $H_k(M)$, which discloses $k = (Q_{-4}, Q_{-3}, Q_{-2}, Q_{-1})$.

Using IV-dependent differential paths. We now present a new type of IV-recovery attacks, that we will apply against MD4. Assume again that we know a specific differential path corresponding to a message difference Δ and with total probability p much larger than 2^{-128}, but assume this time that the path is IV-dependent: it holds only if the IV satisfies a specific condition (SC). In other words, if k satisfies (SC), then a randomly chosen message M will satisfy with probability p:

$$H_k(M) = H_k(M \boxplus \Delta).$$

But if k does not satisfy (SC), the pair $(M, M \boxplus \Delta)$ will drift away from the differential path from the first step, leading us to assume that $H_k(M) = H_k(M \boxplus \Delta)$ will hold with probability only 2^{-128}.

This would lead to the following attack: we would submit approximately $2/p$ pairs $(M, M \boxplus \Delta)$ to the H_k-oracle, and conclude that k satisfies (SC) if and only if $H_k(M) = H_k(M \boxplus \Delta)$ for at least one M. When sufficient information on k has been gathered, the rest of k can be guessed by exhaustive search if we have at least one collision of the form $H_k(M) = H_k(M \boxplus \Delta)$.

Notice that in some sense the differential path of the MD5 pseudo-collision [7] is an example of IV-dependent path where (SC) is the dBB condition, but it does not disclose much information about the IV. We would need to find many IV-dependent paths. Our attack on HMAC/NMAC-MD4 will use 22 such paths, which were found by an automated search.

We note that such attacks require an assumption similar to the previous IV-recovery attack. Namely, we assume that for the same message difference Δ, there is no differential paths with better (or comparable) probability, with or without conditions on the IV. To justify this assumption for our HMAC/NMAC-MD4 attack, we have performed experiments which will be explained in Section 4.

3.3   Subtleties Between the Inner and Outer Keys

Although the recovery of the inner key k_2 and the outer key k_1 both require IV-recovery attacks, we would like to point out subtle differences between the two cases. As mentioned previously, the recovery of k_2 only requires a basic IV-recovery attack, while the recovery of k_1 requires a composite IV-recovery attack. The composite IV-recovery attacks will be explained in Section 4 for MD4, and Section 5 for MD5.

When turning a basic IV-recovery attack into a composite IV-recovery, there are two important restrictions to consider:

- We have no direct control over the input of the outer hash function, it is the result of the inner hash function. So IV-recovery attacks using message modifications will become much less efficient when turned into composite IV-recovery. We will see this in Section 5 when extending the partial key-recovery from [4] into a full key-recovery.
- Since the input of H_{k_1} is a hash, its length is only 128 bits. Any differential path using a message difference Δ with non-zero bits outside these 128 first bits will be useless. This means that the partial key-recovery attacks from [4] against MD4, SHA-0 and reduced SHA-1 can't be extended into a full key-recovery.

  Using related keys one can use a differential path with a difference in the IV and no message difference – such as the one from [7] – and try a given message with both keys. However, if we want to get rid of related keys, we need a differential path with no IV difference and a difference in the beginning of the message.

3.4   Summary

To summarize, our attacks will have essentially the following structure (the MD5 attack will be slightly different because of the related-key setting):

1. Apply an IV-recovery attack to retrieve k_2, repeating sufficiently many times:
   (a) Select many one-block messages M uniformly at random.

(b) Observe if $\text{NMAC}_{k_1,k_2}(M) = \text{NMAC}_{k_1,k_2}(M \boxplus \Delta_1)$ for some M and a well-chosen Δ_1.

(c) Deduce information on k_2.

2. Apply a composite IV-recovery attack to retrieve k_1, repeating sufficiently many times:

(a) Construct carefully many pairs (M_1, M_2).

(b) Observe if $\text{NMAC}_{k_1,k_2}(M_1) = \text{NMAC}_{k_1,k_2}(M_2)$ for some pair (M_1, M_2).

(c) Deduce information on k_1.

4   Attacking HMAC/NMAC-MD4

4.1   Our IV-Recovery Attack Against MD4

In order to find differential paths which leak information about the key, we consider differential paths with a message difference in the first word (eg. $\delta m_0 = 1$). Then in the first steps of the compression function, we have:

$$Q_0 = (Q_{-4} \boxplus \text{IF}(Q_{-1}, Q_{-2}, Q_{-3}) \boxplus m_0) \lll 3$$
$$Q'_0 = (Q_{-4} \boxplus \text{IF}(Q_{-1}, Q_{-2}, Q_{-3}) \boxplus m_0 \boxplus 1) \lll 3$$

So $Q_0^{[3]} \neq Q_0'^{[3]}$. Then

$$Q_1 = (Q_{-3} \boxplus \text{IF}(Q_0, Q_{-1}, Q_{-2}) \boxplus m_1) \lll 7$$
$$Q'_1 = (Q_{-3} \boxplus \text{IF}(Q'_0, Q_{-1}, Q_{-2}) \boxplus m_1) \lll 7$$

Thus, if $Q_{-1}^{[3]} \neq Q_{-2}^{[3]}$, we will have $\text{IF}(Q_0, Q_{-1}, Q_{-2}) \neq \text{IF}(Q'_0, Q_{-1}, Q_{-2})$ and $Q_1 \neq Q'_1$. On the other hand, if $Q_{-1}^{[3]} = Q_{-2}^{[3]}$ and there is no carry when going from Q_0 to Q'_0, then $Q_1 = Q'_1$. Therefore, collision paths where $Q_{-1}^{[3]} = Q_{-2}^{[3]}$ will be significantly different from collision paths where $Q_{-1}^{[3]} \neq Q_{-2}^{[3]}$. This suggests that the collision probability will be correlated with the condition (SC) : $Q_{-1}^{[3]} = Q_{-2}^{[3]}$, and we expect to be able to detect the bias. More precisely we believe that the case $Q_{-1}^{[3]} \neq Q_{-2}^{[3]}$ will give a much smaller collision probability, since it means that an extra difference is introduced in step 1.

To check this intuition experimentally, we ran one cMD4 round (16 steps) with a random IV on message pairs $(M, M \boxplus 1)$ where M was also picked at random, and we looked for pseudo-collisions in $Q_{12}...Q_{15}$ with the following properties:

– The weight of the non-adjacent form of the difference is lower or equal to 4.
– There is no difference on $Q_{12}^{[12]}$.

The second condition is required to eliminate the paths which simply keep the difference introduced in $Q_0^{[3]}$ without modifying it. We ran this with $5 \cdot 10^{11}$ random messages and IVs and found 45624 collisions out of which 45515 respected the condition: this gives a ratio of about 420. This does not prove that we will have such a bias for collisions in the full MD4, but it is a strong evidence.

The same arguments apply when we introduce the message difference in another bit k (ie. $\delta m_0 = 2^k$): we expect to find more collisions if $Q_{-1}^{[k \boxplus s_0]} = Q_{-2}^{[k \boxplus s_0]}$.

We ran a differential path search algorithm to find such paths, and we did find 22 paths for different values of k with $Q_{-1}^{[k \boxplus s_0]} = Q_{-2}^{[k \boxplus s_0]}$. The path for $k = 0$ is given in Appendix B, and the other paths are just a rotation of this one. The corresponding set of sufficient conditions contains 79 conditions on the internal variables Q_i, so we expect that for a random message M:

$$\Pr\left[\text{MD4}(M) = \text{MD4}(M + \Delta)\right] = p \geq 2^{-79} \qquad \text{if } Q_{-1}^{[k \boxplus s_0]} = Q_{-2}^{[k \boxplus s_0]}$$
$$\ll p \qquad \text{if } Q_{-1}^{[k \boxplus s_0]} \neq Q_{-2}^{[k \boxplus s_0]}$$

If we try 2^{82} message pairs per path, we will find a collision for every path whose condition is fulfilled with a probability[3] of more than 99%. Then we know 22 bits of the IV ($Q_{-1}^{[k \boxplus s_0]} = Q_{-2}^{[k \boxplus s_0]}$ or $Q_{-1}^{[k \boxplus s_0]} \neq Q_{-2}^{[k \boxplus s_0]}$), which leaves only 2^{106} IV candidates. To check if a given IV is the correct one, we just check whether it gives a collision on the pairs colliding with the real IV, so we expect to find the IV after computing 2^{105} pairs of hashes in an offline phase.

We show in Appendix A.2 how to reduce the search space to 2^{94} keys by extracting more than one bit of information when a collision is found. This gives an IV-recovery attack against MD4 with a data complexity of 2^{88} MD4 oracle queries, and a time complexity of 2^{94} MD4 evaluations.

4.2 Deriving a Composite IV-Recovery Attack Against MD4

To turn this into a composite IV-recovery attack, we need to efficiently compute message pairs M, M' such that $H_{k_2}(M) = H_{k_2}(M') \boxplus \Delta$ (we will need 2^{82} such pairs). As we know k_2 (in the HMAC attack, we first recover it using the basic IV-recovery attack), we can compute $H_{k_2}(M)$ and find such pairs offline. If we do this naively using the birthday paradox, we need to hash about 2^{106} random messages to have all the pairs we need[4]. Then we can use the IV-recovery attack to get k_1.

Actually, we can do much better: we use the birthday paradox to find one pair of one-block messages (R, R') such that $H_{k_2}(R') = H_{k_2}(R) \boxplus \Delta$, and then we extend it to a family of two-block message pairs such that $H_{k_2}(R'\|Q') = H_{k_2}(R\|Q) \boxplus \Delta$ with very little extra computation. In the end, the cost to generate the messages with $H_{k_2}(M) = H_{k_2}(M') \boxplus \Delta$ will be negligible and the composite IV-recovery attack is as efficient as the basic one. This is the most important part of our work: thanks to our new path, we only need a low level of control on the input of the hash function to extract the IV.

Extending a Pair of Good Messages into a Family of Pairs. Figure 1 shows how we will create many message pairs with $H_{k_2}(M) = H_{k_2}(M') \boxplus \Delta$. We use a pair of one-block message (R, R') such that $H_{k_2}(R') = H_{k_2}(R) \boxplus \Delta$. Then we

---

[3] We have $\left(1 - \left(1 - 2^{-79}\right)^{2^{82}}\right)^{22} > 0.992$.

[4] This gives 2^{210} pairs of messages, and each pair has a probability of 2^{-128} to have the correct difference.

will generate a second block pair (Q, Q') such that $H_{k_2}(R'\|Q') \boxminus H_{k_2}(R\|Q) = \Delta$. Thanks to the Davies-Meyer construction of the compression function, all that we need is a difference path which starts with a Δ difference and ends with a zero difference in the internal state; then the feed-forward of H will keep $\delta H = \Delta$. This path can be found by a differential path search algorithm, or created by hand by slightly modifying a collision path.

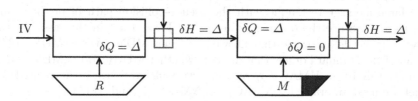

Fig. 1. Generating many pairs of message with a fixed hash difference

In this step, we also have to take care of the padding in MD4. Usually we ignore it because a collision at the end of a block is still a collision after an extra block of padding, but here we want a specific non-zero difference, and this will be broken by an extra block. So we have to adapt our collision finding algorithm to produce a block with a 55-byte message M and the last 9 bytes fixed by the MD4 padding. This can be done with nearly the same complexity as unconstrained MD4 collisions (about 4 MD4 computations per collision) using the technique of Leurent [10]. Thus, the cost of the message generation in the composite IV-recovery attack drops from 2^{106} using the birthday paradox to 2^{90} and becomes negligible in the full attack.

4.3 MD4 Attack Summary

This attack uses the same IV-recovery attack for the inner key and the outer key, with a complexity of 2^{88} online queries and 2^{94} offline computations. We manage to keep the complexity of the composite IV-recovery as low as the basic IV-recovery because we only need to control the hash differences, and we introduce a trick to generate many messages with a fixed hash difference.

In Appendix A.1 we show how to reduce a little bit the query complexity of the attack, and in the end the NMAC full key-recovery attack requires 2^{88} requests to the oracle, and 2×2^{94} offline computations.

5 Attacking NMAC-MD5

In this section, we will describe the attack of Contini and Yin [4], and we extend it to a full key recovery. This improved attack was independently found by Rechberger and Rijmen in [15].

As for MD4, the IV-recovery attack is based on a specific differential path, and assumes that when a collision is found with the given message difference, the Q_i's follow the path. This gives some bits of the internal state already, and a kind of message modification technique to disclose more bits is proposed in [4]. We can learn bits of Q_t using related messages where we fix the first t words.

5.1   The IV-Recovery Attack Against MD5

The IV-recovery attack on MD5 is the same as the one presented in [4]. It uses the related-message technique with the pseudo-collision path of de Boer and Bosselaers [7]. Since the differences are in the IV and not in the message, the IV-recovery needs an oracle that answers whether $\text{MD5}_{\text{IV}}(M) = \text{MD5}_{\text{IV}'}(M)$, instead of the standard oracle that answers whether $\text{MD5}_{\text{IV}}(M) = \text{MD5}_{\text{IV}}(M')$. To apply this to an HMAC key-recovery, we will have to use the related-key model: we need an oracle for NMAC_{k_1,k_2}, NMAC_{k_1',k_2} and $\text{NMAC}_{k_1,k_2'}$.

The IV-recovery attack in the related-key setting requires 2^{47} queries and 2^{45} hash computations, and this translates into a partial key-recovery (we will recover k_2) against NMAC-MD5 in the related-key model with the same complexity.

5.2   Deriving a Composite IV-Recovery Against MD5

To extend this to a composite IV-recovery attack, we run into the problem previously mentioned; to use this attack we need to create many inputs N^* of the hash function related to one input N, but these inputs are the outputs of a first hash function, and we cannot choose them freely: $N = \text{MD5}_{k_2}(M)$. However, we know k_2, so we can compute many $N_R = H_{k_2}(R)$ for random messages R and select those that are related to a particular N; if we want to recover bits of Q_t we will have to choose $32(t + 1)$ bits of N_R. We also run into the problem that any N_R is only 128 bits long; the last 384 bits will be fixed by the padding and there are the same for all messages. Therefore, we can only use the related-message technique to recover bits of the internal state of in the very first steps, whereas in the simple IV-recovery it is more efficient to recover the internal state of later steps (Contini and Yin used step 11 to 14). If we want to recover bits of Q_0 (due to the rotation we can only recover 25 bits of them), we need to produce 24×2^{45} messages N^* with the first 32 bits chosen; this will cost $24 \times 2^{45} \times 2^{32} \approx 2^{82}$ hash computations. Then, we know 25 bits of Q_0, plus the most significant bit of Q_1, Q_2, and Q_3; we still have 100 bits to guess. Thus, we have a related-key composite IV-recovery attack against MD5 with $2 \times 24 \times 2^{45} \approx 2^{51}$ oracle queries and 2^{100} MD5 evaluations.

If we try to guess bits in Q_1, we have to select at least 2^{44} hashes with 64 chosen bits; this costs about 2^{108} MD5, so it does not improve the attack.

5.3   MD5 Attack Summary

Thus, the Contini-Yin NMAC-MD5 attack can be extended into a full key-recovery attack in the related-key setting, with a query complexity of 2^{51}, a time complexity of 2^{100} MD5 operations, and success rate of 2^{-4} (due to the dBB condition for k_1 and k_2).

It is a very simple extension of the attack from Contini and Yin: we apply their technique to recover the outer key, but since we cannot choose the value of $H_{k_2}(M)$, we compute it for many random messages until we find a good one. This requires to change the step in which we extract internal bits, and the complexity become much higher.

Acknowledgement

Part of this work is supported by the Commission of the European Communities through the IST program under contract IST-2002-507932 ECRYPT, and by the French government through the Saphir RNRT project.

References

1. Amirazizi, H.R., Hellman, M.E.: Time-memory-processor trade-offs. IEEE Transactions on Information Theory 34(3), 505–512 (1988)
2. Bellare, M.: New Proofs for NMAC and HMAC: Security Without Collision Resistance. In: Dwork, C. (ed.) CRYPTO 2006. LNCS, vol. 4117, pp. 602–619. Springer, Heidelberg (2006)
3. Bellare, M., Canetti, R., Krawczyk, H.: Keying Hash Functions for Message Authentication. In: Koblitz, N. (ed.) CRYPTO 1996. LNCS, vol. 1109, pp. 1–15. Springer, Heidelberg (1996)
4. Contini, S., Yin, Y.L.: Forgery and Partial Key-Recovery Attacks on HMAC and NMAC Using Hash Collisions. In: Lai, X., Chen, K. (eds.) ASIACRYPT 2006. LNCS, vol. 4284, Springer, Heidelberg (2006)
5. Cramer, R.: In: Cramer, R. (ed.) EUROCRYPT 2005. LNCS, vol. 3494, pp. 22–26. Springer, Heidelberg (2005)
6. Daum, M.: Cryptanalysis of Hash Functions of the MD4-Family. PhD thesis, Ruhr-University of Bochum (2005)
7. den Boer, B., Bosselaers, A.: Collisions for the Compression Function of MD5. In: Helleseth, T. (ed.) EUROCRYPT 1993. LNCS, vol. 765, pp. 293–304. Springer, Heidelberg (1994)
8. Fouque, P.A., Leurent, G., Nguyen, P.: Automatic Search of Differential Path in MD4. ECRYPT Hash Worshop – Cryptology ePrint Archive, Report, 2007/206 (2007), http://eprint.iacr.org/
9. Kim, J., Biryukov, A., Preneel, B., Hong, S.: On the Security of HMAC and NMAC Based on HAVAL, MD4, MD5, SHA-0 and SHA-1. In: De Prisco, R., Yung, M. (eds.) SCN 2006. LNCS, vol. 4116, pp. 242–256. Springer, Heidelberg (2006)
10. Leurent, G.: Message Freedom in MD4 and MD5: Application to APOP Security. In: Biryukov, A. (ed.) FSE. LNCS, Springer, Heidelberg (to appear)
11. Preneel, B., van Oorschot, P.C.: MDx-MAC and Building Fast MACs from Hash Functions. In: Coppersmith, D. (ed.) CRYPTO 1995. LNCS, vol. 963, pp. 1–14. Springer, Heidelberg (1995)
12. Preneel, B., van Oorschot, P.C.: On the Security of Two MAC Algorithms. In: Rueppel, R.A. (ed.) EUROCRYPT 1992. LNCS, vol. 658, pp. 19–32. Springer, Heidelberg (1993)

13. Preneel, B., van Oorschot, P.C.: On the Security of Iterated Message Authentication Codes. IEEE Transactions on Information Theory 45(1), 188–199 (1999)
14. Rechberger, C., Rijmen, V.: Note on Distinguishing, Forgery, and Second Preimage Attacks on HMAC-SHA-1 and a Method to Reduce the Key Entropy of NMAC. Cryptology ePrint Archive, Report, 2006/290 (2006), http://eprint.iacr.org/
15. Rechberger, C., Rijmen, V.: On Authentication with HMAC and Non-Random Properties. In: Dietrich, S. (ed.) Financial Cryptography. LNCS, Springer, Heidelberg (to appear)
16. Shoup, V. (ed.): In: Shoup, V. (ed.) CRYPTO 2005. LNCS, vol. 3621, pp. 14–18. Springer, Heidelberg (2005)
17. Wang, X., Lai, X., Feng, D., Chen, H., Yu, X.: Cryptanalysis of the Hash Functions MD4 and RIPEMD. [5] pp. 1–18
18. Wang, X., Yin, Y.L., Yu, H.: Finding Collisions in the Full SHA-1. [16] pp. 17–36
19. Wang, X., Yu, H.: How to Break MD5 and Other Hash Functions. [5] pp. 19–35
20. Wang, X., Yu, H., Yin, Y.L.: Efficient Collision Search Attacks on SHA-0. [16] pp. 1–16
21. Yu, H., Wang, G., Zhang, G., Wang, X.: The Second-Preimage Attack on MD4. In: Desmedt, Y.G., Wang, H., Mu, Y., Li, Y. (eds.) CANS 2005. LNCS, vol. 3810, pp. 1–12. Springer, Heidelberg (2005)

A   Improving the MD4 IV-Recovery

A.1   Reducing the Online Cost

First, we can easily lower the number of calls to the NMAC-oracle in the first phase of the IV-recovery. Instead of trying 22×2^{82} random message pairs, we will choose the messages more cleverly so that each message belongs to 22 pairs: we first choose 490 bits of the message at random and then use every possibility for the 22 remaining bits. Thus, we only need 2^{83} calls to the oracle instead of 22×2^{83}.

Note that we cannot use this trick in the composite IV-recovery attack, so the number of queries for the full key-recovery will only be halved (the queries for the basic IV-recovery for k_2 become negligible compared to the queries for the composite IV-recovery that will reveal k_2).

A.2   Reducing the Offline Cost

We may also lower the computational cost of the attack, by getting more than one bit of the IV once a collision has been found. This will require the extra assumption the colliding messages follow the differential path in step 1 (previously we only needed step 0), but this seems quite reasonable, for the same reasons. Out of the 22 paths used to learn IV bits, let p be the number of paths for which the condition holds, and a collision is actually found. From each message that collides following the differential path, we can also extract some conditions on the internal states Q_0 and Q_1. These states are not part of the IV, but since we know the message used, we can use these conditions to learn something on the IV. If we have a message pair that collides with $M' \boxminus M = 2^k$, we will call them

$M^{(k)}$ and $M'^{(k)}$ and the condition gives us $Q_0^{[k \boxplus s_0]}(M^{(k)})$ and $Q_1^{[k \boxplus s_0]}(M^{(k)})$. The idea is the following (the symbol '$\blacktriangleright$' summarizes the number of bits to guess at each step):

1. We guess Q_{-1}. Let $n = 32 - |Q_{-1}|$ be the number of 0 bits in Q_{-1} (we use $|x|$ to denote the Hamming weight of x).

2. We compute 22 bits of Q_{-2} using the conditions on the IV, and we guess the others $\blacktriangleright$ 10 bits.

3. We guess the bits of Q_{-3} used to compute Q_0. Since we have $Q_0 = (Q_{-4} \boxplus \mathrm{IF}(Q_{-1}, Q_{-2}, Q_{-3}) \boxplus k_0 \boxplus m_0) \lll s_0$, we only need $Q_{-3}^{[i]}$ when $Q_{-1}^{[i]} = 0$ $\blacktriangleright$ n bits.

4. We have $Q_{-4} = (Q_0 \ggg s_0) \boxminus \mathrm{IF}(Q_{-1}, Q_{-2}, Q_{-3}) \boxminus m_0 \boxminus k_0$. If we use it with the message $M^{(0)}$ and take the equation modulo 2, it becomes: $Q_{-4}^{[0]} = Q_0^{[s_0]}(M^{(0)}) \boxminus (\mathrm{IF}(Q_{-1}, Q_{-2}, Q_{-3}) \boxminus m_0^{(0)} \boxminus k_0) \bmod 2$, and it gives us $Q_{-4}^{[0]}$.

   Then, if we write the equation with $M^{(1)}$ and take it modulo 2, we will learn $Q_0^{[s_0]}(M^{(1)})$ from $Q_{-4}^{[0]}$. Since we know $(Q_0(M^{(1)}) \lll s_0) \bmod 4$ from $Q_0^{[s_0]}(M^{(1)})$ and $Q_0^{[1 \boxplus s_0]}(M^{(1)})$, we can take the equation modulo 4 to learn $Q_{-4}^{[1]}$.

   By repeating this process, we learn the full Q_{-4}, but we need to guess the bit i when we don't have a message pair $M^{(i)}, M'^{(i)}$ $\blacktriangleright$ $32 - p$ bits.

5. We apply the same process to compute the remaining bits of Q_{-3}. We already know n bits and we expect to be able to compute a ratio of $p/32$ of the missing ones. $\blacktriangleright$ $\frac{(32-n)(32-p)}{32}$ bits.

So, for each choice of Q_{-1} in step 1, we have to try a number of choices for the other bits that depends on the Hamming weight $32 - n$ of Q_{-1}. In the end, the number of keys to try is:

$$\sum_{Q_{-1}} 2^{10 + n + 32 - p + (32-n)(32-p)/32} = 2^{74 - 2p} \left(1 + 2^{p/32}\right)^{32}$$

With $p = 11$, this becomes a little less than 2^{90}, but the complexity depends on the number of conditions fulfilled by the key. If we assume that every condition has a probability of one half to hold, we can compute the average number of trials depending on the keys, and we will have to try half of them:

$$\frac{1}{\#k} \sum_k 2^{74 - 2p} \left(1 + 2^{p/32}\right)^{32} < 2^{93.8}$$

Hence, we have an IV-recovery attack requiring less than 2^{88} queries to the NMAC oracle, and less than 2^{94} offline hash computations. See the full version of this paper for a detailed complexity analysis.

B   IV-Dependent Differential Path

Here is one of the 22 IV-dependent paths we found in MD4. The 22 paths can be deduced from this one by rotating all the bit differences and bit conditions:

it works on bit positions 0, 1, 3, 4, 6-8, 12-17, 19-24, 26, 27, and 29, and fails on other positions due to carry expansions.

This path was found using an automated differential paths search algorithm described in [8].

step	s_i	δm_i	$\partial\Phi_i$	∂Q_i	Φ-conditions and $\lll$-conditions
0	3	$\langle\blacktriangle^{[0]}\rangle$		$\langle\blacktriangle^{[3]}\rangle$	
1	7				$Q_{-1}^{[3]} = Q_{-2}^{[3]}$
2	11				$Q_1^{[3]} = 0$
3	19				$Q_2^{[3]} = 1$
4	3			$\langle\blacktriangledown\blacktriangle^{[6,7]}\rangle$	
5	7				$Q_3^{[6]} = Q_2^{[6]}, Q_3^{[7]} = Q_2^{[7]}$
6	11				$Q_5^{[6]} = 0, Q_5^{[7]} = 0$
7	19		$\langle\blacktriangle^{[7]}\rangle$	$\langle\blacktriangle^{[26]}\rangle$	$Q_6^{[6]} = 1, Q_6^{[7]} = 0$
8	3		$\langle\blacktriangledown^{[26]}\rangle$	$\langle\blacktriangle^{[9]}, \blacktriangledown^{[29]}\rangle$	$Q_5^{[26]} = 1, Q_6^{[26]} = 0$
9	7				$Q_7^{[9]} = Q_6^{[9]}, Q_8^{[26]} = 0, Q_7^{[29]} = Q_6^{[29]}$
10	11				$Q_9^{[9]} = 0, Q_9^{[26]} = 1, Q_9^{[29]} = 0$
11	19			$\langle\blacktriangle^{[13]}\rangle$	$Q_{10}^{[9]} = 1, Q_{10}^{[29]} = 1$
12	3			$\langle\blacktriangledown^{[0]}, \blacktriangle^{[12]}\rangle$	$Q_{10}^{[13]} = Q_9^{[13]}$
13	7				$Q_{11}^{[0]} = Q_{10}^{[0]}, Q_{11}^{[12]} = Q_{10}^{[12]}, Q_{12}^{[13]} = 0$
14	11		$\langle\blacktriangledown^{[0]}\rangle$	$\langle\blacktriangle\blacktriangle\blacktriangledown^{[11\ldots13]}\rangle$	$Q_{13}^{[0]} = 1, Q_{13}^{[12]} = 0, Q_{13}^{[13]} = 1$
15	19		$\langle\blacktriangledown^{[13]}\rangle$		$Q_{14}^{[0]} = 1, Q_{13}^{[11]} = Q_{12}^{[11]}, Q_{13}^{[12]} = 0, Q_{13}^{[13]} = 1, Q_{12}^{[13]} = 0$
16	3	$\langle\blacktriangle^{[0]}\rangle$	$\langle\blacktriangle\blacktriangledown^{[12,13]}\rangle$		$Q_{15}^{[11]} = Q_{13}^{[11]}, Q_{15}^{[12]} \neq Q_{13}^{[12]}, Q_{15}^{[13]} \neq Q_{13}^{[13]}$
17	5				$Q_{16}^{[11]} = Q_{15}^{[11]}, Q_{16}^{[12]} = Q_{15}^{[12]}, Q_{16}^{[13]} = Q_{15}^{[13]}$
18	9			$\langle\blacktriangle\blacktriangle\blacktriangle\blacktriangledown^{[20\ldots23]}\rangle$	
19	13				$Q_{17}^{[20]} = Q_{16}^{[20]}, Q_{17}^{[21]} = Q_{16}^{[21]}, Q_{17}^{[22]} = Q_{16}^{[22]}, Q_{17}^{[23]} = Q_{16}^{[23]}$
20	3		$\langle\blacktriangledown^{[23]}\rangle$	$\langle\blacktriangledown^{[26]}\rangle$	$Q_{19}^{[20]} = Q_{17}^{[20]}, Q_{19}^{[21]} = Q_{17}^{[21]}, Q_{19}^{[22]} = Q_{17}^{[22]}, Q_{19}^{[23]} \neq Q_{17}^{[23]}$
21	5				$Q_{20}^{[20]} = Q_{19}^{[20]}, Q_{20}^{[21]} = Q_{19}^{[21]}, Q_{20}^{[22]} = Q_{19}^{[22]}, Q_{20}^{[23]} = Q_{19}^{[23]}, Q_{19}^{[26]} = Q_{18}^{[26]}$
22	9			$\langle\blacktriangledown^{[29]}\rangle$	$Q_{21}^{[26]} = Q_{19}^{[26]}$
23	13				$Q_{22}^{[26]} = Q_{21}^{[26]}, Q_{21}^{[29]} = Q_{20}^{[29]}$
24	3			$\langle\blacktriangle\blacktriangledown^{[29,30]}\rangle$	$Q_{23}^{[29]} = Q_{21}^{[29]}$
25	5				$Q_{23}^{[30]} = Q_{22}^{[30]}$
26	9		$\langle\blacktriangle^{[29]}\rangle$		$Q_{25}^{[29]} \neq Q_{23}^{[29]}, Q_{25}^{[30]} = Q_{23}^{[30]}$
27	13				$Q_{26}^{[29]} = Q_{25}^{[29]}, Q_{26}^{[30]} = Q_{25}^{[30]}$
28	3			$\langle\blacktriangledown^{[0]}\rangle$	
29	5				$Q_{27}^{[0]} = Q_{26}^{[0]}$
30	9				$Q_{29}^{[0]} = Q_{27}^{[0]}$
31	13				$Q_{30}^{[0]} = Q_{29}^{[0]}$
32	3	$\langle\blacktriangle^{[0]}\rangle$			

Path 1. A path with the message difference on the first word

How Should We Solve Search Problems Privately?

Amos Beimel[1], Tal Malkin[2,*], Kobbi Nissim[1,**], and Enav Weinreb[3]

[1] Dept. of Computer Science, Ben-Gurion University, Be'er Sheva, Israel
{beimel,kobbi}@cs.bgu.ac.il
[2] Dept. of Computer Science, Columbia University, New York, NY
tal@cs.columbia.edu
[3] Dept. of Computer Science, Technion, Haifa, Israel
weinreb@cs.technion.ac.il

Abstract. Secure multiparty computation allows a group of distrusting parties to jointly compute a (possibly randomized) *function* of their inputs. However, it is often the case that the parties executing a computation try to solve a *search problem*, where one input may have a multitude of correct answers – such as when the parties compute a shortest path in a graph or find a solution to a set of linear equations.

Picking one output arbitrarily from the solution set has significant implications on the privacy of the algorithm. Beimel et al. [STOC 2006] gave a *minimal* definition for private computation of search problems with focus on proving impossibility result. In this work we aim for stronger definitions of privacy for search problems that provide reasonable privacy. We give two alternative definitions and discuss their privacy guarantees. We also supply algorithmic machinery for designing such protocols for a broad selection of search problems.

1 Introduction

Secure multiparty computation addresses a setting where several distrusting parties want to jointly compute a function $f(x_1, \ldots, x_n)$ of their private inputs $x_1, \ldots, x_n$, while maintaining the privacy of their inputs. One of the most fundamental, and by now well known, achievements in cryptography (initiated by [19,14,8,3], and continued by a long line of research) shows that in fact for any feasible function f, there exists a secure multiparty protocol for f (in a variety of settings). However, in many cases, what the parties wish to compute is not a function with just a single possible output for each input, and not even a randomized function with a well defined output distribution. Rather, in many cases the parties are solving a problem where several correct answers (or *solutions*) may exist for a single instance $x = (x_1, \ldots, x_n)$. For example, the parties may jointly hold a graph and wish to compute a shortest path between two of its

---

* Research partially supported by the NSF (grant No. CCF-0347839).

** Research partially supported by the Israel Science Foundation (grant No. 860/06).

vertices or to find a minimal vertex cover in it.[1] We call such problems *search problems*. In such cases, to apply known results of secure multiparty computation, one has first to decide upon a polynomial-time computable function that solves the search problem.

An approach often taken by designers of secure multiparty protocols for such applications is to arbitrarily choose one of the existing algorithms/heuristics for the search problem, and implement a secure protocol for it. This amounts to choosing an arbitrary (possibly randomized) *function* that provides a solution, and implementing it securely. The privacy implications of such choices have not been analyzed, and it is clear that if the computed function leaks unnecessary information on the parties' private inputs, any protocol realizing it, no matter how secure, will also leak this information. Thus, some privacy requirements should be imposed on the chosen input-output functionality.

To illustrate the necessity of a rigid discussion of secure computation of search problems, consider the following setting. A server holds a database with valuable information, and a client makes queries to this database such that there may be many different answers to a single query. The server is interested in answering the client's queries in a way that reveals the least information possible on the database. However, the strategy the server chooses to answer each query might reveal information. For example, consider a case where the client queries for the name of a person whose details are in the database and satisfies some condition. An arbitrary solution such as answering with the details of the appropriate person whose name is the lexicographically first in the database reveals the fact that every person prior to that person in the lexicographic order does not satisfy the given condition.

In this paper we study the privacy implications of how the output is chosen for search problems, propose suitable privacy requirements, and provide constructions achieving them for several problems (see details below). This generalizes the approach of [1], who introduced the problem of private search algorithms in the context of private approximations. Beimel et al. [1] have put forward what seems to be a minimal requirement of privacy (first coined in the context of private approximation of functions [10], and later extended to search problems):

If two instances x, y have an identical set of possible solutions, their outputs should not be distinguished.

That is, in order for the algorithm to be private, the output must depend only on the solution set, and not on the specific input. In spirit of this requirement, we say that two inputs are equivalent if they have the same set of solutions.

This definition was reasonable in the context of [1] because they provide mostly negative results (so a weaker definition corresponds to stronger infeasibility results), and because in the context of private approximations of functions[2], this

---

[1] Another example is when the parties compute an approximation to a function $f()$ as in [10,16,17]. Again, there is potentially more than one correct answer for an instance.

[2] And similarly for those instances of search problems for which a *unique* solution exists.

turns out to be a significant privacy guarantee (as it implies that no information beyond the original $f(x)$ is leaked). However, in the context of search problems the implication is potentially much weaker – that no information beyond the *entire solution set* of x is leaked. Arguably, for most applications requiring privacy, leaking information up to the entire solution set does not provide a sufficient privacy guarantee. Furthermore, even with this minimal definition of privacy, the notion of private search has so far proved to be very problematic. Search versions of many NP-complete problems do not admit even very weakened notion of private approximation algorithms [1,2], and private search is infeasible even to some problems that do admit polynomial time search algorithms.

We are thus faced with a double challenge: first, strengthen the definition, imposing further requirements on the function in order to provide reasonable privacy guarantees. Second, provide protocols implementing the stronger definition for as wide as possible class of search problems. This is the goal we tackle in this work.

1.1 This Work

As discussed above, the outcome of a private algorithm $\mathcal{A}$ when run on an instance x should only depend on the set of possible solutions to x. Which further requirements should be imposed on this outcome? While the answer to this question may be application dependent, we identify two (incomparable) requirements that are suitable in many situations, and study which problems admit those requirements and which techniques can be used to achieve them. Before elaborating on this, let us start with two naïve proposals that are used to demonstrate essential privacy considerations arising for search problems, and to facilitate our actual proposed definitions and algorithms.

Deterministic vs. Randomized Private Algorithms. Consider first requiring any private algorithm $\mathcal{A}$ to be *deterministic*. As such, it consistently selects one of the solutions, hence subsequent applications of the algorithm on the same (or equivalent) inputs do not reveal further information. A possible choice is to output the lexicographically first solution. This choice is computationally feasible for several polynomially solvable search problems such as the problem of finding a solution for a linear system, and stable marriage (using the stable marriage with restrictions algorithm [9]).[3] Deterministic algorithms, however, leak definite information, that (depending on the application) may turn to be crucial. E.g., the lexicographically first solution rules out all solutions that are ordered below it. Furthermore, deterministic algorithms would enable verifying that the instance x is not equivalent to another instance y, even if x, y have similar solution sets, just by checking the outcome of the algorithm on both instances.

Next, consider a *randomized* algorithm $\mathcal{A}$, which on input x selects from the set of solutions according to a specific distribution (depending only on the solution set).

---

[3] The recent protocols of [15,11] also output a deterministic solution – the outcome of the Gale-Shapley algorithm [12]. However, this is not a private search algorithm.

A natural choice here is to pick a solution uniformly at random. Randomized private algorithms may be advantageous to deterministic private algorithms, as the information they leak is potentially "blurred". For example, if instances x, y have similar solution sets, then the resulting output distributions would be close. On the other hand, when applied repeatedly on the same instance there is a potential for an increased leakage. E.g., for the problem of finding a solution for a linear system of equations, the number of revealed solutions grows exponentially in the number of invocations, until the entire solution space is revealed.

We note that the benefits and disadvantages of deterministic and randomized algorithms are generally incomparable. Moreover, there exist problems for which an algorithm outputting a uniformly selected solutions exist, but no deterministic private algorithm exists (under standard assumptions), and vice versa (see Appendix A).

Framework: Seeded Algorithms. In the following, we restrict our attention to what we call *seeded algorithms*. The idea of seeded algorithms is not new – these are deterministic algorithms that get as input a "random" seed s and an instance x. If the seed s is selected at random the first time the seeded algorithm is invoked, subsequent invocations on the same input may be answered consistently. A seeded algorithm allows selecting a random solution for each instance (separately), while preventing abuse of repeated queries. Arguably, seeded algorithms are less desirable than algorithms that do not need to maintain any state information.[4] However, we note that the state information of seeded algorithms is rather easy to maintain, as they do not need to keep a log of previously answered queries, and hence their state does not grow with the number of queries. In that, the usage of seeded algorithms is similar to that of pseudorandom functions.

Our Results. To focus on the choice of a function for solving a search problem, we abstract out the implementation details of the underlying secure multiparty setting (in analogy to [10,16,1,2]). Our results directly apply to a client-server setup, where the server is willing to let the client learn a solution to a specific search problem applied to its input. They (similarly) directly apply to the setup of a distributed multiparty computation where the parties share an instance x using a secret sharing scheme, as it can be reduced to a client-server setup using secure function evaluation protocols [19,14,8,3]. In the general setup of distributed multiparty computation, however, one may also consider definitions that allow leakage to a party of any information implied by its individual input.

Equivalence Protecting Algorithms. Equivalence protecting algorithms are seeded algorithms that choose a uniformly random answer for each class of equivalent instances. Given the seed, the output is deterministic and respects equivalence of instances – an access to an equivalence protecting algorithm $\mathcal{A}_P$ for a problem

---

[4] In secure multiparty computation, the parties should jointly generate a random seed, and then work with this shared seed in subsequent executions of the algorithm. In a client-server setup, the server should generate the seed the first time it is invoked, and use it in future invocations.

$\mathcal{P}$ simulates an access to a random oracle for $\mathcal{P}$ that answers consistently on inputs with the same solutions.[5]

To some extent, equivalence protecting algorithms enjoy benefits of both the naïve privacy notions discussed above, deterministic and randomized private algorithms: (i) there is a potential for not giving "definite" information; and (ii) leakage is not accumulated with repeated queries. However, equivalence protecting algorithms do allow distinguishing instances even when their solution sets are very close.

In Section 3 we reduce the problem of designing an equivalence protecting algorithm for a search problem, to that of (i) designing a deterministic algorithm for finding a canonical representative of the equivalence class; (ii) designing a randomized private algorithm returning a uniformly chosen solution; and (iii) the existence of pseudorandom functions. We then show how to use this to construct an equivalence protecting algorithm for what we call "monotone search problems", a wide class of functions including perfect matching in bipartite graphs and shortest path in a directed graph. We further demonstrate the power of our general construction by showing an equivalence protecting algorithm for solving a system of linear equations over a finite field.

Resemblance Preserving Algorithms. Our second strengthening of the requirements on a private search algorithm addresses the problem of distinguishing non-equivalent instances with similar solution sets. Similarly to equivalence protecting algorithms, resemblance preserving algorithms choose a random solution for each set of equivalence instances. However, here the choices for non-equivalent instances are highly correlated such that pairs of instances that have close output sets are answered identically with high probability.

In Section 4 we present a generic construction of resemblance preserving algorithms, for any search problem whose output space admits a pairwise independent family of permutations, where the minimum of a permuted solution set can be computed efficiently. Examples of such search problems include finding roots or non-roots of a polynomial, solving a system of linear equations over a finite field, finding a point in a union of rectangles in a fixed dimensional space, and finding a satisfying assignment for a DNF formula. It is interesting to note that for the last problem, finding an efficient equivalence protecting algorithm implies P=NP.

To summarize, we present two definitions (suitable for different applications), provide technical tools to achieve these definitions, and identify generic classes, as well as specific examples, of search problems where our tools can be used to yield private search algorithms with the desired properties. The main conceptual contribution of the paper is in putting forward the need to study private computation of search problems (where a non-private solution is well known), analyzing privacy considerations, and defining equivalence protecting and resemblance preserving algorithms. The main technical contribution of the paper is in the tools and algorithms presented in Section 4 for resemblance preserving algorithms.

---

[5] Such a random oracle can be thought of as an ideal model solution to the problem, which this definition requires to emulate.

2   Definitions

We define a search problem as a function assigning to an instance $x \in \{0,1\}^n$ a solution set $\mathcal{P}_n(x)$. Two instances of a search problem are *equivalent* if they have exactly the same solution set. More formally:

Definition 1 (Search Problem). *A search problem is an ensemble* $\mathcal{P} = \{\mathcal{P}_n\}_{n \in \mathbb{N}}$ *such that* $\mathcal{P}_n : \{0,1\}^n \to 2^{\{0,1\}^{q(n)}}$ *for some positive polynomial* $q(n)$.

Definition 2. *For a search problem* $\mathcal{P}$ *the equivalence relation* $\equiv_{\mathcal{P}}$ *includes all pairs of instances* $x, y \in \{0,1\}^n$ *such that* $\mathcal{P}_n(x) = \mathcal{P}_n(y)$.

We recall the *minimal* definition of private search algorithms from [1]. All our definitions will be stronger – an algorithm that satisfies Definition 7 or Definition 13 trivially satisfies Definition 3.

Definition 3 (Private Search Algorithms [1]). *A probabilistic polynomial time algorithm* $\mathcal{A}_{\mathcal{P}}$ *is a private search algorithm for* $\mathcal{P}$ *if (i)* $\mathcal{A}_{\mathcal{P}}(x) \in \mathcal{P}_n(x)$ *for all* $x \in \{0,1\}^n$, $n \in \mathbb{N}$; *and (ii) for every polynomial-time algorithm* $\mathcal{D}$ *and for every positive polynomial* $q(\cdot)$, *there exists some* $n_0 \in \mathbb{N}$ *such that for every* $x, y \in \{0,1\}^*$ *such that* $x \equiv_{\mathcal{P}} y$ *and* $|x| = |y| \geq n_0$

$$\left| \Pr[\mathcal{D}(\mathcal{A}_{\mathcal{P}}(x), x, y) = 1] - \Pr[\mathcal{D}(\mathcal{A}_{\mathcal{P}}(y), x, y) = 1] \right| \leq \frac{1}{q(|x|)} .$$

That is, when $x \equiv_{\mathcal{P}} y$, *every polynomial time algorithm* $\mathcal{D}$ *cannot distinguish if the input of* $\mathcal{A}_{\mathcal{P}}$ *is* x *or* y.

We proceed to a standard definition of pseudorandom functions from binary strings of size n to binary strings of size $\ell(n)$, where $\ell(\cdot)$ is some fixed polynomial.

Definition 4 (Pseudorandom Functions [13]). *A function ensemble* $F = \{F_n\}_{n \in \mathbb{N}}$ *of functions from* $\{0,1\}^n$ *to* $\{0,1\}^{\ell(n)}$ *is called* pseudorandom *if for every probabilistic polynomial time oracle machine* M, *every polynomial* $p(\cdot)$, *and all sufficiently large* n's,

$$\left| \Pr[M^{F_n}(1^n) = 1] - \Pr[M^{H_n}(1^n) = 1] \right| < \frac{1}{p(n)}$$

where $\ell(\cdot)$ *is some fixed polynomial, and* $H = \{H_n\}_{n \in \mathbb{N}}$ *is the uniform function ensemble over functions from* $\{0,1\}^n$ *to* $\{0,1\}^{\ell(n)}$.

Finally, we define *seeded* algorithms, which are central to our constructions.

Definition 5 (Seeded Algorithms). *A seeded algorithm* $\mathcal{A}$ *is a deterministic polynomial time algorithm taking two inputs* x, s_n *where* $|x| = n$ *and* $|s_n| = p(n)$ *for some polynomial* $p()$. *The distribution induced by a seeded algorithm on an input* x *is the distribution on outcomes* $\mathcal{A}(x, s_n)$ *where* s_n *is chosen uniformly at random from* $\{0,1\}^{p(|x|)}$.

Informally, a seeded algorithm is private if it is a deterministic private algorithm for every choice of the seed s_n, i.e., $\mathcal{A}(x, s_n) = \mathcal{A}(y, s_n)$ for all $s_n \in \{0,1\}^{p(|x|)}$ whenever $x \equiv_{\mathcal{P}} y$.

3 Equivalence Protecting Privacy Definition

In this section we suggest a definition of private algorithm for a search problem and supply efficient algorithms satisfying this definition for a broad class of problems. The privacy guarantee we introduce enjoys the advantages of both deterministic and random algorithms. Based on the existence of pseudorandom functions, it provides solutions that look random but do not leak further information while executed repeatedly on inputs that are equivalent. In order to suggest appropriate privacy definitions for secure computation of a search problem, we need to picture how such a computation would take place in an ideal world. The following two definitions capture random sampling of an answer that depends only on the solution set (and not on the specific input).

Definition 6 (Private Oracle). *Let $\mathcal{P} = \{\mathcal{P}_n\}_{n \in \mathbb{N}}$ be a search problem and p be the polynomial such that $\mathcal{P}_n : \{0,1\}^n \to 2^{\{0,1\}^{p(n)}}$. We say that for a given $n \in \mathbb{N}$ an oracle $O_n : \{0,1\}^n \to \{0,1\}^{p(n)}$ is private with respect to $\mathcal{P}_n$ if*

1. *For every $x \in \{0,1\}^n$ it holds that $O_n(x) \in \mathcal{P}_n(x)$. That is, O_n returns correct answers.*
2. *For every $x, x' \in \{0,1\}^n$ it holds that $x \equiv_{\mathcal{P}} x'$ implies $O_n(x) = O_n(x')$. That is, O_n satisfies the privacy requirement of Definition 3.*

An oracle that is private with respect to $\mathcal{P}$ represents one possible functionality that solves the search problem and protects the equivalence relation. We define an algorithm to be equivalence protecting if it cannot be efficiently distinguished from a random oracle that is private with respect to $\mathcal{P}$.

Definition 7 (Equivalence Protecting Algorithm). *Let $\mathcal{P} = \{\mathcal{P}_n\}_{n \in \mathbb{N}}$ be a search problem. An algorithm $\mathcal{A}(\cdot, \cdot)$ is private with respect to $\equiv_{\mathcal{P}}$, if for every polynomial time oracle machine $\mathcal{D}$, for every polynomial p, and for all sufficiently large n's,*

$$\left| \Pr[\mathcal{D}^{O_n}(1^n) = 1] - \Pr[\mathcal{D}^{\mathcal{A}(\cdot, s_n)}(1^n) = 1] \right| < \frac{1}{p(n)},$$

where the first probability is over the uniform distribution over oracles O_n that are private with respect to $\mathcal{P}$, and the second probability is uniform over the choices of the seed s_n for the algorithm $\mathcal{A}$.

In the above definition we arbitrarily choose the uniform distribution over private oracles. We note that, for some applications, other distributions might be preferred; the definition can be easily adjusted to such scenarios. We note that using the uniform distribution is common in many sampling algorithms, e.g., [18].

The following two definitions will be helpful in constructing equivalence protecting algorithms for various search problems. The first definition discusses algorithms that return a representative element for every equivalence class of the search problem $\mathcal{P}$. The second defines sampling an answer from the output set of a given input.

Definition 8 (Canonical Representative Algorithm). *Let* $\mathcal{P} = \{\mathcal{P}_n\}_{n\in\mathbb{N}}$ *be a search problem. An algorithm $\mathcal{A}$ is a* canonical representative *algorithm for $\mathcal{P}$ if (i) for every $x \in \{0,1\}^n$ it holds that $x \equiv_{\mathcal{P}} \mathcal{A}(x)$; and (ii) for every $x, y \in \{0,1\}^n$, it holds that $\mathcal{A}(x) = \mathcal{A}(y)$ iff $x \equiv_{\mathcal{P}} y$.*

Definition 9 (Output Sampling Algorithm). *Let* $\mathcal{P} = \{\mathcal{P}_n\}_{n\in\mathbb{N}}$ *be a search problem. A randomized algorithm $\mathcal{A}$ is called an* output sampling *algorithm for $\mathcal{P}$ if for every $x \in \{0,1\}^n$ the distribution $\mathcal{A}(x, r)$ is computationally indistinguishable from* $\mathbf{Unif}_{\mathcal{P}(x)}$, *the uniform distribution on the possible outputs on x.*

We reduce the problem of designing an equivalence protecting algorithm for a search problem into designing a canonical representative algorithm and an output sampling algorithm for the problem. The construction is based on the existence of pseudorandom functions. Let $F = \{F_n\}_{n\in\mathbb{N}}$ be an ensemble of pseudorandom functions from $\{0,1\}^n$ to $\{0,1\}^{\ell(n)}$, where $\ell(\cdot)$ is a polynomial that bounds the number of random bits used by the output sampling algorithm. We denote by $f_{s_n}(x)$ the output of the function indexed by s on an input $x \in \{0,1\}^n$. The proof of Theorem 1 is omitted here.

Algorithm General Equivalence Protecting

INPUT: An instance $x \in \{0,1\}^n$ and a seed s_n for a family of pseudorandom functions $F = \{F_n\}_{n\in\mathbb{N}}$.
OUTPUT: A solution sol $\in \mathcal{P}_n(x)$.

1. Compute $y = \mathcal{A}_{\mathrm{rep}}(x)$.
2. Compute $r = F_{s_n}(y)$.
3. Output sol $= \mathcal{A}_{\mathrm{rand}}(y, r)$.

Theorem 1. *Let $\mathcal{P}$ be a search problem. Suppose $\mathcal{P}$ has (i) an efficient output sampling algorithm $\mathcal{A}_{\mathrm{rand}}$; and (ii) an efficient canonical representative algorithm $\mathcal{A}_{\mathrm{rep}}$. Then Algorithm* General Equivalence Protecting *is an efficient equivalence protecting algorithm for $\mathcal{P}$.*

3.1 Private Algorithms for Monotone Search Problems

In view of Theorem 1, the construction of a private algorithm for a given search problem is reduced to finding a canonical representative algorithm and an output sampling algorithm. We focus on search problems in which an output is a subset of the input satisfying some property. We reduce the design of a canonical representative algorithm into deciding whether an input element is contained in some possible output.

Definition 10 (Monotone Search Problem). *Let $\mathcal{P}$ be a search problem and view the inputs to $\mathcal{P}_n$ as subsets of $[n]$. We say that $\mathcal{P}$ is a* monotone search problem *if there exists a set $S \subseteq 2^{[n]}$ such that $\mathcal{P}_n(X) = 2^X \cap S$ for every input*

$X \subseteq [n]$. That is, there is a global set S of solutions and the outputs of X are the solutions that are contained in X.

For example, the problem of finding a perfect matching in a bipartite graph is monotone. The global set of solution consists of all the graphs whose edges form exactly a perfect matching. For every bipartite graph G, the set of solutions on G, is the set of perfect matching graphs whose edges are contained in G.

Definition 11 (Relevant Element). *Let $\mathcal{P}$ be a subset search problem and X be an input to $\mathcal{P}_n$. We say that $i \in X$ is relevant to X if there is an output $Y \in \mathcal{P}_n(X)$ such that $i \in Y$. We denote by $R(X)$ the set of elements relevant to X.*

In the perfect matching example, an edge is relevant if it appears in some perfect matching. The following claim shows that computing $R(X)$ efficiently from X is sufficient to get a representation algorithm.

Claim 1. *Let $\mathcal{P}$ be a monotone search problem and $X, Y \subseteq [n]$ be inputs of $\mathcal{P}_n$. Then (i) $X \equiv_P R(X)$; and (ii) $X \equiv_P Y$ if and only if $R(X) = R(Y)$.*

Proof. (i) We show that X and R(X) have the same sets of solutions. Let Y be a solution to X. Every $i \in Y$ is relevant to X and thus $i \in R(X)$. Hence $Y \subseteq R(X)$ and therefore Y is a solution to $R(X)$. For the other direction let Y be a solution to $R(X)$. Obviously $R(X) \subseteq X$ and thus $Y \subseteq X$ and therefore Y is a solution to X. (ii) Assume $X \equiv_P Y$ and let $i \in R(X)$. Then $i \in Z$ where Z is a solution to X. As $X \equiv_P Y$, we get that Z is also a solution to Y and thus $i \in R(Y)$. The other direction is immediate from (i) and the transitivity of $\equiv_P$. □

3.2 Applications of the Construction

We introduce equivalence protecting algorithms for some well known search problems.

Example 1 (Perfect Matching in Bipartite Graphs). Consider the problem of finding a perfect matching in a bipartite graph $G = \langle G, E \rangle$. To decide whether an input edge $\langle u, v \rangle$ is relevant we do the following: (i) Denote by G' the graph that results from deleting u, v and all the edges adjacent to them from G. (ii) Check whether there is a perfect matching in G'. Evidently, $\langle u, v \rangle$ is relevant to G if and only if G' has a perfect matching. Hence, perfect matching has an efficient canonical representative algorithm.

As an output sampling algorithm, we use the algorithm of Jerrum et al. [18]. The algorithm samples a perfect matching of a bipartite graph from a distribution that is statistically close to uniform. Therefore, we have both a canonical representative and a output sampling algorithm for perfect matching, and thus by Theorem 1, we get that perfect matching has an efficient equivalence protecting algorithm.

Example 2 (Linear Algebra). Let n and m be positive integers, $\mathbb{F}$ be a finite field, M be an $n \times m$ matrix over $\mathbb{F}$, and $v \in \mathbb{F}^n$. Consider the problem of solving the system $My = v$. As this problem is not monotone, we need to design both the canonical representative algorithm and the output sampling algorithm. As a canonical representative algorithm simply perform the Gaussian elimination procedure on the system. Elementary linear algebra argument shows that if two systems have the same sets of solutions, then they have the same structure after performing the Gaussian elimination procedure. We now show a simple output sampling algorithm for the problem: Compute an arbitrary solution $y_0 \in \mathbb{F}^m$ satisfying $My_0 = v$. Compute $k = \text{rank}(M)$ and compute an $m \times (n - k)$ matrix K representing the kernel of the matrix M. Randomly pick a vector $r \in \mathbb{F}^{n-k}$ and output $w = y_0 + Kr$. Again, elementary linear algebra argument shows that w is a random solution to the system $My = v$.

Example 3 (Shortest Path). Consider the problem of finding a shortest path from a vertex s to a vertex t in a directed graph G. In this case there is no global set of solutions, since a path can be an appropriate solution for one graph, while in another graph there may be shorter paths. However, the set of edges that appear in any shortest path in G still form an appropriate solution for the canonical representative algorithm. Checking whether an edge is relevant for G is an easy tasks. To sample a random solution do: (i) Compute for every $v \in V$ the number of shortest paths from v to t. (ii) Starting from s, pick the vertices on the path randomly, where the probabilities are weighted according to the number of paths computed in (i). Hence, by Theorem 1, shortest path has an efficient equivalence protecting algorithm.

Similar ideas are applicable for finding shortest path in a weighted directed graph. Here, however, we do not apply Theorem 1 directly. The equivalence protecting algorithm in this case does the following: (i) Compute the set of edges that appear in at least one shortest path from s to t. (ii) Output a *random* path from s to t in the non-weighted graph computed in (i) (not a shortest path!). The randomness for step (ii) should be extracted like in Theorem 1, by applying a pseudorandom function on the graph computed in stage (i). This example is different in the fact that the canonical input we use in step (ii) is an instance to a problem that is slightly different than the original problem.

4   Resemblance Preserving Algorithms

We now strengthen the requirement on private algorithm in an alternative manner to the definition of equivalence protecting algorithms presented in Section 3. The motivation for the definition in this section is that we want the output of the algorithm will not distinguish between inputs with similar sets of solutions. While this requirement is met by a randomized algorithm that outputs a uniform solution, it cannot be satisfied by a deterministic algorithm for non-trivial search problems (the algorithm would have to output the same "solution" for all inputs contradicting the correctness of the algorithm). As we want an algorithm that does not leak more information on repeated executions, we put forward a definition of resemblance preserving algorithms, which are seeded algorithms that protect inputs with similar sets of outputs.

To measure the similarity between the sets of outputs we use resemblance between sets, a notion used in [5,7,6] and seems to capture well the informal notion of "roughly the same." For example, in [5,7] resemblance between documents was successfully used for clustering documents.

Definition 12 (Resemblance). *Let U be a set, and $A, B \subseteq U$. Then the resemblance between A and B is defined to be*

$$r(A, B) = \frac{|A \cap B|}{|A \cup B|} .$$

For a search problem $\mathcal{P}$ we will consider the resemblance between solution sets of $\mathcal{P}_n(x), \mathcal{P}_n(y)$ of $x, y \in \{0,1\}^n$. Informally, a (perfect) resemblance preserving algorithm is a seeded algorithm that returns the same output for x and y with probability of at least the resemblance between $\mathcal{P}_n(x), \mathcal{P}_n(y)$.

Definition 13 (Resemblance Preserving Algorithm). *An algorithm $\mathcal{A}(\cdot, \cdot)$ is resemblance preserving with respect to $\mathcal{P}$ if:*

1. *For every polynomial-time algorithm $\mathcal{D}$ and every polynomial $p(\cdot)$ there exists some $n_0 \in \mathbb{N}$ such that for every $x \in \{0,1\}^*$ satisfying $|x| > n_0$*

$$\left| \Pr[\mathcal{D}(x, \mathcal{A}(s_n, x)) = 1] - \Pr[\mathcal{D}(x, \mathbf{Unif}(\mathcal{P}_n(x))) = 1] \right| \leq \frac{1}{p(|x|)} .$$

   The probability is taken over the random choice of the seed s_n and the randomness of $\mathcal{D}$. Informally, taking the probability over the seed, the outputs of $\mathcal{A}_\mathcal{P}$ on x is indistinguishable from the uniform distribution on $\mathcal{P}_n(x)$.
2. *There exists a constant $c > 0$ such that for all $x, y \in \{0,1\}^*$ such that $|x| = |y|$*
$$\Pr[\mathcal{A}(s_n, x) = \mathcal{A}(s_n, y)] \geq c \cdot r(\mathcal{P}_n(x), \mathcal{P}_n(y)) .$$

   The probability is taken over the random choice of s_n. That is, the probability that $\mathcal{A}$ returns the same output on two inputs is at least some constant times the resemblance between $\mathcal{P}_n(x)$ and $\mathcal{P}_n(y)$.
3. *If $x \equiv_\mathcal{P} y$ then $\mathcal{A}(s_n, x) = \mathcal{A}(s_n, y)$ for all seeds s_n. That is, if x and y are equivalent then $\mathcal{A}$ always returns the same output on x and on y.*

If $c = 1$ in the above Requirement 2, then $\mathcal{A}(\cdot, \cdot)$ is perfect *resemblance preserving with respect to $\mathcal{P}$.*

Unlike Definition 9, in the definition of resemblance preserving algorithms we do not know how to formulate this privacy using an "ideal world". This difference implies, in particular, that in designing resemblance preserving algorithms we do not need cryptographic assumptions. In our constructions, for example, we only use pairwise independent permutations. Furthermore, Definition 13 does not prevent partial disclosure, or even full disclosure of the seed by the algorithm. This should be considered when using a resemblance preserving algorithm.

Example 4 (Non Roots of a Polynomial). We give an example demonstrating that perfect resemblance preserving algorithms exist. Consider the following problem. The inputs are univariate polynomials of degree $d(n)$ over $\mathbb{F}_{2^n}$, where $d : \mathbb{N} \to \mathbb{N}$ is some fixed increasing function (e.g., $d(n) = n$). The set of solutions of a polynomial Q is the set of all points y which *are not* roots of Q, that is, $\{y \in \mathbb{F}_{2^n} : Q(y) \neq 0\}$. This problem arises, e.g., when we want to find points in which two polynomials disagree.

The seed s_n in the algorithm we construct is a random string of length $(d(n) + 1) \cdot n$ considered as a list of $d(n) + 1$ elements in $\mathbb{F}_{2^n}$. As Q has at most $d(n)$ roots, there is an element in the list s_n that is not a root of Q. The algorithm on input Q returns the first element in s_n that is not a root of Q. We claim that this algorithm is resemblance preserving. First, as the seed is chosen at random, the first element in the list that is not a root is a random non-root of Q. Second, consider two polynomials Q_1 and Q_2 with sets of non-roots Y_1 and Y_2 respectively. The algorithm returns the same non-root on both Q_1 and Q_2 if the first element in the list s_n from Y_1 is also the first element in the list s_n from Y_2. In other words, the algorithm returns the same non-root if the first element in the list s_n which is from the set $Y_1 \cup Y_2$ is from $Y_1 \cap Y_2$. The probability of this event is exactly $r(Y_1, Y_2) = |Y_1 \cap Y_2|/|Y_1 \cup Y_2|$.

4.1 Generic Constructions of Resemblance Preserving Algorithms

We present our main tool for constructing resemblance preserving algorithms – min-wise independent permutations. We will first show a general construction, that (depending on the search problem) may exhibit exponential time complexity. Then, we will present the main contribution of this section – a polynomial-time resemblance preserving algorithm that is applicable for problems for which there is a pairwise independent family of permutations where we can compute the minimum on any set of solutions.

Definition 14 (Family of Min-wise Independent Permutations [6]). *Let U be a set and $\mathcal{F} = \{\pi_s\}_{s \in S}$ be a collection of permutations $\pi_s : U \to U$. The collection $\mathcal{F}$ is a collection of min-wise independent permutations if* $\Pr[\min (\pi_s(A)) = \pi_s(a)] = 1/|A|$ *for all $A \subseteq U$ and all $a \in A$. The probability is taken over the choice of the seed s at uniform from S.*

We will use the following observation that relates min-wise permutations and resemblance:

Observation 1 ([6]). Let $\mathcal{F}$ be a family of min-wise independent permutations $\{\pi_s\}_{s \in S}$ where $\pi_s : U \to U$. Then $\Pr[\min(\pi_s(A)) = \min(\pi_s(B))] = r(A, B)$ for every sets $A, B \subseteq U$. The probability is taken over the choice of the seed s at uniform from S.

In Fig. 1, we describe Algorithm $\mathtt{Minwise}_{\mathcal{P}}$ for a search problem $\mathcal{P}$, where $\mathcal{P}_n : \{0,1\}^n \to \{0,1\}^{q(n)}$. Using Obseration 1 it is easy to see that Algorithm $\mathtt{Minwise}_{\mathcal{P}}$ is perfectly resemblance preserving.

However, Algorithm $\mathtt{Minwise}_{\mathcal{P}}$ maybe inefficient in several aspects:

1. Algorithm Minwise$_\mathcal{P}$ uses a family of min-wise independent permutations. It was shown in [6] that such families are of size $2^{\Omega(|U|)} = 2^{\Omega(2^{q(n)})}$ (where n is the input length), and hence the seed length $|s| = \Omega(2^{q(n)})$. However, for most purposes, the seed length may be reduced to polynomial by using pseudorandom permutations.[6]

2. Algorithm Minwise$_\mathcal{P}$ needs to compute the minimum element, according to π_s in the solution set $\mathcal{P}_n(x)$. This is feasible when it is possible to enumerate in polynomial time the elements of $\mathcal{P}_n(x)$. However, to make Minwise$_\mathcal{P}$ feasible in cases where, for example, $\mathcal{P}_n(x)$ is of super-polynomial size, one needs to carefully use the structure of π_s and the structure of the underlying solution set space.

Algorithm Minwise$_\mathcal{P}$

INPUT: An instance $x \in \{0,1\}^*$, seed s for a family of min-wise independent permutations $\{\pi_s\}_{s \in S}$ where $\pi_s : \{0,1\}^{q(|x|)} \to \{0,1\}^{q(|x|)}$.
OUTPUT: A solution sol $\in \mathcal{P}_n(x)$.

1. Let $A = \mathcal{P}_n(x)$.
2. Output sol $\in A$ such that $\pi_s(\text{sol}) = \min \pi_s(A)$.

Fig. 1. Algorithm Minwise$_\mathcal{P}$

Example 5 (Roots of a Polynomial). As an example for when Algorithm Minwise$_\mathcal{P}$ can be implemented efficiently we consider the problem of finding roots of a polynomial. As in Example 4, the inputs are univariate polynomials of degree $d(n)$ over $\mathbb{F}_{2^n}$, where $d : \mathbb{N} \to \mathbb{N}$ is some fixed increasing function (e.g., $d(n) = n$). The set of solutions of a polynomial Q is the set of all points y which *are* roots of Q, that is, $\{y \in \mathbb{F}_{2^n} : Q(y) = 0\}$. Berlekamp [4] presented an efficient algorithm that finds roots of a polynomial over $\mathbb{F}_{2^n}$. We implement Algorithm Minwise$_\mathcal{P}$, where we use a family of pseudorandom permutations from $\mathbb{F}_{2^n}$ to $\mathbb{F}_{2^n}$ instead of the family of min-wise independent permutations. Furthermore, as the number of roots of a polynomial of degree $d(n)$ is at most $d(n)$, we can use Berlekamp's algorithm to explicitly find all roots of the polynomial, apply the pseudorandom permutation to each roots, and find for which root $\pi_s(y)$ obtains a minimum. The above algorithm can be generalized to any search problems whose entire set of solutions can be generated efficiently.

Observation 2. If for a search problem $\mathcal{P}$ there is an algorithm that generates the set of solutions of an input of $\mathcal{P}$ whose running time in polynomial in the length of the input (and, in particular, the number of solutions in polynomial), then Algorithm Minwise$_\mathcal{P}$ can be efficiently implemented for $\mathcal{P}$.

---

[6] We need the family of pseudorandom permutations to be secure against a non-uniform adversary. Thus, for every long enough inputs x and y a pseudorandom permutation must be min-wise. We omit further details as this is not the approach taken in this study.

4.2   Resemblance Preserving Using Pairwise Independence

To get around the above mentioned problems of implementing Minwise$_\mathcal{P}$ for search problems with super-polynomial number of solutions, we construct a non-perfect resemblance preserving algorithm using pairwise independence permutations instead of min-wise independence.

Definition 15 (Family of Pairwise Independent Permutations). *Let U be a set and $\mathcal{F} = \{\pi_s\}_{s \in S}$ be a collection of permutations $\pi_s : U \to U$. The collection $\mathcal{F}$ is a family of pairwise independent permutations if*

$$\Pr[\pi_s(a) = c \ \wedge \ \pi_s(b) = d] = \frac{1}{|U|(|U| - 1)} \ .$$

for all $a, b \in U$ and $c, d \in U$. The probability is taken over the choice of the seed s at uniform from S.

Theorem 2 ([6]). *Let $\mathcal{F}$ be a family of pairwise independent permutations $\{\pi_s\}_{s \in S}$ where $\pi_s : U \to U$. Then for every set $A \subseteq U$ and every $a \in A$*

$$\frac{1}{2(|A| - 1)} \ \leq \ \Pr[\min(\pi_s(A)) = \pi_s(a)] \ \leq \ \frac{2}{\sqrt{|A| - 1}} \ .$$

The probability is taken over the choice of the seed s at uniform from S.

Lemma 1. *Let $\mathcal{F}$ be a family of pairwise independent permutations $\{\pi_s\}_{s \in S}$ where $\pi_s : U \to U$. Then for every sets $A, B \subseteq U$*

$$\Pr[\min(\pi_s(A)) = \min(\pi_s(B))] \geq \max\left(\frac{r(A, B)}{2}, 1 - \frac{2 \cdot |A \Delta B|}{\sqrt{|A \cup B| - 1}}\right) \ .$$

The probability is taken over the choice of the seed s at uniform from S.

We construct an algorithm Pairwise$_\mathcal{P}$ that is almost identical to Minwise$_\mathcal{P}$ of Fig. 1, where the family of min-wise permutations is replaced with a family of pairwise independent permutations. The following corollary follows directly from Lemma 1:

Corollary 1. *Algorithm Pairwise$_\mathcal{P}$ is resemblance preserving.*

4.3   Applications of the Pairwise Independence Construction

We next show how to apply Algorithm Pairwise$_\mathcal{P}$ to a few search problems. Given a search problem, we need to choose the family of pairwise independent permutations such that the solution minimizing $\pi_s(A)$ can computed efficiently. In our examples we use the following well-known family of pairwise independent permutations from $\mathbb{F}_q^n$ to $\mathbb{F}_q^n$ for some prime-power q:

$$\mathcal{L}_{q,n} \stackrel{\text{def}}{=} \left\{ Hy + b \ : \ H \text{ is an invertible } n \times n \text{ matrix over } \mathbb{F}_q \text{ and } b \in \mathbb{F}_q^n \right\}.$$

Linear Algebra. We show how to construct a resemblance preserving algorithm for finding a solution of a system of equations (as considered in Example 2 for Equivalence Protecting Algorithms).

Linear Algebra over $\mathbb{F}_2$. We assume that the system is over $\mathbb{F}_2$.[7] That is, the input is an $m \times n$ matrix M over $\mathbb{F}_2$ and a vector $v \in \mathbb{F}_2^m$, and a solution is a vector $y \in \mathbb{F}_2^n$ such that $My = v$. We apply Algorithm Pairwise$_\mathcal{P}$ for this problem using the family $\mathcal{L}_{2,n}$. That is, we choose a permutation at random, specified by H and b, and we need to find the lexicographically first z satisfying $z = Hy + b$ for y satisfying $Ay = b$. We view $Ay = b$ and $z = Hy + v$ as a single system of linear equations with $2n$ unknowns, namely, $y = \langle y_1, \ldots, y_n \rangle$ and $z = \langle z_1, \ldots, z_n \rangle$. To find the value of z_1 in the lexicographically first z, we add the equation $z_1 = 0$ to the system of equations. If the new system has a solution, we keep the equation $z_1 = 0$ in the system and continue to find the value of z_2. Otherwise, we understand that $z_1 = 1$ in every solution of the original system of equations, and, in particular, in the lexicographically first z. In this case, we remove the equation $z_1 = 0$ from the system of equations and continue to find the value of z_2. To conclude, we find the lexicographically first z iteratively, where in iteration i we have already found the values of $z_1, \ldots, z_{i-1}$ and we compute the value of z_i in the lexicographically first z as we found z_1. We continue these iterations until we find the lexicographically first z. Recall that $Hy + b$ is a permutation. Thus, once we found z, the solution y is uniquely defined and is easy to compute from the system of equations.

Union of Systems of Equations. We want to use the resemblance preserving algorithm for finding a solution of a system of linear equations to construct resemblance preserving algorithms for other problems. That is, we want to represent the set of solutions of an instance of some search problem as a set of solutions to a system of linear equations. In our applications, we manage to represent the set of solutions of an instance as a *union* of polynomially many systems of linear equations over the same field. We next show how to construct a resemblance preserving algorithm for such a union. That is, the input is a sequence $M_1, v_1, \ldots, M_\ell, v_\ell$ and a solution is a vector y such that $M_i y = v_i$ for at least one i.

Algorithm LinearAlgebraUnion

INPUT: A a sequence $M_1, v_1, \ldots, M_\ell, v_\ell$ and a seed H, b.
OUTPUT: A vector y such that $M_i y = v_i$ for at least one i.

1. Find for each system of equation a solution y_i such that $Hy_i + b$ is minimized amongst all vectors such that $M_i y = v_i$.
2. Output y_j such that $Hy_j + b = \min \{ Hy_i + b : 1 \le i \le \ell \}$.

---

[7] In the full version of this paper we generalize the result to every finite field.

Theorem 3. *There is a resemblance preserving algorithm for finding a solution in a union of polynomially many solution sets of systems of linear equations over the same field.*

Points in a Union of Discrete Rectangles. We show how to use the resemblance preserving algorithm for linear algebra to construct resemblance preserving algorithms for finding a point in a union of discrete rectangles. We construct such algorithms for two cases: (1) unions of rectangles in $[2]^n$, that is, DNF formulae, and (2) unions of rectangles in $[N]^d$ when d is fixed (however, N is not fixed).

Satisfying Assignment for a DNF Formula. We show how to construct a resemblance preserving algorithm for finding a satisfying assignment of a DNF formula. This follows Theorem 3 and the following observations. First, the set of satisfying assignments of a single term is the set of solutions to a system of linear equations over $\mathbb{F}_2$:

- For every variable x_i that appears in the term without negation, add the equation $y_i = 1$.
- For every variable x_i that appears in the term with negation, add the equation $y_i = 0$.

Now, given a DNF formula with ℓ terms, a satisfying assignment to the formula is a assignment satisfying at least one of the terms in the formula, that is, it belongs to the union of solutions of the ℓ systems of linear equations constructed for each of the terms of the formula. Thus, by Theorem 3, we get a resemblance preserving algorithm for finding a satisfying assignment of a DNF formula.

It is interesting to note that, unless P=NP, there is no efficient equivalence protecting algorithm for DNF as an equivalence protecting algorithm for DNF can be used to check if two DNF formulae are equivalent, a problem that is coNP-hard.

Points in a Union of Discrete Rectangles in a d-dimensional Space. We show how to construct a resemblance preserving algorithm for finding a point in a discrete rectangle. That is, for some fixed $d \in \mathbb{N}$ and for an integer $N \in \mathbb{N}$, our inputs are $2d$ elements $a^1, \ldots, a^d, b^1, \ldots, b^d \in [N]$ which represents a rectangle as follows: First, for two points a, b we define the segment $I_{a,b} \stackrel{\text{def}}{=} \{y \in \mathbb{N} : a \le y \le b\}$. Second, we define $R_{a^1, \ldots, a^d, b^1, \ldots, b^d} \stackrel{\text{def}}{=} I_{a^1, b^1} \times I_{a^2, b^2} \times \cdots \times I_{a^d, b^d}$. Let $n \stackrel{\text{def}}{=} \lceil \log N \rceil$, and we represent a number $a \in [N]$ by an n-bit string $a_1, \ldots, a_n$, where $a = \sum_{i=1}^{n} a_i 2^{n-i}$. Note that, in this section, a^i is a string in $\{0,1\}^n$ and a_i is the ith bit of a string a.

We solve the problem of finding a point in a rectangle by representing each rectangle as a union of polynomially many systems of equations over $\mathbb{F}_2$, and then use Theorem 3 to construct the resemblance preserving algorithm.

Let us start with the simple case where $d = 1$ and $b^1 = \langle 1, \ldots, 1 \rangle$ (in words, $b \in \{0,1\}^n$ is the all 1 string). That is, an input is a string a and a solution is a string $y \ge a$.

$$y \ge a \text{ iff } \left(\exists_{i \in [n]} \left(y_i = 1 \land a_i = 0 \right) \land \left(\forall_{1 \le j < i} \ y_i = a_i \right) \right) \ \lor \ \left(\forall_{1 \le j \le n} \left(y_i = a_i \right) \right). (1)$$

For example, $y = \langle y_1, y_2, y_3 \rangle \geq \langle 0, 1, 0 \rangle$ either if $(y_1 = 1)$, or $(y_1 = 0 \wedge y_2 = 1 \wedge y_3 = 1)$, or $(y_1 = 0 \wedge y_2 = 1 \wedge y_3 = 0)$.

Note that, by (1), the set of points $y \geq a$ is a union of solutions of at most $n + 1$ systems of equations. Similarly, the set of points $a \leq y \leq b$ is a union of solutions of at most $2(n + 1)$ systems of equations: Let $a < b$ and i_0 be the minimal index such that $a_i = 0$ and $b_i = 1$ (in particular, $a_j = b_j$ for every $1 \leq j \leq i_0 - 1$).

$$a \leq y \leq b \text{ iff } (\forall_{1 \leq j < i_0} y_j = a_j) \wedge ((y_{i_0} = a_{i_0} \wedge a \leq y) \vee (y_{i_0} = b_{i_0} \wedge y \leq b)). \quad (2)$$

In other words, we partitioned the segment $I_{a,b}$ to at most $2(n+1)$ segments such that the points in each segment are exactly the solutions of a system of linear equations.

Given a rectangle in $(\{0, 1\}^n)^d$, we partition it to $(O(n))^d$ rectangles such that the points in each rectangle correspond to solutions of a system of linear equations, and use Theorem 3 to construct the resemblance preserving algorithm. Notice that given a rectangle $R_{a^1, \ldots, a^d, b^1, \ldots, b^d}$, we can partition each segment I_{a_i, b_i} into $O(n)$ segments $I_{i,1}, \ldots, I_{i,O(n)}$ as in (1) and (2). Thus,

$$\begin{aligned} R_{a^1, \ldots, a^d, b^1, \ldots, b^d} &= I_{a_1, b_1} \times I_{a_2, b_2} \times \cdots \times I_{a_d, b_d} \\ &= (\cup_{j_1} I_{1,j_1}) \times (\cup_{j_2} I_{2,j_2}) \times \cdots \times (\cup_{j_d} I_{d,j_d}) \\ &= \cup_{j_1, \ldots, j_d} I_{1,j_1} \times I_{2,j_2} \times \cdots \times I_{d,j_d}. \end{aligned}$$

Notice that for $i_1 \neq i_2$, the variables of the equations representing $I_{i_1, j_{i_1}}$ and $I_{i_2, j_{i_2}}$ are disjoint, and the points in each rectangle $I_{1,j_1} \times I_{2,j_2} \times \cdots \times I_{d,j_d}$ are solutions to a system of linear equations.

Finally, if our input is a union of ℓ rectangles, we can represent it as a union of $\ell(O(n))^d$ systems of equations, hence:

Theorem 4. *There exists an efficient resemblance preserving algorithm for finding a point in a union of ℓ rectangles in $[N]^d$. The running time of the algorithm is $\text{poly}((\log N)^d, \ell)$.*

The above algorithm is polynomial in ℓ and $(\log N)^d$ while the size of the input is $O(\ell d \log N)$, thus, it is polynomial when d is constant. It would be interesting to construct an efficient algorithm for non-constant d. Notice that a union of ℓ rectangles in $[2]^d$ is equivalent to an ℓ-term DNF formula with n variables. Thus, there is a polynomial resemblance preserving algorithm for union of rectangles in $[2]^d$.

Acknowledgments. We thank the anonymous CYRPTO referees for their useful comments. Part of this research was performed when the authors visited IPAM at UCLA. We thank Rafi Ostrovsky and the IPAM staff for inviting us to IPAM and making our stay pleasant and productive.

References

1. Beimel, A., Carmi, P., Nissim, K., Weinreb, E.: Private approximation of search problems. In: Proc. of the 38th Symp. on the Theory of Comp. pp. 119–128 (2006)
2. Beimel, A., Hallak, R., Nissim, K.: Private approximation of clustering and vertex cover. In: Vadhan, S.P. (ed.) TCC 2007. LNCS, vol. 4392, pp. 383–403. Springer, Heidelberg (2007)
3. Ben-Or, M., Goldwasser, S., Wigderson, A.: Completeness theorems for noncryptographic fault-tolerant distributed computations. In: Proc. of the 20th Symp. on the Theory of Comp. pp. 1–10 (1988)
4. Berlekamp, E.R: Factoring polynomials over large finite fields. Math. Comp. 24, 713–735 (1970)
5. Broder, A.Z.: On the resemblance and containment of documents. In: Compression and Complexity of Sequences 1997, pp. 21–29 (1997)
6. Broder, A.Z., Charikar, M., Frieze, A.M., Mitzenmacher, M.: Min-wise independent permutations. J. of Computer and System Sciences 60(3), 630–659 (2000)
7. Broder, A.Z., Glassman, S.C., Manasse, M.S., Zweig, G.: Syntactic clustering of the web. In: Proc. of World Wide Web conference, pp. 1157–1166 (1997)
8. Chaum, D., Crépeau, C., Damgård, I.: Multiparty unconditionally secure protocols. In: Proc. of the 20th Symp. on the Theory of Comp. pp. 11–19 (1988)
9. Dias, V.M.F., da Fonseca, G.D., de Figueiredo, C.M.H., Szwarcfiter, J.L.: The stable marriage problem with restricted pairs. Theoretical Computer Science 306(1–3), 391–405 (2003)
10. Feigenbaum, J., Ishai, Y., Malkin, T., Nissim, K., Strauss, M.J., Wright, R.N.: Secure multiparty computation of approximations. In: Orejas, F., Spirakis, P.G., van Leeuwen, J. (eds.) ICALP 2001. LNCS, vol. 2076, pp. 927–938. Springer, Heidelberg (2001)
11. Franklin, M., Gondree, M., Mohassel, P.: Improved efficiency for private stable matching. In: Abe, M. (ed.) CT-RSA 2007. LNCS, vol. 4377, pp. 163–177. Springer, Heidelberg (2006)
12. Gale, D., Shapley, L.S.: College admissions and the stability of marriage. American Mathematical Monthly 69, 9–15 (1962)
13. Goldreich, O., Goldwasser, S., Micali, S.: How to construct random functions. J. of the ACM 33(4), 792–807 (1986)
14. Goldreich, O., Micali, S., Wigderson, A.: How to play any mental game. In: Proc. of the 19th Symp. on the Theory of Comp. pp. 218–229 (1987)
15. Golle, P.: A private stable matching algorithm. In: Di Crescenzo, G., Rubin, A. (eds.) FC 2006. LNCS, vol. 4107, pp. 65–80. Springer, Heidelberg (2006)
16. Halevi, S., Krauthgamer, R., Kushilevitz, E., Nissim, K.: Private approximation of NP-hard functions. In: Proc. of the 33th Symp. on the Theory of Comp. pp. 550–559 (2001)
17. Indyk, P., Woodruff, D.: Polylogarithmic private approximations and efficient matching. In: Halevi, S., Rabin, T. (eds.) TCC 2006. LNCS, vol. 3876, pp. 245–264. Springer, Heidelberg (2006)
18. Jerrum, M., Sinclair, A., Vigoda, E.: A polynomial-time approximation algorithm for the permanent of a matrix with nonnegative entries. J. of the ACM 51(4), 671–697 (2004)
19. Yao, A.C.: Protocols for secure computations. In: Proc. of the 23th IEEE Symp. on Foundations of Computer Science, pp. 160–164. IEEE Computer Society Press, Los Alamitos (1982)

A   Deterministic vs. Randomized Private Algorithms

We start with a search problem that admits a randomized private algorithm (outputting a uniformly chosen solution on each instance), but no efficient deterministic one. For $n = p \cdot q$ and $a \in Z_n^*$ with Jacobi Symbol $(\frac{a}{n}) = 1$ define $\mathcal{QR}(n, a) = \{b \in Z_n^* : (\frac{b}{n}) = 1 \ \wedge \ b \in QR_n \Leftrightarrow a \in QR_n\}$.

Claim 2. *The problem* $\mathcal{QR}$ *admits a randomized polynomial time private algorithm, but no efficient deterministic private algorithms, unless quadratic residuosity is decidable in deterministic polynomial time.*

Our second example is of a search problem that admits a deterministic private algorithm but no (non trivial) randomized one.

For a CNF formula ϕ over Boolean variables $x_1, \dots, x_n$ define

$$\mathcal{ZERO} - \mathcal{SAT}(\phi) = \{a \in \{0,1\}^n : a = 0^n \ \vee \ \phi(a)\} \, .$$

If a randomized algorithm for $\mathcal{ZERO} - \mathcal{SAT}$ assigns non-negligible probability to some non-zero assignment whenever ϕ is satisfiable we say it is *non-trivial*.

Claim 3. *The problem* $\mathcal{ZERO} - \mathcal{SAT}$ *admits a deterministic polynomial time private algorithm, but, unless* $NP \subseteq RP$ *no non-trivial randomized private algorithm for* $\mathcal{ZERO} - \mathcal{SAT}$ *exists.*

Public Key Encryption That Allows PIR Queries

Dan Boneh*, Eyal Kushilevitz**,
Rafail Ostrovsky***, and William E. Skeith III†

dabo@cs.stanford.edu, eyalk@cs.technion.ac.il,
rafail@cs.ucla.edu, wskeith@math.ucla.edu

Abstract. Consider the following problem: Alice wishes to maintain her email using a storage-provider Bob (such as a Yahoo! or hotmail e-mail account). This storage-provider should provide for Alice the ability to collect, retrieve, search and delete emails but, at the same time, should learn neither the content of messages sent from the senders to Alice (with Bob as an intermediary), nor the search criteria used by Alice. A trivial solution is that messages will be sent to Bob in encrypted form and Alice, whenever she wants to search for some message, will ask Bob to send her a copy of the entire database of encrypted emails. This however is highly inefficient. We will be interested in solutions that are communication-efficient and, at the same time, respect the privacy of Alice. In this paper, we show how to create a public-key encryption scheme for Alice that allows PIR searching over encrypted documents. Our solution is the first to reveal no partial information regarding the user's search (including the access pattern) in the public-key setting and with non-trivially small communication complexity. This provides a theoretical solution to a problem posed by Boneh, DiCrescenzo, Ostrovsky and Persiano on "Public-key Encryption with Keyword Search." The main technique of our solution also allows for Single-Database PIR writing with sub-linear communication complexity, which we consider of independent interest.

keywords: Searching on encrypted data, Database security, Public-key Encryption with special properties, Private Information Retrieval.

1 Introduction

Problem Overview. Consider the following problem: Alice wishes to maintain her email using a storage-provider Bob (such as Yahoo! or hotmail e-mail account). She

* Stanford Department of Computer Science. Supported by NSF and the Packard foundation.
** Department of Computer Science, Technion. Partially supported by BSF grant 2002-354 and by Israel Science Foundation grant 36/03.
*** Computer Science Department and Department of Mathematics, University of California, Los Angeles, CA 90095. Research partially done while visiting IPAM, and supported in part by IBM Faculty Award, Xerox Innovation Group Award, NSF Cybertrust grant no. 0430254, and U.C. MICRO grant.
† Department of Mathematics, University of California, Los Angeles. Research done in part at IPAM, and supported in part by U.C. Chancellor's Presidential Dissertation Fellowship 2006-2007.

A. Menezes (Ed.): CRYPTO 2007, LNCS 4622, pp. 50–67, 2007.

publishes a public-key for a semantically-secure Public-Key Encryption scheme, and asks all people to send their e-mails, encrypted under her public-key, to the intermediary Bob. Bob (the storage-provider) should allow Alice to collect, retrieve, search and delete emails at her leisure. In known implementations of such services, either the content of the emails is known to the storage-provider Bob (and then the privacy of both Alice and the senders is lost) or the senders can encrypt their messages to Alice, in which case privacy is maintained, but sophisticated services (such as search by keyword) cannot be easily performed and, more importantly leak information, such as Alice's access pattern, to Bob. Of course, Alice can always ask Bob, the storage-provider, to send her a copy of the entire database of emails. This however is highly inefficient in terms of communication, which will be a main focus in this work. In all that follows, we will denote the number of encrypted documents that Bob stores for Alice by the variable n.

We will be interested in solutions that are communication-efficient and, at the same time, respect the complete privacy of Alice. A seemingly related concept is that of *Private Information Retrieval (PIR)* (e.g., [13,23,10], or [27] for a survey). However, existing PIR solutions either allow only for retrieving a (plain or encrypted) record of the database by address, or allow for search by keyword [12,23,25] in a non-encrypted data. The challenge of creating a public-key encryption that allows for keyword search, where keywords are encrypted in a probabilistic manner, remained an open problem prior to this paper.

In our solution, Alice creates a public key that allows arbitrary senders to send her encrypted e-mail messages. Each such message M is accompanied by an "encoded" list of keywords in response to which M should be retrieved. These email messages are collected for Alice by Bob, along with the "encoded" keywords. When Alice wishes to search in the database maintained by Bob for e-mail messages containing certain keywords, she is able to do so in a communication-efficient way and does not allow Bob to learn *anything* about the messages that she wishes to read, download or erase. In particular, Alice is not willing to reveal what particular messages she downloads from the mail database, from which senders these emails are originating and/or what is the search criterion, including the access pattern.

Furthermore, our solution allows the communication from any sender to Bob to be *non-interactive* (i.e. just a single message from the sender to Bob), and allow a single round of communication from Alice to Bob and back to Alice, with total communication complexity sub-linear in n. Furthermore, we show a simple extension that allows honest-but-curious Bob to tolerate malicious senders, who try to corrupt messages that do not belong to them in Bob's database, and reject all such messages with overwhelming probability.

Comparison with Related Work. Recently, there was a lot of work on *searching on encrypted data* (see [7,6] and references therein). However, all previous solutions either revealed some partial information about the data or about the search criterion, or work only in *private-key* settings. In such settings, only entities who have access to the private key can do useful operations; thus, it is inappropriate for our setting, where both the storage-provider and the senders of e-mail messages for Alice have no information on her private key. We emphasize that, in settings that include only a user Alice and a storage-provider, the problem is already solved; for example, one can apply results of [17,29,9,7]. However, the involvement of the senders who are also allowed to *encrypt* data for Alice (but are not allowed to decrypt data encrypted by other senders) requires using public-key encryption. In contrast to the above work, we show how to search, in a communication-efficient manner, on encrypted data in a *public-key setting*, where those who store data (encrypted with a public key of Alice) do not need to know the private key under which this data is encrypted. The only previous results for such a scenario in the public-key setting, is due to Boneh et al. [6] and Abddalla et al. [1] who deal with the same storage-provider setting we describe above; however, their solution *reveals* partial information; namely, the particular keyword that Alice is searching for is given by her, in the clear, to Bob (i.e., only the content of the email messages is kept private while the information that Alice is after is revealed). This, in particular, reveals the *access pattern* of the user. The biggest problem left was creating a scheme that hides the access pattern as well. This is exactly what we achieve in this paper. That is, we show how to hide *all* information in a semantically-secure way.

As mentioned, private information retrieval (PIR) is a related problem that is concerned with communication-efficient retrieval of *public* (i.e., plain) data. Extensions of the basic PIR primitive (such as [12,23], mentioned above, and, more recently, [22,15,25]) allow more powerful keyword search *un-encrypted* data. Therefore, none of those can directly be used to solve the current problem.

It should also be noted that our paper is in some ways only a partial solution to the problem. Specifically, we put the following constraint in our model: the number of total messages associated to each keyword is bounded by a constant. It is an interesting question as to whether this condition can be relaxed, while keeping communication non-trivially small and maintaining the strict notions of security presented here.

Our Techniques. We give a short overview of some of the tools that we use. The right combination of these tools is what allows for our protocol to work.

As a starting point, we examine *Bloom filters* (see Section 2.1 for a definition). Bloom filters allow us to use space which is not proportional to the number of all potential keywords (which is typically huge) but rather to the maximal number of keywords which are in use at any given time (which is typically much smaller). That is, the general approach of our protocols is that the senders will store in the database of the storage-provider some extra information (in encrypted form) that will later allow the efficient search by Alice. *Bloom filters* allow us to keep the space that is used to store this extra information "small". The approach is somewhat similar to Goh's use of Bloom filters [16]; the important difference is that in our case we are looking for a public-key solution, whereas Goh [16] gives a private-key solution. This makes our problem more challenging, and our use of Bloom filter is somewhat different. Furthermore, we require the Bloom filters in our application to encode significantly more information than just set membership. We modify the standard definitions of Bloom filters to accommodate the additional functionality.

Recall that the use of Bloom filters requires the ability to flip bits in the array of extra information. However, the identity of the positions that are flipped should be kept secret from the storage-provider (as they give information about the keywords). This brings us to an important technical challenge: we need a way to specify an encrypted length-n unit vector e_i (i.e., a length n vector with 1 in its i-th position and 0's elsewhere) while keeping the value i secret, and having a representation that is short enough to get communication-efficiency beyond that of the trivial solution. We show that a recent public-key homomorphic-encryption scheme, due to Boneh, Goh and Nissim [5], that supports additions and one multiplication on ciphertexts, allows us to obtain just that. For example, one can specify such a length-n unit vector using communication complexity which is $\sqrt{n}$ times a security parameter. Also, as shown in [26], this is optimal, from an algebraic point of view.

Finally, for Alice to read information from the array of extra information, she applies efficient PIR schemes, e.g. [23,10], that, again, allow keeping the keywords that Alice is after secret.

We emphasize that the communication in the protocol is sub-linear in n. This includes both the communication from the senders to the storage-provider Bob (when sending email messages) and the communication from Alice to Bob (when she retrieves/searches for messages). Furthermore, we allow Alice to *delete* messages from Bob's storage in a way that hides from Bob which messages have been deleted. Our main theorem is as follows:

MAIN THEOREM (informal): There exists Public-Key Encryption schemes that support sending, reading and writing into remote server (honest-but-curious Bob) with the following communication complexity:

- $\mathcal{O}(\sqrt{n}\log^3 n)$ for sending a message from any honest-but-curious Sender to Bob. In case the sender is malicious, the communication complexity for sending a message becomes $\mathcal{O}(\sqrt{n\log n} \cdot \text{polylog}(n))$
- $\mathcal{O}(\text{polylog}(n))$ for reading by Alice from Bob's (encrypted) memory.
- $\mathcal{O}(\sqrt{n}\log^3 n)$ for deleting messages by Alice from Bob's memory.

Organization: In Section 2, we explain and develop the tools needed for our solutions. Section 3 defines the properties we want our protocols to satisfy. Finally, Section 4 gives the construction and its analysis.

1.1   Reference Table of Notation

For the reader's convenience, we provide a table of the most frequently used notation in this work.

- n – size of e-mail database
- s – a security parameter
- k – number of hash functions used in Bloom filter
- m – size of Bloom filter hash table
- $\{h_i\}_{i=1}^{k}$ – Bloom filter hash functions

- H_w – set of hash images for a word $w \in \{0,1\}^*$, i.e. $\{h_i(w) \mid i \in [k]\}$
- B_j – a buffer in a Bloom filter with storage (so, $j \in [m]$)
- σ – size of fixed length buffers in a Bloom filter with storage
- l – size of the associated values in a Bloom filter with storage
- $\mathcal{X}$ – a message sender
- $\mathcal{Y}$ – message receiver (owner of public key)
- $\mathcal{S}$ – owner of remote storage (mail server)
- K – a set of keywords
- M – a message
- $(\mathcal{K}, \mathcal{E}, \mathcal{D})$ – key generation, encryption and decryption, respectively
- c – a constant, greater than 1
- λ – maximum number of messages associated to a specific keyword
- θ – maximum size of a keyword set associated to a specific message

2   Ingredients

We will make use of several basic tools, some of which are being introduced for the first time in this paper. In this section, we define (and create, if needed) these tools, as well as outline their utility in our protocol.

2.1   Bloom Filters

Bloom filters [4] provide a way to probabilistically encode set membership using a small amount of space, even when the universe set is large. The basic idea is as follows. Choose an independent set of hash functions $\{h_i\}_{i=1}^k$, where each function $h_i : \{0,1\}^* \longrightarrow [m]$. Suppose $S = \{a_i\}_{i=1}^l \subset \{0,1\}^*$. We set an array $T = \{t_i\}_{i=1}^m$ such that $t_i = 1 \iff \exists j \in [k]$ and $j' \in [l]$ such that $h_j(a_{j'}) = i$. Now to test the validity of a statement like "$a \in S$", one simply verifies that $t_{h_i(a)} = 1, \forall i \in [k]$. If this does not hold, then certainly $a \notin S$. If the statement does hold, then there is still some probability that $a \notin S$, however this can be shown to be small. Optimal results are obtained by having m proportional to k; in this case, it can be shown that the probability of an inaccurate positive result is negligible as k increases, as will be thoroughly demonstrated in what follows.

This work will use a variation of a Bloom filter, as we require more functionality. We would like our Bloom filters to not just store whether or not a certain element is in the set, but also to store some values $v \in V$ which are associated with the elements in the set (and to preserve those associations).

Definition 1. *Let V be a finite set. A (k, m)-Bloom Filter with Storage is a collection $\{h_i\}_{i=1}^k$ of functions, with $h_i : \{0,1\}^* \longrightarrow [m]$ for all i, together with a collection of sets, $\{B_j\}_{j=1}^m$, where $B_j \subseteq V$. To insert a pair (a, v) into this structure, where $a \in \{0,1\}^*$ and $v \in V$, v is added to $B_{h_i(a)}$ for all $i \in [k]$. Then, to determine whether or not $a \in S$, one examines all of the sets $B_{h_i(a)}$ and returns true if all are non-empty. The set of values associated with $a \in S$ is simply $\bigcap_{i \in [k]} B_{h_i(a)}$. (Note: every inserted value is assumed to have at least one associated value.)*

Next, we analyze the total size of a (k, m)-Bloom filter with storage. For the purpose of analysis, the functions h_i will, as usual, be modeled as uniform, independent randomness. For $w \in \{0, 1\}^*$, define $H_w = \{h_i(w) \mid i \in [k]\}$.

Claim. Let $(\{h_i\}_{i=1}^k, \{B_j\}_{j=1}^m)$ be a (k, m)-Bloom filter with storage. Suppose the filter has been initialized to store some set S of size n and associated values. Suppose also that $m = \lceil cnk \rceil$ where $c > 1$ is a constant. Denote the (binary) relation of element-value associations by $R(\cdot, \cdot)$. Then, for any $a \in \{0, 1\}^*$, the following statements hold true with probability $1 - \text{neg}(k)$, where the probability is over the uniform randomness used to model the h_i:

1. $(a \in S) \iff (B_{h_i(a)} \neq \varnothing \ \forall i \in [k])$
2. $\bigcap_{i \in [k]} B_{h_i(a)} = \{v \mid R(a, v) = 1\}$

Proof. (1., $\Rightarrow$) Certainly if $B_{h_i(a)} = \varnothing$ for some $i \in [k]$, then a was never inserted into the filter, and $a \notin S$. (1., $\Leftarrow$) Suppose that $B_{h_i(a)} \neq \varnothing$ for every $i \in [k]$. We'd like to compute the probability that for an arbitrary $a \in \{0, 1\}^*$, $a \notin S$, we have $H_a \subset \bigcup_{w \in S} H_w$; i.e., that such an element will appear to be in S by our criteria. Recall that we model each evaluation of the functions h_i as independent and uniform randomness. Therefore, a total of nk (not necessarily distinct) random sets are modified to insert the n values of S into the filter. So, we only need to compute the probability that all k functions place a in this subset of the B_j's. By assumption, there are a total of $m = \lceil cnk \rceil$ sets where $c > 1$ is a constant. Let $X_{k,k'}$ denote the random variable that models the experiment of throwing k balls into m bins and counting the number that land in the first k' bins. For a fixed insertion of the elements of S into our filter and letting k' be the number of distinct bins occupied, $X_{k,k'}$ represents how close a random element appears to being in S according to our Bloom filter. More precisely, $\Pr[X_{k,k'} = k]$ is the probability that a random element will appear to be in S for this specific situation. Note that $X_{k,k'}$ is a sum of independent (by assumption) Bernoulli trials, and hence is distributed as a binomial random variable with parameters, $(k, \frac{k'}{cnk})$, where $k' \leq nk$. Hence, $\Pr[X_{k,k'} = k] = \left(\frac{k'}{cnk}\right)^k \leq \left(\frac{1}{c}\right)^k$. So, we've obtained a bound that is negligible in k, independently of k'. Hence, if we let Y_k be the experiment of sampling k' by throwing nk balls into $\lceil cnk \rceil$ bins and counting the distinct number of bins, then taking a random sample from the variable $X_{k,k'}$ and returning 1 if and only if $X_{k,k'} = k$, then Y_k is distributed identically to the variable that describes whether or not a random $a \in \{0, 1\}^*$ will appear to be in S according to our filter. Since we have $\Pr[X_{k,k'} = k] < \text{neg}(k)$ and the bound was independent of k', it is easy to see that $\Pr[Y_k = 1] < \text{neg}(k)$, as needed.

(2.) The argument is quite similar to part 1. ($\supseteq$) If $R(a, v) = 1$, then the value v has been inserted and associated with a and by definition, $v \in B_{h_i(a)}$ for every $i \in [k]$. ($\subseteq$) Suppose $a \in S$ and $v \in B_{h_i(a)}$ for every $i \in [k]$. The probability of this event randomly happening independent of the relation R is maximized if every other element in S is associated with the same value. In this case, the problem reduces to a false positive for set membership with $(n-1)k$ writes if $a \in S$, or the usual nk if $a \notin S$. This has already been shown to be negligible in part 1.

In practice, we will need some data structure to model the sets of our Bloom filter with storage, e.g. a linked list. However, in this work we will be interested in *oblivious* writing to the Bloom filter, in which case a linked list seems quite inappropriate as the dynamic size of the structure would leak information about the writing. So, we would like to briefly analyze the total space required for a Bloom filter with storage if it is implemented with fixed-length buffers to represent the sets. Making some needed assumptions about uniformity of value associations, we can show that with overwhelming probability (exponentially close to 1 as a function of the size of our structure) no buffer will overflow.

Claim. Let $(\{h_i\}_{i=1}^{k}, \{B_j\}_{j=1}^{m})$ be a (k, m)-Bloom filter with storage. Suppose the filter has been initialized to store some set S of size n and associated values. Again, suppose that $m = \lceil cnk \rceil$ where $c > 1$ is a constant, and denote the relation of element-value associations by $R(\cdot, \cdot)$. Let $\lambda > 0$ be any constant. If for every $a \in S$ we have $|\{v \mid R(a, v) = 1\}| \leq \lambda$, then for $\sigma \in \mathbb{N}$ we have that as σ increases, $\Pr\left[\max_{j \in [m]}\{|B_j|\} > \sigma\right] < \mathrm{neg}(\sigma)$. Again, the probability is over the uniform randomness used to model the h_i.

Proof. First, let us analyze the case $\lambda = 1$, so there will be a total of nk values placed randomly into the $\lceil cnk \rceil$ buffers. Let X_j be a random variable that counts the size of B_j after the nk values are randomly placed. X_j has a binomial distribution with parameters $(nk, \frac{1}{cnk})$. Hence $E[X_j] = (1/c)$. If $(1 + \delta) > 2e$, we can apply a Chernoff bound to obtain the following estimation: $\Pr[X_j > (1 + \delta)/c] < 2^{-\delta/c}$. Now, for a given σ we'd like to compute $\Pr[X_j > \sigma]$. So, set $(1 + \delta)/c = \sigma$ and hence $\delta/c = \sigma - 1/c$. Then, $\Pr[X_j > \sigma] < 2^{-\sigma+1/c} = 2^{-\sigma}2^{(1/c)} = \mathrm{neg}(\sigma)$. By the union bound, the probability that *any* X_j is larger than σ is also negligible in σ.

Now, if $\lambda > 1$, what has changed? Our analysis above treated the functions as uniform randomness, but to associate additional values to a specific element of $a \in S$ the same subset of buffers (H_a in our notation) will be written to repeatedly- there is no more randomness to analyze. Each buffer will have at most a factor of λ additional elements in it, so our above bound becomes $\mathrm{neg}(\sigma/\lambda)$ which is still $\mathrm{neg}(\sigma)$ as λ is an independent constant.

So, we can implement a (k, m)-Bloom filter with storage using fixed-length buffers. However, the needed length of such buffers depends on the maximum number of values that could be associated to a specific $a \in S$. A priori, this is bounded only by $|V|$, the size of the value universe: for it could be the case that all values are associated to a particular $a \in S$, and hence the buffers of H_a would need to be as large as this universe. But, since we want to fix the buffers length *ahead of time*, we will enforce a "uniformity" constraint; namely, that the number of values associated to each word is bounded by a constant. We summarize with the following observation.

Observation 1. *One can implement a (k, m)-Bloom filter with storage by using fixed-length arrays to store the sets B_j, with the probability of losing an associated value negligible in the length of the arrays. The total size of such a structure is* *linear in n, k, σ, l and c where n is the maximum number of elements that the filter is designed to store, k is the number of functions (h_i) used (which serves as a correctness parameter), σ is the size of the buffer arrays (which serves as a correctness parameter; note that σ should be chosen to exceed λ, the maximum number of values associated to any single element of the set), l is the storage size of an associated value, and c is any constant greater than 1.*

So, for our application of public-key storage with keyword search, if we assume that there are as many keywords as there are messages, then we have created a structure of size $\mathcal{O}(n \cdot l) = \mathcal{O}(n \log n)$ to hold the keyword set and the message references. However, the correctness parameter σ has logarithmic dependence on n, leaving us with $\mathcal{O}(n \log^2 n)$.

2.2   Oblivious Modification

For our application, we will need message senders to update the contents of a Bloom filter with storage. However, all data is encrypted under a key which neither they, nor the storage provider have. So, they must write to the buffers in an "oblivious" way- they will not (and cannot) know what areas of the buffer are already occupied, as this will reveal information about the user's data, and the message-keyword associations. One model for such a writing protocol has been explored by Ostrovsky and Skeith [25]. They provide a method for obliviously writing to a buffer which, with overwhelming probability in independent correctness parameters, is completely correct: i.e., there is a method for extracting documents from the buffer which outputs exactly the set of documents which were put into it.

In [25], the method for oblivious buffer writing is simply to write messages at uniformly random addresses in a buffer, except to ensure that data is recoverable with very high probability, messages are written repeatedly to an appropriately sized buffer, which has linear dependence on a correctness parameter. To ensure that no additional documents arise from collisions, a "collision detection string" is appended to each document from a special distribution which is designed to not be closed under sums. We can apply the same methods here, which will allow senders to update an encrypted Bloom filter with storage, without knowing anything about what is already contained in the encrypted buffers. For more details on this approach, see [25], or the full version of this work. Another approach to this situation was presented by Bethencourt, Song, and Waters [3], who solve a system of linear equations to recover buffer contents. These methods may also be applicable, but require additional interaction to evaluate a pseudo-random function on appropriate input. So, with an added factor of a correctness parameter to the buffer lengths, one can implement and *obliviously update* an encrypted Bloom filter with storage, using the probabilistic methods of [25], or [3].

As a final note on our Bloom filters with storage, we mention that, in practice, we can replace the functions h_i with pseudo-random functions; in this case our claims about correctness are still valid, only with a computational assumption in place of the assumption about the h_i being truly random, provided that the participating parties are non-adaptive[1].

By now, we have an amicable data structure to work with, but there is a piece of the puzzle missing: this data structure will be held by a central storage provider that we'd like to keep in the dark regarding all operations performed on the data. Next, we give message senders a way to update this data structure without revealing to the storage provider any information about the update, *and using small communication*.

2.3   Modifying Encrypted Data in a Communication-Efficient Way

Our next tool is that of encrypted database modification. This will allow us to privately manipulate our Bloom filters. The situation is as follows:

- A database owner Bob holds an array of ciphertexts $\{c_i\}_{i=1}^n$ where each $c_i = \mathcal{E}(x_i)$ is encrypted using a public-key for which Bob does not have the private key.
- A user would like to modify one plaintext value x_i, without revealing to Bob which value was modified, or how it was modified.

Furthermore, we would like to minimize the communication between the parties beyond the trivial $\mathcal{O}(n)$ solution which could be based on any group homomorphic encryption. Using the cryptosystem of Boneh, Goh, and Nissim [5], we can accomplish this with communication $\mathcal{O}(\sqrt{n})$. The important property of the cryptosystem of [5], for our purposes, is its additional homomorphic property; specifically, in their system, one can compute multivariate polynomials of total degree 2 on ciphertexts; i.e., if $\mathcal{E}$ is the encryption map (and $\mathcal{D}$ is the corresponding decryption) and if $F(X_1, \ldots, X_n) = \sum_{1 \leq i \leq j \leq n} a_{ij} X_i X_j$, then there exists some function $\widetilde{F}$ on ciphertexts (which can be computed using public information alone) such that, for any array of ciphertexts $\{c_l = \mathcal{E}(x_l)\}_{l=1}^n$, it holds that $\mathcal{D}(\widetilde{F}(c_1, \ldots, c_n)) = F(x_1, \ldots, x_n)$.

Applying such a cryptosystem to encrypted database modification is simple. Suppose $\{x_{ij}\}_{i,j=1}^{\sqrt{n}}$ is our database (not encrypted). Then, to increment the value of a particular element at position (i^*, j^*) by some value α, we proceed as

---

[1] In the case of malicious senders, we cannot reveal the seeds for the random functions and still guarantee correctness; however, we can entrust the storage provider (Bob) with the seeds, and have the senders execute a secure two-party computation protocol with Bob to learn the value of the functions. This can be accomplished without Bob learning anything, and with the sender learning only $h_i(w)$ and nothing else. Examples of such a protocol can be found in the work of Katz and Ostrovsky [20], if we disallow concurrency, and the work of Canetti et al. [11], to allow concurrency. Here, the common reference string can be provided as part of the public key. These solutions require additional rounds of communication between the senders and the storage provider Bob, and additional communication. However, the size of the communication is proportional to the security parameter and is independent of the size of the database. We defer this and other extensions to the full version of the paper.

follows: Create two vectors v, w of length $\sqrt{n}$ where, $v_i = \delta_{ii^*}$ and $w_j = \alpha\delta_{jj^*}$ (here $\delta_{k\ell} = 1$ when $k = \ell$ and 0 otherwise). Thus, $v_i w_j = \alpha$ if $(i = i^* \wedge j = j^*)$ and 0 otherwise. Now, we wish to add the value $v_i w_j$ to the (i, j) position of the database. Note that, for each i, j, we are just evaluating a simple polynomial of total degree two on v_i, w_j and the data element x_{ij}. So, if we are given any cryptosystem that allows us to compute multivariate polynomials of total degree two on ciphertexts, then we can simply encrypt every input (the database, and the vectors v, w) and perform the same computation which will give us a private database modification protocol with communication complexity $\mathcal{O}(\sqrt{n})$.

More formally, suppose $(\mathcal{K}, \mathcal{E}, \mathcal{D})$ is a CPA-secure public-key encryption scheme that allows polynomials of total degree two to be computed on ciphertexts, as described above. Suppose also that an array of ciphertexts $\{c_l = \mathcal{E}(x_l)\}_{l=1}^n$ is held by a party $\mathcal{S}$, which have been encrypted under some public key, A_{public}. Suppose that n is a square (if not, it can always be padded by $< 2\sqrt{n} + 1$ extra elements to make it a square). Define $F(X, Y, Z) = X + YZ$. Then by our assumption, there exists some $\widetilde{F}$ such that $\mathcal{D}(\widetilde{F}(\mathcal{E}(x), \mathcal{E}(y), \mathcal{E}(z))) = F(x, y, z)$ for any plaintext values x, y, z. We define a two party protocol $\mathsf{Modify}_{\mathcal{U},\mathcal{S}}(l, \alpha)$ by the following steps, where l and α are private inputs to $\mathcal{U}$:

1. $\mathcal{U}$ computes i^*, j^* as the coordinates of l (i.e., i^* and j^* are the quotient and remainder of l/n, respectively).
2. $\mathcal{U}$ sends $\{\overline{v}_i = \mathcal{E}(\delta_{ii^*})\}_{i=1}^{\sqrt{n}}, \{\overline{w}_j = \mathcal{E}(\alpha\delta_{jj^*})\}_{j=1}^{\sqrt{n}}$ to $\mathcal{S}$ where all values are encrypted under A_{public}.
3. $\mathcal{S}$ computes $\widetilde{F}(c_{ij}, \overline{v}_i, \overline{w}_j)$ for all $i, j \in [\sqrt{n}]$, and replaces each c_{ij} with the corresponding resulting ciphertext.

By our remarks above, this will be a correct database modification protocol. It is also easy to see that it is private, in that it resists a chosen plaintext attack. In a chosen plaintext attack, an adversary would ask many queries consisting of requests for the challenger to execute the protocol to modify positions of the adversary's choice. But all that is exchanged during these protocols is arrays of ciphertexts for which the plaintext is known to the adversary. Distinguishing two different modifications is precisely the problem of distinguishing two finite arrays of ciphertexts, which is easily seen to be infeasible assuming the CPA-security of the underlying cryptosystem and then using a standard hybrid argument.

3   Definitions

In what follows, we will denote message sending parties by $\mathcal{X}$, a message receiving party will be denoted by $\mathcal{Y}$, and a server/storage provider will be denoted by $\mathcal{S}$.

Definition 2. *A* Public Key Storage with Keyword Search *consists of the following probabilistic polynomial time algorithms and protocols:*

- KeyGen(1^s) *outputs public and private keys, A_{public} and $A_{private}$ of length s.*
- Send$_{\mathcal{X},\mathcal{S}}(M, K, A_{public})$ *is (an interactive or non-interactive) two-party*

protocol that allows $\mathcal{X}$ to send the message M to server $\mathcal{S}$, encrypted under A_{public}, and also associates M with each keyword in the set K. The values M, K are private inputs that only the message-sending party $\mathcal{X}$ holds.

- *Retrieve$_{\mathcal{Y},\mathcal{S}}(w, A_{private})$ is a two party protocol between the user $\mathcal{Y}$ and server $\mathcal{S}$ that retrieves all messages associated with the keyword w for $\mathcal{Y}$. The inputs $w, A_{private}$ are held only by $\mathcal{Y}$. This protocol also removes the retrieved messages from the server and properly maintains the keyword references.*

We now describe correctness and privacy for such a system.

Definition 3. *Let $\mathcal{Y}$ be a user, $\mathcal{X}$ be a message sender and $\mathcal{S}$ be a server/storage provider. Let $A_{public}, A_{private} \longleftarrow$ KeyGen(1^s). Fix a finite sequence of messages and keyword sets: $\{(M_i, K_i)\}_{i=1}^m$. Suppose that, for all $i \in [m]$, the protocol* Send$_{\mathcal{X},\mathcal{S}}(M_i, K_i, A_{public})$ *is executed by $\mathcal{X}$ and $\mathcal{S}$. Denote by R_w the set of messages that $\mathcal{Y}$ receives after the execution of* Retrieve$_{\mathcal{Y},\mathcal{S}}(w, A_{private})$. *Then, a Public Key Storage with Keyword Search is said to be* correct *on the sequence $\{(M_i, K_i)\}_{i=1}^m$ if $\Pr\left[R_w = \{M_i \mid w \in K_i\}\right] > 1 - \text{neg}(1^s)$, for every w, where the probability is taken over all internal randomness used in the protocols* Send *and* Retrieve*. A Public Key Storage with Keyword Search is said to be* correct *if it is correct on all such finite sequences.*

Definition 4. *A Public Key Storage with Keyword Search is said to be (n, λ, θ)-* correct *if whenever $\{(M_i, K_i)\}_{i=1}^m$ is a sequence such that (1) $m \leq n$, (2) $|K_i| < \theta$, for every $i \in [m]$, and (3) for every $w \in \bigcup_{i \in [m]} K_i$, at most λ messages are associated with w, then it is correct on $\{(M_i, K_i)\}_{i=1}^m$ in the sense of Definition 3.*

For privacy, there are several parties involved, and hence there will be several definitional components.

Definition 5. *For sender-privacy, consider the following game between an adversary $\mathcal{A}$ and a challenger $\mathcal{C}$. $\mathcal{A}$ will play the role of the storage provider and $\mathcal{C}$ will play the role of a message sender. The game consists of the following steps:*

1. KeyGen(1^s) *is executed by $\mathcal{C}$ who sends the output A_{public} to $\mathcal{A}$.*
2. *$\mathcal{A}$ asks queries of the form (M, K) where M is a message and K is a set of keywords; $\mathcal{C}$ answers by executing the protocol* Send(M, K, A_{public}) *with $\mathcal{A}$.*
3. *$\mathcal{A}$ chooses two pairs $(M_0, K_0), (M_1, K_1)$ and sends this to $\mathcal{C}$, where both the messages and keyword sets are of equal size.*
4. *$\mathcal{C}$ picks a random bit $b \in_R \{0, 1\}$ and executes* Send(M_b, K_b, A_{public}) *with $\mathcal{A}$.*
5. *$\mathcal{A}$ asks more queries of the form (M, K) and $\mathcal{C}$ responds by executing protocol* Send(M, K, A_{public}) *with $\mathcal{A}$.*
6. *$\mathcal{A}$ outputs a bit $b' \in \{0, 1\}$.*

We define the adversary's advantage as $\text{Adv}_{\mathcal{A}}(1^s) = \left|\Pr[b = b'] - \frac{1}{2}\right|$. We say that a Public-Key Storage with Keyword Search is CPA-sender-private if, for all $\mathcal{A} \in$ PPT, we have that $\text{Adv}_{\mathcal{A}}(1^s)$ is a negligible function.[2]

---

[2] "PPT" stands for *Probabilistic Polynomial Time*. We use the notation $\mathcal{A} \in$ PPT to denote that $\mathcal{A}$ is a probabilistic polynomial-time algorithm.

Definition 6. *For* receiver-privacy, *consider the following game between an adversary $\mathcal{A}$ and a challenger $\mathcal{C}$. $\mathcal{A}$ again plays the role of the storage provider, and $\mathcal{C}$ plays the role of a message receiver. The game proceeds as follows:*

1. KeyGen(1^s) *is executed by $\mathcal{C}$ who sends the output A_{public} to $\mathcal{A}$.*
2. *$\mathcal{A}$ asks queries of the form w, where w is a keyword; $\mathcal{C}$ answers by executing the protocol* Retrieve$_{\mathcal{C},\mathcal{A}}(w, A_{private})$ *with $\mathcal{A}$.*
3. *$\mathcal{A}$ chooses two keywords w_0, w_1 and sends both to $\mathcal{C}$.*
4. *$\mathcal{C}$ picks a random bit $b \in_R \{0,1\}$ and executes the protocol* Retrieve$_{\mathcal{C},\mathcal{A}}(w_b, A_{private})$ *with $\mathcal{A}$.*
5. *$\mathcal{A}$ asks more keyword queries w and $\mathcal{C}$ responds by executing the protocol* Retrieve$_{\mathcal{C},\mathcal{A}}$ *$(w, A_{private})$ with $\mathcal{A}$.*
6. *$\mathcal{A}$ outputs a bit $b' \in \{0,1\}$.*

We define the adversary's advantage as $\mathrm{Adv}_{\mathcal{A}}(1^s) = \left| \Pr[b = b'] - \frac{1}{2} \right|$. *We say that a Public Key Storage with Keyword Search is* CPA-receiver-private *if, for all $\mathcal{A} \in$ PPT, we have that* $\mathrm{Adv}_{\mathcal{A}}(1^s)$ *is a negligible function.*

Remark: Note that we could have also included a separate protocol for erasing items. At present, the implementation erases messages as they are retrieved. These processes need not be tied together. We have done so to increase the simplicity of our definitions and exposition.

3.1   Extensions

The reader may have noted that this protocol deviates from the usual view of sending mail in that the process requires interaction between a message sender and a server. For simplicity, this point is not addressed in the main portion of the paper, however, it is quite easy to remedy. The source of the problem is that the mail server must communicate the internal address of the new message back to the sender so that the sender can update the Bloom filter with storage to contain this address at the appropriate locations. However, once again, using probabilistic methods from [25], we can solve this problem. As long as the address space is known (which just requires knowledge of the database size, which could be published) the mail sender can simply instruct the server to write the message to a number of random locations, and simultaneously send modification data which would update the Bloom filter accordingly. There are of course, prices to pay for this, but they will not be so significant. The Bloom filter with storage now has addresses of size $\log^2(n)$, since there will be a logarithmic number of addresses instead of just one and, furthermore, to ensure correctness, the database must also grow by a logarithmic factor. A detailed analysis is given in the full version of this work.

Another potential objection to our construction is that mail senders are somewhat free to access and modify the keyword-message associations. Hence, a malicious message sender could invalidate the message-keyword associations, which is another way that this protocol differs from what one may expect from a mail system. (We stress, however, that a sender has no means of modifying other

senders' mail data - only the keyword association data can be manipulated.)
However, this too can be solved by using "off the shelf" protocols; namely, non-
interactive efficient zero knowledge proof systems of Groth, Ostrovksy and Sahai
[18]. In particular, the receiver publishes a common reference string as in [18]
(based on the same cryptographic assumption that we already use in this paper;
i.e., [5]). The sender is now required to include a NIZK proof that the data for
updating the Bloom filter is correct according to the protocol specification. The
main observation is that the theorem size is $O(\sqrt{n \log n})$ and the circuit that
generates it (and its witness) are $O(\sqrt{n \log n} \cdot \text{polylog}(n))$. The [18] NIZK size
is proportional to the circuit size times the security parameter. Thus, assuming
poly-logarithmic security parameter, the result follows.

4 Main Construction

We present a construction of a public-key storage with keyword search that is
(n, λ, θ)-correct, where the maximum number of messages to store is n, and the
total number of distinct keywords that may be in use at a given time is also n
(however, the keyword universe consists of arbitrary strings of bounded length,
say proportional to the security parameter). Correctness will be proved under a
computational assumption in a "semi-honest" model, and privacy will be proved
based only on a computational assumption. In our context, the term "semi-
honest" refers to a party that correctly executes the protocol, but may collect
information during the protocol's execution. We assume the existence of a semantically secure public-key encryption scheme with homomorphic properties
that allow the computation of polynomials of total degree two on ciphertexts,
e.g., [5]. The key generation, encryption and decryption algorithms of the system
will be denoted by $\mathcal{K}, \mathcal{E}$, and $\mathcal{D}$ respectively. We define the required algorithms
and sub-protocols below. First, let us describe our assumptions about the parties
involved: $\mathcal{X}, \mathcal{Y}$ and $\mathcal{S}$. Recall that $\mathcal{X}$ will always denote a message sender. In general, there could be many senders but, for the purposes of describing the protocol,
we need only to name one. Sender $\mathcal{X}$ is assumed to hold a message, keyword(s)
and the public key. Receiver $\mathcal{Y}$ holds the private key. $\mathcal{S}$ has a storage buffer
for n encrypted messages, and it also has a (k, m)-Bloom filter with storage,
as defined in Definition 1, implemented with fixed-length buffers and encrypted
under the public key distributed by $\mathcal{Y}$. As before, $m = \lceil cnk \rceil$, where $c > 1$ is
a constant; the functions and buffers are denoted by $\{h_i\}_{i=1}^{k}$ and $\{B_j\}_{j=1}^{m}$. The
buffers $\{B_j\}$ will be initialized to 0 in every location. $\mathcal{S}$ maintains in its storage space encryptions of the buffers, and not the buffers themselves. We denote
these encryptions $\{\widehat{B_j}\}_{j=1}^{m}$. The functions h_i are implemented by pseudo-random
functions, which can be published by $\mathcal{Y}$. Recall that for $w \in \{0,1\}^*$, we defined
$H_w = \{h_i(w) \mid i \in [k]\}$.

KeyGen(k): Run $\mathcal{K}(1^s)$, the key generation algorithm of the underlying cryptosystem, to create public and private keys, A_{public} and $A_{private}$ respectively. Private
and public parameters for a PIR scheme are also generated by this algorithm.

$Send_{\mathcal{X},\mathcal{S}}(M, K, A_{public})$: Sender $\mathcal{X}$ holds a message M, keywords K and A_{public} and wishes to send the message to $\mathcal{Y}$ via the server $\mathcal{S}$. The protocol consists of the following steps:

1. $\mathcal{X}$ modifies M to have K appended to it, and then sends $\mathcal{E}(M)$, an encryption of the modified M to $\mathcal{S}$.
2. $\mathcal{S}$ receives $\mathcal{E}(M)$, and stores it at an available address ρ in its message buffer. $\mathcal{S}$ then sends ρ back to $\mathcal{X}$.
3. For every $j \in \bigcup_{w \in K} H_w$, sender $\mathcal{X}$ writes γ copies of the address ρ to $\widehat{B_j}$, using the methods of [25]. However, the information of which buffers were written needs to be hidden from $\mathcal{S}$. For this, $\mathcal{X}$ repeatedly executes protocol $\mathsf{Modify}_{\mathcal{X},\mathcal{S}}(x, \alpha)$ for appropriate (x, α), in order to update the Bloom filter buffers. Writing a single address may take several executions of Modify depending on the size of the plaintext set in the underlying cryptosystem. Also, if $|\bigcup_{w \in K} H_w| < k|K|$, then $\mathcal{X}$ executes additional $\mathsf{Modify}(r, 0)$ invocations (for any random r) so that the total number of times that Modify is invoked is uniform among all keyword sets of equal size.

$Retrieve_{\mathcal{Y},\mathcal{S}}(w, A_{private})$: $\mathcal{Y}$ wishes to retrieve all messages associated with the keyword w, and erase them from the server. The protocol proceeds as follows:

1. $\mathcal{Y}$ repeatedly executes an efficient PIR protocol (e.g., [23,10]) with $\mathcal{S}$ to retrieve the encrypted buffers $\{\widehat{B_j}\}_{j \in H_w}$ which are the Bloom filter contents corresponding to w. If $|H_w| < k$, then $\mathcal{Y}$ executes additional PIR protocols for random locations and discards the results so that the same number of executions are invoked regardless of the keyword w. Recall that $\mathcal{Y}$ possesses the seeds used for the pseudo-random functions h_i, and hence can compute H_w without interacting with $\mathcal{S}$.
2. $\mathcal{Y}$ decrypts the answers for the PIR queries to obtain $\{B_j\}_{j \in H_w}$, using the key $A_{private}$. Receiver $\mathcal{Y}$ then computes $L = \bigcap_{j \in H_w} B_j$, a list of addresses corresponding to w, and then executes PIR protocols again with $\mathcal{S}$ to retrieve the encrypted messages at each address in L. Recall that we have bounded the maximum number of messages associated with a keyword. We refer to this value as λ. Receiver $\mathcal{Y}$ will, as usual, invoke additional random PIR executions so that it appears as if every word has λ messages associated to it. After decrypting the messages, $\mathcal{Y}$ will obtain any other keywords associated to the message(s) (recall that the keywords were appended to the message during the Send protocol). Denote this set of keywords $\overline{K}$.
3. $\mathcal{Y}$ first retrieves the additional buffers $\{\widehat{B_j}\}$, for all $j \in \bigcup_{w' \neq w \in \overline{K}} H_{w'}$, using PIR queries with $\mathcal{S}$. The number of additional buffers is bounded by the constant $\theta \cdot t$. Once again, $\mathcal{Y}$ invokes additional PIR executions with $\mathcal{S}$ so that the number of PIR queries in this step of the protocol is uniform for every w. Next, $\mathcal{Y}$ modifies these buffers, removing any occurrences of any address in L. This is accomplished via repeated execution of $\mathsf{Modify}_{\mathcal{Y},\mathcal{S}}(x, \alpha)$ for appropriate x and α. Additional Modify protocols are invoked to correspond to the maximum $\theta \cdot k$ buffers.

Remark: If one wishes to separate the processes of message retrieval and message erasure, simply modify the retrieval protocol to skip the last step, and then use the current retrieval protocol as the message erasure procedure.

Theorem 2. *The Public-Key Storage with Keyword Search from the preceding construction is (n, λ, θ)-correct according to Definition 4, under the assumption that the functions h_i are pseudo-random.*

Proof sketch: This is a consequence of Claim 2.1, Claim 2.1, and Observation 1. The preceding claims were all proved under the assumption that the functions h_i were uniformly random. In our protocol, they were replaced with pseudo-random functions, but since we are dealing with non-adaptive adversaries, the keywords are chosen before the seeds are generated. Hence they are independent, and if any of the preceding claims failed to be true with pseudo-random functions in place of the h_i, our protocol could be used to distinguish the h_i from the uniform distribution without knowledge of the random seed, violating the assumption of pseudo-randomness. As we mentioned before, we can easily handle adaptive adversaries, by implementing h_i using PRF's, where the seeds are kept by the service provider, and users executing secure two-party computation protocols to get $h_i(w)$ for any w using [20] or, in the case of concurrent users, using [11] and having the common random string required by [11] being part of the public key.   □

We also note that in a model with potentially malicious parties, we can apply additional machinery to force "malicious" behavior using [18] as discussed above.

Theorem 3. *Assuming CPA-security of the underlying cryptosystem[3] (and therefore the security of our* Modify *protocol as well), the Public Key Storage with Keyword Search from the above construction is sender private, according to Definition 5.*

Proof sketch: Suppose that there exists an adversary $\mathcal{A} \in$ PPT that can succeed in breaking the security game, from Definition 5, with some non-negligible advantage. So, under those conditions, $\mathcal{A}$ can distinguish the distribution of Send(M_0, K_0) from the distribution of Send(M_1, K_1), where the word "distribution" refers to the distribution of the transcript of the interaction between the parties. A transcript of Send(M, K) essentially consists of just $\mathcal{E}(M)$ and a transcript of several Modify protocols that update locations of buffers based on K. Label the sequence of Modify protocols used to update the buffer locations for K_i by $\{\text{Modify}(x_{i,j}, \alpha_{i,j})\}_{j=1}^{\nu}$. Note that by our design, if $|K_0| = |K_1|$, then it will take the same number of Modify protocols to update the buffers, so the variable ν does not depend on i in this case. Now consider the following sequence of distributions:

$$
\begin{array}{|llll|}
\hline
\mathcal{E}(M_0) & \text{Modify}(x_{0,0}, \alpha_{0,0}) & \cdots & \text{Modify}(x_{0,\nu}, \alpha_{0,\nu}) \\
\mathcal{E}(M_0) & \text{Modify}(x_{0,0}, \alpha_{0,0}) & \cdots & \text{Modify}(x_{1,\nu}, \alpha_{1,\nu}) \\
\vdots & \vdots & \vdots & \vdots \\
\mathcal{E}(M_0) & \text{Modify}(x_{1,0}, \alpha_{1,0}) & \cdots & \text{Modify}(x_{1,\nu}, \alpha_{1,\nu}) \\
\mathcal{E}(M_1) & \text{Modify}(x_{1,0}, \alpha_{1,0}) & \cdots & \text{Modify}(x_{1,\nu}, \alpha_{1,\nu}) \\
\hline
\end{array}
$$

The first line of distributions in the sequence is the transcript distribution for Send(M_0, K_0) and the last line of distributions is the transcript distribution for Send(M_1, K_1). We assumed that there exists an adversary $\mathcal{A}$ that can distinguish

---

[3] For concreteness, this may be implemented using the cryptosystem of [5], in which case security relies on the subgroup decision problem (see [5]).

these two distributions. Hence, not all of the adjacent intermediate distributions can be computationally indistinguishable since computational indistinguishability is transitive. So, there exists an adversary $\mathcal{A}' \in$ PPT that can distinguish between two adjacent rows in the sequence. If $\mathcal{A}'$ distinguishes within the first $\nu + 1$ rows, then it has distinguished $\mathsf{Modify}(x_{0,j}, \alpha_{0,j})$ from $\mathsf{Modify}(x_{1,j}, \alpha_{1,j})$ for some $j \in [\nu]$ which violates our assumption of the security of Modify. And if $\mathcal{A}'$ distinguishes the last two rows, then it has distinguished $\mathcal{E}(M_0)$ from $\mathcal{E}(M_1)$ which violates our assumption on the security of the underlying cryptosystem. Either way, a contradiction. So we conclude that no such $\mathcal{A}$ exists in the first place, and hence the system is secure according to Definition 5. □

Theorem 4. *Assuming CPA-security of the underlying cryptosystem (and therefore the security of our Modify protocol as well), and assuming that our PIR protocol is semantically secure, the Public Key Storage with Keyword Search from the above construction is receiver private, according to Definition 6.*

Proof sketch: Again, assume that there exists $\mathcal{A} \in$ PPT that can gain a non-negligible advantage in Definition 6. Then, $\mathcal{A}$ can distinguish $\mathsf{Retrieve}(w_0)$ from $\mathsf{Retrieve}(w_1)$ with non-negligible advantage. The transcript of a $\mathsf{Retrieve}$ protocol consists a sequence of PIR protocols from steps 1, 2, and 3, followed by a number of Modify protocols. For a keyword w_i, denote the sequence of PIR protocols that occur in $\mathsf{Retrieve}(w_i)$ by $\{\mathrm{PIR}(z_{i,j})\}_{j=1}^{\zeta}$, and denote the sequence of Modify protocols by $\{\mathsf{Modify}(x_{i,j}, \alpha_{i,j})\}_{j=1}^{\eta}$. Note that by the design of the $\mathsf{Retrieve}$ protocol, there will be equal numbers of these PIR queries and Modify protocols regardless of the keyword w, and hence ζ and η are independent of i. Consider the following sequence of distributions:

$$
\begin{array}{|cccccc|}
\hline
\mathrm{PIR}(z_{0,0}) & \cdots & \mathrm{PIR}(z_{0,\zeta}) & \mathsf{Modify}(x_{0,0}, \alpha_{0,0}) & \cdots & \mathsf{Modify}(x_{0,\eta}, \alpha_{0,\eta}) \\
\mathrm{PIR}(z_{1,0}) & \cdots & \mathrm{PIR}(z_{0,\zeta}) & \mathsf{Modify}(x_{0,0}, \alpha_{0,0}) & \cdots & \mathsf{Modify}(x_{0,\eta}, \alpha_{0,\eta}) \\
\vdots & \ddots & \vdots & \vdots & & \vdots \\
\mathrm{PIR}(z_{1,0}) & \cdots & \mathrm{PIR}(z_{1,\zeta}) & \mathsf{Modify}(x_{0,0}, \alpha_{0,0}) & \cdots & \mathsf{Modify}(x_{0,\eta}, \alpha_{0,\eta}) \\
\mathrm{PIR}(z_{1,0}) & \cdots & \mathrm{PIR}(z_{1,\zeta}) & \mathsf{Modify}(x_{1,0}, \alpha_{1,0}) & \cdots & \mathsf{Modify}(x_{0,\eta}, \alpha_{0,\eta}) \\
\vdots & & \vdots & \vdots & \ddots & \vdots \\
\mathrm{PIR}(z_{1,0}) & \cdots & \mathrm{PIR}(z_{1,\zeta}) & \mathsf{Modify}(x_{1,0}, \alpha_{1,0}) & \cdots & \mathsf{Modify}(x_{1,\eta}, \alpha_{1,\eta}) \\
\hline
\end{array}
$$

The first line is the transcript distribution of $\mathsf{Retrieve}(w_0)$ and the last line is the transcript distribution of $\mathsf{Retrieve}(w_1)$. Since there exists $\mathcal{A} \in$ PPT that can distinguish the first distribution from the last, then there must exist an adversary $\mathcal{A}' \in$ PPT that can distinguish a pair of adjacent distributions in the above sequence, due to the transitivity of computational indistinguishability. Therefore, for some $j \in [\zeta]$ or $j' \in [\eta]$ we have that $\mathcal{A}'$ can distinguish $\mathrm{PIR}(z_{0,j})$ from $\mathrm{PIR}(z_{1,j})$ or $\mathsf{Modify}(x_{0,j'}, \alpha_{0,j'})$ from $\mathsf{Modify}(x_{1,j'}, \alpha_{1,j'})$. In both cases, a contradiction of our initial assumption. Therefore, no such $\mathcal{A} \in$ PPT exists, and hence our construction is secure according to Definition 6. □

Theorem 5. *(Communication Complexity) The Public-Key Storage with Keyword Search from the preceding construction has sub-linear communication complexity in n, the number of documents held by the storage provider $\mathcal{S}$.*

Proof. From Observation 1, a (k, m)-Bloom filter with storage that is designed to store n different keywords is of linear size in n (the maximum number of elements that the filter is designed to store), k (the number of functions h_i used, which serves as a correctness parameter), σ (the size of the buffer arrays, which serves as a correctness parameter; note that σ should be chosen to exceed λ, the maximum number of values associated to any single element of the set), $l = \log n$ (the storage size of an associated value), and c (any constant greater than 1).

However, all the buffers in our construction have been encrypted, giving an extra factor of s, the security parameter. Additionally, there is another correctness parameter, γ coming from our use of the methods of [25], which writes a constant number copies of each document into the buffer. Examining the proof of Theorem 2.1, we see that the parameters k and c are indeed independent of n. However, $\{s, l, \gamma\}$ should have logarithmic dependence on n. So, the total size of the encrypted Bloom filter with storage is $\mathcal{O}(n \cdot k \cdot \sigma \cdot l \cdot c \cdot s \cdot \gamma) = \mathcal{O}(n \log^3 n)$, as all other parameters are constants or correctness parameters independent of n (i.e., their value in preserving correctness does not deteriorate as n grows).

Therefore the communication complexity of the protocol is $\mathcal{O}(\sqrt{n \log^3 n})$ for sending a message assuming honest-but-curious sender; $\mathcal{O}(\sqrt{n \log^3 n} \cdot \text{polylog}(n))$ for any malicious poly-time bounded sender; $\mathcal{O}(\text{polylog}(n))$ for reading using any polylog(n) PIR protocol, e.g. [8,10,24]; and $\mathcal{O}(\sqrt{n \log^3 n})$ for deleting messages.

References

1. Abdalla, M., Bellare, M., Catalano, D., Kiltz, E., Kohno, T., Lange, T., Malone-Lee, J., Neven, G., Paillier, P., Shi, H.: Searchable Encryption Revisited: Consistency Properties, Relation to Anonymous IBE, and Extensions. In: Shoup, V. (ed.) CRYPTO 2005. LNCS, vol. 3621, pp. 205–222. Springer, Heidelberg (2005)
2. Barak, B., Goldreich, O.: Universal Arguments and their Applications. In: IEEE Conference on Computational Complexity, pp. 194–203 (2002)
3. Bethencourt, J., Song, D., Waters, B.: New techniques for private stream searching. Technical Report CMU-CS-06-106, Carnegie Mellon University (March 2006)
4. Bloom, B.: Space/time trade-offs in hash coding with allowable errors. Communications of the ACM 13(7), 422–426 (1970)
5. Boneh, D., Goh, E., Nissim, K.: Evaluating 2-DNF Formulas on Ciphertexts. In: TCC, 325–341 (2005)
6. Boneh, D., Crescenzo, G., Ostrovsky, R., Persiano, G.: Public Key Encryption with Keyword Search. In: Cachin, C., Camenisch, J.L. (eds.) EUROCRYPT 2004. LNCS, vol. 3027, pp. 506–522. Springer, Heidelberg (2004)
7. Curtmola, R., Garay, J., Kamara, S., Ostrovsky, R.: Searchable symmetric encryption: improved definitions and efficient constructions. In: Proc. of CCS-2006, pp. 79–88 (2006)
8. Chang, Y.C.: Single Database Private Information Retrieval with Logarithmic Communication. In: Wang, H., Pieprzyk, J., Varadharajan, V. (eds.) ACISP 2004. LNCS, vol. 3108, Springer, Heidelberg (2004)
9. Chang, Y.C., Mitzenmacher, M.: Privacy Preserving Keyword Searches on Remote Encrypted Data. In: Ioannidis, J., Keromytis, A.D., Yung, M. (eds.) ACNS 2005. LNCS, vol. 3531, pp. 442–455. Springer, Heidelberg (2005)

10. Cachin, C., Micali, S., Stadler, M.: Computationally private information retrieval with polylogarithmic communication. In: Stern, J. (ed.) EUROCRYPT 1999. LNCS, vol. 1592, pp. 402–414. Springer, Heidelberg (1999)
11. Canetti, R., Lindell, Y., Ostrovsky, R., Sahai, A.: Universally composable two-party and multi-party secure computation. In: Proc. of the thiry-fourth annual ACM symposium on Theory of computing, pp. 494–503. ACM Press, New York (2002)
12. Chor, B., Gilboa, N., Naor, M.: Private Information Retrieval by Keywords in Technical Report TR CS0917, Department of Computer Science, Technion (1998)
13. Chor, B., Goldreich, O., Kushilevitz, E., Sudan, M.: Private information retrieval. In: Proc. of the 36th Annu. IEEE Symp. on Foundations of Computer Science, pp. 41–51 (1995). Journal version: J. of the ACM, 45 965–981 (1998)
14. Di Crescenzo, G., Malkin, T., Ostrovsky, R.: Single-database private information retrieval implies oblivious transfer. In: Preneel, B. (ed.) EUROCRYPT 2000. LNCS, vol. 1807, Springer, Heidelberg (2000)
15. Freedman, M., Ishai, Y., Pinkas, B., Reingold, O.: Keyword Search and Oblivious Pseudorandom Functions. In: Kilian, J. (ed.) TCC 2005. LNCS, vol. 3378, Springer, Heidelberg (2005)
16. Goh, E.J.: Secure indexes (2003), available at http://eprint.iacr.org/2003/216
17. Goldreich, O., Ostrovsky, R.: Software Protection and Simulation on Oblivious RAMs. In J. ACM 43(3), 431–473 (1996)
18. Groth, J., Ostrovsky, R., Sahai, A.: Perfect Non-interactive Zero Knowledge for NP. In: Vaudenay, S. (ed.) EUROCRYPT 2006. LNCS, vol. 4004, pp. 339–358. Springer, Heidelberg (2006)
19. Goldwasser, S., Micali, S.: Probabilistic encryption. In J. Comp. Sys. Sci. 28(1), 270–299 (1984)
20. Katz, J., Ostrovsky, R.: Round-Optimal Secure Two-Party Computation. In: Franklin, M. (ed.) CRYPTO 2004. LNCS, vol. 3152, pp. 335–354. Springer, Heidelberg (2004)
21. Kilian, J.: A Note on Efficient Zero-Knowledge Proofs and Arguments (Extended Abstract). In: Proc. of STOC 1992, pp. 723–732 (1992)
22. Kurosawa, K., Ogata, W.: Oblivious Keyword Search. Journal of Complexity (Special issue on coding and cryptography) 20(2-3), 356–371 (2004)
23. Kushilevitz, E., Ostrovsky, R.: Replication is not needed: Single database, computationally-private information retrieval. In: Proc. of the 38th Annu. IEEE Symp. on Foundations of Computer Science, pp. 364–373. IEEE Computer Society Press, Los Alamitos (1997)
24. Lipmaa, H.: An Oblivious Transfer Protocal with Log-Squared Communication. IACR ePrint Cryptology Archive 2004/063
25. Ostrovsky, R., Skeith, W.: Private Searching on Streaming Data. In: Shoup, V. (ed.) CRYPTO 2005. LNCS, vol. 3621, Springer, Heidelberg (2005)
26. Ostrovsky, R., Skeith, W.: Algebraic Lower Bounds for Computing on Encrypted Data. In: Electronic Colloquium on Computational Complexity, ECCC TR07-22
27. Ostrovsky, R., Skeith, W.: A Survey of Single Database PIR: Techniques and Applications. In: Proceedings of Public Key Cryptology (PKC-2007). LNCS, Springer-Verlag/IACR, Heidelberg (2007)
28. Sander, T., Young, A., Yung, M.: Non-Interactive CryptoComputing For NC1. In: FOCS 1999, pp. 554–567 (1999)
29. Song, D.X., Wagner, D., Perrig, A.: Practical Techniques for Searches on Encrypted Data. In: Proc. of IEEE Symposium on Security and Privacy, pp. 44–55. IEEE Computer Society Press, Los Alamitos (2000)

Information Security Economics – and Beyond

Ross Anderson and Tyler Moore

Computer Laboratory, University of Cambridge
15 JJ Thomson Avenue, Cambridge CB3 0FD, United Kingdom
firstname.lastname@cl.cam.ac.uk

Abstract. The economics of information security has recently become a thriving and fast-moving discipline. As distributed systems are assembled from machines belonging to principals with divergent interests, incentives are becoming as important to dependability as technical design. The new field provides valuable insights not just into 'security' topics such as privacy, bugs, spam, and phishing, but into more general areas such as system dependability (the design of peer-to-peer systems and the optimal balance of effort by programmers and testers), and policy (particularly digital rights management). This research program has been starting to spill over into more general security questions (such as law-enforcement strategy), and into the interface between security and sociology. Most recently it has started to interact with psychology, both through the psychology-and-economics tradition and in response to phishing. The promise of this research program is a novel framework for analyzing information security problems – one that is both principled and effective.

1 Introduction

Over the last few years, people have realised that security failure is caused by bad incentives at least as often as by bad design. Systems are particularly prone to failure when the person guarding them does not suffer the full cost of failure. Game theory and microeconomic theory are becoming important to the security engineer, just as as the mathematics of cryptography did a quarter century ago. The growing use of security mechanisms for purposes such as digital rights management and accessory control – which exert power over system owners rather than protecting them from outside enemies – introduces many strategic issues. Where the system owner's interests conflict with those of her machine's designer, economic analysis can shine light on policy options.

We survey recent results and live research challenges in the economics of information security. Our goal is to present several promising applications of economic theory and ideas to practical information security problems. In Section 2, we consider foundational concepts: misaligned incentives in the design and deployment of computer systems, and the impact of externalities. Section 3 discusses information security applications where economic analysis has yielded interesting insights: software vulnerabilities, privacy, and the development of user-control mechanisms to support new business models. Metrics present another challenge:

A. Menezes (Ed.): CRYPTO 2007, LNCS 4622, pp. 68–91, 2007.

risks cannot be managed better until they can be measured better. Most users cannot tell good security from bad, so developers are not compensated for efforts to strengthen their code. Some evaluation schemes are so badly managed that 'approved' products are less secure than random ones. Insurance is also problematic; the local and global correlations exhibited by different attack types largely determine what sort of insurance markets are feasible. Cyber-risk markets are thus generally uncompetitive, underdeveloped or specialised.

Economic factors also explain many challenges to privacy. Price discrimination – which is economically efficient but socially controversial – is simultaneously made more attractive to merchants, and easier to implement, by technological advance. Privacy problems also create many externalities. For example, spam and 'identity theft' impose non-negligible social costs. Information security mechanisms or failures can also create, destroy or distort other markets: digital rights management in online music and software markets provides a topical example. Finally, we look at government policy options for dealing with market failures in Section 4, where we examine regulation and mechanism design.

We conclude by discussing several open research challenges: examining the security impact of network structure on interactions, reliability and robustness.

2   Foundational Concepts

Economic thinkers used to be keenly aware of the interaction between economics and security; wealthy nations could afford large armies and navies. But nowadays a web search on 'economics' and 'security' turns up relatively few articles. The main reason is that, after 1945, economists drifted apart from people working on strategic studies; nuclear weapons were thought to decouple national survival from economic power [1], and a secondary factor may have been that the USA confronted the USSR over security, but Japan and the EU over trade. It has been left to the information security world to re-establish the connection.

2.1   Misaligned Incentives

One of the observations that sparked interest in information security economics came from banking. In the USA, banks are generally liable for the costs of card fraud; when a customer disputes a transaction, the bank must either show she is trying to cheat it, or refund her money. In the UK, the banks had a much easier ride: they generally got away with claiming that their systems were 'secure', and telling customers who complained that they must be mistaken or lying. "Lucky bankers," one might think; yet UK banks spent more on security and suffered more fraud. This may have been what economists call a moral-hazard effect: UK bank staff knew that customer complaints would not be taken seriously, so they became lazy and careless, leading to an epidemic of fraud [2].

In 1997, Ayres and Levitt analysed the Lojack car-theft prevention system and found that once a threshold of car owners in a city had installed it, auto theft plummeted, as the stolen car trade became too hazardous [3]. This is a classic example of an *externality*, a side-effect of an economic transaction that may have positive or negative effects on third parties. Camp and Wolfram built on this in 2000 to analyze information security vulnerabilities as negative externalities, like air pollution: someone who connects an insecure PC to the Internet does not face the full economic costs of that, any more than someone burning a coal fire. They proposed trading vulnerability credits in the same way as carbon credits [4].

Also in 2000, Varian looked at the anti-virus software market. People did not spend as much on protecting their computers as they logically should have. At that time, a typical virus payload was a service-denial attack against the website of Amazon or Microsoft. While a rational consumer might well spend $20 to stop a virus trashing her hard disk, she will be less likely to do so just to protect a wealthy corporation [5].

Legal theorists have long known that liability should be assigned to the party that can best manage the risk. Yet everywhere we look, we see online risks allocated poorly, resulting in privacy failures and protracted regulatory tussles. For instance, medical record systems are bought by hospital directors and insurance companies, whose interests in account management, cost control and research are not well aligned with the patients' interests in privacy; this mismatch of incentives led in the USA to HIPAA, a law that sets standards for privacy in health IT. Bohm et al. [6] documented how many banks used online banking as a means of dumping on their customers many of the transaction risks that they previously bore in the days of cheque-based banking; for a recent update on liability in payment systems, see [7].

Asymmetric information plays a large role in information security. Moore showed that we can classify many problems as hidden-information or hidden-action problems [8]. The classic case of hidden information is the 'market for lemons' [26]. Akerlof won a Nobel prize for the following simple yet profound insight: suppose that there are 100 used cars for sale in a town: 50 well-maintained cars worth $2000 each, and 50 'lemons' worth $1000. The sellers know which is which, but the buyers don't. What is the market price of a used car? You might think $1500; but at that price no good cars will be offered for sale. So the market price will be close to $1000. Hidden information, about product quality, is one reason poor security products predominate. When users can't tell good from bad, they might as well buy a cheap antivirus product for $10 as a better one for $20, and we may expect a race to the bottom on price.

Hidden-action problems arise when two parties wish to transact, but one party's unobservable actions can impact the outcome. The classic example is insurance, where a policyholder may behave recklessly without the insurance company observing this. Network nodes can hide malicious or antisocial behavior from their peers; routers can quietly drop selected packets or falsify responses to routing requests; nodes can redirect network traffic to eavesdrop on conversations; and players in file-sharing systems can hide whether they share with others, so some may 'free-ride' rather than to help sustain the system. Once the problem is seen in this light, designers can minimise the capacity for hidden action, or to make it easy to enforce suitable contracts.

This helps explain the evolution of peer-to-peer systems. Early systems proposed by academics, such as Eternity, Freenet, Chord, Pastry and OceanStore, required users to serve a random selection of other users' files [9]. These systems were never widely adopted. Later systems that did attract large numbers of users, like Gnutella and Kazaa, instead allow peer nodes to serve only the content they have downloaded for their own use, rather than burdening them with others' files. The comparison between these architectures originally focused on purely technical aspects: the cost of search, retrieval, communications and storage. However, analysing incentives turned out to be fruitful too.

First, a system structured as an association of clubs reduces the potential for hidden action; club members are more able to assess which members are contributing. Second, clubs might have quite divergent interests. Though peer-to-peer systems are now seen as mechanisms for sharing music, early systems were designed for censorship resistance. A system might serve a number of quite different groups – maybe Chinese dissidents, critics of Scientology, or aficionados of sado-masochistic imagery that is legal in California but banned in Tennessee. Early peer-to-peer systems required such users to serve each other's files, so that they ended up protecting each others' free speech. But might such groups not fight harder to defend their own colleagues, rather than people involved in struggles in which they have no interest?

Danezis and Anderson introduced the Red-Blue model to analyze this [10]. Each node has a preference among resource types, for instance left-leaning versus right-leaning political texts, while a censor will try to impose his own preference. His action will suit some nodes but not others. The model proceeds as a multi-round game in which nodes set defense budgets that affect the probability that they will defeat the censor or be overwhelmed by him. Under reasonable assumptions, the authors show that diversity (with each node storing its preferred resource mix) performs better under attack than solidarity (where each node stores the same resource mix). Diversity makes nodes willing to allocate higher defense budgets; the greater the diversity, the more quickly will solidarity crumble in the face of attack. This model was an early venture on the boundary between economics and sociology; it sheds light on the general problem of diversity versus solidarity, which has had a high profile recently because of the question whether the growing diversity of modern societies is in tension with the solidarity on which modern welfare systems are founded [11].

2.2 Security as an Externality

Information industries have many different types of externality. They tend to have dominant firms for three reasons. First, there are often network externalities, whereby the value of a network grows more than linearly in the number of users; for example, anyone wanting to auction some goods will usually go to the largest auction house, as it will attract more bidders. Second, there is often technical lock-in stemming from interoperability, and markets can be two-sided: software firms develop for Windows to access more customers, and users buy Windows machines to get access to more software. Third, information industries tend to combine high fixed and low marginal costs: the first copy of a software program (or a music download ot even a DVD) may cost millions to produce, while subsequent copies are almost free. These three features separately can lead to industries with dominant firms; together, they are even more likely to.

This not only helps explain the rise and dominance of operating systems, from System/360 through DOS and Windows to Symbian; it also helps explain patterns of security flaws. While a platform vendor is building market dominance, it has to appeal to vendors of software as well as to users, and security could get in their way. So vendors start off with minimal protection; once they have become dominant, they add security to lock their customers in more tightly [12]. We'll discuss this in more detail later.

Further externalities affect security investment, as protection often depends on the efforts of many principals. Hirshleifer told the story of Anarchia, an island whose flood defences were constructed by individual families and whose defence depends on the weakest link, that is, the laziest family; he compared this with a city whose defences against ICBM attack depend on the single best defensive shot [13]. Varian extended this to three cases of interest to the dependability of information systems – where performance depends on the minimum effort, the best effort, or the sum-of-efforts [14].

Program correctness can depend on minimum effort (the most careless programmer introducing a vulnerability) while software vulnerability testing may depend on the sum of everyone's efforts. Security may also depend on the best effort – the actions taken by an individual champion such as a security architect. When it depends on the sum of individual efforts, the burden will tend to be shouldered by the agents with the highest benefit-cost ratio, while the others free-ride. In the minimum-effort case, the agent with the lowest benefit-cost ratio dominates. As more agents are added, systems become more reliable in the total-effort case but less reliable in the weakest-link case. What are the implications? Well, software companies should hire more software testers and fewer but more competent programmers. (Of course, measuring programmer competence can be hard, which brings us back to hidden information.)

This work inspired other researchers to consider interdependent risk. A recent influential model by Kunreuther and Heal notes that an individual taking protective measures creates positive externalities for others that in turn may discourage them from investment [15]. This insight has implications far beyond information security. The decision by one apartment owner to install a sprinkler system will decrease his neighbours' fire risk and make them less likely to do the same; airlines may decide not to screen luggage transferred from other carriers who are believed to be careful with security; and people thinking of vaccinating their children may choose to free-ride off the herd immunity instead. In each case, several widely varying equilibria are possible, from complete adoption to total refusal, depending on the levels of coordination between principals.

Katz and Shapiro famously analyzed how network externalities influenced the adoption of technology: they lead to the classic S-shaped adoption curve in which slow early adoption gives way to rapid deployment once the number of users reaches some critical mass [16]. Network effects can also influence the initial deployment of security technology, whose benefit may depend on the number of users who adopt it. The cost may exceed the benefit until a minimum number adopt; so everyone might wait for others to go first, and the technology never gets deployed. Recently, Ozment and Schechter have analyzed different approaches for overcoming such bootstrapping problems [17].

This challenge is particularly topical. A number of core Internet protocols, such as DNS and routing, are considered insecure. Better protocols exist (e.g., DNSSEC, S-BGP); the challenge is to get them adopted. Two widely-deployed security protocols, SSH and IPsec, both overcame the bootstrapping problem by providing significant internal benefits to adopting firms, with the result that they could be adopted one firm at a time, rather than needing everyone to move at once. The deployment of fax machines was similar: many companies initially bought fax machines to connect their own offices.

3   Applications

3.1   Economics of Vulnerabilities

There has been much debate about 'open source security', and more generally whether actively seeking and disclosing vulnerabilities is socially desirable. Anderson showed in 2002 that, under standard assumptions of reliability growth, open systems and proprietary systems are just as secure as each other; opening up a system helps the attackers and defenders equally [18]. Thus the open-security question may be an empirical one, turning on the extent to which a given real system follows the standard model.

Rescorla argued in 2004 that for software with many latent vulnerabilities, removing one bug makes little difference to the likelihood of an attacker finding another one later [19]. Since exploits are often based on vulnerabilities inferred from patches, he argued against disclosure and frequent patching unless the same vulnerabilities are likely to be rediscovered. This also raised the question of whether software follows the standard dependability model, of independent vulnerabilities. Ozment found that for FreeBSD, vulnerabilities are correlated in that they are likely to be rediscovered [20]. Ozment and Schechter also found that the rate at which unique vulnerabilities were disclosed for the core and unchanged FreeBSD operating system has decreased over a six-year period [21]. These findings suggest that vulnerability disclosure can improve system security over the long term. Vulnerability disclosure also helps motivate vendors to fix bugs [22]. Arora et al. showed that public disclosure made vendors respond with fixes more quickly; attacks increased to begin with, but reported vulnerabilities declined over time [23].

This discussion begs a deeper question: why do so many vulnerabilities exist in the first place? A useful analogy might come from considering large software project failures: it has been known for years that perhaps 30% of large development projects fail [24], and this figure does not seem to change despite improvements in tools and training: people just built much bigger disasters nowadays than they did in the 1970s. This suggests that project failure is not fundamentally about technical risk but about the surrounding socio-economic factors (a point to which we will return later). Similarly, when considering security, software writers have better tools and training than ten years ago, and are capable of creating more secure software, yet the economics of the software industry provide them with little incentive to do so.

In many markets, the attitude of 'ship it Tuesday and get it right by version 3' is perfectly rational behaviour. Many software markets have dominant firms thanks to the combination of high fixed and low marginal costs, network externalities and client lock-in noted above [25], so winning market races is all-important. In such races, competitors must appeal to complementers, such as application developers, for whom security gets in the way; and security tends to be a lemons market anyway. So platform vendors start off with too little security, and such as they provide tends to be designed so that the compliance costs are dumped on the end users [12]. Once a dominant position has been established, the vendor may add more security than is needed, but engineered in such a way as to maximise customer lock-in [27].

In some cases, security is even worse than a lemons market: even the vendor does not know how secure its software is. So buyers have no reason to pay more for protection, and vendors are disinclined to invest in it.

How can this be tackled? Economics has suggested two novel approaches to software security metrics: vulnerability markets and insurance.

Vulnerability markets help buyers and sellers establish the actual cost of finding a vulnerability in software. To begin with, some standards specified a minimum cost of various kinds of technical compromise; one example is banking standards for point-of-sale terminals [28]. Camp and Wolfram suggested in 2000 that markets might work better here than central planning [4]. Schechter developed this into a proposal for open markets in reports of previously undiscovered vulnerabilities [29]. Two firms, iDefense and Tipping Point, are now openly buying vulnerabilities, so the market actually exists (unfortunately, the prices are not published). Their business model is to provide vulnerability data simultaneously to their customers and to the affected vendor, so that their customers can update their firewalls before anyone else. However, the incentives here are suboptimal: bug-market organisations might increase the value of their product by leaking vulnerability information to harm non-subscribers [30].

Several variations on vulnerability markets have been proposed. Böhme has argued that software derivatives might be better [31]. Contracts for software would be issued in pairs: the first pays a fixed value if no vulnerability is found in a program by a specific date, and the second pays another value if one is found. If these contracts can be traded, then their price should reflect the consensus on software quality. Software vendors, software company investors, and insurance companies could use such derivatives to hedge risks. A third possibility, due to Ozment, is to design a vulnerability market as an auction [32].

One criticism of all market-based approaches is that they might increase the number of identified vulnerabilities by motivating more people to search flaws. Thus some care must be exercised in designing them.

An alternative approach is insurance. Underwriters often use expert assessors to look at a client firm's IT infrastructure and management; this provides data to both the insured and the insurer. Over the long run, insurers learn to value risks more accurately. Right now, however, the cyber-insurance market is both under-developed and underutilised. One reason, according to Böhme and Kataria [33], is the interdependence of risk, which takes both local and global forms. Firms' IT infrastructure is connected to other entities – so their efforts may be undermined by failures elsewhere. Cyber-attacks often exploit a vulnerability in a program used by many firms. Interdependence can make some cyber-risks unattractive to insurers – particularly those risks that are globally rather than locally correlated, such as worm and virus attacks, and systemic risks such as Y2K.

Many writers have called for software risks to be transferred to the vendors; but if this were the law, it is unlikely that Microsoft would be able to buy insurance. So far, vendors have succeeded in dumping most software risks; but this outcome is also far from being socially optimal. Even at the level of customer firms, correlated risk makes firms under-invest in both security technology and cyber-insurance [34]. Cyber-insurance markets may in any case lack the volume and liquidity to become efficient.

3.2 Economics of Privacy

The persistent erosion of personal privacy has frustrated policy makers and practitioners alike. People say that they value privacy, yet act otherwise. Privacy-enhancing technologies have been offered for sale, yet most have failed in the marketplace. Why should this be?

Privacy is one aspect of information security that interested economists before 2000. In 1978, Posner defined privacy in terms of secrecy [35], and the following year extended this to seclusion [36]. In 1980, Hirshleifer published a seminal paper in which he argued that rather than being about withdrawing from society, privacy was a means of organising society, arising from evolved territorial behaviour; internalised respect for property is what allows autonomy to persist in society. These privacy debates in the 1970s led in Europe to generic data-protection laws, while the USA limited itself to a few sector-specific laws such as HIPAA. Economists' appetite for work on privacy was further whetted recently by the Internet, the dotcom boom, and the exploding trade in personal information about online shoppers.

An early modern view of privacy can be found in a 1996 paper by Varian who analysed privacy in terms of information markets [38]. Consumers want to not be annoyed by irrelevant marketing calls while marketers do not want to waste effort. Yet both are frustrated, because of search costs, externalities and other factors. Varian suggested giving consumers rights in information about themselves, and letting them lease it to marketers with the proviso that it not be resold without permission.

The recent proliferation of complex, information-intensive business models demand a broader approach. Odlyzko argued in 2003 that privacy erosion is a consequence of the desire to charge different prices for similar services [39]. Technology is simultaneously increasing both the incentives and the opportunities for price discrimination. Companies can mine online purchases and interactions for data revealing individuals' willingness to pay. From airline yield-management systems to complex and ever-changing software and telecommunications prices, differential pricing is economically efficient – but increasingly resented. Acquisti and Varian analyzed the market conditions under which personalised price discrimination is profitable [40]: it may thrive in industries with wide variation in consumer valuation for services, where services can be personalised at low marginal cost, and where repeated purchases are likely.

Acquisti and Grossklags tackled the specific problem of why people express a high preference for privacy when interviewed but reveal a much lower preference through their behaviour both online and offline [41]. They find that people mostly lack sufficient information to make informed choices, and even when they do they often trade long-term privacy for short-term benefits. Vila et al. characterised privacy economics as a lemons market [42], arguing that consumers disregard future price discrimination when giving information to merchants.

Swire argued that we should measure the costs of privacy intrusion more broadly [43]. If a telesales operator calls 100 prospects, sells three of them insurance, and annoys 80, then the conventional analysis considers only the benefit to the three and to the insurer. However, persistent annoyance causes millions of people to go ex-directory, to not answer the phone during dinner, or to screen calls through an answering machine. The long-run societal harm can be considerable. Several empirical studies have backed this up by examining people's privacy valuations.

So much for the factors that make privacy intrusions more likely. What factors make them less so? Campbell et al. found that the stock price of companies reporting a security breach is more likely to fall if the breach leaked confidential information [44]. Acquisti, Friedman and Telang conducted a similar analysis for privacy breaches [45]. Their initial results are less conclusive but still point to a negative impact on stock price followed by an eventual recovery.

Regulatory responses (pioneered in Europe) have largely centred on requiring companies to allow consumers to either 'opt-in' or 'opt-out' of data collection. While privacy advocates typically support opt-in policies as they result in lower rates of data collection, Bouckaert and Degryse argue for opt-out on competition grounds [46]: the availability of information about the buying habits of most customers, rather than a few customers, may help competitors to enter a market.

Empirically, there is wide variation in 'opt-out' rates between different types of consumer, but their motives are not always clear. Varian et al. analyzed the FCC's telephone-sales blacklist by district [47]. They found that educated people are more likely to sign up: but is that because rich households get more calls, because they value their time more, or because they understand the risks better?

Incentives also affect the design of privacy technology. Builders of anonymity systems know they depend on network externalities: more users mean more cover traffic to hide activities from the enemy [48]. An interesting case is Tor [49], which anonymises web traffic and emphasises usability to increase adoption rates. It developed from a US Navy communications system, but eventually all internet users were invited to participate in order to build network size, and it is now the largest anonymous communication system known.

3.3   Incentives and the Deployment of Security Mechanisms

Insurance is not the only market affected by information security. Some very high-profile debates have centred on DRM; record companies have pushed for years for DRM to be incorporated into computers and consumer electronics, while digital-rights activists have opposed them. What light can security economics shed on this debate?

Many researchers have set the debate in a much wider context than just record companies versus downloaders. Varian pointed out in 2002 that DRM and similar mechanisms were also about tying, bundling and price discrimination; and that their unfettered use could damage competition [50]. A paper by Samuelson and Scotchmer studied what might go wrong if technical and legal restraints were to undermine the right to reverse engineer software products for compatibility. It provided the scholarly underpinnings for much of the work on the anti-competitive effects of the DMCA, copyright control mechanisms, and information security mechanisms applied to new business models.

'Trusted Computing' (TC) mechanisms have come in for significant analysis and criticism. Von Hippel showed how most of the innovations that spur economic growth are not anticipated by the manufacturers of the platforms on which they are based; the PC, for example, was conceived as an engine for running spreadsheets, and if IBM had been able to limit it to doing that, a huge opportunity would have been lost. Furthermore, technological change in IT markets is usually cumulative. If security technology can be abused by incumbent firms to make life harder for innovators, this will create all sorts of traps and perverse incentives [52]. Anderson pointed out the potential for competitive abuse of the TC mechanisms; for example, by transferring control of user data from the owner of the machine on which it is stored to the creator of the file in which it is stored, the potential for lock-in is hugely increased [27]. Lookabaugh and Sicker discussed an existing case history of an industry crippled by security-related technical lock-in [53]. US cable industry operators are locked in to their set-top-box vendors; and although they largely negotiated away the direct costs of this when choosing a suppler, the indirect costs were large and unmanageable. Innovation suffered and cable fell behind other platforms, such as the Internet, as the two platform vendors did not individually have the incentive to invest in improving their platforms.

Economic research has been applied to the record industry itself, with results it found disturbing. In 2004, Oberholzer and Strumpf published a now-famous paper, in which they examined how music downloads and record sales were correlated [54].

They showed that downloads do not do significant harm to the music industry. Even in the most pessimistic interpretation, five thousand downloads are needed to displace a single album sale, while high-selling albums actually benefit from file sharing.

In January 2005, Varian presented a surprising result [55]: that stronger DRM would help system vendors more than the music industry, because the computer industry is more concentrated (with only three serious suppliers of DRM platforms – Microsoft, Sony, and the dominant firm, Apple). The content industry scoffed, but by the end of that year music publishers were protesting that Apple was getting too large a share of the cash from online music sales. As power in the supply chain moved from the music majors to the platform vendors, so power in the music industry appears to be shifting from the majors to the independents, just as airline deregulation favoured aircraft makers and low-cost airlines. This is a striking demonstration of the predictive power of economic analysis. By fighting a non-existent threat, the record industry had helped the computer industry forge a weapon that may be its undoing.

3.4   Protecting Computer Systems from Rational Adversaries

Information security practitioners traditionally assumed two types of user: honest ones who always behave as directed, and malicious ones intent on wreaking havoc at any cost. But systems are often undermined by what economists call *strategic* users: users who act out of self-interest rather than malice. Many file-sharing systems suffer from 'free-riding', where users download files without uploading their own. This is perfectly rational behaviour, given that upload bandwidth is typically more scarce and file uploaders are at higher risk of getting sued. The cumulative effect is degraded performance.

Another nuisance caused by selfish users is spam. The cost per transmission to the spammer is so low that a tiny success rate is acceptable [56]. Furthermore, while spam imposes significant costs on recipients, these costs are not felt by the spammers. Böhme and Holz examined stock spam and identified statistically significant increases in the price of touted stocks [57]. Frieder and Zittrain independently find a similar effect [58].

Several network protocols may be exploited by selfish users at the expense of system-wide performance. In TCP, the protocol used to transmit most Internet data, Akella et al. find that selfish provision of congestion control mechanisms can lead to suboptimal performance [59].

Researchers have used game theory to study the negative effects of selfish behaviour on systems more generally. Koutsoupias and Papadimitriou termed the 'price of anarchy' as the ratio of the utilities of the worst-case Nash equilibrium to the social optimum [60]. The price of anarchy has become a standard measurement of the inefficiency of selfish behaviour in computer networks. Roughgarden and Tardos studied selfish routing in a congested network, comparing congestion levels in a network where users choose the shortest path available to congestion when a network planner chooses paths to maximise flow [61]. They established an upper bound of $\frac{4}{3}$ for the price of anarchy when congestion costs are linear;

furthermore, in general, the total latency of a selfish network is at most the same as an optimal flow routing twice as much traffic.

Other topics hindered by selfish activity include network creation, where users decide whether to create costly links to shorten paths or free-ride over longer, indirect connections [62,63,64]; wireless spectrum sharing, where service providers compete to acquire channels from access points [65]; and computer virus inoculation, where users incur a high cost for inoculating themselves and the benefits accrue to unprotected nodes [66].

To account for user self-interest, computer scientists have proposed several mechanisms with an informal notion of 'fairness' in mind. To address spam, Dwork and Naor propose attaching to emails a 'proof-of-work' that is easy to do for a few emails but impractical for a flood [67]. Laurie and Clayton criticise 'proof-of-work' schemes, demonstrating that the additional burden may be cumbersome for many legitimate users while spam senders could use botnets to perform the computations [68]. Furthermore, ISPs may not be prepared to block traffic from these compromised machines. Serjantov and Clayton analyse the incentives on ISPs to block traffic from other ISPs with many infected machines, and back this up with data [69]. They also show how a number of existing spam-blocking strategies are irrational and counterproductive.

Reputation systems have been widely proposed to overcome free-riding in peer-to-peer networks. The best-known fielded example may be feedback on eBay's online auctions. Dellarocas argues that leniency in the feedback mechanism (only 1% of ratings are negative) encourages stability in the marketplace [71]. Serjantov and Anderson use social choice theory to recommend improvements to reputation system proposals [72]. Feldman et al model such systems as an iterated prisoner's dilemma game, where users in each round alternate between roles as client and server [70]. Recently, researchers have begun to consider more formally how to construct fair systems using mechanism design. We discuss these developments in Section 5.1.

4 The Role of Governments

The information security world has been regulated from the beginning, although initially government concerns had nothing to do with competition policy. The first driver was a non-proliferation concern. Governments used export licenses and manipulated research funding to restrict access to cryptography for as long as possible. This effort was largely abandoned in 2000. The second driver was the difficulty that even the US government had over many years in procuring systems for its own use, once information security came to encompass software security too. Thus, during the 80s and 90s, it was policy to promote research in security while hindering research in cryptography.

Landwehr describes the efforts of the US government from the mid-1980s to tackle a the lemons problem in the security software business [73]. The first attempted fix was a government evaluation scheme – the Orange Book – but that brought its own problems. Managers' desire for the latest software eased certification requirements: vendors had to simply show that they had initiated the certification process, which often was never completed. Evaluations were also conducted at government expense by NSA civil servants, who being risk-averse took their time; evaluated products were often unusably out of date. There were also problems interworking with allies' systems, as countries such as the UK and Germany had their own incompatible schemes.

This led the NATO governments to establish the 'Common Criteria' as a successor to the Orange Book. Most evaluations are carried out by commercial laboratories and are paid for by the vendor who is supposed to be motivated by the cachet of a successful evaluation. The Common Criteria suffer from different problems, most notably adverse selection: vendors shop around for the evaluator who will give them the easiest ride, and the national agencies who certify the evaluation labs are very reluctant to revoke a license, even following scandal, because of fears that confidence in the scheme will be undermined [74].

Regulation is increasingly justified by perceived market failures in the information security industry. The European Union has proposed a Network Security Policy that sets out a common European response to attacks on information systems [75]. This starts using economic arguments about market failure to justify government action in this sector. The proposed solutions are familiar, involving everything from consciousness raising to more Common Criteria evaluations.

Another explicit use of security economics in policymaking was the German government's comments on Trusted Computing [76]. These set out concerns about issues from certification and trapdoors through data protection to economic policy matters. They were hugely influential in persuading the Trusted Computing Group to incorporate and adopt membership rules that mitigated the risk of its program discriminating against small-to-medium sized enterprises. Recently the European Commission's DG Competition has been considering the economic implications of the security mechanisms of Vista.

Among academic scholars of regulation, Barnes studies the incentives facing the virus writers, software vendors and computer users [77], and contemplates various policy initiatives to make computers less liable to infection, from rewarding those who discover vulnerabilities to penalising users who do not adopt minimal security standards. Garcia and Horowitz observe that the gap between the social value of internet service providers, and the revenue at stake associated with their insecurity, is continuing to increase [78]. If this continues, they argue, mandatory security standards may become likely.

Moore presents an interesting regulatory question from forensics. While PCs use standard disc formats, mobile phones use proprietary interfaces, which make data recovery from handsets difficult; recovery tools exist only for the most common models. So criminals should buy unfashionable phones, while the police should push for open standards [79].

Heavy-handed regulation can introduce high costs – whether directly, or as a result of agency issues and other secondary factors. Ghose and Rajan discuss how three US laws – Sarbanes-Oxley, Gramm-Leach-Bliley and HIPAA – place a disproportionate burden on small and medium sized businesses, largely through a one-model-fits-all approach to compliance by the big accounting firms [80]. They show how mandatory investment in security compliance can create unintended consequences from distorting security markets to reducing competition.

Given the high costs and doubtful effectiveness of regulation, self-regulation has been tried in a number of contexts, but some attempts failed spectacularly. For example, a number of organisations have set up certification services to vouch for the quality of software products or web sites. Their aim was twofold: to overcome public wariness about electronic commerce, and to forestall more expensive regulation by the government. But (as with the Common Criteria) certification markets can easily be ruined by a race to the bottom; dubious companies are more likely to buy certificates than reputable ones, and even ordinary companies may shop around for the easiest deal. In the absence of a capable motivated regulator, ruin can arrive quickly.

Edelman analysed this 'adverse selection' in the case of website approvals and online advertising [81]: while about 3% of websites are malicious, some 8% of websites with certification from one large vendor are malicious. He also compared ordinary web search results and those from paid advertising, finding that while 2.73% of companies ranked top in a web search were bad, 4.44% of companies who had bought ads from the search engine were bad. His conclusion – 'Don't click on ads' – could be bad news for the search industry.

Self-regulation has fared somewhat better for patch management. Analysis by Arora et al. shows that competition in software markets hastens patch release even more than the threat of vulnerability disclosure in two out of three studied strategies [83]. Beattie et al. found that pioneers who apply patches quickly end up discovering problems that break their systems, but laggards are more vulnerable to attack [82].

Governments also facilitate the sharing of security information between private companies. Two papers analyse the incentives that firms have to share information on security breaches within the Information Sharing and Analysis Centers (ISACs) set up after 9/11 by the US government [84,85]. Theoretical tools developed to model trade associations and research joint ventures can be applied to work out optimal membership fees and other incentives.

5 Open Problems

There are many active areas of security-economics research. Here we highlight just four live problems. Each lies not just at the boundary between security and economics, but also at the boundary between economics and some other discipline – respectively algorithmic mechanism design, network science, organisational theory and psychology.

5.1 Algorithmic Mechanism Design

Given the largely unsatisfactory impact of information security regulation, a complementary approach based on mechanism design is emerging. Researchers are beginning to design network protocols and interfaces that are 'strategy-proof': that is, designed so that no-one can gain by cheating [86]. Designing bad behavior out of systems may be cheaper than policing it afterwards.

One key challenge is to allocate scare digital resources fairly. Nisan and Segal show that although one can solve the allocation problem using strategy-proof mechanisms, the number of bits that must be communicated grows exponentially; thus in many cases the best practical mechanism will be a simple bundled auction [87]. They also suggest that if arbitrary valuations are allowed, players can submit bids that will cause communications complexity problems for all but the smallest auctions.

Some promising initial results look at mechanism design and protocols. Feigenbaum et al. show how combinatorial auction techniques can be used to provide distributed strategy-proof routing mechanisms [88]. Schneidman et al. compare the incentive mechanisms in BitTorrent, a popular peer-to-peer file-sharing application, to theoretical guarantees of faithfulness [89].

5.2   Network Topology and Information Security

There has been an interesting collaboration recently between physicists and sociologists in analyzing the topology of complex networks and its effect on social interactions. Computer networks, like social networks, are complex but emerge from ad-hoc interactions of many entities using simple ground rules. The new discipline of network analysis takes ideas from sociology, condensed-matter physics and graph theory, and in turn provides tools for modelling and investigating such networks (see [90] for a recent survey). Some economists have also recognised the impact of network structure on a range of activities, from crime [91,92] to the diffusion of new technologies [93]. Other researchers have focused on why networks are formed, where the individual costs of establishing links between agents is weighed against the overall benefit of improved connectivity [94]. Economic models are well-suited to comparing the social efficiency of different network types and predicting which structures are likely to emerge when agents act selfishly. See [95] for a collection of recent work.

Network topology can strongly influence conflict dynamics. Often an attacker tries to disconnect a network or increase its diameter by destroying nodes or edges, while the defender counters using various resilience mechanisms. Examples include a music industry body attempting to close down a peer-to-peer file-sharing network; a police force trying to decapitate a terrorist organisation; and a totalitarian government harrassing political activists. Police forces have been curious for some years about whether network science might be of practical use in covert conflicts – whether to insurgents or to counterinsurgency forces.

Different topologies have different robustness properties. Albert, Jeong and Barabási showed that certain real world networks with scale-free degree distributions resist random attacks much better than targeted attacks [96]. This is because scale-free networks – like many real-world networks – get much of their connectivity from a few nodes with high vertex order. This resilience makes them highly robust against random upsets; but remove the 'kingpin' nodes, and connectivity collapses.

This is the static case – for example, when a police force becomes aware of a criminal or terrorist network, and sets out to disrupt it by finding and arresting its key people. Nagaraja and Anderson extend this to the dynamic case. In their model, the attacker can remove a certain number of nodes at each round, after which the defenders recruit other nodes to replace them [97]. They studied how attack and defence interact using multi-round simulations, and found that forming localised clique structures at key network points works reasonably well while defences based on rings did not work well at all. This helps explain why peer-to-peer systems with ring architectures turned out to be rather fragile – and why revolutionaries have tended to organise themselves in cells.

An open challenge is how to reconcile the differences between generated network models and computer networks. Degree distribution is only one factor in the structure of a network. Li et al. closely examined the topology of computer networks [98] and found that degree-centrality attacks on the Internet do not work well since edge routers that connect to homes have much higher degree than backbone routers at major IPSs. For attacks on privacy, however, topological analysis has proven quite effective. When Danezis and Wittneben applied these network analysis ideas to privacy [99], they found that doing traffic analysis against just a few well-connected organisers can draw a surprising number of members of a dissident organisation into the surveillance net.

5.3 Large Project Management

As well as extending into system design, crime, and covert conflict, security economics may help the student of information systems management. Perhaps the largest issue here is the risk of large software project failures, which can cost billions and threaten the survival of organisations.

We noted above that perhaps 30% of large development projects fail [24], and this figure seems impervious to technological progress: better tools help engineers make larger systems, the same proportion of which still fail as before. This suggests that project failure is not technical but down to socio-economic factors such as the way decisions are taken in firms. There is thus a temptation to place what we now know about the economics of dependability alongside institutional economics and perform a gap analysis.

One interesting question is whether public-sector organisations are particularly prone to large software project failure. The CIO of the UK's Department of Work and Pensions recently admitted that only 30% of government IT projects succeed [100]. There are many possible reasons. The dependability literature teaches that large software project failures are mostly due to overambitious, vague or changing specifications, coupled with poor communications and an inability to acknowledge the signs of failure early enough to take corrective action. Good industrial project managers try to close down options fast, and get the customer to take the hard decisions upfront. Elected politicians, on the other hand, are in the business of mediating conflicts between different interests and groups in society, and as many of these conflicts are transient, avoiding or delaying hard choices is a virtue. Furthermore, at equilibrium, systems have too many features because the marginal benefit of the typical feature accrues to a small vocal group, while the cost is distributed across a large user base as a slightly increased risk of failure. This equilibrium may be even further from the optimum when design decisions are taken by elected officials: the well-known incentives to dump liability, to discount consequences that will arrive after the next election or reshuffle, and to avoid ever admitting error, surely add their share. The economics of dependability may thus be an interesting topic for researchers in schools of government.

5.4   Psychology and Security

Security engineers have so far had at least three points of contact with psychology. First, three famous experiments in social psychology showed the ease with which people could be bullied by authority figures, or persuaded by peers, to behave inappropriately. In 1951, Solomon Asch showed that most people could be induced to deny the evidence of their own eyes in order to conform to a group [101]; in 1961, Milgram showed that most people would administer severe electric shocks to an actor playing the role of a 'learner' at the behest of an experimenter playing the role of the 'teacher' – even when the 'learner' appeared to be in severe pain and begged the subject to stop [102]; and in 1971, the Stanford Prisoner Experiment showed that normal people can egg each other on to behave wickedly even in the absence of orders. There, students playing the role of warders so brutalised students playing the role of prisoners that the experiment had to be stopped [103].

Inappropriate obedience is a live problem: card thieves call up cardholders, pretend to be from the bank, and demand the PIN [2,74]. Worse, in 1995-2005, a hoaxer calling himself 'Officer Scott' ordered the managers of dozens of US stores and restaurants to detain some young employee on suspicion of theft and strip-search her or him. Various other degradations were ordered, including beatings and sexual assaults. At least 13 people who obeyed the caller and did searches were charged with crimes, and seven were convicted [104].

The second point of contact has been security usability, which has become a growth area recently; early results are collected in [105]. The third has been the study of deception – a somewhat less well-defined field, but which extends from conjuring to camouflage to the study of fraud, and which is interesting the security usability community more as phishing becomes a serious problem.

There is a potentially valuable interface with economics here too. Economic analysis traditionally assumed that the principals are rational and act out of pure self-interest. Real people depart in a number of ways from this ideal, and there has arisen in recent years a vigorous school of economic psychology or behavioural economics, which studies the effects that human social and cognitive biases have on economic decision-making. The Nobel prize was recently awarded to Kahnemann and Tversky for their seminal role in establishing this field, and particularly in decision-making under risk and uncertainty. Our mental accounting rules are not really rational; for example, most people are disproportionately reluctant to risk money they already have, and to write off money that they have wasted.

Schneier has discussed cognitive biases as the root cause of our societies' vulnerability to terrorism [106]. The psychologist Daniel Gilbert, in an article provocatively entitled 'If only gay sex caused global warming', also discusses why we are much more afraid of terrorism than of climate change [107]. We have many built-in biases that made perfect evolutionary sense on the plains of Africa half a million years ago, but may now be maladaptive. For example, we are more sensitive to risks involving intentionality, whether of a person or animal, as the common causes of violent death back then included hungry lions and enemies with sharp sticks. We are also more afraid of uncertainty; of rare or unfamiliar risks; of risks controlled by others, particularly 'outsiders' or other people we don't trust or find morally offensive. A number of these biases tie in with defects in our mental accounting.

The study of cognitive biases may also help illuminate fraud and phishing. The *fundamental attribution error* – that people often err by trying to explain things by intentionality when their causes are in fact impersonal – undermines efforts to curb phishing by teaching users about the gory design details of the Internet – for example, by telling them to parse URLs in emails that seem to come from a bank. As soon as users get confused, they will revent to judging a website by its 'look and feel'.

One potential area of research is gender. Recently people have realised that software can create barriers to females, and this has led to research work on 'gender HCI' – on how software should be designed so that women as well as men can use it effectively. The psychologist Simon Baron-Cohen classifies human brains into type S (systematizers) and type E (empathizers) [108]. Type S people are better at geometry and some kinds of symbolic reasoning, while type Es are better at language and multiprocessing. Most men are type S, while most women are type E. Of course, innate abilities can be modulated by many developmental and social factors. Yet, even at a casual reading, this material raises a suspicion that many security mechanisms are far from gender-neutral. Is it unlawful sex discrimination for a bank to expect its customers to detect phishing attacks by parsing URLs?

Another interesting insight from Baron-Cohen's work is that humans are most distinct from other primates in that we have a theory of mind; our brains are wired so that we can imagine others as being like ourselves, to empathise with them better. A side-effect is that we are much better at deception. Chimps learn to 'hack' each other, and learn defences against such exploits, more or less at random; humans can plan and execute complex deceptions. We are also equipped to detect detection by others, and no doubt our capabilities co-evolved over many generations of lies, social manipulation, sexual infidelities and revenge. The hominids who left the most descendants were those who were best at cheating, at detecting cheating by others, or both. So we finish this section with the following provocative thought: if we are really not 'homo sapiens sapiens' so much as 'homo sapiens deceptor', then perhaps the design and analysis of system security mechanisms can be seen as one of the culminations of human intellectual development.

6   Conclusions

Over the last few years, a research program on the economics of security has built many cross-disciplinary links and has produced many useful (and indeed delightful) insights from unexpected places. Many perverse things, long known to security practitioners but just dismissed as 'bad weather', turn out to be quite explicable in terms of the incentives facing individuals and organisations, and in terms of different kinds of market failure.

As for the future, the work of the hundred or so researchers active in this field has started to spill over into at least four new domains. The first is the technical question of how we can design better systems by making protocols strategy-proof so that the incentives for strategic or malicious behaviour are removed a priori.

The second is the economics of security generally, where there is convergence with economists studying topics such as crime and warfare. The causes of insurgency, and tools for understanding and dealing with insurgent networks, are an obvious attractor.

The third is the economics of dependability. Large system failures cost industry billions, and the problems seem even more intractable in the public sector. We need a better understanding of what sort of institutions can best evolve and manage large complex interconnected systems.

Finally, the border between economics and psychology seems particularly fruitful, both as a source of practical ideas for designing more usable secure systems, and as a source of deeper insights into foundational issues.

Acknowledgments. Tyler Moore is supported by the UK Marshall Aid Commemoration Commission and the US National Science Foundation.

References

1. Mastanduno, M.: Economics and Security in Statecraft and Scholarship. International Organization 52(4) (1998)
2. Anderson, R.J.: Why Cryptosystems Fail. Communications of the ACM 37(11), 32–40 (1994)
3. Ayres, I., Levitt, S.: Measuring Positive Externalities from Unobservable Victim Precaution: An Empirical Analysis of Lojack, NBER Working Paper no W5928; also in The Quarterly Journal of Economics. 113, 43–77
4. Camp, J., Wolfram, C.: Pricing Security. In: Proceedings of the CERT Information Survivability Workshop, October 24-26, 2000, pp. 31–39 (2000)
5. Varian, H.: Managing Online Security Risks, Economic Science Column, The New York Times (June 1, 2000)
6. Bohm, N., Brown, I., Gladman, B.: Electronic Commerce: Who Carries the Risk of Fraud? Journal of Information, Law and Technology 3 (2000)
7. Anderson, R.J.: Closing the Phishing Hole – Fraud, Risk and Nonbanks. In: Nonbanks in the Payment System, Santa Fe (May 2007)
8. Moore, T.: Countering Hidden-Action Attacks on Networked Systems. In: Fourth Workshop on the Economics of Information Security, Harvard (2005)
9. Anderson, R.J.: The Eternity Service. In: Pragocrypt 96 (1996)

10. Danezis, G., Anderson, R.J.: The Economics of Resisting Censorship. IEEE Security & Privacy 3(1), 45–50 (2005)
11. Goodhart, D.: Too Diverse? In: Prospect (February 2004) at
    http://www.guardian.co.uk/race/story/0,11374,1154684,00.html
12. Anderson, R.J.: Why Information Security is Hard – An Economic Perspective. In: 17th Annual Computer Security Applications Conference, December 2001 (2001), and at http://www.cl.cam.ac.uk/users/rja14/Papers/econ.pdf
13. Hirshleifer, J.: From weakest-link to best-shot: the voluntary provision of public goods. Public Choice 41, 371–386 (1983)
14. Varian, H.: System Reliability and Free Riding. In: Economics of Information Security, pp. 1–15. Kluwer, Dordrecht (2004)
15. Kunreuther, H., Heal, G.: Interdependent Security. Journal of Risk and Uncertainty 26(2–3), 231–249 (2003)
16. Katz, M., Shapiro, C.: Network Externalities, Competition, and Compatibility. The American Economic Review 75(3), 424–440 (1985)
17. Ozment, J.A., Schechter, S.E.: Bootstrapping the Adoption of Internet Security Protocols. In: Fifth Workshop on the Economics of Information Security, Cambridge, UK, June 26–28,
18. Anderson, R.J.: Open and Closed Systems are Equivalent (that is, in an ideal world. In: Perspectives on Free and Open Source Software, pp. 127–142. MIT Press, Cambridge (2005)
19. Rescorla, E.: Is Finding Security Holes a Good Idea? In: Third Workshop on the Economics of Information Security (2004)
20. Ozment, J.A.: The Likelihood of Vulnerability Rediscovery and the Social Utility of Vulnerability Hunting. In: Fourth Workshop on the Economics of Information Security (2005)
21. Ozment, J.A., Schechter, S.E.: Milk or Wine: Does Software Security Improve with Age? In: 15th Usenix Security Symposium (2006)
22. Arora, A., Telang, R., Xu, H.: Optimal Policy for Software Vulnerability Disclosure. In: Third Workshop on the Economics of Information Security, Minneapolis, MN, May 2004 (2004)
23. Arora, A., Krishnan, R., Nandkumar, A., Telang, R., Yang, Y.: Impact of Vulnerability Disclosure and Patch Availability – An Empirical Analysis. In: Third Workshop on the Economics of Information Security (2004)
24. Curtis, B., Krasner, H., Iscoe, N.: A Field Study of the Software Design Process for Large Systems. Communications of the ACM 31(11), 1268–1287 (1988)
25. Shapiro, C., Varian, H.: Information Rules. Harvard Business School Press (1998)
26. Akerlof, G.: The Market for 'Lemons: Quality Uncertainty and the Market Mechanism. The Quarterly Journal of Economics 84(3), 488–500 (1970)
27. Anderson, R.J.: Cryptography and Competition Policy – Issues with Trusted Computing. In: Second Workshop on Economics and Information Security (2003)
28. VISA, PIN Management Requirements: PIN Entry Device Security Requirements Manual (2004)
29. Schechter, S.E.: Computer Security Strength & Risk: A Quantitative Approach. Harvard University (May 2004)
30. Kannan, K., Telang, R.: Economic Analysis of Market for Software Vulnerabilities. In: Third Workshop on the Economics of Information Security (2004)
31. Böhme, R.: A Comparison of Market Approaches to Software Vulnerability Disclosure. In: Müller, G. (ed.) ETRICS 2006. LNCS, vol. 3995, pp. 298–311. Springer, Heidelberg (2006)

32. Ozment, J.A.: Bug Auctions: Vulnerability Markets Reconsidered. In: Third Workshop on the Economics of Information Security (2004)
33. Böhme, R., Kataria, G.: Models and Measures for Correlation in Cyber-Insurance. In: Fifth Workshop on the Economics of Information Security (2006)
34. Ogut, H., Menon, N., Raghunathan, S.: Cyber Insurance and IT Security Investment: Impact of Interdependent Risk. In: Fourth Workshop on the Economics of Information Security (2005)
35. Posner, R.: An Economic Theory of Privacy. Regulation, 19–26 (1978)
36. Posner, R.: Privacy, Secrecy and Reputation. Buffalo Law Review 28(1) (1979)
37. Hirshleifer, J.: Privacy: its Origin, Function and Future. Journal of Legal Studies 9, 649–664 (1980)
38. Varian, H.: Economic Apects of Personal Privacy. In: Privacy and Self-Regulation in the Information Age, National Telecommunications and Information Administration report (1996)
39. Odlyzko, A.M.: Privacy, economics, and price discrimination on the Internet. In: ICEC '03: Proceedings of the 5th international conference on Electronic commerce, pp. 355–366
40. Acquisti, A., Varian, H.: Conditioning Prices on Purchase History. Marketing Science 24(3) (2005)
41. Acquisti, A., Grossklags, J.: Privacy and Rationality: Preliminary Evidence from Pilot Data. In: Third Workshop on the Economics of Information Security, Minneapolis, Mn (2004)
42. Vila, T., Greenstadt, R., Molnar, D.: Why we can't be bothered to read privacy policies. In: Economics of Information Security, pp. 143–154. Kluwer, Dordrecht (2004)
43. Swire, P.: Efficient Confidentiality for Privacy, Security, and Confidential Business Information. Brookings-Wharton Papers on Financial Services Brookings (2003)
44. Campbell, K., Gordon, L.A., Loeb, M., Zhou, L.: The economic cost of publicly announced information security breaches: empirical evidence from the stock market. Journal of Computer Security 11(3), 431–448 (2003)
45. Acquisti, A., Friedman, A., Telang, R.: Is There a Cost to Privacy Breaches? In: Fifth Workshop on the Economics of Information Security (2006)
46. Bouckaert, J., Degryse, H.: Opt In Versus Opt Out: A Free-Entry Analysis of Privacy Policies. In: Fifth Workshop on the Economics of Information Security (2006)
47. Varian, H., Wallenberg, F., Woroch, G.: The Demographics of the Do-Not-Call List. IEEE Security & Privacy 3(1), 34–39 (2005)
48. Dingledine, R., Matthewson, N.: Anonymity Loves Company: Usability and the Network Effect. In: Workshop on Usable Privacy and Security Software (2004)
49. http://tor.eff.org
50. Varian, H.: New chips and keep a tight rein on consumers, even after they buy a product. New York Times (July 4, 2002)
51. Samuelson, P., Scotchmer, S.: The Law and Economics of Reverse Engineering. Yale Law Journal (2002)
52. von Hippel, E.: Open Source Software Projects as User Innovation Networks. Open Source Software Economics (Toulouse) (2002)
53. Lookabaugh, T., Sicker, D.: Security and Lock-In: The Case of the U.S. Cable Industry. In: Workshop on the Economics of Information Security, also in Economics of Information Security. Advances in Information Security, vol. 12, pp. 225–246. Kluwer, Dordrecht (2003)

54. Oberholzer, F., Strumpf, K.: The Effect of File Sharing on Record Sales – An Empirical Analysis. Cambridge, Ma (2004)
55. Varian, H.: Keynote address to the Third Digital Rights Management Conference, Berlin, Germany (January 13, 2005)
56. Cobb, S.: The Economics of Spam. ePrivacy Group (2003), http://www.spamhelp.org/articles/economics_of_spam.pdf
57. Böhme, R., Holz, T.: The Effect of Stock Spam on Financial Markets. In: Workshop on the Economics of Information Security (2006)
58. Frieder, L., Zittrain, J.: Spam Works: Evidence from Stock Touts and Corresponding Market Activity. Berkman Center Research Publication No. 2006-11 (2006)
59. Akella, A., Seshan, S., Karp, R., Shenker, S., Papadimitriou, C.: Selfish Behavior and Stability of the Internet: A Game-Theoretic Analysis of TCP. ACM SIG-COMM, 117–130
60. Koutsoupias, E., Papadimitriou, C.: Worst-case equilibria. In: Meinel, C., Tison, S. (eds.) STACS 99. LNCS, vol. 1563, pp. 387–396. Springer, Heidelberg (1999)
61. Roughgarden, T., Tardos, É.: How bad is selfish routing? Journal of the ACM 49(2), 236–259 (2002)
62. Fabrikant, A., Luthra, A., Maneva, E., Papadimitriou, C., Shenker, S.: On a network creation game. In: 22nd PODC, pp. 347–351 (2003)
63. Anshelevich, E., Dasgupta, A., Tardos, É., Wexler, T.: Near-optimal network design with selfish agents. In: 35th STOC, pp. 511–520 (2003)
64. Anshelevich, E., Dasgupta, A., Kleinberg, J., Tardos, É., Wexler, T., Roughgarden, T.: The price of stability for network design with fair cost allocation. In: 45th FOCS, pp. 295–304 (2004)
65. Halldórsson, M.M., Halpern, J., Li, L., Mirrokni, V.: On spectrum sharing games. In: 23rd PODC, pp. 107–114 (2004)
66. Aspnes, J., Chang, K., Yampolskiy, A.: Inoculation strategies for victims of viruses and the sum-of-squares partition problem. In: 16th ACM-SIAM Symposium on Discrete Algorithms, pp. 43–52 (2005)
67. Dwork, C., Naor, M.: Pricing via processing or combatting junk mail. In: Crypto 92, pp. 139–147.
68. Laurie, B., Clayton, R.: Proof-of-Work' Proves Not to Work. In: Third Workshop on the Economics of Information Security (2004)
69. Serjantov, A., Clayton, R.: Modeling Incentives for Email Blocking Strategies. In: Fourth Workshop on the Economics of Information Security (2005)
70. Feldman, M., Lai, K., Stoica, I., Chuang, J.: Robust Incentive Techniques for Peer-to-Peer Networks. In: Fifth ACM Conference on Electronic Commerce (2004)
71. Dellarocas, C.: Analyzing the economic efficiency of eBay-like online reputation mechanisms. In: Third ACM Conference on Electronic Commerce (2001)
72. Serjantov, A., Anderson, R.J.: On dealing with adversaries fairly. In: Third Workshop on the Economics of Information Security (2004)
73. Landwehr, C.: Improving Information Flow in the Information Security Market. In: Economics of Information Security, pp. 155–164. Kluwer, Dordrecht (2004)
74. Anderson, R.J.: Security Engineering. Wiley, Chichester (2001)
75. European Commission proposal for a Council framework decision on attacks against information systems (April 2002)
76. German Federal Government's Comments on the TCG and NGSCB in the Field of Trusted Computing (2004), at http://www.bsi.bund.de/sichere_plattformen/index.htm
77. Barnes, D.: Deworming the Internet. Texas Law Journal 83(279), 279–329 (2004)

78. Garcia, A., Horowitz, B.: The Potential for Underinvestment in Internet Security: Implications for Regulatory Policy. In: Fifth Workshop on the Economics of Information Security (2006)
79. Moore, T.: The Economics of Digital Forensics. In: Fifth Workshop on the Economics of Information Security (2006)
80. Ghose, A., Rajan, U.: The Economic Impact of Regulatory Information Disclosure on Information Security Investments, Competition, and Social Welfare. In: Fifth Workshop on the Economics of Information Security (2006)
81. Edelman, B.: Adverse Selection in Online 'Trust' Certificates. In: Fifth Workshop on the Economics of Information Security (2006)
82. Beattie, S., Arnold, S., Cowan, C., Wagle, P., Wright, C., Shostack, A.: Timing the Application of Security Patches for Optimal Uptime. In: LISA 2002, pp. 233–242 (2002)
83. Arora, A., Forman, C., Nandkumar, A., Telang, R.: Competitive and Strategic Effects in the Timing of Patch Release. In: Fifth Workshop on the Economics of Information Security (2006)
84. Gal-Or, E., Ghose, A.: Economic Consequences of Sharing Security Information. In: Information System Research, pp. 186–208 (2005)
85. Gordon, L.A., Loeb, M., Lucyshyn, W.: An Economics Perspective on the Sharing of Information Related to Security Breaches. In: First Workshop on the Economics of Information Security, Berkeley, CA, May 16-17 2002, pp. 16–17 (2002)
86. Nisan, N., Ronen, A.: Algorithmic mechanism design (extended abstract). In: STOC '99, pp. 129–140 (1999)
87. Nisan, N., Segal, I.: The communication complexity of efficient allocation problems. Draft. Second version (March 5, 2002)
88. Feigenbaum, J., Papadimitriou, C., Sami, R., Shenker, S.: A BGP-based mechanism for lowest-cost routing. In: PODC '02, pp. 173–182 (2002)
89. Shneidman, J., Parkes, D.C., Massouli, L.: Faithfulness in internet algorithms. In: PINS '04: Proceedings of the ACM SIGCOMM workshop on Practice and theory of Incentives in Networked Systems (2004)
90. Newman, M.: The structure and function of complex networks. SIAM Review 45, 167–256
91. Sah, R.: Social osmosis and patterns of crime. Journal of Political Economy 99(6), 1272–1295 (1991)
92. Ballester, C., Calvó-Armengol, A., Zenou, Y.: 'Who's, who in crime networks? Wanted – The Key Player, No 617, Working Paper Series from Research Institute of Industrial Economics
93. Bramoulle, Y., Kranton, R.: Strategic experimentation in networks. NajEcon Working Paper no. 784828000000000417 from http://www.najecon.org
94. Jackson, M.: The economics of social networks. CalTech Division of the Humanities and Social Sciences Working Paper 1237. In: Proceedings of the 9th World Congress of the Econometric Society CUP (2006)
95. Demange, G., Wooders, M.: Group formation in economics: networks, clubs and coalitions. Cambridge University Press, Cambridge (2005)
96. Albert, R., Jeong, H., Barabsi, A.-L.: Error and attack tolerance of complex networks. Nature 406(1), 387–482 (2000)
97. Nagaraja, S., Anderson, R.J.: The Topology of Covert Conflict. In: Fifth Workshop on the Economics of Information Security, UK (2006)
98. Li, L., Alderson, D., Willinger, W., Doyle, J.: A first-principles approach to understanding the internet's router-level topology. In: SIGCOMM 2004, pp. 3–14 (2004)

99. Danezis, G., Wittneben, B.: The Economics of Mass Surveillance. In: Fifth Workshop on the Economics of Information Security (2006)
100. Harley, J.: keynote talk, Government UK IT Summit, (May 2007)
101. Asch, S.E.: 'Social Psychology', OUP (1952)
102. Milgram, S.: 'Obedience to Authority: An Experimental View', HarperCollins (1974, reprinted 2004)
103. Zimbardo, P.: 'The Lucifer Effect', Random House (2007)
104. Wolfson, A.: A hoax most cruel. The Courier-Journal (2005)
105. Cranor, L.: 'Security Usability', O'Reilly (2005)
106. Schneier, B.: The Psychology of Security. In: RSA (2007), at http://www.schneier.com
107. Gilbert, D.: If only gay sex caused global warming, LA Times (July 2, 2006)
108. Baron-Cohen, S.: The Essential Difference: Men, Women, and the Extreme Male Brain, Penguin (2003)

Cryptography with Constant Input Locality*
(Extended Abstract)

Benny Applebaum, Yuval Ishai**, and Eyal Kushilevitz***

Computer Science Department, Technion, Haifa 32000, Israel
{abenny,yuvali,eyalk}@cs.technion.ac.il

Abstract. We study the following natural question: Which cryptographic primitives (if any) can be realized by functions with constant input locality, namely functions in which every bit of the *input* influences only a constant number of bits of the output? This continues the study of cryptography in low complexity classes. It was recently shown (Applebaum et al., FOCS 2004) that, under standard cryptographic assumptions, most cryptographic primitives can be realized by functions with constant *output* locality, namely ones in which every bit of the *output* is influenced by a constant number of bits from the input.

We (almost) characterize what cryptographic tasks can be performed with constant input locality. On the negative side, we show that primitives which require some form of non-malleability (such as digital signatures, message authentication, or non-malleable encryption) *cannot* be realized with constant input locality. On the positive side, assuming the intractability of certain problems from the domain of error correcting codes (namely, hardness of decoding a random linear code or the security of the McEliece cryptosystem), we obtain new constructions of one-way functions, pseudorandom generators, commitments, and semantically-secure public-key encryption schemes whose input locality is constant. Moreover, these constructions also enjoy constant *output locality*. Therefore, they give rise to cryptographic hardware that has constant-depth, constant fan-in and constant *fan-out*. As a byproduct, we obtain a pseudorandom generator whose output and input locality are both optimal (namely, 3).

1 Introduction

The question of minimizing the complexity of cryptographic primitives has been the subject of an extensive body of research (see [23,3] and references therein). On one extreme, it is natural to ask whether one can implement cryptographic primitives in NC^0, i.e., by functions in which each output bit depends on a

---

* Research supported by grant 1310/06 from the Israel Science Foundation.
** Supported by grant 2004361 from the U.S.-Israel Binational Science Foundation.
*** Supported by grant 2002354 from the U.S.-Israel Binational Science Foundation.

A. Menezes (Ed.): CRYPTO 2007, LNCS 4622, pp. 92–110, 2007.

constant number of input bits.[1] Few primitives, including pseudorandom *functions* [12], cannot even be realized in AC^0 [20]; no similar negative results are known for other primitives. However, it was shown recently [3,2] that, under standard assumptions, most cryptographic primitives can be realized by functions with output locality 4, namely by NC^0 functions in which each bit of the output depends on at most 4 bits of the input.

Another possible extreme is the complementary question of implementing cryptographic primitives by functions in which each *input* bit affects only a constant number of output bits. This was not settled by [3], and was suggested as an open problem. This natural question can be motivated from several distinct perspectives:

- (Theoretical examination of a common practice) A well known design principle for practical cryptosystems asserts that each input bit must affect many output bits. This principle is sometimes referred to as Confusion/Diffusion or Avalanche property. It is easy to justify this principle in the context of block-ciphers (which are theoretically modeled as pseudorandom functions or permutations), but is it also necessary in other cryptographic applications (e.g., stream ciphers)?
- (Hardware perspective) Unlike NC^0 functions, functions with both constant input locality and constant output locality can be computed by constant depth circuits with bounded fan-in and *bounded fan-out*. Hence, the parallel time complexity of such functions is constant in a wider class of implementation scenarios.
- (Complexity theoretic perspective) One can state the existence of cryptography in NC^0 in terms of average-case hardness of Constraint Satisfaction Problems in which each constraint involves a constant number of variables (k-CSPs). The new question can therefore be formulated in terms of k-CSPs with bounded occurrences of each variable. It is known that NP hardness and inapproximability results can be carried from the CSP setting to this setting [24,6], hence it is interesting to ask whether the same phenomenon occurs with respect to cryptographic hardness as well.

Motivated by the above, we would like to understand which cryptographic tasks (if any) can be realized with constant input *and* output locality, or even with constant input locality alone.

Another question considered in this work, which was also posed in [3], is that of closing the (small) gap between positive results for cryptography with locality 4 and the impossibility of cryptography with locality 2. It was shown in [3] that the existence of a OWF with locality 3 follows from the intractability of decoding a random linear code. The possibility of closing this gap for other primitives remained open.

---

[1] Equivalently, NC^0 is the class of functions computed by boolean circuits of polynomial size, constant depth, and bounded fan-in gates. We will also mention the classes AC^0 and NC^1 which extend this class. Specifically, in AC^0 we allow unbounded fan-in AND and OR gates, and in NC^1 the circuit depth is logarithmic.

1.1 Our Results

We provide an almost full characterization of the cryptographic tasks that can be realized by functions with constant input locality. On the negative side, we show that primitives which require some form of non-malleability (e.g., signatures, MACs, non-malleable encryption schemes) *cannot* be realized with constant (or, in some cases, even logarithmic) input locality.

On the positive side, assuming the intractability of some problems from the domain of error correcting codes, we obtain constructions of pseudorandom generators, commitments, and semantically-secure public-key encryption schemes with constant input locality and constant output locality. In particular, we obtain the following results:

- For PRGs, we answer simultaneously both of the above questions. Namely, we construct a collection[2] of PRGs whose output locality and input locality are both 3. We show that this is optimal in both output locality and input locality. Our construction is based on the intractability of decoding a random linear code. Previous constructions of PRGs (or even OWFs) [4,9] which enjoyed constant input locality and constant output locality at the same time, were based on non-standard intractability assumptions.
- We construct a non-interactive commitment scheme, in the common reference string model, in which the output locality of the commitment function is 4, and its input locality is 3. The security of this scheme also follows from the intractability of decoding a random linear code. (We can also get a non-interactive commitment scheme in the standard model under the assumption that there exists an explicit binary linear code that has a large minimal distance but is hard to decode.)
- We construct a semantically secure public-key encryption scheme whose encryption algorithm has input locality 3. This scheme is based on the security of the McEliece cryptosystem [21], an assumption which is related to the intractability of decoding a random linear code, but is seemingly stronger. Our encryption function also has constant output locality, if the security of the McEliece cryptosystem holds when it is instantiated with some error correcting code whose relative distance is constant.
- We show that MACs, signatures and non-malleable symmetric or public-key encryption schemes cannot be realized by functions whose input locality is constant or, in some cases, even logarithmic in the input length. In fact, we prove that even the weakest versions of these primitives (e.g., one-time secure MACs) cannot be constructed in this model.

1.2 Our Techniques

Our constructions rely on the machinery of *randomized encoding*, which was was explicitly introduced in [16] (under the algebraic framework of *randomizing*

---

[2] All of our collections are indexed by a public random key. That is, $\{G_z\}_{z\in\{0,1\}^*}$ is a collection of PRGs if for every z the function G_z expands its input and the pair $(z, G_z(x))$ is pseudorandom for random x and z.

polynomials) and was implicitly used, in weaker forms, in the context of secure multiparty computation (e.g., [19,8]). A randomized encoding of a function $f(x)$ is a randomized mapping $\hat{f}(x, r)$ whose output distribution depends only on the output of f. Specifically, it is required that: (1) there exists a decoder algorithm that recovers $f(x)$ from $\hat{f}(x, r)$, and (2) there exists a simulator algorithm that given $f(x)$ samples from the distribution $\hat{f}(x, r)$ induced by a uniform choice of r. That is, the distribution $\hat{f}(x, r)$ hides all the information about x except for the value $f(x)$.

In [3] it was shown that the security of most cryptographic primitives is inherited by their randomized encoding. Suppose that we want to construct some cryptographic primitive $\mathcal{P}$ in some low complexity class WEAK. Then, we can try to encode functions from a higher complexity class STRONG by functions from WEAK. Now, if we have an implementation f of the primitive $\mathcal{P}$ in STRONG, we can replace f by its encoding $\hat{f} \in$ WEAK and obtain a low-complexity implementation of $\mathcal{P}$. This paradigm was used in [3,2]. For example, it was shown that STRONG can be NC^1 and WEAK can be the class of functions whose output locality is 4.

However, it seems hard to adapt this approach to the current setting, since it is not clear whether there are non-trivial functions that can be encoded by functions with constant input locality. (In fact, we show that some very simple NC^0 functions cannot be encoded in this class.) We solve this problem by introducing a new construction of randomized encodings. Our construction shows that there exists a complexity class $\mathcal{C}$ of simple (but non-trivial) functions that can be encoded by functions with constant input locality. Roughly speaking, a function f is in $\mathcal{C}$ if each of its output bits can be written as a sum of terms over $\mathbb{F}_2$ such that each input variable of f participates in a constant number of *distinct* terms, ranging over all outputs of f. Moreover, if the algebraic degree of theses terms is constant, then f can be encoded by a function with constant input locality as well as constant output locality. (In particular, all linear functions over $\mathbb{F}_2$ admit such an encoding.)

By relying on the nice algebraic structure of intractability assumptions related to decoding random linear codes, and using techniques from [4], we construct PRGs, commitments and public-key encryption schemes in $\mathcal{C}$ whose algebraic degree is constant. Then, we use the new construction to encode these primitives, and obtain implementations whose input locality and output locality are both constant.

Interestingly, unlike previous constructions of randomized encodings, the new encoding does not have a universal simulator nor a universal decoder; that is, one should use different decoders and simulators for different functions in C. This phenomenon is inherent to the setting of constant input locality and is closely related to the fact that MACs cannot be realized in this model. See Section 6.2 for a discussion.

1.3   Previous Work

The existence of cryptographic primitives in NC^0 has been recently studied in [7,22,3]. Goldreich observed that a function whose output locality is 2 cannot even be one-way [9]. Cryan and Miltersen [7] proved that a PRG whose output locality is 3 cannot achieve a superlinear stretch; namely, it can only stretch n bits to $n + O(n)$ bits. Mossel et al. [22] extended this impossibility to functions whose output locality is 4.

On the positive side, Goldreich [9] suggested an approach for constructing OWFs based on expander graphs, an approach whose conjectured security does not follow from any well-known assumption. This general construction can be instantiated by functions with constant output locality and constant input locality. Mossel et al. [22] constructed (non-cryptographic) ε-biased generators with (non-optimal) constant input and output locality. Applebaum et al. [3,2] subsequently showed that: (1) the existence of many cryptographic primitives (including OWFs, PRGs, encryptions, signatures and hash functions) in NC^1 implies their existence with output locality 4; and (2) the existence of these primitives in NC^1 is implied by most standard cryptographic assumptions such as the intractability of factoring, discrete logarithms and lattice problems. They also constructed a OWF with (optimal) output locality 3 based on the intractability of decoding a random linear code. However, all these constructions did not achieve constant input locality. The constructions in [3] were also limited to PRGs with small (sub-linear) stretch, namely, one that stretches a seed of length n to a pseudorandom string of length $n + o(n)$. This problem was addressed by [4], who gave a construction of a linear-stretch PRG with (large) constant output locality under a non-standard assumption taken from [1]. In fact, the construction of [4] can also give an NC^0 PRG with (large) constant input locality (under the same non-standard assumption).

2   Preliminaries

Notation. All logarithms in this paper are to the base 2. We use U_n to denote a random variable uniformly distributed over $\{0,1\}^n$. We let $\text{H}_2(\cdot)$ denote the binary entropy function, i.e., for $0 < p < 1$, $\text{H}_2(p) \overset{\text{def}}{=} -p \log(p) - (1-p)\log(1-p)$. The *statistical distance* between discrete probability distributions Y and Y', denoted $\text{SD}(Y, Y')$, is defined as the maximum, over all functions A, of the *distinguishing advantage* $|\Pr[A(Y) = 1] - \Pr[A(Y') = 1]|$.

A function $\varepsilon(\cdot)$ is said to be *negligible* if $\varepsilon(n) < n^{-c}$ for any constant $c > 0$ and sufficiently large n. We will sometimes use $\text{neg}(\cdot)$ to denote an unspecified negligible function. For two distribution ensembles $\{X_n\}_{n \in \mathbb{N}}$ and $\{Y_n\}_{n \in \mathbb{N}}$, we write $X_n \equiv Y_n$ if X_n and Y_n are identically distributed, and $X_n \overset{s}{\equiv} Y_n$ if the two ensembles are *statistically indistinguishable*; namely, $\text{SD}(X_n, Y_n)$ is negligible in n. A weaker notion of closeness between distributions is that of *computational* indistinguishability: We write $X_n \overset{c}{\equiv} Y_n$ if for every (non-uniform) polynomial-size circuit family $\{A_n\}$, the distinguishing advantage $|\Pr[A_n(X_n) = 1] - \Pr[A_n(Y_n) = 1]|$ is negligible. A distribution ensemble $\{X_n\}_{n \in \mathbb{N}}$ is said to be *pseudorandom* if $X_n \overset{c}{\equiv} U_{m(n)}$ where $m(n)$ is the length of strings over which X_n is distributed.

Locality. Let $f : \{0,1\}^n \rightarrow \{0,1\}^s$ be a function. The *output locality* of f is c if each of its output bits depends on at most c input bits. The *locality* of an input variable x_i in f is c if at most c output bits depend on x_i. The *input locality* of f is c if the input locality of all the input variables of f is bounded by c. The output locality (resp. input locality) of a function family $f : \{0,1\}^* \to \{0,1\}^*$ is c if for every n the restriction of f to n-bit inputs has output locality (resp. input locality) c. We envision circuits as having their inputs at the bottom and their outputs at the top. Hence, for functions $l(n), m(n)$, we let $\text{Local}_{l(n)}^{m(n)}$ (resp. $\text{Local}_{l(n)}$, $\text{Local}^{m(n)}$) denote the non-uniform class which includes all functions $f : \{0,1\}^* \to \{0,1\}^*$ whose input locality is $l(n)$ and output locality is $m(n)$ (resp. whose input locality is $l(n)$, whose output locality is $m(n)$). The uniform versions of these classes contain only functions that can be computed in polynomial time. (All of our positive results are indeed uniform.) Note that $\text{Local}^{O(1)}$ is equivalent to the class NC^0 which is the class of functions that can be computed by constant depth circuits with bounded fan-in. Also, the class $\text{Local}_{O(1)}^{O(1)}$ is equivalent to the class of functions that can be computed by constant depth circuits with bounded fan-in and bounded fan-out.

2.1 Randomized Encoding

We review the notions of randomized encoding and randomizing polynomials from [16,17,3].

Definition 1. (Perfect randomized encoding [3]) *Let $f : \{0,1\}^n \to \{0,1\}^l$ be a function. We say that a function $\hat{f} : \{0,1\}^n \times \{0,1\}^m \to \{0,1\}^s$ is a perfect randomized encoding of f, if there exist an algorithm B, called a decoder, and a randomized algorithm S, called a simulator, for which the following hold:*

- **perfect correctness.** $B(\hat{f}(x,r)) = f(x)$ *for any input* $x \in \{0,1\}^n, r \in \{0,1\}^m$.
- **perfect privacy.** $S(f(x)) \equiv \hat{f}(x, U_m)$ *for any* $x \in \{0,1\}^n$.
- **balance.** $S(U_l) \equiv U_s$.
- **stretch preservation.** $s - (n+m) = l - n$, *or equivalently* $m = s - l$.

We refer to the second input of $\hat{f}$ as its *random input*, and to m and s as the *randomness complexity* and the *output complexity* of $\hat{f}$, respectively. The *overall complexity* (or complexity) of $\hat{f}$ is defined to be $m + s$.

Definition 1 naturally extends to infinite functions $f : \{0,1\}^* \to \{0,1\}^*$. In this case, the parameters l, m, s are all viewed as functions of the input length n, and the algorithms B, S receive 1^n as an additional input. By default, we require $\hat{f}$ to be computable in poly(n) time whenever f is. In particular, both $m(n)$ and $s(n)$ are polynomially bounded. We also require both the decoder and the simulator to be efficient.

We will rely on the following composition property of randomized encodings.

Lemma 1 (Lemma 4.6 in [3]). **(Composition)** *Let $g(x, r_g)$ be a perfect encoding of $f(x)$ and $h((x, r_g), r_h)$ be a perfect encoding of $g((x, r_g))$ (viewed as a single-argument function). Then, the function $\hat{f}(x, (r_g, r_h)) \stackrel{def}{=} h((x, r_g), r_h)$ is a perfect encoding of f.*

3   Randomized Encoding with Constant Input Locality

In this section we will show that functions with a "simple" algebraic structure (and in particular *linear* functions over $\mathbb{F}_2$) can be encoded by functions with constant input locality. We begin with the following construction that shows how to reduce the input locality of a function which is represented as a sum of functions.

Construction 1. (Basic input locality construction) *Let*

$$f(x) = (a(x) + b_1(x), a(x) + b_2(x), \ldots, a(x) + b_k(x), c_1(x), \ldots, c_l(x)),$$

where $f : \mathbb{F}_2^n \to \mathbb{F}_2^{k+l}$ *and* $a, b_1, \ldots, b_k, c_1, \ldots, c_l : \mathbb{F}_2^n \to \mathbb{F}_2$. *The encoding* $\hat{f} :$ $\mathbb{F}_2^{n+k} \to \mathbb{F}_2^{2k+l}$ *is defined by:*

$$\hat{f}(x, (r_1, \ldots, r_k)) \stackrel{def}{=} (r_1 + b_1(x), r_2 + b_2(x), \ldots, r_k + b_k(x),$$
$$a(x) - r_1, r_1 - r_2, \ldots, r_{k-1} - r_k, c_1(x), \ldots, c_l(x)) .$$

Note that after the transformation the function $a(x)$ appears only once and therefore the locality of the input variables that appear in a is reduced. In addition, the locality of all the other original input variables does not increase.

Lemma 2. (Input locality lemma) *Let* f *and* $\hat{f}$ *be as in Construction 1. Then,* $\hat{f}$ *is a perfect randomized encoding of* f.

Proof. The encoding $\hat{f}$ is stretch-preserving since the number of random inputs equals the number of additional outputs (i.e., k). Moreover, given a string $\hat{y} = \hat{f}(x, r)$ we can decode the value of $f(x)$ as follows: To recover $a(x) + b_i(x)$, compute the sum $y_i + y_{k+1} + y_{k+2} + \ldots + y_{k+i}$; To compute $c_i(x)$, simply take y_{2k+i}. This decoder never errs.

Fix some $x \in \{0, 1\}^n$. Let $y = f(x)$ and let $\hat{y}$ denote the distribution $\hat{f}(x, U_k)$. To prove perfect privacy, note that: (1) the last l bits of $\hat{y}$ are fixed and equal to $y_{[k+1 \ldots k+l]}$; (2) the first k bits of $\hat{y}$ are independently uniformly distributed; (3) the remaining bits of $\hat{y}$ are uniquely determined by y and $\hat{y}_1, \ldots, \hat{y}_k$. To see (3), observe that, by the definition of $\hat{f}$, we have $\hat{y}_{k+1} = y_1 - \hat{y}_1$; and for every $1 < i \leq k$, we also have $\hat{y}_{k+i} = y_i - \hat{y}_i - \sum_{j=1}^{i-1} \hat{y}_{k+j}$.

Hence, define a perfect simulator as follows. Given $y \in \{0, 1\}^{k+l}$, the simulator S chooses a random string r of length k, and outputs $(r, s, y_{[k+1 \ldots k+l]})$, where $s_1 = y_1 - r_1$ and $s_i = y_i - r_i - \sum_{j=0}^{i-1} s_j$ for $1 < i \leq k$. This simulator is also balanced as each of its outputs is a linear function that contains a fresh random bit. (Namely, the output bit $S(y; r)_i$ depends on: (1) r_i if $1 \leq i \leq k$; or (2) y_{i-k} if $k + 1 \leq i \leq 2k + l$.) $\square$

An additive representation of a function $f : \mathbb{F}_2^n \to \mathbb{F}_2^l$ is a representation in which each output bit is written as as a sum (over $\mathbb{F}_2$) of functions of the input x. That is, each output bit f_i can be written as $f_i(x) = \sum_{a \in T_i} a(x)$, where T_i is a set of boolean functions over n variables. We specify such an additive representation by an l-tuple $(T_1, \ldots, T_l)$ where T_i is a set of boolean functions

$a : \mathbb{F}_2^n \to \mathbb{F}_2$. We assume, without loss of generality, that none of the T_i's contains the constant functions 0 or 1. The following measures are defined with respect to a given additive representation of f. For a function $a : \mathbb{F}_2^n \to \mathbb{F}_2$, define the *multiplicity* of a to be the number of T_i's in which a appears, i.e., $\#a = |\{T_i \mid a \in T_i\}|$. For a variable x_j, we define the *rank* of x_j to be the number of different boolean functions a which depend on x_j and appear in some T_i. That is, $\mathrm{rank}(x_j) = |\{a : \mathbb{F}_2^n \to \mathbb{F}_2 \mid a \text{ depends on } x_j, a \in T_1 \bigcup \ldots \bigcup T_l\}|$.

Theorem 2. *Let $f : \mathbb{F}_2^n \to \mathbb{F}_2^l$ be a function, and fix some additive representation $(T_1, \ldots, T_l)$ for f. Then f can be perfectly encoded by a function $\hat{f} : \mathbb{F}_2^n \times \mathbb{F}_2^m \to \mathbb{F}_2^s$ such that the following hold:*

1. *The input locality of every x_j in $\hat{f}$ is at most $\mathrm{rank}(x_j)$, and the input locality of the random inputs r_i of $\hat{f}$ is at most 3.*
2. *If the output locality of f is i, then the output locality of $\hat{f}$ is $\max(i, 2)$.*
3. *The randomness complexity of $\hat{f}$ is $m = \sum_{a \in T} \#a$, where $T = \bigcup_{i=1}^{l} T_i$.*

Proof. We will use the following convention. The additive representation of a function $\hat{g}$ resulting from applying Construction 1 to a function g is the (natural) representation induced by the original additive representation of g. We construct $\hat{f}$ iteratively via the following process. (1) Let $f^{(0)} = f, i = 0$. (2) For $j = 1, \ldots, n$ do the following: (2a) while there exists a function a in $f^{(i)}$ that depends on x_j, whose multiplicity is greater than 1, apply Construction 1 to $f^{(i)}$, let $f^{(i+1)}$ be the resulting encoding and let $i = i + 1$. (3) Let $\hat{f} = f^{(i)}$. By Lemma 2, the function $f^{(i)}$ perfectly encodes the function $f^{(i-1)}$, hence by the composition property of randomized encodings (Lemma 1), the final function $\hat{f}$ perfectly encodes f. The first item of the theorem follows from the following observations: (1) In each iteration the input locality and the rank of each original variable x_j do not increase. (2) The multiplicity in $\hat{f}$ of every function a that depends on some original input variable x_j is 1. (3) The input locality of the random inputs which are introduced by the locality construction is at most 3. The last two items of the theorem follow directly from the definition of Construction 1 and the construction of $\hat{f}$. □

Remarks on Theorem 2.

1. By Theorem 2, every linear function admits an encoding of constant input locality, since each output bit can be written as a sum of degree 1 monomials. More generally, every function f whose canonic representation as a sum of monomials (i.e., each output bit is written as a sum of monomials) includes a constant number of monomials per input bit can be encoded by a function of constant input locality.
2. Interestingly, Construction 1 does not provide a universal encoding for any natural class of functions (e.g., the class of linear functions mapping n bits into l bits). This is contrasted with previous constructions of randomized encoding with constant output locality (cf. [16,17,3]). In fact, in Section 6.1 we prove that there is no universal encoding with constant input locality for the class of linear function $L : \mathbb{F}_2^n \to \mathbb{F}_2$.

3. When Theorem 2 is applied to a function family $f_n : \{0,1\}^n \to \{0,1\}^{l(n)}$ then the resulting encoding is uniform whenever the additive representation $(T_1, \ldots, T_l)$ is polynomial-time computable.
4. In Section 6.1, we show that Theorem 2 is tight in the sense that for each integer i we can construct a function f in which the rank of x_1 is i, and in every encoding $\hat{f}$ of f the input locality of x_1 is at least i.

In some cases we can combine Theorem 2 and the output-locality construction from [3, Construction 4.11] to derive an encoding which enjoys low input locality and output locality at the same time. In particular, we will use the following lemma which is implicit in [3].

Lemma 3 (implicit in [3]). *Let $f : \mathbb{F}_2^n \to \mathbb{F}_2^l$ be a function such that each of its output bits can be written as sum of monomials of degree d. Then, we can perfectly encode f by a function $\hat{f}$ such that: (1) The output locality of $\hat{f}$ is $d+1$; (2) The rank of every original variable x_i in $\hat{f}$ is equal to the rank of x_i in f; (3) The new variables introduced by $\hat{f}$ appear only in monomials of degree 1; hence their rank is 1.*

By combining Lemma 3 with Theorem 2 we get:

Corollary 1. *Let $f : \mathbb{F}_2^n \to \mathbb{F}_2^l$ be a function. Fix some additive representation for f in which each output bit is written as a sum of monomials of degree (at most) d and the* rank *of each variable is at most ρ. Then, f can be perfectly encoded by a function $\hat{f}$ of input locality $\max(\rho, 3)$ and output locality $d + 1$. Moreover, the resulting encoding is uniform whenever the additive representation is polynomial-time computable.*

Proof. First, by Lemma 3, we can perfectly encode f by a function $f' \in \mathrm{Local}^{d+1}$ without increasing the rank of the input variables of f. Next, we apply Theorem 2 and perfectly encode f' by a function $\hat{f} \in \mathrm{Local}^{d+1}_{\max(\rho,3)}$. By the composition property of randomized encodings (Lemma 1), the resulting function $\hat{f}$ perfectly encodes f. Finally, the proofs of Theorem 2 and Lemma 3 both allow to efficiently transform an additive representation of the function f into an encoding $\hat{f}$ in $\mathrm{Local}^{d+1}_{\max(\rho,3)}$. Hence, the uniformity of f is inherited by $\hat{f}$. □

We remark that Theorem 2 as well as Lemma 3 generalize to any finite field $\mathbb{F}$. Hence, so does Corollary 1.

4 Primitives with Constant Input Locality and Output Locality

4.1 Main Assumption: Intractability of Decoding Random Linear Code

Our positive results are based on the intractability of decoding a random linear code. In the following we introduce and formalize this assumption.

An (m, n, δ) *binary linear code* is a n-dimensional linear subspace of $\mathbb{F}_2^m$ in which the Hamming distance between each two distinct vectors (codewords) is at least δm. We refer to the ratio n/m as the *rate* of the code and to δ as its (relative) *distance*. Such a code can be defined by an $m \times n$ *generator matrix* whose columns span the space of codewords. It follows from the Gilbert–Varshamov bound that whenever $n/m < 1 - H_2(\delta) - \varepsilon$, almost all $m \times n$ generator matrices form (m, n, δ)-linear codes. Formally,

Fact 3 ([26]). *Let $0 < \delta < 1/2$ and $\varepsilon > 0$. Let $n/m \le 1 - H_2(\delta) - \varepsilon$. Then, a randomly chosen $m \times n$ generator matrix generates an (m, n, δ) code with probability $1 - 2^{-(\varepsilon/2)m}$.*

A proof of the above version of the Gilbert–Varshamov bound can be found in [25, Lecture 5].

Definition 2. *Let $m(n) \le \mathrm{poly}(n)$ be a code length parameter, and $0 < \mu(n) < 1/2$ be a noise parameter. We say that $\mathrm{CODE}(m, \mu)$ is intractable if for every polynomial-time adversary A,*

$$\Pr[A(C, Cx + e) = x] \le \mathrm{neg}(n),$$

where C is an $m(n) \times n$ random binary generator matrix, $x \leftarrow U_n$, and $e \in \{0, 1\}^m$ is a random error vector in which each entry is chosen to be 1 with probability μ (independently of other entries), and arithmetic is over $\mathbb{F}_2$.

Typically, we let $m(n) = O(n)$ and μ be a constant such that $n/m(n) < 1 - H_2(\mu + \varepsilon)$ where $\varepsilon > 0$ is a constant. Hence, by Fact 3, the random code C is, with overwhelming probability, an $(m, n, \mu + \varepsilon)$ code. Note that, except with negligible probability, the noise vector flips less than $\mu + \varepsilon$ of the bits of y. In this case, the fact that the noise is random (rather than adversarial) guarantees, by Shannon's coding theorem (for random linear codes), that x will be unique with overwhelming probability. That is, roughly speaking, we assume that it is intractable to correct μn *random* errors in a random linear code of relative distance $\mu + \varepsilon > \mu$. The plausibility of such an assumption is supported by the fact that a successful adversary would imply a major breakthrough in coding theory. Similar assumptions were put forward in [13,5,10].

We will rely on the following Lemma of [5].

Lemma 4. *Let $m(n)$ be a code length parameter, and $\mu(n)$ be a noise parameter. If $\mathrm{CODE}(m, \mu)$ is intractable then the distribution $(C, Cx + e)$ is pseudorandom, where C, x and e are as in Definition 2.*

4.2 Pseudorandom Generator in Local_3^3

A pseudorandom generator (PRG) is an efficiently computable function G which expands its input and its output distribution $G(U_n)$ is pseudorandom. An efficiently computable collection of functions $\{G_z\}_{z \in \{0,1\}^*}$ is a PRG collection if for every z, the function G_z expands its input and the pair $(z, G_z(x))$ is pseudorandom for random x and z. We show that pseudorandom generators (and therefore also one-way functions and one-time symmetric encryption schemes) can be realized by $\text{Local}_{O(1)}^{O(1)}$ functions. Specifically, we get a PRG in Local_3^3. In the full version we also show that such a PRG has optimal output locality and optimal input locality. We rely on the following assumption.

Assumption 4. *The problem* $\text{CODE}(13n, 1/4)$ *is intractable.*

Note that the code considered here is of rate $n/m = 1/13$ which is strictly smaller than $1 - H_2(\frac{1}{3})$. Therefore, except with negligible probability, its relative distance is at least $\frac{1}{3}$. Hence the above assumption roughly says that it is intractable to correct $n/4$ random errors in a random linear code of relative distance $\frac{1}{3}$. (We did not attempt to optimize the constant 13 in the above.)

Let $m(n) = 13n$. Let $C \leftarrow U_{m(n) \times n}$, $x \leftarrow U_n$ and $e \in \{0,1\}^m$ be a random error vector of rate $1/4$, that is, each of the entries of e is 1 with probability $1/4$ (independently of the other entries). By Lemma 4, the distribution $(C, Cx + e)$ is pseudorandom under the above assumption. Since the noise rate is $1/4$, it is natural to sample the noise distribution e by using $2m$ random bits $r_1, \ldots, r_{2m}$ and letting the i-th bit of e be the product of two fresh random bits, i.e., $e_i = r_{2i-1} \cdot r_{2i}$. We can now define the mapping $f(C, x, r) = (C, Cx + e(r))$ where $e(r) = (r_{2i-1} \cdot r_{2i})_{i=1}^m$. The output distribution of f is pseudorandom, however, f is not a PRG since it does not expand its input. In [4], it was shown how to bypass this problem by applying a randomness extractor. Namely, the following function was shown to be a PRG: $G(C, x, r, s) = (C, Cx + e(r), \text{Ext}(r, s))$. Although the setting of parameters in [4] is different than ours, a similar solution works here as well. We rely on the leftover hashing lemma of [15] and base our extractor on a family of pairwise independent hash functions (which is realized by the mapping $x \mapsto Ax + b$ where A is a random matrix and b is a random vector).[3]

Construction 5. *Let* $m = 13n$ *and let* $t = \lceil 1.1 \cdot m \rceil$. *Define the function*

$$G(x, C, r, A, b) \overset{def}{=} (C, Cx + e(r), Ar + b, A, b),$$

where $x \in \{0,1\}^n$, $C \in \{0,1\}^{m \times n}$, $r \in \{0,1\}^{2m}$, $A \in \{0,1\}^{t \times 2m}$, *and* $b \in \{0,1\}^t$.

Theorem 6. *Under Assumption 4, the function* G *defined in Construction 5 is a PRG.*

The proof of the above theorem is deferred to the full version of this paper. From now on, we fix the parameters m, t according to Construction 5. We can redefine the above construction as a collection of PRGs by letting C, A, b be the keys of the collection. Namely,

$$G_{C,A,b}(x, r) = (Cx + e(r), Ar + b).$$

We can now prove the main theorem of this section.

---

[3] We remark that in [4] one had to rely on a specially made extractor in order to maintain the large stretch of the PRG. In particular, the leftover hashing lemma could not be used there.

Theorem 7. *Under Assumption 4, there exists a collection of pseudorandom generators $\{G_z\}_{z \in \{0,1\}^{p(n)}}$ in Local_3^3. Namely, for every $z \in \{0,1\}^{p(n)}$, it holds that $G_z \in \text{Local}_3^3$.*

Proof. Fix C, A, b and write each output bit of $G_{C,A,b}(x,r)$ as a sum of monomials. Note that in this case, each variable x_i appears only in degree 1 monomials, and each variable r_i appears only in the monomial $r_{2i-1}r_{2i}$ and also in degree 1 monomials. Hence, the rank of each variable is at most 2. Moreover, the (algebraic) degree of each output bit of $G_{C,A,b}$ is at most 2. Therefore, by Corollary 1, we can perfectly encode the function $G_{C,A,b}$ by a function $\hat{G}_{C,A,b}$ in Local_3^3. In [3, Lemma 6.1] it was shown that a uniform perfect encoding of a PRG is also a PRG. Thus, we get a collection of PRGs in Local_3^3. $\qquad\square$

We can rely on Theorem 7 to obtain a one-time semantically-secure symmetric encryption scheme (E, D) whose encryption algorithm is in Local_3^3 (see [2, Construction 4.3]). (This scheme allows to encrypt an arbitrary polynomially long message with a short key.) A similar approach can be also used to give multiple message security, at the price of requiring the encryption and decryption algorithms to maintain a synchronized *state*. The results of Section 4.4 give a direct construction of public-key encryption (hence also symmetric encryption) with constant input locality under the stronger assumption that the McEliece cryptosystem is one-way secure.

4.3   Commitment in Local_3^4

We will consider a non-interactive commitment scheme in the common reference string (CRS) model. In such a scheme, the sender and the receiver share a common public random key k (that can be selected once and be used in many invocations of the scheme). To commit to a bit b, the sender computes the commitment function $\text{COM}_k(b, r)$ that outputs a commitment c using the randomness r, and sends the output to the receiver. To open the commitment, the sender sends the randomness r and the committed bit b to the receiver who checks whether the opening is valid by computing the function $\text{REC}_k(c, b, r)$. The scheme should be both (computationally) hiding and (statistically) binding. Hiding requires that $c = \text{COM}_k(b, r)$ keep b computationally secret. Binding means that, except with negligible probability over the choice of the random public key, it is impossible for the sender to open its commitment in two different ways.

We construct a commitment scheme in Local_3^4, i.e., a commitment of input locality 3 and output locality 4. Let c be a constant that satisfies $c > \frac{1}{1-H_2(1/4)}$. Let $m = m(n) = \lceil cn \rceil$. Then, by Fact 3, a random $m \times n$ generator matrix generates, except with negligible probability (i.e., $2^{-\Omega(m)} = 2^{-\Omega(n)}$), a code whose relative distance is $1/4 + \varepsilon$, for some constant $\varepsilon > 0$. The public key of our scheme will be a random $m(n) \times n$ generator matrix C. To commit to a bit b, we first choose a random information word $x \in \{0,1\}^n$ and hide it by computing $Cx + e$, where $e \in \{0,1\}^m$ is a noise vector of rate $1/8$, and then take the exclusive-or of b with a hardcore bit $\beta(x)$ of the above function. That is, we send the receiver the value $(Cx + e, b + \beta(x))$. In particular, we can use the Goldreich-Levin [14] hardcore bit and get

$$\text{COM}_C(b, (x, r, s)) = (Cx + e(r), s, b + \langle x, s \rangle),$$

where r is a random $3m$-bit string, $e(r) = (r_1 r_2 r_3, r_4 r_5 r_6, \ldots, r_{3m-2} r_{3m-1} r_{3m})$, s is a random n-bit string and $\langle \cdot, \cdot \rangle$ denotes inner product (over $\mathbb{F}_2$). Assuming that $\text{CODE}(m, 1/8)$ is intractable, this commitment hides the committed bit b. (This is so because $\langle x, s \rangle$ is unpredictable given $(C, Cx + e, s)$, cf. [10, Construction 4.4.2].) Suppose that the relative distance of C is indeed $1/4 + \varepsilon$. Then, if e contains no more than $1/8 + \varepsilon/2$ ones, x is uniquely determined by $Cx + e$. Of course, the sender might try to cheat and open the commitment ambiguously by claiming that the weight of the error vector is larger than $1/8 + \varepsilon/2$. Hence, we let the receiver verify that the Hamming weight of the noise vector e given to him by the sender in the opening phase is indeed smaller than $1/8 + \varepsilon/2$. This way, the receiver will always catch a cheating sender (assuming that C is indeed a good code). Note that an honest sender will be rejected only if its randomly chosen noise vector is heavier than $1/8 + \varepsilon/2$, which, by a Chernoff bound, happens with negligible probability (i.e., $e^{-\Omega(m)} = e^{-\Omega(n)}$) as the noise rate is $1/8$. Hence, the pair (COM, REC) defined above is indeed a commitment scheme. When C is fixed, the rank and algebraic degree of the function COM_C are 2 and 3 (with respect to the natural representation as a sum of monomials). Hence, by Corollary 1, we can encode COM_C by a function $\hat{\text{COM}}_C \in \text{Local}_3^4$. By [3], this encoding is also a commitment scheme. Summarizing, we have:

Theorem 8. *Let c be a constant that satisfies $c > \frac{1}{1 - H_2(1/4)}$, and $m = m(n) = \lceil cn \rceil$. If $\text{CODE}(m, 1/8)$ is intractable, then there exists a commitment scheme (COM, REC) in Local_3^4; i.e., for every public key C, we have $\text{COM}_C \in \text{Local}_3^4$.*

We remark that we can eliminate the use of the CRS by letting C be a generator matrix of some fixed error correcting error whose relative distance is large (i.e., $1/4$ or any other constant) in which decoding is intractable. For example, one might use the dual of a BCH code.

4.4   Semantically Secure Public-Key Encryption in $\text{Local}_3^{O(1)}$

We construct a semantically-secure public-key encryption scheme (PKE) whose encryption algorithm is in $\text{Local}_{O(1)}^{O(1)}$. Our scheme is based on the McEliece cryptosystem [21]. We begin by reviewing the general scheme proposed by McEliece.

- **System parameters:** Let $m(n) : \mathbf{N} \to \mathbf{N}$, where $m(n) > n$, and $\mu(n) : \mathbf{N} \to (0, 1)$. For every $n \in \mathbf{N}$, let $\mathcal{C}_n$ be a set of generating matrices of $(m(n), n, 2(\mu(n) + \varepsilon))$ codes that have a (universal) efficient decoding algorithm D that, given a generating matrix from $\mathcal{C}_n$, can correct up to $(\mu(n) + \varepsilon) \cdot m(n)$ errors, where $\varepsilon > 0$ is some constant. We also assume that there exists an efficient sampling algorithm that samples a generator matrix of a random code from $\mathcal{C}_n$.

- **Key Generation:** Given a security parameter 1^n, use the sampling algorithm to choose a random code from C_n and let C be its generating matrix. Let $m = m(n)$ and $\mu = \mu(n)$. Choose a random $n \times n$ non-singular matrix S over $\mathbb{F}_2$, and a random $m \times m$ permutation matrix P. Let $C' = P \cdot C \cdot S$ be the public key and P, S, D_C be the private key where D_C is the efficient decoding algorithm of C.
- **Encryption:** To encrypt $x \in \{0,1\}^n$ compute $c = C'x + e$ where $e \in \{0,1\}^m$ is an error vector of noise rate μ.
- **Decryption:** To decrypt a ciphertext c, compute $P^{-1}y = P^{-1}(C'x + e) = CSx + P^{-1}e = CSx + e'$ where e' is a vector whose weight equals to the weight of e (since P^{-1} is also a permutation matrix). Now, use the decoding algorithm D to recover the information word Sx (i.e., $D(C, CSx + P^{-1}e) = Sx$). Finally, to get x multiply Sx on the left by S^{-1}.

By Chernoff bound, the weight of the error vector e is, except with negligible probability, smaller than $(\mu + \varepsilon) \cdot m$ and so the decryption algorithm almost never errs.[4] As for the security of the scheme, it is not hard to see that the scheme is *not* semantically secure. (For example, it is easy to verify that a ciphertext c is an encryption of a given plaintext x by checking whether the weight of $c - Cx$ is approximately μn.)

However, the scheme is conjectured to be a one-way cryptosystem; namely, it is widely believed that, for proper choice of parameters, any efficient adversary fails with probability $1 - \text{neg}(n)$ to recover x from $(c = C'x + e, C')$ where x is a random n-bit string.

Suppose that the scheme is indeed one-way with respect to the parameters $m(n), \mu(n)$ and C_n. Then, we can convert it into a semantically secure public-key encryption scheme by extracting a hardcore predicate and xoring it with a 1-bit plaintext b (this transformation is similar to the one used for commitments in the previous section). That is, we encrypt the bit b by the ciphertext $(C'x + e, s, \langle s, x \rangle + b)$ where x, s are random n-bit strings, and e is a noise vector of rate μ. (Again, we use the Goldreich-Levin hardcore predicate [14].) To decrypt the message, we first compute x, by invoking the McEliece decryption algorithm, and then compute $\langle s, x \rangle$ and xor it with the last entry of the ciphertext. We refer to this scheme as the *modified* McEliece public-key encryption scheme. If the McEliece cryptosystem is indeed one-way, then $\langle s, x \rangle$ is pseudorandom given $(C', C'x + e, s)$, and thus the modified McEliece public-key is semantically secure. Formally,

Lemma 5. *If the McEliece cryptosystem is one-way with respect to the parameters $m(n), \mu(n)$ and C_n, then the modified McEliece PKE is semantically secure with respect to the same parameters.*

The proof of this lemma is essentially the same as the proof of [11, Prop. 5.3.14].

---

[4] In fact, we may allow ε to decrease with n. In such case, we might get a non-negligible decryption error. This can be fixed (without increasing the rank or the degree of the encryption function) by repeating the encryption with independent fresh randomness. Details omitted.

Let $\mu(n) = 2^{-t(n)}$. Then, we can sample the noise vector e by using the function $e(r) = \left(\prod_{j=1}^{t} r_{t \cdot (i-1)+j} \right)_{i=1}^{m(n)}$ where r is a $t(n) \cdot m(n)$ bit string. In this case, we can write the encryption function of the modified McEliece as $E_{C'}(b, x, r, s) = (C'x + e(r), s, \langle x, s \rangle + b)$.

The rank of each variable of this function is at most 2, and its algebraic degree is at most $t(n)$. Hence, by Corollary 1, we can encode it by a function $\hat{E} \in \mathrm{Local}_3^{t(n)+1}$, i.e., the output locality of $\hat{E}$ is $t(n) + 1$ and its input locality is 3. In [3, Lem. 7.5] it was shown that randomized encoding preserves the security of PKE. Namely, if (G, E, D) is a semantically secure PKE then $(G, \hat{E}, \hat{D})$ is also an encryption scheme where $\hat{E}$ is an encoding of E, $\hat{D}(c) = D(B(c))$ and B is the decoder of the encoding. Hence we have,

Theorem 9. *If the McEliece cryptosystem is one-way with respect to to the parameters* $m(n), \mu(n) = 2^{-t(n)}$ *and* $\mathcal{C}_n$, *then there exists a semantically secure PKE whose encryption algorithm is in* $\mathrm{Local}_3^{t(n)}$.

The scheme we construct encrypts a single bit, however we can use concatenation to derive a PKE for messages of arbitrary (polynomial) length without increasing the input and output locality. Theorem 9 gives a PKE with constant output locality whenever the noise rate μ is constant. Unfortunately, the binary classical Goppa Codes, which are commonly used with the McEliece scheme [21], are known to have an efficient decoding only for *subconstant* noise rate. Hence, we cannot use them for the purpose of achieving constant output locality and constant input locality simultaneously. Instead, we suggest using algebraic-geometric (AG) codes which generalize the classical Goppa Codes and enjoy an efficient decoding algorithm for constant noise rate. It seems that the use of such codes does not decrease the security of the McEliece cryptosystem [18].

5   Negative Results for Cryptographic Primitives

In this section we show that cryptographic tasks which require some form of "non-malleability" cannot be performed by functions with low input locality. This includes MACs, signatures and non-malleable encryption schemes (e.g., CCA2 secure encryptions). We prove our results in the private-key setting (i.e., for MAC and symmetric encryption). This makes them stronger as any construction that gains security in the public-key setting is also secure in the private-key setting.

We will use the following simple observation.

Lemma 6. *Let* $f : \{0,1\}^n \rightarrow \{0,1\}^{s(n)}$ *be a function in* $\mathrm{Local}_{l(n)}$. *Then, there exist a (probabilistic) polynomial-size circuit family* $\{A_n\}$ *such that for every* $x \in \{0,1\}^n$ *and* $i \in [n]$, *the output of* A_n *on* $(y = f(x), i, 1^n)$ *equals, with probability* $2^{-l(n)}$, *to the string* $y' = f(x')$ *where* x' *differs from* x *only in the* i-*th bit. In particular, when* $l(n) = O(\log(n))$, *the success probability of* A_n *is* $1/\mathrm{poly}(n)$.

Proof. Since f is in $\mathrm{Local}_{l(n)}$, the input variable x_i affects at most $l(n)$ output bits. Hence, y and y' differ in at most $l(n)$ bits. Thus, we can randomly choose y' from a set of strings whose size is at most $2^{l(n)}$. (We assume that the set of output bits which are affected by the i-th input bit is hardwired into A_n.) $\square$

In the full version we show how to get rid of the non-uniformity when f is polynomial-time computable. We now sketch the impossibility results.

5.1   MACs and Signatures

Let (S, V) be a MAC scheme, where the randomized signing function $S(k, \alpha, r)$ computes a signature β on the document α using the (random and secret) key k and randomness r, and the verification algorithm $V(k, \alpha, \beta)$ verifies that β is a valid signature on α using the key k. The scheme is secure (unforgeable) if it is infeasible to forge a signature in a chosen message attack. Namely, any efficient adversary that gets an oracle access to the signing process $S(s, \cdot)$ fails to produce a valid signature β on a document α (with respect to the corresponding key k) for which it has not requested a signature from the oracle.[5] The scheme is one-time secure if the adversary is allowed to query the signing oracle only once.

Suppose that the signature function $S(k, \alpha, r)$ has logarithmic input locality (i.e., $S(k, \alpha, r) \in \mathrm{Local}_{O(\log(|k|))}$). Then, by Lemma 6, we can break the scheme by transforming, with noticeable probability, a valid pair (α, β) of document and signature into a valid pair (α', β') for which α' and α differ in, say, their first bit. Since we used a single oracle call, such a scheme cannot be even one-time secure.

Now, suppose that for each *fixed* key k the signature function $S_k(\alpha, r) = S(k, \alpha, r)$ has input locality $\ell(n)$. In this case we cannot use Lemma 6 directly as we do not know which output bits are affected by the i-th input bit. When $\ell(n) = c$ is constant, we can easily overcome this problem. We guess which bits are affected by, say, the first input bit and then guess their value as in Lemma 6. This attack succeeds with probability $1/(m^c \cdot 2^c) = 1/\mathrm{poly}(n)$ where m is the length of the message (and so is polynomial in n). Again, this shows that the scheme is not even one-time secure. To summarize:

Theorem 10. *Let (S, V) be a MAC scheme. If $S(k, \alpha, r) \in \mathrm{Local}_{O(\log(|k|))}$ or $S_k(\alpha, r) \in \mathrm{Local}_{O(1)}$ for every k, then the scheme is not one-time secure.*

5.2   Non-malleable Encryption

Let (E, D) be a private-key encryption scheme, where the encryption function $E(k, m, r)$ computes a ciphertext c encrypting the message m using the (random and secret) key k and randomness r, and the decryption algorithm $D(k, c, r)$ decrypts the ciphertext c that was encrypted under the key k. Roughly speaking, non-malleability of an encryption scheme guarantees that it is infeasible to modify a ciphertext c into a ciphertext c' of a message related to the decryption of c. In the full version we prove the following theorem:

---

[5] When querying the signing oracle, the adversary chooses only the message and is not allowed to choose the randomness which the oracle uses to produce the signature.

Theorem 11. *Let (E, D) be a private-key encryption scheme. If $E(k, m, r) \in$ $\text{Local}_{O(\log(|k|))}$ or $E_k(m, r) \in \text{Local}_{O(1)}$ for every k, then the scheme is malleable with respect to an adversary that has no access to neither the encryption oracle nor the decryption oracle. If (G, E, D) is a public-key encryption scheme and $E_k(m, r) \in \text{Local}_{O(\log(|k|))}$ for every k, then the scheme is malleable.*

6   Negative Results for Randomized Encodings

In the following, we prove some negative results regarding randomized encoding with low input locality. In Section 6.1, we provide a necessary condition for a function to have such an encoding. We use this condition to prove that some simple (NC^0) functions cannot be encoded by functions having sub-linear input locality (regardless of the complexity of the encoding). This is contrasted with the case of constant output locality, where it is known [17,3] that *every* function f can be encoded by a function $\hat{f}$ whose output locality is 4 (and whose complexity is polynomial in the size of the branching program that computes f). In Section 6.2 we show that, although linear functions do admit efficient constant-input encoding, they do not admit an efficient *universal* constant-input encoding. That is, one should use different decoders and simulators for each linear function.

6.1   A Necessary Condition for Encoding with Low Input Locality

Let $f : \{0, 1\}^n \to \{0, 1\}^l$ be a function. For a string $x \in \{0, 1\}^n$, let $x_{\oplus i}$ denote the string x with the i-th bit flipped. Define an undirected graph G_i over $\text{Im}(f)$ such that there is an edge between the strings y and y' if there exists $x \in \{0, 1\}^n$ such that $f(x) = y$ and $f(x_{\oplus i}) = y'$. Let $\hat{f} : \{0, 1\}^n \times \{0, 1\}^m \to \{0, 1\}^s$ be a (perfectly correct and private) randomized encoding of f with decoder B and simulator S. Let t_i be the number of output bits in $\hat{f}$ which are affected by the input variable x_i. We rely on the following lemma whose proof is omitted and deferred to the full version of this paper.

Lemma 7. *The size of each connected component of G_i is at most 2^{t_i}.*

We conclude that a function $f : \{0, 1\}^n \to \{0, 1\}^l$ can be perfectly encoded by a function $\hat{f} : \{0, 1\}^n \times \{0, 1\}^m \to \{0, 1\}^s$ in Local_t only if for every $1 \le i \le n$ the size of the connected components of G_i is at most 2^t. This shows that even some very simple functions do not admit an encoding with constant input locality. Consider, for example, the function

$$f(x_1, \ldots, x_n) = x_1 \cdot (x_2, \ldots, x_n) = (x_1 \cdot x_2, x_1 \cdot x_3, \ldots, x_1 \cdot x_n).$$

For every $y \in \text{Im}(f) = \{0, 1\}^{n-1}$ it holds that $f(1, y) = y$ and $f(0, y) = 0^{n-1}$. Hence, every vertex in G_1 is a neighbor of 0^{n-1} and the size of the connected component of G_1 is 2^{n-1}. Thus, the input locality of x_1 in any perfect encoding of this function is $n - 1$. (Note that this matches the results of Section 3 since $\text{rank}(x_1) = n - 1$.)

6.2   Impossibility of Universal Encoding for Linear Functions

For a class C of functions that map n-bits into l-bits, we say that C has a universal encoding in the class $\hat{C}$ if there exists a universal simulator S and a universal decoder B such that, for every function $f_z \in C$, there is an encoding $\hat{f}_z \in \hat{C}$ which is private and correct with respect to the simulator S and the decoder B.

We show that, although linear functions do admit encodings with constant input locality, they do not admit such a *universal* encoding. Suppose that the class of linear (equivalently affine) functions had a universal encoding with constant input locality. Then, by the results of [3], we would have a one-time secure MACs (S, V) whose signing algorithm has constant input locality for every fixed key; i.e., $S_k(\alpha, r) \in \text{Local}_{O(1)}$ for every fixed key k. However, the results of Section 5.1 rule out the existence of such a scheme. In the full version of this paper, we give a more direct proof to the impossibility of obtaining a universal constant-input encoding for linear functions. This proof is based on the notions presented in Section 6.1.

Acknowledgments. We thank Ronny Roth for helpful discussions.

References

1. Alekhnovich, M.: More on average case vs approximation complexity. In: Proc. 44th FOCS, pp. 298–307 (2003)
2. Applebaum, B., Ishai, Y., Kushilevitz, E.: Computationally private randomizing polynomials and their applications. Computional Complexity 15(2), 115–162 (2006)
3. Applebaum, B., Ishai, Y., Kushilevitz, E.: Cryptography in NC^0. SIAM J. Comput. 36(4), 845–888 (2006)
4. Applebaum, B., Ishai, Y., Kushilevitz, E.: On pseudorandom generators with linear stretch in NC^0. In: Proc. 10th Random (2006)
5. Blum, A., Furst, M., Kearns, M., Lipton, R.J.: Cryptographic primitives based on hard learning problems. In: Stinson, D.R. (ed.) CRYPTO 1993. LNCS, vol. 773, pp. 278–291. Springer, Heidelberg (1994)
6. Cook, S.A.: The complexity of theorem-proving procedures. In: STOC '71: Proceedings of the third annual ACM symposium on Theory of computing, pp. 151–158. ACM Press, New York (1971)
7. Cryan, M., Miltersen, P.B.: On pseudorandom generators in NC^0. In: Sgall, J., Pultr, A., Kolman, P. (eds.) MFCS 2001. LNCS, vol. 2136, pp. 272–284. Springer, Heidelberg (2001)
8. Feige, U., Killian, J., Naor, M.: A minimal model for secure computation (extended abstract). In: Proc. of the 26th STOC, pp. 554–563 (1994)
9. Goldreich, O.: Candidate one-way functions based on expander graphs. Electronic Colloquium on Computational Complexity (ECCC), 7(090) (2000)
10. Goldreich, O.: Foundations of Cryptography: Basic Tools. Cambridge University Press, Cambridge (2001)
11. Goldreich, O.: Foundations of Cryptography: Basic Applications. Cambridge University Press, Cambridge (2004)
12. Goldreich, O., Goldwasser, S., Micali, S.: How to construct random functions. J. of the ACM. 33, 792–807 (1986)

13. Goldreich, O., Krawczyk, H., Luby, M.: On the existence of pseudorandom generators. SIAM J. Comput. 22(6), 1163–1175 (1993)
14. Goldreich, O., Levin, L.: A hard-core predicate for all one-way functions. In: Proc. 21st STOC, pp. 25–32 (1989)
15. Impagliazzo, R., Levin, L.A., Luby, M.: Pseudorandom generation from one-way functions. In: Proc. 21st STOC, pp. 12–24 (1989)
16. Ishai, Y., Kushilevitz, E.: Randomizing polynomials: A new representation with applications to round-efficient secure computation. In: Proc. 41st FOCS, pp. 294–304 (2000)
17. Ishai, Y., Kushilevitz, E.: Perfect constant-round secure computation via perfect randomizing polynomials. In: Widmayer, P., Triguero, F., Morales, R., Hennessy, M., Eidenbenz, S., Conejo, R. (eds.) ICALP 2002. LNCS, vol. 2380, pp. 244–256. Springer, Heidelberg (2002)
18. Janwa, H., Moreno, O.: Mceliece public key cryptosystems using algebraic-geometric codes. Des. Codes Cryptography 8(3), 293–307 (1996)
19. Kilian, J.: Founding cryptography on oblivious transfer. In: Proc. 20th STOC, pp. 20–31 (1988)
20. Linial, N., Mansour, Y., Nisan, N.: Constant depth circuits, fourier transform, and learnability. J. ACM 40(3), 607–620 (1993)
21. McEliece, R.J.: A public-key cryptosystem based on algebraic coding theory. Technical Report DSN PR 42-44, Jet Prop. Lab (1978)
22. Mossel, E., Shpilka, A., Trevisan, L.: On ϵ-biased generators in NC^0. In: Proc. 44th FOCS, pp. 136–145 (2003)
23. Naor, M., Reingold, O.: Synthesizers and their application to the parallel construction of pseudo-random functions. J. of Computer and Systems Sciences 58(2), 336–375 (1999)
24. Papadimitriou, C., Yannakakis, M.: Optimization, approximation, and complexity classes. J. of Computer and Systems Sciences 43, 425–440 (1991)
25. Sudan, M.: Algorithmic introduction to coding theory - lecture notes (2002), http://theory.csail.mit.edu/~madhu/FT01/
26. Varshamov, R.: Estimate of the number of signals in error correcting codes. Doklady Akademii Nauk SSSR 117, 739–741 (1957)

Universally-Composable Two-Party Computation in Two Rounds

Omer Horvitz* and Jonathan Katz**

Dept. of Computer Science, University of Maryland
{horvitz,jkatz}@cs.umd.edu

Abstract. Round complexity is a central measure of efficiency, and characterizing the round complexity of various cryptographic tasks is of both theoretical and practical importance. We show here a universally-composable (UC) protocol (in the common reference string model) for two-party computation of any functionality, where *both* parties receive output, using only two rounds. (This assumes honest parties are allowed to transmit messages simultaneously in any given round; we obtain a three-round protocol when parties are required to alternate messages.) Our results match the obvious lower bounds for the round complexity of secure two-party computation under any reasonable definition of security, regardless of what setup is used. Thus, our results establish that secure two-party computation can be obtained under a commonly-used setup assumption with *maximal* security (i.e., security under general composition) in a *minimal* number of rounds.

To give but one example of the power of our general result, we observe that as an almost immediate corollary we obtain a two-round UC blind signature scheme, matching a result by Fischlin at Crypto 2006 (though, in contrast to Fischlin, we use specific number-theoretic assumptions).

1   Introduction

Round complexity is an important measure of efficiency for cryptographic protocols, and much research has focused on trying to characterize the round complexity of various tasks such as zero knowledge [GK96a, GK96b], Byzantine agreement [PSL80, FL82, FM97, GM98], Verifiable Secret-Sharing [GIKR01, FGG$^+$06], and secure two-party/multi-party computation [Yao86, BMR90, IK00, Lin01, GIKR02, KOS03, KO04]. (Needless to say, this list is not exhaustive.) Here, we focus on the goal of secure two-party computation. Feasibility results in this case are clearly of theoretical importance, both in their own right and because two-party computation may be viewed as the "base case" for secure computation without honest majority. Results in this case are also of potential practical importance since many interesting cryptographic problems (zero knowledge, commitment, and — as we

---

* Research supported by the U.S. Army Research Laboratory and the U.K. Ministry of Defence under agreement #W911NF-06-3-0001.

** Research supported by NSF CAREER award #0447075 and U.S.-Israel Binational Science Foundation grant #2004240. A portion of this work was done at IPAM.

A. Menezes (Ed.): CRYPTO 2007, LNCS 4622, pp. 111–129, 2007.

will see — blind signatures) can be solved by casting them as specific instances of secure two-party computation.

The round complexity of secure two-party computation in the *stand-alone* setting has been studied extensively. Yao [Yao86] gave a constant-round protocol for the case when parties are honest-but-curious. Goldreich, Micali, and Wigderson [GMW87, Gol04] showed how to obtain a protocol tolerating malicious adversaries; however, their protocol does not run in a constant number of rounds. Lindell [Lin01] gave the first constant-round protocol for secure two-party computation in the presence of malicious adversaries. Katz and Ostrovsky [KO04] showed a five-round protocol for malicious adversaries, and proved a lower bound showing that five rounds are necessary (for black-box proofs of security) when no setup is assumed. (Both the upper and lower bound assume parties talk in alternating rounds.) Two-round protocols for secure two-party computation, where only a single player receives output, have been studied in, e.g., [SYY99,CCKM00]; in particular, Cachin et al. [CCKM00] show a two-round protocol for computing arbitrary functionalities in this case assuming a common reference string (CRS) available to all participating parties.

It is by now well known that protocols secure when run in a stand-alone setting may no longer be secure when many copies of the protocol are run concurrently in an arbitrary manner (possibly among different parties), or when run alongside other protocols in a larger network. To address this issue, researchers have proposed models and definitions that would guarantee security in exactly such settings [PW00, Can01]. In this work, we adopt the model of *universal composability* (UC) introduced by Canetti [Can01].

The initial work of Canetti showed broad feasibility results for UC multiparty computation in the presence of a strict majority of honest players. Unfortunately, subsequent work of Canetti and Fischlin [CF01] showed that even for the case of two parties, one of whom may be malicious, there exist functionalities that cannot be securely computed within the UC framework. Further characterization of all such "impossible-to-realize" two-party functionalities is given by [CKL06]. These impossibility results hold for the "plain" model; in contrast, it is known that these negative results can be bypassed if one is willing to assume some sort of "trusted setup". Various forms of trusted setup have been explored [CF01,BCNP04,HMU05,CDPW07,Katz07], the most common of which is the availability of a CRS to all parties in the network. Under this assumption, universally composable multi-party computation of any (well-formed) functionality is possible for any number of corrupted parties [CLOS02].

The round complexity of UC two-party computation has not been explored in detail. The two-party protocol given in [CLOS02] does not run in a constant number of rounds, though this may be due at least in part to the fact that the goal of their work was security under *adaptive* corruptions (where corruptions may happen at any point during the execution of the protocol, and not necessarily at its outset, as is the case with *passive* corruptions). Indeed, it is a long-standing open question to construct a constant-round protocol for adaptively-secure two-party computation even in the stand-alone setting. Jarecki and Shmatikov [JS07] recently showed a four-round protocol, assuming a CRS, for functionalities that generate output for only one of the parties; they also show a two-round protocol in the random oracle model. Using a standard transformation [Gol04], their protocols can be used to compute two-output functionalities at the cost of an additional round.

Our Results. We show a protocol for securely realizing any (well-formed) two-party functionality in the UC framework using only *two* rounds of communication; we stress that *both parties* may receive output. In our work, we allow both parties to simultaneously send a message in any given round (i.e., when both parties are honest), but prove security against a *rushing* adversary who may observe the other party's message in a given round before sending his own. Although this communication model is non-standard in the two-party setting, it matches the convention used in the study of multi-party protocols and allows for a more accurate characterization of the round complexity. Our result holds under any one of various standard number-theoretic assumptions, and does not rely on random oracles. We assume a CRS but, as we have seen, some form of setup is necessary for two-party computation to be possible. We consider static corruptions only; again, recall that even in the stand-alone setting it is not known how to achieve adaptive security in constant rounds.

We achieve our result via the following steps:

- We first show a two-round protocol (where only one party speaks in each round) for secure computation of any single-output functionality. This protocol is similar to that of Cachin et al. [CCKM00], though our protocol is secure in the UC framework. The protocol relies on Yao's "garbled circuit" technique [Yao86], the two-round oblivious transfer protocol of Tauman [Tau05], and the non-interactive zero-knowledge proofs of De Santis et al. [DDO+01]. Using standard techniques [Gol04, Propositions 7.2.11 and 7.4.4], this immediately implies a three-round protocol (where only one party speaks in each round) for any two-output functionality.
- As our main result, we show how two instances of our initial protocol can be run "in parallel" so as to obtain a two-round protocol (where now both parties speak[1] in each round) *even if both parties are to receive output*. The challenging aspect here is to "bind" the two executions so that each party uses the same input in each of the two protocol instances.

It is not hard to see that one-round secure computation, even if both parties are allowed to speak simultaneously, is impossible under any reasonable definition of security and regardless of any global setup assumption; a similar observation holds for two-round protocols when parties speak in alternate rounds. (It may be possible, however, to obtain such protocols given some preprocessing phase run by the two parties.) Thus, interestingly, the round complexity of our protocols is optimal *for any setting of secure computation* and not "just" for the setting of universal composability with a CRS.

---

[1] We stress again that our security analysis takes into account a *rushing* adversary.

The low round complexity of our protocol implies round-efficient solutions for various cryptographic tasks. To give an example, we show that blind signatures [Cha82] can be reduced to secure computation of a particular functionality (here, we simplify the prior result of [JL97] to the same effect); thus, as almost an immediate corollary of our result we obtain a two-round blind signature protocol, matching a recent result by Fischlin [Fis06]. Our result has certain technical advantages as compared to Fischlin's work: our scheme can be applied to *any* underlying signature scheme and achieves strong unforgeability "for free" (as long as the underlying signature scheme does); in contrast, Fischlin's result applies to a specific signature scheme and achieves strong unforgeability only with significant additional complications. On the other hand, Fishlin's result holds under more general assumptions.

As a second example, we observe that the evaluation of a trust policy, held by a server, on a set of credentials, held by a client, can be cast as an instance of two-party computation. Applying our protocol yields a solution that provides input privacy to both the client and the server in a minimal number of rounds while preserving security under general composition, a combination of traits not present in current solutions (see [BHS04, LDB03, NT05, LL06, BMC06, FAL06] and references therein). The full version of this work contains a more detailed discussion [Hor07].

2   Framework, Tools, and Assumptions

Preliminaries. Let $X = \{X(k, z)\}_{k \in \mathbb{N}, z \in \{0,1\}^*}$ denote an ensemble of binary distributions, where $X(k, z)$ represents the output of a probabilistic, polynomial time (PPT) algorithm on a *security parameter* k and advice z (the ensemble may be parameterized by additional variables, and the algorithm may take additional inputs). We say that ensembles X, Y are *computationally indistinguishable*, and write $X \overset{c}{\approx} Y$, if for any $a \in \mathbb{N}$ there exists $k_a \in \mathbb{N}$ such that for all $k > k_a$, for all z (and for all values any additional variables parameterizing the ensemble may take), we have $|\Pr[X(k, z) = 1] - \Pr[Y(k, z) = 1]| < k^{-a}$.

Universally Composable Security. We work in the Universal Composability (UC) framework of [Can01]. Our focus is on the two-party, static corruption setting. We highlight a few features of the definition we use that are standard but not universal: (1) The real model offers authenticated communication and universal access to a common reference string. Formally, this corresponds to the $(\mathcal{F}_{\text{AUTH}}, \mathcal{F}_{\text{CRS}})$-hybrid model of [Can01]. (2) Message delivery in both the real and ideal models is carried out by the adversary (contrast with [Can01], where messages between the dummy parties and the ideal functionality in the ideal model are delivered immediately). (3) The ideal functionality is not informed of party corruption by the ideal adversary. We make this choice purely to simplify the exposition; our results extend to the more general setting by the same means employed in [CLOS02] (see section 3.3 there).

Universally Composable Zero Knowledge. We use a standard definition of the *ideal zero-knowledge* functionality $\mathcal{F}_{ZK}$, following the treatment of [CLOS02]. The functionality, parameterized by a relation R, accepts a statement x to be proven, along with a witness w, from a *prover*; it then forwards x to a *verifier* if and only if $R(x, w) = 1$ (i.e., if and only if it is a correct statement). Looking ahead, our constructions will be presented in the $\mathcal{F}_{ZK}$-hybrid model.

For the case of static adversaries, De Santis et. al. [DDO+01] give a non-interactive protocol (i.e., consists of a single message from the prover to the verifier) that UC realizes $\mathcal{F}_{ZK}$ for any NP relation (see also a discussion in [CLOS02, Section 6]); the protocol is given in the CRS model and assumes the existence of enhanced trapdoor-permutations (see [Gol04, Appendix C.1] for a discussion of this assumption).

The Decisional Diffie-Hellman (DDH) Assumption. We use a two-round *oblivious transfer (OT)* protocol as a building block in our constructions; any OT protocol based on *smooth projective hashing for hard subset-membership problems* per Tauman's framework [Tau05] will do. To simplify the exposition, we describe our constructions in terms of a protocol based on the Decisional Diffie-Hellman (DDH) assumption [DH76] which we recall here.

A *group generator* GroupGen is a PPT which on input $k \in \mathbb{N}$ outputs a description of a cyclic group $\mathcal{G}$ of prime order q, the order q with $|q| \geq k$, and a generator $g \in \mathcal{G}$. Looking ahead, we will want to associate messages of length k with group elements; for simplicity we thus assume that $|q| \geq k$ (alternatively, we could use hashing). We say that the DDH problem *is hard for* GroupGen if for any PPT algorithm A, the following ensembles are computationally indistinguishable:

(1) $\left\{ (\mathcal{G}, q, g) \xleftarrow{R} \mathsf{GroupGen}(k); a, b \xleftarrow{R} \mathbb{Z}_q : A(k, z, \mathcal{G}, q, g, g^a, g^b, g^{ab}) \right\}_{k \in \mathbb{N}, z \in \{0,1\}^*}$

(2) $\left\{ (\mathcal{G}, q, g) \xleftarrow{R} \mathsf{GroupGen}(k); a, b, c \xleftarrow{R} \mathbb{Z}_q : A(k, z, \mathcal{G}, q, g, g^a, g^b, g^c) \right\}_{k \in \mathbb{N}, z \in \{0,1\}^*}.$

Yao's "Garbled Circuit" Technique. Our protocols use the "garbled-circuit" technique of Yao [Yao86, LP04]; we follow [KO04] in abstracting the technique, and refer the reader to [LP04] for a full account. Let F_k be a description of a two-input/single-output circuit whose inputs and output are of length k (the technique easily extends to lengths polynomial in k). Yao's results provide two PPT algorithms:

1. Yao_1 is a randomized algorithm which takes as input a security parameter $k \in \mathbb{N}$, a circuit F_k, and a string $y \in \{0,1\}^k$. It outputs a *garbled circuit* Circuit and *input-wire labels* $\{Z_{i,\sigma}\}_{i \in \{1,\ldots,k\}, \sigma \in \{0,1\}}$.
2. Yao_2 is a deterministic algorithm which takes as input a security parameter $k \in \mathbb{N}$, a "garbled-circuit" Circuit and values $\{Z_i\}_{i \in \{1,\ldots,k\}}$ where $Z_i \in \{0,1\}^k$. It outputs either an invalid symbol $\perp$, or a value $v \in \{0,1\}^k$.

We informally describe how the above algorithms may be used for secure computation when the participating parties are honest-but-curious. Let P_1 hold input

$x = x_1 \ldots x_k \in \{0,1\}^k$, P_2 hold input $y \in \{0,1\}^k$, and assume P_1 is to obtain the output $F_k(x,y)$. First, P_2 computes $(\mathsf{Circuit}, \{Z_{i,\sigma}\}_{i,\sigma}) \xleftarrow{R} \mathsf{Yao}_1(k, F_k, y)$ and sends Circuit to P_1. Then the players engage in k instances of *Oblivious Transfer*: in the ith instance, P_1 enters with input x_i, P_2 enters with input $(Z_{i,0}, Z_{i,1})$, and P_1 obtains $Z_i \overset{\text{def}}{=} Z_{i,x_i}$ (P_2 learns "nothing" about x_i, and P_1 learns "nothing" about $Z_{i,1-x_i}$). P_1 then computes $v \leftarrow \mathsf{Yao}_2(\mathsf{Circuit}, \{Z_i\}_i)$, and outputs v.

With the above in mind, we describe the properties required of $\mathsf{Yao}_1, \mathsf{Yao}_2$. We first require *correctness*: for any F_k, y, any output $(\mathsf{Circuit}, \{Z_{i,\sigma}\}_{i,\sigma})$ of $\mathsf{Yao}_1(k, F_k, y)$ and any x, we have $F_k(x,y) = \mathsf{Yao}_2(k, \mathsf{Circuit}, \{Z_{i,x_i}\}_i)$. The algorithms also satisfy the following notion of *security*: there exists a PPT *simulator* $\mathsf{Yao\text{-}Sim}$ which takes k, F_k, x, v as inputs, and outputs Circuit and a set of k input-wire labels $\{Z_i\}_i$; furthermore, for any PPT A, the following two ensembles are computationally indistinguishable:

(1) $\left\{ (\mathsf{Circuit}, \{Z_{i,\sigma}\}_{i,\sigma}) \xleftarrow{R} \mathsf{Yao}_1(k, F_k, y) : A(k, z, x, y, \mathsf{Circuit}, \{Z_{i,x_i}\}_i) \right\}_{\substack{k \in \mathbb{N}, z \in \{0,1\}^* \\ x,y \in \{0,1\}^k}}$

(2) $\left\{ v = F_k(x,y) : A(k, z, x, y, \mathsf{Yao\text{-}Sim}(k, F_k, x, v)) \right\}_{\substack{k \in \mathbb{N}, z \in \{0,1\}^* \\ x,y \in \{0,1\}^k}}.$

3   Round-Efficient UC Two-Party Computation

We begin by describing a two-round (where parties take turns in speaking), UC protocol for computing functionalities that provide output for only one of the parties. The protocol may be compiled into one that UC computes functionalities providing output to both parties at the cost of an additional round, using standard tools. We then show how to bind two instances of the initial protocol so as to obtain a two-round (where both parties may speak at any given round), UC protocol for computing functionalities that provide output to both parties. We conclude by showing that two rounds are necessary.

Our constructions use UC zero-knowledge, Yao's garbled circuit technique, and two-message oblivious transfer (OT) as building blocks. As mentioned earlier, any OT protocol based on *smooth projective hashing for a hard subset-membership problem* per Tauman's framework [Tau05] will do. We stress that such OT protocols satisfy a weaker notion of security than the one needed here; we use zero-knowledge to lift the security guarantees to the level we need. To simplify the exposition, we use a protocol from the framework based on the DDH assumption, simplifying a construction due to Naor and Pinkas [NP01]. We remark that other protocols conforming to Tauman's framework are known to exist under the DDH assumption [AIR01], under the Nth-residuosity assumption and under both the Quadratic-Residuosity assumption and the Extended Riemann hypothesis [Tau05].

3.1   A Two-Round Protocol for Single-Output Functionalities

Let $\mathcal{F} = \{F_k\}_{k \in \mathbb{N}}$ be a non-reactive, polynomial-sized, two-party functionality that provides output to a single party, say P_1. To simplify matters, we assume that $\mathcal{F}$ is deterministic; randomized functionalities can be handled using standard tools [Gol04, Prop. 7.4.4]. Without loss of generality, assume that F_k takes two k-bit inputs and produces a k-bit output (the protocol easily extends to input/output lengths polynomial in k). Let GroupGen be a group generator as in Sect. 2.

Informally, the first round of our protocol is used to set up k instances of oblivious transfer. The second round is used to communicate a "garbled circuit" per Yao's construction, and for completing the oblivious-transfer of circuit input-wire labels that correspond to P_1's input (cf. Sect. 2). To gain more intuition, we sketch a single oblivious transfer instance, assuming both parties are honest (the actual construction accounts for possibly malicious behavior by the parties with the aid of zero-knowledge). Let $\mathcal{G}$ be a group and g a generator, provided by GroupGen. To obtain the label corresponding to an input x_i for wire i, P_1 picks elements a, b uniformly at random from $\mathcal{G}$ and sends P_2 a tuple $(u = g^a, v = g^b, w = g^c)$, where c is set to ab if $x_i = 0$, to $(ab - 1)$ otherwise. Note that if the DDH problem is hard for GroupGen, P_2 will not be able to tell a tuple generated for $x_i = 0$ from one generated for $x_i = 1$, preserving P_1's privacy. Let $Z_{i,\sigma}$ be the label corresponding to input bit σ for wire i. P_2 selects r_0, s_0, r_1, s_1 uniformly at random from $\mathcal{G}$, and sends P_1 two pairs as follows:

$$(K_0 = u^{r_0} \cdot g^{s_0} \ , \ C_0 = w^{r_0} \cdot v^{s_0} \cdot Z_{i,0}) \ ; \text{ and}$$
$$(K_1 = u^{r_1} \cdot g^{s_1} \ , \ C_1 = (g \cdot w)^{r_1} \cdot v^{s_1} \cdot Z_{i,1}).$$

It is easy to verify that P_1 can obtain Z_{i,x_i} by computing $K_{x_i}^{-b} \cdot C_{x_i}$. Moreover, it can be shown that the tuple (K_{1-x_i}, C_{1-x_i}) is uniformly distributed (over the choice of r_{1-x_i}, s_{1-x_i}), and therefore P_1 "learns nothing" (information-theoretically) about the label corresponding to input $(1-x_i)$ for wire i, preserving P_2's privacy.

In the following, we describe our two-round protocol $\pi_{\mathcal{F}}$ for UC realizing $\mathcal{F}$ in the $\mathcal{F}_{ZK}$-hybrid model. In our description, we always let i range from 1 to k and σ range from 0 to 1.

Common Reference String: On security parameter $k \in \mathbb{N}$, the CRS is $(\mathcal{G}, q, g) \xleftarrow{R}$ GroupGen(k).

First Round: P_1 on inputs $k \in \mathbb{N}$, $x = x_1 \ldots x_k \in \{0,1\}^k$ and sid, proceeds as follows:

1. For every i, chooses a_i, b_i uniformly at random from $\mathbb{Z}_q$, sets:

$$c_i = \begin{cases} a_i b_i & x_i = 0 \\ a_i b_i - 1 & \text{otherwise,} \end{cases}$$

   and lets $u_i = g^{a_i}, v_i = g^{b_i}, w_i = g^{c_i}$.
2. P_1 sends

$$(\mathsf{ZK\text{-}prover}, sid \circ 1, (\{u_i, v_i, w_i\}_i, (\mathcal{G}, q, g), k), (x, \{a_i, b_i\}_i))$$

to $\mathcal{F}_{ZK}^1$, where $\mathcal{F}_{ZK}^1$ is parameterized by the relation:

$$R_1 = \left\{ \left((\{u_i, v_i, w_i\}_i, (\mathcal{G}, q, g), k), (x, \{a_i, b_i\}_i) \right) \,\middle|\, \begin{array}{l} \forall i, u_i = g^{a_i}, v_i = g^{b_i}, w_i = g^{c_i}, \\ \text{where } c_i = \left\{ \begin{array}{ll} a_i b_i & x_i = 0 \\ a_i b_i - 1 & \text{otherwise} \end{array} \right. \end{array} \right\}$$

and is set up such that P_1 is the prover and P_2 is the verifier.

Second Round: P_2, on inputs $k \in \mathbb{N}$, $y = y_1 \dots y_k \in \{0,1\}^k$ and sid, and upon receiving

$$(\text{ZK-proof}, sid \circ 1, (\{u_i, v_i, w_i\}_i, (\mathcal{G}', q', g'), k'))$$

from $\mathcal{F}_{ZK}^1$, first verifies that $\mathcal{G}' = \mathcal{G}, q' = q, g' = g$ and $k' = k$. If any of these conditions fail, P_2 ignores the message. Otherwise, it proceeds as follows:

1. Generates a "garbled circuit" (cf. Sect. 2) for F_k, based on its own input y. This involves choosing random coins Ω and computing $(\text{Circuit}, \{Z_{i,\sigma}\}_{i,\sigma}) \leftarrow \text{Yao}_1(k, F_k, y; \Omega)$.
2. For every i and σ, chooses $r_{i,\sigma}, s_{i,\sigma}$ uniformly at random from $\mathbb{Z}_q$, and sets:

$$K_{i,0} = u_i^{r_{i,0}} \cdot g^{s_{i,0}}, \; C_{i,0} = w_i^{r_{i,0}} \cdot v_i^{s_{i,0}} \cdot Z_{i,0};$$
$$K_{i,1} = u_i^{r_{i,1}} \cdot g^{s_{i,1}}, \; C_{i,1} = (g \cdot w_i)^{r_{i,1}} \cdot v_i^{s_{i,1}} \cdot Z_{i,1}.$$

3. Sends

$$\left(\text{ZK-prover}, sid \circ 2, \left(\begin{array}{l} \text{Circuit}, \{K_{i,\sigma}, C_{i,\sigma}\}_{i,\sigma} \\ (\mathcal{G}, q, g), k, \{u_i, v_i, w_i\}_i \end{array} \right), \left(\begin{array}{l} y, \Omega, \{Z_{i,\sigma}\}_{i,\sigma} \\ \{r_{i,\sigma}, s_{i,\sigma}\}_{i,\sigma} \end{array} \right) \right)$$

to $\mathcal{F}_{ZK}^2$, where $\mathcal{F}_{ZK}^2$ is parameterized by the relation:

$$R_2 = \left\{ \left(\left(\begin{array}{l} \text{Circuit} \\ \{K_{i,\sigma}, C_{i,\sigma}\}_{i,\sigma} \\ (\mathcal{G}, q, g), k \\ \{u_i, v_i, w_i\}_i \end{array} \right. , \begin{array}{l} y, \Omega \\ \{Z_{i,\sigma}\}_{i,\sigma} \\ \{r_{i,\sigma}, s_{i,\sigma}\}_{i,\sigma} \end{array} \right) \,\middle|\, \begin{array}{l} (\text{Circuit}, \{Z_{i,\sigma}\}_{i,\sigma}) = \text{Yao}_1(k, F_k, y; \Omega) \\ \wedge \forall i, \\ K_{i,0} = u_i^{r_{i,0}} \cdot g^{s_{i,0}}, C_{i,0} = w_i^{r_{i,0}} \cdot v_i^{s_{i,0}} \cdot Z_{i,0}; \\ K_{i,1} = u_i^{r_{i,1}} \cdot g^{s_{i,1}}, C_{i,1} = (g \cdot w_i)^{r_{i,1}} \cdot v_i^{s_{i,1}} \cdot Z_{i,1} \end{array} \right\}$$

and is set up such that P_2 is the prover and P_1 is the verifier.

Output Computation: P_1, upon receipt of message

$$(\text{ZK-proof}, sid \circ 2, (\text{Circuit}, \{K_{i,\sigma}, C_{i,\sigma}\}_{i,\sigma}, (\mathcal{G}', q', g'), k', \{u_i', v_i', w_i'\}_i))$$

from $\mathcal{F}_{ZK}^2$, first verifies that $\mathcal{G}' = \mathcal{G}, q' = q, g' = g, k' = k$ and $\{u_i', v_i', w_i'\}_i = \{u_i, v_i, w_i\}_i$. If any of these conditions fail, P_1 ignores the message. Otherwise, it completes the protocol by computing $Z_i \stackrel{\text{def}}{=} K_{i,x_i}^{-b_i} \cdot C_{i,x_i}$, computing $v \leftarrow \text{Yao}_2(k, \text{Circuit}, \{Z_i\}_i)$ and reporting v as output if $v \neq \perp$.

Concrete Round Complexity. When composed with the non-interactive protocol of De Santis et al. [DDO+01] UC-realizing F_{ZK}, our protocol takes two communication rounds. Its security now additionally rests on the existence of enhanced trapdoor permutations.

Security. The protocol may be viewed as a degenerate version of the construction we present next, and its security follows in a straightforward manner from security of the latter.

3.2   A Two-Round Protocol for Two-Output Functionalities

Let $\mathcal{F} = \left\{ F_k \stackrel{\text{def}}{=} (F_k^1, F_k^2) \right\}_{k \in \mathbb{N}}$ be a non-reactive, polynomial-sized, two-party functionality such that P_1 wishes to obtain $F_k^1(x, y)$ and P_2 wishes to obtain $F_k^2(x, y)$ when P_1 holds x and P_2 holds y. Without loss of generality, assume once more that $\mathcal{F}$ is deterministic; that x, y and the outputs of F_k^1, F_k^2 are k-bit strings; and that GroupGen is as in Sect. 2.

The protocol of the preceding section provides means to securely compute a functionality that provides output to *one* of the parties, in two rounds. To securely-compute our two-output functionality $F_k = (F_k^1, F_k^2)$, we run one instance of that protocol such that P_1 receives F_k^1 (with a first-round message originating from P_1 and a second-round message from P_2), and a second instance such that P_2 receives F_k^2 (with a first-round message originating from P_2 and a second-round message from P_1); if we allow the parties to transmit messages *simultaneously* in any given round, this yields a two-round protocol. All that's left to ensure is that each party enters both instances of the protocol with the same input. Here, we have the relation parameterizing the second round zero-knowledge functionality enforce this condition[2].

Below, we describe our two-round protocol $\pi_{\mathcal{F}}$ for UC realizing $\mathcal{F}$ in the $\mathcal{F}_{\text{ZK}}$-hybrid model when parties are allowed to send messages simultaneously in any given round. We describe our protocol from the perspective of P_1; P_2 behaves analogously (i.e., the protocol is symmetric). In the description, we always let i range from 1 to k and σ range from 0 to 1.

Common Reference String: On security parameter $k \in \mathbb{N}$, the CRS is $(\mathcal{G}, q, g) \xleftarrow{R}$ GroupGen(k).

First Round: P_1 on inputs $k \in \mathbb{N}$, $x = x_1 \ldots x_k \in \{0,1\}^k$ and *sid*, proceeds as follows:

1. For every i, chooses a_i, b_i uniformly at random from $\mathbb{Z}_q$, sets:

$$c_i = \begin{cases} a_i b_i & x_i = 0 \\ a_i b_i - 1 & \text{otherwise,} \end{cases}$$

---

[2] Alternatively, we can make the following modifications to the initial protocol: each party will add a commitment to its input to its original protocol message, and modify its zero-knowledge assertion to reflect that it has constructed its initial message with an input that is consistent with the commitment. Two instances of this protocol can now be run in parallel as above without further modifications (note that the second-round commitments become redundant). We omit the details here.

and lets $u_i = g^{a_i}, v_i = g^{b_i}, w_i = g^{c_i}$.

2. Sends

$$(\text{ZK-prover}, sid \circ 1 \circ P_1, (\{u_i, v_i, w_i\}_i, (\mathcal{G}, q, g), k), (x, \{a_i, b_i\}_i))$$

to $\mathcal{F}_{\text{ZK}}^{1, P_1 \to P_2}$, where $\mathcal{F}_{\text{ZK}}^{1, P_1 \to P_2}$ is parameterized by the relation:

$$R_1 = \left\{ ((\{u_i, v_i, w_i\}_i, (\mathcal{G}, q, g), k), (x, \{a_i, b_i\}_i)) \left| \begin{array}{l} \forall i, u_i = g^{a_i}, v_i = g^{b_i}, w_i = g^{c_i}, \\ \text{where } c_i = \begin{cases} a_i b_i & x_i = 0 \\ a_i b_i - 1 & \text{otherwise} \end{cases} \end{array} \right. \right\}$$

and is set up such that P_1 is the prover and P_2 is the verifier.

Second Round: Upon receiving the symmetric first-round message

$$(\text{ZK-proof}, sid \circ 1 \circ P_2, (\{\bar{u}_i, \bar{v}_i, \bar{w}_i\}_i, (\mathcal{G}', q', g'), k'))$$

from $\mathcal{F}_{\text{ZK}}^{1, P_2 \to P_1}$ (defined analogously to $\mathcal{F}_{\text{ZK}}^{1, P_1 \to P_2}$ using the relation R_1, but set up such that P_2 is the prover and P_1 is the verifier), P_1 verifies that $\mathcal{G}' = \mathcal{G}, q' = q, g' = g$ and $k' = k$. If any of these conditions fail, P_1 ignores the message. Otherwise, it proceeds as follows:

1. Generates a "garbled circuit" (cf. Sect. 2) for F_k^2, based on its own input x. This involves choosing random coins Ω and computing $(\text{Circuit}, \{Z_{i,\sigma}\}_{i,\sigma}) \leftarrow \text{Yao}_1(k, F_k^2, x; \Omega)$.

2. For every i and σ, chooses $r_{i,\sigma}, s_{i,\sigma}$ uniformly at random from $\mathbb{Z}_q$, and sets:

$$K_{i,0} = \bar{u}_i^{r_{i,0}} \cdot g^{s_{i,0}}, \quad C_{i,0} = \bar{w}_i^{r_{i,0}} \cdot \bar{v}_i^{s_{i,0}} \cdot Z_{i,0};$$
$$K_{i,1} = \bar{u}_i^{r_{i,1}} \cdot g^{s_{i,1}}, \quad C_{i,1} = (g \cdot \bar{w}_i)^{r_{i,1}} \cdot \bar{v}_i^{s_{i,1}} \cdot Z_{i,1}.$$

3. Sends

$$\left(\text{ZK-prover}, sid \circ 2 \circ P_1, \begin{pmatrix} \text{Circuit}, \{K_{i,\sigma}, C_{i,\sigma}\}_{i,\sigma} \\ (\mathcal{G}, q, g), k, \{\bar{u}_i, \bar{v}_i, \bar{w}_i\}_i \\ \{u_i, v_i, w_i\}_i \end{pmatrix}, \begin{pmatrix} x, \Omega, \{Z_{i,\sigma}\}_{i,\sigma} \\ \{r_{i,\sigma}, s_{i,\sigma}\}_{i,\sigma} \\ \{a_i, b_i\}_i \end{pmatrix} \right)$$

to $\mathcal{F}_{\text{ZK}}^{2, P_1 \to P_2}$, where $\mathcal{F}_{\text{ZK}}^{2, P_1 \to P_2}$ is parameterized by the relation:

$$R_2 = \left\{ \begin{pmatrix} \begin{array}{l} \text{Circuit} \\ \{K_{i,\sigma}, C_{i,\sigma}\}_{i,\sigma} \\ (\mathcal{G}, q, g), k \\ \{\bar{u}_i, \bar{v}_i, \bar{w}_i\}_i \\ \{u_i, v_i, w_i\}_i \end{array}, \begin{array}{l} x, \Omega \\ \{Z_{i,\sigma}\}_{i,\sigma} \\ \{r_{i,\sigma}, s_{i,\sigma}\}_{i,\sigma} \\ \{a_i, b_i\}_i \end{array} \end{pmatrix} \left| \begin{array}{l} (\text{Circuit}, \{Z_{i,\sigma}\}_{i,\sigma}) = \text{Yao}_1(k, F_k^2, x; \Omega) \\ \wedge \forall i, \\ K_{i,0} = \bar{u}_i^{r_{i,0}} \cdot g^{s_{i,0}}, \ C_{i,0} = \bar{w}_i^{r_{i,0}} \cdot \bar{v}_i^{s_{i,0}} \cdot Z_{i,0} \\ K_{i,1} = \bar{u}_i^{r_{i,1}} \cdot g^{s_{i,1}}, \ C_{i,1} = (g \cdot \bar{w}_i)^{r_{i,1}} \cdot \bar{v}_i^{s_{i,1}} \cdot Z_{i,1} \\ \wedge \forall i, u_i = g^{a_i}, v_i = g^{b_i}, w_i = g^{c_i}, \\ \text{where } c_i = \begin{cases} a_i b_i & x_i = 0 \\ a_i b_i - 1 & \text{otherwise} \end{cases} \end{array} \right. \right\}$$

and is set up such that P_1 is the prover and P_2 is the verifier.

Output Computation: Upon receiving the symmetric second-round message

$$(\text{ZK-proof}, sid \circ 2 \circ P_2, (\overline{\text{Circuit}}, \{\bar{K}_{i,\sigma}, \bar{C}_{i,\sigma}\}_{i,\sigma}, (\mathcal{G}', q', g'), k', \{u_i', v_i', w_i'\}_i, \{\bar{u}_i', \bar{v}_i', \bar{w}_i'\}_i))$$

from $\mathcal{F}_{\text{ZK}}^{2,P_2 \to P_1}$ (defined analogously to $\mathcal{F}_{\text{ZK}}^{2,P_1 \to P_2}$ using the relation R_2, but set up such that P_2 is the prover and P_1 is the verifier), P_1 verifies that $\mathcal{G}' = \mathcal{G}, q' = q, g' = g, k' = k$, that $\{u_i', v_i', w_i'\}_i = \{u_i, v_i, w_i\}_i$ and that $\{\bar{u}_i', \bar{v}_i', \bar{w}_i'\}_i = \{\bar{u}_i, \bar{v}_i, \bar{w}_i\}_i$. If any of these conditions fail, P_1 ignores the message. Otherwise, it completes the protocol by computing $\bar{Z}_i \stackrel{\text{def}}{=} \bar{K}_{i,x_i}^{-b_i} \cdot \bar{C}_{i,x_i}$, computing $v \leftarrow \text{Yao}_2(k, \overline{\text{Circuit}}, \{\bar{Z}_i\}_i)$ and reporting v as output if $v \neq \perp$.

Concrete Round Complexity. As in our first protocol, this takes two rounds when composed with the protocol of De Santis et al. [DDO+01] realizing $\mathcal{F}_{\text{ZK}}$; the security of our protocols now additionally relies on the existence of enhanced trapdoor permutations.

Theorem 1. *Assuming that the DDH problem is hard for* GroupGen*, the above protocol UC-realizes $\mathcal{F}$ in the $\mathcal{F}_{\text{ZK}}$-hybrid model (in the presence of static adversaries).*

Let $\mathcal{A}$ be a (static) adversary operating against $\pi_{\mathcal{F}}$ in the $\mathcal{F}_{\text{ZK}}$-hybrid model. To prove the theorem, we construct a simulator $\mathcal{S}$ such that no environment $\mathcal{Z}$ can tell with a non-negligible probability whether it is interacting with $\mathcal{A}$ and P_1, P_2 running $\pi_{\mathcal{F}}$ in the $\mathcal{F}_{\text{ZK}}$-hybrid model or with $\mathcal{S}$ and $\tilde{P}_1, \tilde{P}_2$ in the ideal process for $\mathcal{F}$. $\mathcal{S}$ will internally run a copy of $\mathcal{A}$, "simulating" for it an execution of $\pi_{\mathcal{F}}$ in the $\mathcal{F}_{\text{ZK}}$-hybrid model (by simulating an environment, a CRS, ideal $\mathcal{F}_{\text{ZK}}$ functionalities and parties P_1, P_2) that matches $\mathcal{S}$'s view of the ideal process; $\mathcal{S}$ will use $\mathcal{A}$'s actions to guide its own in the ideal process. We refer to an event as occurring in the *internal simulation* if it happens within the execution environment that $\mathcal{S}$ simulates for A. We refer to an event as occurring in the *external process* if it happens within the ideal process, in which $\mathcal{S}$ is participating. $\mathcal{S}$ proceeds as follows:

Initial Activation. $\mathcal{S}$ sets the (simulated) CRS to be $(\mathcal{G}, q, g) \stackrel{R}{\leftarrow} \text{GroupGen}(k)$. It copies the input value written by $\mathcal{Z}$ on its own input tape onto $\mathcal{A}$'s input tape and activates $\mathcal{A}$. If $\mathcal{A}$ corrupts party P_i (in the internal simulation), $\mathcal{S}$ corrupts $\tilde{P}_i$ (in the external process). When $\mathcal{A}$ completes its activation, $\mathcal{S}$ copies the output value written by $\mathcal{A}$ on its output tape to $\mathcal{S}$'s own output tape, and ends its activation.

P_2 Only is Corrupted. Upon activation, $\mathcal{S}$ copies the input value written by $\mathcal{Z}$ on its own input tape onto $\mathcal{A}$'s input tape. In addition, if $\tilde{P}_1$ has added a message $(\mathcal{F}\text{-input}_1, sid, \cdot)$ for $\mathcal{F}$ to its outgoing communication tape (in the external process; recall that $\mathcal{S}$ can only read the *public headers* of messages on the outgoing communication tapes of uncorrupted dummy parties), $\mathcal{S}$, for every i, chooses a_i, b_i uniformly at random from $\mathbb{Z}_q$, sets $u_i = g^{a_i}, v_i = g^{b_i}, w_i = g^{a_i b_i}$ for future use, and adds a message $(\text{ZK-prover}, sid \circ 1 \circ P_1, \perp, \perp)$ for $\mathcal{F}_{\text{ZK}}^{1,P_1 \to P_2}$

to P_1's outgoing communication tape (in the internal simulation; recall that A will only be able to read the *public header* of a message intended for $\mathcal{F}_{\text{ZK}}$ on the outgoing communication tape of an uncorrupted party in the $\mathcal{F}_{\text{ZK}}$-hybrid model). S then activates A.

Upon completion of A's activation, S acts as follows:

1. If A delivered the message (ZK-prover, $sid \circ 1 \circ P_1, \perp, \perp$) from P_1 to $\mathcal{F}_{\text{ZK}}^{1,P_1 \to P_2}$ (in the internal simulation), S adds the message

$$(\text{ZK-proof}, sid \circ 1 \circ P_1, (\{u_i, v_i, w_i\}_i, (\mathcal{G}, q, g), k))$$

for P_2 and A to $\mathcal{F}_{\text{ZK}}^{1,P_1 \to P_2}$'s outgoing communication tape (in the internal simulation). Informally, S constructs the message from $\mathcal{F}_{\text{ZK}}^{1,P_1 \to P_2}$ to P_2 and A (in the internal simulation) in accordance with $\pi_{\mathcal{F}}$, except that it always lets w_i be $g^{a_i b_i}$.

2. If A delivered a message

$$(\text{ZK-prover}, sid \circ 1 \circ P_2, (\{\bar{u}_i, \bar{v}_i, \bar{w}_i\}_i, (\mathcal{G}', q', g'), k'), (y, \{\bar{a}_i, \bar{b}_i\}_i))$$

from P_2 to $\mathcal{F}_{\text{ZK}}^{1,P_2 \to P_1}$ (in the internal simulation), S verifies that

$$((\{\bar{u}_i, \bar{v}_i, \bar{w}_i\}_i, (\mathcal{G}', q', g'), k'), (y, \{\bar{a}_i, \bar{b}_i\}_i)) \in R_1.$$

If the verification fails, S does nothing. Otherwise, S adds the message

$$(\text{ZK-proof}, sid \circ 1 \circ P_2, (\{\bar{u}_i, \bar{v}_i, \bar{w}_i\}_i, (\mathcal{G}', q', g'), k'))$$

for P_1 and A to $\mathcal{F}_{\text{ZK}}^{1,P_2 \to P_1}$'s outgoing communication tape (in the internal simulation), and delivers the message ($\mathcal{F}$-input$_2$, sid, y) from (the corrupted) $\tilde{P}_2$ to $\mathcal{F}$ (in the external simulation). S records the values y and $\{\bar{u}_i, \bar{v}_i, \bar{w}_i\}_i$.

3. If A delivered the message

$$(\text{ZK-proof}, sid \circ 1 \circ P_2, (\{\bar{u}_i, \bar{v}_i, \bar{w}_i\}_i, (\mathcal{G}', q', g'), k'))$$

from $\mathcal{F}_{\text{ZK}}^{1,P_2 \to P_1}$ to P_1 (in the internal simulation), S first verifies that $\tilde{P}_1$ has a message ($\mathcal{F}$-input$_1$, $sid, \cdot$) for $\mathcal{F}$ on its outgoing communication tape (in the external process) and that $\mathcal{G}' = \mathcal{G}, q' = q, g' = g$ and $k' = k$. If any of these fail, S does nothing. Otherwise, it adds the message (ZK-prover, $sid \circ 2 \circ P_1, \perp, \perp$) for $\mathcal{F}_{\text{ZK}}^{2,P_1 \to P_2}$ to P_1's outgoing communication tape (in the internal simulation), delivers ($\mathcal{F}$-input$_1$, $sid, \cdot$) from $\tilde{P}_1$ to $\mathcal{F}$ (in the external process), and notes to itself that the Round-1 message from $\mathcal{F}_{\text{ZK}}^{1,P_2 \to P_1}$ to P_1 (in the internal simulation) has been delivered. Note that once the activation of S will be complete, $\mathcal{F}$ will be in possession of both its inputs and will be activated next (in the external process).

4. If A delivered the message (ZK-prover, $sid \circ 2 \circ P_1, \perp, \perp$) from P_1 to $\mathcal{F}_{\text{ZK}}^{2,P_1 \to P_2}$, S proceeds as follows. First note that at this point, we are guaranteed that two inputs were delivered to $\mathcal{F}$ and that $\mathcal{F}$ has been activated subsequently (in the external process); therefore, $\mathcal{F}$ has written a message

($\mathcal{F}$-output$_2$, sid, v) for $\tilde{P}_2$ on its outgoing communication tape (in the external process; note that $\mathcal{S}$ may read the contents of a message from $\mathcal{F}$ to a corrupted party). Also note that at this point, $\mathcal{S}$ has recorded values y and $\{\bar{u}_i, \bar{v}_i, \bar{w}_i\}_i$ sent by (the corrupted) P_2 in its first-round message to $\mathcal{F}_{\mathrm{ZK}}^{1, P_2 \to P_1}$. $\mathcal{S}$ produces a simulated "garbled circuit" and input-wire labels using F_k^2, y and v (cf. Sect. 2) by computing $(\text{Circuit}, \{Z_i\}_i) \xleftarrow{R}$ Yao-Sim(k, F_k^2, y, v). For every i, it chooses r_{i,y_i}, s_{i,y_i} uniformly at random from $\mathbb{Z}_q$, sets:

$$K_{i,y_i} = \bar{u}_i^{r_{i,y_i}} \cdot g^{s_{i,y_i}}$$

$$C_{i,y_i} = \begin{cases} \bar{w}_i^{r_{i,y_i}} \cdot \bar{v}_i^{s_{i,y_i}} \cdot Z_i & \text{if } y_i = 0 \\ (g \cdot \bar{w}_i)^{r_{i,y_i}} \cdot \bar{v}_i^{s_{i,y_i}} \cdot Z_i & \text{otherwise,} \end{cases}$$

and sets $K_{i,1-y_i}, C_{i,1-y_i}$ to be elements selected uniformly at random from $\mathcal{G}$. It then adds the message

$$\begin{pmatrix} & \text{Circuit}, \{K_{i,\sigma}, C_{i,\sigma}\}_{i,\sigma} \\ \text{ZK-proof}, sid \circ 2 \circ P_1, & (\mathcal{G}, q, g), k, \{\bar{u}_i, \bar{v}_i, \bar{w}_i\}_i \\ & \{u_i, v_i, w_i\}_i \end{pmatrix}$$

for P_2 and $\mathcal{A}$ to the outgoing communication tape of $\mathcal{F}_{\mathrm{ZK}}^{2, P_1 \to P_2}$. Informally, $\mathcal{S}$ constructs the message in accordance with $\pi_{\mathcal{F}}$, except that it uses simulated circuit and input wire labels, and sets $\{K_{i,1-y_i}, C_{i,1-y_i}\}_i$ to be uniform elements in $\mathcal{G}$.

5. If $\mathcal{A}$ delivered a message

$$\begin{pmatrix} \overline{\text{Circuit}}, \{\bar{K}_{i,\sigma}, \bar{C}_{i,\sigma}\}_{i,\sigma} & y', \bar{\Omega}, \{\bar{Z}_{i,\sigma}\}_{i,\sigma} \\ \text{ZK-prover}, sid \circ 2 \circ P_2 \ , \ (\mathcal{G}', q', g'), k', \{u_i', v_i', w_i'\}_i, & \{\bar{r}_{i,\sigma}, \bar{s}_{i,\sigma}\}_{i,\sigma} \\ \{\bar{u}_i', \bar{v}_i', \bar{w}_i'\}_i & \{\bar{a}_i', \bar{b}_i'\}_i \end{pmatrix}$$

from $\mathcal{A}$ to $\mathcal{F}_{\mathrm{ZK}}^{2, P_2 \to P_1}$ (in the internal simulation), $\mathcal{S}$ verifies that

$$\begin{pmatrix} \overline{\text{Circuit}}, \{\bar{K}_{i,\sigma}, \bar{C}_{i,\sigma}\}_{i,\sigma} & y', \bar{\Omega}, \{\bar{Z}_{i,\sigma}\}_{i,\sigma} \\ (\mathcal{G}', q', g'), k', \{u_i', v_i', w_i'\}_i, & \{\bar{r}_{i,\sigma}, \bar{s}_{i,\sigma}\}_{i,\sigma} \\ \{\bar{u}_i', \bar{v}_i', \bar{w}_i'\}_i & \{\bar{a}_i', \bar{b}_i'\}_i \end{pmatrix} \in R_2.$$

If the verification fails, $\mathcal{S}$ does nothing. Otherwise, $\mathcal{S}$ adds the message

$$\begin{pmatrix} & \overline{\text{Circuit}}, \{\bar{K}_{i,\sigma}, \bar{C}_{i,\sigma}\}_{i,\sigma} \\ \text{ZK-proof}, sid \circ 2 \circ P_2, (\mathcal{G}', q', g'), k', \{u_i', v_i', w_i'\}_i \\ & \{\bar{u}_i', \bar{v}_i', \bar{w}_i'\}_i \end{pmatrix}$$

for P_1 and $\mathcal{A}$ to $\mathcal{F}_{\mathrm{ZK}}^{2, P_2 \to P_1}$'s outgoing communication tape (in the internal simulation).

6. If $\mathcal{A}$ delivered the message

$$\begin{pmatrix} & \overline{\text{Circuit}}, \{\bar{K}_{i,\sigma}, \bar{C}_{i,\sigma}\}_{i,\sigma} \\ \text{ZK-proof}, sid \circ 2 \circ P_2, (\mathcal{G}', q', g'), k', \{u_i', v_i', w_i'\}_i \\ & \{\bar{u}_i', \bar{v}_i', \bar{w}_i'\}_i \end{pmatrix}$$

from $\mathcal{F}_{\text{ZK}}^{2,P_2 \to P_1}$ to P_1 (in the internal simulation), $\mathcal{S}$ first checks whether a Round-1 message from $\mathcal{F}_{\text{ZK}}^{1,P_2 \to P_1}$ to P_1 (in the internal simulation) has been delivered, per Item 3 above; if not, $\mathcal{S}$ does nothing. Otherwise, we are guaranteed that two inputs were delivered to $\mathcal{F}$, and that $\mathcal{F}$ has subsequently been activated and written a message $(\mathcal{F}\text{-output}_1, sid, \cdot)$ for $\tilde{P}_1$ on its outgoing communication tape (in the external process). $\mathcal{S}$ verifies that $\mathcal{G}' = \mathcal{G}, q' = q, g' = g, k' = k$, that $\{u_i', v_i', w_i'\}_i = \{u_i, v_i, w_i\}_i$ and that $\{\bar{u}_i', \bar{v}_i', \bar{w}_i'\}_i = \{\bar{u}_i, \bar{v}_i, \bar{w}_i\}_i$ (intuitively, the checks, along those performed by $\mathcal{S}$ on behalf of $\mathcal{F}_{\text{ZK}}^{2,P_2 \to P_1}$ per Item 5 above, guarantee that (the corrupted) P_2 has used the same input consistently in both rounds, i.e., that $y' = y$); if so, $\mathcal{S}$ delivers the message $(\mathcal{F}\text{-output}_1, sid, \cdot)$ from $\mathcal{F}$ to $\tilde{P}_1$ (in the external process).

After performing one of the above (if any), $\mathcal{S}$ copies the output value written by $\mathcal{A}$ on its output tape to $\mathcal{S}$'s own output tape, and ends its activation.

Other Corruption Scenarios. $\mathcal{S}$'s actions for the case where P_1 is corrupted are symmetric to the above; its actions for the case where both parties are corrupted and for the case where neither is, are straightforward.

This concludes the description of $\mathcal{S}$. We claim that for any $\mathcal{Z}$:

$$\text{EXEC}_{\pi_{\mathcal{F}}, \mathcal{A}, \mathcal{Z}}^{\mathcal{F}_{\text{ZK}}} \stackrel{c}{\approx} \text{IDEAL}_{\mathcal{F}, \mathcal{S}, \mathcal{Z}}, \tag{1}$$

We prove the above in cases corresponding to the parties $\mathcal{A}$ corrupts. We give here an informal description of the case where P_2 only is corrupted (the case where P_1 only is corrupted is symmetric, and the cases where either or neither parties are corrupted are straightforward); refer to [Hor07] for the complete proof.

Loosely speaking, when P_2 only is corrupted, the following differences between a real-life execution of $\pi_{\mathcal{F}}$ among P_1, P_2 in the $\mathcal{F}_{\text{ZK}}$-hybrid model and the ideal process for $\mathcal{F}$ among $\tilde{P}_1, \tilde{P}_2$ may be noted: (1) in the former, P_1 computes its output based on a "garbled circuit" and obliviously-transferred input-wire labels corresponding to its input, received in the second round of the protocol, while in the latter, $\tilde{P}_1$ receives its output from $\mathcal{F}$ based on the value y that $\mathcal{S}$ obtained while simulating $\mathcal{F}_{\text{ZK}}^{1,P_2 \to P_1}$ for the first round of the protocol; (2) in the former, the first round message from $F_{\text{ZK}}^{1,P_1 \to P_2}$ to P_2 contains values $w_i = g^{c_i}$ where $c_i = a_i b_i$ when $x_i = 0$, $c_i = a_i b_i - 1$ when $x_i = 1$, while in the latter, the message (in the internal simulation) contains $w_i = g^{a_i b_i}$ for all i; (3) in the former, the second-round message from $\mathcal{F}_{\text{ZK}}^{2,P_1 \to P_2}$ to P_2 contains values $K_{i,(1-y_i)}, C_{i,(1-y_i)}$ computed as in the specification of the protocol, while in the latter, those values (in the internal simulation) are chosen uniformly at random from $\mathcal{G}$; and (4) in the former, Yao_1 is used to compute the "garbled circuit" and input-wire labels for the second-round message from $\mathcal{F}_{\text{ZK}}^{2,P_1 \to P_2}$ to P_2, while in the latter, Yao-Sim is used for that purpose, based on P_2's output from $\mathcal{F}(x, y)$, where y was obtained by $\mathcal{S}$ while simulating $F_{\text{ZK}}^{1,P_2 \to P_1}$ for the first round of the protocol.

Nevertheless, we claim that Eq. 1 holds, based on (1) the correctness of Yao's "garbled circuit" technique, the correctness of the oblivious transfer protocol and the enforcement of parties entering the two rounds of the protocol with a consistent input; (2) the hardness of the DDH assumption for GroupGen; (3) the uniformity of $K_{i,(1-y_i)}, C_{i,(1-y_i)}$ per $\pi_{\mathcal{F}}$ in $\mathcal{G}$; and (4) the security Yao's construction.

3.3   Two Rounds Are Necessary

It is almost immediate that two rounds are necessary for two-party computation under any reasonable definition of security. Loosely speaking, consider a candidate single-round protocol for a functionality that provides output to one of the parties, say P_2. Since (an honest) P_1 sends its message independently of P_2's input, P_2 can (honestly) run its output-computation side of the protocol on the incoming message multiple times using inputs of its choice, and learn the output of the functionality on each. This clearly violates security except for functions that do not depend on P_2's input.

More formally and in the context of UC security, consider the functionality $\mathcal{F}_=$, which on input a pair of two-bit strings $x, y \in \{0, 1\}^2$, provides P_2 with output 1 if $x = y$, 0 otherwise. Assume π UC realizes $\mathcal{F}_=$ in a single round. Let π^{P_1} be the procedure in π that takes P_1's input x and a security parameter k and outputs P_1's outgoing message m; let π^{P_2} be the procedure in π that takes P_2's input y, an incoming message m and a security parameter k, and computes P_2's output value v. As π UC realizes $\mathcal{F}_=$, it must be the case that for any x, y and with all but negligible probability in k, if $m \xleftarrow{R} \pi^{P_1}(x, k)$ and $v \xleftarrow{R} \pi^{P_2}(y, m, k)$, then $v = \mathcal{F}_=(x, y)$ (by considering a benign adversary that does not corrupt any party and delivers all messages as prescribed by π).

Consider an environment $\mathcal{Z}$ which picks x uniformly at random from $\{0, 1\}^2$ and provides x as input to P_1. Consider an adversary $\mathcal{A}$, participating in a real-life execution of π, that acts as follows. $\mathcal{A}$ corrupts P_2 on the onset of the execution. On an incoming message m from P_1, $\mathcal{A}$ computes $\pi^{P_2}(y, m, k)$ on *all* four strings $y \in \{0, 1\}^2$, and outputs (the lexicographically first) y on which the computation produces 1. Note that by the above, with all but negligible probability, $\mathcal{A}$ outputs x. We claim that for any ideal-process adversary $\mathcal{S}$, $\mathcal{Z}$ may distinguish a real-life execution of π in the presence of $\mathcal{A}$ from the ideal process involving $\mathcal{S}$ and $\mathcal{F}_=$. To see this, observe that $\mathcal{S}$'s probability of outputting x is at most $1/4$, as its view in the ideal process is independent of x.

4   Two-Round Universally-Composable Blind Signatures

In this section, we briefly discuss how our work can be used to construct a round-optimal (i.e., two-round) UC-secure blind signature scheme in the CRS model. We begin with a quick recap of the definitions. Roughly speaking, a blind signature scheme should guarantee *unforgeability* and *blindness*. The first requires that if a malicious user interacts with the honest signer for a total of ℓ executions of the protocol (in an arbitrarily-interleaved fashion), then the user should be unable to output valid signatures on $\ell + 1$ distinct messages. (A stronger requirement called *strong unforgeability* requires that the user cannot even output $\ell + 1$

distinct signatures on $\ell+1$ possibly-repeating messages.) Blindness requires, very informally, that a malicious signer cannot "link" a particular execution of the protocol to a particular user *even after observing the signature obtained by the user*. This is formalized (see, e.g., [Fis06]) by a game in which the signer interacts with two users in an order determined by a randomly-chosen selector bit b, and should be unable to guess the value of b (with probability significantly better than $1/2$) even after being given the signatures computed by these two users. This definition also allows the malicious signer to generate its public key in any manner (and not necessarily following the legitimate key-generation algorithm).

The above represent the "classical" definitions of security for blind signatures. Fischlin [Fis06] formally defines a *blind signature functionality* in the UC framework. He also gives a two-round protocol realizing this functionality. Interestingly, one of the motivations cited in [Fis06] for *not* relying on the generic results of [CLOS02] is the desire to obtain a round-optimal protocol.

Assume we have a (standard) signature scheme (Gen, Sign, Vrfy), and consider the (randomized) functionality $f_{\mathsf{sign}}(SK, m) = \mathsf{Sign}_{SK}(m)$. Contrary to what might be a naive first impression, secure computation of this functionality does *not* (in general) yield a secure blind signature scheme! (See also [JL97].) Specifically, the problem is that the signer may use *different* secret keys SK, SK' in different executions of the protocol. Furthermore, the public key may be set up in such a way that each secret key yields a valid signature. Then, upon observing the signatures computed by the users, the signer may be able to tell which key was used to generate each signature, thereby violating the users' anonymity.

Juels, Luby, and Ostrovsky [JL97] suggest a relatively complex method for handling this issue. We observe that a much simpler solution is possible by simply forcing the signer to use *a fixed signing key in every execution of the protocol*. This is done in the following way: To generate a public/secret key, the signer first computes $(PK, SK) \leftarrow \mathsf{Gen}(1^k)$. It then computes a (perfectly-binding) commitment $\mathsf{com} = \mathsf{Com}(SK; \omega)$ to SK using randomness ω. The public key is PK, com and the secret key contains SK and ω.

Define functionality $f^*_{\mathsf{sign}}((SK, \omega), (\mathsf{com}, m))$ as follows: if $\mathsf{Com}(SK; \omega) = \mathsf{com}$, then the second party receives output $\mathsf{Sign}_{SK}(m)$ (when Sign is randomized, the functionality chooses a uniform random tape for computing this signature). Otherwise, the second party receives output $\perp$. The first party receives no output in either case.

It is not hard to see that a protocol for secure computation of f^*_{sign} yields a secure blind signature scheme (a proof is omitted); using a UC two-party computation protocol for f^*_{sign} gives a UC blind signature scheme. Using the simple two-round protocol constructed in Sect. 3.1, and noticing that only one party receives output here, we thus obtain a two-round UC blind signature scheme.

Acknowledgments

We are extremely grateful to the anonymous reviewers from Eurocrypt 2007 who suggested a way to recast our result in the setting where both parties communicate in a given round.

References

[AIR01]   Aiello, W., Ishai, Y., Reingold, O.: Priced Oblivious Transfer: How to Sell Digital Goods. In: Pfitzmann, B. (ed.) EUROCRYPT 2001. LNCS, vol. 2045, pp. 119–135. Springer, Heidelberg (2001)

[BCNP04]   Barak, B., Canetti, R., Nielsen, J.B., Pass, R.: Universally Composable Protocols with Relaxed Set-Up Assumptions. In: Proceedings of the 45th IEEE Symposium on Foundations of Computer Science FOCS, pp. 186–195. IEEE, Los Alamitos (2004)

[BHS04]   Bradshaw, R., Holt, J., Seamons, K.: Concealing complex policies with hidden credentials. In: Proceedings of the 11th ACM Conference on Computer and Communications Security (CCS), pp. 146–157. ACM Press, New York (2004)

[BMC06]   Bagga, W., Molva, R., Crosta, S.: Policy-Based Encryption Schemes from Bilinear Pairings. In: Proceedings of the ACM Symposium on Information, Computer and Communications Security (ASIACCS), p. 368 (2006)

[BMR90]   Beaver, D., Micali, S., Rogaway, P.: The Round Complexity of Secure Protocols. In: Procedings of the 22nd ACM Symposium on Theory of Computing (STOC), pp. 503–513. ACM, New York (1990)

[Can01]   Canetti, R.: Universally Composable Security: A New Paradigm for Cryptographic Protocols. In: Proceedings of the 42nd IEEE Symposium on foundations of computer Science (FOCS), pp. 136–145. IEEE, Los Alamitos (2001), http://eprint.iacr.org/2000/067

[Cha82]   Chaum, D.: Blind Signatures for Untraceable Payments. In: McCurley, K.S., Ziegler, C.D. (eds.) Advances in Cryptology 1981 - 1997. LNCS, vol. 1440, pp. 199–203. Springer, Heidelberg (1982)

[CCKM00]   Cachin, C., Camenisch, J., Kilian, J., Müller, J.: One-Round Secure Computation and Secure Autonomous Mobile Agents. In: Welzl, E., Montanari, U., Rolim, J.D.P. (eds.) ICALP 2000. LNCS, vol. 1853, pp. 512–523. Springer, Heidelberg (2000)

[CDPW07]   Canetti, R., Dodis, Y., Pass, R., Walfish, S.: Universally Composable Security with Global Setup. In: Vadhan, S.P. (ed.) TCC 2007. LNCS, vol. 4392, pp. 61–85. Springer, Heidelberg (2007)

[CF01]   Canetti, R., Fischlin, M.: Universally Composable Commitments. In: Kilian, J. (ed.) CRYPTO 2001. LNCS, vol. 2139, pp. 19–40. Springer, Heidelberg (2001)

[CKL06]   Canetti, R., Kushilevitz, E., Lindell, Y.: On the Limitations of Universally Composable Two-Party Computation Without Set-Up Assumptions. Journal of Cryptology 19(2), 135–167 (2006)

[CLOS02]   Canetti, R., Lindell, Y., Ostrovsky, R., Sahai, A.: Universally Composable Two-Party and Multi-Party Secure Computation. In: Proceedings of the 34th ACM Symposium on Theory of Computing (STOC), pp. 494–503. ACM, New York (2002), http://eprint.iacr.org/2002/140

[DDO+01]   De Santis, A., Di Crescenzo, G., Ostrovsky, R., Persiano, G., Sahai, A.: Robust Non-Interactive Zero-Knowledge. In: Kilian, J. (ed.) CRYPTO 2001. LNCS, vol. 2139, pp. 566–598. Springer, Heidelberg (2001)

[DH76]   Diffie, W., Hellman, M.: New Directions in Cryptography. IEEE Transactions on Information Theory 22(6), 644–654 (1976)

[Fis06]   Fischlin, M.: Round-Optimal Composable Blind Signatures in the Common Reference String Model. In: Dwork, C. (ed.) CRYPTO 2006. LNCS, vol. 4117, pp. 60–77. Springer, Heidelberg (2006)

[FAL06]   Frikken, K.B., Atallah, M.J., Li, J.: Attribute-Based Access Control with Hidden Policies and Hidden Credentials. IEEE Transactions on Computers 55(10), 1259–1270 (2006)

[FGG$^+$06]   Fitzi, M., Garay, J., Gollakota, S., Rangan, C.P., Srinathan, K.: Round-Optimal and Efficient Verifiable Secret Sharing. In: Halevi, S., Rabin, T. (eds.) TCC 2006. LNCS, vol. 3876, pp. 329–342. Springer, Heidelberg (2006)

[FL82]   Fischer, M.J., Lynch, N.A.: A Lower Bound for the Time to Assure Interactive Consistency. Information Processing Letters 14(4), 183–186 (1982)

[FM97]   Feldman, P., Micali, S.: An Optimal Probabilistic Protocol for Synchronous Byzantine Agreement. SIAM Journal of Computing 26(4), 873–933 (1997)

[Gol01]   Goldreich, O.: Foundations of Cryptography: – Basic Tools, vol. 1. Cambridge University Press, Cambridge (2001)

[Gol04]   Goldreich, O.: Foundations of Cryptography: – Basic Applications, vol. 2. Cambridge University Press, Cambridge (2004)

[GIKR01]   Gennaro, R., Ishai, Y., Kushilevitz, E., Rabin, T.: The Round Complexity of Verifiable Secret Sharing and Secure Multicast. In: Proceedings of the 33rd ACM Symposium on Theory of Computing STOC, pp. 580–589. ACM, New York (2001)

[GIKR02]   Gennaro, R., Ishai, Y., Kushilevitz, E., Rabin, T.: On 2-Round Secure Multiparty Computation. In: Yung, M. (ed.) CRYPTO 2002. LNCS, vol. 2442, pp. 178–193. Springer, Heidelberg (2002)

[GK96a]   Goldreich, O., Kahan, A.: How to Construct Constant-Round Zero-Knowledge Proof Systems for NP. Journal of Cryptology 9(3), 167–190 (1996)

[GK96b]   Goldreich, O., Krawczyk, H.: On the Composition of Zero-Knowledge Proof Systems. SIAM Journal of Computing 25(1), 169–192 (1996)

[GM98]   Garay, J., Moses, Y.: Fully Polynomial Byzantine Agreement for $n > 3t$ Processors in $t + 1$ Rounds. SIAM Journal of Computing 27(1), 247–290 (1998)

[GMR89]   Goldwasser, S., Micali, S., Rackoff, C.: The Knowledge Complexity of Interactive Proof Systems. SIAM J. of Computing 18(1), 186–208 (1989)

[GMW87]   Goldreich, O., Micali, S., Wigderson, A.: How to Play any Mental Game, or A Completeness Theorem for Protocols with Honest Majority. In: Proceedings of the 19th ACM Symposium on Theory Computing (STOC), pp. 218–229. ACM, New York (1987)

[Hor07]   Horvitz, O.: Expressiveness of Definitions and Efficiency of Constructions in Computational Cryptography. Ph.D. thesis, University of Maryland (2007)

[HMU05]   Hofheinz, D., Müller-Quade, J., Unruh, D.: Universally Composable Zero-Knowledge Arguments and Commitments from Signature Cards. In: Proceedings of the 5th Central European Conference on Cryptology — MoraviaCrypt (2005)

[IK00]   Ishai, Y., Kushilevitz, E.: Randomizing Polynomials: A New Representation with Applications to Round-Efficient Secure Computation. In: Proceedings of the 41st IEEE Symposium on Foundations of Computer Science (FOCS), pp. 294–304. IEEE, Los Alamitos (2000)

[JL97]   Juels, A., Luby, M., Ostrovsky, R.: Security of Blind Digital Signatures. In: Kaliski Jr., B.S. (ed.) CRYPTO 1997. LNCS, vol. 1294, pp. 150–164. Springer, Heidelberg (1997)

[JS07]   Jarecki, S., Shmatikov, V.: Efficient Two-Party Secure Computation on
           Committed Inputs. In: Naor, M. (ed.) EUROCRYPT 2007, vol. 4515,
           Springer, Heidelberg (2007)
[Katz07]   Katz, J.: Universally Composable Multi-Party Computation using
           Tamper-Proof Hardware. In: Naor, M. (ed.) EUROCRYPT 2007,
           vol. 4515, Springer, Heidelberg (2007)
[KO04]   Katz, J., Ostrovsky, R.: Round-Optimal Secure Two-Party Computation.
           In: Franklin, M. (ed.) CRYPTO 2004. LNCS, vol. 3152, Springer, Heidel-
           berg (2004)
[KOS03]   Katz, J., Ostrovsky, R., Smith, A.: Round Efficiency of Multi-Party Com-
           putation with a Dishonest Majority. In: Biham, E. (ed.) EUROCRPYT
           2003. LNCS, vol. 2656, Springer, Heidelberg (2003)
[Lin01]   Lindell, Y.: Parallel Coin-Tossing and Constant-Round Secure Two-Party
           Computation. Journal of Crypto 16(3), 143–184 (2003)
[LDB03]   Li, N., Du, W., Boneh, D.: Oblivious Signature-Based Envelope. In: Pro-
           ceedings of the 22nd ACM Symposium on Principles of Distributed Com-
           puting (PODC), pp. 182–189. ACM Press, New York (2003)
[LL06]   Li, J., Li., N.: A Construction for General and Efficient Oblivious Com-
           mitment Based Envelope Protocols. In: Ning, P., Qing, S., Li, N. (eds.)
           ICICS 2006. LNCS, vol. 4307, pp. 122–138. Springer, Heidelberg (2006)
[LP04]   Lindell, Y., Pinkas, B.: A Proof of Yao's Protocol for Secure Two-
           Party Computation. Journal of Cryptology, Full version available at
           http://eprint.iacr.org/,/175 (toappear)
[NP01]   Naor, M., Pinkas, B.: Efficient Oblivious Transfer Protocols. In: Proceed-
           ings of the 12th Symposium on Discrete Algorithms (SODA), pp. 448–457
           (2001)
[NT05]   Nasserian, S., Tsudik, G.: Revisiting Oblivious Signature-Based Envelopes
           Available at, http://eprint.iacr.org/2005/283
[PSL80]   Pease, M., Shostak, R., Lamport, L.: Reaching Agreement in the Presence
           of Faults. Journal of the ACM 27(2), 228–234 (1980)
[PW00]   Pfitzmann, B., Waidner, M.: Composition and Integrity Preservation of
           Secure Reactive Systems. In: Proceedings of the 7th ACM Conference
           on Computer and Communications Security (CCS), pp. 245–254. ACM
           Press, New York (2000)
[SYY99]   Sander, T., Young, A., Yung, M.: Non-Interactive CryptoComputing For
           NC[1]. In: Proceedings of the 40th IEEE Symposium on Foundations of
           Computer Science (FOCS), pp. 554–567. IEEE, Los Alamitos (1999)
[Tau05]   Tauman Kalai, Y.: Smooth Projective Hashing and Two-Message Obliv-
           ious Transfer. In: Cramer, R.J.F. (ed.) EUROCRYPT 2005. LNCS,
           vol. 3494, pp. 78–95. Springer, Heidelberg (2005)
[Yao86]   Yao, A.C.-C.: How to Generate and Exchange secrets. In: Proceedings of
           the 27 th IEEE Symposium on Foundations of Computer Science (FOCS),
           pp. 162–167. IEEE, Los Alamitos (1986)

Indistinguishability Amplification

Ueli Maurer[1], Krzysztof Pietrzak[2], and Renato Renner[3]

[1] Department of Computer Science, ETH Zurich
maurer@inf.ethz.ch
[2] CWI Amsterdam
pietrzak@cwi.nl
[3] University of Cambridge
r.renner@damtp.cam.ac.uk

Abstract. Many aspects of cryptographic security proofs can be seen as the proof that a certain system (e.g. a block cipher) is indistinguishable from an ideal system (e.g. a random permutation), for different types of distinguishers.

This paper presents a new generic approach to proving upper bounds on the information-theoretic distinguishing advantage (from an ideal system) for a combined system, assuming upper bounds of certain types for the component systems. For a general type of combination operation of systems, including the XOR of functions or the cascade of permutations, we prove two amplification theorems. The first is a product theorem, in the spirit of XOR-lemmas: The distinguishing advantage of the combination of two systems is at most twice the product of the individual distinguishing advantages. This bound is optimal. The second theorem states that the combination of systems is secure against some strong class of distinguishers, assuming only that the components are secure against some weaker class of distinguishers.

A key technical tool of the paper is the proof of a tight two-way correspondence, previously only known to hold in one direction, between the distinguishing advantage of two systems and the probability of winning an appropriately defined game.

1 Introduction

1.1 Indistinguishability Amplification for Random Variables

This paper is concerned with the indistinguishability of systems that interact with their environment. As a motivation for this paper, we consider an indistinguishability amplification result for random variables. A random variable can be understood as the special case of a system, which is non-interactive. Lemma 1 below states that the distance from uniform, of random variables, can be amplified by combining two or more (independent) moderately uniform random variables.

To state the lemma, we recall the following definitions.

Definition 1. The *statistical distance* of two random variables X and X' over $\mathcal{X}$ is defined as

A. Menezes (Ed.): CRYPTO 2007, LNCS 4622, pp. 130–149, 2007.

$$\delta(X, X') := \|P_X - P_{X'}\| = \tfrac{1}{2}\sum_{x \in \mathcal{X}} |P_X(x) - P_{X'}(x)|.$$

The *distance* of a random variable X *from uniform* is $d(X) := \delta(X, U)$, where U is a uniform random variable on $\mathcal{X}$.

The advantage of the best distinguisher for X and X' is $\delta(X, X')$.

Lemma 1. *For any two independent random variables X and Y over a finite domain $\mathcal{X}$ and any quasi-group operation[1] $\star$ on $\mathcal{X}$,*

$$d(X \star Y) \leq 2\, d(X)\, d(Y).$$

This bound is tight, as the following example illustrates.

Example 1. Consider two independent biased bits, X with a 40/60-bias and Y with a 30/70-bias. Then $d(X) = 0.1$, $d(Y) = 0.2$, and $d(X \oplus Y) = 0.04$ ($= 2 \cdot 0.1 \cdot 0.2$), since $X \oplus Y$ is 54/46-biased.

Corollary 2 of this paper can be seen as a natural generalization of Lemma 1. It states (for example) that if **F** and **G** are systems, for each of which the best distinguisher's advantage in distinguishing it from a uniform random function is bounded by ϵ and ϵ', respectively, then the system $\mathbf{F} \star \mathbf{G}$ obtained by using **F** and **G** in parallel and combining their outputs with $\star$, can be distinguished with advantage at most $2\epsilon\epsilon'$ from a uniform random function (for the same number of queries issued by the distinguisher). Actually, the proof of Corollary 2, restricted to random variables, appears to be a natural proof for Lemma 1.

As the abstraction underlying the quasi-group operation we introduce the concept of a *neutralizing combination* of two systems, which means that if any one (or both) of the systems is an ideal system (e.g. a uniform random function), then the combined system is also ideal. This is for example true for $X \star Y$: If either X or Y is uniform, then so is $X \star Y$.

1.2 Contributions of This Paper

The *amplification* of security properties is an important theme in cryptography. Examples of amplification results are XOR-lemmas, Vaudenay's product theorem for random permutations [Vau99], and the theorems proving adaptive security from non-adaptive security assumptions of [MP04] and [MOPS06].

This paper generalizes, strengthens, and unifies these results and provides a framework for proving such amplification results. We explore the general problem of proving various indistinguishability amplification results for systems. In contrast to earlier works, we do not restrict ourselves to *stateless* systems. The term "amplification" is used with two different meanings:

---

[1] A quasi-group operation $\star$ on a set $\mathcal{X}$ is a function $\mathcal{X}^2 \to \mathcal{X} : (a, b) \mapsto c = a \star b$ such given a and c (b and c), b (a) is uniquely determined. An important example is the bit-wise XOR of bit-strings. Any group operation is also a quasi-group operation.

- **Reduction of the distinguishing advantage.** We prove a general theorem (Theorem 1), in the spirit of Lemma 1, which states that the distinguishing advantage of a neutralizing combination of two systems is at most twice the product of the individual distinguishing advantages.
- **Attack strengthening.** We prove a general theorem (Theorem 2), which states that the adaptive distinguishing advantage of a neutralizing combination of two systems is bounded by the sum of the individual distinguishing advantages for a weaker distinguisher class (e.g. non-adaptive, or for permutations, one-sided instead of two-sided queries).

Our results are stated in the random systems framework of [Mau02] (see Section 2). They hold in the information-theoretic setting, with computationally unbounded distinguishers. In practice one is often interested in *computational* indistinguishability. Although the results from this paper do not directly translate to the computational setting[2], they have implications in the computational setting as well.

A main technical tool of this paper is a tight relation between the distinguishing advantage and the game-winning probability, discussed in the following section.

1.3   Discrete Systems, Indistinguishability, and Game-Winning

Many cryptographic systems (e.g. a block cipher, the CBC-MAC construction, or more complex games) can be modeled as *discrete systems*. A discrete system interacts with its environment by taking a sequence of inputs and producing, for each new input, an output (for a single, a fixed, or an unbounded number of such interactions).

Two major paradigms for cryptographic security definitions are:

- **Indistinguishability:** An ideal-world system is indistinguishable from a real-world system. For example, a secure encryption scheme can be seen as realizing a secure channel (ideal world) from an authenticated channel (real world).
- **Game-winning:** Breaking a system means that the adversary must achieve a certain goal, i.e., win a certain game. For example, a MAC is secure if the adversary cannot generate a fresh message together with the correct MAC, even if he can query the system arbitrarily.

The first type of security definition requires to prove that the distinguishing advantage of a certain class of distinguishers for two systems is very small. The second type of security definition requires to prove that no adversary of a certain type can win the game, except with very small probability.

In this paper we establish a tight relation between the above two problems in the information-theoretic setting. More precisely, game-winning can be modeled as an internal monotone condition in a system. Indeed, an important paradigm

---

[2] Actually, some results from this paper are known to be false in the computational case under standard assumptions [Pie05].

in indistinguishability proofs is the definition of such an internal monotone condition in a system (sometimes also called a "bad event") such that for any distinguisher **D** the distinguishing advantage can be shown to be upper bounded by the probability that **D** provokes this condition. A key technical tool of the paper (Lemma 5) is to show that this holds also in the other direction: For two systems **S** and **T** one can always define new systems $\hat{\mathbf{S}}$ and $\hat{\mathbf{T}}$, which are equivalent to **S** and **T**, respectively, but have an additional monotone binary output (MBO), such that

(i) for any distinguisher **D** the distinguishing advantage for **F** and **G** is *equal* to the probability that **D** sets the MBO to 1 in $\hat{\mathbf{S}}$ (or $\hat{\mathbf{T}}$), and

(ii) the systems $\hat{\mathbf{S}}$ and $\hat{\mathbf{T}}$ are equivalent as long as the respective MBOs are 0.

1.4   Related Work and Applications

This section is perhaps best read after reading the technical part of the paper.

Lemma 5 from this paper improves on Lemma 9 of [MP04] where a relation between distinguishing advantage and monotone binary outputs (there called conditions) was introduced, but which was not tight by a logarithmic factor and whose proof was quite technical, based on martingales. This paper settles a main open problem from [MP04], as Lemma 5 is tight.

The product theorem for sequential composition of stateless permutations, implied by Corollary 3, was proved earlier by Vaudenay within his decorrelation framework (see [Vau98] for the non-adaptive and [Vau99] for the adaptive case). Vaudenay's proofs, which use matrix norms, are tailored to the construction and attack at hand (i.e. sequential composition and stateless permutations), and do not extend to our general setting. While Vaudenay's decorrelation theory [Vau03] is purely information-theoretic, its application is for the design of actual (computationally secure) block-ciphers. In the same sense, our results have applications in the computational setting, where one considers computationally bounded adversaries.

In the computational setting, a product theorem for the sequential composition of permutations was proved by Luby and Rackoff [LR86]. Myers [Mye03] proved a product theorem[3] for a construction which is basically the parallel composition but with some extra random values XOR-ed to the inputs.

Our stronger results (compared to [MP04]) on adaptive security by composition, namely Corollaries 4 and 5, immediately apply to all results that made use of the bounds of [MP04]. For example, the construction of Kaplan, Naor and Reingold [KNR05] of randomness-efficient constructions of almost k-wise independent permutations, achieve *a priori* only non-adaptive security, but the authors observe that one can apply the results from [MP04] in order to obtain adaptive security. This paper allows to improve the bound of [KNR05]. Another application of Corollary 5 is in the already mentioned decorrelation theory

---

[3] Which in some sense is stronger than the amplification from [LR86], see [Mye03] for a discussion.

where it implies better security against adaptive attacks, even if the considered block-cipher only satisfies a non-adaptive notion of decorrelation.

The question whether composition implies adaptive security also in the computational setting (i.e. for pseudorandom systems) has been investigated in [Mye04, Pie05]. Unlike for the product amplification results, these attack-strengthening results do not hold for pseudorandom systems in general, though some positive results have also been achieved in this setting [Pie06].

Theorem 2 can be used to prove the adaptive security of more complicated constructions than the sequential and parallel composition considered in this paper. In [MOPS06], (a generalization of) Theorem 2 is used to prove that the four-round Feistel network with non-adaptively secure round functions is adaptively secure. That paper also shows that in the computational setting this is no longer true.

A result using Lemma 5 of a completely different vain than the problems considered in this paper is given in [PS07], where the security of some constructions for range extension of weak random functions is proven in the information theoretic setting (again, in the computational setting those results no longer hold).

2   Random Systems

This section follows and extends [Mau02], in slightly different notation.

2.1   Random Systems

Essentially every kind of discrete system (say $\mathbf{S}$), in particular a cryptographic system, can be described as follows. It takes inputs $X_1, X_2, \ldots$ (from some alphabet[4] $\mathcal{X}$) and generates, for each new input X_i, an output Y_i (from some alphabet $\mathcal{Y}$). The output Y_i depends (possibly probabilistically) on the current input X_i and on the internal state. Such a system is called an $(\mathcal{X}, \mathcal{Y})$-system.

In most contexts, only the *observable* input-output behavior, but not the internal state representation, is of interest. For example, if one considers the distinguishing advantage of a certain distinguisher $\mathbf{D}$ for two systems $\mathbf{S}$ and $\mathbf{T}$, then all that matters is the input-output behavior of the systems $\mathbf{D}$, $\mathbf{S}$ and $\mathbf{T}$. Hence the input-output behavior is the abstraction of a system that needs to be captured. This is analogous, for example, to a memoryless channel $\mathbf{C}$ in communication theory whose abstraction is captured by a conditional probability distribution $\mathsf{p}^{\mathbf{C}}_{Y|X}$ of the output Y, given the input X, independently of the physical description of the channel. A system is more complex than a channel; what is the abstraction of a (discrete) system?

A system is described exactly by the conditional probability distributions of the ith output Y_i, given $X_1, \ldots, X_i$ and $Y_1, \ldots, Y_{i-1}$, for all i. We use the shorthand notation $X^i := [X_1, \ldots, X_i]$. This is captured by the following definition from [Mau02].

---

[4] It is not a restriction to consider fixed input and output alphabets. This allows to model also systems where inputs and outputs come from different alphabets for different i.

Definition 2. An $(\mathcal{X}, \mathcal{Y})$-*random*[5] *system* **S** is a (generally infinite) sequence of conditional probability distributions[6] $\mathsf{p}^{\mathbf{S}}_{Y_i|X^iY^{i-1}}$ for $i \geq 1$.[7]

This description of a system is exact and minimal in the sense that two systems with different input-output behavior correspond to two different random systems, and two different random systems have different input-output behavior.

Note that the name **S** is used interchangeably for a system **S** (which can be described arbitrarily, for example by its internal workings) and the corresponding random system. This should cause no confusion. It is therefore also meaningful to say that two systems are equivalent if they have the same behavior, even though their internal structure may be different.

Definition 3. Two systems **S** and **T** are *equivalent*, denoted **S** $\equiv$ **T**, if they correspond to the same random system, i.e., if for all $i \geq 1$[8]

$$\mathsf{p}^{\mathbf{S}}_{Y_i|X^iY^{i-1}} = \mathsf{p}^{\mathbf{T}}_{Y_i|X^iY^{i-1}}.$$

The results of this paper are stated for random systems, but we emphasize that they hold for arbitrary systems, as the only property of a system that is relevant here is the input-output behavior. When several random systems appear in the same random experiment, they are (tacitly) assumed to be independent. In a more general theory, random systems could be dependent.

A random system **S** can be characterized equivalently by the sequence $\mathsf{p}^{\mathbf{S}}_{Y^i|X^i}$, for $i \geq 1$, of conditional probability distributions. This description is often convenient, but is not minimal.[9] The conversion between the two forms is given by

$$\mathsf{p}^{\mathbf{S}}_{Y^i|X^i} = \prod_{j=1}^{i} \mathsf{p}^{\mathbf{S}}_{Y_j|X^jY^{j-1}} \quad \text{and} \quad \mathsf{p}^{\mathbf{S}}_{Y_i|X^iY^{i-1}} = \frac{\mathsf{p}^{\mathbf{S}}_{Y^i|X^i}}{\mathsf{p}^{\mathbf{S}}_{Y^{i-1}|X^{i-1}}}. \qquad (1)$$

S and **T** are equivalent if and only if $\mathsf{p}^{\mathbf{S}}_{Y^i|X^i} = \mathsf{p}^{\mathbf{T}}_{Y^i|X^i}$ for $i \geq 1$.

2.2 Special Random Systems

Definition 4. A *random function* $\mathcal{X} \to \mathcal{Y}$ is a random system which answers consistently in the sense that $X_i = X_j \implies Y_i = Y_j$. A random function is *stateless* if it corresponds to a random variable taking on as values function tables $\mathcal{X} \to \mathcal{Y}$. A *random permutation* on $\mathcal{X}$ is a random function $\mathcal{X} \to \mathcal{X}$ mapping distinct inputs to distinct outputs: $X_i \neq X_j \implies Y_i \neq Y_j$.

---

[5] Throughout the paper, the term "random" is used in the same sense as it is used in the term "random variable", without implying uniformity of a distribution.

[6] We use a lower-case p to stress the fact that these conditional distributions by themselves do not define a random experiment in which probabilities are defined.

[7] For arguments x^{i-1} and y^{i-1} such that $\mathsf{p}^{\mathbf{S}}_{Y^{i-1}|X^{i-1}}(y^{i-1}, x^{i-1}) = 0$, $\mathsf{p}^{\mathbf{S}}_{Y_i|X^iY^{i-1}}$ need not be defined.

[8] This equality is an equality of (partial) functions, where two conditional probability distributions are considered to be equal if they are equal for all arguments for which both are defined.

[9] The distributions $\mathsf{p}^{\mathbf{S}}_{Y^i|X^i}$ must satisfy a consistency condition for the different i.

Note that in general a random function is not stateless. For example, a system defined by $Y_i = X_1$ for all i is not stateless.

We discuss a few examples of random systems.

Example 2. A $\mathcal{Y}$-*beacon*, usually denoted as **B**, is a random system which outputs a new independent uniformly (over $\mathcal{Y}$) output Y_i for every new input X_i: $\mathsf{p}^{\mathbf{B}}_{Y_i|X^iY^{i-1}} = 1/|\mathcal{Y}|$ for all choices of the arguments.

Example 3. A *uniform random function*, usually denoted as **R**, from some domain $\mathcal{X}$ to some finite range $\mathcal{Y}$. Typically $\mathcal{X} = \{0,1\}^m$ for some m or $\mathcal{X} = \{0,1\}^*$, and $\mathcal{Y} = \{0,1\}^n$ for some n. If $\mathcal{X}$ is finite, then this corresponds to a randomly selected function table. We have

$$\mathsf{p}^{\mathbf{R}}_{Y_i|X^iY^{i-1}}(y_i, x^i, y^{i-1}) = \begin{cases} 1 & \text{if } x_i = x_j \text{ for some } j < i \text{ and } y_i = y_j \\ 0 & \text{if } x_i = x_j \text{ for some } j < i \text{ and } y_i \neq y_j \\ 1/|\mathcal{Y}| & \text{else.} \end{cases}$$

$\mathsf{p}^{\mathbf{R}}_{Y_i|X^iY^{i-1}}(y_i, x^i, y^{i-1})$ is undefined if $x_j = x_k$ and $y_j \neq y_k$ for $j < k < i$.

We point out that when analyzing constructions involving uniform random functions (or other random systems), there is no need to resort to this apparently complex description. Any complete description is fine. Using the concept of random systems buys precision and simplicity, without requiring technical complexity of the arguments.

Example 4. A *uniform random permutation*, usually denoted as **P**, for domain and range $\mathcal{X}$, is a function randomly selected from all bijective functions $\mathcal{X} \to \mathcal{X}$.

2.3 Distinguishing Random Systems

We are interested in distinguishing two systems **S** and **T** by means of a distinguisher **D**. In the sequel, we will usually tacitly assume that the two systems are compatible, i.e., have the same input and output alphabets.

A distinguisher **D** for distinguishing two $(\mathcal{X}, \mathcal{Y})$-systems generates X_1 as an input, receives the output Y_1, then generates X_2, receives Y_2, etc. Finally, after receiving Y_k, it outputs a binary decision bit, say W. More formally:

Definition 5. A *distinguisher* **D** *for* $(\mathcal{X}, \mathcal{Y})$-*random systems* is a $(\mathcal{Y}, \mathcal{X})$-random system, which is one query ahead, meaning that it is defined by $\mathsf{p}^{\mathbf{D}}_{X_i|Y^{i-1}X^{i-1}}$ (instead of $\mathsf{p}^{\mathbf{D}}_{X_i|Y^iX^{i-1}}$) for all i.[10] **D** outputs a bit W after a certain number k of queries, based on the transcript (X^k, Y^k).

When a distinguisher **D** is connected to a system **S**, which we denote simply as **DS**, this defines a random experiment. The probabilities of an event $\mathcal{E}$ in this

---

[10] In particular the first output $\mathsf{p}^{\mathbf{D}}_{X_1}$ is defined before **D** is fed with any input.

experiment will be denoted as $\mathsf{P}^{\mathbf{DS}}(\mathcal{E})$. We note that the probability distribution $\mathsf{p}^{\mathbf{DS}}_{X^k Y^k}$ can be expressed by

$$\mathsf{p}^{\mathbf{DS}}_{X^k Y^k}(x^k, y^k) = \prod_{i=1}^{k} \mathsf{p}^{\mathbf{D}}_{X_i | X^{i-1} Y^{i-1}}(x_i, x^{i-1}, y^{i-1}) \, \mathsf{p}^{\mathbf{S}}_{Y_i | X^i Y^{i-1}}(y_i, x^i, y^{i-1})$$

$$= \mathsf{p}^{\mathbf{D}}_{X^k | Y^{k-1}}(x^k, y^{k-1}) \, \mathsf{p}^{\mathbf{S}}_{Y^k | X^k}(y^k, x^k) \, , \tag{2}$$

where the last equality follows from (1).

The performance of a distinguisher, called the advantage, can be defined in two equivalent ways, both of which will be useful for us. We first state the standard definition.

Definition 6. The *advantage* of distinguisher $\mathbf{D}$ for random systems $\mathbf{S}$ and $\mathbf{T}$, for k queries, denoted $\Delta^{\mathbf{D}}_k(\mathbf{S}, \mathbf{T})$, is defined as

$$\Delta^{\mathbf{D}}_k(\mathbf{S}, \mathbf{T}) := \left| \mathsf{P}^{\mathbf{DS}}(W = 1) - \mathsf{P}^{\mathbf{DT}}(W = 1) \right|.$$

For a class $\mathcal{D}$ of distinguishers, the advantage of the best $\mathbf{D}$ in $\mathcal{D}$, asking at most k queries, is denoted as

$$\Delta^{\mathcal{D}}_k(\mathbf{S}, \mathbf{T}) := \max_{\mathbf{D} \in \mathcal{D}} \Delta^{\mathbf{D}}_k(\mathbf{S}, \mathbf{T}).$$

For the class of *all* distinguishers we simply write $\Delta_k(\mathbf{S}, \mathbf{T})$.

To state an equivalent definition of the advantage we need the following definition.

Definition 7. For two compatible systems $\mathbf{S}$ and $\mathbf{T}$, $\langle \mathbf{S}/\mathbf{T} \rangle$ denotes the random system which is equal to system $\mathbf{S}$ or $\mathbf{T}$ with probability $\frac{1}{2}$ each. To make the independent unbiased binary random variable, say Z, selecting between $\mathbf{S}$ (for $Z = 0$) and $\mathbf{T}$ (for $Z = 1$) explicit, we write $\langle \mathbf{S}/\mathbf{T} \rangle_Z$.[11]

The advantage $\Delta^{\mathbf{D}}_k(\mathbf{S}, \mathbf{T})$ can be defined equivalently in terms of the probability that $\mathbf{D}$, interacting with the mixed system $\langle \mathbf{S}/\mathbf{T} \rangle_Z$, guesses Z correctly:

Lemma 2. *For every distinguisher* $\mathbf{D}$,[12]

$$\Delta^{\mathbf{D}}_k(\mathbf{S}, \mathbf{T}) = 2 \left| \mathsf{P}^{\mathbf{D} \langle \mathbf{S}/\mathbf{T} \rangle_Z}(W = Z) - \tfrac{1}{2} \right|.$$

Proof. Let p_z for $z \in \{0, 1\}$ denote the probability that $W = 1$ if $Z = z$. Then $\Delta^{\mathbf{D}}_k(\mathbf{S}, \mathbf{T}) = |p_0 - p_1|$ and $\mathsf{P}^{\mathbf{D} \langle \mathbf{S}/\mathbf{T} \rangle_Z}(W = Z) = \frac{1}{2}(1 - p_0 + p_1)$, hence $2 \left| \mathsf{P}^{\mathbf{D} \langle \mathbf{S}/\mathbf{T} \rangle_Z}(W = Z) - \frac{1}{2} \right| = |p_0 - p_1|$. $\square$

The following distinguisher classes are usually of special interest:

Definition 8. By NA we denote the class of computationally unbounded non-adaptive distinguishers which select all queries $X_1, \ldots, X_k$ in advance (i.e.,

---

[11] It is helpful to think of Z as the position of a switch selecting between the systems $\mathbf{S}$ and $\mathbf{T}$.

[12] The normalization factor 2 assures that the advantage is between 0 and 1. The absolute value in $|\mathsf{P}^{\mathbf{D} \langle \mathbf{S}/\mathbf{T} \rangle_Z}(W = Z) - \frac{1}{2}|$ takes into account the fact that one can always invert the output of a distinguisher whose success probability is below $\frac{1}{2}$.

independent of the outputs Y_i).[13] By RI we denote the class of computation-ally unbounded distinguishers which (cannot select the queries but) are given uniformly random values $X_1, \ldots, X_k$ (and the corresponding outputs $Y_1, \ldots, Y_k$).

Clearly, RI $\subseteq$ NA. The class NA is sometimes called nCPA (non-adaptive chosen-plaintext attack) in the literature and the class RI is sometimes called KPA (known-plaintext attack).

The following lemma captures the simple fact that if one has to distinguish the systems $\mathbf{S}$ and $\langle \mathbf{S}/\mathbf{T} \rangle_Z$, then the advantage is only half of the advantage when distinguishing $\mathbf{S}$ and $\mathbf{T}$. In a sense, $\langle \mathbf{S}/\mathbf{T} \rangle_Z$ is half-way between $\mathbf{S}$ and $\mathbf{T}$.

Lemma 3. *For every* $\mathbf{D}$, $\Delta_k^{\mathbf{D}}(\mathbf{S}, \langle \mathbf{S}/\mathbf{T} \rangle_Z) = \frac{1}{2}\Delta_k^{\mathbf{D}}(\mathbf{S}, \mathbf{T})$.

Proof. This follows from the linearity of the probability of $\mathbf{D}$ outputting a 1: we have $\mathsf{P}^{\mathbf{D}\langle \mathbf{S}/\mathbf{T} \rangle_Z}(W = 1) = \frac{1}{2}(\mathsf{P}^{\mathbf{D}\mathbf{S}}(W = 1) + \mathsf{P}^{\mathbf{D}\mathbf{T}}(W = 1))$.

2.4 Game-Winning and Monotone Binary Outputs

An important paradigm in certain security definitions is the notion of winning a game. Without loss of generality, a game with one player (e.g. the adversary) can be described by an $(\mathcal{X}, \mathcal{Y})$-system which interacts with its environment by taking inputs $X_1, X_2, \ldots$ (considered as moves) and answering with outputs $Y_1, Y_2, \ldots$. In addition, after every input it also outputs a bit indicating whether the game has been won. This bit is monotone in the sense that it is initially set to 0 and that, once it has turned to 1 (the game is won), it can not turn back to 0. This motivates the following definition, which captures the notion of game-winning.

Definition 9. For a $(\mathcal{X}, \mathcal{Y} \times \{0, 1\})$-system $\mathbf{S}$ the binary component A_i of the output (Y_i, A_i) is called a *monotone binary output (MBO)* if $A_i = 1$ implies $A_j = 1$ for $j \geq i$. For such a system $\mathbf{S}$ with MBO we define two derived systems:

(i) $\mathbf{S}^-$ is the $(\mathcal{X}, \mathcal{Y})$-system resulting from $\mathbf{S}$ by ignoring the MBO.
(ii) $\mathbf{S}^{\dashv}$ is the $(\mathcal{X}, \mathcal{Y} \times \{0, 1\})$-system which masks the $\mathcal{Y}$-output to a dummy symbol ($\perp$) as soon as the MBO turns to 1. More precisely, the following function is applied to the outputs of $\mathbf{S}$:

$$(y, a) \mapsto (y', a) \quad \text{where} \quad y' = \begin{cases} y & \text{if } a = 0 \\ \perp & \text{if } a = 1. \end{cases}$$

Definition 10. Two systems $\mathbf{S}$ and $\mathbf{T}$ with MBOs are called *restricted equivalent* if $\mathbf{S}^{\dashv} \equiv \mathbf{T}^{\dashv}$, i.e., if they are equivalent as long as the MBO is 0.

A system (or player) $\mathbf{D}$ interacting with $\mathbf{S}$, trying to win the game defined by $\mathbf{S}$, is like a distinguisher, except that it need not have a binary output W. Whether or not $\mathbf{D}$ "sees" the MBO is irrelevant; one can think of $\mathbf{D}$ interacting with $\mathbf{S}^-$ instead of $\mathbf{S}$. One could call such a $\mathbf{D}$ a "player" or a "provoker", as it tries to provoke the MBO to become 1, but for consistency we will continue to call a $\mathbf{D}$ distinguisher.

---

[13] One can view such a distinguisher as making a single (compound) query $(x_1, \ldots, x_k)$.

Definition 11. For a $(\mathcal{X}, \mathcal{Y} \times \{0,1\})$-random system $\mathbf{S}$ with an MBO (called A_i) and for a distinguisher $\mathbf{D}$, we denote with $\nu_k^{\mathbf{D}}(\mathbf{S})$ the probability that $\mathbf{D}$ wins the game within k queries:

$$\nu_k^{\mathbf{D}}(\mathbf{S}) := \mathsf{P}^{\mathbf{DS}}(A_k = 1).$$

For a class $\mathcal{D}$ of distinguishers, the winning probability of the best $\mathbf{D}$ in $\mathcal{D}$ within k queries is denoted as

$$\nu_k^{\mathcal{D}}(\mathbf{S}) := \max_{\mathbf{D} \in \mathcal{D}} \nu_k^{\mathbf{D}}(\mathbf{S}).$$

For the class of *all* distinguishers we simply write $\nu_k(\mathbf{S})$.

3   Relating Indistinguishability and Game-Winning

3.1   From Game-Winning to Indistinguishability

The following lemma was proved in [Mau02]. Versions of this lemma for special types of systems appeared subsequently.

Lemma 4. *Let* $\mathbf{S}$ *and* $\mathbf{T}$ *be two* $(\mathcal{X}, \mathcal{Y} \times \{0,1\})$-*random systems with MBOs. If* $\mathbf{S}^{\dashv} \equiv \mathbf{T}^{\dashv}$, *then*

$$\Delta_k^{\mathbf{D}}(\mathbf{S}^-, \mathbf{T}^-) \leq \nu_k^{\mathbf{D}}(\mathbf{S}) = \nu_k^{\mathbf{D}}(\mathbf{T})$$

for all distinguishers $\mathbf{D}$ *for* $(\mathcal{X}, \mathcal{Y})$-*random systems.*[14] *In particular, for any distinguisher class* $\mathcal{D}$, $\Delta_k^{\mathcal{D}}(\mathbf{S}^-, \mathbf{T}^-) \leq \nu_k^{\mathcal{D}}(\mathbf{S})$, *hence* $\Delta_k(\mathbf{S}^-, \mathbf{T}^-) \leq \nu_k(\mathbf{S})$ *and* $\Delta_k^{\mathsf{NA}}(\mathbf{S}^-, \mathbf{T}^-) \leq \nu_k^{\mathsf{NA}}(\mathbf{S})$.

Proof. According to Lemma 2, $\Delta_k^{\mathbf{D}}(\mathbf{S}, \mathbf{T})$ can be computed in terms of the probability that $\mathbf{D}$ guesses the switch Z in $\langle \mathbf{S}/\mathbf{T} \rangle_Z$ correctly. The condition $\mathbf{S}^{\dashv} \equiv \mathbf{T}^{\dashv}$ implies that if the MBO of $\langle \mathbf{S}/\mathbf{T} \rangle_Z$ is 0, then the output of $\langle \mathbf{S}/\mathbf{T} \rangle_Z$ is independent of Z, and therefore in this case $\mathbf{D}$ cannot do better than guess randomly. (If the MBO is 1, the success probability is bounded by 1.) Hence, if we denote by p the probability that $\mathbf{D}$ sets the MBO to 1, the probability that $\mathbf{D}$ guesses Z correctly is bounded by $\frac{1}{2}(1-p) + p = \frac{1}{2} + \frac{1}{2}p$, where $p = \nu_k^{\mathbf{D}}(\langle \mathbf{S}/\mathbf{T} \rangle_Z) = \nu_k^{\mathbf{D}}(\mathbf{S}) = \nu_k^{\mathbf{D}}(\mathbf{T})$. Applying Lemma 2 completes the proof. $\qquad\square$

3.2   From Indistinguishability to Game-Winning

The following lemma states, in a certain sense, a converse to Lemma 4, and is a key tool for the proofs of the main results. While Lemma 4 holds for every distinguisher, whether computationally bounded or not, and whether or not its binary output is determined optimally based on the transcript, the converse only holds in the information-theoretic setting and if we assume that the decision bit is computed optimally. More precisely, it is a statement about the statistical distance of transcripts.

---

[14] Recall that it is well-defined what it means for such a distinguisher to play the game for $\mathbf{S}$ which is defined *with* an MBO.

Definition 12. Let

$$\delta_k^{\mathbf{D}}(\mathbf{S}, \mathbf{T}) := \| \mathsf{P}_{X^k Y^k}^{\mathbf{DS}} - \mathsf{P}_{X^k Y^k}^{\mathbf{DT}} \|$$

be the statistical distance of the transcripts $(X^k Y^k)$ when $\mathbf{D}$ interacts with $\mathbf{S}$ and $\mathbf{T}$, respectively. For a class $\mathcal{D}$ of distinguishers we define[15]

$$\delta_k^{\mathcal{D}}(\mathbf{S}, \mathbf{T}) := \max_{\mathbf{D} \in \mathcal{D}} \delta_k^{\mathbf{D}}(\mathbf{S}, \mathbf{T}).$$

Note that in general we have $\Delta_k^{\mathbf{D}}(\mathbf{S}, \mathbf{T}) \leq \delta_k^{\mathbf{D}}(\mathbf{S}, \mathbf{T})$, but for a computationally unbounded distinguisher $\mathbf{D}$ that chooses the output bit optimally, we have $\Delta_k^{\mathbf{D}}(\mathbf{S}, \mathbf{T}) = \delta_k^{\mathbf{D}}(\mathbf{S}, \mathbf{T})$. In particular,

$$\Delta_k(\mathbf{S}, \mathbf{T}) = \delta_k(\mathbf{S}, \mathbf{T}) \quad \text{and} \quad \Delta_k^{\mathsf{NA}}(\mathbf{S}, \mathbf{T}) = \delta_k^{\mathsf{NA}}(\mathbf{S}, \mathbf{T}).$$

Lemma 5. *For any two* $(\mathcal{X}, \mathcal{Y})$-*systems* $\mathbf{S}$ *and* $\mathbf{T}$ *there exist* $(\mathcal{X}, \mathcal{Y} \times \{0, 1\})$-*random systems* $\hat{\mathbf{S}}$ *and* $\hat{\mathbf{T}}$ *with MBOs such that*

(i) $\hat{\mathbf{S}}^- \equiv \mathbf{S}$,
(ii) $\hat{\mathbf{T}}^- \equiv \mathbf{T}$,
(iii) $\hat{\mathbf{S}}^{\dashv} \equiv \hat{\mathbf{T}}^{\dashv}$, and
(iv) $\delta_k^{\mathbf{D}}(\mathbf{S}, \mathbf{T}) = \nu_k^{\mathbf{D}}(\hat{\mathbf{S}}) = \nu_k^{\mathbf{D}}(\hat{\mathbf{T}})$ *for all* $\mathbf{D}$.[16]

To illustrate the idea of the proof of Lemma 5, we consider an analogous statement (in fact, a special case) where probability distributions P_X and Q_X (over some alphabet $\mathcal{X}$) take the place of the random systems $\hat{\mathbf{S}}$ and $\hat{\mathbf{T}}$. In this case, the systems with MBO can be replaced by joint distributions $\hat{P}_{XA}$ and $\hat{Q}_{XA}$, where A is binary. Indeed, if we define these distributions by

$$\hat{P}_{XA}(x, 0) = \hat{Q}_{XA}(x, 0) = \min(P_X(x), Q_X(x))$$
$$\hat{P}_{XA}(x, 1) = P_X(x) - \min(P_X(x), Q_X(x))$$
$$\hat{Q}_{XA}(x, 1) = Q_X(x) - \min(P_X(x), Q_X(x))$$

(for any $x \in \mathcal{X}$) it is easy to verify that $\hat{P}_X = P_X$ and $\hat{Q}_X = Q_X$, which corresponds to (i) and (ii), respectively. Furthermore, and trivially, $\hat{P}_{XA}(\cdot, 0) = \hat{Q}_{XA}(\cdot, 0)$, which is (iii). Finally, because the statistical distance can be written as

$$\delta(P_X, Q_X) = 1 - \sum_x \min(P_X(x), Q_X(x)), \tag{3}$$

the equivalent of (iv) follows from the fact that the right-hand side of (3) equals $\hat{P}_A(1) = \hat{Q}_A(1)$.

---

[15] For the class of *all* distinguishers we simply write $\delta_k(\mathbf{S}, \mathbf{T})$.

[16] This also implies, for example, $\Delta_k(\mathbf{S}, \mathbf{T}) = \nu_k(\hat{\mathbf{S}})$ and $\Delta_k^{\mathsf{NA}}(\mathbf{S}, \mathbf{T}) = \nu_k^{\mathsf{NA}}(\hat{\mathbf{S}})$.

Proof (of Lemma 5). The idea is to define the system $\hat{\mathbf{S}}$ with MBO A_i (and, likewise, $\hat{\mathbf{T}}$) such that, for all $i \geq 1$,

$$
\begin{aligned}
\mathsf{p}^{\hat{\mathbf{S}}}_{Y^i A_i | X^i}(y^i, 0, x^i) &:= m_{x^i, y^i} \\
\mathsf{p}^{\hat{\mathbf{S}}}_{Y^i A_i | X^i}(y^i, 1, x^i) &:= \mathsf{p}^{\mathbf{S}}_{Y^i | X^i}(y^i, x^i) - m_{x^i, y^i},
\end{aligned}
\tag{4}
$$

where

$$
m_{x^i, y^i} := \min(\mathsf{p}^{\mathbf{S}}_{Y^i | X^i}(y^i, x^i), \mathsf{p}^{\mathbf{T}}_{Y^i | X^i}(y^i, x^i)) .
$$

We will verify below that this can always be done consistently.

Note that properties (i), (ii), and (iii) follow immediately from these equations (similarly to the above argument for random variables). To verify (iv), we recall that the probabilities of $\mathsf{P}^{\mathbf{DS}}_{X^k Y^k}$ (and, likewise, $\mathsf{P}^{\mathbf{DT}}_{X^k Y^k}$) can be expressed by equation (2). Using formula (3) for the statistical distance we find

$$
\begin{aligned}
\delta^{\mathbf{D}}_k(\mathbf{S}, \mathbf{T}) &= \| \mathsf{P}^{\mathbf{DS}}_{X^k Y^k} - \mathsf{P}^{\mathbf{DT}}_{X^k Y^k} \| \\
&= 1 - \sum_{x^k, y^k} \min\big(\mathsf{P}^{\mathbf{DS}}_{X^k Y^k}(x^k, y^k), \mathsf{P}^{\mathbf{DT}}_{X^k Y^k}(x^k, y^k)\big) \\
&= 1 - \sum_{x^k, y^k} \mathsf{p}^{\mathbf{D}}_{X^k | Y^{k-1}}(x^k, y^{k-1}) \min\big(\mathsf{p}^{\mathbf{S}}_{Y^k | X^k}(y^k, x^k), \mathsf{p}^{\mathbf{T}}_{Y^k | X^k}(y^k, x^k)\big) .
\end{aligned}
$$

Property (iv) then follows because the probability that the MBO A_k of $\hat{\mathbf{S}}$ (and, likewise, $\hat{\mathbf{T}}$) equals 1 after k steps is given by

$$
\begin{aligned}
\nu^{\mathbf{D}}_k(\hat{\mathbf{S}}) &= 1 - \sum_{x^k, y^k} \mathsf{P}^{\mathbf{D}\hat{\mathbf{S}}}_{X^k Y^k A_k}(x^k, y^k, 0) \\
&= 1 - \sum_{x^k, y^k} \mathsf{p}^{\mathbf{D}}_{X^k | Y^{k-1}}(x^k, y^{k-1}) \mathsf{p}^{\hat{\mathbf{S}}}_{Y^k A_k | X^k}(y^k, 0, x^k) ,
\end{aligned}
$$

which equals the above expression for $\delta^{\mathbf{D}}_k(\mathbf{S}, \mathbf{T})$.

It remains to verify that there exists a system $\hat{\mathbf{S}}$ satisfying (4) (the argument for $\hat{\mathbf{T}}$ follows by symmetry).

Note that (4) only determines the interrelation between the system's output Y_i and the value A_i of the MBO at the same step, but it does not specify the dependency on previous values A^{i-1}. In fact, there are various degrees of freedom in the definition of $\hat{\mathbf{S}}$, for instance in the choice of the probabilities $r_{x^i, y^i} := \mathsf{p}^{\hat{\mathbf{S}}}_{Y^i A^{i-1} | X^i}(y^i, 0^{i-1}, x^i)$. Most generally, the probabilities defining $\hat{\mathbf{S}}$, conditioned on the event that the previous MBO equals 0, can be written as[17]

$$
\mathsf{p}^{\hat{\mathbf{S}}}_{Y_i A_i | X^i Y^{i-1} A^{i-1}}(y_i, a_i, x^i, y^{i-1}, 0^{i-1}) := \begin{cases} \dfrac{m_{x^i, y^i}}{m_{x^{i-1}, y^{i-1}}} & \text{if } a_i = 0 \\[2mm] \dfrac{r_{x^i, y^i} - m_{x^i, y^i}}{m_{x^{i-1}, y^{i-1}}} & \text{if } a_i = 1, \end{cases}
$$

---

[17] We use the convention $\mathsf{p}^{\mathbf{S}}_{Y^0 | X^0} \equiv 1$ and, in particular, $m_{x^0, y^0} = 1$.

for any $i \geq 1$, where $r_{x_i,y_i} \in [m_{x^i,y^i}, \mathsf{p}^{\mathbf{S}}_{Y^i|X^i}(y^i,x^i)]$ are parameters. To make sure that the conditional probabilities sum up to 1, we require

$$\sum_{y_i} r_{x^i,y^i} = m_{x^{i-1},y^{i-1}} \, , \tag{5}$$

for any fixed x^i and y^{i-1}. Note that such a choice of r_{x^i,y^i} always exists because the right side of (5) lies in the interval

$$m_{x^{i-1},y^{i-1}} \in \left[\sum_{y_i} m_{x^i,y^i}, \sum_{y_i} \mathsf{p}^{\mathbf{S}}_{Y^i|X^i}(y^i,x^i)\right] \, .$$

To complete the definition of $\hat{\mathbf{S}}$, we set, for any $i > 1$ and $a^{i-1} \neq 0$,

$$\mathsf{p}^{\hat{\mathbf{S}}}_{Y_i A_i|X^i Y^{i-1} A^{i-1}}(y_i, 1, x^i, y^{i-1}, a^{i-1}) := \frac{\mathsf{p}^{\mathbf{S}}_{Y^i|X^i}(y^i, x^i) - r_{x^i,y^i}}{\mathsf{p}^{\mathbf{S}}_{Y^{i-1}|X^{i-1}}(y^{i-1}, x^{i-1}) - m_{x^{i-1},y^{i-1}}} \, .$$

Again, the conditional probabilities are well-defined because all values are non-negative and, by (5), sum up to 1. Furthermore, it is easy to see that the outputs A_i of $\hat{\mathbf{S}}$ are indeed monotone. Finally, by induction over i, it is straightforward to verify that $\hat{\mathbf{S}}$ satisfies (4), which concludes the proof. $\qquad\square$

We give another interpretation of Lemma 5. If two probability distributions P_X and Q_X have statistical distance δ then there exists a (common) random experiment with two random variables X' and X'', distributed according to P_X and Q_X, respectively, such that $X' = X''$ with probability $1 - \delta$. Lemma 5 can be interpreted as the generalization of this statement to random systems. For any distinguisher $\mathbf{D}$, two random systems $\mathbf{S}$ and $\mathbf{T}$ are equal with probability $1 - \delta$, where δ is $\mathbf{D}$'s distinguishing advantage.

4 Amplification of the Distinguishing Advantage

4.1 Neutralizing Constructions

Throughout the rest of the paper we let $\mathbf{C}(\cdot, \cdot)$ be a construction invoking two systems. For example $\mathbf{C}(\mathbf{F}, \mathbf{G})$ denotes the system obtained when $\mathbf{C}(\cdot, \cdot)$ invokes the two systems $\mathbf{F}$ and $\mathbf{G}$.

Definition 13. A construction $\mathbf{C}(\cdot, \cdot)$ is called *neutralizing* for the pairs $(\mathbf{F}, \mathbf{I})$ and $(\mathbf{G}, \mathbf{J})$ of (independent) systems if

$$\mathbf{C}(\mathbf{F}, \mathbf{J}) \equiv \mathbf{C}(\mathbf{I}, \mathbf{G}) \equiv \mathbf{C}(\mathbf{I}, \mathbf{J}) \equiv \mathbf{Q} \tag{6}$$

(for some $\mathbf{Q}$). Moreover, we denote by k' and k'' the maximal number of queries made to the first and the second subsystem, respectively, when the number of queries to $\mathbf{C}(\cdot, \cdot)$ is k.

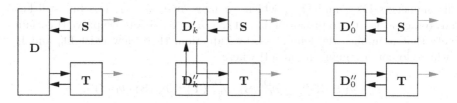

Fig. 1. Illustration for the proof of Lemma 6. **D** can be seen as a pair $(\mathbf{D}'_k, \mathbf{D}''_k)$ of distinguishers which can exchange up to $k = 2k''$ messages (simply set $\mathbf{D}'_k \equiv \mathbf{D}$ and $\mathbf{D}''_k$ to be the trivial system which only passes messages). The gray arrows indicate the MBOs.

4.2   Winning Independent Games

The following lemma states that the best combined strategy for winning two independent games is not better than applying the individually best strategies separately. We note that this is (of course) also true for real games, like playing black jack, but we phrase the result at an abstract (and hence very general) level.

We need some new notation: For two systems **S** and **T** with MBOs let $[\mathbf{S} \, \| \, \mathbf{T}]^\wedge$ be the system consisting of **S** and **T** being accessible independently, with an MBO which is 1 if and only if the MBOs of **S** and **T** are *both* 1. Let $\nu^{\mathcal{D}}_{k',k''}([\mathbf{S} \, \| \, \mathbf{T}]^\wedge)$ denote the advantage of the best distinguisher in $\mathcal{D}$, making k' and k'' (arbitrarily scheduled) queries to **S** and **T**, respectively, in setting the MBO to 1 (we simply write $\nu_{k',k''}([\mathbf{S} \, \| \, \mathbf{T}]^\wedge)$ if $\mathcal{D}$ is the class of all distinguishers).

Lemma 6. *For any random systems* **S** *and* **T** *with MBOs, and any k' and k'',*

$$\nu_{k',k''}([\mathbf{S} \, \| \, \mathbf{T}]^\wedge) \; = \; \nu_{k'}(\mathbf{S}) \, \nu_{k''}(\mathbf{T}), \tag{7}$$

and

$$\nu^{\mathsf{NA}}_{k',k''}([\mathbf{S} \, \| \, \mathbf{T}]^\wedge) \; = \; \nu^{\mathsf{NA}}_{k'}(\mathbf{S}) \, \nu^{\mathsf{NA}}_{k''}(\mathbf{T}). \tag{8}$$

Proof. The non-adaptive case (8) follows from the adaptive case (7) by viewing the non-adaptive queries as a single adaptive query. To prove (7), let **D** be an optimal distinguisher for the task considered, i.e.

$$\nu_{k',k''}([\mathbf{S} \, \| \, \mathbf{T}]^\wedge) \; = \; \nu^{\mathbf{D}}_{k',k''}([\mathbf{S} \, \| \, \mathbf{T}]^\wedge).$$

Let $A_1, \ldots, A_{k'}$ and $B_1, \ldots, B_{k''}$ denote the MBOs of **S** and **T**, respectively. We can interpret **D** as a pair $(\mathbf{D}'_k, \mathbf{D}''_k)$ of distinguishers which can exchange up to $k = 2k''$ messages with each other, as shown in Figure 1. As this is just a conceptual change, the advantage of setting both MBOs to 1 is exactly the same for **D** as for the pair $(\mathbf{D}'_k, \mathbf{D}''_k)$.

Now assume that there is a pair of distinguishers $\mathbf{D}'_\ell$ and $\mathbf{D}''_\ell$ which can exchange up to ℓ messages and have advantage ϵ to provoke $(A_{k'} = 1) \wedge (B_{k''} = 1)$ when querying **S** and **T**, respectively. We claim that then there also exist distinguishers $\mathbf{D}'_{\ell-1}$ and $\mathbf{D}''_{\ell-1}$ which exchange one message less but still have advantage at least ϵ to provoke $(A_k = 1) \wedge (B_k = 1)$. Before we prove this claim, note that it implies the lemma as, by induction, there now exist $\mathbf{D}'_0$ and $\mathbf{D}''_0$ (which do not communicate at all) where

$$\nu_{k',k''}([\mathbf{S} \| \mathbf{T}]^{\wedge}) \le \nu_{k'}^{\mathbf{D}'_0}(\mathbf{S}) \cdot \nu_{k''}^{\mathbf{D}''_0}(\mathbf{T}) \le \nu_{k'}(\mathbf{S}) \cdot \nu_{k''}(\mathbf{T}).$$

We actually have equality above as the other direction ($\ge$) is trivial. To prove the claim, assume that the (last) ℓ-th message is sent from $\mathbf{D}'_{\ell}$ to $\mathbf{D}''_{\ell}$. Let the random variable M denote this last message, and let V be the "view" of $\mathbf{D}''_{\ell}$ just before receiving the message. Let $\mathcal{E}$ denote this random experiment where $\mathbf{D}'_{\ell}$ and $\mathbf{D}''_{\ell}$ are querying $\mathbf{S}$ and $\mathbf{T}$ respectively. The probability that we have $A_{k'} = 1 \wedge B_{k''} = 1$ is

$$\sum_{m,v} \mathsf{P}^{\mathcal{E}}[A_{k'} = 1 \wedge M = m \wedge V = v] \cdot \mathsf{P}^{\mathcal{E}}[B_{k''} = 1 | M = m \wedge V = v]. \qquad (9)$$

We used $\mathsf{P}^{\mathcal{E}}[B_{k''} = 1 | A_{k'} = 1 \wedge M = m \wedge V = v] = \mathsf{P}^{\mathcal{E}}[B_{k''} = 1 | M = m \wedge V = v]$ which holds as $\mathbf{S}$ is independent of $\mathbf{T}$ and the whole interaction between these systems is captured by M and V. Now consider a new system $\mathbf{D}''_{\ell-1}$ which simulates $\mathbf{D}''_{\ell}$ but does not expect the (last) ℓ-th message M and instead replaces it with a message m' which maximizes the probability of $B_{k''} = 1$ (given the view V). Also, let $\mathbf{D}'_{\ell-1}$ be the system $\mathbf{D}'_{\ell}$, but where the last message is not sent (note that this change does not affect the probability of $A_{k'} = 1$ or the distribution of V). The probability that the pair $(\mathbf{D}'_{\ell-1}, \mathbf{D}''_{\ell-1})$ can provoke $A_{k'} = 1 \wedge B_{k''} = 1$ is thus

$$\sum_{m,v} \mathsf{P}^{\mathcal{E}}[A_{k'} = 1 \wedge M = m \wedge V = v] \cdot \max_{m'} \mathsf{P}^{\mathcal{E}}[B_{k''} = 1 | M = m' \wedge V = v]$$

which is at least equal to (9). $\qquad \square$

4.3   The Product Theorem

We can now state the first main result of the paper. Recall Definition 13.

Theorem 1. *If* $\mathbf{C}(\cdot, \cdot)$ *is neutralizing for the pairs* $(\mathbf{F}, \mathbf{I})$ *and* $(\mathbf{G}, \mathbf{J})$ *of systems, then, for all* k,

$$\Delta_k(\mathbf{C}(\mathbf{F}, \mathbf{G}), \mathbf{C}(\mathbf{I}, \mathbf{J})) \le 2\, \Delta_{k'}(\mathbf{F}, \mathbf{I})\, \Delta_{k''}(\mathbf{G}, \mathbf{J}).$$

Proof. We consider the systems $\mathbf{H}_{Z,Z'} := \mathbf{C}(\langle \mathbf{I}/\mathbf{F} \rangle_Z, \langle \mathbf{J}/\mathbf{G} \rangle_{Z'})$, indexed by Z and Z', where Z and Z' are independent unbiased bits. Due to (6) we have $\mathbf{H}_{11} \equiv \mathbf{C}(\mathbf{F}, \mathbf{G})$ and $\mathbf{H}_{00} \equiv \mathbf{H}_{01} \equiv \mathbf{H}_{10} \equiv \mathbf{Q} \equiv \mathbf{C}(\mathbf{I}, \mathbf{J})$. One can hence easily verify that

$$\mathbf{H}_{Z,Z'} \equiv \langle\langle \mathbf{Q}/\mathbf{C}(\mathbf{F}, \mathbf{G}) \rangle_{Z'}/\mathbf{Q} \rangle_{Z \oplus Z'},$$

by checking the equivalence for all four values of the pair (Z, Z').

Lemma 3 implies that $\Delta_k(\mathbf{C}(\mathbf{F}, \mathbf{G}), \mathbf{Q}) = 2 \cdot \Delta_k(\langle \mathbf{Q}/\mathbf{C}(\mathbf{F}, \mathbf{G}) \rangle_{Z'}, \mathbf{Q})$, where, according to Lemma 2, $\Delta_k(\langle \mathbf{Q}/\mathbf{C}(\mathbf{F}, \mathbf{G}) \rangle_{Z'}, \mathbf{Q})$ is equal to the optimal advantage in guessing $Z \oplus Z'$ with k queries to $\mathbf{H}_{Z,Z'}$, since Z' and $Z \oplus Z'$ are independent unbiased bits. For the analysis of this advantage we consider the form $\mathbf{H}_{Z,Z'} = \mathbf{C}(\langle \mathbf{I}/\mathbf{F} \rangle_Z, \langle \mathbf{J}/\mathbf{G} \rangle_{Z'})$.

Let $\hat{\mathbf{F}}$ and $\hat{\mathbf{I}}$ be defined as guaranteed by Lemma 5, where $\hat{\mathbf{F}}^- \equiv \mathbf{F}$, $\hat{\mathbf{I}}^- \equiv \mathbf{I}$, $\hat{\mathbf{F}}^{\dashv} \equiv \hat{\mathbf{I}}^{\dashv}$, and $\delta_{k'}(\mathbf{F}, \mathbf{I}) = \Delta_{k'}(\mathbf{F}, \mathbf{I}) = \nu_{k'}(\hat{\mathbf{F}})$. Similarly, let $\hat{\mathbf{G}}$ and $\hat{\mathbf{J}}$ be defined such that $\hat{\mathbf{G}}^- \equiv \mathbf{G}$, $\hat{\mathbf{J}}^- \equiv \mathbf{J}$, $\hat{\mathbf{G}}^{\dashv} \equiv \hat{\mathbf{J}}^{\dashv}$, and $\delta_{k''}(\mathbf{G}, \mathbf{J}) = \Delta_{k''}(\mathbf{G}, \mathbf{J}) = \nu_{k''}(\hat{\mathbf{G}})$. We define the system

$$\hat{\mathbf{H}}_{Z,Z'} := \mathbf{C}(\langle \hat{\mathbf{I}}/\hat{\mathbf{F}} \rangle_Z, \langle \hat{\mathbf{J}}/\hat{\mathbf{G}} \rangle_{Z'})$$

with two MBOs. If we define $\hat{\mathbf{H}}_{Z,Z'}^-$ as $\hat{\mathbf{H}}_{Z,Z'}$ with *both* MBOs ignored, then $\hat{\mathbf{H}}_{Z,Z'}^- \equiv \mathbf{H}_{Z,Z'}$.

Since the MBOs can always be ignored, guessing $Z \oplus Z'$ can only become easier in $\hat{\mathbf{H}}_{Z,Z'}$ (compared to $\mathbf{H}_{Z,Z'}$.) If we assume further that whenever an MBO turns to 1, the corresponding bit (Z or Z') is also output (i.e., given to the distinguisher for free), this can only improve the advantage further.

If either MBO is 0, the advantage in guessing that bit (Z or Z') is 0, and hence also the advantage in guessing $Z \oplus Z'$ is 0. Thus the optimal strategy for guessing $Z \oplus Z'$ is to provoke *both* MBOs (i.e., win both games), and the probability that this succeeds is the advantage in guessing $Z \oplus Z'$.

We can now consider making the distinguisher's task even easier. Instead of having to provoke the two MBOs in the system $\hat{\mathbf{H}}_{Z,Z'}$, we give the distinguisher direct access to the systems $\langle \hat{\mathbf{I}}/\hat{\mathbf{F}} \rangle_Z$ and $\langle \hat{\mathbf{J}}/\hat{\mathbf{G}} \rangle_{Z'}$, allowing k' and k'' queries, respectively. Lemma 6 implies that in this setting, using individual optimal strategies is optimal. The probabilities of provoking the MBOs by individually optimal strategies are $\nu_{k'}(\hat{\mathbf{F}}) = \Delta_{k'}(\mathbf{F}, \mathbf{I})$ and $\nu_{k''}(\hat{\mathbf{G}}) = \Delta_{k''}(\mathbf{G}, \mathbf{J})$, respectively, hence the advantage in guessing $Z \oplus Z'$ is $\Delta_{k'}(\mathbf{F}, \mathbf{I})\Delta_{k''}(\mathbf{G}, \mathbf{J})$. Taking into account the factor 2 from above (due to Lemma 3) this completes the proof. $\qquad\square$

We say that a construction $\mathbf{C}(\cdot, \cdot)$ is *feed-forward* if, within the evaluation of a single query to $\mathbf{C}(\mathbf{F}, \mathbf{G})$, no input to $\mathbf{F}$ (or $\mathbf{G}$) depends on a previous output of $\mathbf{F}$ (or $\mathbf{G}$) of the same evaluation of $\mathbf{C}(\mathbf{F}, \mathbf{G})$. We will only consider constructions $\mathbf{C}(\cdot, \cdot)$ that make a single call to the invoked systems per invocation of $\mathbf{C}(\cdot, \cdot)$, and such constructions are always feed-forward. The proof of the following result is omitted.

Corollary 1. *Consider the setting of Theorem 1. If $\mathbf{C}(\cdot, \cdot)$ is a feed-forward construction, then the inequality also holds for non-adaptive strategies:*

$$\Delta_k^{\mathsf{NA}}(\mathbf{C}(\mathbf{F}, \mathbf{G}), \mathbf{C}(\mathbf{I}, \mathbf{J})) \leq 2\, \Delta_{k'}^{\mathsf{NA}}(\mathbf{F}, \mathbf{I})\, \Delta_{k''}^{\mathsf{NA}}(\mathbf{G}, \mathbf{J}).$$

4.4 Implications of the Product Theorem

Recall that $\mathbf{R}$ ($\mathbf{P}$) denotes a uniform random function (permutation).

Definition 14. For two $(\mathcal{X}, \mathcal{Y})$-systems $\mathbf{F}$ and $\mathbf{G}$ and a quasi-group operation $\star$ on $\mathcal{Y}$, we define $\mathbf{F} \star \mathbf{G}$ as the system obtained by feeding each input to both systems and combining the outputs using $\star$.

Corollary 2. *For any random functions $\mathbf{F}$ and $\mathbf{G}$, any quasi-group operation $\star$, and for all k,*

$$\Delta_k(\mathbf{F} \star \mathbf{G}, \mathbf{R}) \leq 2\, \Delta_k(\mathbf{F}, \mathbf{R})\, \Delta_k(\mathbf{G}, \mathbf{R})$$

and

$$\Delta_k^{\mathsf{NA}}(\mathbf{F} \star \mathbf{G}, \mathbf{R}) \leq 2\, \Delta_k^{\mathsf{NA}}(\mathbf{F}, \mathbf{R})\, \Delta_k^{\mathsf{NA}}(\mathbf{G}, \mathbf{R}).$$

The same statements hold for general random systems $\mathbf{F}$ and $\mathbf{G}$ when $\mathbf{R}$ is replaced by (a beacon) $\mathbf{B}$.

Proof. Let $\mathbf{I} := \mathbf{R}$ and $\mathbf{J} := \mathbf{R}$ in Theorem 1 and $\mathbf{C}(\mathbf{F}, \mathbf{G}) := \mathbf{F} \star \mathbf{G}$. Condition (6) is satisfied since $\mathbf{F} \star \mathbf{R} \equiv \mathbf{R}$, $\mathbf{R} \star \mathbf{G} \equiv \mathbf{R}$, and $\mathbf{R} \star \mathbf{R} \equiv \mathbf{R}$. This proves the first inequality. The second inequality follows from Corollary 1 since $\mathbf{F} \star \mathbf{G}$ is clearly a feed-forward construction. The proof of the last statement is analogous.   □

Definition 15. For two $(\mathcal{X}, \mathcal{X})$-random permutations $\mathbf{F}$ and $\mathbf{G}$ we define $\mathbf{F} \triangleright \mathbf{G}$ as the system obtained by cascading $\mathbf{F}$ and $\mathbf{G}$, i.e., the input to $\mathbf{F} \triangleright \mathbf{G}$ is fed to $\mathbf{F}$, its output is fed to $\mathbf{G}$, and $\mathbf{G}$'s output is the output of $\mathbf{F} \triangleright \mathbf{G}$. Moreover, for a random permutation $\mathbf{F}$, we denote by $\langle \mathbf{F} \rangle$ the random permutation which can be queried from "both sides", i.e., one can also provide an output and receive the corresponding input.[18]

Corollary 3. *For any compatible random permutations $\mathbf{F}$ and $\mathbf{G}$, where $\mathbf{G}$ is stateless, for all k,*

$$\Delta_k(\mathbf{F} \triangleright \mathbf{G}, \mathbf{P}) \leq 2\, \Delta_k(\mathbf{F}, \mathbf{P})\, \Delta_k(\mathbf{G}, \mathbf{P})$$

and

$$\Delta_k^{\mathsf{NA}}(\mathbf{F} \triangleright \mathbf{G}, \mathbf{P}) \leq 2\, \Delta_k^{\mathsf{NA}}(\mathbf{F}, \mathbf{P})\, \Delta_k^{\mathsf{NA}}(\mathbf{G}, \mathbf{P}).$$

If also $\mathbf{F}$ is stateless, then the corresponding two inequalities also hold when bi-directional permutations are considered.[19]

Proof. Let $\mathbf{I} := \mathbf{P}$ and $\mathbf{J} := \mathbf{P}$ in Theorem 1 and $\mathbf{C}(\mathbf{F}, \mathbf{G}) := \mathbf{F} \triangleright \mathbf{G}$. Condition (6) is satisfied since $\mathbf{F} \triangleright \mathbf{P} \equiv \mathbf{P}$, $\mathbf{P} \triangleright \mathbf{G} \equiv \mathbf{P}$, and $\mathbf{P} \triangleright \mathbf{P} \equiv \mathbf{P}$. Note that $\mathbf{P} \triangleright \mathbf{G} \equiv \mathbf{P}$ is only guaranteed to hold if $\mathbf{G}$ is stateless.[20] No restriction applies to $\mathbf{F}$. This proves the first inequality. The second inequality follows from Corollary 1 since the cascade construction is feed-forward. The proof of the last statement is similar but omitted.   □

---

[18] This definition is motivated by considering chosen-plaintext and chosen-ciphertext attacks against a block-cipher. One-sided and two-sided attacks are sometimes also called CCA and nCCA, for the adaptive and the non-adaptive version.

[19] E.g., $\Delta_k(\langle \mathbf{F} \rangle \triangleright \langle \mathbf{G} \rangle, \langle \mathbf{P} \rangle) \leq 2\, \Delta_k(\langle \mathbf{F} \rangle, \langle \mathbf{P} \rangle)\, \Delta_k(\langle \mathbf{G} \rangle, \langle \mathbf{P} \rangle)$.

[20] As an example, consider a stateful random permutation $\mathbf{G}$ which internally builds a permutation function table by always taking the least unused element.

5   Amplification of the Distinguisher Class

The second main result of this paper states that if subsystems of a neutralizing construction are only indistinguishable from ideal systems by a *weak* distinguisher class, then the construction is indistinguishable for a *stronger* distinguisher class. Recall Definition 13.

Theorem 2. *If* $\mathbf{C}(\cdot,\cdot)$ *is neutralizing for the pairs* $(\mathbf{F},\mathbf{I})$ *and* $(\mathbf{G},\mathbf{J})$ *of systems, then, for all k and all distinguishers* $\mathbf{D}$,[21]

$$\delta_k^{\mathbf{D}}(\mathbf{C}(\mathbf{F},\mathbf{G}),\mathbf{C}(\mathbf{I},\mathbf{J})) \;\leq\; \delta_{k'}^{\mathbf{DC}(\cdot,\mathbf{J})}(\mathbf{F},\mathbf{I}) + \delta_{k''}^{\mathbf{DC}(\mathbf{I},\cdot)}(\mathbf{G},\mathbf{J}).$$

Proof. As in the proof of Theorem 1, let $\hat{\mathbf{F}}$ and $\hat{\mathbf{I}}$ be defined as guaranteed by Lemma 5, where $\hat{\mathbf{F}}^- \equiv \mathbf{F}$, $\hat{\mathbf{I}}^- \equiv \mathbf{I}$, $\hat{\mathbf{F}}^{\dashv} \equiv \hat{\mathbf{I}}^{\dashv}$, and $\delta_k^{\mathbf{D}}(\mathbf{F},\mathbf{I}) = \nu_k^{\mathbf{D}}(\hat{\mathbf{F}}) = \nu_k^{\mathbf{D}}(\hat{\mathbf{I}})$ for all $\mathbf{D}$. (Note that this $\mathbf{D}$ is different from that in the theorem.) Similarly, let $\hat{\mathbf{G}}$ and $\hat{\mathbf{J}}$ be defined such that $\hat{\mathbf{G}}^- \equiv \mathbf{G}$, $\hat{\mathbf{J}}^- \equiv \mathbf{J}$, $\hat{\mathbf{G}}^{\dashv} \equiv \hat{\mathbf{J}}^{\dashv}$, and $\delta_k^{\mathbf{D}}(\mathbf{G},\mathbf{J}) = \nu_k^{\mathbf{D}}(\hat{\mathbf{G}}) = \nu_k^{\mathbf{D}}(\hat{\mathbf{J}})$ for all $\mathbf{D}$.

We can consider the following two systems with MBO: $\hat{\mathbf{H}}_{00} := \mathbf{C}(\hat{\mathbf{I}},\hat{\mathbf{J}})$ and $\hat{\mathbf{H}}_{11} := \mathbf{C}(\hat{\mathbf{F}},\hat{\mathbf{G}})$, where for each system the MBO is defined as the OR of the two internal MBOs. We have $\hat{\mathbf{H}}_{00}^{\dashv} \equiv \hat{\mathbf{H}}_{11}^{\dashv}$ because $\hat{\mathbf{F}}^{\dashv} \equiv \hat{\mathbf{I}}^{\dashv}$ and $\hat{\mathbf{G}}^{\dashv} \equiv \hat{\mathbf{J}}^{\dashv}$. Therefore, since $\hat{\mathbf{H}}_{00}^- \equiv \mathbf{C}(\mathbf{I},\mathbf{J})$ and $\hat{\mathbf{H}}_{11}^- \equiv \mathbf{C}(\mathbf{F},\mathbf{G})$, Lemma 4 implies that

$$\delta_k^{\mathbf{D}}(\mathbf{C}(\mathbf{F},\mathbf{G}),\mathbf{C}(\mathbf{I},\mathbf{J})) \;\leq\; \nu_k^{\mathbf{D}}(\hat{\mathbf{H}}_{00}).$$

It remains to determine a bound on $\nu_k^{\mathbf{D}}(\hat{\mathbf{H}}_{00})$. The MBO in $\hat{\mathbf{H}}_{00}$ (i.e., in $\mathbf{C}(\hat{\mathbf{I}},\hat{\mathbf{J}})$) is provoked if either of the two internal MBOs is provoked. We can apply the union bound and consider the provocation of each MBO separately. More precisely, we consider the following systems with MBO: $\mathbf{C}(\hat{\mathbf{I}},\mathbf{J})$ and $\mathbf{C}(\mathbf{I},\hat{\mathbf{J}})$. Then $\nu_k^{\mathbf{D}}(\hat{\mathbf{H}}_{00})$ is bounded by the sum of the probabilities that $\mathbf{D}$ provokes the MBO in each of these systems, i.e.,

$$\nu_k^{\mathbf{D}}(\hat{\mathbf{H}}_{00}) \;\leq\; \nu_k^{\mathbf{D}}(\mathbf{C}(\hat{\mathbf{I}},\mathbf{J})) + \nu_k^{\mathbf{D}}(\mathbf{C}(\mathbf{I},\hat{\mathbf{J}})).$$

The proof is completed, using Lemma 5, by noting that $\nu_k^{\mathbf{D}}(\mathbf{C}(\hat{\mathbf{I}},\mathbf{J})) = \nu_{k'}^{\mathbf{DC}(\cdot,\mathbf{J})}(\hat{\mathbf{I}})$ $= \delta_{k'}^{\mathbf{DC}(\cdot,\mathbf{J})}(\mathbf{F},\mathbf{I})$ and $\nu_k^{\mathbf{D}}(\mathbf{C}(\mathbf{I},\hat{\mathbf{J}})) = \nu_{k''}^{\mathbf{DC}(\mathbf{I},\cdot)}(\hat{\mathbf{J}}) = \delta_{k''}^{\mathbf{DC}(\mathbf{I},\cdot)}(\mathbf{G},\mathbf{J})$. $\square$

Note that since Theorem 2 applies to every distinguisher, it also applies to any distinguisher class $\mathcal{D}$, for instance the class of all distinguishers. Recalling that $\Delta_k(\mathbf{S},\mathbf{T}) = \delta_k(\mathbf{S},\mathbf{T})$ and $\Delta_k^{\mathsf{NA}}(\mathbf{S},\mathbf{T}) = \delta_k^{\mathsf{NA}}(\mathbf{S},\mathbf{T})$, we obtain:

Corollary 4. *For any compatible random functions* $\mathbf{F}$ *and* $\mathbf{G}$ *and any quasigroup operation* $\star$, *and all k,*

$$\Delta_k(\mathbf{F}\star\mathbf{G},\mathbf{R}) \;\leq\; \Delta_k^{\mathsf{NA}}(\mathbf{F},\mathbf{R}) + \Delta_k^{\mathsf{NA}}(\mathbf{G},\mathbf{R}).$$

---

[21] Here, for example, $\mathbf{DC}(\cdot,\mathbf{J})$ denotes the distinguisher consisting of $\mathbf{D}$ connected to $\mathbf{C}(\cdot,\cdot)$ where the second subsystem is simulated as $\mathbf{J}$ and the system to be distinguished is placed as the first subsystem.

Proof. We recall that the $\star$-combination is neutralizing: $\mathbf{F}\star\mathbf{R} \equiv \mathbf{R}\star\mathbf{G} \equiv \mathbf{R}\star\mathbf{R} \equiv \mathbf{R}$. It remains to show that the distinguisher classes correspond to the class of non-adaptive distinguishers.

For any $\mathbf{D}$, the distinguisher $\mathbf{DC}(\cdot, \mathbf{J})$ (i.e., the distinguisher $\mathbf{D}(\cdot \star \mathbf{R})$) for provoking the MBO in $\hat{\mathbf{F}}$ obtains only random outputs, independently of $\hat{\mathbf{F}}$. A distinguisher could simulate these random outputs itself, ignoring the output of $\mathbf{F} \star \mathbf{R}$, and hence corresponds to a non-adaptive distinguisher. The same argument also applies to the distinguisher $\mathbf{DC}(\mathbf{I}, \cdot)$ for provoking the MBO in $\hat{\mathbf{G}}$. $\qquad\square$

Corollary 5. *For any compatible random permutations $\mathbf{F}$ and $\mathbf{G}$, where $\mathbf{G}$ is stateless, for all k,*

$$\Delta_k(\mathbf{F} \triangleright \mathbf{G}, \mathbf{P}) \;\leq\; \Delta_k^{\mathsf{NA}}(\mathbf{F}, \mathbf{P}) + \Delta_k^{\mathsf{RI}}(\mathbf{G}, \mathbf{P}).$$

If also $\mathbf{F}$ is stateless, then[22]

$$\Delta_k(\langle\mathbf{F} \triangleright \mathbf{G}^{-1}\rangle, \langle\mathbf{P}\rangle) \;\leq\; \Delta_k^{\mathsf{NA}}(\mathbf{F}, \mathbf{P}) + \Delta_k^{\mathsf{NA}}(\mathbf{G}, \mathbf{P}).$$

The last statement means that $\langle\mathbf{F} \triangleright \mathbf{G}^{-1}\rangle$ is adaptively indistinguishable (from both sides) if $\mathbf{F}$ and $\mathbf{G}$ are only non-adaptively indistinguishable (from one side).

Proof. We recall that the $\triangleright$-combination is neutralizing: $\mathbf{F}\triangleright\mathbf{P} \equiv \mathbf{P}\triangleright\mathbf{G} \equiv \mathbf{P}\triangleright\mathbf{P} \equiv \mathbf{P}$. It remains to show that the distinguisher classes correspond to the class NA of non-adaptive distinguishers and the class RI of random-input distinguishers, respectively.

For any $\mathbf{D}$, the distinguisher $\mathbf{DC}(\cdot, \mathbf{J})$, i.e., the distinguisher $\mathbf{D}(\cdot \triangleright \mathbf{P})$, obtains only random outputs, independently of $\mathbf{F}$. A distinguisher could simulate these random outputs itself, ignoring the output of $\mathbf{F} \triangleright \mathbf{P}$, and hence corresponds to a non-adaptive distinguisher.

Similarly, the distinguisher $\mathbf{DC}(\mathbf{I}, \cdot)$, i.e., the distinguisher $\mathbf{D}(\mathbf{P} \triangleright \cdot)$, can only produce random inputs to $\mathbf{G}$, with the possibility of repeating a previous input. Because $\mathbf{G}$ is stateless, repeating an input does not help in provoking the MBO in $\mathbf{G}$.

The proof of the second statement is omitted. $\qquad\square$

Acknowledgments

It is a pleasure to thank Yevgeniy Dodis, Ghislain Fourny, Thomas Holenstein, Dominik Raub, Johan Sjödin, and Stefano Tessaro for discussions about random systems.

References

[KNR05]   Kaplan, E., Naor, M., Reingold, O.: Derandomized constructions of k-wise (almost) independent permutations. In: Chekuri, C., Jansen, K., Rolim, J.D.P., Trevisan, L. (eds.) APPROX 2005 and RANDOM 2005. LNCS, vol. 3624, pp. 354–365. Springer, Heidelberg (2005)

---

[22] Here $\mathbf{G}^{-1}$ is the inverse of $\mathbf{G}$, which is well defined as $\mathbf{G}$ is a stateless random permutation.

[LR86]   Luby, M., Rackoff, C.: Pseudo-random permutation generators and cryp-
            tographic composition. In: Proc, 18th ACM Symposium on the Theory of
            Computing (STOC), pp. 356–363 (1986)

[Mau02]   Maurer, U.: Indistinguishability of random systems. In: Knudsen, L.R.
            (ed.) EUROCRYPT 2002. LNCS, vol. 2332, pp. 110–132. Springer, Hei-
            delberg (2002)

[MOPS06]   Maurer, U., Oswald, Y.A., Pietrzak, K., Sjödin, J.: Luby-Rackoff ciphers
            with weak round functions. In: Vaudenay, S. (ed.) EUROCRYPT 2006.
            LNCS, vol. 4004, pp. 391–408. Springer, Heidelberg (2006)

[MP04]   Maurer, U., Pietrzak, K.: Composition of random systems: When two weak
            make one strong. In: Naor, M. (ed.) TCC 2004. LNCS, vol. 2951, pp. 410–
            427. Springer, Heidelberg (2004)

[Mye03]   Myers, S.: Efficient amplification of the security of weak pseudo-random
            function generators. Journal of Cryptology 16(1), 1–24 (2003)

[Mye04]   Myers, S.: Black-box composition does not imply adaptive security. In:
            Cachin, C., Camenisch, J.L. (eds.) EUROCRYPT 2004. LNCS, vol. 3027,
            pp. 189–206. Springer, Heidelberg (2004)

[Pie05]   Pietrzak, K.: Composition does not imply adaptive security. In: Shoup, V.
            (ed.) CRYPTO 2005. LNCS, vol. 3621, pp. 55–65. Springer, Heidelberg
            (2005)

[Pie06]   Pietrzak, K.: Composition implies adaptive security in minicrypt. In:
            Vaudenay, S. (ed.) EUROCRYPT 2006. LNCS, vol. 4004, pp. 328–338.
            Springer, Heidelberg (2006)

[PS07]   Pietrzak, K., Sjödin, J.: Domain extension for weak PRFs; the good, the
            bad, and the ugly. In: Yung, M. (ed.) CRYPTO 2002. LNCS, vol. 2442,
            pp. 517–533. Springer, Heidelberg (2002)

[Vau98]   Vaudenay, S.: Provable security for block ciphers by decorrelation. In:
            Meinel, C., Morvan, M. (eds.) STACS 98. LNCS, vol. 1373, pp. 249–275.
            Springer, Heidelberg (1998)

[Vau99]   Vaudenay, S.: Adaptive-attack norm for decorrelation and super-
            pseudorandomness. In: Heys, H.M., Adams, C.M. (eds.) SAC 1999. LNCS,
            vol. 1758, pp. 49–61. Springer, Heidelberg (2000)

[Vau03]   Vaudenay, S.: Decorrelation: A theory for block cipher security. J. Cryp-
            tology 16(4), 249–286 (2003)

A Hybrid Lattice-Reduction and Meet-in-the-Middle Attack Against NTRU

Nick Howgrave-Graham

NTRU Cryptosystems, Inc.
nhowgravegraham@ntru.com

Abstract. To date the NTRUEncrypt security parameters have been based on the existence of two types of attack: a meet-in-the-middle attack due to Odlyzko, and a conservative extrapolation of the running times of the best (known) lattice reduction schemes to recover the private key. We show that there is in fact a continuum of more efficient attacks between these two attacks. We show that by combining lattice reduction and a meet-in-the-middle strategy one can reduce the number of loops in attacking the NTRUEncrypt private key from $2^{84.2}$ to $2^{60.3}$, for the $k = 80$ parameter set. In practice the attack is still expensive (dependent on ones choice of cost-metric), although there are certain space/time trade-offs that can be applied. Asymptotically our attack remains exponential in the security parameter k, but it dictates that NTRUEncrypt parameters must be chosen so that the meet-in-the-middle attack has complexity 2^k even after an initial lattice basis reduction of complexity 2^k.

1 Introduction

It is well known that the closest vector problem (CVP) can be solved efficiently in the case that the given point in space is very close to a lattice vector [7,18]. If this CVP algorithm takes time t and a set S has the property that it includes at least one point $v_0 \in S$ which is very close to a lattice vector, then clearly v_0 can be found in time $O(|S|t)$ by exhaustively enumerating the set S. We show that if the points of S can be represented as $S = S' \oplus S'$, i.e. for every $(v, v') \in S \times S'$ there exists an $v'' \in S'$ such that $v = v' + v''$, then there are conditions under which there is actually an efficient meet-in-the-middle algorithm on this space to find the point v_0 in time $O(|S|^{1/2}t)$.

We can translate this CVP result to a result about lattice basis reduction by defining the set S to be some linear combinations of the last $n - m$ rows of a given basis $\{b_1 \ldots, b_n\}$, and then using the CVP algorithm on the elements of S and the basis $\{b_1, \ldots, b_m\}$. We note that a similar approach is taken by Schnorr in [21] for reducing generic lattices with the SHORT algorithm. Schnorr also suggests that "birthday" improvements might be possible for his method (generalizing results from [24]) but concludes that, in general, storage requirements may be prohibitive.

In this paper we show that, in the case of searching for the NTRUEncrypt private key, meet-in-the-middle techniques are indeed possible. We show that

A. Menezes (Ed.): CRYPTO 2007, LNCS 4622, pp. 150–169, 2007.

Odlyzko's storage ideas may be generalized to remain efficient even when used after lattice reduction, and we optimize the set S for the structure of the NTRU-Encrypt private key.

1.1   Roadmap

In section 2 we describe the key recovery problem behind NTRUEncrypt, and we explain the best known attacks against it. We introduce the following question regarding its parameters: "for a given N and q and security parameter k, how low can d_f be?": this is the fundamental mathematical question that our paper addresses, and it ultimately shows that d_f cannot be as low as previously thought. This is important because d_f is one of the factors that govern the efficiency of NTRUEncrypt, and so from a parameter generation point of view there is a practical desire to keep it as low as possible.

In section 3 we give a brief summary of the theory lattices including the usefulness of triangularization of lattice bases. In section 3.1 we discuss the practical consequences of running lattice reduction schemes on the NTRU public basis.

In section 4 we explain the mathematics behind the new hybrid technique, and in section 5 we analyze the cost of the technique in theory and in practice.

As with many meet-in-the-middle techniques the storage requirements of our technique are considerable. In section 6 we discuss methods to lessen these requirements at the cost of increasing the running time.

In section 7 we discuss possible generalizations of our work in more generic lattice situations, and give conclusions in section 8.

2   The NTRU Cryptosystem

NTRUEncrypt was invented in 1996, and was first published in [10]. It is based in the ring $\mathcal{R} = \mathbb{Z}[X]/(X^N - 1, q)$ whose elements can be represented by vectors of length N with integer entries modulo q. To aid exposition we will differentiate between a vector representation $a \in V_N(\mathbb{Z})$ and a ring representation $\mathsf{a} \in \mathcal{R}$ by the use of the LaTeX fonts shown. The NTRUEncrypt private key is two "binary" vectors $f, g \in V_N(\{0, 1\})$ with d_f and d_g ones respectively, and the remaining entries zero[1]. The NTRUEncrypt public key is $\mathsf{h} = \mathsf{g}/\mathsf{f}$ in the ring $\mathcal{R}$, where h is typically viewed as h, a vector of length N with integer entries modulo q.

There are many good descriptions of how the NTRU cryptosystem works [9,10,11,14], but in this paper we directly take on the problem of recovering the private key from the public information, so we do not need to delve into details of encryption and decryption. Out of interest we note that the encryption and decryption algorithms are both very efficient operations (both encryption and decryption are $O(k^2)$ in the security parameter k), and all known attacks against NTRUEncrypt are exponential in the security parameter k (including

---

[1] Other sets of small vectors are possible for the set of NTRUEncrypt private keys, but this is the one we will initially concentrate on.

the one demonstrated in this paper). Another potential upside of NTRUEncrypt is its apparent resistance to attack by quantum computers. The downsides to NTRUEncrypt are that the public key-size and ciphertext size are both slightly large, and that there is expansion in encryption (a raw N-bit plaintext (after padding) is encrypted to a $(N \log_2 q)$-bit ciphertext) so NTRUEncrypt lacks some of the nice properties that an encryption-permutation allows.

The parameter choices for N, q, d_f, d_g have undergone several changes since the invention of NTRUEncrypt due to both progress in cryptanalysis [11], and fine tuning of the parameters for efficiency reasons [14]. The currently recommended choices for $k = 80$ bit security are $N = 251$, $q = 197$, $d_f = 48$, $d_g = 125$. This parameter set is known as ees251ep6 in the IEEE P1363.1 draft standard [16].

The attack demonstrated in this paper is applicable, to some degree, to all the NTRUEncrypt parameter sets since its invention. Unfortunately it is most effective on the currently recommended parameter sets because d_f has been lowered considerably for efficiency reasons.

2.1   Lattice Attacks Against NTRU

The recovery of the NTRUEncrypt private key from public information can be posed as a lattice problem. This was known by the inventors of NTRU [10], and further explored in [5].

From the definition of $\mathsf{h} = \mathsf{g}/\mathsf{f}$, it is clear that there is an length-$(2N)$ integer vector (k, f) such that

$$(k, f) \begin{pmatrix} qI & 0 \\ H & I \end{pmatrix} = (g, f), \tag{1}$$

where H is a circulant matrix generated from h, i.e. $H_{i,j} = h_{i+j \bmod N}$. Note that the vector/matrix multiplication fH respects the multiplication fh in the ring $\mathbb{Z}[X]/(X^N - 1)$, and the k part of the vector corresponds to the reduction of each coefficient modulo q. The $(2N) \times (2N)$ basis $((qI, 0), (H, I))$ is referred to as the *NTRU public lattice basis*.

The discriminant of the NTRU public lattice basis is clearly q^N, whilst the (g, f) vector has size $(d_f + d_g)^{1/2}$. The Gaussian heuristic therefore suggests that there are no smaller vectors in the lattice than (g, f) and so lattice reduction might be used to find it. We note that the NTRU public lattice basis does not contain just one small vector (g, f) but all N "rotations" $(g^{(i)}, f^{(i)})$ where $g^{(i)}$, $f^{(i)}$ correspond to $\mathsf{f}^{(i)} = \mathsf{f}X^i$ and $\mathsf{g}^{(i)} = \mathsf{g}X^i$ respectively, for $i = 0, \ldots, N - 1$, since $(\mathsf{f}X^i)\mathsf{h} = \mathsf{g}X^i$ in the ring $\mathcal{R}$.

Although the rotations of the (g, f) vectors are the smallest vectors in the NTRU lattice, the best (known) direct lattice reduction techniques find it hard to recover any of these vectors in practice. Indeed typically lattice reduction methods appear to be fully exponential in the security parameter k. For the lattice family to which the ees251ep6 parameter set belongs, it is stated in [14,16] that lattice reduction has a complexity of at least

$$R = 2^{0.4245N - 3.44} \tag{2}$$

for $N > 120$, to find any vector smaller that a q-vector. For $N = 251$ this corresponds to time of $2^{103.1}$ to directly find a (g, f) rotation from the public basis.

2.2   Odlyzko's Meet-in-the-Middle Attack on NTRU

NTRU parameter sets have always been secure against a meet-in-the-middle attack discovered by Odlyzko, which is described in [15].

The idea is that if f_1 and f_2 are such that $f = f_1 + f_2$ then the entries of $x_1 = f_1 h$ and $x_2 = -f_2 h$ differ only by 0 or 1 mod q, since $(f_1 + f_2)h = g$ and g is binary.

Assuming f has d_f ones and d_f is even, then the attack progresses by sampling a binary ring element f_1 with $d_f/2$ ones, and computing $x_1 = f_1 h$.

The vector x_1 corresponding to x_1 is of length N with entries satisfying $-q/2 < (x_1)_i \leq q/2$. For each index i of x_1 we determine a bit β_i, where $\beta_i = 1$ if $(x_1)_i > 0$ and 0 otherwise. We can therefore determine an N-bit string from x_1, namely $a_1 = \beta_1 : \beta_2 : \ldots : \beta_N$, which we call an "address" or "label'. Let $\overline{\beta} = 1 - \beta$ denote the complement of a bit β, and let $\overline{a}$ denote the component-wise complement of a bit-string a. The element f_1 is stored in two "boxes": one with address a_1, and one with address $\overline{a_1}$.

The meet-in-the-middle technique carries on sampling f_1 as above, and storing them in boxes dependent on the x_1. If two binary elements f_1 and f_2 are sampled such that $f_1 + f_2 = f$ then one can hope that the a_1 corresponding to $x_1 = f_1 h$ is the same as the $\overline{a_2}$ corresponding to $x_2 = f_2 h$, since $x_1 = -x_2 + g$. This will only be the case if the entries of g do not cause af the entries of x_1 to "change sign", but this technicality can be dealt with by either simply accepting the probability of the occurrence, or by storing the f_1 in more boxes if the x_1 have coefficients that may change sign. These approaches are discussed further in [15] and later in this report.

In this introduction we will assume that whenever $f_1 + f_2 = f$ then with certainty $a_1 = \overline{a_2}$, i.e. sampling f_1, f_2 such that $f_1 + f_2 = f$ can be detected by a collision in a box. For any collisions we can retrieve the f_1, f_2 stored in the box, and check if $(f_1 + f_2)h$ is binary: if so we have found a very small vector in the NTRU public basis; undoubtedly[2] one of the rotations of (g, f).

To estimate the complexity of this attack, let V denote the set of f_1 which are actually a subset of the ones of some rotation of f. Assuming the rotations have a small number of intersections we see that $|V| \approx N \binom{d_f}{d_f/2}$, and we can expect a collision in the set of such f_1 after $O(|V|^{1/2})$ samples. The probability of sampling from this set is $|V|/\binom{N}{d_f/2}$, so the expected number of loops of the algorithm before a collision is

$$L = \frac{1}{\sqrt{N}} \binom{N}{d_f/2} \binom{d_f}{d_f/2}^{-1/2}. \tag{3}$$

For the ees251ep6 parameter set this turns out to be $2^{84.2}$.

---

[2] It will almost certainly be a (g, f) rotation because of the way we performed the search, although it is worth noting that discovering any short enough vector is tantamount to breaking NTRUEncrypt, as observed in [5].

2.3   Choosing NTRUEncrypt Parameters

NTRU are typically conservative when choosing parameters, so the true lattice security (when considering BKZ attacks only) is probably significantly higher than equation 2 suggests, due to the upward concavity of the observed running times. Similarly, ensuring that the number of loops, L, given by equation 3 is greater than 2^{80} is conservative for two reasons:

- There are hidden computational costs per loop, e.g. Odlyzko's attack requires summing together $d_f/2$ vectors of length N and reducing their coefficients modulo q. If we count "one addition modulo q" as an "intrinsic operation" then this cost could arguably add $\log_2(Nd_f/2)$ bits of security.
- The storage requirements of Odlyzko's attack is slightly greater than the number loops given by equation 2, since we may need to store the f_1's in several boxes per loop (on average 8, say). Also the f_1's take at least $\log_2 \binom{N}{d_f/2}$ bits to store.

Thus one might conclude that, in practice, Odlyzko's attack on the ees251ep6 parameter set will require too many operations ($2^{95.8}$ modular additions) and/or too much storage (2^{94} bits) to be feasible, and hence the parameter set is more than adequate for a $k = 80$ security level. Of these two constraints the storage requirement is by far the larger obstacle given today's hardware.

Although NTRU have been conservative in their parameter choices, this is with respect to the best known attacks. In this paper we demonstrate a new class of attack that may cause NTRU to re-evaluate their parameter sets. Indeed, as a piece of mathematics, the contribution of this paper can be summed up as an improved answer to the question "for a fixed N and q and a security level k, how low can d_f go?": we show that d_f cannot be as low as previously thought.

To gauge the practicality of our attack, we examine how it changes the running time and storage requirements of Odlyzko's attack in section 5, and discuss methods to make the storage requirements more feasible (at the cost of extra computation) in section 6.

This paper is not about suggesting new parameter sets which would require a large amount of analysis to justify well, however we do mention techniques that can mitigate our attack in section 8. When it does come to choosing new NTRU parameters we advocate the methodology outlined in [19], i.e. for a fixed PC architecture working out how much time it takes to break a symmetric key algorithm (e.g. DES), and how much time it takes to break a small NTRUEncrypt example, and then extrapolating the two to work out when NTRUEncrypt will require the same amount of work as an 80-bit symmetric algorithm (and similarly for higher k-bit security levels).

The above methodology (of comparison with an exhaustive search on a DES symmetric key) essentially benchmarks the PC in a standard way, and for example, allows one to argue how much security $2^{95.8}$ modular additions truly gives (it is certainly less than a bit security level of $k = 95.8$).

3   Lattice Basis Representation and Lattice Reduction

We take a row-oriented view of matrices and allow some flexibility between basis representations and matrix representations, e.g. we call a *matrix* BKZ-reduced if the *rows* of the matrix form a BKZ-reduced basis [22].

For a thorough grounding on lattices see [3,4], however for our purposes the following will suffice: for a given basis $\mathcal{B} = \{b_1, \ldots, b_n\}$ of $\mathbb{R}^n$ a lattice is defined to be the set of points

$$\mathcal{L} = \left\{ y \in \mathbb{R}^n \;\middle|\; y = \sum_{i=1}^{n} a_i b_i, \; a_i \in \mathbb{Z} \right\}$$

Clearly many bases will generate the same set of lattice points; indeed if we represent a basis $\mathcal{B}$ by a matrix B with rows $\{b_1, \ldots, b_n\}$ then it is exactly the rows of UB for any $U \in GL_n(\mathbb{Z})$ that generate these points.

However it is often convenient to give ourselves even more freedom with matrix representations of bases in that one can consider bases of isomorphic lattices too[3].

Definition 1. *Two lattices $\mathcal{L}, \mathcal{L}'$ are called isomorphic if there is a length-preserving bijection $\phi : \mathcal{L} \to \mathcal{L}'$ satisfying $\phi(x + y) = \phi(x) + \phi(y)$.*

In terms of matrix representations this means that if the rows of B form a basis for a lattice $\mathcal{L}$ then the rows of $B' = UBY$ where $U \in GL_n(\mathbb{Z})$ and Y is orthonormal, form a basis for an isomorphic lattice $\mathcal{L}'$, even though the rows of B' do not necessarily generate the same *points* of $\mathcal{L}$.

The point of allowing the extra freedom of post-multiplying by an orthonormal matrix is that if (for some reason) one can find an integer vector u such that uB' is small, then $uU^{-1}B$ is also small, i.e. solving lattice problems in an isomorphic lattice can help solve them in the original lattice. It is worth noting that this freedom also allows one to always consider lower triangular lattice bases by forming Y from the Gram-Schmidt procedure[4]. Explicitly $T_{i,j} = \mu_{i,j}|b_j^*|$ where $\mu_{i,j} = \langle b_i, b_j^* \rangle / |b_j^*|^2$ for $1 \leq j < i \leq n$ and $\mu_{i,i} = 1$.

Given that there are many bases of the same lattice $\mathcal{L}$, there is a significant amount of research around defining which bases are "more reduced" than others, and generating efficient algorithms to produce such bases [22,8,21]. The most commonly used reduction scheme in cryptography is BKZ [22] and its efficient implementation in the number theory library NTL [23].

Lattice reduction typically transforms a basis $\{b_1, \ldots, b_n\}$ so that the size of the Gram-Schmidt vectors b_i^* do not decrease "too quickly". This allows one to

---

[3] This phenomenon is usually explained through the language of quadratic forms, but such a presentation typically misses the concreteness of the isomorphic lattice bases, which we prefer in this report.

[4] It is worth saying that mathematicians do not always apply this transformation because some non-lower triangular lattice bases naturally have integer entries (as opposed to general real entries), and putting a lattice in lower triangular form can force the use of square roots of rational numbers (or real approximations) in this case.

prove an approximation factor between the size of the first vector b_1 and the size of the smallest vector in the lattice λ_1 (which is normally bounded by the size of b_n^*). Thus lattice reduction can be used to solve the (approximate) shortest vector problem (SVP).

Another well-studied lattice problem is the (approximate) closest vector problem (CVP): one is given an arbitrary point in space $y \in \mathbb{R}^n$ and the problem is to find the closest lattice point to this point (or more generally a lattice point within a radius of a multiple of λ_1). We make use of the following simple CVP-algorithm when the lattice basis is given by the rows of a lower triangular $(n) \times (n)$ matrix T (as explained above a basis can always be represented this way, and this avoids explicit use of b_n^*). We remark that this algorithm has a long history; it is sometimes called "weak reduction" or "size reduction" of the vector y against the basis T and is an essential component of lattice reduction techniques, however is usually referred to as Babai's nearest plane algorithm from the analysis in [2].

Algorithm 1. weakly reducing y against T

1: $x \leftarrow y$
2: **for** $i = n$ down to 1 **do**
3:     let u_i to be the nearest integer to $x_i/T_{i,i}$
4:     $x \leftarrow x - u_i T_i$
5: **end for**
6: **return** the reduced vector x

The following lemma (first shown in [7]) shows that if a point in space is "particularly close" to a lattice vector then it can be recovered by algorithm 1.

Lemma 1 (Furst, Kannan). *Assume* $y = uT + x$ *for some* $u \in V_n(\mathbb{Z})$, $x \in V_n(\mathbb{R})$ *and a lower triangular* $T \in M_n(\mathbb{R})$. *If the entries of* x *satisfy*

$$-T_{i,i}/2 < \quad x_i \quad \leq T_{i,i}/2 \qquad (4)$$

for $1 \leq i \leq n$, *then* x *can be recovered by algorithm 1.*

Proof. It is simple to confirm that the "error" vector x does not change any of the "rounding" computations of u_i in step 3 of algorithm 1.

3.1   Reducing the NTRU Public Basis

The NTRU public basis is given[5] in equation 1. The state of a partially-reduced NTRU lattice can be expressed well by plotting the $\log_q |b_i^*|$ for $i = 1, \ldots, 2N$, as done in [9]. Figure 1 shows the various states of reduction of an NTRU public basis with the ees251ep6 parameter set.

---

[5] We note that this basis description is slightly different from the original description in [10], but we prefer putting the small q-vectors first.

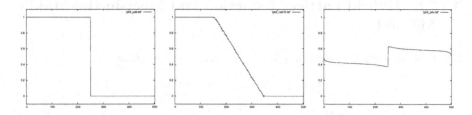

Fig. 1. A representation of lattice reduction on NTRU lattices via plotting $\log_q |b_i^*|$. The left figure is the public basis, the middle figure is after reduction with BKZ with blocksize 15, the right is the fully reduced private basis.

As can be seen from the middle graph of figure 1, the first few b_i^* vectors of a partially-reduced NTRU basis typically remain q-vectors, whilst the last b_i^* vectors typically satisfy $|b_i^*| = 1$. We note the (approximate) symmetry of these graphs can be made exact by using the symplectic lattice reduction techniques of [9]. Notice the central region of a partially-reduced NTRU basis is approximately linear in the log scale, i.e. it obeys the geometric series assumption (GSA) as defined in [21].

Given that the first and last vectors are untouched, reducing an NTRU basis can be speeded up by extracting a suitable lower triangular submatrix B', reducing this, and putting the basis back in a lower triangular form[6]. If the public basis is represented by B, then this transformation can be written $UBY = T$, where the structure of U, B, Y and T are shown below:

$$\left(\begin{array}{c|c|c} I_r & 0 & 0 \\ \hline 0 & U' & 0 \\ \hline 0 & 0 & I_{r'} \end{array} \right) \left(\begin{array}{c|c|c} qI_r & 0 & 0 \\ \hline * & B' & 0 \\ \hline * & * & I_{r'} \end{array} \right) \left(\begin{array}{c|c|c} I_r & 0 & 0 \\ \hline 0 & Y' & 0 \\ \hline 0 & 0 & I_{r'} \end{array} \right) = \left(\begin{array}{c|c|c} qI_r & 0 & 0 \\ \hline * & T' & 0 \\ \hline * & * & I_{r'} \end{array} \right). \tag{5}$$

The (partially-reduced) matrix T also has N small vectors given by $(g^{(i)}, f^{(i)}) Y$, where $(g^{(i)}, f^{(i)})$ are the original small vectors corresponding to $\mathsf{f} = \mathsf{f} X^i$, $\mathsf{g} = \mathsf{g} X^i$ for $i = 1, \ldots, N$. From the structure of Y we see that these small vectors have binary entries for the first r entries and the last r' entries, and only the middle entries are affected by Y'.

Let $m = 2N - r'$ denote the number of vectors in the first two "blocks", and let $\{b'_1, \ldots, b'_{m-r}\}$ denote the rows of B'. Our strategy to recover the NTRUEncrypt private key is to pick a submatrix B' such that $(b'_{m-r})^*$ can be made reasonably large (so that lemma 1 may be usefully employed), whilst at the same time making m reasonably large so that the last r' entries of $(g^{(i)}, f^{(i)}) Y$ can either be guessed, or (less-restrictively) have a meet-in-the-middle attack mounted on them.

We remark that the standard way to ensure $(b'_{m-r})^*$ is large, is to try to minimize the first vector in the dual matrix of B', as described in [6,12,13,20].

---

[6] This is completely akin to the treatment of blocks in a block reduction scheme.

4   The Hybrid Lattice-Reduction and Meet-in-the-Middle Method

Let T be as defined in equation 5, and let u, v, s be such that

$$(u|v)T = (s|v) = (g^{(i)}, f^{(i)})Y,$$

for some $i = 1, \ldots, 2N$ and where u, s are of length m, and v is of length $r' = 2N - m$.

We start by showing that an algorithm that enumerates all possible v is enough to recover $(u|v)$, and then show that there is actually a meet-in-the-middle algorithm to recover the same information. We note that knowledge of $(u|v)$ is clearly equivalent to knowledge of $(s|v)$ and (g, f).

Lemma 2. *The vector $(0|v)T - (0|v)$ is a distance of $|s|$ away from a lattice point of T.*

Proof. We know

$$(0|v)T - (0|v) = (u|v)T - (u|0)T - (0|v)$$
$$= (s|0) - (u|0)T.$$

Corollary 1. *If s if "small enough" to satisfy the conditions of lemma 1 then it can be found by algorithm 1.*

In our analysis we always ensure that s satisfies the conditions of lemma 1 for a large proportion[7] of the rotations $(g^{(i)}, f^{(i)})$. In principle this condition could be slightly relaxed by using the methods of [18,21], at the cost of doing some extra "searching".

We use the output of algorithm 1 to determine a number of "addresses" for "boxes" to store meet-in-the-middle data in to. As mentioned in section 2.2 there are slight complications in working out which boxes to store information in to increase the probability of good collisions. In our analysis we always ensure that r (the number of initial q-vectors untouched) is large enough so that the storage requirements of the meet-in-the-middle attack is less that 2^r. This way we can use Odlyzko's storage strategy directly without having to consider i for which $T_{i,i} < q$. We note that the following definition could be generalized to handle the case when $|x_i| \leq T_{i,i}/2$ and the error on the x_i is non-constant, unknown but small (rather than the fixed value 1), but this is presently unnecessary.

Definition 2. *For a fixed integer r, and any vector $(x|0)$ with entries satisfying $-q/2 < x_i \leq q/2$ for $1 \leq i \leq r$ we define an associated set, $\mathcal{A}_x^{(r)}$, of r-bit integer "addresses" where $\mathcal{A}_x^{(r)}$ contains every r-bit integer a satisfying both of the following properties:*

---

[7] This probability depends on the form of T and the effect of Y so can be checked easily.

- bit $a_i = 1$ for all indices i, $1 \leq i \leq r$, such that $x_i > 1$, and
- bit $a_i = 0$ for all indices i, $1 \leq i \leq r$, such that $x_i \leq 0$.

Example 1. To help explain definition 2 we do a simple example with $r = 10$, $q = 11$, and

$$x = (2, 3, -4, -1, 1, 5, -3, -2, 0, 1).$$

In this case there are 2 entries satisfying $x_i = 1$, so

$$\mathcal{A}_x^{(r)} = \{1100010000_2, 1100010001_2$$
$$1100110000_2, 1100110001_2\}.$$

Lemma 3. *When the first r entries of $(x|0)$ are random integers modulo q and independent of each other, then the expected size of $\mathcal{A}_x^{(r)}$ is 2^z where*

$$z = \sum_{j=0}^{r} j \binom{r}{j} \frac{(q-1)^{r-j}}{q^r}.$$

Proof. Each $x_i = 1$ doubles the entries of $\mathcal{A}_x^{(r)}$ (one with bit $a_i = 0$ and the other with bit $a_i = 1$), so the number of expected entries in $\mathcal{A}_x^{(r)}$ is 2^z where z is the number of expected 1's in the first r entries of x.

Assuming the entries of x are random modulo q and independent of each other, the probability that x has j ones in its first r entries is

$$p_j = \binom{r}{j} \frac{(q-1)^{r-j}}{q^r},$$

and the expected value is therefore given by $z = \sum_{j=0}^{r} j p_j$.

Example 2. In the case $r = 159$ and $q = 197$ then $p_0 \approx 0.45, p_1 \approx 0.36, p_2 \approx 0.14, p_3 \approx 0.04$, so $z \approx 0.36 + 2(0.14) + 3(0.04) \approx 0.76$, and the probability that $z > 3$ is very low (so in practice we discard such x since they are costly to store).

Lemma 4. *If the first r entries of a vector s are binary, and $-q/2 < x_i - s_i \leq q/2$ for $1 \leq i \leq r$, then the set $\mathcal{A}_x^{(r)} \cap \mathcal{A}_{x-s}^{(r)}$ is non-empty.*

Proof. If the first r entries of a vector s are binary then the sign of the first r entries of $x - s$ are unchanged whenever $x_i > 1$ or $x_i \leq 0$. If $0 < x_i \leq 1$ then the sign does change but in that case $\mathcal{A}_x^{(r)}$ contained addresses with both choices of bit a_i.

The meet-in-the-middle attack is described in algorithm 2. To analyze its properties we create the following definition.

Definition 3. *A vector v_1 of length $r' = 2N - m$ is called s-admissible if the x_1, u_1 gotten from algorithm 1 satisfy:*

$$(0|v_1)T - (0|v_1) = (x_1|0) - (u_1|0)T, \quad \text{and} \tag{6}$$
$$(0|v_1)T - (s|v_1) = (x_1 - s|0) - (u_1|0)T,$$

i.e. the subtraction of $(s|0)$ does not affect the multiple of T taken away during algorithm 1.

Algorithm 2. meet-in-the-middle on v

1: **loop**
2:     guess a binary vector v_1 of length $r' = 2N - m$ with c ones
3:     use algorithm 1 to calculate x_1, u_1 such that $(0|v_1)T - (0|v_1) = (x_1|0) - (u_1|0)T$

4:     store v_1 in the boxes addressed by a, for every $a \in \mathcal{A}_{x_1}^{(r)} \cup \mathcal{A}_{-x_1}^{(r)}$
5:     **if** there is already a value v_2 stored in any of the above boxes **then**
6:         let $v = v_1 + v_2$ and use algorithm 1 to calculate x, u such that $(0|v)T - (0|v) = (x|0) - (u|0)T$
7:         **if** $(g|f) = (x|v)Y^{-1}$ is binary **then**
8:             **return** f, g
9:         **end if**
10:    **end if**
11: **end loop**

Lemma 5. *If a vector v_1 is s-admissible, then the vector $v_2 = v - v_1$ is also s-admissible.*

Proof. We have

$$(0|v - v_1)T - (0|v - v_1) = (0|v)T - (0|v) - (0|v_1)T + (0|v_1)$$
$$= (s|0) - (u|0)T - (x_1|0) + (u_1|0)T$$
$$= (s - x_1|0) - (u - u_1|0)T$$

and

$$(0|v - v_1)T - (s|v - v_1) = (-x_1|0) - (u - u_1|0)T.$$

Theorem 1. *Let v_1, v_2 be two s-admissible vectors such that $v_1 + v_2 = v$. If x_i, u_i are gotten from applying algorithm 1 to $(0|v_i)T - (0|v_i)$ for $i = 1, 2$, then $x_1 + x_2 = s$.*

Proof. We know

$$(0|v_1)T - (s|v_1) = (x_1 - s|0) - (u_1|0)T$$
$$(0|v_2)T - (0|v_2) = (x_2|0) - (u_2|0)T,$$

where $-T_{i,i}/2 < (x_1)_i - s_i \leq T_{i,i}/2$, $-T_{i,i}/2 < (x_2)_i \leq T_{i,i}/2$, so summing these equations yields

$$(0|v)T - (s|v) = (x_1 + x_2 - s|0) - (u_1 + u_2|0)T$$
$$(u|0)T = (x_1 + x_2 - s|0) - (u_1 + u_2|0)T$$
$$(u - u_1 - u_2|0)T = (x_1 + x_2 - s|0).$$

Thus $(x_1)_m + (x_2)_m - s_m = 0$ modulo $T_{m,m}$, but in fact we can deduce $(x_1)_m + (x_2)_m - s_m = 0$ over the integers because of the size restrictions on $(x_1)_m$ and $(x_2)_m - s_m$ (two real numbers modulo $T_{m,m}$ cannot be as large as $2T_{m,m}$). This implies $u_m = (u_1)_m + (u_2)_m$, and given that one can then re-apply a similar argument to coefficients $(m - 1), \ldots, 1$ to realize $x_1 + x_2 = s$, and $u_1 + u_2 = u$.

Theorem 2. *If v_1, v_2 are s-admissible such that $v_1 + v_2 = v$ and they are chosen in separate loops of algorithm 2 then there exists a box which contains both v_1 and v_2.*

Proof. Since v_1 and v_2 are both admissible and $v = v_1 + v_2$ then by theorem 1 we know $x_1 + x_2 = s$. We know v_2 is contained in all the boxes addressed by $a \in \mathcal{A}_{-x_2}^{(r)} = \mathcal{A}_{x_1-s}^{(r)}$. But v_1 is stored in all the boxes addressed by $a \in \mathcal{A}_{x_1}^{(r)}$ so by lemma 4 there is at least one box which contains both v_1 and v_2.

Remark 1. The problem of estimating the probability that a vector v_1 chosen in step 2 of algorithm 2 is s-admissible can be modelled by the problem of calculating the probability distribution of a coordinate ϵ of a point obtained by multiplying a binary vector times an orthonormal matrix[8]. In particular, the square of a coordinate of the image of a binary vector with d 1's and $m-r-d$ 0's after multiplication by an orthonormal matrix will have $|\epsilon| < \sqrt{d}$ and expected value $E(\epsilon^2) = d/(m-r)$. Denote by $p_d(\delta, \delta')$ the probability that $\delta \leq |\epsilon| < \delta'$ and choose $0 = \delta_1 < \delta_2 < \cdots < \delta_K = \sqrt{d}$. Let $T_{m,m} = q^\alpha$ and assume that the GSA holds and that the density function associated to p_d is decreasing for $\delta \geq \delta_2$. When $i > r$, the ith coefficient of v_1 is $x_i + s_i$ where x_i is uniformly distributed in $[-T_{i,i}/2, T_{i,i}/2]$, where $T_{i,i} = q^{e_i}$. Let's approximate the density function of s_i by a step function on intervals $[\delta_k, \delta_{k+1}]$. In practice, the probabilities $p_d(\delta_k, \delta_{k+1})$ are obtained experimentally, for the choices of the partition by δ_k, but the precise values aren't relevant for the validity of the formula below. More precisely, for each i, $i \geq r+1$ the factors on the right hand side below represent lower and upper approximations to an integral which involves the convolution of the density functions of x_i and s_i. The factor $(1 - 1/q)^{r/2}$ is present because for $i \leq r$, x_i is a uniform random variable in $[-q/2, q/2)$ and s_i takes on the values $\{0, 1\}$ each with probability $1/2$. Under these assumptions, and given these approximations, the following "theoretical" computation provides a reality check for a probability determined experimentally. This is the probability p_s that a vector v_1 chosen in step 2 of algorithm 2 is s-admissible. In particular, we have

$$p_s > \left(1 - \frac{1}{q}\right)^{r/2} \prod_{i=r+1}^{m} \left(\sum_{k=1}^{K-1} \left(1 - \frac{\delta_{k+1}}{q^{e_i}}\right) p_d(\delta_k, \delta_{k+1})\right)$$

$$p_s < \left(1 - \frac{1}{q}\right)^{r/2} \prod_{i=r+1}^{m} \left(\sum_{k=1}^{K} \left(1 - \frac{\delta_k}{q^{e_i}}\right) p_d(\delta_k, \delta_{k+1})\right)$$

Here $e_i = ((\alpha - 1)i + (m - \alpha r))/(m - r)$.

Remark 2. The interval for p_s given by the above inequality comes reasonably close to calculations of p_s obtained by direct sampling and testing (given T and Y). For example, taking the parameters of the $N = 251$ example in the table,

---

[8] That is, we are modelling Y' as a random orthonormal matrix, which can be approximated by applying the Gram-Schmidt procedure (with normalization) to a random matrix.

with $\alpha = 0.3$, computations show that $2^{-5.95} < p_s < 2^{-6.82}$, while sampling directly gave $p_s = 2^{-6.7}$.

Lemma 6. *The probability that a vector v_1 sampled in step 2 of algorithm 2 is such that $v = v_1 + v_2$, for some v_2 with c ones, is given by*

$$p_h = w \binom{2c}{c} \binom{2N - m}{c}^{-1},$$

where w is the number of rotations of $(g|f)$ resulting in $2c$ distinct ones in v.

Proof. This is just the ratio of the sizes of the respective sets, assuming no intersections of the rotations.

Theorem 3. *The expected number of loops of algorithm 2 before (f, g) is returned is estimated by*

$$L^* = \binom{2N - m}{c} \left(p_s w \binom{2c}{c} \right)^{-1/2},$$

where w is the number of rotations of $(g|f)$ resulting in $2c$ ones in v.

Proof. Let V denote the set of s-admissible vectors v_1 with c ones, such that $v - v_1$ also has c ones. Assuming independence the probability of choosing an element of V in step 2 of algorithm 2 is $p_s p_h$ so we can expect to draw from V about 1 in every $(p_s p_h)^{-1}$ samples. Again assuming independence the size of the set V is $|V| = p_s w \binom{2c}{c}$, and after about $|V|^{1/2}$ samples of the set V we can expect to have sampled a v_1 and a v_2 such that $v_1 + v_2 = v$.

Remark 3. Although the number L^* seems the most natural measure of the cost of the hybrid method, it does ignore the change in the cost per loop in going from Odlyzko's attack to the hybrid attack. We previously estimated the inner-loop cost of Odlyzko's attack as $N d_f / 2$, whereas the corresponding cost of the hybrid scheme is $m^2 c / 2$ modular additions.

5 Results

To determine the practicality of our attack we have implemented it fully on a small example, and have done a thorough analysis for the **ees251ep6** parameter set, namely: examples with $m = 302$ and $m = 325$ based on actual lattice reduction data, and an example with $m = 344$ based on extrapolated data.

5.1 A Small Example

Algorithm 2 has been fully implemented for a small example with $N = 53$, $q = 37$, $d_f = d_g = 16$. In this example Odlyzko's meet-in-the-middle attack should have taken $2^{20.1}$ loops whereas algorithm 2 has $2^{13.1}$ loops.

With such a small example lattice reduction can often recover the NTRU-Encrypt private key, so care was taken to avoid this. We extracted the lower triangular submatrix from rows/columns 24 to 76 inclusive, and LLL-reduced this basis (it took a few seconds using NTL on a 2GHz laptop with 1GB of RAM running Cygwin on a Windows XP platform).

This left the last $r' = 30$ to launch the meet-in-the-middle attack on. We assumed there were 8 ones in these last 30 entries, which was true for 11 of the 57 rotations of (g, f).

We chose many combinations of $c = 4$ ones from these last $r' = 30$, storing the choices in boxes dependent on the output of algorithm 1. After $2^{13.1}$ loops we successfully found a rotation of the (g, f) vector.

On average we stored information in roughly 4 boxes per loop, so the storage complexity was $2^{15.1}$.

5.2   ees251ep6 with $m = 302$

In the first of the experiments of the ees251ep6 parameter set we extracted the lower triangular submatrix from rows/columns 160 to 300 inclusive, and BKZ-reduced this basis with blocksize 15. The form of the matrix T is shown in the left side of figure 2.

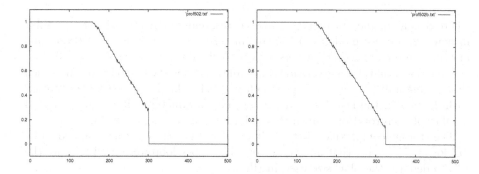

Fig. 2. $\log_{197} |b_i^*|$ for $i = 1, \ldots, 502$. Left: $m = 302$, Right: $m = 325$.

This left the last $r' = 200$ to launch the meet-in-the-middle attack on. We assumed there were 34 ones in these last 200 entries, which was true for 2 of the 251 rotations of (g, f).

We chose many combinations of $c = 17$ ones from these last $r' = 200$, storing the choices in boxes dependent on the output of algorithm 1. Our analysis predicts algorithm 2 will find the NTRUEncrypt private key after $2^{66.9}$ loops.

5.3   A Table of Results with an Extrapolation

From an initial analysis of BKZ lattice running times, the connection between running time and number of q-vectors removed appears to be

$$y = 2378.28 - 132.652x + 23.7371x \log x \tag{7}$$

where it takes 2^y time to remove x q-vectors, and we assume $x > 98$. Out of interest we note that the accuracy of this equation would imply that equation 2 is a severe underestimate of the lattice security of the ees251ep6 parameter set against BKZ attacks.

Given equation 7 it seems reasonable to assume that in less time than it takes to do $2^{76.2}$ modular additions, one can hope to perform lattice reduction with $r = 136$ and $m = 344$. If this is indeed the case then the security of the ees251ep6 parameter set is at most that given by $2^{76.2}$ modular additions and a storage requirement of $2^{65.6}$. A stronger initial lattice reduction would improve both of these figures.

N	q	d_f	m	β	t (secs)	r	c	α	p_s	w	L	L^*	#adds	#store
53	37	16	76	2	5	23	4	0.35	$2^{-6.3}$	11	$2^{20.1}$	$2^{13.1}$	$2^{26.6}$	2^{19}
107	67	32	151	15	360	36	7	0.235	$2^{-8.6}$	2	$2^{44.0}$	$2^{28.3}$	$2^{44.6}$	$2^{36.2}$
251	197	48	302	15	780	159	17	0.287	$2^{-6.8}$	9	$2^{84.2}$	$2^{66.9}$	$2^{86.4}$	$2^{76.2}$
251	197	48	325	22	48727	144	14	0.182	2^{-13}	4	$2^{84.2}$	$2^{60.3}$	$2^{79.8}$	$2^{69.4}$
251	197	48	344	*	*	136	12	0.106	$2^{-20.4}$	4	$2^{84.2}$	$2^{56.7}$	$2^{76.2}$	$2^{65.6}$

The extrapolation was done assuming the GSA [21], which gives the relation

$$\alpha = \frac{2N - m - r}{m - r},$$

where α is as defined in remark 1 and corresponds to the "height of the cliffs" in figure 2, and we modelled Y' as an $(m - r) \times (m - r)$ random orthonormal matrix (these assumptions fit very well with the data from real examples).

We remark that, for a given m, the parameter c was chosen to be minimal such that a randomly chosen f has probability ≥ 0.4 of having at least one rotation with $2c$ ones in the last $2N - m$ entries. The parameter w holds the expected number of such rotations, given that f has at least one such rotation.

The storage complexity "#store" is measured in bits and was assumed to be $8L^* \log_2 \binom{2N-m}{c}$, i.e. that an average of 8 boxes per loop were used to store the v_1's. β denotes the blocksize used in BKZ.

Note that "#adds" count the number of *modular additions* and does not correspond to *bit-security* (see section 2.3 for a further discussion on this distinction). The usefulness of this measure is that it shows the factor of improvement over existing attacks.

6   Lessening Storage Requirements

The storage requirements of the extrapolated data point for the ees251ep6 parameter set corresponds to a total of $2^{65.6}$ bits. Although this is significantly better than the storage requirements for Odlyzko's attack (2^{94} bits), it is still very expensive given today's hardware. In this section we discuss how to reduce this requirement to a more manageable figure, e.g. a total of $2^{53.6}$ bits of storage[9].

---

[9] At the time of writing a 1TB hard drive costs about \$400, so this amount of storage can be had for approximately \$500,000.

The idea is to perform the attack as before, but now we assume we know more about the structure of f. For example in the `ees251ep6` parameter set after reducing the first $m = 344$ rows, we assume there is a rotation of f satisfying:

$\leftarrow 93 \rightarrow$	$\leftarrow 108 \rightarrow$	$\leftarrow D = 50 \rightarrow$
22 ones	$2c' = 20$ ones	$c'' = 6$ ones

When f is randomly chosen with 48 ones, there is a probability of 0.4 that there will be a rotation of f satisfying this pattern (if f is not of this structure the method will end in failure, and another common form should be chosen). Given that there is at least one rotation of f of this form, the expected number of rotations of this form is $w = 3$.

The attacker[10] chooses a fixed vector v_0 of length $r' = 2N - m$ with c'' of the last D entries set to 1, and the remaining entries 0.

We now explain how to slightly modify algorithm 2 to solve CVP with the point $(0|v_0)T$ rather than SVP. In step 2 the algorithm should guess a vector v_1' of length $r' = 2N - m$ such that $v_1 = v_1' + v_0$ has c ones (so v_1' has $c' = 11$ ones in the first $r' - D$ entries, assuming the above form of f).

In step 3 we then calculate the $x_1, u_1,$ and x_1', u_1' corresponding to both v_1 and v_1' respectively. In step 4 we store v_1' in the boxes addressed by a for every $a \in \mathcal{A}_{x_1}^{(r)} \cup \mathcal{A}_{-x_1}^{(r)} \cup \mathcal{A}_{x_1'}^{(r)} \cup \mathcal{A}_{-x_1'}^{(r)}$.

In step 6 we let $v = v_1' + v_2' + v_0$ and apply algorithm 1 as before. It is easy to confirm that theorem 2 still holds where we now have $v_1 = v_1' + v_0$ and $v_2 = v_2'$, so if these are s-admissible then there will be a collision in a box and v can be recovered.

The cost per loop of the modified algorithm is roughly twice the running time and storage of algorithm 2.

The number of loops of the modified algorithm is

$$L^\# = \binom{2N - m - D}{c'} \left(p_s w \binom{2c'}{c'} \right)^{-1/2},$$

and the cost per loop is $m^2 c'$ modular additions, and there are $\binom{D}{c''}$ choices for v_0. Thus the total work done is $2^{89.2}$ modular additions, which is still better than Odlyzko's attack ($2^{95.8}$) but now we have brought the storage down to $8L^\# \log_2 \binom{2N-m-D}{c'} = 2^{53.6}$ bits, which is a factor of $2^{40.4}$ less than Odlyzko's attack.

7   Generalizations

As mentioned in the introduction, the idea of this paper is really about achieving a meet-in-the-middle technique when one has a set $S = S' \oplus S'$ which contains a vector v_0 which is close to a lattice point of a well-reduced lattice basis. We

---

[10] Or multiple attackers, since this part is totally parallelizable.

have seen how the idea can be applied to NTRUEncrypt, but there is also hope it can be applied in more generic lattice situations.

To place the approximate-SVP problem on a basis $\mathcal{B} = \{b_1, \ldots, b_n\}$ in to the above framework one can split the basis in to two parts: $\mathcal{B}_1 = \{b_1, \ldots, b_m\}$ and $\mathcal{B}_2 = \{b_{m+1} \ldots, b_n\}$. The set S' can then be generated by linear combinations of $\mathcal{B}_2$ which are small in the space orthogonal to $\mathcal{B}_1$. In [21] Schnorr proposes that $\mathcal{B}_2$ be sampled with the SHORT algorithm[11], but many other approaches are possible[12], and indeed other approaches may result in shorter vectors (in the space orthogonal to $\mathcal{B}_1$).

There are several competing criteria dictating what value of m, $1 \leq m \leq n$ one should use: m should be small enough to ensure:

- S' is large enough: An important criteria for the technique to work is that there should be some vector $v_0 \in S = S' \oplus S'$ which, when projected in to the space generated by $\mathcal{B}_1$, is close to a lattice point of $\mathcal{B}_1$. In the case of NTRUEncrypt the structure of the private key guarantees this, but in a more generic lattice situation one must ensure that S' is large enough for there to be a reasonable probability of this being true.
- b_m^* is large enough: For the technique to work m must be chosen such that lattice reduction is likely to b_m^* large enough (with respect to the closeness of v_0 to a lattice point of $\mathcal{B}_1$) for the s-admissible probability to be non-negligible.

However there are also reasons for making m large:

- There need to be enough boxes to store the multiple of $\mathcal{B}_2$ in to: If m is too low there will be too many collisions and the technique will not work. Interestingly this means that lattices resulting from subset sum problems do not seem good candidates for this approach (they typically only contain one non-trivial column).
- Sampling from $\mathcal{B}_2$ should not take too long.

There is hope that the parameter m, and the amount of initial lattice reduction can be fine tuned to optimize this meet-in-the-middle approach, and possibly improve on Kannan's exhaustive search algorithm [17]. Asymptotically we know that Kannan's algorithm will be beaten by sieving techniques [1], but in relatively low dimensions there may be a space for a hybrid algorithm out-performing both existing techniques. We leave the investigation of this idea to future research.

As a final generalization we also note that algorithm 1 is not essential to the method, indeed the s-admissible probability can be improved by using a better CVP algorithm than Babai's closest plane algorithm (e.g. mixing Babai's CVP (which is essentially blocksize 1) with searching in higher blocksizes $2, 3, \ldots$), but this means a more costly CVP approach has to be performed for each loop, so

---

[11] This can be seen as a slightly modified version of Babai's nearest plane algorithm which takes a binary auxiliary "error" vector as input, and makes a slight error in rounding wherever this vector has a non-zero entry.

[12] For example an exhaustive search of small vectors within some bound.

care should be taken to keep it relatively efficient[13]. When using such a higher blocksize CVP approach it is possible to calculate the addresses for the boxes from the multiple of the rows take away in the CVP process rather than the sign of the x_i.

8 Conclusions

We have demonstrated a new class of attack on the NTRU cryptosystem: one where there is an initial amount of lattice reduction, followed by a generalized meet-in-the-middle procedure.

One way this result can be viewed is as a large strengthening of the result in [5]. In that paper it was shown that lattice reduction sufficient to retrieve a vector of size less that q could be used to break NTRUEncrypt; in this paper we show that far less lattice reduction is needed to mount a successful attack.

With regards to the ees251ep6 parameter set, we have performed lattice reduction to a sufficient degree to make our method 2^{16} times quicker than Odlyzko's attack, whilst at the same time requiring a factor of $2^{24.6}$ less storage. However, due to the original conservative choice of NTRUEncrypt parameters there still remains a substantially hard problem to recover the NTRUEncrypt private key (primarily due to the storage requirements).

We extrapolated lattice running times to make our method $2^{19.6}$ times quicker than Odlyzko's attack, and requiring a factor of $2^{28.4}$ less storage, but the storage requirements were still substantial.

We have thus modified the attack to require a factor of $2^{40.4}$ less memory, and take time about one hundredth of that of Odlyzko's attack. This is therefore the most practical attack on NTRUEncrypt since its inception in 1996. Progress in lattice reduction will improve our results (both the running time and storage requirements), and so should be factored in if choosing new parameters.

Our attack is still exponential in the security parameter k, so it does not "break" NTRUEncrypt in an asymptotic sense. However to avoid this attack it is imperative to chose parameters so that the meet-in-the-middle attack has complexity 2^k even after an initial lattice basis reduction of complexity 2^k. We also note that when choosing parameters it seems overly-cautious to allow the attacker up to 2^k storage, especially for security levels $k > 80$, but some realistic model of the attackers' storage capabilities should be made.

We observe that in order to defend against this attack it is probably a good idea to "thicken" the NTRUEncrypt private vector (g, f), i.e. to set $d_f = d_g \approx N/2$, or preferably to use a "trinary" vector (g, f) with -1's, 0's, and 1's, to make meet-in-the-middle attacks substantially harder without increasing N considerably.

Acknowledgements. The author would very much like to thank Jeff Hoffstein and Phil Hirschhorn for verifying the analysis in this paper, and Jill Pipher and William Whyte for several useful and stimulating conversations. Jeff and Jill also gave great help with remark 1.

---

[13] For example, it could be only used where the b_i^* are small.

References

1. Ajtai, M., Kumar, R., Sivakumar, D.: A sieve algorithm for the shortest lattice vector problem. In: Proc. of 29th STOC, pp. 284–293. ACM Press, New York (1997)
2. Babai, L.: On Lovász lattice reduction and the nearest lattice point problem. Combinatorica 6, 1–13 (1986)
3. Cassels, J.W.S.: An introduction to the geometry of numbers, Springer-Verlag, Reprint of the 1st ed. Berlin Heidelberg New York, Corr. 2nd printing 1971, 1997, VIII (1959)
4. Conway, J.H., Sloane, N.J.A.: Sphere Packings, Lattices and Groups. Grundlehren der mathematischen Wissenschaften, 290 (1993)
5. Coppersmith, D., Shamir, A.: Lattice Attacks on NTRU. In: Fumy, W. (ed.) EUROCRYPT 1997. LNCS, vol. 1233, pp. 52–61. Springer, Heidelberg (1997)
6. Fincke, U., Pohst, M.: Improved methods for calculating vectors of short length in a lattice, including a complexity analysis. Math. Comp. 44, 463–471 (1985)
7. Furst, M.L., Kannan, R.: Succinct certificates for almost all subset sum problems. SIAM Journal on Computing 1989, 550–558
8. Gama, N., Howgrave-Graham, N., Nguyen, P.Q.: Rankin's Constant and Blockwise Lattice Reduction. In: Dwork, C. (ed.) CRYPTO 2006. LNCS, vol. 4117, pp. 112–130. Springer, Heidelberg (2006)
9. Gama, N., Howgrave-Graham, N., Nguyen, P.Q.: Symplectic Lattice Reduction and NTRU. In: Vaudenay, S. (ed.) EUROCRYPT 2006. LNCS, vol. 4004, pp. 233–253. Springer, Heidelberg (2006)
10. Hoffstein, J., Pipher, J., Silverman, J.H.: NTRU: A Ring-Based Public Key Cryptosystem. In: Buhler, J.P. (ed.) Algorithmic Number Theory. LNCS, vol. 1423, pp. 267–288. Springer, Heidelberg (1998)
11. Howgrave-Graham, N., Nguyen, P.Q., Pointcheval, D., Proos, J., Silverman, J.H., Singer, A., Whyte, W.: The Impact of Decryption Failures on the Security of NTRU Encryption. In: Boneh, D. (ed.) CRYPTO 2003. LNCS, vol. 2729, pp. 226–246. Springer, Heidelberg (2003)
12. Howgrave-Graham, N.: Computational Mathematics Inspired by RSA, PhD. Thesis, University of Bath (1998)
13. Howgrave-Graham, N.: Finding Small Roots of Univariate Modular Equations Revisited. IMA Int. Conf. pp. 131–142 (1997)
14. Howgrave-Graham, N., Silverman, J.H., Whyte, W.: Choosing Parameter Sets for NTRUEncrypt with NAEP and SVES-3. In: Menezes, A.J. (ed.) CT-RSA 2005. LNCS, vol. 3376, pp. 118–135. Springer, Heidelberg (2005), http://www.ntru.com/cryptolab/articles.htm#2005_1
15. Howgrave-Graham, N., Silverman, J.H., Whyte, W.: A Meet-In-The-Middle Attack on an NTRU Private Key, http://www.ntru.com/cryptolab/tech_notes.htm#004!
16. W. Whyte, (ed.) IEEE P1363, 1/D9 Draft Standard for Public-Key Cryptographic Techniques Based on Hard Problems over Lattices
17. Kannan, R.: Improved algorithms for integer programming and related lattice problems. In: Proc. of the 15th Symposium on the Theory of Computing (STOC 1983), pp. 99–108. ACM Press, New York (1983)
18. Philip, N.: Finding the closest lattice vector when it's unusually close. In: Proceedings, ACM-SIAM Symposium on Discrete Algorithms, pp. 937–941. ACM, New York (2000)

19. Lenstra, A., Verheul, E.: Selecting Cryptographic Key Sizes. Journal of Cryptology 14(4), 255–293 (2001)
20. Pohst, M.: On the computation of lattice vectors of minimal length, successive minima and reduced bases with applications. ACM SIGSAM Bull. 15, 37–44 (1981)
21. Schnorr, C.P.: Lattice Reduction by Random Sampling and Birthday Methods. In: Alt, H., Habib, M. (eds.) STACS 2003. LNCS, vol. 2607, pp. 145–156. Springer, Heidelberg (2003)
22. Schnorr, C.P., Euchner, M.: Lattice Basis Reduction: Improved Practical Algorithms and Solving Subset Sum Problems. Mathematical Programming 66, 181–191 (1994)
23. Shoup, V.: NTL: A Library for doing Number Theory, Version 5.4, http://www.shoup.net/ntl
24. Wagner, D.: A Generalized Birthday Problem. In: Yung, M. (ed.) CRYPTO 2002. LNCS, vol. 2442, pp. 288–303. Springer, Heidelberg (2002), http://www.cs.berkeley.edu/daw/papers

Improved Analysis of Kannan's Shortest Lattice Vector Algorithm

(Extended Abstract)

Guillaume Hanrot[1,*] and Damien Stehlé[2]

[1] LORIA/INRIA Lorraine, Technopôle de Nancy-Brabois,
615 rue du jardin botanique, F-54602 Villers-lès-Nancy Cedex, France
hanrot@loria.fr
http://www.loria.fr/~hanrot
[2] CNRS and ÉNS Lyon/ LIP, 46 allée d'Italie, 69364 Lyon Cedex 07, France
damien.stehle@ens-lyon.fr
http://perso.ens-lyon.fr/damien.stehle

Abstract. The security of lattice-based cryptosystems such as NTRU, GGH and Ajtai-Dwork essentially relies upon the intractability of computing a shortest non-zero lattice vector and a closest lattice vector to a given target vector in high dimensions. The best algorithms for these tasks are due to Kannan, and, though remarkably simple, their complexity estimates have not been improved since over twenty years. Kannan's algorithm for solving the shortest vector problem (SVP) is in particular crucial in Schnorr's celebrated block reduction algorithm, on which rely the best known generic attacks against the lattice-based encryption schemes mentioned above. In this paper we improve the complexity upper-bounds of Kannan's algorithms. The analysis provides new insight on the practical cost of solving SVP, and helps progressing towards providing meaningful key-sizes.

1 Introduction

A lattice L is a discrete subgroup of some $\mathbb{R}^n$. Such an object can always be represented as the set of integer linear combinations of at most n vectors $\boldsymbol{b}_1, \ldots, \boldsymbol{b}_d$. These vectors can be chosen linearly independent, and in that case, we say that they are a basis of the lattice L. The most famous algorithmic problem associated with lattices is the so-called shortest vector problem (SVP). Its computational variant is to find a non-zero lattice vector of smallest Euclidean length — this length being the minimum $\lambda(L)$ of the lattice — given a basis of the lattice. Its decisional variant is known to be NP-hard under randomised reductions [2], even if one only asks for a vector whose length is no more than $2^{(\log d)^{1-\epsilon}}$ times the length of a shortest vector [12] (for any $\epsilon > 0$).

SVP is of prime importance in cryptography since a now quite large family of public-key cryptosystems relies more or less on it. The Ajtai-Dwork cryptosystem [4] relies on d^c-SVP for some $c > 0$, where $f(d)$-SVP is the problem of finding

---

* Work partially supported by CNRS GDR 2251 "Réseau de théorie des nombres".

A. Menezes (Ed.): CRYPTO 2007, LNCS 4622, pp. 170–186, 2007.

the shortest non-zero vector in the lattice L, under the promise that any vector of length less than $f(d) \cdot \lambda(L)$ is parallel to it. The GGH cryptosystem [11] relies on special instances of the Closest Vector Problem (CVP), a non-homogeneous version of SVP. Both the Ajtai-Dwork and GGH cryptosystems have been shown impractical for real-life parameters [25,23] (the initial GGH containing a major theoretical flaw as well). Finally, one strongly suspects that in NTRU [15] the private key can be read on the coordinates of a shortest vector of the Coppersmith-Shamir lattice [8]. The best known generic attacks against these encryption schemes are based on solving SVP. It is therefore highly important to know precisely what complexity is achievable, both in theory and practice, in particular to select meaningful key-sizes. Most often, for cryptanalysing lattice-based cryptosystems, one considers Schnorr's block-based algorithms [28, 30], such as BKZ. These algorithms internally solve instances of SVP in much lower dimensions (related to the size of the block). They help solving relaxed variants of SVP in high dimensions. Increasing the dimensions up to which one can solve SVP helps decreasing the relaxation factors that are achievable in higher dimensions. Solving the instances of SVP is the computationally expensive part of the block-based reduction algorithms.

Two main algorithms are known for solving SVP. The first one is based on the deterministic exhaustive enumeration of lattice points within a small convex body. It is known as Fincke-Pohst's enumeration algorithm [9] in the algorithmic number theory community. Cryptographers know it as Kannan's algorithm [16]. There are two main differences between both: firstly, in Kannan's algorithm, a long pre-computation on the basis is performed before starting the enumeration process; secondly, Kannan enumerates integer points in a hyper-parallelepiped whereas Fincke and Pohst consider an hyper-ellipsoid which is strictly contained in Kannan's hyper-parallelepiped – though Kannan may have chosen the hyper-parallelepiped in order to simplify the complexity analysis. Kannan obtained a $d^{d+o(d)}$ complexity bound (in the complexity bounds mentioned in the introduction, there is an implicit factor that is polynomial in the bit-size of the input). In 1985, Helfrich [13] refined Kannan's analysis, and obtained a $d^{d/2+o(d)}$ complexity bound. On the other hand, Ajtai, Kumar and Sivakumar [5] designed a probabilistic algorithm of complexity $2^{O(d)}$. The best exponent constant is likely to be small, as suggested by some recent progress [26]. A major drawback of this algorithm is that it requires an exponential space, whereas Kannan's requires a polynomial space.

Our main result is to lower Helfrich's complexity bound on Kannan's algorithm, from $d^{\frac{d}{2}+o(d)} \approx d^{0.5 \cdot d}$ to $d^{\frac{d}{2e}+o(d)} \approx d^{0.184 \cdot d + o(d)}$. This may explain why Kannan's algorithm is tractable even in moderate dimensions. Our analysis can also be adapted to Kannan's algorithm for CVP: it decreases Helfrich's complexity bound from $d^{d+o(d)}$ to $d^{d/2+o(d)}$. The complexity improvement for SVP provides better worst-case efficiency/quality trade-offs for Schnorr's block-based algorithms [28, 30, 10].

It must be noted that if one follows our analysis step by step, the derived $o(d)$ may be large when evaluated for some practical d. The hidden constants can be improved (for some of them it may be easy, for others it is probably much harder).

No attempt was made to improve them and we believe that it would have complicated the proof with irrelevant details. In fact, most of our analysis consists in estimating the number of lattice points within convex bodies and showing that the approximations by the volumes are almost valid. By replacing this discretisation by heuristic volume estimates, one obtains very small hidden constants.

Our complexity improvement is based on a fairly simple idea. It is equivalent to generate all lattice points within a ball and to generate all integer points within an ellipsoid (consider the ellipsoid defined by the quadratic form naturally associated with the given lattice basis). Fincke and Pohst noticed that it was more efficient to work with the ellipsoid than to consider a parallelepiped containing it: indeed, when the dimension increases, the ratio between the two volumes tends to 0 very quickly. In his analysis, instead of considering the ellipsoid, Kannan bounds the volume of the parallelepiped. Using rather involved technicalities, we bound the number of points within related ellipsoids. Some parts of our proof could be of independent interest. For example, we show that for any Hermite-Korkine-Zolotarev-reduced (HKZ-reduced for short) lattice basis $(b_1, \ldots, b_d)$, and any subset I of $\{1, \ldots, d\}$, we have:

$$\frac{\|b_1\|^{|I|}}{\prod_{i \in I} \|b_i^*\|} \leq \sqrt{d}^{|I|(1 + \log \frac{d}{|I|})},$$

where $(b_i^*)_{i \leq d}$ is the Gram-Schmidt orthogonalisation of the b_i's. This generalises the results of [28] on the quality of HKZ-reduced bases.

PRACTICAL IMPLICATIONS. We do not change Kannan's algorithm, but only improve its complexity upper-bound. As a consequence, the running-time of Kannan's algorithm remains the same. Nevertheless, our work may still have some important practical impact. First of all, it revives the interest on Kannan's algorithm. Surprisingly, although it has the best complexity upper-bound, it is not the one implemented in the usual number theory libraries (e.g., NTL [32] and Magma [18] implement Schnorr-Euchner's variant [30]): we show that by using Kannan's principle (i.e., pre-processing the basis before starting the enumeration), one can solve SVP in larger dimensions. This might point a problem in NTRU's security estimates, since they are derived from experimentations with NTL. Secondly, our analysis helps providing a heuristic measure of the (practical) cost of solving SVP for a particular instance, which is both efficiently computable and reliable: given a lattice basis, it provides very quickly a heuristic upper bound on the cost of finding a shortest vector.

ROAD-MAP OF THE PAPER. In Section 2, we recall some basic definitions and properties on lattice reduction. Section 3 is devoted to the description of Kannan's algorithm and Section 4 to its complexity analysis. In Section 5, we give without much detail our sibling result on CVP, as well as direct consequences of our result for block-based algorithms. In Section 6, we discuss the practical implications of our work.

NOTATION. All logarithms are natural logarithms, i.e., $\log(e) = 1$. Let $\|\cdot\|$ and $\langle \cdot, \cdot \rangle$ be the Euclidean norm and inner product of $\mathbb{R}^n$. Bold variables are vectors. We use the bit complexity model. The notation $\mathcal{P}(n_1, \ldots, n_i)$ means $(n_1 \cdot \ldots \cdot n_i)^c$ for some constant $c > 0$. If x is real, we denote by $\lfloor x \rceil$ a closest integer to it (with any convention for making it unique) and we define the centred fractional part $\{x\}$ as $x - \lfloor x \rceil$. Finally, for any integers a and b, we define $[\![a, b]\!]$ as $[a, b] \cap \mathbb{Z}$.

2   Background on Lattice Reduction

We assume that the reader is familiar with the geometry of numbers and its algorithmic aspects. Introductions may be found in [21] and [27].

Lattice Invariants. Let $\boldsymbol{b}_1, \ldots, \boldsymbol{b}_d$ be linearly independent vectors. Their *Gram-Schmidt orthogonalisation* (GSO) $\boldsymbol{b}_1^*, \ldots, \boldsymbol{b}_d^*$ is the orthogonal family defined recursively as follows: the vector $\boldsymbol{b}_i^*$ is the component of $\boldsymbol{b}_i$ which is orthogonal to the span of the vectors $\boldsymbol{b}_1, \ldots, \boldsymbol{b}_{i-1}$. We have $\boldsymbol{b}_i^* = \boldsymbol{b}_i - \sum_{j=1}^{i-1} \mu_{i,j} \boldsymbol{b}_j^*$ where $\mu_{i,j} = \frac{\langle \boldsymbol{b}_i, \boldsymbol{b}_j^* \rangle}{\|\boldsymbol{b}_j^*\|^2}$. For $i \leq d$ we let $\mu_{i,i} = 1$. Notice that the GSO family depends on the order of the vectors. If the $\boldsymbol{b}_i$'s are integer vectors, the $\boldsymbol{b}_i^*$'s and the $\mu_{i,j}$'s are rational. The *volume* of a lattice L is defined as $\det(L) = \prod_{i=1}^{d} \|\boldsymbol{b}_i^*\|$, where the $\boldsymbol{b}_i$'s are any basis of L. It does not depend on the choice of the basis of L and can be interpreted as the geometric volume of the parallelepiped naturally spanned by the basis vectors. Another important lattice invariant is the minimum. The *minimum* $\lambda(L)$ is the length of a shortest non-zero lattice vector.

The most famous lattice problem is the *shortest vector problem* (SVP). Here is its computational variant: given a basis of a lattice L, find a lattice vector whose norm is exactly $\lambda(L)$. The *closest vector problem* (CVP) is a non-homogeneous variant of SVP. We give here its computational variant: given a basis of a lattice L and a target vector in the real span of L, find a vector of L which is closest to the target vector.

The volume and the minimum of a lattice cannot behave independently. Hermite [14] was the first to bound the ratio $\frac{\lambda(L)}{(\det L)^{1/d}}$ as a function of the dimension only. His bound was later on greatly improved by Minkowski in his *Geometrie der Zahlen* [22]. *Hermite's constant* γ_d is defined as the supremum over d-dimensional lattices L of $\frac{\lambda(L)^2}{(\det L)^{2/d}}$. We have $\gamma_d \leq \frac{d+4}{4}$ (see [19]), which we will refer to as *Minkowski's theorem*.

Lattice Reduction. In order to solve lattice problems, a classical strategy consists in considering a lattice basis and trying to improve its quality (e.g., the slow decrease of the $\|\boldsymbol{b}_i^*\|$'s). This is called *lattice reduction*. The most usual notions of reduction are probably L^3 and HKZ. HKZ-reduction is very strong, but expensive to compute. On the contrary, L^3-reduction is fairly cheap, but an L^3-reduced basis is of much lower quality.

A basis $(\boldsymbol{b}_1, \ldots, \boldsymbol{b}_d)$ is *size-reduced* if its GSO family satisfies $|\mu_{i,j}| \leq 1/2$ for all $1 \leq j < i \leq d$. A basis $(\boldsymbol{b}_1, \ldots, \boldsymbol{b}_d)$ is said to be *Hermite-Korkine-Zolotarev-reduced* if it is size-reduced, the vector $\boldsymbol{b}_1$ reaches the lattice minimum, and the projections of the $(\boldsymbol{b}_i)_{i \geq 2}$'s orthogonally to the vector $\boldsymbol{b}_1$ are themselves an

HKZ-reduced basis. Lemma 1 immediately follows from this definition and Minkowski's theorem. It is the sole property on HKZ-reduced bases that we will use.

Lemma 1. *If* $(b_1, \ldots, b_d)$ *is HKZ-reduced, then for any* $i \leq d$, *we have:*

$$\|b_i^*\| \leq \sqrt{\frac{d-i+5}{4}} \cdot \left(\prod_{j \geq i} \|b_j^*\| \right)^{\frac{1}{d-i+1}}.$$

A basis $(b_1, \ldots, b_d)$ is L^3-*reduced* [17] if it is size-reduced and if its GSO satisfies the $(d-1)$ Lovász conditions: $\frac{3}{4} \cdot \left\|b_{\kappa-1}^*\right\|^2 \leq \left\|b_\kappa^* + \mu_{\kappa,\kappa-1}b_{\kappa-1}^*\right\|^2$. The L^3-reduction implies that the norms of the GSO vectors never drop too fast: intuitively, the vectors are not far from being orthogonal. Such bases have useful properties, like providing exponential approximations to SVP and CVP. In particular, their first vector is relatively short.

Theorem 1 ([17]). *Let* $(b_1, \ldots, b_d)$ *be an* L^3-*reduced basis of a lattice* L. *Then we have* $\|b_1\| \leq 2^{\frac{d-1}{4}} \cdot (\det L)^{1/d}$. *Moreover, there exists an algorithm that takes as input any set of integer vectors and outputs in deterministic polynomial time an* L^3-*reduced basis of the lattice they span.*

In the following, we will also need the fact that if the set of vectors given as input to the L^3 algorithm starts with a shortest non-zero lattice vector, then this vector is not changed during the execution of the algorithm: the output basis starts with the same vector.

3   Kannan's SVP Algorithm

Kannan's SVP algorithm [16] relies on multiple calls to the so-called short lattice points enumeration procedure. The latter finds all vectors of a given lattice that are in the sphere centred in $\mathbf{0}$ and of some prescribed radius. Variants of the enumeration procedure are described in [1].

3.1   Short Lattice Points Enumeration

Let $(b_1, \ldots, b_d)$ be a basis of a lattice $L \subset \mathbb{Z}^n$ and let $A \in \mathbb{Z}$. Our goal is to find all lattice vectors $\sum_{i=1}^d x_i b_i$ of squared Euclidean norm $\leq A$. The enumeration works as follows. Suppose that $\left\|\sum_i x_i b_i\right\|^2 \leq A$ for some integers x_i's. Then, by considering the components of the vector $\sum_i x_i b_i$ on each of the b_i^*'s, we obtain d equations:

$$(x_d)^2 \cdot \|b_d^*\|^2 \leq A,$$
$$(x_{d-1} + \mu_{d,d-1}x_d)^2 \cdot \|b_{d-1}^*\|^2 \leq A - (x_d)^2 \cdot \|b_d^*\|^2,$$

$$\cdots$$

$$\left(x_i + \sum_{j=i+1}^{d} \mu_{j,i} x_j\right)^2 \cdot \|b_i^*\|^2 \leq A - \sum_{j=i+1}^{d} l_j,$$

$$\cdots$$

where $l_i = (x_i + \sum_{j>i} x_j \mu_{j,i})^2 \cdot \|b_i^*\|^2$. The algorithm of Figure 1 mimics the equations above. It can be shown that the bit-cost of this algorithm is bounded by the number of loop iterations times a polynomial in the bit-size of the input. We will prove that if the input basis $(b_1, \ldots, b_d)$ is sufficiently reduced and if $A = \|b_1\|^2$, there are $\leq d^{\frac{d}{2e} + o(d)}$ loop iterations.

Input: An integer lattice basis $(b_1, \ldots, b_d)$, a bound $A \in \mathbb{Z}$.
Output: All vectors in $L(b_1, \ldots, b_d)$ that are of squared norm $\leq A$.
1. Compute the rational $\mu_{i,j}$'s and $\|b_i^*\|^2$'s.
2. $x := 0, l := 0, S := \emptyset$.
3. $i := 1$. While $i \leq d$, do
4. $\quad l_i := (x_i + \sum_{j>i} x_j \mu_{j,i})^2 \|b_i^*\|^2$.
5. $\quad$ If $i = 1$ and $\sum_{j=1}^{d} l_j \leq A$, then $S := S \cup \{\sum_{j=1}^{d} x_j b_j\}$, $x_1 := x_1 + 1$.
6. $\quad$ If $i \neq 1$ and $\sum_{j \geq i} l_j \leq A$, then
7. $\qquad i := i - 1, x_i := \left\lceil -\sum_{j>i}(x_j \mu_{j,i}) - \sqrt{\frac{A - \sum_{j>i} l_j}{\|b_i^*\|^2}} \right\rceil$.
8. $\quad$ If $\sum_{j \geq i} l_j > A$, then $i := i + 1, x_i := x_i + 1$.
9. Return S.

Fig. 1. The enumeration algorithm

3.2 Solving SVP

To solve SVP, Kannan provides an algorithm that computes HKZ-reduced bases, see Figure 2. The cost of the enumeration procedure dominates the overall cost and mostly depends on the quality of the input basis. The main idea of Kannan's algorithm is to spend a lot of time pre-computing a basis of excellent quality before calling the enumeration procedure. More precisely, it pre-computes a so-called quasi-HKZ-reduced basis.

Definition 1 (Quasi-HKZ-reduction). *A basis $(b_1, \ldots, b_d)$ is quasi-HKZ-reduced if it is size-reduced, if $\|b_2^*\| \geq \|b_1^*\|/2$ and if once projected orthogonally to b_1, the other b_i's are HKZ-reduced.*

A few comments need to be made on the algorithm of Figure 2. Steps 3 and 9 are recursive calls. However, the b_i''s may be rational vectors, whereas the input of the algorithm must be integral. These vectors may be scaled by a common factor. Steps 4 and 10 may be performed by expressing the reduced basis vectors as integer linear combinations of the initial ones, using these coefficients to recover lattice vectors and subtracting a correct multiple of the vector b_1. In Step 6, it is possible to choose such a vector b_0, since this enumeration always provides non-zero solutions (the vector b_1 is one of them).

Input: An integer lattice basis $(b_1, \ldots, b_d)$.
Output: An HKZ-reduced basis of the same lattice.
1. L^3-reduce the basis $(b_1, \ldots, b_d)$.
2. Compute the projections $(b'_i)_{i \geq 2}$ of the b_i's orthogonally to b_1.
3. HKZ-reduce the $(d-1)$-dimensional basis $(b'_2, \ldots, b'_d)$.
4. Extend the obtained $(b'_i)_{i \geq 2}$'s into vectors of L by adding to them rational multiples of b_1, in such a way that we have $|\mu_{i,1}| \leq 1/2$ for any $i > 1$.
5. If $(b_1, \ldots, b_d)$ is not quasi-HKZ-reduced, swap b_1 and b_2 and go to Step 2.
6. Call the enumeration procedure to find all lattice vectors of length $\leq \|b_1\|$. Let b_0 be a shortest non-zero vector among them.
7. $(b_1, \ldots, b_d) := L^3(b_0, \ldots, b_d)$.
8. Compute the projections $(b'_i)_{i \geq 2}$'s of the b_i's orthogonally to the vector b_1.
9. HKZ-reduce the $(d-1)$-dimensional basis $(b'_2, \ldots, b'_d)$.
10. Extend the obtained $(b'_i)_{i \geq 2}$'s into vectors of L by adding to them rational multiples of b_1, in such a way that we have $|\mu_{i,1}| \leq 1/2$ for any $i > 1$.

Fig. 2. Kannan's SVP algorithm

3.3   Cost of Kannan's SVP Solver

We recall briefly Helfrich's analysis [13] of Kannan's algorithm and explain our complexity improvement. Let $C(d, n, B)$ be the worst-case complexity of the algorithm of Figure 2 when given as input a d-dimensional basis which is embedded in $\mathbb{Z}^n$ and whose coefficients are smaller than B in absolute value. The following properties hold:

- Kannan's algorithm computes an HKZ-reduced basis of the lattice spanned by the input vectors.
- All arithmetic operations performed during the execution are of cost $\mathcal{P}(d, n, \log B)$. This implies that $C(d, n, B)$ can be bounded by $C(d) \cdot \mathcal{P}(\log B, n)$ for some function $C(d)$.
- There are fewer than $O(1) + \log d$ iterations of the loop of Steps 2–5.
- The cost of the call to the enumeration procedure at Step 6 is bounded by $\mathcal{P}(\log B, n) \cdot d^{d/2 + o(d)}$.

From these properties and those of the L^3 algorithm as recalled in the previous section, it is easy to obtain the following equation:

$$C(d) \leq (O(1) + \log d)(C(d-1) + \mathcal{P}(d)) + \mathcal{P}(d) + d^{\frac{d}{2} + o(d)}.$$

One can then derive the bound $C(d, B, n) \leq \mathcal{P}(\log B, n) \cdot d^{\frac{d}{2} + o(d)}$.

The main result of the present paper is to improve this complexity upper bound to $\mathcal{P}(\log B, n) \cdot d^{\frac{d}{2e} + o(d)}$. In fact, we show the following:

Theorem 2. *Given as inputs a quasi-HKZ-reduced basis $(b_1, \ldots, b_d)$ and $A = \|b_1\|^2$, there are $2^{O(d)} \cdot d^{\frac{d}{2e}}$ loop iterations during the execution of the enumeration* *algorithm as described in Figure 1. As a consequence, given a d-dimensional basis of n-dimensional vectors whose entries are integers with absolute values $\leq B$, one can compute an HKZ-reduced basis of the spanned lattice in deterministic time $\mathcal{P}(\log B, n) \cdot d^{\frac{d}{2e}+o(d)}$.*

4   Complexity of the Enumeration Procedure

This section is devoted to proving Theorem 2. The previous section has shown that the cost of Kannan's algorithm is dominated by the time for enumerating the integer points in the hyper-ellipsoids $(\mathcal{E}_i)_{1 \leq i \leq d}$ defined by $\mathcal{E}_i = \left\{ (y_i, \ldots, y_d) \in \mathbb{R}^{d-i+1}, \| \sum_{j \geq i} y_j \boldsymbol{b}_j^{(i)} \| \leq \|\boldsymbol{b}_1\| \right\}$, where $\boldsymbol{b}_j^{(i)} = \boldsymbol{b}_j - \sum_{k < i} \mu_{j,k} \boldsymbol{b}_k^*$ is the vector $\boldsymbol{b}_j$ once projected orthogonally to $\boldsymbol{b}_1^*, \ldots, \boldsymbol{b}_{i-1}^*$. Classically, the number of integer points in a body of some $\mathbb{R}^n$ is heuristically estimated by the n-dimensional volume of the body. This yields the following heuristic complexity upper-bound for Kannan's algorithm:

$$\max_{i \leq d} \frac{V_i \|\boldsymbol{b}_1\|^i}{\prod_{j \geq d-i+1} \|\boldsymbol{b}_j^*\|} \lesssim \max_{i \leq d} \frac{\|\boldsymbol{b}_1\|^i}{(\sqrt{i})^i \cdot \prod_{j \geq d-i+1} \|\boldsymbol{b}_j^*\|}, \tag{1}$$

where V_i is the volume of the i-dimensional unit ball.

Here, such an estimate may be too optimistic since the hyper-ellipsoids might be too flat for the approximation by the volume to be valid. The first step of our analysis is to prove a slight modification of this heuristic estimate. This is essentially an adaptation of a method due to Mazo and Odlyzko [20] to bound the number of integer points in hyper-spheres. We prove the weaker upper bound $\max_{I \subset [\![1,d]\!]} \frac{\|\boldsymbol{b}_1\|^{|I|}}{\sqrt{d}^{|I|} \prod_{i \in I} \|\boldsymbol{b}_i^*\|}$, for quasi-HKZ-reduced bases (Subsections 4.1 and 4.2).

In the second step of our analysis (Subsection 4.3), we bound the above quantity. This involves a rather precise study of the geometry of HKZ-reduced bases. The only available tool is Minkowski's inequality, which is used numerous times. For the intuition, the reader should consider the typical case where $(\boldsymbol{b}_i)_{1 \leq i \leq d}$ is an HKZ-reduced basis for which $(\|\boldsymbol{b}_i^*\|)_i$ is a non-increasing sequence. In that case, the first part of the analysis shows that one has to consider a set I of much simpler shape: it is an interval $[\![i, d]\!]$ starting at some index i. Lemmata 2 and 3 (which should thus be considered as the core of the proof) and the fact that $x \log x \geq -1/e$ for $x \in [0, 1]$ are sufficient to deal with such sets.

Non-connex sets I are harder to handle. We split the HKZ-reduced basis into *blocks* (defined by the expression of I as a union of intervals), i.e., groups of consecutive vectors $\boldsymbol{b}_i, \ldots, \boldsymbol{b}_{j-1}$ such that $i, \ldots, k-1 \notin I$ and $k, \ldots, j-1 \in I$. The former vectors will be the "large ones" and the latter the "small ones". Over each block, Lemma 3 relates the average size of the small vectors to the average size of the whole block. We consider the blocks by decreasing indices and use an amortised analysis to combine the local behaviours on blocks to obtain a global bound (Lemma 4). A final convexity argument gives the result (Lemma 5).

4.1   Integer Points in Hyper-Ellipsoids

In this subsection, we do not assume anything on the input basis vectors $b_1, \dots, b_d$ and on the input bound A. Up to some polynomial in d and $\log B$, the complexity of the enumeration procedure of Figure 1 is the number of loop iterations. This number of iterations is itself bounded by $3 \sum_{i=1}^{d} |\mathcal{E}_i|$. Indeed, the truncated coordinate $(x_i, \dots, x_d)$ is either a valid one, i.e., we have $\|\sum_{j=i}^{d} x_j b_j^{(i)}\|^2 \leq A$, or $(x_i - 1, \dots, x_d)$ is a valid one, or $(x_{i+1}, \dots, x_d)$ is a valid one. In fact, if $(x_i, \dots, x_d)$ is a valid truncated coordinate, at most two non-valid ones related to that one may be considered during the execution of the algorithm: $(x_i + 1, \dots, x_d)$ and $(x_{i-1}, x_i \dots, x_d)$ for at most one integer x_{i-1}. We now fix some $i \leq d$. By applying the change of variable $x_j \leftarrow x_j - \lfloor \sum_{k>j} \mu_{k,j} x_k \rceil$, we obtain:

$$
\begin{aligned}
|\mathcal{E}_{d-i+1}| &\leq \left| \left\{ (x_j)_{i \leq j \leq d} \in \mathbb{Z}^{d-i+1}, \sum_{j \geq i} (x_j + \sum_{k>j} \mu_{k,j} x_k)^2 \cdot \|b_j^*\|^2 \leq A \right\} \right| \\
&\leq \left| \left\{ (x_j)_{i \leq j \leq d} \in \mathbb{Z}^{d-i+1}, \sum_{j \geq i} (x_j + \{ \sum_{k>j} \mu_{k,j} x_k \})^2 \cdot \|b_j^*\|^2 \leq A \right\} \right|.
\end{aligned}
$$

If x is an integer and $\epsilon \in [-1/2, 1/2]$, then we have $(x + \epsilon)^2 \geq x^2/4$ (it suffices to use the inequality $|\epsilon| \leq 1/2 \leq |x|/2$, which is valid for a non-zero x). As a consequence, up to a polynomial factor, the complexity of the enumeration is bounded by $\sum_{i \leq d} N_i$, where $N_i = |\mathcal{E}_i' \cap \mathbb{Z}^{d-i+1}|$ and $\mathcal{E}_i' = \left\{ (y_i, \dots, y_d) \in \mathbb{R}^{d-i+1}, \sum_{j \geq i} y_j^2 \|b_j^*\|^2 \leq 4A \right\}$, for any $i \leq d$.

We again fix some index i. The following sequence of relations is inspired from [20, Lemma 1].

$$
\begin{aligned}
N_i &= \sum_{(x_i, \dots, x_d) \in \mathbb{Z}^{d-i+1}} 1_{\mathcal{E}_i'}(x_i, \dots, x_d) \leq \exp\left(d\left(1 - \sum_{j \geq i} x_j^2 \frac{\|b_j^*\|^2}{4A} \right) \right) \\
&\leq e^d \cdot \prod_{j \geq i} \sum_{x \in \mathbb{Z}} \exp\left(-x^2 \frac{d\|b_j^*\|^2}{4A} \right) = e^d \cdot \prod_{j \geq i} \Theta\left(\frac{d\|b_j^*\|^2}{4A} \right),
\end{aligned}
$$

where $\Theta(t) = \sum_{x \in \mathbb{Z}} \exp(-tx^2)$ is defined for $t > 0$. Notice that $\Theta(t) = 1 + 2\sum_{x \geq 1} \exp(-tx^2) \leq 1 + 2\int_0^\infty \exp(-tx^2)dx = 1 + \sqrt{\frac{\pi}{t}}$. Hence $\Theta(t) \leq \frac{1+\sqrt{\pi}}{\sqrt{t}}$ for $t \leq 1$ and $\Theta(t) \leq 1 + \sqrt{\pi}$ for $t \geq 1$. As a consequence, we have:

$$
N_i \leq (4e(1 + \sqrt{\pi}))^d \cdot \prod_{j \geq i} \max\left(1, \frac{\sqrt{A}}{\sqrt{d}\|b_j^*\|} \right). \tag{2}
$$

One thus concludes that the cost of the enumeration is bounded by:

$$
\mathcal{P}(n, \log A, \log B) \cdot 2^{O(d)} \cdot \max_{I \subset [\![1,d]\!]} \left(\frac{(\sqrt{A})^{|I|}}{(\sqrt{d})^{|I|} \prod_{i \in I} \|b_i^*\|} \right).
$$

4.2   The Case of Quasi-HKZ-Reduced Bases

We now suppose that $A = \|\boldsymbol{b}_1\|^2$ and that the input basis $(\boldsymbol{b}_1, \ldots, \boldsymbol{b}_d)$ is quasi-HKZ-reduced. We are to strengthen the quasi-HKZ-reducedness hypothesis into an HKZ-reducedness hypothesis. Let $I \subset [\![1, d]\!]$. If $1 \notin I$, then, because of the quasi-HKZ-reducedness assumption:

$$\frac{\|\boldsymbol{b}_1\|^{|I|}}{(\sqrt{d})^{|I|} \prod_{i \in I} \|\boldsymbol{b}_i^*\|} \leq 2^d \frac{\|\boldsymbol{b}_2^*\|^{|I|}}{(\sqrt{d})^{|I|} \prod_{i \in I} \|\boldsymbol{b}_i^*\|}.$$

If $1 \in I$, we have, by removing $\|\boldsymbol{b}_1^*\|$ from the product $\prod_{i \in I - \{1\}} \|\boldsymbol{b}_i^*\|$:

$$\frac{\|\boldsymbol{b}_1\|^{|I|}}{(\sqrt{d})^{|I|} \prod_{i \in I} \|\boldsymbol{b}_i^*\|} \leq 2^d \frac{\|\boldsymbol{b}_2^*\|^{|I|-1}}{(\sqrt{d})^{|I|-1} \prod_{i \in I - \{1\}} \|\boldsymbol{b}_i^*\|}.$$

As a consequence, Theorem 2 follows from the following:

Theorem 3. *Let* $(\boldsymbol{b}_1, \ldots, \boldsymbol{b}_d)$ *be HKZ-reduced and* $I \subset [\![1, d]\!]$. *Then*

$$\frac{\|\boldsymbol{b}_1\|^{|I|}}{\prod_{i \in I} \|\boldsymbol{b}_i^*\|} \leq (\sqrt{d})^{|I| \left(1 + \log \frac{d}{|I|}\right)} \leq (\sqrt{d})^{\frac{d}{e} + |I|}.$$

By applying Theorem 3 the HKZ-reduced basis $(\boldsymbol{b}_1, \ldots, \boldsymbol{b}_i)$ and $I = \{i\}$, we recover the result of [28]: $\|\boldsymbol{b}_i^*\| \geq (\sqrt{i})^{-\log i - 1} \cdot \|\boldsymbol{b}_1\|$.

4.3   A Property on the Geometry of HKZ-Reduced Bases

In this section, we prove Theorem 3, which is the last missing part to obtain the claimed result. The proofs of the following lemmata will be contained in the full version of this paper. In the sequel, $(\boldsymbol{b}_i)_{i \leq d}$ is an HKZ-reduced basis of a lattice L of dimension $d \geq 2$.

Definition 2. *For any* $I \subset [\![1, d]\!]$, *we define* $\pi_I = \left(\prod_{i \in I} \|\boldsymbol{b}_i^*\|\right)^{\frac{1}{|I|}}$. *Moreover, if* $k \in [\![1, d-1]\!]$, *we define* $\Gamma_d(k) = \prod_{i=d-k}^{d-1} (\gamma_{i+1})^{\frac{1}{2i}}$.

We need upper bounds on $\Gamma_d(k)$ and a technical lemma allowing us to finely recombine such bounds. Intuitively, the following lemma is a rigorous version of the identity:

$$\log \Gamma_d(k) \approx \int_{x=d-k}^{d} \frac{1}{2x} \log x \, \mathrm{d}x \approx \frac{\log^2(d) - \log^2(d-k)}{4} \lesssim \frac{\log d}{2} \log \frac{d}{d-k}.$$

Lemma 2. *For all* $1 \leq k < d$, *we have* $\Gamma_d(k) \leq \sqrt{d}^{\log \frac{d}{d-k}}$.

We now give an "averaged" version of [28, Lemma 4], deriving from Lemma 2. This provides the result claimed in Theorem 3 for any set I of the shape $[\![i, j]\!]$, for any $i \leq j \leq d$.

Lemma 3. *For all* $k \in [\![0, d-1]\!]$, *we have* $\pi_{[\![1,k]\!]} \leq (\Gamma_d(k))^{d/k} \cdot \pi_{[\![k+1,d]\!]}$ *and*
$\pi_{[\![k+1,d]\!]} \geq (\Gamma_d(k))^{-1} \cdot (\det L)^{1/d} \geq \sqrt{d}^{\log \frac{d-k}{d}} (\det L)^{1/d}$.

We prove Theorem 3 by induction on the number of intervals occurring in the expression of the set I as a union of intervals. The following lemma is the induction step. This is a recombination step, where we join one block (between the indices 1 and v, the "small vectors" being those between $u+1$ and v) to one or more already considered blocks on its right. An important point is to ensure that the densities δ_i defined below actually decrease when their indices increase. Its proof is based on Lemma 3.

Lemma 4. *Let* $(\boldsymbol{b}_1, \ldots, \boldsymbol{b}_d)$ *be an HKZ-reduced basis. Let* $v \in [\![2, d]\!]$, $I \subset [\![v+1, d]\!]$ *and* $u \in [\![1, v]\!]$. *Assume that:*

$$\pi_I^{|I|} \geq \prod_{i < t} \left(\pi_{[\![\alpha_i+1, \alpha_{i+1}]\!]}^{|I_i|} \cdot \sqrt{d}^{|I_i| \log \delta_i} \right),$$

where $I_i = I \cap [\![\alpha_i+1, \alpha_{i+1}]\!]$, $\delta_i = \frac{|I_i|}{\alpha_{i+1}-\alpha_i}$ *is the density of the set* I *in* $[\![\alpha_i+1, \alpha_{i+1}]\!]$, *and the integers* t *and* α_i's, *and the densities* δ_i's *satisfy* $t \geq 1$, $v = \alpha_1 < \ldots < \alpha_t \leq d$ *and* $1 \geq \delta_1 > \ldots > \delta_{t-1} > 0$. *Then, we have*

$$\pi_{I'}^{|I'|} \geq \prod_{i < t'} \left(\pi_{[\![\alpha_i'+1, \alpha_{i+1}']\!]}^{|I_i'|} \cdot \sqrt{d}^{|I_i'| \log \delta_i'} \right),$$

where $I' = [\![u+1, v]\!] \cup I$, $I_i' = I' \cap [\![\alpha_i'+1, \alpha_{i+1}']\!]$, $\delta_i' = \frac{|I_i'|}{\alpha_{i+1}'-\alpha_i'}$ *and the integers* t' *and* α_i''s, *and the densities* δ_i' *satisfy* $t' \geq 1$, $0 = \alpha_1' < \ldots < \alpha_{t'}' \leq d$ *and* $1 \geq \delta_1' > \ldots > \delta_{t'-1}' > 0$.

The last ingredient to the proof of Theorem 3 is the following, which derives from the convexity of the function $x \mapsto x \log x$.

Lemma 5. *Let* $\Delta \geq 1$, *and define* $F_\Delta(k, d) = \Delta^{-k \log \frac{k}{d}}$. *We have, for any* $t \in \mathbb{Z}$, *for any* $k_1, \ldots, k_t \in \mathbb{Z}$ *and* $d_1, \ldots, d_t \in \mathbb{Z}$ *such that* $1 \leq k_i < d_i$ *for all* $i \leq t$,

$$\prod_{i \leq t} F_\Delta(k_i, d_i) \leq F_\Delta \left(\sum_{i \leq t} k_i, \sum_{i \leq t} d_i \right).$$

Finally, Theorem 3 follows from Lemmata 4 and 5.

Proof of Theorem 3. Lemma 4 gives us, by induction on the size of the considered set I, that for all $I \subset [\![1, d]\!]$:

$$\pi_I^{|I|} \geq \prod_{i < t} \left(\pi_{[\![\alpha_i+1, \alpha_{i+1}]\!]}^{|I_i|} \cdot \sqrt{d}^{|I_i| \log \delta_i} \right),$$

where $I_i = I \cap [\![\alpha_i+1, \alpha_{i+1}]\!]$, and t, the α_i's, and the densities $\delta_i = \frac{|I_i|}{\alpha_{i+1}-\alpha_i}$ satisfy $t \geq 1$, $0 = \alpha_1 < \ldots < \alpha_t \leq d$ and $1 \geq \delta_1 > \ldots > \delta_{t-1} > 0$. By using

Lemma 5 with $\Delta := \sqrt{d}, k_i := |I_i|$ and $d_i := \alpha_{i+1} - \alpha_i$, we obtain:

$$\pi_I^{|I|} \geq \left(\sqrt{d}^{|I| \log \frac{|I|}{\alpha_t - \alpha_1}}\right) \cdot \left(\prod_{i<t} \pi_{[\![\alpha_i+1, \alpha_{i+1}]\!]}^{|I_i|}\right).$$

We define $\delta_t = 0$. Because of the definition of the α_i's, we have:

$$\prod_{i<t} \pi_{[\![\alpha_i+1, \alpha_{i+1}]\!]}^{|I_i|} = \prod_{i<t} \left(\pi_{[\![\alpha_i+1, \alpha_{i+1}]\!]}^{\alpha_{i+1}-\alpha_i}\right)^{\delta_i} = \prod_{i<t} \prod_{i \leq j < t} \left(\pi_{[\![\alpha_i+1, \alpha_{i+1}]\!]}^{\alpha_{i+1}-\alpha_i}\right)^{\delta_j - \delta_{j+1}}$$

$$= \prod_{j<t} \left(\prod_{i \leq j} \pi_{[\![\alpha_i+1, \alpha_{i+1}]\!]}^{\alpha_{i+1}-\alpha_i}\right)^{\delta_j - \delta_{j+1}} = \prod_{j<t} \left(\pi_{[\![1, \alpha_{j+1}]\!]}^{\alpha_{j+1}}\right)^{\delta_j - \delta_{j+1}}.$$

By using $t - 1$ times Minkowski's theorem, we obtain that:

$$\frac{\pi_I^{|I|}}{\sqrt{d}^{|I| \log \frac{|I|}{d}}} \geq \left(\frac{\|b_1\|}{\sqrt{d}}\right)^{\sum_{j<t} \alpha_{j+1}(\delta_j - \delta_{j+1})} \geq \left(\frac{\|b_1\|}{\sqrt{d}}\right)^{|I|}.$$

The final inequality of the theorem comes from the fact that the function $x \mapsto x \log(d/x)$ is maximal for $x = d/e$.   □

5   CVP and Other Related Problems

Our improved analysis of Kannan's algorithm can be adapted to the Closest Vector Problem and other problems related to strong lattice reduction.

In CVP, we are given a basis $(b_1, \ldots, b_d)$ and a target vector t, and we look for a lattice vector that is closest to t. Kannan's CVP algorithm starts by HKZ-reducing the b_i's. Then it runs a slight modification of the enumeration algorithm of Figure 1. For the sake of simplicity, we assume that $\|b_1^*\|$ is the largest of the $\|b_i^*\|$'s (we refer to Kannan's proof [16] for the general case). By using Babai's nearest hyperplane strategy [6], we see that there is a lattice vector b at distance less than $\sqrt{d} \cdot \|b_1\|$ of the target vector t. As a consequence, if we take $A = d \cdot \|b_1\|^2$ in the modified enumeration procedure, we will find all solutions. The analysis then reduces (at the level of Equation (2)) to bound the ratio $\frac{\|b_1\|^d}{\prod_{i \leq d} \|b_i^*\|}$, which can be done with Minkowski's theorem.

Theorem 4. *Given a basis $(b_1, \ldots, b_d)$ and a target vector t, all of them in $\mathbb{Z}^n$ and with coordinates whose absolute values are smaller than some B, one can compute all vectors in the lattice spanned by the b_i's that are closest to t in deterministic time $\mathcal{P}(\log B, n) \cdot d^{d/2+o(d)}$.*

The best deterministic complexity upper bound previously known for this problem was $\mathcal{P}(\log B, n) \cdot d^{d+o(d)}$ (see [13, 7]).

Our result can also be adapted to the enumeration of all vectors of a given lattice that are of length below a prescribed bound, which is in particular useful in the context of computing lattice theta series. Another important consequence of our analysis is a significant worst-case bound improvement of Schnorr's block-based strategy [28] to compute relatively short vectors in high-dimensional lattices. More precisely, if we take the bounds given in [10] for the quality of Schnorr's semi-$2k$ reduction and for the transference reduction, we obtain the table of Figure 3. Each entry of the table gives the upper bound of the quantity $\frac{\|b_1\|}{(\det L)^{1/d}}$ which is reachable for a computational effort of 2^t, for t growing to infinity. To sum up, the exponent constant is divided by $e \approx 2.7$. The table upper bounds may be adapted to the quantity $\frac{\|b_1\|}{\lambda_1(L)}$ by squaring them.

	Semi-$2k$ reduction	Transference reduction
Using [13]	$\lesssim 2^{\frac{\log 2}{2} \frac{d \log^2 t}{t}} \approx 2^{0.347 \frac{d \log^2 t}{t}}$	$\lesssim 2^{\frac{1}{4} \frac{d \log^2 t}{t}} \approx 2^{0.250 \frac{d \log^2 t}{t}}$
Using Theorem 2	$\lesssim 2^{\frac{\log 2}{2e} \frac{d \log^2 t}{t}} \approx 2^{0.128 \frac{d \log^2 t}{t}}$	$\lesssim 2^{\frac{1}{4e} \frac{d \log^2 t}{t}} \approx 2^{0.092 \frac{d \log^2 t}{t}}$

Fig. 3. Worst-case bounds for block-based reduction algorithms

6   Practical Implications

As mentioned in the introduction, the main contribution of the present paper is to improve the worst-case complexity analysis of an already known algorithm, namely, Kannan's HKZ-reduction algorithm. Our improvement has no direct impact on the practical capabilities of lattice reduction algorithms. However, our work may have two indirect consequences: popularising Kannan's principle and providing easily computable cost estimates for SVP instances.

6.1   Pre-processing Before Enumerating

In the main libraries containing lattice reduction routines, the shortest vector problem is solved with the enumeration routine, but starting from only L^3-reduced bases. This is the case for the BKZ routines of Victor Shoup's NTL [32], which, depending on a parameter k, compute strongly reduced bases in high dimensions (the quality being quantified by k). This is also the case in Magma's ShortestVectors routine [18], which computes the shortest vectors of a given lattice. Both rely on the enumeration of Schnorr and Euchner [30]. On the theoretical side, this strategy is worse than using Kannan's algorithm, the worst-case complexity being $2^{O(d^2)}$ instead of $d^{O(d)}$. To justify this choice, one might argue that L^3 computes much better bases in practice than guaranteed by the worst-case bounds, in particular in low dimensions (see [24] for more details), and that the asymptotically superior algorithm of Kannan may overtake the L^3-based enumeration only for large dimensions (in particular too large to be tractable).

It may be that the genuine Kannan algorithm is expensive. However, the general principle of enumerating from a more than L^3-reduced basis works, as the following experiments tend to show. For a given dimension d, we consider the lattice spanned by the columns of the following matrix:

$$\begin{pmatrix} x_1 & x_2 & \dots & x_d \\ 1 & 0 & \dots & 0 \\ 0 & 1 & \dots & 0 \\ \vdots & \vdots & \ddots & \vdots \\ 0 & 0 & \dots & 1 \end{pmatrix},$$

where the x_i's are chosen uniformly and independently in $[\![0, 2^{100 \cdot d}]\!]$. The basis is then L^3-reduced with a close to optimal parameter ($\delta = 0.99$). For the same lattice, we compute more reduced bases, namely BKZ_k-reduced for different parameters k, using NTL's BKZ_FP routine without pruning and close to optimal factor ($\delta = 0.99$). We run the same enumeration routine starting from these different bases and compare the timings. The results of the experiments are given in Figure 4. The enumeration is a non-optimised C-code, which updates the norm upper bound during the enumeration [30]. All timings are given in seconds and include the BKZ-reduction (unless we start from the L^3-reduced basis). Each point corresponds to the average over at least 10 samples. The experiments were performed on 2.4 GHz AMD Opterons. The enumeration from an L^3-reduced basis is clearly outperformed. BKZ-reducing the basis with larger block-sizes becomes more interesting when the dimension increases: it seems that in moderate dimension, a BKZ_k reduced basis is close to being HKZ-reduced, even when k is small with respect to the dimension.

pre-processing	$d = 40$	$d = 43$	$d = 46$	$d = 49$	$d = 52$	$d = 55$	$d = 58$
L^3	1.8	15	110	990	$5.0 \cdot 10^3$	–	–
BKZ_{10}	0.36	1.6	6.7	36	160	–	–
BKZ_{20}	0.40	1.3	4.7	21	96	800	$2.5 \cdot 10^3$
BKZ_{30}	0.57	1.7	5.2	19	68	660	$1.6 \cdot 10^3$

Fig. 4. Comparison between various pre-processings

6.2 Estimating the Cost of Solving SVP

The cost of solving SVP on a particular instance with the enumeration routine is essentially dominated by the cost of the highest-dimensional enumeration. Up to a polynomial factor, the cost of the enumeration as described in Figure 1 can be estimated with Equation (1):

$$E(\boldsymbol{b}_1, \dots, \boldsymbol{b}_d) := \max_{i \leq d} \frac{\pi^{i/2} \cdot \|\boldsymbol{b}_1\|^i}{\Gamma(i/2 + 1) \cdot \prod_{j \geq d-i+1} \|\boldsymbol{b}_j^*\|}.$$

This estimate is simply the application of the Gaussian heuristic, stating that the number of integer points within a body is essentially the volume of the body. It can be computed in polynomial time from the basis from which the enumeration will be started. We computed $E(\boldsymbol{b}_1, \dots, \boldsymbol{b}_d)$ for random bases generated as above

pre-processing	$d = 40$	$d = 45$	$d = 50$	$d = 55$	$d = 60$	$d = 65$	$d = 70$	$d = 75$
L^3	$1.0 \cdot 10^8$	$4.4 \cdot 10^9$	$1.5 \cdot 10^{14}$	$9.6 \cdot 10^{16}$	$3.0 \cdot 10^{18}$	$6.1 \cdot 10^{21}$	$2.8 \cdot 10^{27}$	$1.6 \cdot 10^{30}$
BKZ_{10}	$4.6 \cdot 10^5$	$1.2 \cdot 10^7$	$1.1 \cdot 10^8$	$1.3 \cdot 10^{10}$	$7.6 \cdot 10^{11}$	$1.7 \cdot 10^{14}$	$4.3 \cdot 10^{16}$	$1.9 \cdot 10^{19}$
BKZ_{20}	$2.4 \cdot 10^5$	$2.7 \cdot 10^6$	$3.1 \cdot 10^7$	$1.3 \cdot 10^9$	$4.1 \cdot 10^{10}$	$3.7 \cdot 10^{12}$	$6.4 \cdot 10^{13}$	$2.1 \cdot 10^{16}$
BKZ_{30}	$1.9 \cdot 10^5$	$1.6 \cdot 10^6$	$1.8 \cdot 10^7$	$3.0 \cdot 10^8$	$4.3 \cdot 10^9$	$1.1 \cdot 10^{11}$	$3.7 \cdot 10^{12}$	$1.9 \cdot 10^{14}$

Fig. 5. Value of $E(b_1, \ldots, b_d)$ for randomly generated $(b_1, \ldots, b_d)$

and obtained the table of Figure 5. It confirms that a strong pre-processing should help increasing the dimension up to which SVP may be solved completely.

If one is looking for vectors smaller than some prescribed B (for example if the existence of an unusually short vector is promised), then $\|b_1\|$ may be replaced by B in the estimate. Overall, these estimates are rather crude since factors that are polynomial in the dimension should be considered as well. Furthermore, it does not take into account more elaborate techniques such as updating the norm during the enumeration, pruning [30, 31] and random sampling [29].

OPEN PROBLEM. One may wonder if the complexity upper bound for Kannan's SVP algorithm can be decreased further. Work under progress seems to show, by using a technique due to Ajtai [3], that it is sharp, in the sense that for all $\epsilon > 0$, we can build HKZ-reduced bases for which the number of steps of Kannan's algorithm would be at least $d^{d(\frac{1}{2e} - \epsilon)}$.

Acknowledgements. We thank Frederik Vercauteren for helpful discussions, as well as John Cannon and the University of Sydney for having hosted the second author while a large part of this work was completed.

References

1. Agrell, E., Eriksson, T., Vardy, A., Zeger, K.: Closest point search in lattices. IEEE Trans. Inform. Theory 48(8), 2201–2214 (2002)
2. Ajtai, M.: The shortest vector problem in l_2 is NP-hard for randomized reductions (extended abstract). In: Proc. of STOC 1998, pp. 284–293. ACM Press, New York (1998)
3. Ajtai, M.: The worst-case behavior of Schnorr's algorithm approximating the shortest nonzero vector in a lattice. In: Proc. of STOC 2003, pp. 396–406. ACM Press, New York (2003)
4. Ajtai, M., Dwork, C.: A public-key cryptosystem with worst-case/average-case equivalence. In: Proc. of STOC 1997, pp. 284–293. ACM Press, New York (1997)
5. Ajtai, M., Kumar, R., Sivakumar, D.: A sieve algorithm for the shortest lattice vector problem. In: Proc of STOC 2001, pp. 601–610. ACM Press, New York (2001)
6. Babai, L.: On Lovász lattice reduction and the nearest lattice point problem. Combinatorica 6, 1–13 (1986)
7. Blömer, J.: Closest vectors, successive minima and dual-HKZ bases of lattices. In: Welzl, E., Montanari, U., Rolim, J.D.P. (eds.) ICALP 2000. LNCS, vol. 1853, pp. 248–259. Springer, Heidelberg (2000)

8. Coppersmith, D., Shamir, A.: Lattice attacks on NTRU. In: Fumy, W. (ed.) EUROCRYPT 1997. LNCS, vol. 1233, pp. 52–61. Springer, Heidelberg (1997)

9. Fincke, U., Pohst, M.: A procedure for determining algebraic integers of given norm. In: van Hulzen, J.A. (ed.) ISSAC 1983 and EUROCAL 1983. LNCS, vol. 162, pp. 194–202. Springer, Heidelberg (1983)

10. Gama, N., Howgrave-Graham, N., Koy, H., Nguyen, P.: Rankin's constant and blockwise lattice reduction. In: Dwork, C. (ed.) CRYPTO 2006. LNCS, vol. 4117, pp. 112–130. Springer, Heidelberg (2006)

11. Goldreich, O., Goldwasser, S., Halevi, S.: Public-key cryptosystems from lattice reduction problems. In: Kaliski Jr., B.S. (ed.) CRYPTO 1997. LNCS, vol. 1294, pp. 112–131. Springer, Heidelberg (1997)

12. Haviv, I., Regev, O.: Tensor-based hardness of the shortest vector problem to within almost polynomial factors. In: Proc. of STOC 2007 (2007)

13. Helfrich, B.: Algorithms to construct Minkowski reduced and Hermite reduced lattice bases. Theoret. Comput. Sci. 41, 125–139 (1985)

14. Hermite, C.: Extraits de lettres de M. Hermite à M. Jacobi sur différents objets de la théorie des nombres, deuxième lettre. J. Reine Angew. Math. 40, 279–290 (1850)

15. Hoffstein, J., Pipher, J., Silverman, J.H.: NTRU : a ring based public key cryptosystem. In: Buhler, J.P. (ed.) Algorithmic Number Theory. LNCS, vol. 1423, pp. 267–288. Springer, Heidelberg (1998)

16. Kannan, R.: Improved algorithms for integer programming and related lattice problems. In: Proc. of STOC 1983, pp. 99–108. ACM Press, New York (1983)

17. Lenstra, A.K., Lenstra Jr., H.W., Lovász, L.: Factoring polynomials with rational coefficients. Math. Ann. 261, 513–534 (1982)

18. Magma. The Magma computational algebra system for algebra, number theory and geometry. Available at http://magma.maths.usyd.edu.au/magma/

19. Martinet, J.: Perfect Lattices in Euclidean Spaces. Springer, Heidelberg (2002)

20. Mazo, J., Odlyzko, A.: Lattice points in high-dimensional spheres. Monatsh. Math. 110, 47–61 (1990)

21. Micciancio, D., Goldwasser, S.: Complexity of lattice problems : a cryptographic perspective. Kluwer Academic Publishers, Dordrecht (2002)

22. Minkowski, H.: Geometrie der Zahlen. Teubner-V (1896)

23. Nguyen, P.: Cryptanalysis of the Goldreich-Goldwasser-Halevi cryptosystem from Crypto'97. In: Wiener, M.J. (ed.) CRYPTO 1999. LNCS, vol. 1666, pp. 288–304. Springer, Heidelberg (1999)

24. Nguyen, P., Stehlé, D.: LLL on the average. In: Hess, F., Pauli, S., Pohst, M. (eds.) ANTSVII. LNCS, vol. 4076, pp. 238–256. Springer, Heidelberg (2006)

25. Nguyen, P., Stern, J.: Cryptanalysis of the Ajtai-Dwork cryptosystem. In: Krawczyk, H. (ed.) CRYPTO 1998. LNCS, vol. 1462, pp. 223–242. Springer, Heidelberg (1998)

26. Nguyen, P., Vidick, T.: Assessing sieve algorithms for the shortest vector problem. Draft (2007)

27. Regev, O.: Lecture notes of lattices in computer science, taught at the Computer Science Tel Aviv University. Available at http://www.cs.tau.il/~odedr

28. Schnorr, C.P.: A hierarchy of polynomial lattice basis reduction algorithms. Theoret. Comput. Sci. 53, 201–224 (1987)

29. Schnorr, C.P.: Lattice reduction by random sampling and birthday methods. In: Alt, H., Habib, M. (eds.) STACS 2003. LNCS, vol. 2607, pp. 145–156. Springer, Heidelberg (2003)
30. Schnorr, C.P., Euchner, M.: Lattice basis reduction : improved practical algorithms and solving subset sum problems. Math. Programming 66, 181–199 (1994)
31. Schnorr, C.P., Hörner, H.H.: Attacking the Chor-Rivest cryptosystem by improved lattice reduction. In: Guillou, L.C., Quisquater, J.-J. (eds.) EUROCRYPT 1995. LNCS, vol. 921, pp. 1–12. Springer, Heidelberg (1995)
32. Shoup, V.: NTL, Number Theory Library. Available, at http://www.shoup.net/ntl/

Domain Extension of Public Random Functions: Beyond the Birthday Barrier[*]

Ueli Maurer and Stefano Tessaro

Department of Computer Science, ETH Zurich, 8092 Zurich, Switzerland
{maurer,tessaros}@inf.ethz.ch

Abstract. A *public* random function is a random function that is accessible by all parties, including the adversary. For example, a (public) random oracle is a public random function $\{0,1\}^* \to \{0,1\}^n$. The natural problem of constructing a public random oracle from a public random function $\{0,1\}^m \to \{0,1\}^n$ (for some $m > n$) was first considered at Crypto 2005 by Coron et al. who proved the security of variants of the Merkle-Damgård construction against adversaries issuing up to $\mathcal{O}(2^{n/2})$ queries to the construction and to the underlying compression function. This bound is less than the square root of $n2^m$, the number of random bits contained in the underlying random function.

In this paper, we investigate domain extenders for public random functions approaching optimal security. In particular, for all $\epsilon \in (0, 1)$ and all functions m and ℓ (polynomial in n), we provide a construction $\mathbf{C}_{\epsilon,m,\ell}(\cdot)$ which extends a public random function $\mathbf{R} : \{0,1\}^n \to \{0,1\}^n$ to a function $\mathbf{C}_{\epsilon,m,\ell}(\mathbf{R}) : \{0,1\}^{m(n)} \to \{0,1\}^{\ell(n)}$ with time-complexity polynomial in n and $1/\epsilon$ and which is secure against adversaries which make up to $\Theta(2^{n(1-\epsilon)})$ queries. A central tool for achieving high security are special classes of unbalanced bipartite expander graphs with small degree. The achievability of practical (as opposed to complexity-theoretic) efficiency is proved by a non-constructive existence proof.

Combined with the iterated constructions of Coron et al., our result leads to the first iterated construction of a hash function $\{0,1\}^* \to \{0,1\}^n$ from a component function $\{0,1\}^n \to \{0,1\}^n$ that withstands all recently proposed generic attacks against iterated hash functions, like Joux's multi-collision attack, Kelsey and Schneier's second-preimage attack, and Kelsey and Kohno's herding attacks.

1 Introduction

1.1 Secret vs. Public Random Functions

Primitives that provide some form of randomness are of central importance in cryptography, both as a primitive assumed to be given (e.g. a secret key), and as a primitive constructed from a weaker one to "behave like" a certain ideal random primitive (e.g. a random function), according to some security notion.

---

[*] This research was partially supported by the Swiss National Science Foundation (SNF), project no. 200020-113700/1.

A. Menezes (Ed.): CRYPTO 2007, LNCS 4622, pp. 187–204, 2007.

An adversary may have different types of access to a random primitive. The two extreme cases are that the adversary has *no access* and that he has *complete access*[1] to it. For example, the adversary is assumed to have no access to a secret key, and a pseudo-random function (PRF) is a (computationally-secure) realization from such a secret key of a secret random function to which the adversary has no access. In contrast, a (public) random oracle, as used in the so-called random-oracle model [7], is a function $\{0,1\}^* \to \{0,1\}^n$ to which the adversary has *complete* access, like the legitimate parties. Similarly, a *public parameter* (e.g. the parameter selecting a hash function from a class) is a finite random string to which the adversary has complete access. It is natural to also consider finite-domain public random functions.

In this paper we are interested in such *public* random primitives and reductions among them. The question whether (and how) a certain primitive can be securely realized from another primitive is substantially more complex in the public setting, compared to the secret setting, and even the security notion is more involved. For example, while the CBC-construction can be seen as the secure realization of a secret random function $\{0,1\}^* \to \{0,1\}^n$ from a secret random function $\{0,1\}^n \to \{0,1\}^n$ [5,19], the same statement is false if public functions (accessible to the adversary) are considered. Another famous example of a reduction problem for public primitives is the realization of a (public) random oracle from a public parameter. This was shown to be impossible [8,21].

1.2 Domain Extension and the Birthday Barrier

A random primitive (both secret or public) can be characterized by the number of random bits it contains. An ℓ-bit key is a string (or table) containing ℓ random bits, a random function $\{0,1\}^m \to \{0,1\}^n$ corresponds to a table of $n2^m$ random bits which can be accessed efficiently, and a random oracle $\{0,1\}^* \to \{0,1\}^n$ corresponds to a countably infinite table of random bits.[2] Of course, a random table of N bits can be interpreted as a random function $\{0,1\}^m \to \{0,1\}^n$ for any m and n with $n2^m \leq N$. For example, n can be doubled at the apparently minor expense of reducing m by 1.

An important topic in cryptography is the secure expansion of such a table, considered as an ideal system. This is referred to as *domain extension*, say from $\{0,1\}^m$ to $\{0,1\}^{2m}$ (or to $\{0,1\}^*$), which corresponds to an exponential (or even infinite) blow-up of the table size. (In contrast, increasing the range, say from $\{0,1\}^n$ to $\{0,1\}^{2n}$, corresponds to merely a doubling of the table size.)

---

[1] Side-channel attack analyses, where part of the secret key is assumed to leak, are examples of intermediate scenarios.

[2] Each bit can be accessed in time logarithmic in its position in the table, which is optimal since the specification of the position requires logarithmically many bits. In this paper we only consider such random primitives where the bits can be accessed efficiently, but there are also more complicated primitives, like an ideal cipher, which on one hand has a special permutation structure and also allows on the other hand a special additional type of access, namely inverse queries.

In [21] a generalization of indistinguishability to systems with public access, called *indifferentiability*, was proposed. Like for indistinguishability, there is a computational and a stronger, information-theoretic, version of indifferentiability. This general notion allows to discuss the secure realization of a *public* random primitive from another public random primitive. In [21] also a simple general framework was proposed, based on entropy arguments, for proving impossibility results like that of [8]. One can easily show that not even a single-bit extension of a public parameter, from ℓ to $\ell+1$ bits, is possible, let alone to an exponentially large table (corresponding to a public random function $\{0,1\}^m \rightarrow \{0,1\}^n$) or even to an infinite table (corresponding to the impossibility of realizing a random oracle [8,21]).

However, the situation is different if one starts from a public random function (as opposed to just a public random string). Coron et al. [11] considered the problem of constructing a random oracle $\{0,1\}^* \rightarrow \{0,1\}^n$ from a public random function $\{0,1\}^m \rightarrow \{0,1\}^n$ (where $m > n$) and showed that a modified Merkle-Damgård construction [24,12] works, with information-theoretic security (i.e., indifferentiability) up to about $\mathcal{O}(2^{n/2})$ queries. This bound, only the square root of $\mathcal{O}(2^n)$, is usually called the "birthday barrier". The term "birthday" is used because the birthday paradox applies (as soon as two different inputs to the function occur which produce the same output, security is lost) and the term "barrier" is used because breaking it is non-trivial if at all possible.

For *secret* random functions, many constructions in the literature, also those based on universal hashing [9,26] and the CBC-construction [5,19], suffer from the birthday problem, and hence several researchers [1,4,19] considered the problem of achieving security beyond the birthday barrier. The goal of this paper is to solve the corresponding problem for public random functions. Namely, we want to achieve essentially maximal security, i.e., up to $\Theta(2^{n(1-\epsilon)})$ queries for any $\epsilon > 0$ (where the construction may depend on ϵ). Like for other problems (see e.g. [13]), going from the "secret case" to the "public case" appears to involve substantial new construction elements and analysis techniques.

1.3   Significance of Domain Extension for Public Random Functions

The domain extension problem for public random functions has important implications for the design of cryptographic functions, in addition to being of general theoretical interest. We also refer to [11] for a discussion of the significance of this problem.

Cryptographic functions with arbitrary input-length are of crucial importance in cryptography. Desirable properties for such functions are collision-resistance, second-preimage resistance, multi-collision resistance, being pseudo-random, or being a secure MAC, etc. A general paradigm for constructing a cryptographic function $\{0,1\}^* \rightarrow \{0,1\}^n$, both in the secret and the public case, is to make use of a component function $\mathbf{F} : \{0,1\}^m \rightarrow \{0,1\}^n$ and to embed it into an iterated construction $\mathbf{C}(\cdot)$ (e.g. the CBC or the Merkle-Damgård construction), resulting in the overall function $\mathbf{C}(\mathbf{F}) : \{0,1\}^* \rightarrow \{0,1\}^n$.

It is important to be able to separate the reasoning about the component function **F** and about the construction **C**(·). Typically, **F** is simply assumed to have some property, like being collision-resistant, second-preimage resistant, a secure MAC, etc. In contrast, the construction **C**(·) is (or should be!) designed in a way that one can *prove* certain properties.

There are two types of such proofs for **C**(·). The first type is a complexity-theoretic reduction proof showing that if there exists an adversary breaking a certain property of **C**(**F**), then there exists a comparably efficient adversary breaking a property (the same or a different one) of **F**. For example, using such an argument one can prove that the Merkle-Damgård [24,12] construction is collision-resistant if the component function is. Similarly, one can prove that the CBC construction is a PRF if the component function is [5], or that certain constructions [2,22] are secure MACs if the component function is.

A second type of proof, which is the subject of [11] and of this paper, is the proof that if **F** is a public random function, then so is **C**(**F**), up to a certain number B of queries. Such a proof implies the absence of a generic (black-box) attack against **C**(**F**), i.e., an attack which does not exploit specific properties of **F**, but uses it merely as a black-box.[3] Such a generic proof is not an ultimate security proof for **C**(**F**), but it proves that the construction **C**(·) itself has no weakness. A main advantage of such a proof is that it applies to *every* cryptographic property of interest (which a random function has), not just to specific properties like collision-resistance.

The number B of queries up to which security is guaranteed is a crucial parameter of such a proof, especially in view of several surprises of the past years regarding weaknesses of iterated constructions. Joux [15] showed that the security of the Merkle-Damgård construction (with compression function with n-bit output) against finding multi-collisions is not much higher than the security against normal collision attacks, namely the birthday barrier $\mathcal{O}(2^{n/2})$, which is surprising because for a random function, finding an r-multi-collision requires $\Theta(2^{\frac{r-1}{r}n})$ queries. Joux's attack has been generalized to a wider class of constructions [14]. Other attacks in a similar spirit against iterated constructions are the second-preimage attack by Kelsey and Schneier [17], and herding attacks [16]. One possibility to overcome these issues is to rely on a compression function with input domain much larger than the size of the output of the construction (cf. for example the constructions in [18] and the double block-length construction of [10]), but this does not seem to be the best possible approach, both from a theoretical and from a practical viewpoint, as explained below.

A proof, like that of [11], for a construction **C**(·) of a public random function, implies that **C**(·) is secure against all possible attacks, up to the bound B on the number of queries stated in the proof. Since the bound in [11] is the birthday barrier, this implies nothing (beyond the birthday barrier) for attacks that require more queries, like the attacks of [15,17,10] mentioned above, and indeed the constructions of [11] also suffer from the same attacks.

---

[3] This is analogous to security proofs in the generic group model [27,20] which show that no attack exists that does not exploit the particular representation of group elements.

The bound B is also of importance since it determines the input and output sizes of $\mathbf{F}$. For example, because collision-resistance is a property that can hold only up to $2^{n/2}$ queries (due to the birthday paradox), n must be chosen twice as large as one might expect to be feasible in a naïve security analysis. Moreover, since the function must be compressing to be useful in a construction $\mathbf{C}(\cdot)$, the input size m must be larger than the output size n. However, if collision-resistance is not required, but instead for example second-preimage resistance, then the input size m of $\mathbf{F}$ can potentially be smaller or, turning the argument around, security for a given m can be much higher.

The input size m of $\mathbf{F}$ is relevant for two more reasons. First, if one considers the (perhaps not very realistic) possibility of finding a random function in Nature (say, by scanning the surface of the moon or by appropriately accessing the WWW), then m is a crucial parameter since the table size $n2^m$ is exponential in m. Second, for a given maximal computing time for $\mathbf{F}$, the difficulty of designing a concrete cryptographic function $\mathbf{F} : \{0,1\}^m \rightarrow \{0,1\}^n$ that is supposed to "look random" increases significantly if m is large. This can be seen as follows. Such a function $\mathbf{F}$ for large m could be modified in many different ways to reduce m to $m' < m$ (e.g. set $m - m'$ input bits to 0 or to any fixed value, or repeat an input of size m' until a block of length m is filled, etc.), and for each of these modifications it would still have to be secure.[4] Hence simply designing a new function with doubled m is not a very reasonable solution for the birthday barrier problem. Rather, one should find a construction that doubles (or multiplies) the input size but at the same time preserves the security almost optimally.

1.4   Contributions and Outline of This Paper

The main contribution of this paper is a construction paradigm for breaking the birthday barrier for domain extension of public random functions. More precisely, in Section 3 we prove that for every $\epsilon \in (0,1)$, m and ℓ, there exists an efficient construction $\mathbf{C}_{\epsilon,m,\ell}(\cdot)$ which extends a public random function $\{0,1\}^n \rightarrow \{0,1\}^n$ to a public random function $\{0,1\}^m \rightarrow \{0,1\}^\ell$, and which guarantees security for up to $\Theta(2^{n(1-\epsilon)})$ queries.

A central tool in our approach is a new combinatorial object, which we call an *input-restricting function family*. Section 4 discusses constructions of such families from highly-unbalanced bipartite expander graphs. While current expander constructions only allow our paradigm to be efficient in a complexity-theoretic sense (i.e. polynomial-time), an existence proof shows that very efficient constructions exist which would be of real practical interest if such graphs could be made explicit. We hope this paper provides additional motivation to investigate explicit constructions of unbalanced bipartite expanders for parameters ranges which have not received much attention so far.

Finally, our techniques allow to use a public random function $\{0,1\}^n \rightarrow \{0,1\}^n$ to construct a compression function with sufficiently large domain and

---

[4] This argument applies even though we know that a public random function is not securely realizable from a public random parameter.

range and to plug it into the construction of [11] to achieve the first iterated construction of a public random oracle $\{0,1\}^* \to \{0,1\}^n$ from a public random function $\{0,1\}^n \to \{0,1\}^n$ with security above the birthday barrier. We discuss this in Section 5.

2   Preliminaries

2.1   Notation and Probabilities

Throughout this paper, calligraphic letters (e.g. $\mathcal{U}$) denote sets. A k-tuple is denoted as $u^k = [u_1, \ldots, u_k]$, and the set of k-tuples of elements of $\mathcal{U}$ is denoted as $\mathcal{U}^k$. We use capital letters (e.g. U) to name random variables, whereas their concrete values are often denoted by the corresponding lower-case letters (e.g. u). Also, we write P_U for the probability distribution of U, and we use the shorthand $\mathsf{P}_U(u)$ for $\mathsf{P}(U = u)$. Given random variables U and V, as well as events $\mathcal{A}$ and $\mathcal{B}$, $\mathsf{P}_{U\mathcal{A}|V\mathcal{B}}$ denotes the corresponding conditional probability distribution, which is interpreted as a function $\mathcal{U} \times \mathcal{V} \to \mathbb{R}_{\geq 0}$, where the value $\mathsf{P}_{U\mathcal{A}|V\mathcal{B}}(u, v)$ is well-defined for all $u \in \mathcal{U}$ and $v \in \mathcal{V}$ such that $\mathsf{P}_{V\mathcal{B}}(v) > 0$ (and undefined otherwise). Two probability distributions P_U and $\mathsf{P}_{U'}$ on the same set $\mathcal{U}$ are equal, denoted $\mathsf{P}_U = \mathsf{P}_{U'}$, if $\mathsf{P}_U(u) = \mathsf{P}_{U'}(u)$ for all $u \in \mathcal{U}$. Also, for conditional probability distributions, equality holds if it holds for all inputs for which *both* are defined. We often need to deal with distinct random experiments where equally-named random variables and/or events appear. To avoid confusion, we add superscripts to probability distributions (e.g. $\mathsf{P}_{U|V}^{\mathcal{E}}(u, v)$) to make the random experiment explicit. Finally, we denote by $s \| s'$ the concatenation of two binary strings $s, s' \in \{0,1\}^*$.

2.2   Indistinguishability of Random Systems

In this section, we review basic definitions and facts from the framework of *random systems* of [19]. A random system is the abstraction of the input-output behavior of any discrete system.

Definition 1. *An $(\mathcal{X}, \mathcal{Y})$-random system $\mathbf{F}$ is a (generally infinite) sequence of conditional probability distributions[5] $\mathsf{p}_{Y_i|X^iY^{i-1}}^{\mathbf{F}}$ for all $i \geq 1$. Two random systems $\mathbf{F}$ and $\mathbf{G}$ are equivalent, denoted $\mathbf{F} \equiv \mathbf{G}$, if $\mathsf{p}_{Y_i|X^iY^{i-1}}^{\mathbf{F}} = \mathsf{p}_{Y_i|X^iY^{i-1}}^{\mathbf{G}}$ for all $i \geq 1$.*

The system is described by the conditional probabilities $\mathsf{p}_{Y_i|X^iY^{i-1}}^{\mathbf{F}}(y_i, x^i, y^{i-1})$ (for $i \geq 1$) of obtaining the output $y_i \in \mathcal{Y}$ on query $x_i \in \mathcal{X}$ given the previous $i - 1$ queries $x^{i-1} = [x_1, \ldots, x_{i-1}] \in \mathcal{X}^{i-1}$ and their corresponding outputs $y^{i-1} = [y_1, \ldots, y_{i-1}] \in \mathcal{Y}^{i-1}$. An example of a random system that we consider in the following is a *random function* $\mathbf{R} : \{0,1\}^m \to \{0,1\}^n$, which

---

[5] We use a lower-case p to stress the fact that these conditional distributions by themselves do not define a random experiment.

returns for every distinct input value $x \in \{0,1\}^m$ an independent and uniformly-distributed n-bit value. Moreover, a *random oracle* $\mathbf{O} : \{0,1\}^* \rightarrow \{0,1\}^n$ is a random function taking inputs of arbitrary length.

A *distinguisher* $\mathbf{D}$ for an $(\mathcal{X}, \mathcal{Y})$-random system is a $(\mathcal{Y}, \mathcal{X})$-random system which is one query ahead, i.e. it is defined by the conditional probability distributions $\mathsf{p}^{\mathbf{D}}_{X_i|X^{i-1}Y^{i-1}}$ for all $i \geq 1$. In particular, $\mathsf{p}^{\mathbf{D}}_{X_1}$ is the probability distribution of the first value queried by $\mathbf{D}$. Finally, the distinguisher outputs a bit after a certain number (say k) of queries depending on the transcript (X^k, Y^k). For an $(\mathcal{X}, \mathcal{Y})$-random system $\mathbf{F}$ and a distinguisher $\mathbf{D}$, we denote by $\mathbf{D} \circ \mathbf{F}$ the random experiment[6] where $\mathbf{D}$ interacts with $\mathbf{F}$. Furthermore, given an additional $(\mathcal{X}, \mathcal{Y})$-random system $\mathbf{G}$, the *distinguishing advantage* of $\mathbf{D}$ in distinguishing systems $\mathbf{F}$ and $\mathbf{G}$ is defined as $\varDelta^{\mathbf{D}}(\mathbf{F}, \mathbf{G}) := \left| \mathsf{P}^{\mathbf{D} \circ \mathbf{F}}(1) - \mathsf{P}^{\mathbf{D} \circ \mathbf{G}}(1) \right|$, where $\mathsf{P}^{\mathbf{D} \circ \mathbf{F}}(1)$ and $\mathsf{P}^{\mathbf{D} \circ \mathbf{G}}(1)$ denote the probabilities that $\mathbf{D}$ outputs 1 after its k queries when interacting with $\mathbf{F}$ and $\mathbf{G}$, respectively.

We are interested in considering an internal *monotone condition* defined on a random system $\mathbf{F}$. Such a condition is initially true, and once it fails, it cannot become true any more. In particular, a *system* $\mathbf{F}^{\mathcal{A}}$ *with a monotone condition* $\mathcal{A}$ is an $(\mathcal{X}, \mathcal{Y} \times \{0,1\})$-random system, where the additional output bit indicates whether the condition $\mathcal{A}$ holds after the i'th query has been answered. In general, we characterize such a condition by a sequence of events $\mathcal{A} = A_0, A_1, \ldots$, where A_0 always holds, and A_i holds if the condition holds after query i. The condition *fails* at query i if $A_{i-1} \wedge \overline{A_i}$ occurs. For a system with a monotone condition $\mathbf{F}^{\mathcal{A}}$, we write $\mathbf{F}$ for the system where the additional output bit is ignored. Generally, we are interested in considering the behavior of systems only as long as a certain monotone condition holds: Given two systems $\mathbf{F}^{\mathcal{A}}$ and $\mathbf{G}^{\mathcal{B}}$ with monotone conditions $\mathcal{A}$ and $\mathcal{B}$, respectively, they are *equivalent*, denoted $\mathbf{F}^{\mathcal{A}} \equiv \mathbf{G}^{\mathcal{B}}$, if $\mathsf{p}^{\mathbf{F}}_{A_iY_i|X^iY^{i-1}A_{i-1}} = \mathsf{p}^{\mathbf{G}}_{B_iY_i|X^iY^{i-1}B_{i-1}}$ holds for all $i \geq 1$.

The probability that a distinguisher $\mathbf{D}$ issuing k queries makes a monotone condition $\mathcal{A}$ fail in the random experiment $\mathbf{D} \circ \mathbf{F}$ is defined as $\nu^{\mathbf{D}}(\mathbf{F}^{\mathcal{A}}) := \mathsf{P}^{\mathbf{D} \circ \mathbf{F}}_{\overline{A_k}}$. The following lemma from [19] relates this probability with the distinguishing advantage.

Lemma 1. *If* $\mathbf{F}^{\mathcal{A}} \equiv \mathbf{G}^{\mathcal{B}}$ *holds, then* $\varDelta^{\mathbf{D}}(\mathbf{F}, \mathbf{G}) \leq \nu^{\mathbf{D}}(\mathbf{F}^{\mathcal{A}}) = \nu^{\mathbf{D}}(\mathbf{G}^{\mathcal{B}})$ *for all distinguishers* $\mathbf{D}$.

One can use a random system $\mathbf{F}$ as a component of a larger system: In particular, we are interested in *constructions* $\mathbf{C}(\cdot)$ such that the resulting random system $\mathbf{C}(\mathbf{F})$ invokes $\mathbf{F}$ as a subsystem. (Note that $\mathbf{C}(\cdot)$ itself is not a random system, while $\mathbf{C}(\mathbf{F})$ is a random system.)

Finally, we remark that in general when we mention that a construction (or a distinguisher) is *efficient* we mean that there exists a probabilistic interactive Turing machine implementing the same input-output behavior and with polynomial running time (in the understood security parameter).

---

[6] In particular, in this random experiment, the joint distribution $\mathsf{P}^{\mathbf{D} \circ \mathbf{F}}_{X^k Y^k}$ is well-defined as $\prod_{i=1}^{k} \mathsf{p}^{\mathbf{D}}_{X_i|X^{i-1}Y^{i-1}} \cdot \mathsf{p}^{\mathbf{F}}_{Y_i|X^iY^{i-1}}$.

2.3   Indifferentiability, Reductions, and Public Random Primitives

The notion of *indifferentiability* [21] naturally extends the concept of indistinguishability to systems with a *public* and a *private* interface[7] adopting a simulation-based approach. The public interface can be used by all parties, including the adversary, whereas the legitimate parties have exclusive access to the private interface. Generally, we denote such a system as an ordered pair $\mathbf{F} = [\mathbf{F}_{pub}, \mathbf{F}_{priv}]$. Furthermore, given constructions $\mathbf{S}(\cdot)$ and $\mathbf{C}(\cdot)$ leaving, respectively, private and public queries unmodified, we simply write $\mathbf{S}(\mathbf{F}) = [\mathbf{S}(\mathbf{F}_{pub}), \mathbf{F}_{priv}]$ and $\mathbf{C}(\mathbf{F}) = [\mathbf{F}_{pub}, \mathbf{C}(\mathbf{F}_{priv})]$.

Public random primitives are a special case of such systems. A *public random function (puRF)* $\mathbf{R} : \{0,1\}^m \rightarrow \{0,1\}^n$ is a system with a public and a private interface which behaves as the *same* random function at *both* interfaces.[8] In particular, both interfaces answer consistently. Furthermore, a *public random oracle (puRO)* $\mathbf{O} : \{0,1\}^* \rightarrow \{0,1\}^n$ is a public random function which takes inputs of arbitrary bit-length.

The following definition refines the notion of (information-theoretic) indifferentiability from [21] to deal with concrete parameters.

Definition 2. *Let $\alpha : \mathbb{N} \rightarrow \mathbb{R}_{\geq 0}$ and $\sigma : \mathbb{N} \rightarrow \mathbb{N}$ be functions. A system $\mathbf{F}$ is (α, σ)-indifferentiable from $\mathbf{G}$, denoted $\mathbf{F} \overset{\alpha,\sigma}{\sqsubset} \mathbf{G}$, if there exists a simulator $\mathbf{S}$ such that $\Delta^{\mathbf{D}}([\mathbf{F}_{pub}, \mathbf{F}_{priv}], [\mathbf{S}(\mathbf{G}_{pub}), \mathbf{G}_{priv}]) \leq \alpha(k)$ for all distinguishers $\mathbf{D}$ making at most k queries, and $\mathbf{S}$ makes at most $\sigma(k)$ queries to $\mathbf{G}_{pub}$ when interacting with $\mathbf{D}$.*

The purpose of the simulator is to mimic $\mathbf{F}_{pub}$ by querying $\mathbf{G}_{pub}$, but without seeing the queries made to $\mathbf{G}_{priv}$. Indifferentiability directly implies a notion of reducibility.

Definition 3. *A system $\mathbf{G}$ is (α, σ)-reducible to a system $\mathbf{F}$ if there exists an efficient, deterministic, and stateless construction $\mathbf{C}(\cdot)$ such that $[\mathbf{F}_{pub}, \mathbf{C}(\mathbf{F}_{priv})] \overset{\alpha,\sigma}{\sqsubset} \mathbf{G}$. The construction $\mathbf{C}(\cdot)$ is called an (α, σ)-reduction.*

Note that if a random primitive is (α, σ)-reducible to a further random primitive with an N-bit table, then $\alpha(k) \geq \frac{1}{2}$ for all $k > N$, and hence security can only be achieved with respect to distinguishers issuing at most N queries. (We refer the reader to the full version [23] for a proof.) The following lemma states that reducibility is transitive. We omit its simple proof.

Lemma 2. *Let $\mathbf{E}, \mathbf{F}$, and $\mathbf{G}$ be systems. If $\mathbf{C}(\cdot)$ is a (α, σ)-reduction of $\mathbf{F}$ to $\mathbf{E}$, and $\mathbf{C}'(\cdot)$ is an (α', σ') reduction of $\mathbf{G}$ to $\mathbf{F}$ that makes at most $k_{\mathbf{C}'}(k)$ queries to $\mathbf{F}_{priv}$ when queried k times, then $\mathbf{C}'(\mathbf{C}(\cdot))$ is an $(\overline{\alpha}, \overline{\sigma})$-reduction of $\mathbf{G}$ to $\mathbf{E}$, where $\overline{\alpha}(k) = \alpha(k + k_{\mathbf{C}'}(k)) + \alpha'(k + \sigma(k))$ and $\overline{\sigma}(k) = \sigma'(\sigma(k))$.*

---

[7] Formally, this can be seen as a random system with a single interface and two types of queries.

[8] For this reason, we generally write both $\mathbf{R}_{pub}$ and $\mathbf{R}_{priv}$ as $\mathbf{R}$.

The computational variant of indifferentiability is obtained by requiring $\mathbf{S}$ to be efficient and the advantage $\Delta^{\mathbf{D}}([\mathbf{F}_{\text{pub}}, \mathbf{F}_{\text{priv}}], [\mathbf{S}(\mathbf{G}_{\text{pub}}), \mathbf{G}_{\text{priv}}])$ to be negligible for all efficient distinguishers $\mathbf{D}$. A *computational reduction* is defined accordingly. In the information theoretic case, it is sometimes desirable to prove that the simulator is efficient when queried by an efficient distinguisher, as this then implies the corresponding complexity-theoretic statement. We refer the reader to [21,11] for the implications of computational indifferentiability.

In contrast, as long as we are only interested in excluding generic attacks against security properties of a random function, the running time of the simulator is irrelevant. If $\mathbf{C}(\cdot)$ is an (α, σ)-reduction of a puRO $\mathbf{O} : \{0,1\}^* \rightarrow \{0,1\}^n$ (or of a puRF $\mathbf{R}' : \{0,1\}^m \rightarrow \{0,1\}^{\ell}$) to a puRF $\mathbf{R} : \{0,1\}^n \rightarrow \{0,1\}^n$, then $\mathbf{C}(\mathbf{R})$ inherits *all* the security properties of the truly-random oracle $\mathbf{O}$ (or of $\mathbf{R}'$), as long as the number of queries keeps $\alpha(k)$ small: Any adversary A making k queries (to both $\mathbf{R}$ and $\mathbf{C}(\mathbf{R})$) and breaking some property of $\mathbf{C}(\mathbf{R})$ with probability $\pi(k)$ can be transformed (combining it with the simulator) into an adversary A' making at most $k + \sigma(k)$ queries to $\mathbf{O}$ and breaking the same property for $\mathbf{O}$ with probability at least $\pi(k) - \alpha(k)$, and if no such A' can exist, then also no adversary A exists. The actual running time of A' is irrelevant, as the security of a random function (or oracle) with respect to a certain property is determined by the number of queries of the adversary, and not by its running time. For example, if $\sigma(k) = \Theta(k)$, then, given a random element $s \in \{0,1\}^m$, no adversary can find a second preimage $s' \in \{0,1\}^m$ with $s' \neq s$ and $\mathbf{C}(\mathbf{R})(s) = \mathbf{C}(\mathbf{R})(s')$ with probability higher than $\Theta(k \cdot 2^{-n}) + \alpha(k)$.

3 Beyond-Birthday Domain Extension for Public Random Functions

3.1 The Construction

We first discuss at an abstract level the main construction of this paper (represented in Figure 1), which implements a function mapping m-bit strings to ℓ-bit strings from $r + t$ independent puRF's $\mathbf{F}_1, \ldots, \mathbf{F}_r : \{0,1\}^n \rightarrow \{0,1\}^{t\rho n}$ and $\mathbf{G}_1, \ldots, \mathbf{G}_t : \{0,1\}^n \rightarrow \{0,1\}^{\ell}$ (for given parameters r, t, and ρ). Let $E_1, \ldots, E_r : \{0,1\}^m \rightarrow \{0,1\}^n$ be efficiently-computable functions (to be instantiated below). On input $s \in \{0,1\}^m$, the construction operates in three stages:

1. The values $\mathbf{F}_p(E_p(s)) = \mathbf{F}_p^{(1)}(E_p(s)) \| \cdots \| \mathbf{F}_p^{(t)}(E_p(s)) \in \{0,1\}^{t\rho n}$ are computed for all $p = 1, \ldots, r$, where $\mathbf{F}_p^{(q)}(E_p(s)) \in \{0,1\}^{\rho n}$ for all $q = 1, \ldots, t$;
2. The value $w(s) = w^{(1)}(s) \| \cdots \| w^{(t)}(s)$ is computed, where $w^{(q)}(s)$ equals (for all $q = 1, \ldots, t$) the first n bits of the product $\bigodot_{p=1}^{r} \mathbf{F}_p^{(q)}(E_p(s))$, and $\odot$ denotes multiplication in $GF(2^{\rho n})$ with ρn-bit strings interpreted as elements of the finite field $GF(2^{\rho n})$;
3. Finally, the value $\bigoplus_{q=1}^{t} \mathbf{G}_q(w^{(q)}(s))$ is output.

Our approach relies on the observation that if for each new query to the construction with input $s \in \{0,1\}^m$ there exists an index $q \in \{1, \ldots, t\}$ for which $\mathbf{G}_q$

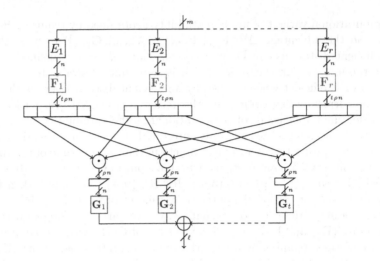

Fig. 1. Main construction, where $\mathbf{F}_1, \ldots, \mathbf{F}_r$ and $\mathbf{G}_1, \ldots, \mathbf{G}_t$ are independent puRF's and $E_1, \ldots, E_r : \{0,1\}^m \to \{0,1\}^n$ are efficiently-computable functions

has not been queried yet at the value $w^{(q)}(s)$, either directly at its public interface or by the construction at the private interface, the resulting output value is uniformly distributed and independent from all previously-returned values. This resembles the approach taken to extend the domain of (secret) random functions [1,4,19]. However, we stress that the role of the first two stages (including the functions $E_1, \ldots, E_r$) is crucial here: Not only they have to guarantee that such an index q always exists, but they must also permit simulation of the puRF's $\mathbf{F}_1, \ldots, \mathbf{F}_r$ and $\mathbf{G}_1, \ldots, \mathbf{G}_t$ given only access to the public interface of an (ideal) puRF $\mathbf{R} : \{0,1\}^m \to \{0,1\}^\ell$, without seeing the queries made to the private interface of $\mathbf{R}$. Also, the probability that the simulation fails must be small enough to allow security beyond the birthday barrier.

3.2   Input-Restricting Functions

For every $s \in \{0,1\}^m$ one can always learn the value $w(s)$ by querying the public interfaces of $\mathbf{F}_1, \ldots, \mathbf{F}_r$ with appropriate inputs $E_1(s), \ldots, E_p(s)$, respectively. For every such s, the sum $\bigoplus_{q=1}^t \mathbf{G}_q(w^{(q)}(s))$ equals the output of the construction on input s. The simulator must ensure that its answers for queries to the functions $\mathbf{G}_1, \ldots, \mathbf{G}_t$ are consistent with these constraints. However, if $E_1, \ldots, E_r$ allow a relatively small number of queries to the functions $\mathbf{F}_1, \ldots, \mathbf{F}_t$ to reveal a too large number of values $w(s)$, then the simulator possibly fails to satisfy all constraints. For example, the *Benes* construction [1] adopts an approach similar to the one of our construction, but suffers from this problem and its security in the setting of puRF's is inherently bounded by the birthday barrier. (We provide a concrete attack in the full version [23].) To overcome this problem, we introduce the following combinatorial notion.

Definition 4. *Let $\epsilon \in (0,1)$, and let $m > n$. A family $\mathcal{E}$ of functions $E_1, \ldots, E_r$: $\{0,1\}^m \rightarrow \{0,1\}^n$ is called (m, δ, ϵ)-input restricting if it satisfies the following two properties:*

Injective. *For all $s \neq s' \in \{0,1\}^m$, there exists $p \in \{1, \ldots, r\}$ such that $E_p(s) \neq E_p(s')$.*

Input-Restricting. *For all subsets $\mathcal{U}_1, \ldots, \mathcal{U}_r \subseteq \{0,1\}^n$ such that $|\mathcal{U}_1| + \cdots + |\mathcal{U}_r| \leq 2^{n(1-\epsilon)}$, we have*

$$\left| \{s \in \{0,1\}^m \mid E_p(s) \in \mathcal{U}_p \text{ for all } p = 1, \ldots, r\} \right| \leq \delta \cdot (|\mathcal{U}_1| + \cdots + |\mathcal{U}_r|).$$

It is easy to see that $\delta \geq 1/r$ must hold. Furthermore, we need $r \cdot n \geq m$ for the family to be injective. When talking about efficiency, we can naturally extend the notion to asymptotic families $\mathcal{E} = \{\mathcal{E}_n\}_{n \in \mathbb{N}}$ of function families by letting m, δ, ϵ, and r be functions of n, and $\mathcal{E}_n = \{E_1^n, \ldots, E_{r(n)}^n\}$, with $E_p^n : \{0,1\}^{m(n)} \rightarrow \{0,1\}^n$. In particular, note that we allow the size of the family to grow with the security parameter. The family $\mathcal{E}_n$ is called *explicit* if $r = r(n)$ is polynomial in n and if there exists a (uniform) polynomial-time (in n) algorithm E that outputs $E_p^n(s) \in \{0,1\}^n$ on input $n \in \mathbb{N}$, $s \in \{0,1\}^{m(n)}$, and $p \in \{1, \ldots, r(n)\}$. The family is additionally called *invertible* if there exists an algorithm which on input the sets $\mathcal{U}_1, \ldots, \mathcal{U}_r \subseteq \{0,1\}^n$ and n returns the set of all $s \in \{0,1\}^m$ for which $E_p(s) \in \mathcal{U}_p$ for all $p = 1, \ldots, r$ in time polynomial in $|\mathcal{U}_1| + \cdots + |\mathcal{U}_r|$ and in n. We will not, however, stress the asymptotic point of view in the following, as long as it is clear from the context that the statements can be also formalized in this sense.

We postpone the discussion of the existence of explicit function families to Section 4, where we construct (for all constants ϵ) explicit families of (m, δ, ϵ)-input-restricting functions for all polynomials m and sufficiently-small δ using highly unbalanced expander graphs with polynomial-degree.

3.3 Main Result

Let $\epsilon \in (0,1)$. The concrete construction $\mathbf{C}_{\epsilon,m,\ell}^{\mathcal{E}}(\cdot)$ is obtained from the description in Section 3.1 by instantiating the functions $E_1, \ldots, E_r$ with an explicit family $\mathcal{E} = \{E_1, \ldots, E_r\}$ of (m, δ, ϵ)-input restricting functions with n-bit output. Also, we let $\rho := \lceil \frac{m}{n} + 2 - \epsilon \rceil$ and $t := \lceil 2/\epsilon - 1 \rceil$. Note that underlying $r + t$ puRF's can be seen as a single puRF $\mathbf{R}' : \{0,1\}^{n+\phi(n)} \rightarrow \{0,1\}^n$, where $\phi(n) = \lceil \log(r \cdot t\rho + t\ell/n) \rceil$. If m, ℓ, and $1/\epsilon$ are polynomial in n, then in particular $\phi(n) = \mathcal{O}(\log n)$. Also, it is easy to see that $\mathbf{C}_{\epsilon,m,\ell}^{\mathcal{E}}(\cdot)$ is efficient, as long as the function family $\mathcal{E}$ is explicit. The following is the main theorem of this paper and it is proved in the next section.

Theorem 1. *The construction $\mathbf{C}_{\epsilon,m,\ell}^{\mathcal{E}}(\cdot)$ is an (α, σ)-reduction of the puRF $\mathbf{R}$: $\{0,1\}^m \rightarrow \{0,1\}^\ell$ to the puRF's $\mathbf{F}_1, \ldots, \mathbf{F}_r : \{0,1\}^n \rightarrow \{0,1\}^{t \cdot \rho n}$ and $\mathbf{G}_1, \ldots, \mathbf{G}_t$: $\{0,1\}^n \rightarrow \{0,1\}^\ell$, where for all $k \leq 2^{n(1-\epsilon)} - r$,*

$$\alpha(k) \leq 2r^t(\delta+1)^{t+1} \cdot k^{t+2} \cdot 2^{-nt} + \frac{1}{2}t(\delta+1) \cdot k \cdot (k + 2r + 1) \cdot 2^{m-\rho n}$$

and $\sigma(k) \leq \delta(n) \cdot k$. If the family $\mathcal{E}$ is invertible, the simulator runs in time poly-nomial in k and n, and in particular $\mathbf{C}_{\epsilon,m,\ell}^{\mathcal{E}}(\cdot)$ is also a computational reduction.

We remark the following two important consequences of Theorem 1. First, if ϵ is constant and r, δ polynomial in n, the above advantage $\alpha(k)$ is negligible for all parameters k up to $k = 2^{n(1-\epsilon)} - r$. In particular, choosing $\epsilon < \frac{1}{2}$ leads to secu-rity beyond the birthday barrier,[9] and we are going to provide input-restricting families of functions with appropriate parameters in Section 4. Second, the re-sult can be used to extend the domain of a puRF $\mathbf{R}' : \{0,1\}^n \to \{0,1\}^n$ with security up to $2^{n(1-\mu)}$ queries: One chooses any $\epsilon < \mu$ and n' maximal such that $n' + \phi(n') \leq n$, and interprets the function $\mathbf{R}'$ as a puRF $\{0,1\}^{n'+\phi(n')} \to \{0,1\}^{n'}$ by dropping approximately $\phi(n')$ bits of the output. The above advan-tage is still negligible for all $k \leq 2^{n'(1-\epsilon)} - r$, and hence for all $k \leq 2^{n(1-\mu)}$ for n large enough, since $n - n' = o(n)$.

3.4   Proof of Theorem 1

We prove that there exists a simulator $\mathbf{S}$ such that $\Delta^{\mathbf{D}}(\mathbf{H}_1, \mathbf{H}_2)$ is bounded by the above expression for all distinguishers $\mathbf{D}$ making at most $k \leq 2^{n(1-\epsilon)} - r$ queries, where for notational convenience $\mathbf{H}_1$ and $\mathbf{H}_2$ are defined as

$$\mathbf{H}_1 := [\mathbf{F}_1, \dots, \mathbf{F}_r, \mathbf{G}_1, \dots, \mathbf{G}_t, \mathbf{C}_{\epsilon,m,\ell}^{\mathcal{E}}(\mathbf{F}_1, \dots, \mathbf{F}_r, \mathbf{G}_1, \dots, \mathbf{G}_t)]$$
$$\mathbf{H}_2 := [\mathbf{S}(\mathbf{R}), \mathbf{R}].$$

There are three types of queries to the systems $\mathbf{H}_1$ and $\mathbf{H}_2$: The first two types are $\mathbf{F}$-queries, denoted $(\mathbf{F}, p, u)$ for $p \in \{1, \dots, r\}$ and $u \in \{0,1\}^n$, and $\mathbf{G}$-queries, denoted $(\mathbf{G}, q, v)$, for $v \in \{0,1\}^n$ and $q \in \{1, \dots, t\}$. In $\mathbf{H}_1$, a query $(\mathbf{F}, p, u)$ returns the value $\mathbf{F}_p(u)$ and a query $(\mathbf{G}, q, v)$ returns the value $\mathbf{G}_q(v)$, while in $\mathbf{H}_2$ both query-types are answered by the simulator $\mathbf{S}$. The third type of queries, called $\mathbf{R}$-queries, are denoted $(\mathbf{R}, s)$ for $s \in \{0,1\}^m$ and are answered by the construction $\mathbf{C}_{\epsilon,m,\ell}^{\mathcal{E}}(\cdot)$ in $\mathbf{H}_1$, and by the private interface of the ran-dom function $\mathbf{R}$ in $\mathbf{H}_2$. Given the first i queries $x^i = [x_1, \dots, x_i]$, where $x_j \in \{(\mathbf{F}, p, u), (\mathbf{G}, q, v), (\mathbf{R}, s)\}$ for all $j = 1, \dots, i$, we define for all indices p and q the sets $\mathcal{F}_{p,i}$ and $\mathcal{G}_{q,i}$ that contain, respectively, all values $u \in \{0,1\}^n$ for which a query $(\mathbf{F}, p, u)$ and all $v \in \{0,1\}^n$ for which a query $(\mathbf{G}, q, v)$ appears in x^i. Also, we let $\mathcal{R}_i$ be the set of values $s \in \{0,1\}^m$ for which a query $(\mathbf{R}, s)$ appears in x^i, and we let $\mathcal{S}_i$ consist of all the values $s \in \{0,1\}^m$ such that $E_p(s) \in \mathcal{F}_{p,i}$ for all $p = 1, \dots, r$. Furthermore, let $\Delta\mathcal{S}_i := \mathcal{S}_i \setminus \mathcal{S}_{i-1}$. Notice that the set $\mathcal{S}_i$ contains all inputs for which the values returned by the first i queries allow to compute the value $w(s)$. Clearly, $|\mathcal{S}_i| = \sum_{j=1}^{i} |\Delta\mathcal{S}_j| \leq \delta \cdot i$ for all $i \leq 2^{n(1-\epsilon)}$, since the family $\mathcal{E}$ is input-restricting. For $s \in \mathcal{S}_i$, we define $w(s) = w^{(1)}(s)\| \cdots \|w^{(t)}(s)$ as in the description of $\mathbf{C}_{\epsilon,m,\ell}^{\mathcal{E}}(\cdot)$ according to the answers of the first queries, and for a set $\mathcal{S} \subseteq \mathcal{S}_i$ we use the shorthand $w^{(q)}(\mathcal{S}) := \{w^{(q)}(s) \mid s \in \mathcal{S}\}$.

---

[9] Note that ϵ could even be some function going (slowly) towards zero, even though this may require setting t differently.

The simulator $\mathbf{S}$ defines the function tables of $\mathbf{F}_1, \ldots, \mathbf{F}_r$ and of $\mathbf{G}_1, \ldots, \mathbf{G}_t$ *dynamically*. That is, all values $\mathbf{F}_p(u)$ and $\mathbf{G}_q(v)$ are initially *undefined* for all $u, v \in \{0,1\}^n$ and indices p and q. Upon processing a new $\mathbf{F}$-query $x_i = (\mathrm{F}, p, u)$, the simulator sets the value $\mathbf{F}_p(u)$ to a fresh random value and computes the set $\Delta\mathcal{S}_i$: The simulator knows this set, as it processes all $\mathbf{F}$-queries. For each $s \in \Delta\mathcal{S}_i$, the equality $\bigoplus_{q=1}^t \mathbf{G}_q(w^{(q)}(s)) = \mathbf{R}(s)$ must be satisfied, and hence $\mathbf{S}$ tries to satisfy these constraints by appropriately setting the values of the functions $\mathbf{G}_1, \ldots, \mathbf{G}_t$. More precisely, it looks for an ordering of $\Delta\mathcal{S}_i = \{s_1, \ldots, s_{|\Delta\mathcal{S}_i|}\}$ with the property that for all $j = 1, \ldots, |\Delta\mathcal{S}_i|$ there exists $q_j \in \{1, \ldots, t\}$ such that $w^{(q_j)}(s_j) \notin \{w^{(q_j)}(s_1), \ldots, w^{(q_j)}(s_{j-1})\} \cup \mathcal{G}_{q,i-1}$, and sets $\mathbf{G}_{q_j}(w^{(q_j)}(s_j)) := \mathbf{R}(s_j) \oplus \bigoplus_{q \neq q_j} \mathbf{G}_q(w^{(q)}(s_j))$ for $j = 1, \ldots, |\Delta\mathcal{S}_i|$, where each undefined value in the sums is set to an independent random value. A query to the public interface of $\mathbf{R}$ is issued in order to learn $\mathbf{R}(s_j)$. If no such ordering exists, then the simulator aborts.[10] Finally, the value $\mathbf{F}_p(u)$ is returned. For a query $x_i = (\mathrm{G}, q, v)$, the simulator returns $\mathbf{G}_q(v)$, defining it to a random value if undefined. In the full version of this paper [23], we provide a detailed pseudo-code description of the simulator $\mathbf{S}$. The number of $\mathbf{R}$-queries made by the simulator after $i \leq 2^{n(1-\epsilon)}$ queries is $|\mathcal{S}_i| \leq \delta \cdot i$. Also, as long as the family $\mathcal{E}$ is invertible and an appropriate ordering can be efficiently found, its running time is efficient in k and n. In fact, we show that with very high probability *any* ordering can be used.

Without loss of generality, it is convenient to advance the generation of the random functions $\mathbf{F}_1, \ldots, \mathbf{F}_r$ to the initialization phase, that is, their *entire* function tables are generated once uniformly at random in both $\mathbf{H}_1$ and $\mathbf{H}_2$. Subsequently, all queries (F, p, u) are answered according to the initial choice. In particular, this means that in $\mathbf{H}_2$ the simulator $\mathbf{S}$ uses the value $\mathbf{F}_p(u)$ already defined instead of generating a new fresh random value. It is clear that the behavior of both systems is unchanged. This also allows us to define the value $w(s) = w^{(1)}(s) \| \cdots \| w^{(t)}(s)$ for *all* $s \in \{0,1\}^m$ and each such value induces a constraint, namely the answer of an $\mathbf{R}$-query (R, s) must equal $\bigoplus_{q=1}^t \mathbf{G}_q(w^{(q)}(s))$. Such a constraint remains hidden until $s \in \Delta\mathcal{S}_i$ from some i, and in this case the simulator attempts to fill the function tables of $\mathbf{G}_1, \ldots, \mathbf{G}_t$ consistently. To avoid possible problems, we have to account for two things captured by the two following monotone conditions which we define on both $\mathbf{H}_1$ and $\mathbf{H}_2$:

(a) The monotone condition $\mathcal{A} = A_0, A_1, \ldots$ fails at query i if there exists an $s \in \Delta\mathcal{S}_i$ such that $w^{(q)}(s) \in w^{(q)}(\mathcal{S}_i \setminus \{s\}) \cup \mathcal{G}_{q,i-1}$ for all $q = 1, \ldots, t$.
(b) The monotone condition $\mathcal{B} = B_0, B_1, \ldots$ fails at query i if there exists $s \in \mathcal{R}_i \setminus \mathcal{S}_i$ such that $w^{(q)}(s) \in w^{(q)}(\mathcal{S}_i \cup \mathcal{R}_i \setminus \{s\}) \cup \mathcal{G}_{q,i}$ for all $q = 1, \ldots, t$.

As long as $\mathcal{A}$ does not fail, the simulator never aborts. This in particular implies that $\mathbf{R}$-queries (R, s) for $s \in \mathcal{S}_i$ in $\mathbf{H}_2$ are consistent with $\mathbf{G}$-queries answered by the simulator. However, all $\mathbf{R}$-queries (R, s) for $s \notin \mathcal{S}_i$ are answered independently and uniformly at random in $\mathbf{H}_2$, and $\mathcal{B}$ ensures that this happens in $\mathbf{H}_1$

---

[10] Note that there is no need to formalize the exact meaning of abortion, since whenever the simulator fails to find such an ordering, then the distinguisher is assumed to win.

as well. In the full version [23], we prove the following lemma, which formalizes this argument and states that as long as neither $\mathcal{A}$ nor $\mathcal{B}$ fail, then $\mathbf{H}_1$ and $\mathbf{H}_2$ behave identically.

Lemma 3. $\mathbf{H}_1^{\mathcal{A} \wedge \mathcal{B}} \equiv \mathbf{H}_2^{\mathcal{A} \wedge \mathcal{B}}$.

To provide some intuition as to why the probability that a distinguisher $\mathbf{D}$ makes $\mathcal{A} \wedge \mathcal{B}$ fail is small, let us assume first that for any two distinct $s, s' \in \{0,1\}^m$ (such that at least one of them is not in $\mathcal{S}_i$) and for all $q = 1, \ldots, t$, the probability (conditioned on the answers to the previous queries) that $w^{(q)}(s) = w^{(q)}(s')$ is bounded by some small value φ (say $\varphi \approx 2^{-n}$). In order to upper bound the probability of $\mathcal{A}$ failing after query i, combining the union bound with the above assumption we see that $\mathsf{P}(w^{(q)}(s) \in w^{(q)}(\mathcal{S}_i \setminus \{s\}) \cup \mathcal{G}_{q,i-1}) \leq |w^{(q)}(\mathcal{S}_i \setminus \{s\}) \cup \mathcal{G}_{q,i-1}| \cdot \varphi \leq (\delta + 1) \cdot i \cdot \varphi$ for all $s \in \Delta\mathcal{S}_i$, since $\mathcal{E}$ is input-restricting. Furthermore, for all distinct $q, q' \in \{1, \ldots, t\}$ and $s, s' \in \{0,1\}^n$ (possibly $s = s'$), the structure of the first two stages of $\mathbf{C}_{\epsilon,m,\ell}^{\mathcal{E}}(\cdot)$ ensures that the values $w^{(q)}(s)$ and $w^{(q')}(s')$ are statistically independent, and hence

$$\mathsf{P}(\forall q : w^{(q)}(s) \in w^{(q)}(\mathcal{S}_i \setminus \{s\}) \cup \mathcal{G}_{q,i-1}) \leq (\delta + 1)^t \cdot i^t \cdot \varphi^t.$$

Therefore, the probability $\mathsf{p}_{A_i | X^i Y^{i-1} A_{i-1}}^{\mathbf{H}_1}(x^i, y^{i-1}) = \mathsf{p}_{A_i | X^i Y^{i-1} A_{i-1}}^{\mathbf{H}_2}(x^i, y^{i-1})$ that there exists an $s \in \Delta\mathcal{S}_i$ making $\mathcal{A}$ fail after query i is bounded by $|\Delta\mathcal{S}_i| \cdot (\delta + 1)^t \cdot i^t \cdot \varphi^t$, where $|\Delta\mathcal{S}_i|$ is small for all $i \leq 2^{n(1-\epsilon)}$.

Nevertheless, turning this intuition into a formal proof (and extending it to the monotone condition $\mathcal{B}$) requires additional care. The probability that $w^{(q)}(s)$ equals $w^{(q)}(s')$ happens to be small only with overwhelming probability (taken over the answers to the previous queries): This fact follows from the use of multiplication in $GF(2^{\rho n})$ and the choice of a sufficiently large parameter ρ.

In particular, a complete proof of the following lemma appears in the full version of this paper [23].

Lemma 4. For all distinguishers $\mathbf{D}$ making at most $k \leq 2^{n(1-\epsilon)} - r$ queries we have $\nu^{\mathbf{D}}(\mathbf{H}_1^{\mathcal{A} \wedge \mathcal{B}}) = \nu^{\mathbf{D}}(\mathbf{H}_2^{\mathcal{A} \wedge \mathcal{B}}) \leq 2r^t(\delta + 1)^{t+1} \cdot k^{t+2} \cdot 2^{-nt} + \frac{1}{2}t(\delta + 1) \cdot k \cdot (k + 2r + 1) \cdot 2^{m - \rho n}$.

By combining Lemmas 3 and 4, Theorem 1 follows making use of Lemma 1.

4   Existence of Input-Restricting Function Families

In this section, we prove the existence of input-restricting function families according to Definition 4, and we study their relationship to *highly unbalanced* bipartite expander graphs. First, we recall the following definition.

Definition 5. *A bipartite graph* $G = (V_1, V_2, E)$ *is* (K, γ)-expanding *if* $|\Gamma(X)| \geq \gamma \cdot |X|$ *for all subsets* $X \subset V_1$ *such that* $|X| \leq K$, *where* $\Gamma(X) \subseteq V_2$ *is the set of neighbors of* X. *Furthermore, such a graph has* left-degree D *if the degree of all* $v \in V_1$ *is bounded by* D.

A family of graphs $G = (V_1, V_2, E)$ with $V_1 := \{0,1\}^{m(n)}$, $V_2 := \{0,1\}^n$ (parameterized by the security parameter n) with left-degree $D = D(n)$ is called *explicit* if there exists a (uniform) algorithm which, on input 1^n, $v \in \{0,1\}^{m(n)}$ and $i \in \{1, \ldots, D(n)\}$ outputs the i'th neighbor of v in time polynomial in n. (The ordering of the neighbors is arbitrary.)

Given a bipartite graph $G = (V_1, V_2, E)$ with $V_1 = \{0,1\}^m$, $V_2 = \{0,1\}^n$, and left-degree D, we construct the family of functions $\mathcal{E} = \{E_1, \ldots, E_r\}$, where $r = D + \lceil m/n \rceil$, and the functions $E_1, \ldots, E_D : \{0,1\}^m \to \{0,1\}^n$ are such that $E_p(s)$ is the p'th neighbor of s in G for all $p = 1, \ldots, D$. Furthermore, the functions $E_{D+1}, \ldots, E_{D+\lceil m/n \rceil}$ are defined as $E_{D+p}(s) := s^{(p)}$ for $p = 1, \ldots, \lceil m/n \rceil$, where extra zeros are appended to s to make its length a multiple of n. Clearly, this family is injective. Furthermore, it turns out that good expanding properties for G imply that the family $\mathcal{E}$ is input-restricting. We refer the reader to the full version [23] for a proof of the following lemma.

Lemma 5. *Let $m \geq n$. Assume that there exists an explicit family of bipartite (K, γ)-expander graphs $G = (V_1, V_2, E)$ with polynomially-bounded left-degree D where $V_1 = \{0,1\}^m$ and $V_2 = \{0,1\}^n$. Then, for all $\epsilon > 0$ such that $\epsilon > 1 - \frac{\log(K\gamma)}{n}$ for n large enough, there exists an explicit (m, δ, ϵ)-input-restricting family of functions with $\delta = \gamma^{-1}$ and cardinality $r := D + \lceil m/n \rceil$. Furthermore, if $\lceil m/n \rceil$ is constant, then the family is invertible.*

For example, if a family exists with $K = 2^{n(1-\eta)}$ and constant expansion factor $\gamma > 1$, then $1 - \frac{\log K\gamma}{n} = \eta - o(1)$, and hence the family is (m, γ^{-1}, η)-input restricting. It remains to be shown that an explicit family of unbalanced expander graphs with sufficiently small (i.e. polynomially-bounded) left-degree exists. Much work in this area has been devoted to *lossless* unbalanced expanders, i.e., with $\gamma \approx D$, but the best known constructions (cf. e.g. [28,25]) for this case lead to either super-polynomial degree or a much too small bound K for our choice of parameters. However, we are satisfied even if the expansion factor is much smaller than the left-degree, as long as the latter stays small, and it is possible to obtain such graphs by appropriately composing known constructions. We discuss the following result in the full version of this paper [23].[11]

Theorem 2. *For all polynomials γ and constants $\eta \in (0,1)$, and all functions m (polynomially-bounded in n), there exists an explicit family of expander graphs $G = (V_1, V_2, E)$ with $V_1 = \{0,1\}^m$, $V_2 = \{0,1\}^n$ which is $(2^{n(1-\eta)}, \gamma)$-expanding and has left-degree polynomially-bounded in n.*

Note that these techniques even allow to obtain slightly stronger results, for instance allowing η to be a moderately vanishing function. Combining this with Lemma 5 we see that for all constants $\epsilon \in (0,1)$ there exist explicit (m, δ, ϵ)-input-restricting families with δ^{-1} polynomial in n. However, by dropping the explicitness requirement, families with much better parameters exist. In particular, the following result is a simple application of the probabilistic method.

---

[11] Also note that a very similar result appears in unpublished work by Baltz et al.[3].

Lemma 6. *Let K and γ be arbitrary such that $K \cdot \gamma \leq 2^n$, and let m be such that $m \geq n$. There exists a graph $G = (V_1, V_2, E)$ where $V_1 = \{0,1\}^m$ and $V_2 = \{0,1\}^n$ which is (K, γ)-expanding and with left-degree $D = \left\lceil \frac{1+\gamma \log e + m}{n - \log(K\gamma)} + \gamma \right\rceil$.*

For example, setting $m = \ell = 2n$, $\gamma = 1$ and $K = 2^{n(1-\epsilon)}$, we obtain left-degree $D = 1 + \frac{2}{\epsilon} + (\log e + 1)/(\epsilon \cdot n)$. For $\epsilon = \frac{1}{4}$ and $n = 128$, this leads to a family of size 12 by Lemma 5. Furthermore in this case $t = 7$ and $\rho = 4$, and all these values do not grow with n. (And a similar reasoning applies to all constants $\epsilon > 0$.) With these parameters, the construction is of practical interest, as it only relies on the design of a secure component function $\{0,1\}^n \to \{0,1\}^n$ which may be very efficient. We hope this motivates further research on de-randomizing families of unbalanced expander graphs for a wider range of parameters.

5 Constructing Public Random Oracles

We first review a slightly generalized version of the *prefix-free Merkle-Damgård* construction [11]. Let n be the given output size, and let $\ell \geq n$. We are given both a compression function $f : \{0,1\}^{b+\ell} \to \{0,1\}^\ell$ and a *prefix-free padding scheme*, that is, a mapping $\mathsf{pad} : \{0,1\}^* \to (\{0,1\}^b)^+$ such that $\mathsf{pad}(s)$ is not a prefix of $\mathsf{pad}(s')$ for all distinct $s, s' \in \{0,1\}^*$. The *prefix-free Merkle-Damgård construction* $\mathbf{pfMD}_{b,\ell,n}(f)$ proceeds as follows. On input $s \in \{0,1\}^*$, it computes $s_1 \| \cdots \| s_l = \mathsf{pad}(s)$ (with $s_i \in \{0,1\}^b$) and the chaining values $v_i := f(s_i, v_{i-1})$ for all $1 \leq i \leq l$, where v_0 is set to some initialization vector $IV \in \{0,1\}^\ell$. Finally, the construction outputs the first n bits of v_l. The following theorem easily[12] follows from Theorem 2 in [11].

Theorem 3. *Let $\mathbf{F} : \{0,1\}^{\ell+b} \to \{0,1\}^\ell$ be a puRF and let $\mathbf{O} : \{0,1\}^* \to \{0,1\}^n$ be a puRO. The construction $\mathbf{pfMD}_{b,\ell,n}(\cdot)$ is an (α', σ')-reduction of $\mathbf{O}$ to $\mathbf{F}$ with $\alpha'(k) = \mathcal{O}((l_{\max} \cdot k)^2 \cdot 2^{-\ell})$ and $\sigma'(k) = k$, where $l_{\max}$ is the maximal length (of the padding) of a message input to the construction.*

We note that there exists a trade-off between the number of queries and the length of the queries to the construction.[13] This issue is inevitable in all iterated constructions. We take now $\ell, b > 0$ as in the above explanation, and some $\epsilon > 0$. We set $m := \ell + b$, and we let $\mathcal{E}$ be an explicit (m, δ, ϵ)-input restricting family of functions. If given only a compression function $\mathbf{R}' : \{0,1\}^{n+\phi(n)} \to \{0,1\}^n$ (for $\phi(n)$ defined as in Section 3.3), we obtain a construction $\mathbf{pfMD}_{b,\ell,n}(\mathbf{C}_{\epsilon,m,\ell}^{\mathcal{E}}(\cdot))$ which replaces calls to the compression functions by calls to the construction $\mathbf{C}_{\epsilon,m,\ell}^{\mathcal{E}}(\cdot)$. We obtain the following theorem using Lemma 2.

---

[12] The only difference with respect to the original result is that we allow the chaining value to be larger than the output value, i.e. $\ell > n$.

[13] A possible distinguishing strategy would consist of doing few very long queries, instead of many queries, and security is guaranteed only as long as $l_{\max} \cdot k < 2^{\ell/2}$.

Theorem 4. *The construction* $\mathbf{pfMD}_{b,\ell,n}(\mathbf{C}^{\mathcal{E}}_{\epsilon,m,\ell}(\cdot))$ *is an* $(\overline{\alpha}, \overline{\sigma})$*-reduction of a puRO* $\mathbf{O} : \{0,1\}^* \to \{0,1\}^n$ *to* $\mathbf{R}'$*, where* $\overline{\alpha}(k) = \alpha((l_{\max} + 1)k) + \alpha'((\delta + 1)k)$ *and* $\overline{\sigma}(k) = \delta \cdot k$*, with* α *and* α' *as in Theorems 1 and 3, respectively.*

Setting $\ell > 2n(1 - \epsilon)$ leads to security for all distinguishers such that $l_{\max} \cdot k \leq \Theta(2^{n(1-\epsilon)})$. We finally note that our approach also works with all other known constructions of a public random oracle from a public compression function, as for example the constructions of [6,10], or other constructions discussed in [11].

Setting ϵ small enough provides high levels of security for properties like preimage resistance, second preimage resistance, multicollision resistance, or CTFP preimage resistance [16], and also excludes the existence of attacks for these properties (up to the obtained bound), that is, even with respect to adversaries which perform enough queries to find collisions for the component function $f : \{0,1\}^n \to \{0,1\}^n$.

References

1. Aiello, W., Venkatesan, R.: Foiling birthday attacks in length-doubling transformations. In: Maurer, U.M. (ed.) EUROCRYPT 1996. LNCS, vol. 1070, pp. 307–320. Springer, Heidelberg (1996)
2. An, J.H., Bellare, M.: Constructing VIL-MACs from FIL-MACs: Message authentication under weakened assumptions. In: Wiener, M.J. (ed.) CRYPTO 1999. LNCS, vol. 1666, pp. 252–269. Springer, Heidelberg (1999)
3. Baltz, A., Jäger, G., Srivastav, A., Ta-Shma, A.: An explicit construction of sparse asymmetric connectors. Manuscript (2003)
4. Bellare, M., Goldreich, O., Krawczyk, H.: Stateless evaluation of pseudorandom functions: Security beyond the birthday barrier. In: Wiener, M.J. (ed.) CRYPTO 1999. LNCS, vol. 1666, pp. 270–287. Springer, Heidelberg (1999)
5. Bellare, M., Kilian, J., Rogaway, P.: The security of the cipher block chaining message authentication code. Journal of Computer and System Sciences 61(3), 362–399 (2000)
6. Bellare, M., Ristenpart, T.: Multi-property-preserving hash domain extension and the EMD transform. In: Lai, X., Chen, K. (eds.) ASIACRYPT 2006. LNCS, vol. 4284, pp. 299–314. Springer, Heidelberg (2006)
7. Bellare, M., Rogaway, P.: Random oracles are practical: A paradigm for designing efficient protocols. In: CCS '93: Proceedings of the 1st ACM conference on Computer and Communications Security, pp. 62–73. ACM Press, New York (1993)
8. Canetti, R., Goldreich, O., Halevi, S.: The random oracle methodology, revisited. Journal of the ACM 51(4), 557–594 (2004)
9. Carter, J.L., Wegman, M.N.: Universal classes of hash functions. Journal of Computer and System Sciences 18(2), 143–154 (1979)
10. Chang, D., Lee, S., Nandi, M., Yung, M.: Indifferentiable security analysis of popular hash functions with prefix-free padding. In: Lai, X., Chen, K. (eds.) ASIACRYPT 2006. LNCS, vol. 4284, pp. 283–298. Springer, Heidelberg (2006)
11. Coron, J.-S., Dodis, Y., Malinaud, C., Puniya, P.: Merkle–Damgård revisited: How to construct a hash function. In: Shoup, V. (ed.) CRYPTO 2005. LNCS, vol. 3621, pp. 430–448. Springer, Heidelberg (2005)
12. Damgård, I.B.: A design principle for hash functions. In: Brassard, G. (ed.) CRYPTO 1989. LNCS, vol. 435, pp. 416–427. Springer, Heidelberg (1989)

13. Dodis, Y., Puniya, P.: On the relation between the ideal cipher and the random oracle models. In: Halevi, S., Rabin, T. (eds.) TCC 2006. LNCS, vol. 3876, pp. 184–206. Springer, Heidelberg (2006)

14. Hoch, J.J., Shamir, A.: Breaking the ICE — finding multicollisions in iterated concatenated and expanded (ICE) hash functions. In: Robshaw, M. (ed.) FSE 2006. LNCS, vol. 4047, pp. 179–194. Springer, Heidelberg (2006)

15. Joux, A.: Multicollisions in iterated hash functions. Application to cascaded constructions. In: Franklin, M. (ed.) CRYPTO 2004. LNCS, vol. 3152, pp. 306–316. Springer, Heidelberg (2004)

16. Kelsey, J., Kohno, T.: Herding hash functions and the Nostradamus attack. In: Vaudenay, S. (ed.) EUROCRYPT 2006. LNCS, vol. 4004, pp. 183–200. Springer, Heidelberg (2006)

17. Kelsey, J., Schneier, B.: Second preimages on n-bit hash functions for much less than 2^n work. In: Cramer, R.J.F. (ed.) EUROCRYPT 2005. LNCS, vol. 3494, pp. 474–490. Springer, Heidelberg (2005)

18. Lucks, S.: A failure-friendly design principle for hash functions. In: Roy, B. (ed.) ASIACRYPT 2005. LNCS, vol. 3788, pp. 474–494. Springer, Heidelberg (2005)

19. Maurer, U.: Indistinguishability of random systems. In: Knudsen, L.R. (ed.) EUROCRYPT 2002. LNCS, vol. 2332, pp. 110–132. Springer, Heidelberg (2002)

20. Maurer, U.: Abstract models of computation in cryptography. In: Smart, N.P. (ed.) Cryptography and Coding. LNCS, vol. 3796, pp. 1–12. Springer, Heidelberg (2005)

21. Maurer, U., Renner, R., Holenstein, C.: Indifferentiability, impossibility results on reductions, and applications to the random oracle methodology. In: Kilian, J. (ed.) TCC 2005. LNCS, vol. 3378, pp. 21–39. Springer, Heidelberg (2005)

22. Maurer, U., Sjödin, J.: Single-key AIL-MACs from any FIL-MAC. In: Caires, L., Italiano, G.F., Monteiro, L., Palamidessi, C., Yung, M. (eds.) ICALP 2005. LNCS, vol. 3580, pp. 472–484. Springer, Heidelberg (2005)

23. Maurer, U., Tessaro, S.: Full version of this paper. Cryptology ePrint Archive, Report 2007/229, http://eprint.iacr.org/

24. Merkle, R.C.: A certified digital signature. In: Brassard, G. (ed.) CRYPTO 1989. LNCS, vol. 435, pp. 218–238. Springer, Heidelberg (1989)

25. Moran, T., Shaltiel, R., Ta-Shma, A.: Non-interactive timestamping in the bounded storage model. In: Franklin, M. (ed.) CRYPTO 2004. LNCS, vol. 3152, pp. 460–476. Springer, Heidelberg (2004)

26. Shoup, V.: On fast and provably secure message authentication based on universal hashing. In: Koblitz, N. (ed.) CRYPTO 1996. LNCS, vol. 1109, pp. 313–328. Springer, Heidelberg (1996)

27. Shoup, V.: Lower bounds for discrete logarithms and related problems. In: Fumy, W. (ed.) EUROCRYPT 1997. LNCS, vol. 1233, pp. 256–266. Springer, Heidelberg (1997)

28. Ta-Shma, A., Umans, C., Zuckerman, D.: Lossless condensers, unbalanced expanders, and extractors. In: STOC '01: Proceedings of the 33rd Annual ACM Symposium on Theory of Computing, pp. 143–152. ACM Press, New York (2001)

Random Oracles and Auxiliary Input

Dominique Unruh*

Saarland University, Saarbrücken, Germany
unruh@cs.uni-sb.de

Abstract. We introduce a variant of the random oracle model where
oracle-dependent auxiliary input is allowed. In this setting, the adversary
gets an auxiliary input that can contain information about the random
oracle. Using simple examples we show that this model should be preferred over the classical variant where the auxiliary input is independent
of the random oracle.

In the presence of oracle-dependent auxiliary input, the most important proof technique in the random oracle model—lazy sampling—does
not apply directly. We present a theorem and a variant of the lazy sampling technique that allows one to perform proofs in the new model
almost as easily as in the old one.

As an application of our approach and to illustrate how existing proofs
can be adapted, we prove that RSA-OAEP is IND-CCA2 secure in the
random oracle model with oracle-dependent auxiliary input.

Keywords: Random oracles, auxiliary input, proof techniques, foundations.

1 Introduction

In [3] the following heuristic was advocated as a practical way to design cryptographic protocols:[1] To prove the security of a cryptographic scheme, one first
introduces a *random oracle* $\mathcal{O}$, i.e., a randomly chosen function to which all
parties including the adversary have access. One then proves the security of the
scheme that uses the random oracle and subsequently replaces the random oracle by a suitably chosen function (or family of functions) H. The *random oracle
heuristic* now states that if the scheme using $\mathcal{O}$ is secure, the scheme using H is
secure as well.

Unfortunately, a counter-example to this heuristic has been given in [6]. It
was shown that there exist public key encryption and signature schemes that are
secure in the random oracle model but lose their security when instantiated with
any function or family of functions. Nonetheless, the random oracle heuristic still
is an important design guideline for implementing cryptographic schemes.

Furthermore, [15] pointed out that zero-knowledge proofs in the random oracle
model can lose their deniability when instantiated with a fixed function. In contrast

---

* Part of this work was done while the author was at the IAKS, University of Karlsruhe,
  Germany.
[1] However, the basic idea seems to have already appeared earlier.

A. Menezes (Ed.): CRYPTO 2007, LNCS 4622, pp. 205–223, 2007.

to the result of [6], this happens even for natural protocols. However, [15] was able to give conditions under which this effect does not occur and gave a protocol that fulfilled these conditions.

Although the heuristic is known not to be sound in general, no practical scheme is known where it fails, and schemes that are proven to be secure using this heuristic tend to be simpler and more efficient than schemes that are shown to be secure in the standard model. As a consequence, schemes used in practise are often based on the random oracle heuristic, e.g., the RSA-OAEP encryption scheme, introduced in [2] and standardised in [17], is one of the most widely used public-key encryption schemes, and its security is based on the random oracle heuristic.

In the light of the results of [6] and [15], and of the practical importance of the random oracle heuristic, it is important to try and learn what the exact limitations of the heuristic are, and, if possible, give criteria to distinguish those protocols in the random oracle model that become insecure when instantiated due to those limitations, and those protocols where we can at least hope—if not prove—that their instantiations are secure. The augmented definition of zero-knowledge by [15] is an example of such a criterion.

In this paper, we uncover another such limitation of the random oracle world. We will see that there are natural schemes secure in the random oracle model that become insecure *with respect to auxiliary input* (or equivalently, with respect to nonuniform adversaries) when instantiated. As [15] did for the deniability, we give augmented definitions for the random oracle model with auxiliary input that allow one to distinguish protocols that fail upon instantiation from those that do not (at least not due to the abovementioned limitation).

Although such a result does not imply the soundness of the random oracle model, it helps to better understand which protocol can reasonably be expected to be secure when instantiated with a fixed function.

We will now investigate the problem of auxiliary input in the random oracle model in more detail. An important concept in cryptology is the *auxiliary input*. The auxiliary input is a string that is given to the adversary at the beginning of the execution of some cryptographic protocol. This string is usually chosen nonuniformly and depends on all protocol inputs. In other words, the auxiliary input models the possibility that the adversary has some additional knowledge concerning the situation at the beginning of the protocol. This additional knowledge may, e.g., represent information acquired in prior protocol runs. It turns out that in many cases the presence of an auxiliary input is an essential concept for proving secure sequential composition. Therefore, most modern cryptographic schemes are designed to be secure even in the presence of an auxiliary input (given that the underlying complexity assumptions hold against nonuniform adversaries).

However, when we try to combine these two concepts, the random oracle model and the auxiliary input, undesirable effects may occur. We will demonstrate this by studying the definitions of two simple security notions: one-wayness and collision-resistance. First, consider the notion of a one-way function. We construct a function $f := \mathcal{O}$ in the random oracle model and ask whether it is one-way. For this, we substitute $\mathcal{O}$ for f in the definition of one-wayness with respect to auxiliary input, and get the following definition: The function f is one-way if for any polynomial-time adversary A and any auxiliary input z, the following probability is negligible in the security parameter k:

$$P\big(x \xleftarrow{\$} \{0,1\}^k, \ x' \leftarrow A^{\mathcal{O}}(1^k, z, \mathcal{O}(x)) : \ \mathcal{O}(x') = \mathcal{O}(x)\big). \tag{1}$$

Here $\mathcal{O}$ is a randomly chosen function (with some given domain and range), and the adversary is given black-box access to $\mathcal{O}$. It is now easy to see that $f := \mathcal{O}$ is indeed secure in the above sense: The adversary can make at most a polynomial number of queries, and each query except $\mathcal{O}(x)$ returns a uniformly random image (exploiting this latter fact is later called the lazy sampling technique). From this fact one can conclude that the adversary must make an exponential number of queries to find a preimage of $\mathcal{O}(x)$, hence f is secure. The presence of the auxiliary input does not have noticeable impact on the proof. The random oracle heuristic now claims that $f := H$ is oneway for a sufficiently unstructured function H, even in the presence of auxiliary input. So far, nothing out of the ordinary has happened.

We now try to use the same approach for another security property: collision-resistance. Again, we set $f := \mathcal{O}$, and then collision-resistance of f means that for any polynomial-time adversary A and any auxiliary input z, the following probability is negligible:

$$P\big((x_1, x_2) \leftarrow A^{\mathcal{O}}(1^k, z) : \ x_1 \neq x_2 \text{ and } \mathcal{O}(x') = \mathcal{O}(x)\big). \tag{2}$$

This can again easily be proven using the lazy sampling technique: the answers to the adversary's queries are independent random values, and finding a collision requires two of these random values to be identical which happens only with negligible probability. Again, the auxiliary input does not help the adversary, since it does not contain any information on where a collision might be. We now use the random oracle heuristic, replace $\mathcal{O}$ by some sufficiently unstructured function H, so that $f = H$, and then claim that f is collision-resistant in the presence of an auxiliary input. But this of course is impossible, since the auxiliary input may simply contain a collision of H, since H is a fixed function.[2]

Hence, the random oracle heuristic should not be applied to collision-resistance. On the other hand, we would like to prove the one-wayness of $f := \mathcal{O}$ in the random oracle model. We hence need a stronger variant of the random oracle heuristic that does not allow one to prove the collision-resistance of f, but still allows one to prove its one-wayness. An inspection of our proof above reveals the mistake we made: In the random oracle model, the auxiliary input was chosen before the random oracle,

---

[2] If we replace $\mathcal{O}$ by a family of functions, i.e., some parameter i is chosen at the beginning of the protocol, and then a funcion H_i is used, then the problem described here does not occur. Unfortunately, one is not always free to use such a family of functions. On one hand, the index has in some way to be chosen, and we do not want to leave that choice to the corrupted parties. On the other hand, practical applications usually instantiate the random oracle using a fixed function like SHA-1 or SHA-256 [8].

so it could not contain a collision. After instantiation, the function H was fixed, so the auxiliary input did depend on H and therefore could provide a collision. The random oracle heuristic should hence be recast as follows in the case of auxiliary input: When a scheme is secure in the random oracle model with *oracle-dependent* auxiliary input, it is still secure after replacing the random oracle by a sufficiently unstructured fixed function H, even in the presence of auxiliary input.

It remains to clarify the formal meaning of oracle-dependent auxiliary input. Unfortunately, we cannot simply say: "for randomly chosen $\mathcal{O}$ and every z". At least, the semantics underlying constructions like (1) and (2) get highly nontrivial in this case. Fortunately, there is another possibility. By an oracle function z we mean a function that returns a string $z^{\mathcal{O}}$ for each possible value of the random oracle $\mathcal{O}$. So formally, z is simply a function that maps functions to strings. Then a scheme is called secure in the random oracle model with oracle-dependent auxiliary input if for any polynomial-time adversary A and for any oracle function z into strings of polynomial length (that may depend on k, of course), the adversary cannot break the scheme *even when given $z^{\mathcal{O}}$ as auxiliary input*.

As with the traditional random oracle model, the exact form these definitions take depends on the security notion under consideration. For example, the one-wayness and the collision-resistance of $f := \mathcal{O}$ take the following form: for any polynomial-time adversary A and any oracle function z into strings of polynomial-length, we have

$$P\big(x \xleftarrow{\$} \{0,1\}^k,\ x' \leftarrow A^{\mathcal{O}}(1^k, z^{\mathcal{O}}, \mathcal{O}(x)) :\ \mathcal{O}(x') = \mathcal{O}(x)\big) \qquad (3)$$

or

$$P\big((x_1, x_2) \leftarrow A^{\mathcal{O}}(1^k, z^{\mathcal{O}}) :\ x_1 \neq x_2 \text{ and } \mathcal{O}(x') = \mathcal{O}(x)\big), \qquad (4)$$

respectively. However, we can give a simple guideline on how to transform a security definition in the random oracle model with an oracle-*independent* auxiliary input z into a security definition with oracle-*dependent* auxiliary input. First, one quantifies over oracle functions z instead of strings z. And then one replaces all occurrences of the string z by $z^{\mathcal{O}}$.

It is now easy to see that (4) is not negligible: let $z^{\mathcal{O}}$ encode a collision x_1, x_2, and let the adversary output that collision. Since such a collision always exists (assuming a *length-reducing* f), this breaks the collision-resistance of f in the presence of oracle-dependent auxiliary input, as we would have expected.

On the other hand, we expect (3) to be negligible in the presence of oracle-dependent auxiliary input. However, it is not so easy to see whether there may not be some possibility to encode information about the random oracle in a string of polynomial length that allows one to find a preimage with non-negligible probability. Although one-wayness is one of the weakest conceivable security notions, proving its security with respect to oracle-dependent auxiliary input is quite difficult. (We encourage the reader to try and find an elementary proof for the one-wayness of f.[3]) The reason for this difficulty lies in the fact that it is not possible any more to apply the lazy sampling technique: given some

---

[3] We give a proof using the techniques from this paper in Lemma 10.

information $z^{\mathcal{O}}$ on the random oracle $\mathcal{O}$, the images under the random oracle are not independently nor uniformly distributed any more. We therefore need new techniques if we want to be able to cope with oracle-dependent auxiliary input and to prove more complex cryptographic schemes secure in this model. Such techniques will be presented in this paper.

On Nonuniform and Uniform Auxiliary Input. In this work, we always consider nonuniform auxiliary inputs, that is, the auxiliary input is not required to be the result of an efficient computation. This is the most common modelling of auxiliary input in the cryptographic community. However, it is also possible to consider uniform auxiliary inputs: In this case, the auxiliary input is not an arbitrary sequence of strings, but is instead the output of a *uniform* probabilistic algorithm. The main motivation of the auxiliary input, namely to model information gained from prior executions of cryptographic protocols on the same data, and thus to allow for composability, is preserved by this uniform approach. (See [11] for a detailed analysis.) The main disadvantage of the uniform approach is that definitions and proofs get more complicated due to the presence of another machine. This is why the nonuniform auxiliary input is more commonly used.

Applying the uniform approach to our setting, a uniform oracle-dependent auxiliary input would be the output of a polynomial-time oracle Turing machine Z with access to the random oracle. Since that Turing machine could only make a polynomial number of queries, using the lazy sampling technique would be easy: all positions of the random oracle that have not been queried by Z can be considered random.

However, if we use the random oracle heuristic to motivate the security of a protocol with respect to *uniform* auxiliary input, the result is incompatible with existing theorems and definitions in the *nonuniform* auxiliary input model. So to use the random oracle heuristic together with existing results, we either have to reprove all existing results for the uniform case, or we have to use *nonuniform* oracle-dependent auxiliary input. It is the latter approach we follow in this work.

Instantiating the Random Oracle with Keyed Families of Functions. Above, we showed that the random oracle is (unsurprisingly) not collision-resistant in the presence of auxiliary input. It follows that we may not instantiate the random oracle with a fixed function if we need collision-resistance. On the other hand, replacing the random-oracle by a keyed family of functions may be secure, since the auxiliary input cannot encode a collision for each function. We do not claim that it is necessary to use oracle-dependent auxiliary input when instantiating with families of functions. Rather, oracle-dependent auxiliary input provides a tool for distinguishing the cases where the use of a single function[4] is sufficient (e.g., in the case where we require only one-wayness) and where a

---

[4] Here, by a single function we mean that the function is not parametrised by a key that has to be known by all parties. However, the function may depend on the security parameter. Otherwise a property like collision-resistance trivially cannot be fulfilled by a single function, even against uniform adversaries. See also [16] in this context.

keyed family of functions is necessary (e.g., in the case that we require collision-resistance). Since instantiating with a single function is much simpler (e.g., we do not have to worry about who chooses the key), and is the usual practice in real-world protocols, examining random oracle based protocols with respect to oracle-dependent auxiliary input may give additional insight into when instantiation with single functions is permitted and when we have to use keyed families. Another disadvantage of using a family of functions is that we have to ensure that the key is honestly generated, which may introduce additional difficulties if no trusted party is available for this task.

Designing Special Protocols. An alternative to the approach in this paper would be to systematically construct or transform a protocol so that it is secure with respect to oracle-dependent auxiliary input (instead of verifying a given protocol). However, here the same arguments as in the previous paragraph apply. First, we might not be interested in a new protocol, but might want to examine the security of an existing protocol (that possibly even has already been implemented). Further, efficiency considerations might prevent the use of more elaborate constructions.

1.1   Our Results

We introduce and motivate the random oracle model with *oracle-dependent auxiliary input* (preceding section). In this model, the auxiliary input given to the adversary may depend on the random oracle.

In order to be able to prove security in the new model, we introduce a new variant of the lazy sampling technique that is applicable even in the presence of oracle-dependent auxiliary input. We show that one can replace the random-oracle $\mathcal{O}$ by a new random oracle $\mathcal{P}$ that is independent of the auxiliary input, except for a *presampling*. That is, a small fraction of the total random oracle $\mathcal{P}$ is fixed (and dependent on the auxiliary input), while all other images are chosen independently and uniformly at random (and in particular are independent of the auxiliary input). In this new setting, lazy sampling is possible again: an oracle query that is not in the presampled set is given a random answer.

This also gives some insight into why some schemes are secure and some fail in the presence of oracle-dependent auxiliary input: Intuitively the protocols that fail are those for which you can have a "reason for a failure" (e.g., a collision) contained in a few entries of the random oracle.

As a technical tool, we also formulate a *security amplification* technique: for many security notions, security with respect to nonuniform polynomial-time adversaries implies security with respect to nonuniform adversaries whose running time is bounded by some suitable superpolynomial function f. This technique is useful in the context of oracle-dependent auxiliary input, since some reduction proofs with presampling tend to introduce superpolynomial adversaries.

As an application of our techniques, we show that RSA-OAEP is IND-CCA2 secure in the random oracle model *with oracle-dependent auxiliary input*. Our proof closely follows the proof of [9] where the security of RSA-OAEP was shown in the classical random oracle model. This allows the reader to better compare the differences in the proof introduced by the oracle-dependent auxiliary input. However, we believe that the result does not only exemplify our techniques but is worthwhile in its own light: it gives the first evidence that RSA-OAEP as used in practical application (i.e., with the random oracle instantiated with a fixed function H), is secure *even in the presence of an auxiliary input.*

1.2 Related Work

In [19], the problem of composition of zero-knowledge proofs in the random-oracle model is investigated. It is shown that to guarantee sequential composition, oracle-dependent auxiliary input is necessary. Their definition of oracle-dependent auxiliary input is somewhat weaker than ours in that the machine z generating the auxiliary input is allowed only a polynomial number of queries to the random oracle (it is similar to uniform oracle-dependent auxiliary input in that respect). They give protocols that are secure with respect to that notion. It would be interesting to know whether the techniques developed here allow to show their protocols to be secure even with respect to our stronger notion of oracle-dependent auxiliary input.

In [10], it was shown that a random *permutation* is one-way with respect to oracle-dependent auxiliary input. They showed that the advantage of the adversary is in $2^{-\Omega(k)}$ which is essentially the same bound as we achieve for random *functions* in Section 3. However, their proof is specific to the property of one-wayness and does not generalise to our setting. According to [10], a similar result was shown for random *functions* in [12]. However, their proofs apply only to the one-wayness of the random oracle, while our results imply that many more cryptographic properties of the random oracle are preserved in the presence of oracle-dependent auxiliary input.

In [14,5,4,7], unconditional security proofs in the bounded-storage model were investigated. In this model, one assumes that the adversary is computationally unlimited, but that it may only store a limited amount of data. One assumes that at the beginning of the protocol some large source of randomness (e.g., a random oracle) is available to all parties. The security of the protocol then roughly hinges on the following idea: The honest parties store some (small) random part of the source. Since the adversary does not know which part has been chosen, and since it may not store the whole source, with high probability the honest parties will find some part of the random source they both have information about, but that is unknown to the adversary. To prove the security in this model, it is crucial to show that the adversary cannot compress the source in a manner that contains enough information to break the protocol. This is very similar to the scenario investigated here, since the oracle-dependent auxiliary input can be seen as compressed information on the random oracle. Our results differ from those in the bounded-storage model in two ways: first, our results cover a more general case, since we consider the effect of auxiliary input on arbitrary protocols, while in the bounded-storage model a single protocol is analysed that is specially designed to extract information from the random source that cannot be extracted given only a part of the source. On the other hand, precisely due to the specialised nature of the protocols, the bounds achieved in the bounded-storage model are better than those presented here. In particular, there are protocols in the bounded storage model that are secure given a random source of polynomial size [7], while our results are—at least with the present bounds—only useful if the domain of the random oracle has superpolynomial size (cf. the exact bounds given by Theorem 2). It would be interesting to know whether our techniques can be used in the context of the bounded-storage model, and to what extent the techniques developed in the bounded-storage model can be applied to improve our bounds.

1.3   Further Applications

Besides the application described above, namely to be able to use the random-oracle heuristic in the case of auxiliary input, our main result (the lazy sampling technique) may also be useful in other situations.

In [10] it was shown that a random permutation is one-way in the presence of oracle-dependent auxiliary input. This was the main ingredient for several lower bounds on black-box constructions using one-way permutations. Using our techniques, we might find lower bounds on black-box constructions based on other cryptographic primitives: namely, we would show that the random oracle (or a protocol using the random oracle) has a given security property X even in the presence of oracle-dependent auxiliary input. Then using techniques from [10], lower bounds on black-box constructions based on cryptographic primitives fulfilling X might be derived.

1.4   Organisation

In Section 1 we introduce and motivate the concept of oracle-dependent auxiliary input. In Section 2 we present the main result of this paper: a theorem that allows one to use the lazy sampling technique even in the presence of oracle-dependent auxiliary input. In Section 3 we give a simple example to show how to use the lazy sampling technique. In Section 4 we present the security amplification technique. This technique allows one to use superpolynomial adversaries in reduction proofs, which sometimes is needed when using the lazy sampling technique. In Section 5 we prove that RSA-OAEP is IND-CCA2 secure in the random oracle model with oracle-dependent auxiliary input. Details and proofs left out in this paper are given in the full version [18].

1.5   Notation

For random variables A and B, we denote the Shannon-entropy of A by $H(A)$, and the conditional entropy of A given B by $H(A|B)$. The statistical distance between A and B is denoted $\Delta(A; B)$. The operator log means the logarithm base 2. The variable k always denotes the security parameter. In asymptotic statements of theorems or definitions, some variables implicitly depend on the security parameter k. These variables are then listed at the end of the theorem/definition. We call a nonnegative function in k negligible, if it lies in $k^{-\omega(1)}$. We call a nonnegative function non-negligible if it is not negligible.

Let $\mathcal{O}$ always denote the random oracle. Let $Domain$ be the domain and $Range$ the range of the random oracle, i.e., $\mathcal{O}$ is a uniformly random function from $Domain \rightarrow Range$. In an asymptotic setting, $\mathcal{O}$, $Domain$ and $Range$ implicitly depend on the security parameter k. In this case we always assume $\#Domain$ and $\#Range$ to grow at least exponentially in k.

An *oracle function* g *into* X is a mapping from $Domain \rightarrow Range$ into X. We write the image of some function $\mathcal{O}$ under g as $g^{\mathcal{O}}$.

An *assignment* S is a list $S = (x_1 \rightarrow y_1, \ldots, x_n \rightarrow y_n)$ with $x_i \in Domain$ and $y_i \in Range$ and with $x_i \neq x_j$ for $i \neq j$. The length of S is n. We call y_i the image of x_i under S. We write $x \in S$ if $x_i = x$ for some i. The image $\operatorname{im} S$ is defined as $\operatorname{im} S = \{y_1, \ldots, y_n\}$.

2 Lazy Sampling with Auxiliary Input

The main result of this paper is the following theorem which guarantees that we can replace a random oracle with oracle-dependent auxiliary input by a new random oracle that is independent of the auxiliary input with the exception of some fraction of its domain (which is *presampled*). In order to formulate the theorem, we first need to state what exactly we mean by an oracle with presampling:

Definition 1 (Random oracle with presampling). *Let* $S = (x_1 \rightarrow y_1, \ldots, x_n \rightarrow y_n)$ *be an assignment. Then the* random oracle $\mathcal{P}$ with presampling S *is defined as follows:*

When queried $x \in Domain$ *with* $x = x_i$ *for some* $i \in \{1, \ldots, n\}$*, the oracle returns* y_i*. If* x *has already been queried, the same answer is given again. Otherwise, a uniformly random element* y *is chosen from* $Range$ *and returned.*

We can now state the main theorem.

Theorem 2 (Lazy sampling with auxiliary input). *Let* $f \geq 1$ *and* $q \geq 0$ *be integers. Let* z *be an oracle function with finite range* Z *and* $p := \log \#Z$.

Then there is an oracle function S*, such that* $S^{\mathcal{O}}$ *is an assignment of length at most* f*, so that the following holds: For any probabilistic oracle Turing machine* A *that makes at most* q *queries to the random oracle, it is*

$$\Delta\big(A^{\mathcal{O}}(z^{\mathcal{O}}); \ A^{\mathcal{P}}(z^{\mathcal{O}})\big) \leq \sqrt{\frac{pq}{2f}}$$

where $\mathcal{P}$ *is the random oracle with presampling* $S^{\mathcal{O}}$.

Before presenting the actual proof, we give a short sketch that is intended to serve as a guide through the rest of the proof. To ease comparison with the details given later, we provide forward references to the lemmas of the actual proof.

For any i, let J_i be the maximum amount of information that a sequence of i queries to the random oracle $\mathcal{O}$ gives about the auxiliary input $z^{\mathcal{O}}$. Since $|z| = p$, $J_i \leq p$ for all i. Let F_i be the sequence of i queries that achieves this bound, that is, the mutual information between $z^{\mathcal{O}}$ and the oracle's answers to F_i is J_i.

Assume that the queries F_i have already been performed. Let G be a sequence of q queries. Then the answers to the queries F_i and G together contain at most J_{q+i} bits of information about z. Thus the answers to G contain at most $J_{q+i} - J_i$ bits of information about z beyond what is already known from the answers to F_i.

Consider the quantities $J_0, J_q, J_{2q}, \ldots, J_{f+q}$ (assuming that q divides f). Since $J_0 \geq 0$ and $J_{f+q} \leq p$, there must be some $f' \leq f$ such that the $J_{f'+q} - J_{f'} \leq \frac{pq}{f}$. Thus, given the answers to $F := F_{f'}$, any sequence G of q queries reveals at most $\frac{pq}{f}$ bits about the auxiliary input z. In other words, the answers to G are almost independent of z (assuming that $\frac{pq}{f}$ is sufficiently small). Thus, if we fix the oracle $\mathcal{P}$ to match the answers to F, but choose $\mathcal{P}$ independently of z everywhere else, with q queries we cannot distinguish between $\mathcal{P}$ and the original oracle $\mathcal{O}$. This gives Theorem 2 (except for the concrete bound $\sqrt{pq/2f}$).

In reality, however, the queries performed by A are adaptive, i.e., they depend on z and on the answer to prior queries. So we cannot talk about a fixed sequence G of queries made by A. To overcome this problem, we introduce the concept of an adaptive list (Definitions 3 and 4), which is a generalisation of a sequence of queries where the queries are allowed to be adaptive. When considering adaptive lists, it does not make immediate sense to speak about the mutual information between the answers to an adaptive list G and the auxiliary input $z^{\mathcal{O}}$. In Definition 5 we therefore introduce quantities $J(G)$ and $J(G|F)$ denoting the information that the answers to the adaptive list G contains about $z^{\mathcal{O}}$ (given the answers to the adaptive list F in the case of $J(G|F)$). For this quantity, we show that $J(F) \leq p$ (Lemma 7) and give a chain rule for the information contained in the concatenation of adaptive lists (Lemma 6). Then we can construct the sequence F as in the proof sketch above (Lemma 8). However, F is now an adaptive list. Finally, Theorem 2 is proven (page 216) by showing that the adversary A can be considered as an adaptive list G of length q, and therefore cannot distinguish the answers to queries outside F from uniform randomness. For convenience, in Corollary 9 we formulate an asymptotic version of Theorem 2.

We now give the details of the proof, broken down to several lemmas. First we have to define the concept of an adaptive list. To capture the possibility of adaptive queries, an adaptive list is formally just a deterministic oracle Turing machine. An adaptive list of length n makes n queries to the oracle and outputs an assignment containing the queries and the results of these queries. To be able to talk about the concatenation of adaptive lists, we slightly extend this idea. An adaptive list takes an auxiliary input z, but also an assignment X. This assignment can be thought of as the queries made by an adaptive list executed earlier. So in a concatenation of two adaptive lists, the queries of the second adaptive list can depend on the results of the queries made by the first adaptive list. For definitional convenience, an adaptive list does not only output its queries, but also the queries received as input. An adaptive list expecting a

queries as input and then making $b - a$ queries, we call an $a \to b$ adaptive list. We require that an adaptive list never repeats a query. Note that an adaptive list is indeed a generalisation of a non-adaptive sequence of queries: a sequence $(x_1, \ldots, x_n)$ corresponds to the $0 \to n$ adaptive list querying the positions x_1 to x_n and returning the results.

Definition 3 (Adaptive list). *Let $\#Domain \geq b \geq a \geq 0$. An $a \to b$ adaptive list M is defined as a deterministic oracle Turing machine that takes an assignment $X = (x_1 \to y_1, \ldots, x_a \to y_a)$ and a string $z \in \Sigma^*$ as input and satisfies the following properties*

- *$M = M(X, z)$ does not query the oracle at positions $x_1, \ldots, x_a$.*
- *M never queries the oracle twice at the same position.*
- *M queries the oracle exactly $b - a$ times.*
- *Let $x'_1, \ldots, x'_{b-a}$ be the positions of the oracle calls made by M (in that order). Let $y'_i := \mathcal{O}(x'_i)$ be the corresponding oracle answers.*
- *Then M outputs the assignment $(x_1 \to y_1, \ldots, x_a \to y_a, x'_1 \to y'_1, \ldots, x'_{b-a} \to y'_{b-a})$.*

We can now define simple operations on adaptive lists. The length of an adaptive list is the number of queries it makes, and the composition of two adaptive lists is the adaptive list that first queries the first list, and then executes the second, which gets the queries made by the first as input.

Definition 4 (Operations on adaptive lists). *Let M be an $a \to b$ adaptive list. Then the* length *$|M|$ is defined as $|M| := b - a$.*

Let N be an $a \to b$ adaptive list, and M some $b \to c$ adaptive list. Then the composition *$M \circ N$ is defined as the oracle Turing machine that upon input of an assignment X and a string $z \in \Sigma^*$ outputs $M^{\mathcal{O}}(N^{\mathcal{O}}(X, z), z)$.*

Obviously, $M \circ N$ is an $a \to c$ adaptive list, and $|M \circ N| = |M| + |N|$.

We can now define the quantity $J(M|N)$ for adaptive lists M, N. Intuitively, $J(M|N)$ denotes the information that the queries made by M (when executed after N) contain about the auxiliary input $z^{\mathcal{O}}$ beyond what is already known from the queries made by N. Since the results to the queries made by M should be uniformly random if they are independent of $z^{\mathcal{O}}$, we define $J(M|N)$ as the quantity by which the conditional entropy of M's queries given N's queries and $z^{\mathcal{O}}$ is lower than the hypothetical value of $|M| \cdot \log \#Range$.

Definition 5 (Information of an adaptive list). *Let N be some $0 \to b$ adaptive list, and M some $b \to c$ adaptive list. Let further $\mathcal{O}$ be the random oracle and z a random variable (where z does not need to be independent of $\mathcal{O}$). Then the* information *$J(M|N)$ is defined by*

$$J(M|N) := |M| \cdot \log \#Range - H(M \circ N^{\mathcal{O}}(z)|N^{\mathcal{O}}(z), z).$$

(Note that $J(M|N)$ implicitly depends on the joint distribution of $\mathcal{O}$ and z.)
We write short $J(M)$ for $J(M|\varnothing)$ where $\varnothing$ is the adaptive list making no queries.

We now give two simple properties of the information $J(M|N)$: a chain rule and an upper bound in terms of the auxiliary input's length.

Lemma 6 (Chain rule for the information). *Let N be some $0 \to b$ adaptive list, M_2 some $b \to c$ adaptive list, and M_1 some $c \to d$ adaptive list. Then*

$$J(M_1 \circ M_2|N) \geq J(M_1|M_2 \circ N) + J(M_2|N).$$

Lemma 7 (Bounds for the information). *Let z be a random variable with finite range Z and $p := \log \#Z$. Let F be some $0 \to b$ adaptive list. Then $J(F) \leq p$.*

The proofs of these lemmas as well as of the subsequent ones are given in the full version [18].

Let $J_i := \max_F J(F)$ where F ranges over all adaptive lists of length $|F| = i$. Choose F_i such that $J(F_i) = J_i$. Consider the quantities $J_0, J_q, J_{2q}, \ldots, J_{f+q}$ (assuming that q divides f). Since $J_0 \geq 0$ and $J_{f+q} \leq p$ by Lemma 7, there must be some $f' \leq f$ such that $J_{f'+q} - J_{f'} \leq \frac{pq}{f} =: \varepsilon$. Defining $F := F_{f'}$ we get $J(G|F) \leq J(G \circ F_{f'}) - J(F_{f'}) \leq J_{q+f'} - J_{f'} \leq \varepsilon$ by Lemma 6.

By definition of $J(G|F)$, this implies that the results of the queries made by G are only ε away from the maximum possible entropy $|G| \cdot \log \#Range$. This implies using a result from [13] that the statistical distance between those query-results and the uniform distribution is bounded by $\sqrt{\varepsilon/2}$, even when given the results of the queries made by F and the auxiliary input $z^{\mathcal{O}}$. This is formally captured by the following lemma which is the core of the proof of Theorem 2.

Lemma 8 (The adaptive list F). *Let $f, q \geq 1$ be integers. Let z be a random variable with finite range Z (z may depend on the random oracle $\mathcal{O}$) and $p := \log \#Z$. Let U_n denote the uniform distributions on n-tuples over $\#Range$ (independent of z and $\mathcal{O}$).*

Then there is an adaptive list F with $|F| \leq f$, such that for any $|F| \to \min\{|F| + q, \#Domain\}$ adaptive list G, it is

$$\Delta\left(\nabla G \circ F^{\mathcal{O}}(z), F^{\mathcal{O}}(z), z; \ U_{|G|}, F^{\mathcal{O}}(z), z\right) \leq \sqrt{\frac{pq}{2f}}.$$

Here $\nabla G \circ F$ denotes the oracle Turing machine that behaves as $G \circ F$ but only outputs the oracle answers that G got (instead of also outputting G's input and G's queries). More formally, if $G \circ F^{\mathcal{O}}(z) = (x_1 \to y_1, \ldots, x_{|F|+|G|} \to y_{|F|+|G|})$, we have $\nabla G \circ F^{\mathcal{O}}(z) = (y_{|F|+1}, \ldots, y_{|F|+|G|})$.

Using Lemma 8, proving the main Theorem 2 is easy. For some adversary A let $\mu := \Delta\left(A^{\mathcal{O}}(z^{\mathcal{O}}); A^P(z^{\mathcal{O}})\right)$. By fixing the worst-case random-tape, we can make the adversary A deterministic. Then A's output depends only on its input $z^{\mathcal{O}}$ and the answers to its oracle queries. So if we let A just output the queries it made, the statistical distance μ does not diminish. Further, if we give the presampled queries $S^{\mathcal{O}}$ as an additional input to A, we can assume A to make exactly q distinct queries, and not to query any $x \in S^{\mathcal{O}}$. But then A fulfils the definition of an adaptive list, so by Lemma 8 we have $\mu \leq \sqrt{\frac{pq}{2f}}$, which proves Theorem 2.

We give the full details of the proof in the full version [18].

An interesting question is whether the bound $\sqrt{pq/2f}$ on the statistical distance Δ achieved by Theorem 2 is tight. In particular, the bound falls only sublinearly with f, while we were unable to find a counterexample where Δ did not fall exponentially with f. So a tighter bound may be possible. However, this would need to use new proof techniques, since the approach in this paper uses an averaging argument that will at best give a bound that falls polynomially in f (cf. the computation of $J_{f'+q} - J_{f'}$ below Lemma 7 above.)

Finally, for convenience we state an asymptotic version of Theorem 2 that hides the exact bounds achieved there:

Corollary 9 (Lazy sampling with auxiliary input, asymptotic version).
For any superpolynomial function f and any polynomial q and oracle function z into strings of polynomial length, there is an oracle function S, such that $S^{\mathcal{O}}$ is an assignment of length at most f, so that for any probabilistic oracle Turing machine A making at most q queries, the following random variables are statistically indistinguishable:

$$A^{\mathcal{O}}(1^k, z^{\mathcal{O}}) \quad and \quad A^{\mathcal{P}}(1^k, z^{\mathcal{O}}).$$

Here $\mathcal{P}$ is the random oracle with presampling $S^{\mathcal{O}}$.

(In this corollary, $\mathcal{O}$, z, and S depend implicitly on the security parameter k.)

Proof. Immediate from Theorem 2.   □

3   Example: One-Wayness of the Random Oracle

To give a first impression on how the lazy sampling technique is used in the random oracle model with oracle-dependent auxiliary input, we show a very simple result: If we let $f := \mathcal{O}$, then f is a one-way function.

In the full version [18], as a second example we show that $f := \mathcal{O}$ is given-preimage collision-resistant.

Lemma 10 (The random oracle is one-way). *Let $g := \mathcal{O}$ where $\mathcal{O}$ denotes the random oracle. Then g is a one-way function in the random oracle model with oracle-dependent auxiliary input.*

More formally, for any probabilistic polynomial-time oracle Turing machine A and any oracle function z into strings of polynomial length, the following probability is negligible (in k):

$$\mathrm{Adv}_A := P\big(x \overset{\$}{\leftarrow} Domain,\ x' \leftarrow A^{\mathcal{O}}(1^k, z^{\mathcal{O}}, \mathcal{O}(x)):\ \mathcal{O}(x') = \mathcal{O}(x)\big)$$

(In this lemma, $\mathcal{O}$, Domain, f, and z depend implicitly on the security parameter k.)

We present this proof in some detail, to illustrate how Theorem 2 or Corollary 9 can be used. Since these steps are almost identical in most situations, knowledge of this proof facilitates understanding of the proofs given later on.

Proof. Let $f := \min\{\sqrt{\#Range}, \sqrt{\#Domain}\}$. Let $\tilde{A}$ be the oracle Turing machine that chooses a random x from $Domain_k$, then let $A(1^k, z^{\mathcal{O}}, \mathcal{O}(x))$ choose x', and outputs 1 if and only if $\mathcal{O}(x') = \mathcal{O}(x)$. Then $\mathrm{Adv}_A = P(\tilde{A}^{\mathcal{O}}(1^k, z^{\mathcal{O}}) = 1)$.

Since A is polynomial-time, $\tilde{A}$ makes only a polynomial number of queries, so Corollary 9 applies to $\tilde{A}$, hence $\tilde{A}^{\mathcal{O}}(1^k, z^{\mathcal{O}})$ and $\tilde{A}^{\mathcal{P}}(1^k, z^{\mathcal{O}})$ are statistically indistinguishable (where $\mathcal{P}$ is the random oracle with presampling $S^{\mathcal{O}}$, and S is as in Corollary 9). Then consider the following game:

Game 1: $x \xleftarrow{\$} Domain, \ x' \leftarrow A^{\mathcal{P}}(1^k, z^{\mathcal{O}}, \mathcal{P}(x)) : \ \mathcal{P}(x') = \mathcal{P}(x)$.

We call the probability that the last expression evaluates to true (i.e., that $\mathcal{P}(x') = \mathcal{P}(x)$) the advantage Adv_1 of the game. Since Adv_1 is the probability that $\tilde{A}^{\mathcal{P}}(1^k, z^{\mathcal{O}})$ outputs 1, $|\mathrm{Adv}_A - \mathrm{Adv}_1|$ is negligible.

(This step probably occurs at the beginning of virtually all proofs that use Theorem 2 or Corollary 9. We are now in the situation that with at most f exceptions, the oracle query $\mathcal{P}(x)$ returns a fresh random value, and can use standard techniques based on lazy sampling.)

We now modify A in the following way resulting in a machine A_2: A_2 expects an assignment S as an additional argument. Whenever A would query the random oracle $\mathcal{P}$ with a value x, A_2 first checks if $x \in S$. If so, A returns the image of x under S. Otherwise, A_2 queries its oracle. Then consider the following game:

Game 2: $x \xleftarrow{\$} Domain, \ y \leftarrow \mathcal{P}(x), \ x' \leftarrow A_2^{\mathcal{P}}(1^k, z^{\mathcal{O}}, y, S^{\mathcal{O}}) : \ y = \mathcal{P}(x')$

Obviously, $\mathrm{Adv}_1 = \mathrm{Adv}_2$.

Since for some $x \notin S^{\mathcal{O}}$ (which happens with probability at least $1 - f/\#Domain$), the oracle $\mathcal{P}$ returns a random $y \in \#Range$, the probability that $y \in \mathrm{im}\, S^{\mathcal{O}}$ is at most $f/\#Domain + f/\#Range$. Furthermore, if $x' \in S^{\mathcal{O}}$ but $y \notin \mathrm{im}\, S^{\mathcal{O}}$, the predicate $y = \mathcal{P}(x')$ will be false.

So $|\mathrm{Adv}_2 - \mathrm{Adv}_3| \leq P(y \in \mathrm{im}\, S^{\mathcal{O}})$ is negligible for the following game 3:

Game 3: $x \xleftarrow{\$} Domain, \ y \leftarrow \mathcal{P}(x), \ x' \leftarrow A_2^{\mathcal{P}}(1^k, z^{\mathcal{O}}, y, S^{\mathcal{O}}) :$
$\qquad\qquad$ if $(x' \in S^{\mathcal{O}})$ then false else $y = \mathcal{P}(x')$.

Note that in game 3, A_2 never queries $\mathcal{P}$ at a position in $S^{\mathcal{O}}$. Furthermore, the query $\mathcal{P}(x')$ is only executed if $x' \notin S$. So $\mathcal{P}$ is only queried at a position in $S^{\mathcal{O}}$, if $x \in S^{\mathcal{O}}$, which has probability at most $f/\#Domain$. But when queried at positions outside $S^{\mathcal{O}}$, $\mathcal{P}$ behaves like a normal random oracle (i.e., without presampling). We can therefore replace the oracle $\mathcal{P}$ by a random oracle $\mathcal{R}$ (independent of $\mathcal{O}$):

Game 4: $x \xleftarrow{\$} Domain, \ y \leftarrow \mathcal{R}(x), \ x' \leftarrow A_2^{\mathcal{R}}(1^k, z^{\mathcal{O}}, y, S^{\mathcal{O}}) :$
$\qquad\qquad$ if $(x' \in S^{\mathcal{O}})$ then false else $y = \mathcal{R}(x')$.

Then $|\mathrm{Adv}_3 - \mathrm{Adv}_4| \leq P(x \in S^{\mathcal{O}})$ is negligible.

(We have now succeeded in completely separating the oracle from the auxiliary input; $\mathcal{R}$ is independent from $(z^{\mathcal{O}}, S^{\mathcal{O}})$. From here on, the proof is a standard proof of one-wayness of the random oracle. Note however, that $S^{\mathcal{O}}$ has a length that may be superpolynomial, so A_2 is not polynomially bounded any more. In our case, this does not pose a problem, since we only use the fact that A_2 uses a polynomial number of queries. In proof that additionally need computational assumptions, one might need additional tools which we present in Section 4.)

Consider the following game:

Game 5: $x \xleftarrow{\$} Domain$, $y \xleftarrow{\$} Range$, $x' \leftarrow A_2^{\mathcal{R}}(1^k, z^{\mathcal{O}}, y, S^{\mathcal{O}})$:

if $(x' \in S^{\mathcal{O}})$ then false else $y = \mathcal{R}(x')$.

Since A was polynomially bounded, there is a polynomial q bounding the number of oracle queries of A_2. The probability that A_2 queries $\mathcal{R}$ at position x is therefore at most $q/\#Domain$ (since x is randomly chosen and never used). Furthermore, game 4 and 5 only differ if A_2 queries $\mathcal{R}$ at position x. So $|\text{Adv}_4 - \text{Adv}_5| \leq q/\#Domain$ is negligible.

Since A_2 makes at most q queries, the probability that one of these returns y is at most $q/\#Domain$. If x' returns a value x' it has not queried before, the probability that $y = \mathcal{R}(x')$ is at most $1/\#Domain$. So $\text{Adv}_5 \leq (q-1)/\#Domain$ is negligible.

Collecting the bounds shown so far, we see that Adv_A is negligible.   □

In the preceding proof, we have only verified that the advantage of the adversary is negligible. By using Theorem 2 instead of Corollary 9 and computing the exact bounds, we even get $\text{Adv}_A \in 2^{-\Omega(k)}$ which is essentially the same bound as given in [12] and [10].

4   Security Amplification

When using a random oracle with presampling, reduction proofs sometimes run into situations where the adversary gets the presampling $S^{\mathcal{O}}$ as an input. Unfortunately, this presampling is usually of superpolynomial size, so the resulting adversary is not polynomial-time any more and a reduction to complexity assumptions relative to polynomial-time adversaries is bound to fail. (E.g., in the proof of Lemma 10 the adversary A_2 was not polynomial-time in the security parameter any more. In that case however, this did not matter since we only used the polynomial bound on the number of queries made by A_2.) An example of a situation where superpolynomial adversaries occur, and do pose a problem, is the proof that RSA-OAEP is secure with respect to oracle-dependent auxiliary input, cf. Section 5. One possibility is simply to assume a stronger security notion; in the case of RSA-OAEP one could use, e.g., the RSA-assumption against quasi-polynomial adversaries.

Fortunately, there is another way which allows to use standard assumptions (i.e., with respect to polynomial-time adversaries) in many cases. We show that for some kinds of security notions, security against polynomial-time adversaries implies security against adversaries with f-bounded runtime, where f is a suitably chosen *superpolynomial* function. Using this fact we can finish our reduction proof: Corollary 9 guarantees that for any superpolynomial function f', we can replace the random oracle by a random oracle with presampling of length f'. We then choose f' to be the largest function such that all adversaries constructed in our proof are still f-bounded. Such an f' is still superpolynomial, so Corollary 9 applies. On the other hand, the resulting adversaries are efficient enough for the reduction to go through. This proof method is applied in Section 5 to show the security of RSA-OAEP.

Instead of giving a general proof of our *security amplification* technique, we give here a proof for the security notion of partial-domain one-wayness (Definition 11). The proof can easily be adapted to other security notions (in particular, our proof does not exploit how the advantage Adv is defined for this particular notion). In the full version [18] we give a more general characterisation of the security notions for which security amplification is possible.[5]

Definition 11 (Partial-domain one-way). *A family of 1-1 functions f_{pk} : $B \times C \to D$ is* partial-domain one-way, *if for any nonuniform polynomial-time adversary A, the following advantage is negligible:*

$$\mathrm{Adv}_{A,k} := P\big(pk \leftarrow K(1^k), \ (s,t) \xleftarrow{\$} B \times C, \ y \leftarrow f_{pk}(s,t), \ s' \leftarrow A(1^k, y) : \ s = s'\big).$$

Here K denotes the index generation algorithm for the family f_{pk} of functions. Partial-domain one-way against f-bounded adversaries for some function f is defined analogously.

(In this definition, $B, C,$ and D depend implicitly on the security parameter k.)

Lemma 12 (Security amplification for partial-domain one-wayness). *Let the family f_{pk} be partial-domain one-way (against polynomial-time nonuniform adversaries). Then there exists a superpolynomial function f such that f_{pk} is partial-domain one-way against f-bounded nonuniform adversaries.*

Proof. For $n \in \mathbb{N}$ let $\mu_n(k) := \max_{|A| \leq n}(\mathrm{Adv}_{A,k})$ where A goes over all circuits of size at most n. Assume there was a polynomial p with integer coefficients (an integer polynomial for short) such that $\mu_{p(k)}(k)$ is not negligible in k. Then there is a nonuniform adversary A consisting of circuits A_k with $|A_k| \leq p(k)$ such that $\mathrm{Adv}_{A,k} \geq \mu_{p(k)}(k)$ is non-negligible. Since A is polynomial-time, this contradicts the assumption that the f_{pk} are partial-domain one-way. Hence $\mu_p := \mu_{p(k)}(k)$ is negligible for all integer polynomials p.

We say that a function μ asymptotically dominates a function ν if for all sufficiently large k we have $\mu(k) \geq \nu(k)$. [1] proves that for any countable set S of negligible functions, there is a negligible function μ^* that asymptotically dominates all $\mu \in S$.

---

[5] This includes one-wayness, partial-domain one-wayness, IND-CPA, IND-CCA2, black-box stand-alone security of function evaluations, UC (where the amplification concerns the running time of the environment), black-box zero-knowledge, arguments, black-box arguments of knowledge.

Therefore, there is a negligible function μ^*, that asymptotically dominates μ_p for every integer polynomial p.

Let $f(k) := \max\{p \in \mathbb{N} : \mu_p(k) \leq \mu^*(k)\}$. Then $\mu_{f^\times(k)}(k) \leq \mu^*(k)$ is negligible. So for any nonuniform f-bounded adversary A the advantage $\mathrm{Adv}_{A,k}$ is negligible. Furthermore, we can show that f is superpolynomial. Assume this is not the case. Then there an integer polynomial p such that $p > f$ infinitely often. But then $\mu_p > \mu^*$ holds infinitely often, in contradiction to the choice of μ^* (by definition of f). Thus f is superpolynomial.   □

5   OAEP Encryption

In [9] it was shown that RSA-OAEP (introduced by [2]) is secure in the random oracle model under the RSA-assumption. However, their proof only covers the case that no auxiliary input is given (or at least that the auxiliary input is not oracle-dependent). In this section, we extend this result to encompass the case of oracle-dependent auxiliary input. On one hand, this gives a nontrivial example of the application of the lazy sampling technique in combination with the security amplification technique. On the other hand, this result is important in its own light, since it gives evidence that RSA-OAEP may be secure with respect to an auxiliary input, even when the random oracle has been instantiated with a fixed function.

To read this section, it is helpful to have at least basic knowledge of the OAEP construction and its proof from [9]. We recommend [9] as an introduction.

Theorem 13 (OAEP is secure with respect to oracle-dependent auxiliary input). *Let f_{pk} be a family of partial-domain one-way trapdoor 1-1 functions (with the property, that the elements of the domain of f_{pk} consist of two components each of superlogarithmic length).*

Then the OAEP encryption scheme based on f_{pk} is IND-CCA2 secure in the random oracle model with oracle-dependent auxiliary input.

This theorem implies that RSA-OAEP is IND-CCA2 secure under the RSA-assumption with respect to oracle-dependent auxiliary input, since in [9] it is shown that the RSA family of functions is partial-domain one-way.

At this point, we only describe on a high level, in what points our proof differs from the proof in [9]. In the full version [18], we reproduce the full proof of [9] and highlight our changes for comparison.

In [9], the proof has roughly the following outer form: First, the IND-CCA2 game is formulated for the special case of the OAEP encryption scheme. Then the game is rewritten in a series of small changes, to finally yield a plaintext extractor. If the first game had a non-negligible success probability (i.e., the OAEP encryption scheme was not IND-CCA2 secure), the plaintext extractor had, for some random ciphertext $f_{pk}(s, t)$, a non-negligible probability of outputting s. This breaks the assumption that f_{pk} is partial-domain one-way.

Our proof starts with the same game, except that the adversary now has access to an oracle-dependent auxiliary input $z^{\mathcal{O}}$. Then we can use Corollary 9 to replace the random oracle $\mathcal{O}$ by a random oracle $\mathcal{P}$ with presampling $S^{\mathcal{O}}$ of a yet to determine superpolynomial subexponential length $\mathfrak{f}$ (similar to the first step in the proof of Lemma 10).[6] In this new situation, for randomly chosen $x \in Domain$, with overwhelming probability, the oracle response $\mathcal{P}(x)$ is uniformly distributed. Using this fact, most of the rewriting steps in the sequence of games are the same as in [9], sometimes with slightly larger errors to account for the possibility of randomly choosing an $x \in S^{\mathcal{O}}$. Only in the construction of the plaintext extractor additional care has to be taken. Here the original argument uses that the answer to an oracle query can be assumed to be random if the adversary has not yet queried it. From this they conclude any ciphertext the decryption oracle would accept can also be decrypted by encrypting and comparing all oracles queries that have been made by the adversary so far. This does not hold any more since the auxiliary input $z^{\mathcal{O}}$ can supply additional information on the presampled queries $S^{\mathcal{O}}$. We thus have to change the plaintext extractor not only to encrypt all oracle queries but also all presampled queries $S^{\mathcal{O}}$. Therefore the plaintext extractor is not polynomial-time anymore, but instead a nonuniform machine with running time $p(\mathfrak{f})$ for some polynomial p. We consequently do not directly obtain a contradiction to the partial-domain one-wayness, since therefore the plaintext extractor would have to be polynomial-time.

However, we can use the security amplification technique. By Lemma 12, there is a superpolynomial function f' such that f_{pk} is partial-domain one-way even against nonuniform f'-bounded adversaries. By choosing $\mathfrak{f}$ small enough (but still superpolynomial), it is $p(\mathfrak{f}) \leq f'$, so the plaintext extractor is f'-bounded, and the fact that the plaintext extractor returns s for some $f_{pk}(s,t)$ with non-negligible probability is a contradiction.

6   Open Questions

We have shown how to apply the lazy sampling technique to the case of oracle-dependent auxiliary input. Going further, the following open problems come to mind:

- Polynomial presampling: In Corollary 9, we require the length f of the pre-sampling to be superpolynomial. This makes reduction proofs more difficult, in particular it necessitates the use of the security amplification technique. It would be preferable to be able to use a polynomial length f (in this case, the length would of course have to depend on the length of auxiliary input and the number of queries made by the adversary).
- The random oracle as considered here is only a specific example of the class of random objects that are given as oracle to the parties. Other examples include random permutations (with or without access to the inverse), the generic group model, ideal ciphers, or just random oracles with a skewed distribution. When using these to motivate security results, the same arguments apply as in the case of random oracles, and oracle-dependent auxiliary

---

[6] The actual proof uses Theorem 2, but the asymptotic version is sufficient.

input should be considered. It is then necessary to extend the lazy sampling technique to these constructions as well.

In the full version [18], we discuss these open questions in slightly more detail.

Acknowledgements. We thank Michael Backes, Dennis Hofheinz, Yuval Ishai, Jörn Müller-Quade, Hoeteck Wee, and Jürg Wullschleger for valuable discussions. We further thank the anonymous referees for helpful comments.

References

1. Bellare, M.: A note on negligible functions. Journal of Cryptology 15(4), 271–284 (2002)
2. Bellare, M., Rogaway, P.: Optimal asymmetric encryption—how to encrypt with RSA. In: De Santis, A. (ed.) EUROCRYPT 1994. LNCS, vol. 950, pp. 92–111. Springer, Heidelberg (1995)
3. Bellare, M., Rogaway, P.: Random oracles are practical: A paradigm for designing efficient protocols. In: Proceedings of CCS 1993, pp. 62–73 (1993)
4. Cachin, C., Crépeau, C., Marcil, J.: Oblivious transfer with a memory-bounded receiver. In: Proceedings of STOC 2002, pp. 493–502.
5. Cachin, C., Maurer, U.: Unconditional security against memory-bounded adversaries. In: Kaliski Jr., B.S. (ed.) CRYPTO 1997. LNCS, vol. 1294, pp. 292–306. Springer, Heidelberg (1997)
6. Canetti, R., Goldreich, O., Halevi, S.: The random oracle methodology, revisited. In: Proceedings of STOC 1998, pp. 209–218 (1998)
7. Dziembowski, S., Maurer, U.: Tight security proofs for the bounded-storage model. In: Proceedings of STOC 2002, pp. 341–350 (2002)
8. Federal Information Processing Standards Publications. FIBS PUB 180-2: Secure Hash Standard (SHS) (August 2002)
9. Fujisaki, E., Okamoto, T., Pointcheval, D., Stern, J.: RSA-OAEP is secure under the RSA assumption. Journal of Cryptology 17(2), 81–104 (2004)
10. Gennaro, R., Gertner, Y., Katz, J., Trevisan, L.: Bounds on the efficiency of generic cryptographic constructions. SIAM J Computing 35(1), 217–246 (2005)
11. Goldreich, O.: A uniform-complexity treatment of encryption and zero-knowledge. Journal of Cryptology 6(1), 21–53 (1993)
12. Impagliazzo, R.: Very strong one-way functions and pseudo-random generators exist relative to a random oracle. Manuscript (1996)
13. Kullback, S.: A lower bound for discrimination information in terms of variation (corresp.). IEEE Transactions on Information Theory 13(1), 126–127 (1967)
14. Maurer, U.: Conditionally-perfect secrecy and a provably-secure randomized cipher. Journal of Cryptology 5(1), 53–66 (1992)
15. Pass, R.: On deniability in the common reference string and random oracle model. In: Boneh, D. (ed.) CRYPTO 2003. LNCS, vol. 2729, pp. 316–337. Springer, Heidelberg (2003)
16. Rogaway, P.: Formalizing human ignorance: Collision-resistant hashing without the keys. In: Nguyen, P.Q. (ed.) VIETCRYPT 2006. LNCS, vol. 4341, pp. 221–228. Springer, Heidelberg (2006)
17. RSA Laboratories. PKCS #1: RSA Cryptography Standard, Version 2.1, 2002.
18. Unruh, D.: Random oracles and auxiliary input. IACR ePrint, 2007/168. Full version of this paper
19. Wee, H.: Zero knowledge in the random oracle model, revisited. Manuscript (2006)

Security-Amplifying Combiners for Collision-Resistant Hash Functions

Marc Fischlin and Anja Lehmann

Darmstadt University of Technology, Germany
www.minicrypt.de

Abstract. The classical combiner $\mathsf{Comb}_{\mathrm{class}}^{H_0,H_1}(M) = H_0(M)\|H_1(M)$ for hash functions H_0, H_1 provides collision-resistance as long as at least one of the two underlying hash functions is secure. This statement is complemented by the multi-collision attack of Joux (Crypto 2004) for iterated hash functions H_0, H_1 with n-bit outputs. He shows that one can break the classical combiner in $\frac{n}{2} \cdot T_0 + T_1$ steps if one can find collisions for H_0 and H_1 in time T_0 and T_1, respectively. Here we address the question if there are *security-amplifying* combiners where the security of the building blocks increases the security of the combined hash function, thus beating the bound of Joux. We discuss that one can indeed have such combiners and, somewhat surprisingly in light of results of Nandi and Stinson (ePrint 2004) and of Hoch and Shamir (FSE 2006), our solution is essentially as efficient as the classical combiner.

1 Introduction

A hash function combiner [6] takes two hash functions H_0 and H_1 and combines them into a single, failure-resistant hash function. That is, collision-resistance of the combined function is granted, given that at least one of the starting hash functions H_0, H_1 is secure. A classical example of a secure combiner is $\mathsf{Comb}_{\mathrm{class}}^{H_0,H_1}(M) = H_0(M)\|H_1(M)$, concatenating the outputs of the two hash functions. For this combiner any collision $M \neq M'$ immediately gives collisions for both hash functions H_0 and H_1.

From a more quantitative viewpoint, the classical combiner provides the following security guarantee: If breaking H_0 and H_1 requires T_0 and T_1 steps, respectively, finding a collision for the classical combiner takes at least $T_0 + T_1$ steps. This almost matches an upper bound by Joux [8], showing that for Merkle-Damgård hash functions H_0, H_1 with n-bit outputs the classical combiner can be broken in $\frac{n}{2} \cdot T_0 + T_1$ steps. This means that if the security level of each hash function is degraded only moderately through a new attack method, e.g., from 2^{80} to 2^{60}, then the classical combiner, too, merely warrants a reduced security level of $T_0 + T_1 = 2 \cdot 2^{60}$. Ideally, we would like to have a better security bound for combiners and such moderate degradations, going beyond the $T_0 + T_1$ limit and the bound due to Joux.

Our Results. Here we introduce the notion of security-amplifying combiners for collision-resistant hash functions. Such combiners guarantee a security level $\alpha \cdot$

A. Menezes (Ed.): CRYPTO 2007, LNCS 4622, pp. 224–243, 2007.

$(T_0 + T_1)$ for some $\alpha > 1$ and, in a sense, are therefore stronger than the sum of their components. Note that the classical combiner (and similar proposals) are *not* security amplifying according to the previous discussion, indicating that constructing such security-amplifying combiners is far from trivial.

We next discuss how to achieve security amplification. Consider two Merkle-Damgård hash functions H_0, H_1 (given by compression functions f_0, f_1) and the classical combiner, but limited to input messages $M = m_0 || \ldots || m_{t-1}$ of $t < \frac{n}{4}$ blocks exactly:

$$\mathsf{Comb}_{\mathrm{amp},t}^{H_0,H_1}(M) = H_0(m_0 || \ldots || m_{t-1}) \, || \, H_1(m_0 || \ldots || m_{t-1})$$

This is clearly a secure combiner in the traditional sense, guaranteeing collision resistance if at least one of both hash functions is collision-resistant. But we show that it is even a security-amplifying combiner, assuming that the underlying compression functions behave ideally. More precisely, we consider an attack model in which the compression functions f_0, f_1 are given by random functions, but where the adversary against the combiner can use subroutines $\mathcal{C}_0, \mathcal{C}_1$ to generate collisions for the corresponding compression function. Intuitively, these collision finder oracles implement the best known strategy to find collisions, and each time the adversary calls $\mathcal{C}_b$ to get a collision for f_b, we charge T_b steps. The adversary's task is now to turn such collisions derived through $\mathcal{C}_0, \mathcal{C}_1$ into one against the combiner.

We note that the adversary against the combiner in our model is quite powerful. For each query to the collision finders the adversary can significantly bias the outcome, e.g., by presetting parts of the colliding messages. To give further support of the significance of our model, we show that we can implement the attack of Joux on the classical combiner $\mathsf{Comb}_{\mathrm{class}}$ in our model. We can also realize similar attacks for more advanced combiners like $\mathsf{Comb}^{H_0,H_1}(M) = H_0(M) || H_1(H_0(M) \oplus M)$.

Our main result is to certify the security amplification of our combiner $\mathsf{Comb}_{\mathrm{amp},t}$. The proof is basically split into two parts: one covering general statements about our model (such as pre-image resistance, even in presence of the collision finders), and the other part uses the basic facts to prove our specific combiner $\mathsf{Comb}_{\mathrm{amp},t}$ to be security-amplifying. In our security proof we show that calling each collision finder $\mathcal{C}_0, \mathcal{C}_1$ only polynomially many times does not help to find a collision for $\mathsf{Comb}_{\mathrm{amp},t}$. Therefore, successful attacks on the combiner require more than $\mathrm{poly}(n) \cdot (T_0 + T_1)$ steps.

Viewed from a different perspective we can think of our result as a supplementary lower bound to the attack of Joux. His attack breaks the classical combiner in $\frac{n}{2} \cdot T_0 + T_1$ steps if the hash functions allow to process $t \geq \frac{n}{2}$ message blocks. Our result indicates that restricting the input to $t < \frac{n}{4}$ many blocks suffices to make the combiner security-amplifying and to overcome the bound by Joux. The situation for t in between $\frac{n}{4}$ and $\frac{n}{2}$ remains open.

Finally, recall that our proposal at this point only allows to hash messages of $t < \frac{n}{4}$ blocks. To extend the combiner to handle arbitrarily long messages one can use hash trees in a straightforward way (with our combiner placed at every node of the tree). Since finding collisions in such hash trees requires to come up with collisions in one of the nodes, our security amplification result carries over instantaneously. For messages of k blocks the classical combiner takes about $2k$ applications of the compression functions, compared to roughly $\frac{t}{t-1} \cdot 2k$ applications for our tree-based combiner (but coming with the stronger security amplification guarantee).

Limitations of the Model. Our hash combiner guarantees security amplification in an idealized world where the underlying compression functions behave like random functions. In this model only generic attacks on the hash function are allowed, in the sense that the adversary cannot take advantage of weaknesses of the compression functions beyond the ability to generate collisions (albeit the collision finders are quite flexible). It remains open if similar results can be obtained in a non-idealized setting at all.

Currently, our collision finders return two values mapping to the same compression function output. A recent work of Yu and Wang [14], however, shows that very weak compression functions as in MD4 may allow K-multi-collision attacks, where one is able to find K instead of 2 simultaneous collisions for the compression functions. We expect our results to transfer to this case, when restriciting the number of message blocks further to $t < \frac{n}{4\log_2 K}$. This will be addressed in the full version of the paper.

Related Work. The idea of cryptographic combiners has been considered explicitly by Herzberg [6]. Among others, he analyzes the classical combiner $\mathsf{Comb}_{\mathrm{class}}$ concatenating the hash function values. As for hash function combiners, Boneh and Boyen [1] and subsequently Pietrzak [12] show that collision-resistant combiners cannot do better than the classical combiner in terms of the length, i.e., the output length of a secure combiner must essentially equal the sum of the output lengths of the hash functions (as in our construction).

Interestingly, the idea of security amplification for cryptographic combiners already appears implicitly in Yao's work [13]. He shows that the existence of weak one-way functions —where inversion may succeed with probability $1 - 1/\mathrm{poly}(n)$— can be turned into strong one-way functions where inversion almost surely fails. The construction can be viewed as a security-amplifying self-combiner for one-way functions. See also [5] for improvements and [9] for related results.

Other relevant works are the upper bounds of Nandi and Stinson [11] and of Hoch and Shamir [7]. They extend the attack of Joux to arbitrary combiners for iterated hash functions, where each message block is possibly processed via the compression function more than once but at most a constant number of times. They also transfer their results to tree-based constructions. However, in their model the output of one compression function must not serve as an input to the other compression function, thus disallowing mixes of intermediate hash values. By this, the hash-tree based extension of our combiner circumvents their bounds.

Finally we remark that, in a concurrent work, Canetti et al. [3] also consider amplification of collision resistance. In contrast to our idealized setting they use a complexity-theoretic approach.

Organization. We start by defining our model and security amplifying combiners (Section 2). Next, in Section 3, we discuss that the classical combiner and similar proposals are not security amplifying. Section 4 present some general conclusions in our model. The main result appears in Section 5 and the proof of this result is given in Section 6. Some proofs in this version have been moved to the Appendix.

2 Preliminaries

2.1 Hash Functions and Combiners

A hash function $\mathcal{H} = (\mathsf{HKGen}, \mathsf{H})$ is a pair of efficient algorithms such that HKGen for input 1^n returns (the description of) a hash function H, and H for input H and $M \in \{0,1\}^*$ deterministically outputs a digest $H(M)$. The hash function is called *collision-resistant* if for any efficient algorithm $\mathcal{A}$ the probability that for $H \leftarrow \mathsf{HKGen}(1^n)$ and $(M, M') \leftarrow \mathcal{A}(H)$ we have $M \neq M'$ but $\mathsf{H}(H, M) = \mathsf{H}(H, M')$, is negligible (as a function of n).

Definition 1. *A hash function combiner* Comb *for hash functions* $\mathcal{H}_0, \mathcal{H}_1$ *is an efficient deterministic algorithm such that, for input* $H_0 \leftarrow \mathsf{HKGen}_0(1^n)$, $H_1 \leftarrow \mathsf{HKGen}_1(1^n)$ *and* $M \in \{0,1\}^*$, *it returns a digest* $\mathsf{Comb}(H_0, H_1, M)$. *In addition, the pair* $(\mathsf{CKGen}, \mathsf{Comb})$, *where* $\mathsf{CKGen}(1^n)$ *generates* $H_0 \leftarrow \mathsf{HKGen}_0(1^n)$ *and* $H_1 \leftarrow \mathsf{HKGen}_1(1^n)$ *and outputs* (H_0, H_1), *is a collision-resistant hash function as long as* $\mathcal{H}_0$ *or* $\mathcal{H}_1$ *is collision-resistant.*

The popular Merkle-Damgård construction [10,4] of a hash function takes any collision-resistant compression function $f : \{0,1\}^{l+n} \to \{0,1\}^n$ and an initial vector IV. To compute a digest one divides (and possibly pads) the message $M = m_0 m_1 \ldots m_{k-1}$ into blocks m_i of l bits and computes the digest $H(M) = \mathrm{iv}_k$ as

$$\mathrm{iv}_0 = \mathrm{IV}, \qquad \mathrm{iv}_{i+1} = f(\mathrm{iv}_i, m_i) \qquad \text{for } i = 0, 1, \ldots, k-1.$$

In this case the description of the hash function simply consists of the pair (f, IV). We note that, in order to make this construction collision-resistant for messages of arbitrary length, one still needs to apply the compression function once more to the bit length of the message.

In the *idealized* Merkle-Damgård construction we assume that the compression function f behaves like a random function (drawn from the set of all functions mapping $(l + n)$-bit strings to n-bit strings). In particular, if an algorithm now gets as input the description of such an idealized MD-hash function then it is understood that this algorithms gets IV as input string and oracle access to the random function f. This holds also for a combiner Comb of such idealized MD hash function, i.e., Comb gets oracle access to f_0, f_1 and receives the strings $\mathrm{IV}_0, \mathrm{IV}_1$ as input. We then often write $\mathsf{Comb}^{H_0, H_1}(\cdot)$ instead of $\mathsf{Comb}^{f_0, f_1}(\mathrm{IV}_0, \mathrm{IV}_1, \cdot)$. *We emphasize that the combiner may assemble a solution from the compression functions and the initial vectors which is not necessarily an iterated hash function.*

2.2 Our Model

To analyze the security amplification of a combiner for two idealized MD hash functions (f_0, IV_0) and (f_1, IV_1) we consider an adversary $\mathcal{A}$ with oracle access to f_0, f_1 and input $\mathrm{IV}_0, \mathrm{IV}_1$. The task of this algorithm is to find a collision for the combiner. Since finding collisions for the random compression function directly is restricted to the birthday attack, we allow $\mathcal{A}$ oracle access to two

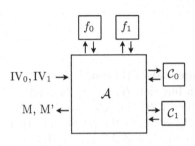

collision finder oracles $\mathcal{C}_0, \mathcal{C}_1$ generating collisions for each compression function (both oracles themselves have access to f_0, f_1). These collision finders can be viewed as the best known algorithm to generate collision for the compression function. See Figure 1. In its most simple form algorithm $\mathcal{A}$ can query the collision finder $\mathcal{C}_b$ by forwarding values $\mathrm{iv}_b, \mathrm{iv}'_b$ and getting a collision (m_b, m'_b) with $f_b(\mathrm{iv}_b, m_b) = f_b(\mathrm{iv}'_b, m'_b)$ from $\mathcal{C}_b$. More gener-

Fig. 1. Attack Model

ally, the adversary may want to influence the colliding messages or enforce dependencies between the initial values $\mathrm{iv}_b, \mathrm{iv}'_b$ and the messages m_b, m'_b. To model such advanced collision finding strategies we allow the adversary to pass (the description of) a circuit $C_b : \{0,1\}^i \to \{0,1\}^{l+n}$ (possibly containing f_0- and f_1-gates) to $\mathcal{C}_b$ instead of $\mathrm{iv}_b, \mathrm{iv}'_b$ only. The collision finder then applies an internal stateful source $S = S(C_b)$ to continuously generate i-bit strings $s \leftarrow S$ and successively provides each s as input to the circuit C_b. See Figure 2(a).[1]

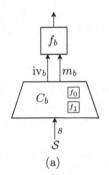

SAMPLES$_b(C_b)$ contains all tested pairs $(C_b(s), f_b(C_b(s)))$ in $\mathcal{C}_b$'s collision search for input circuit C_b

CVAL$_b$ contains all collisions returned by collision finder $\mathcal{C}_b$

FVAL$_b$ contains all pairs $(x, f_b(x))$ appearing in direct f_b-box queries of $\mathcal{A}$ or in an evaluation of a circuit C_b

(a)   (b)

Fig. 2. Operation of collision finder $\mathcal{C}_b$ (a), Sets of function values (b)

For the circuit's output $(\mathrm{iv}_b, m_b) = C_b(s)$ to the next input value s the finder computes $f_b(\mathrm{iv}_b, m_b)$ and checks if for some previously computed value (iv'_b, m'_b) a

---

[1] The source S can be thought of the collision finder's strategy to generate collisions for the input circuit, and is possibly even known by $\mathcal{A}$. Since we will later quantify over all collision finders we do not specify this distribution; the reader may for now think of S sequentially outputting the values $0, 1, 2, \ldots$ in binary.

collision $f_b(\text{iv}_b, m_b) = f_b(\text{iv}'_b, m'_b)$ occurs. If so, $\mathcal{C}_b$ immediately stops and outputs the collision $((\text{iv}_b, m_b), f_b(\text{iv}_b, m_b), s)$ and $((\text{iv}'_b, m'_b), f_b(\text{iv}'_b, m'_b), s')$. Otherwise it stores the new triple $((\text{iv}_b, m_b), f_b(\text{iv}_b, m_b), s)$ and continues its computations. If $\mathcal{C}_b$ does not find a collision among all i-bit inputs s to the circuit it returns $\perp$. We assume that the adversary implicitly gets to know all consulted input values s, gathered in an ordered set $\text{SVAL}(\mathcal{C}_b)$. Note that we leave it essentially up to the adversary and his choice for $\mathcal{C}_b$ to minimize the likelihood of undefined outputs or trivial collisions (i.e., for the same pre-image).

2.3 Lucky Collisions

The collision finders should be the only possibility to derive collisions, i.e., we exclude accidental collisions (say, $\mathcal{A}$ ignoring the collision finders and finding an f_0-collision by querying the f_0-oracle many times). To capture such *lucky collisions* we assume that each answer $((\text{iv}_b, m_b), f_b(\text{iv}_b, m_b), s)$, $((\text{iv}'_b, m'_b), f_b(\text{iv}'_b, m'_b), s')$ of $\mathcal{C}_b$ is augmented by all pre-image/image pairs (x, y) of f_0- and f_1-gate evaluations in the circuit computations during the search. We stress that this excludes all samples $(C_b(s), f_b(C_b(s)))$ which the collision finder probes to find the collision, unless the sample also appears in one of the circuit evaluations (see also the discussion below).

For a query C_b to $\mathcal{C}_b$ we denote the set of the pre-image/image pairs returned to $\mathcal{A}$ by $\text{FVAL}_b^{\text{cf}}(C_b)$ and by $\text{FVAL}_b^{\text{cf}}$ we denote the union of $\text{FVAL}_b^{\text{cf}}(C_b)$ over all queries C_b made to $\mathcal{C}_b$ during $\mathcal{A}$'s computation. Here we assume that the set $\text{FVAL}_b^{\text{cf}}$ is updated immediately after each function gate evaluation during a circuit evaluation. Similarly, $\text{FVAL}_b^{\text{box}}$ stands for the pre-image/image pairs generated by $\mathcal{A}$ as queries and answers to the f_b-box directly. We now set FVAL as the union of $\text{FVAL}_b^{\text{cf}}$ and $\text{FVAL}_b^{\text{box}}$ for both $b = 0, 1$.

Definition 2 (Lucky Collision). *A pair (x, x') is called a* lucky collision *if for an execution we have $x \neq x'$ and $(x, y), (x', y) \in \text{FVAL}$ for some y.*

In the definition below $\mathcal{A}$ will not be considered successful if a lucky collision occurs during an execution. It therefore lies in $\mathcal{A}$'s responsibility to prevent lucky collisions when querying f-boxes or the collision finders.

For notational convenience we collect the pre-image/image pairs of collisions generated by the collision-finders in the set CVAL, which is the union of all answers $\text{CVAL}_b(C_b)$ of collision-finder $\mathcal{C}_b$ for query C_b, over all queries C_b and $b = 0, 1$. We also let $\text{SAMPLES}_b(C_b)$ denote all samples $(C_b(s), f_b(C_b(s)))$ which the collision finder $\mathcal{C}_b$ collects to find a collision for query C_b, and SAMPLES stands for the union over all $\text{SAMPLES}_b(C_b)$ for all queries C_b and $b \in \{0, 1\}$. Clearly, $\text{CVAL}_b(C_b) \subseteq \text{SAMPLES}_b(C_b)$. An informal overview about the sets is given in Figure 2(b).

We remark that we do not include the pairs $(C_b(s), f_b(C_b(s)))$ which the collision finder probes in FVAL_b (unless they appear in the circuit's evaluations). This is in order to not punish the adversary for the collision finder's search and strengthens the model, as lucky collisions become less likely. However, for an answer of the collision finder the adversary $\mathcal{A}$ can re-compute all or some of those values by browsing through the ordered set SVAL(C_b), containing all inspected s-values, and submitting $C_b(s)$ to the f_b-oracle. This value is then added to the set FVAL$_b$, of course.

2.4 Security Amplification

As for the costs of each oracle call to collision finder C_b we charge the adversary $\mathcal{A}$ a pre-determined number T_b of steps for each call (e.g., $T_b = 2^{n/2}$ if C_b implements the birthday attack, ignoring the fact that the collision finder may even fail with some probability in this case). We do not charge the adversary for other steps than these calls. In the definition below we make no restriction on the number of calls to the collision finders, yet one might often want to limit this number in some non-trivial way, e.g., for our main result we assume that the adversary makes at most a polynomial number of calls.

Definition 3. *A hash function combiner* Comb *for idealized Merkle-Damgård hash functions* $\mathcal{H}_0, \mathcal{H}_1$ *is called* $\alpha(n)$-*security amplifying if for any oracles* C_0, C_1 *(with running times* $T_0(n)$ *and* $T_1(n)$, *respectively) and any algorithm* $\mathcal{A}$ *making at most* $\alpha(n) \cdot (T_0(n) + \cdot T_1(n))$ *steps we have*

$$\mathrm{Prob}\left[\mathbf{Exp}^{amp,\mathsf{Comb}}_{\mathcal{A},\mathcal{H}_0,\mathcal{H}_1,C_0,C_1}(n) = 1 \right] \approx 0$$

where

> *Experiment* $\mathbf{Exp}^{amp,\mathsf{Comb}}_{\mathcal{A},\mathcal{H}_0,\mathcal{H}_1,C_0,C_1}(n)$:
> *initialize* $(f_0, IV_0) \leftarrow \mathsf{HKGen}_0(1^n)$, $(f_1, IV_1) \leftarrow \mathsf{HKGen}_1(1^n)$
> *let* $(M, M') \leftarrow \mathcal{A}^{f_0,f_1,C_0,C_1}(IV_0, IV_1)$
> *output* 1 *iff*
> > $M \neq M'$, *and*
> > $\mathsf{Comb}^{f_0,f_1}(IV_0, IV_1, M) = \mathsf{Comb}^{f_0,f_1}(IV_0, IV_1, M')$, *and*
> > *no lucky collisions during* $\mathcal{A}$'s *computation occured.*

The combiner is called security amplifying *if it is* $\alpha(n)$-*security amplifying for some function* $\alpha(n)$ *with* $\alpha(n) > 1$ *for all sufficiently large* n's.

Our definition allows $\alpha(n)$ to converge to 1 rapidly, e.g., $\alpha(n) = 1 + 2^{-n}$. We do not exclude such cases explicitly, but merely remark that, as long as $T_0(n)$ and $T_1(n)$ are polynomially related and the combiner is security-amplifying, one can always find a suitable function $\alpha(n)$ bounded away from 1 by a polynomial fraction. For simplicity we have defined compression functions f_0, f_1 of equal output length n (which is also the security parameter). We remark that all our definitions and results remain valid for different output lengths n_0, n_1 by considering $n = \min\{n_0, n_1\}$.

3 Warming Up: Attack on the Classical Combiner

In this section, to get accustomed to our model, we first present the attack of Joux on the classical combiner, showing that this one is not security amplifying (even though it is a secure combiner in the traditional sense). This also proves that finding such security-amplifying is far from trivial. Recall that the classical combiner is given by

$$\mathsf{Comb}_{\mathrm{class}}^{H_0 H_1}(M) := H_0(M)\|H_1(M)$$

for idealized Merkle-Damgård hash functions. Obviously this combiner is collision-resistant as long as at least one of the hash functions has this property. Yet, it does not have the desired security-amplification property, because an adversary $\mathcal{A}$ can use the strategy of Joux [8] to find a collision rapidly. The idea is to build a multi-collision set of size $2^{\frac{n}{2}}$ for H_0 by calling $\mathcal{C}_0$ only $\frac{n}{2}$ times, and then to let $\mathcal{C}_1$ search for a pair among those messages in the multi-collision set which also constitutes a collision under H_1.

Adversary $\mathcal{A}^{f_0,f_1,\mathcal{C}_0,\mathcal{C}_1}(\mathrm{IV}_0,\mathrm{IV}_1)$:

for $i = 0, 1, \ldots, k := \frac{n}{2} - 1$:
    let $C_{0,i} : \{0,1\}^l \to \{0,1\}^{l+n}$ be the circuit $C_{0,i}(s) = (\mathrm{iv}_{0,i}, s)$, where $\mathrm{iv}_{0,0} = \mathrm{IV}_0$
    get $((\mathrm{iv}_{0,i}, m_i), y_i, s), ((\mathrm{iv}_{0,i}, m_i'), y_i, s') \leftarrow \mathcal{C}_0(C_{0,i})$
        where $m_i \neq m_i'$ by the choice of $C_{0,i}$
    set $\mathrm{iv}_{0,i+1} = y_i$
end of for

construct circuit $C_1 : \{0,1\}^{n/2} \to \{0,1\}^{l+n}$, containing all received collisions (m_i, m_i')
from the first stage, as follows:
    for $i = 0, 1, \ldots, k = \frac{n}{2} - 1$:
        for the i-th input bit s_i let $\hat{m}_i = m_i$ if $s_i = 0$, and $\hat{m}_i = m_i'$ otherwise
        except for the last round, compute $\mathrm{iv}_{1,i+1} = f_1(\mathrm{iv}_{1,i}, \hat{m}_i)$, where $\mathrm{iv}_{1,0} = \mathrm{IV}_1$
    end of for
    let the circuit output $(\mathrm{iv}_{1,k}, \hat{m}_k)$

get $((\mathrm{iv}_{1,k}, \hat{m}_k), y_k, s), ((\mathrm{iv}_{1,k}', \hat{m}_k'), y_k, s') \leftarrow \mathcal{C}_1(C_1)$

reconstruct the successful combination M, M' of $\mathcal{C}_1$ by using the values s, s'
for the pairs (m_i, m_i') as above, and output M, M'

First, the collision finder $\mathcal{C}_0$ is called $\frac{n}{2}$ times by the adversary to derive $\frac{n}{2}$ pairs of colliding message blocks (m_i, m_i') where $f_0(\mathrm{iv}_{0,i}, m_i) = f_0(\mathrm{iv}_{0,i}, m_i')$ for $i = 0, 1, \ldots, k$. Since the circuit $C_{0,i}$ passed to $\mathcal{C}_0$ does not evaluate the functions f_0, f_1, no lucky collision can occur in this stage. The query to collision finder $\mathcal{C}_1$ then requires $\frac{n}{2}$ compression function evaluations in the circuit C_1 for each input $s \in \{0,1\}^{n/2}$, which selects one of the $2^{\frac{n}{2}}$ multi-collisions derived from $\mathcal{C}_0$'s answers. Yet, for each common prefix of the s-values the same function evaluations are repeated, and the set $\mathrm{FVAL}_1^{\mathrm{cf}}$ therefore contains at most $2^{\frac{n}{2}}$ pre-image/image pairs (x, y) from the circuit evaluations. This implies that the probability for a lucky collision is at most $\frac{1}{2}$.

On the other hand, given that no collision in FVAL_1 occurs, all circuit outputs are distinct and the set of probed values of the collision finder is at least $2^{\frac{n}{2}}$. But then, $\mathcal{C}_0$ will find a collision among the values with constant probability (which is roughly equal to $1 - e^{-1/2}$ for the Euler constant e). Hence, the adversary succeeds with constant probability, taking only $\frac{n}{2} \cdot T_0(n) + T_1(n)$ steps. This implies that the classical combiner is *not* security amplifying, because no appropriate function $\alpha(n) > 1$ exists.

Our model allows to implement attacks on more sophisticated hash combiners such as $\mathsf{Comb}^{H_0,H_1}(M) = H_0(M)\|H_1(H_0(M)\oplus M)$, which may seem to be more secure than the classical combiner at first glance due to the dependency of both hash functions. However, by using the circuit C_1 to compute valid inputs for H_1 we can realize a similiar attack as the one for $\mathsf{Comb}_{\mathrm{class}}$.

4 Basic Conclusions

In this section we provide some basic conclusions in our model, e.g., that the functions f_0, f_1 are still pre-image resistant in presence of the collision finders. These results will also be useful when proving our combiner to be security amplifying.

The first lemma basically restates the well-known birthday paradox that, if the adversary $\mathcal{A}$ in experiment $\mathbf{Exp}^{amp,\mathsf{Comb}}_{\mathcal{A},\mathcal{H}_0,\mathcal{H}_1,\mathcal{C}_0,\mathcal{C}_1}(n)$ makes too many f_0- and f_1-queries (either directly or through the collision-finders), then most likely a lucky collision will occur and $\mathcal{A}$ cannot succeed anymore. This result —like all results in this section— hold for arbitrary combiners (based on the idealized Merkle-Damgård model):

Lemma 1 (Birthday Paradox). *Consider experiment* $\mathbf{Exp}^{amp,\mathsf{Comb}}_{\mathcal{A},\mathcal{H}_0,\mathcal{H}_1,\mathcal{C}_0,\mathcal{C}_1}(n)$ *and assume that* $|\mathrm{FVAL}_b| > 2^{dn}$ *for* $b \in \{0,1\}$ *and a constant* $d > \frac{1}{2}$. *Then the probability that no lucky collisions occur is negligible (and, in particular, the probability that the experiment returns 1 is negligible, too).*

Proof. Suppose $|\mathrm{FVAL}_b| > 2^{dn}$ for some b. Then the birthday paradox implies that with probability at most $\exp(-\binom{2^{dn}+1}{2}/2^n) \leq \exp(-2^{(2d-1)n-1})$ there would be *no* lucky collision. Since $d > \frac{1}{2}$ the term $2^{(2d-1)n-1}$ grows exponentially in n. But if a lucky collision occurs, then the experiment outputs 0.   □

We next show that the images of sample values $\mathrm{SAMPLES}\setminus\mathrm{CVAL}$ appearing during the search of the collision finder (but which are not returned to $\mathcal{A}$) are essentially uniformly distributed from $\mathcal{A}$'s viewpoint (i.e., given the sets $\mathrm{FVAL}, \mathrm{CVAL}$). This holds at any point in the execution and even if $\mathcal{A}$ does not win:

Lemma 2 (Image Uncertainty). *Assume that $\mathcal{A}$ in experiment* $\mathbf{Exp}^{amp,\mathsf{Comb}}_{\mathcal{A},\mathcal{H}_0,\mathcal{H}_1,\mathcal{C}_0,\mathcal{C}_1}(n)$ *makes at most 2^{cn} calls to each collision-finder $\mathcal{C}_0, \mathcal{C}_1$ and that $\mathrm{FVAL}_0, \mathrm{FVAL}_1$ each contain at most 2^{cn} elements for a constant $c < 1$. Then for any $(iv,m), y$ and $b \in \{0,1\}$ such that $((iv,m), f_b(iv,m)) \notin \mathrm{FVAL}_b \cup \mathrm{CVAL}_b$, we have* $\mathrm{Prob}[f_b(iv,m) = y \mid \mathrm{FVAL}, \mathrm{CVAL}] \leq 2 \cdot 2^{-n}$ *(for sufficiently large n's).*

Proof. Consider the information about the image of a value (iv,m) (not appearing in $\mathrm{FVAL} \cup \mathrm{CVAL}$) available through $\mathrm{FVAL}, \mathrm{CVAL}$. Suppose that this value (iv,m) appears in the course of a collision search —else the claim already follows because the image is completely undetermined— and thus the image belongs to $\mathrm{SAMPLES} \setminus (\mathrm{FVAL} \cup \mathrm{CVAL})$. This only leaks the information that the image of (iv, m) must be distinct from other images in such a collision search, or else the collision finder would have output (iv, m) as part of the collision. Hence, the information available through FVAL, CVAL only exclude the images in SAMPLES $\cap$ (FVAL$_b \cup$ CVAL$_b$) —values for the other bit $\bar{b}$ are not relevant— which is a set of size at most $|\text{FVAL}_b \cup \text{CVAL}_b| \leq 3 \cdot 2^{cn}$ (since each of the 2^{cn} calls to $\mathcal{C}_b$ yields at most two entries in CVAL$_b$). Thus, for large n's there are at least $2^n - 3 \cdot 2^{cn} \geq \frac{1}{2} \cdot 2^n$ candidate images left, each one being equally like. $\square$

The next lemma says that the collision-finders cannot be used to break pre-image resistance, i.e., despite the ability to find collisions via $\mathcal{C}_0, \mathcal{C}_1$, searching for a pre-image to a chosen value is still infeasible. Below we formalize this by executing an adversary $\mathcal{B}$ in mode challenge first, in which $\mathcal{B}$ explicitly determines an image y for which a pre-image should be found under f_b. To avoid trivial attacks we also presume that no (iv, m) with $f_b(iv, m) = y$ has been found up to this point. Then, we continue $\mathcal{B}$'s execution in mode find in which $\mathcal{B}$ tries to find a suitable pre-image (iv, m). This assumes that $\mathcal{B}$ cannot try out too many collision-finder replies (i.e., at most 2^{cn} many for some constant $c < \frac{1}{2}$):

Lemma 3 (Chosen Pre-Image Resistance). *For any algorithm $\mathcal{B}$ and any constant $c < \frac{1}{2}$ the following experiment $\mathbf{Exp}^{pre,\text{Comb}}_{\mathcal{B},\mathcal{H}_0,\mathcal{H}_1,\mathcal{C}_0,\mathcal{C}_1}(n)$ has negligible probability of returning 1:*

> *Experiment $\mathbf{Exp}^{pre,\text{Comb}}_{\mathcal{B},\mathcal{H}_0,\mathcal{H}_1,\mathcal{C}_0,\mathcal{C}_1}(n)$:*
> *initialize $(f_0, IV_0) \leftarrow \text{HKGen}_0(1^n)$, $(f_1, IV_1) \leftarrow \text{HKGen}_1(1^n)$*
> *let $(y, b, state) \leftarrow \mathcal{B}^{f_0,f_1,\mathcal{C}_0,\mathcal{C}_1}(\text{challenge}, IV_0, IV_1)$*
> *let $\text{VAL}_b^{ch} = \text{FVAL}_b \cup \text{CVAL}_b$ at this point*
> *let $(iv, m) \leftarrow \mathcal{B}^{f_0,f_1,\mathcal{C}_0,\mathcal{C}_1}(\text{find}, state)$*
> *return 1 iff*
> > *$f_b(iv, m) = y$ and $((iv, m), y) \notin \text{VAL}_b^{ch}$, and*
> > *$\mathcal{B}$ made at most 2^{cn} calls to collision-finder $\mathcal{C}_b$ (in both phases together), and*
> > *no lucky collisions occured during $\mathcal{B}$'s computation (in both phases together)*

The proof is delegated to Appendix A. The proof idea is as follows. For any value appearing in FVAL$_b \setminus$ CVAL$_b$ during the find phase the probability of matching y is at most $2 \cdot 2^{-n}$ by the image uncertainty. Furthermore, according to the Birthday Lemma 1 the set FVAL$_b$ cannot contain more than 2^{dn} elements for some $d > \frac{1}{2}$ (or else a lucky collision is very likely). But then the probability of finding another pre-image among those values is negligible.

The harder part is to show that $\mathcal{B}$ cannot significantly influence the collision finder $\mathcal{C}_b$ to search for a collision with image y (which would then appear in CVAL$_b$ and could be output by $\mathcal{B}$). Here we use the property of our model saying that the circuit's output $C_b(s)$ for each sample is essentially determined by $\mathcal{B}$ (or, to be precise, by the previous values in FVAL and CVAL). But then the Image Uncertainty Lemma applies again, and each sample $C_b(s)$ yields y with probability at most $2 \cdot 2^{-n}$. The final step is to note that each collision search most likely requires approximately $2^{\frac{n}{2}}$ or less samples, and $\mathcal{B}$ initiates at most

2^{cn} many searches for $c < \frac{1}{2}$. Hence, with overwhelming probability there is no value with image y in SAMPLES in the find phase at all, and thus no such value in CVAL$_b$. This shows Chosen Pre-Image Resistance.

For the final conclusions about our model, we prove that, given a collision (iv, m), (iv', m') produced by a collision finder C_b, generating another pre-image also mapping to $f_b(iv, m) = f_b(iv', m')$, is infeasible. The proof is in two steps, first showing that one cannot use the f_b-boxes to find such an additional value, and the second lemma shows that this remains true if one tries to use the collision finder (if one does not call the collision finder more than a polynomial number of times). We remark that this aspect refers to collisions *for the compression functions* only; given a collision generated by the finders one can of course extend this to further collisions *for the iterated hash function* by appending message blocks:

Lemma 4 (f-Replication Resistance). *Assume adversary $\mathcal{A}$ in* $\text{Exp}_{\mathcal{A}, \mathcal{H}_0, \mathcal{H}_1, C_0, C_1}^{amp, \text{Comb}}(n)$ *makes at most 2^{cn} calls to each collision-finder C_0, C_1 and that each set FVAL$_0$, FVAL$_1$ contains at most 2^{dn} elements for constants c, d with $c + d < 1$. Then the probability that there exist values $((iv, m), y) \in$ CVAL$_b$ and $((iv', m'), y) \in$ FVAL$_b \setminus$ CVAL$_b$ for $b \in \{0, 1\}$, is negligible.*

Proof. Fix a bit b. Since $\mathcal{A}$ makes at most 2^{cn} calls to C_b and each reply returns two elements, the set CVAL$_b$ is of size at most $2 \cdot 2^{cn}$. Consider any value $((iv, m), y) \in$ CVAL$_b$ and any value $((iv', m'), y') \in$ FVAL$_b \setminus$ CVAL$_b$. Then, because $((iv', m'), y') \notin$ CVAL$_b$, we must have $y' \neq y$ or $(iv, m) \neq (iv', m')$. In the first case we have no match, in the second case a match can occur with probability at most $2 \cdot 2^{-n}$ by the image uncertainty (considering the point in the execution where the the second of the two values appears for the first time).

Now sum over all $2 \cdot 2^{cn} \cdot 2^{dn} = 2 \cdot 2^{(c+d)n}$ combinations, such that the probability of finding any match is at most $4 \cdot 2^{(c+d-1)n}$. Since $c + d < 1$ this is negligible, and stays negligible if we sum over both choices for b. $\square$

Note that the fact above indicates that, after having generated collisions through the finder, finding other matching function values through the f-boxes is infeasible. This holds at any point in the execution, i.e., $\mathcal{A}$ may not even successfully produce a collision but rather stop prematurely. Next, we use this fact (together with pre-image resistance) to prove replication resistance with respect to the collision finders:

Lemma 5 (C-Replication Resistance). *Assume adversary $\mathcal{A}$ in* $\text{Exp}_{\mathcal{A}, \mathcal{H}_0, \mathcal{H}_1, C_0, C_1}^{amp, \text{Comb}}(n)$ *makes at most $\text{poly}(n)$ calls to each collision-finder C_0, C_1 and that FVAL$_0$, FVAL$_1$ each contain at most 2^{dn} elements for a constant $d < 1$. Then the probability that there exist values $((iv, m), y), ((iv', m'), y), ((iv^*, m^*), y) \in$ CVAL$_b$ for $b \in \{0, 1\}$ with pairwise distinct $(iv, m), (iv', m'), (iv^*, m^*)$, is negligible.*

The proof is in Appendix B. The basic idea is that, at some point in the execution, there must be at most two of the three values in CVAL$_b$ and then another call adds the third value with the same image. But then this contradicts the chosen pre-image resistance, because the right call to the collision finder among the polynomially many ones can be guessed with probability $1/\text{poly}(n)$. We note that the full argument needs to take care of the case that the third value appears in FVAL$_b$ before.

5   A Security-Amplifying Combiner

Our (input-restricted) security-amplifying combiner takes messages $M = m_0$ $||\ldots||m_{t-1}$ of exactly t blocks with $t \leq en$ for some constant $e < \frac{1}{4}$ and applies each of the two hash functions H_0, H_1 to the message $m_0||\ldots||m_t$ and outputs the concatenation:

Theorem 1. *Let $\mathcal{H}_0, \mathcal{H}_1$ be idealized Merkle-Damgård hash functions. Let $e < \frac{1}{4}$ be a constant and assume that $t \leq en$. Then the combiner*

$$\text{Comb}_{amp,t}^{H_0 H_1}(M) = H_0(m_0||\ldots||m_{t-1}) \; || \; H_1(m_0||\ldots||m_{t-1})$$

of $\mathcal{H}_0$ and $\mathcal{H}_1$ is $\alpha(n)$-security-amplifying for $\alpha(n) = \text{poly}(n)$ if the adversary in experiment $\mathbf{Exp}_{A,\mathcal{H}_0,\mathcal{H}_1,C_0,C_1}^{amp,\text{Comb}_{amp,t}}(n)$ makes at most $\alpha(n) = \text{poly}(n)$ calls to each collision finder.

We also remark that our combiner is obviously a (classically) secure combiner in the non-idealized setting. The theorem shows that we get the improved security-amplification guarantee against attacks in the idealized world.

For the proof idea it is instructive to investigate why the straightforward application of the attack of Joux for the case of at most $t \leq \frac{n}{4}$ message blocks fails. In this case one would again build a multi-collision set for either hash function of size at most $2^t \leq 2^{\frac{n}{4}}$. But this time the probability that any of the $2^{2t} < 2^{\frac{n}{2}}$ pairs in such a multi-collision set also collides under the other hash function, should be approximately $2^{\frac{n}{2}} \cdot 2^{-n} = 2^{-\frac{n}{2}}$. Most likely, even approximately $2^{\frac{n}{2}}$ multi-collision sets should therefore not help to find a collision under both hash functions. Our proof follows these lines of reasoning, i.e., bounding the size of multi-collision sets and the probability that message pairs in such a multi-collision set also collide under the other hash function. We stress, however, that a full proof in our model still needs to deal with more general adversaries, possibly taking advantage of the collision finders through "clever" queries.

To process messages of arbitrary length without losing the security-amplification property we apply a hash-tree construction [10] to our combiner. Since the construction is somewhat standard we merely give an example for $t = 2$ in Figure 3. For a similar and more formal treatment see for instance [2]. In general the input restriction t of the hash combiner gives us an t-ary tree, processing k message blocks $m_0 \ldots m_{k-1}$.

If two messages $M \neq M'$ lead to a collision in the root of the hash tree, it can be either the result of a non-trivial collision in the final application of the combiner for different message lengths $|M| \neq |M'|$ (in which case we get a

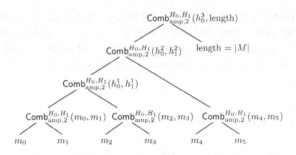

Fig. 3. Example of a hash tree construction for our combiner ($t = 2, k = 6$)

non-trivial collision for the basic combiner), or else the tree structures must be identical. In the latter case the collision can always be traced back to a collision for an earlier application of the combiner. Hence, in both cases the reason for the tree collision is at least one collision for the basic combiner.

As for the efficiency, for a full t-ary tree (with $k = t^r$, the number of message blocks, being a power of t) we apply our basic combiner $\frac{k-1}{t-1} + 1$ times. Each time we need $2t$ applications of the compression functions, making our solution about $\frac{t}{t-1}$ times slower than the classical combiner with $2k$ applications (but with the advantage of security amplification for our combiner).

6   Proof of Security Amplification

Before giving the proof we first show a technical conclusions stating that the adversary against our (input-restricted) combiner essentially cannot win if the function values of the output are undetermined (the proof of this first lemma follows from the image uncertainty and appears in Appendix C):

Lemma 6 (Output Knowledge). *Assume that $\mathcal{A}$ in experiment* $\mathbf{Exp}_{\mathcal{A},\mathcal{H}_0,\mathcal{H}_1,\mathcal{C}_0,\mathcal{C}_1}^{amp,\mathrm{Comb}_{amp,t}}(n)$ *makes at most 2^{cn} calls to each collision-finder $\mathcal{C}_0, \mathcal{C}_1$ for some constant $c < 1$. Assume that $\mathcal{A}$ eventually outputs $M = m_0||\dots||m_{t-1} \neq M' = m'_0||\dots||m'_{t-1}$ such that*

$$iv_{b,0} = iv'_{b,0} = IV_b, \quad iv_{b,i+1} = f_b(iv_{b,i}, m_i),$$

$$iv'_{b,i+1} = f_b(iv'_{b,i}, m'_i) \text{ for } b \in \{0,1\} \text{ and } i \in \{0,1,\dots,t-1\}$$

Suppose further that $((iv_{b,i}, m_i), iv_{b,i+1})$ or $((iv'_{b,i}, m'_i), iv'_{b,i+1}))$ does not belong to $\mathrm{FVAL}_b \cup \mathrm{CVAL}_b$ for some $b \in \{0,1\}$ and some $i \in \{0,1,\dots,t-1\}$. Then the probability that the experiment returns 1 is negligible.

The following lemma proves that, for t message blocks there can only be 2^t multi-collisions, as long as each collision finder is only called a polynomial number of times:

Lemma 7 (Multi-Collisions). *Assume attacker $\mathcal{A}$ in experiment* $\mathbf{Exp}_{\mathcal{A},\mathcal{H}_0,\mathcal{H}_1,\mathcal{C}_0,\mathcal{C}_1}^{amp,\mathrm{Comb}_{amp,t}}(n)$ *makes at most $poly(n)$ calls to each collision-finder $\mathcal{C}_0, \mathcal{C}_1$*

and that the experiment returns 1. *Then, the probability that for some* $b \in \{0, 1\}$ *and some* $iv_{b,t}$, *the set*

$$\text{MULTI}_b(iv_{b,t}) := \left\{ M = m_0 || \dots || m_{t-1} : \begin{array}{l} iv_{b,i+1} = f_b(iv_{b,i}, m_i) \in \text{FVAL}_b \cup \text{CVAL}_b \\ for\ i = 0, 1, \dots, t-1,\ where\ iv_{b,0} = IV_b \end{array} \right\}$$

contains more than 2^t *elements, is negligible.*

The lemma holds because if there was a multi-collision set with more than 2^t elements, then there must be distinct values $(iv_{b,i}, m_i)$, $(iv'_{b,i}, m'_i)$ and $(iv^*_{b,i}, m^*_i)$ mapping to the same image under f_b. According to the previous lemma we can assume that all of them belong to $\text{FVAL}_b \cup \text{CVAL}_b$, but then they would either be a lucky collision (two or three values in FVAL_b), refute f-replication resistance (one value in FVAL_b) or contradict $\mathcal{C}$-replication resistance (no value in FVAL_b).

With these two lemmas we can now prove that our combiner is security-amplifying. The full proof appears in Appendix C. For an outline consider the multi-collision sets defined in the previous lemma. Lemma 6 implies that, in order to win, the adversary must know the images of the final output $M \neq M'$. Hence, each of the two messages must appear in some multi-collision set, and to constitute a collision under hash function H_b, they must appear in the same multi-collision set $\text{MULTI}_b(y_b)$ for some y_b. Moreover, since the messages must collide under both hash functions simultaneously they must belong to an intersection $\text{MULTI}_0(y_0) \cap \text{MULTI}_1(y_1)$ for some y_0, y_1.

Lemma 7 now says that each multi-collision set has at most 2^t elements. Thus, there are at most $2^{2t} \leq 2^{2en}$ such pairs in each multi-collision set. Furthermore, we can bound the number of multi-collision sets by the number of elements in $\text{FVAL}_b \cup \text{CVAL}_b$, and therefore by $3 \cdot 2^{dn}$ for a constant $d > \frac{1}{2}$ with $d + 2e < 1$ (here we use the fact that $e < \frac{1}{4}$). We therefore have at most $3 \cdot 2^{(d+2e)n}$ possible pairs $M \neq M'$. The proof then shows that, by the image uncertainty, any of the pairs M, M' in some multi-collision set $\text{MULTI}_b(y_b)$ also collides under the other hash function $H_{\overline{b}}$, with probability at most $6 \cdot 2^{(d+2e-1)n}$ which is negligible. Put differently, with overwhelming probability the intersections of mulit-collision sets for both hash functions are empty and the adversary cannot find appropriate messages M, M'.

Acknowledgments

We thank the anonymous reviewers for valuable comments. Both authors are supported by the Emmy Noether Program Fi 940/2-1 of the German Research Foundation (DFG).

References

1. Boneh, D., Boyen, X.: On the Impossibility of Efficiently Combining Collision Resistant Hash Functions. In: Dwork, C. (ed.) CRYPTO 2006. LNCS, vol. 4117, pp. 570–583. Springer, Heidelberg (2006)

2. Bellare, M., Rogaway, P.: Collision-Resistant Hashing: Towards Making UOWHFs Practical. In: Kaliski Jr., B.S. (ed.) CRYPTO 1997. LNCS, vol. 1294, pp. 470–484. Springer, Heidelberg (1997)
3. Canetti, R., Rivest, R., Sudan, M., Trevisan, L., Vadhan, S., Wee, H.: Amplifying Collision Resistance: A Complexity-Theoretic Treatment. In: Menezes, A. (ed.) CRYPTO 2007. LNCS, vol. 4622, Springer, Heidelberg (2007)
4. Damgård, I.: A Design Principle for Hash Functions. In:Brassard, G. (ed.) CRYPTO 1989. LNCS, vol. 435, pp. 416–427. Springer, Heidelberg (1990)
5. Goldreich, O., Impagliazzo, R., Levin, L., Venkatesan, R., Zuckerman, D.: Security Preserving Amplification of Hardness. In: Proceedings of the Annual Symposium on Foundations of Computer Science (FOCS)'90, pp. 318–326. IEEE Computer Society Press, Los Alamitos (1990)
6. Herzberg, A.: On Tolerant Cryptographic Constructions. In: Menezes, A.J. (ed.) CT-RSA 2005. LNCS, vol. 3376, pp. 172–190. Springer, Heidelberg (2005)
7. Hoch, J., Shamir, A.: Breaking the ICE — Finding Multicollisions in Iterated Concatenated and Expanded (ICE) Hash Functions. In: Robshaw, M. (ed.) FSE 2006. LNCS, vol. 4047, Springer, Heidelberg (2006)
8. Joux, A.: Multicollisions in Iterated Hash Functions. In: Franklin, M. (ed.) CRYPTO 2004. LNCS, vol. 3152, Springer, Heidelberg (2004)
9. Lin, H., Trevisan, L., Wee, H.: On Hardness Amplification of One-Way Functions. In: Kilian, J. (ed.) TCC 2005. LNCS, vol. 3378, pp. 34–49. Springer, Heidelberg (2005)
10. Merkle, R.: One Way Hash Functions and DES. In: Brassard, G. (ed.) CRYPTO 1989. LNCS, vol. 435, pp. 428–446. Springer, Heidelberg (1990)
11. Nandi, M., Stinson, D.: Multicollision Attacks on a Class of Hash Functions. Number 2004/330 in Cryptology eprint archive (2004), eprint.iacr.org
12. Pietrzak, K.: Non-Trivial Black-Box Combiners for Collision-Resistant Hash-Functions don't Exist. In: Naor, M. (ed.) EUROCRYPT 2007. LNCS, vol. 4515, Springer, Heidelberg (2007)
13. Yao, A.: Theory and Applications of Trapdoor Functions. In: Proceedings of the Annual Symposium on Foundations of Computer Science (FOCS), IEEE Computer Society Press, Los Alamitos (1982)
14. Yu, H., Wang, X.: MultiCollision Attack on the Compression Functions of MD4 and 3-Pass HAVAL. Number 2007/085 in Cryptology eprint archive (2007) eprint.iacr.org

A   Proof of Chosen Pre-image Resistance (Lemma 3)

In this section we prove Lemma 3, showing that no adversary $\mathcal{B}$ can determine an image y and later find another pre-image to this value.

Proof. Let d be a constant with $\frac{1}{2} < d < 1$. Assume that FVAL_b contains more than 2^{dn} elements at the end. Then Lemma 1 implies that such executions can only contribute with negligible probability to $\mathcal{B}$'s success. From now on we can therefore condition on this bound 2^{dn} of number on elements in FVAL_b.

By the image uncertainty we conclude that the probability that any $((\mathrm{iv}, m), f_b(\mathrm{iv}, m)) \in \mathrm{FVAL}_b \backslash \mathrm{CVAL}_b$ in $\mathcal{B}$'s find phase yields y, is at most $2 \cdot 2^{-n}$. Here we use the fact that any function evaluation adding to $\mathrm{FVAL}_b \backslash \mathrm{CVAL}_b$ is either via a direct call to the f_b-box, or via an f_b-gate evaluation in the computation of a circuit $C(s)$, carried out through one of the collision finders. In any case, the input to the function only depends on the values in FVAL and CVAL before the corresponding query; for f_b-box queries this is clear and for circuit computations it follows as the circuit is chosen by $\mathcal{B}$ and all previous function evaluations immediately appear in FVAL$_b$. Therefore, the uncertainty bound applies. Summing over all at most 2^{dn} many values in FVAL$_b$ shows that the probability of hitting y is bounded from above by $2 \cdot 2^{(d-1)n}$ and is thus negligible. In the sequel we therefore presume that no $((\mathrm{iv}, m), y) \in \mathrm{FVAL}_b \setminus \mathrm{CVAL}_b$ appears (unless it has been in $\mathrm{VAL}_b^{\mathrm{ch}}$ before, in which case $\mathcal{B}$ cannot use it anymore for a successful run).

We next investigate the effect of collision finder calls on CVAL$_b$, addressing the question if $\mathcal{B}$ can force the collision finder to bias collisions towards y in some way. Recall that the collision finder makes at most 2^{cn} many runs for $c < \frac{1}{2}$. Let $e = \frac{3}{4} - \frac{c}{2} > \frac{1}{2}$. Then we can assume that each run probes at most 2^{en} new elements previously not in SAMPLES. This is so since, for a single run, the probability of finding no collisions after 2^{en} many trials for fresh values, is double-exponentially small (see Lemma 1 and note that this remains true for a slightly larger probability of $2 \cdot 2^{-n}$). The probability that any of the 2^{cn} calls would require more fresh samples, is therefore still negligible. From now on we thus presume that each call adds at most 2^{en} new entries to SAMPLES.

Consider the j-th call C_b to the collision finder $\mathcal{C}_b$ in the find stage. Let $\mathrm{CVAL}_{b,j}^{\mathrm{before}}$ be the set CVAL_b before this call, such that $\mathrm{CVAL}_{b,1}^{\mathrm{before}}$ denotes the set CVAL_b at the beginning of the find phase. Note that $\mathrm{CVAL}_{b,j}^{\mathrm{before}}$ does not change during the collision search, but only when the finder returns the collision. Suppose further that $\mathrm{CVAL}_{b,j}^{\mathrm{before}}$ does not contain any element $((\mathrm{iv}, m), y)$ which is not already in $\mathrm{VAL}_b^{\mathrm{ch}}$. This is obviously true for $\mathrm{CVAL}_{b,1}^{\mathrm{before}}$.

A crucial aspect in our consideration is that all circuit values $C_b(s)$ during the collision search are fully determined given FVAL$_b$ (containing the pairs of the entire execution but whose images are distinct from y by assumption) as well as $\mathrm{CVAL}_{b,j}^{\mathrm{before}}$. Hence, the uncertainty bound applies again, and the probability that a specific sample $C_b(s)$ gives a new pair $(C_b(s), y) \notin \mathrm{CVAL}_{b,j}^{\mathrm{before}} \cup \mathrm{VAL}_b^{\mathrm{ch}}$, is at most $2 \cdot 2^{-n}$ (noting that any entry $(C_b(s), f_b(C_b(s))) \in (\mathrm{FVAL}_b \cup \mathrm{CVAL}_{b,j}^{\mathrm{before}}) \setminus \mathrm{VAL}_b^{\mathrm{ch}}$ has an image different from y by assumption). Since there are at most 2^{en} new samples, only with probability at most $2 \cdot 2^{(e-1)n}$ some new sample $C_b(s)$ in $\mathcal{C}_b$'s search yields y. It follows that, except with probability $2 \cdot 2^{(e-1)n}$, the set $\mathrm{CVAL}_{b,j+1}^{\mathrm{before}}$ including the new collisions will not contain a suitable entry.

Finally, sum over all at most 2^{cn} many calls to $\mathcal{C}_b$ to derive that CVAL$_b$ does not contain a new entry $((\mathrm{iv}, m), y) \in \mathrm{CVAL}_b \setminus \mathrm{VAL}_b^{\mathrm{ch}}$, except with probability $2 \cdot 2^{(c+e-1)n}$ for $c + e = \frac{3}{4} + \frac{c}{2} < 1$ which is negligible. Since the same holds for $\mathrm{FVAL}_b \setminus \mathrm{CVAL}_b$ the overall probability of finding a suitable pre-image (iv_0, m), including possibly the final output which is not a member in $\mathrm{FVAL}_b \cup \mathrm{CVAL}_b$, is negligible. $\square$

B Proof of $\mathcal{C}$-Replication Resistance (Lemma 5)

In this section we prove that no adversary can find three values in CVAL mapping to the same image.

Proof. We discuss that if $\mathcal{A}$ could find three (or more) of those values then this would contradict either f-replication resistance or chosen pre-image resistance. Consider adversary $\mathcal{B}$ against the chosen pre-image resistance which basically runs a black-box simulation of $\mathcal{A}$. In the challenge-phase, $\mathcal{B}$ initially makes a guess for a specific call j adversary $\mathcal{A}$ makes to one of the collision finders. Then $\mathcal{B}$ runs $\mathcal{A}$ up to the point where $\mathcal{A}$ receives the answer $((iv, m), y), ((\widehat{iv}, \widehat{m}), y)$ of $\mathcal{C}_b$ for this j-th call. Then $\mathcal{B}$ outputs y, b (and all internal information of $\mathcal{A}$ as state) and concludes this stage. In the find-phase $\mathcal{B}$ continues $\mathcal{A}$'s simulation and waits to see a value $((iv^*, m^*), y)$ in the execution, and then outputs (iv^*, m^*) and stops.

We next analyze $\mathcal{B}$'s success probability. Since each call to the collision-finders adds at most two new values to CVAL_b, there must be a point in $\mathcal{A}$'s execution where there is $(iv, m) \in \text{CVAL}_b$ (and possibly $(iv', m') \in \text{CVAL}_b$) and only the next call to $\mathcal{C}_b$ adds the value (iv^*, m^*) to CVAL_b, i.e., so far $(iv^*, m^*) \notin \text{CVAL}_b$. Suppose that the conditional probability (given such a value (iv^*, m^*) with the same image really appears in the execution) that this value belongs to FVAL_b after the corresponding call to $\mathcal{C}_b$, was noticeable. Then this would clearly contradict the f-replication resistance (bounding the polynomial number of calls by 2^{cn} for the constant $c = \frac{1}{2} - \frac{d}{2}$ with $c + d < 1$). We may therefore assume that $(iv^*, m^*) \notin \text{VAL}_b^{\text{ch}} = \text{CVAL}_b \cup \text{FVAL}_b$ at this point. But then $\mathcal{B}$ guesses the right call j with probability $1/\text{poly}(n)$, and thus predicts a function value with noticeable probability. This, however, contradicts the chosen pre-image resistance. $\square$

C Proof of Security Amplification (Theorem 1)

In this section we provide the proofs of the claims in Section 6 and of the theorem. First we prove that an adversary must essentially know the function values of the output (Lemma 6):

Proof (of Lemma 6). Suppose $\mathcal{A}$ outputs such values M, M' and succeeds with noticeable probability. Assume for simplicity that $((iv_{b,i}, m_i), iv_{b,i+1}) \notin \text{FVAL}_b \cup \text{CVAL}_b$; the case $((iv'_{b,i}, m'_i), iv'_{b,i+1})$ is treated analogously. Let i be maximal and fix the bit b.

By Lemma 1 we can assume $|\text{FVAL}_b| \leq 2^{dn}$ for $d = \max\{\frac{3}{4}, c\}$, except with negligible probability. Hence, from now on we can condition on $|\text{FVAL}_b \cup \text{CVAL}_b| \leq 3 \cdot 2^{dn}$. For a success the messages M and M' must collide under H_b. If $i = t - 1$ then $f_b(iv_{b,i}, m_i) = iv_{b,i+1}$ is the output of the hash function, and since this value does not appear in $\text{FVAL}_b \cup \text{CVAL}_b$, the probability of matching $iv'_{b,i+1}$ is bounded from above by $2 \cdot 2^{-n}$ by the image uncertainty.

If $i < t - 1$ then there must exist an entry $((iv_{b,i+1}, m_{i+1}), iv_{b,i+2}) \in \text{FVAL}_b \cup \text{CVAL}_b$ (because i is chosen to be maximal). However, the probability that the

value $f_b(\mathrm{iv}_{b,i}, m_i)$ appears as a prefix in any of the $3 \cdot 2^{dn}$ values in $\mathrm{FVAL}_b \cup \mathrm{CVAL}_b$, is at most $6 \cdot 2^{(d-1)n}$ and thus negligible. On the other hand, if the prefix $f_b(\mathrm{iv}_{b,i}, m_i)$ does not appear in $\mathrm{FVAL}_b \cup \mathrm{CVAL}_b$, then this contradicts the maximal choice of i. Doubling the probability for both choices of b concludes the proof.
□

We next prove Lemma 7, bounding the number of messages in a multi-collision set by 2^t:

Proof (of Lemma 7). Assume that the experiment returns 1 (such that, except with negligible probability, $\mathrm{FVAL}_0, \mathrm{FVAL}_1$ are of size at most 2^{dn} each, for some constant $d < 1$). If some set $\mathrm{MULTI}_b(\mathrm{iv}_{b,t})$ contains more than 2^t elements then there must be an index i such that there are (at least) three distinct values $(\mathrm{iv}_{b,i}, m_i)$, $(\mathrm{iv}'_{b,i}, m'_i)$ and $(\mathrm{iv}^*_{b,i}, m^*_i)$ mapping to the same image under f_b. If two or more of those values belong to $\mathrm{FVAL}_b \setminus \mathrm{CVAL}_b$ then this constitutes a lucky collision and refutes the fact that the experiment returns 1. If one of the values lies in $\mathrm{FVAL}_b \setminus \mathrm{CVAL}_b$, whereas the other two values belong to CVAL_b, then this contradicts the f-replication resistance and this can only happen with negligible probability. Finally, the case that all three values belong to CVAL_b can only happen with negligible probability, too, under the $\mathcal{C}$-replication resistance.   □

Finally, we give the full proof that our combiner is security amplyifing:

Proof (of Theorem 1). According to our definition a combiner is called security-amplifying if for any algorithm $\mathcal{A}$ making at most $\alpha(n) \cdot (T_0(n) + T_1(n))$ steps the probability of finding a collision is negligible (for some $\alpha(n) > 1$). Hence we will show that, with overwhelming probability, no collisions for $\mathsf{Comb}_{\mathrm{amp},t}$ (with $t < en$ for constant $e < \frac{1}{4}$) can be computed for any $\alpha(n) = \mathrm{poly}(n)$ when calling each collision finders at most $\alpha(n) = \mathrm{poly}(n)$ many times.

Let $d = \frac{3}{4} - e$ such that the constant d is at larger than $\frac{1}{2}$ and $d + 2e < 1$. Then we can assume that $\mathrm{FVAL}_0, \mathrm{FVAL}_1$ in $\mathcal{A}$'s attack each contain at most 2^{dn} elements, otherwise the probability of winning would be negligible. Also assume that the number of collision finder calls is bounded by $2 \cdot \mathrm{poly}(n) \leq 2^{dn}$ (for sufficiently large n's). Hence, in the following, we can assume that $\mathrm{FVAL}_b \cup \mathrm{CVAL}_b$ contains at most $3 \cdot 2^{dn}$ many elements for $b \in \{0,1\}$.

For any $b \in \{0,1\}$ and any $\mathrm{iv}_{b,t}$ we again consider all sets of multi-collisions,

$$\mathrm{MULTI}_b(\mathrm{iv}_{b,t}) = \left\{ M = m_0 || \ldots || m_{t-1} \; : \; \begin{array}{l} \mathrm{iv}_{b,i+1} = f_b(\mathrm{iv}_{b,i}, m_i) \in \mathrm{FVAL}_b \cup \mathrm{CVAL}_b \\ \text{for } i = 0, 1, \ldots, t-1, \text{ where } \mathrm{iv}_{b,0} = \mathrm{IV}_b \end{array} \right\}$$

but this time we divide them into different stages (depending on the calls to the collision finders). We denote by $\mathrm{MULTI}^{\mathrm{before}}_{b,j}(y)$ the set of multi-collisions *before* the j-th call to one of the two collision finders. The transition to the next phase therefore adds all messages with respect to the new function values from the collision finder's reply as well as all subsequent function evaluations through the f-boxes. Clearly, $\mathrm{MULTI}^{\mathrm{before}}_{b,j}(y) \subseteq \mathrm{MULTI}^{\mathrm{before}}_{b,j+1}(y)$ for all j and $\mathrm{MULTI}^{\mathrm{before}}_{b, 2 \cdot \mathrm{poly}(n)+1}(y)$

—which we denote by $\text{MULTI}_b^{\text{end}}(y)$— contains all multi-collisions for y under H_b at the end of the execution.

By Lemma 6 adversary $\mathcal{A}$ must "know" all function values in the final output, i.e., they must belong to $\text{FVAL}_b \cup \text{CVAL}_b$ for some $b \in \{0,1\}$. Hence, both messages of the collision $M \neq M'$ for H_b output by $\mathcal{A}$ must also appear in the same set $\text{MULTI}_b^{\text{end}}(y_b)$ for some y_b. This basically reduces the task of showing that $\mathcal{A}$ fails, to the proof that no $M \neq M'$ and y_0, y_1 with $M, M' \in \text{MULTI}_0^{\text{end}}(y_0) \cap \text{MULTI}_1^{\text{end}}(y_1)$ exist (except with some very small probability or if one of the success requirements such as the absence of lucky collisions is violated).

We will show that, given that no success requirements are violated, with overwhelming probability the intersection of multi-collision sets for $b = 0, 1$ will be empty in the course of the execution. This is done by a careful inductive argument, where we use the invariant that for no y_b the set $\text{MULTI}_{b,j}^{\text{before}}(y_b)$ contains $M \neq M'$ such that they collide under $H_{\bar{b}}$. This is clearly true for $\text{MULTI}_{b,1}^{\text{before}}(y_b)$ because up to the point where the first collision finder is called, only f-queries have been made, and each set $\text{MULTI}_{b,1}^{\text{before}}(y_b)$ can contain only one element (or a lucky collision already occurs).

We also use that, according to the Multi-Collision Lemma 7, each set $\text{MULTI}_{b,j}^{\text{before}}(y_b)$ can contain at most 2^t elements (with overwhelming probability). Additionally, we always have at most $3 \cdot 2^{dn}$ non-empty multi-collision sets, because there can only be an element in a such set if there is at least one value from $\text{FVAL}_b \cup \text{CVAL}_b$. Hence, at any point there are at most $2^{2t} \cdot 3 \cdot 2^{dn} \leq 3 \cdot 2^{(d+2e)n}$ many collision pairs (M, M') appearing together in one of the multi-collision sets, for the constant $d + 2e < 1$.

Now suppose we make the j-th call to one of the collision finders, $\mathcal{C}_b$. After this call (and all subsequent f-function evaluations) take any pair $M \neq M'$ belonging to the same set $\text{MULTI}_{b,j+1}^{\text{before}}(y_b)$ for some y_b. The next step is to note that, most likely, this pair M, M' cannot belong to some $\text{MULTI}_{\bar{b},j+1}^{\text{before}}(y_{\bar{b}})$. Note that if M and M' lie in multi-collision sets $\text{MULTI}_{\bar{b},j+1}^{\text{before}}(y_{\bar{b}})$ and $\text{MULTI}_{\bar{b},j+1}^{\text{before}}(y'_{\bar{b}})$ for $y_{\bar{b}} \neq y'_{\bar{b}}$ then they clearly do not collide under $H_{\bar{b}}$ as those sets must be disjoint.

Assume, towards contradiction, that M, M' appear in a single multi-collision set for $\bar{b}$. We already know that M, M' cannot belong to some $\text{MULTI}_{\bar{b},j}^{\text{before}}(y_{\bar{b}})$ of the previous stage, because none of these pairs constitutes a collision under $H_{\bar{b}}$, except with negligible probability. Hence, at least one of the two messages (say, M) must have been added to $\text{MULTI}_{\bar{b},j+1}^{\text{before}}(y_{\bar{b}})$ because of an $f_{\bar{b}}$-function evaluation of $\mathcal{C}_b$ or via a direct evaluation of $f_{\bar{b}}$, taking into account that $\text{CVAL}_{\bar{b}}$ does not change between the two points in time.

Suppose that M is added to some set $\text{MULTI}_{\bar{b},j+1}^{\text{before}}(y_{\bar{b}})$ via a new $f_{\bar{b}}$-value (which has not been in $\text{CVAL}_{\bar{b}}$), and assume that either M' is added only now or has already been in this set before the call. Consider the maximal i for which a new function value is added (when one would process the blocks m_i of message M through the iterated hash function). If the final value $\text{iv}_{\bar{b},t} = f_{\bar{b}}(\text{iv}_{\bar{b},t-1}, m_{t-1})$ is added ($i = t - 1$) then, if for M' processing the final message block $\text{iv}_{\bar{b},t} = f_{\bar{b}}(\text{iv}'_{\bar{b},t-1}, m'_{t-1})$ has been in $\text{FVAL}_{\bar{b}}$ before or is added to $\text{FVAL}_{\bar{b}}$ now, we would have a lucky collision. So $iv_{\bar{b},t} = f_{\bar{b}}(iv'_{\bar{b},t-1}, m'_{t-1})$ must have been in $\text{CVAL}_{\bar{b}}$ before. But then this would contradict the f-replication resistance. For any other $i < t-1$ we note that, if $f_{\bar{b}}(iv_{\bar{b},j}, m_i)$ has not been determined before by $\mathcal{A}$, the probability that it matches any prefix of the at most $3 \cdot 2^{dn}$ previous values in $\text{FVAL}_{\bar{b}} \cup \text{CVAL}_{\bar{b}}$, is negligible (namely, at most $6 \cdot 2^{(d-1)n}$ by the image uncertainty). But this would contradict the maximal choice of i.

In conclusion, for any of the pairs M, M' there must still be an $f_{\bar{b}}$-value not in $\text{FVAL}_{\bar{b}} \cup \text{CVAL}_{\bar{b}}$ at this point, and the probability that the pair M, M' collides under $H_{\bar{b}}$ at all, is thus at most $2 \cdot 2^{-n}$. Therefore, the probability that any of the at most $3 \cdot 2^{(d+2e)n}$ pairs M, M' for $d + 2e < 1$ constitutes a collision under $H_{\bar{b}}$, is negligible. The same argument applies now vice versa, no pair M, M' from a set $\text{MULTI}^{\text{before}}_{\bar{b},j+1}(y_{\bar{b}})$ yields a collision under H_b, except for some negligible error. This gives us the invariant.

The argument can now be set forth to the at most $2 \cdot \text{poly}(n) + 1$ many phases, showing that the final multi-collision sets for $b = 0, 1$ never intersect in more than one element. This proves the theorem. □

Hash Functions and the (Amplified) Boomerang Attack

Antoine Joux[1,3] and Thomas Peyrin[2,3]

[1] DGA
[2] France Télécom R&D
thomas.peyrin@orange-ftgroup.com
[3] Université de Versailles Saint-Quentin-en-Yvelines
antoine.joux@prism.uvsq.fr

Abstract. Since Crypto 2004, hash functions have been the target of many attacks which showed that several well-known functions such as SHA-0 or MD5 can no longer be considered secure collision free hash functions. These attacks use classical cryptographic techniques from block cipher analysis such as differential cryptanalysis together with some specific methods. Among those, we can cite the neutral bits of Biham and Chen or the message modification techniques of Wang *et al*. In this paper, we show that another tool of block cipher analysis, the boomerang attack, can also be used in this context. In particular, we show that using this boomerang attack as a neutral bits tool, it becomes possible to lower the complexity of the attacks on SHA-1.

Keywords: hash functions, boomerang attack, SHA-1.

1 Introduction

The most famous design principle for dedicated hash functions is indisputably the MD-SHA family, firstly introduced by R. Rivest with MD4 [16] in 1990 and its improved version MD5 [15] in 1991. Two years after, the NIST publishes [12] a very similar hash function, SHA-0, that will be patched [13] in 1995 to give birth to SHA-1. This family is still very active, as NIST recently proposed [14] a 256-bit new version SHA-256 in order to anticipate the potential cryptanalysis results and also to increase its security with regard to the fast growth of the computation power. Basically, MD-SHA family hash functions use the Merkle-Damgård extension domain and their compression function is build upon a block cipher in Davies-Meyer mode: the output of the compression function is the output of the block cipher with a feed-forward of the chaining variable.

The first cryptanalysis of a member of this family dates from Dobbertin [7] with a collision attack against MD4. Then, Chabaud-Joux [5] provided the first theoretical collision attack against SHA-0 and Biham-Chen [1] introduced the idea of neutral bits, which led to the computation of a real collision with four blocks of message [2]. Later on, a novel framework of collision attack, using

A. Menezes (Ed.): CRYPTO 2007, LNCS 4622, pp. 244–263, 2007.

modular difference and message modification techniques, surprised the cryptography community [19,23,24,22]. Those devastating attacks broke a lot of hash functions, such as MD4, MD5, SHA-0, SHA-1, RIPEMD or HAVAL-128.

Even if SHA-1 is theoretically broken (with 2^{69} message modifications), the computational power needed in practice is too important and the question arise that when will someone be able to come up with a real collision. Recently [20,21], it has been claimed that the complexity of this attack can be improved up to 2^{63} message modifications.

In this article we study the application of boomerang attacks, originally introduced by D. Wagner [18] for block ciphers, to the case of hash functions. In particular, we show that this very generic method may improve the already known collision attacks against various hash functions when used with classic improvements such as neutral bits or message modification. Although this method is generic, some aspects are closely related to the particular hash function one is planning to attack. Thus, we give a practical proof of concept by applying this improvement to SHA-1. We provide here the detailed constraints and advantages of this particular case. Finally, we are able to present a novel attack against SHA-1, dividing the work factor by 32 from the previous attacks.

An independent work by Klima, describing tunnels in MD5 was posted on ePrint [10], shortly before our first public presentation of the boomerang attack [8] applied to hash function. Each tunnel in Klima's work can be decomposed into a collection of auxiliary differential in our attack. Note that due to the simple message expansion in MD5, the tunnel can be directly observed in a preexisting attack. In our SHA-1 application, a specific differential attack must be constructed to accommodate the auxiliary differentials.

The paper is structured as follows. In Section 2, we recall the concept of boomerang attack for block ciphers and in Section 3 we show how this concept can be applied to hash functions. In particular, we give two different possible approaches for using this method. Then, in Section 4, we treat a practical example with the case of SHA-1. We explain all the specific aspects of the application of boomerang attacks for SHA-1 and show that this method leads to improvements for a collision attack. Finally, we draw conclusions and give future works in Section 5.

Notations. In the following, + will stand for the addition on 32-bit words (modulo 2^{32}) and $\oplus$ will represent the bitwise exclusive-OR. The left (resp. right) bit rotation will be denoted $\ll$ (resp. $\gg$), and $\wedge$ (resp. $\vee$) is the bitwise AND (resp. OR). The j-th bit (modulo 32) of a 32-bit word X is denoted X^j and the bitwise complementary of X will be denoted $\overline{X}$.

2   The Boomerang Attack

The boomerang attack was proposed by D. Wagner as a tool for the cryptanalysis of block ciphers in [18]. It allows to weave two partial and independent differential characteristics together into a global attack on the block cipher. The basic idea

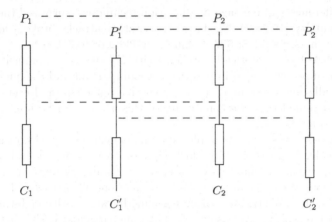

Fig. 1. Schematic view of the boomerang attack on block ciphers

is quite simple. Assume that we are given a first differential characteristic D_1 on the first half of the block cipher which predicts that an input difference Δ leads to an output difference Δ^* with probability p_1. Then, assume a second differential on the second half which predicts that an input difference ∇^* leads to an output difference ∇ with probability p_2. Using these two differentials, we can draw a diagram (see Figure 1) that involves four plaintext/ciphertext pairs.

This diagram can be turned into an attack as follows. First, the attacker choses a random plaintext and asks for the encryption of both this plaintext P_1 and of the plaintext P_2 obtained by xoring P_1 with Δ. The resulting ciphertexts are denoted by C_1 and C_2. After that, the attacker computes C_1' by xoring C_1 with ∇ and C_2' by xoring C_2 with ∇. Then, he asks for the decrypted plaintext P_1' and P_2'. The key idea of the attack is to remark that when the pair (P_1, P_2) follows the Δ differential path and both decryptions follow the ∇ differential path, then the intermediate values corresponding to P_1' and P_2' have the correct difference Δ^*. If in addition (P_1', P_2') is also a correct pair for Δ then the attacker finds that $P_1' \oplus P_2'$ is Δ.

Assuming independence between the four instances of differential paths, we obtain a probability of success $p_1^2 p_2^2$. Basically, this yields a distinguisher that allows us to make the difference between the block cipher and a random permutation.

3 Adapting the Boomerang Attack to Hash Functions

At first, since many hash functions are based on block ciphers, it seems tempting to directly apply the boomerang attack to these hash functions, however several obstructions are quickly encountered and prevent this straightforward approach from working. In particular, the need for decryption, which is an essential part of the boomerang attack, can not be available in the context of hash functions.

Yet, we now show that the boomerang attack, and more specifically its chosen plaintext variant (so-called amplified boomerang attack [9]), can be adapted to the hash function setting and yields improvements compared to previously known differential attacks. The basic idea to adapt the boomerang attack is to use, in addition to the good global differential path used in the now classical differential attacks, several partial differential paths which are very good on a limited number of steps but fail to cover the complete compression function. In order to combine these differential paths together, we use the same basic diagram as with the boomerang attack against block ciphers. However, some specific obstructions appear and need to be removed. The first problem, that we already described when considering the direct application, is the fact that in order to obtain collisions, we cannot use the compression function in the backward direction. The second problem is that we no longer have a nice symmetry with two characteristics playing almost the same role. Instead, there is a main differential path which is our target and some auxiliary paths which help in applying the main one.

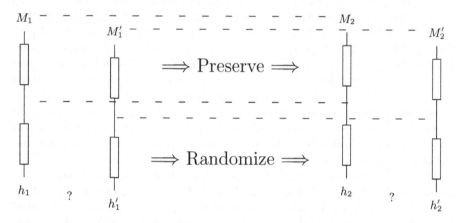

Fig. 2. Schematic view of the boomerang attack on hash functions

Our adapted boomerang attack on iterated hash functions is based on a simple basic block, which we now describe. We start from a basic differential path on an iterated hash function. For the sake of simplicity, we assume that this differential path is of the simple type which yields a collision after a single iteration. Generalizing this description to near-collisions or multiple iterations is a straightforward matter. The basic differential path consists in a message difference Δ, possibly completed by a list of restrictions on acceptable messages, such that the two single block messages M and $M \oplus \Delta$ collide, with probability p_Δ. As usual, this probability do not take into account the so-called early steps where parts of the message M can be chosen independently of each others. From now on, we split the rest of the steps into two main parts, the middle steps and the late steps[1]. To each part, we associate a corresponding probability p_M for

---

[1] In some multi-block attacks, some of the final steps can be treated specifically, ignoring partial misbehaviors which can be corrected in the subsequent blocks.

the middle steps and p_F for the late steps. Under a classical step independence assumption, we have $p_\Delta = p_M \cdot p_F$. The goal of the boomerang based attack is to improve p_M and thus the total complexity of the process. For this, we use an auxiliary differential path that covers both the early and the middle steps as a tool. Assume such an auxiliary differential path, that predicts that, with probability p_δ two messages M and $M \oplus \delta$ yield, after the middle steps, two intermediate internal states with some prescribed (not necessarily null) difference. Take a message pair M and $M' = M \oplus \Delta$ that conforms to the main differential path on the early and middle steps. Assume that both $(M, M \oplus \delta)$ and $(M', M' \oplus \delta)$ conform to the auxiliary differential path. Then, we see that the internal states differences cancel out, and that the pair $(M \oplus \delta, M' \oplus \delta)$ also conforms to the main differential up to the beginning of the late steps (see Figure 2). Assuming independence, this pair yields a collision with probability $p_\delta^2 \cdot P_F$.

The basic block we just described is quite promising. Indeed, when $p_\delta^2 < p_M$ we can expect an improved attack. However, matters are not that simple. Indeed, unless we are given a first pair (M, M'), we cannot construct the second pair. Thus, the basic block, by itself, at best doubles the number of candidate pairs. Luckily, when a large number of auxiliary differential paths can be found, which is a reasonable hypothesis since we are dealing with a small number of steps, we can apply the basic block many times. Assuming that $p_\delta = 1$, for each of t auxiliary differentials, we amplify a single candidate pair into 2^t pairs. Of course, we need to arrange the auxiliary differentials to make sure that they do not overlap or present other similar incompatibilities. When p_δ is smaller than 1 (but not too small), we still amplify a single pair into many.

After this overview of our adapted boomerang attack, the reader may rise two important objections. The first one is the fact that the independence hypothesis is extremely unnatural, because all these messages pairs are extremely correlated. Experimentally, this hypothesis is *false*, however, we remarked that for well-chosen differential characteristic, the bias induced by the dependencies is playing for the attacker and not against him. The main gain is that, since M and $M \oplus \Delta$ gives similar computations, the overall success probability of the two copies of each auxiliary differential is usually nearer to p_δ than p_δ^2. The second objection is that, at first, the early steps do not seem to come for free for the auxiliary differentials. This would be a major problem, since we want p_δ to be much better than p_M. In fact, we propose two different ways of putting together the message construction and the auxiliary differentials choice in order to effectively overcome this objection. Depending on the hash function under consideration and the properties of the differential characteristics in use, each has its own advantages.

3.1 Neutral Bits Approach

The first way to use the adapted boomerang attack is to note its similarity with the neutral bit technique proposed by Biham and Chen [1] at Crypto'04. There, the authors remarked in the case of SHA-0 that given a differential path, corresponding to our main path, it is possible to find so-called *neutral bits*. For a message pair that conforms to the differential characteristic up to some reference step, a neutral bit[2] is a bit of the message which when its value is flipped yields a new message pair that still conforms to the main path up to the reference step. In [1], the neutral bits are found using a guided exhaustive search technique. We argue that using auxiliary differential paths in place of or in addition to these neutral bits, leads to a better attack. Otherwise, this way of implementing our attack closely follows the method of Biham and Chen. The first step is to identify among a large list of candidate auxiliary differential paths those which works for the current message pair. Once this is done, we check, one pair at a time, whether the acceptable differentials are mutually compatible. Even without writing down the explicit algebraic conditions which need be satisfied for each differential, it is clear that this pairwise compatibility check only works for pairs of differential which do not strongly interact[3]. Then, build a large clique of mutually compatible differentials in the graph of pairwise compatible ones.

Once this clique is build, assume that it contains t auxiliary differentials and, using the basic technique presented above, construct the 2^t pairs of messages obtained by adding any subset of these differentials to the original message. We expect that a good proportion of the derived pairs conforms to the main characteristic up to the start of the final steps.

The main drawback of this technique is that the auxiliary differentials do not take advantage of the free early steps. Indeed, the original message pair is chosen independently of them, thus some probability must be paid for the early steps. This prevents us from using auxiliary differentials which are very good in the middle range but have a low probability of success in the early steps. It can be improved by trying to use the free steps both on the main characteristic path and on the auxiliary paths. However, if too many auxiliary paths are considered during a single step, the probability of making a correct choice becomes too low and no initial message pair can be constructed. The second approach given below gives a way out of this dilemma.

3.2 Explicit Conditions Approach

In order to get a good set of auxiliary characteristics, it is preferable to construct the first message pair carefully, forcing it to conform both to the main differential path and to the chosen auxiliary paths in the early steps. In order to do this, we should write down explicit conditions on bit values that are sufficient for each auxiliary characteristic to hold (or at least such that p_δ is increased). Once this is done, we can check whether the condition of the various characteristics are mutually compatible and, if so, we can choose the message values for each of the early steps, except the last one or two, in order to satisfy these explicit

---

[2] Here the term bit is taken in its information theoretic sense and may be a group of several elementary message bits which are all flipped simultaneously.

[3] Some long range interaction, such as carry propagation over several bits may be overlooked. However, they rarely occur anyway and can be ignored in a first approximation.

conditions. In the sequel, we call the partial message resulting from these choices a message seed. Note that, while simple as a principle, this approach requires a lot of specific work for each hash function in order to find a good way of writing and satisfying these explicit conditions.

After building a message seed, we can complete it in many ways on the one or two missing blocks to get an initial message pair. If the pair conforms to the main differential path far enough, we can use the neutral bit technique described above on the message pair, using as neutral bits the set of auxiliary paths that we have forced into the message. Compared to the straight neutral bit approach, the resulting auxiliary paths on the message pair are much more effective. Moreover, with this approach, we may be able to build auxiliary differential paths remaining conformant for more steps than in the case of neutral bits. In other words, the final steps will contain less steps than in the classical attacks such as neutral bits or message modification, and the total complexity will therefore decrease. To resume, while more complicated to set up in practice, this approach yields much better attacks.

For this method to succeed, we have to be able to build a main differential path containing all the sufficient conditions needed for every auxiliary differential paths we are planing to use. In order to make the approach efficient in practice, it would be very useful to have an automated tool that generates a main differential path satisfying those conditions. The availability and efficiency of such a tool greatly depend on the hash function we are considering. In the sequel, we show that the path generator proposed by De Cannière and Rechberger in [3] for SHA-1 can be used together with our boomerang approach.

4   Application to SHA-1

In this section, we show how our new attack applies for the case of SHA-1. After a short description of the algorithm and the state-of-the-art attacks, we explain how to build auxiliary differential paths, place them in a main differential path and use them during the collision search.

4.1   A Short Description of SHA-1

SHA-1 is a 160-bit dedicated hash function based on the design principle of MD4. Like most hash functions, SHA-1 uses the Merkle-Damgård paradigm [6,11] and thus only specifies a compression function. After a padding process, the message is divided into k blocks of 512 bits. At each iteration of the compression function h, a 160-bit chaining variable cv_i is updated using one message block m_{i+1}, i.e. $cv_{i+1} = h(cv_i, m_{i+1})$. The initial value cv_0 (also called IV) is predefined and cv_k is the output of the hash function.

The SHA-1 compression function is build upon the Davies-Meyer construction. It uses a function E as a block cipher with cv_i for the message input and m_{i+1} for the key input, a feed-forward is then needed in order to break the invertibility of the process: $cv_{i+1} = E(cv_i, m_{i+1}) \oplus cv_i$. This function is composed of 80 steps

(4 rounds of 20 steps), each processing a 32-bit message word W_i to update 5 32-bit internal registers $(A_i, B_i, C_i, D_i, E_i)$. Since more message bits than available are utilized, a message expansion is therefore defined.

Message expansion. First, m_i is split into 16 32-bit words $M_0, \ldots, M_{15}$. These 16 words are then expanded linearly into 80 32-bit words W_i, as follows:

$$W_i = \begin{cases} M_i, & \text{for } 0 \leq i \leq 15 \\ (W_{i-3} \oplus W_{i-8} \oplus W_{i-14} \oplus W_{i-16}) \lll 1, & \text{for } 16 \leq i \leq 79 \end{cases}$$

State update. First, the chaining variable cv_i is divided into 5 32-bit words to fill the 5 registers $(A_i, B_i, C_i, D_i, E_i)$. Then we apply 80 times the following transformation:

$$STEP_{i+1} := \begin{cases} A_{i+1} = (A_i \lll 5) + f_i(B_i, C_i, D_i) + E_i + K_i + W_i, \\ B_{i+1} = A_i, \\ C_{i+1} = B_i \ggg 2, \\ D_{i+1} = C_i, \\ E_{i+1} = D_i. \end{cases}$$

where K_i are predetermined constants and f_i are boolean functions defined in Table 1:

Table 1. Boolean function and constants in SHA-1

round	step i	$f_i(B, C, D)$	K_i
1	$1 \leq i \leq 20$	$f_{IF} = (B \wedge C) \oplus (\overline{B} \wedge D)$	0x5a827999
2	$21 \leq i \leq 40$	$f_{XOR} = B \oplus C \oplus D$	0x6ed6eba1
3	$41 \leq i \leq 60$	$f_{MAJ} = (B \wedge C) \oplus (B \wedge D) \oplus (C \wedge D)$	0x8fabbcdc
4	$61 \leq i \leq 80$	$f_{XOR} = B \oplus C \oplus D$	0xca62c1d6

We refer to [13] for a more exhaustive description. Note that all updated registers but A_{i+1} are just rotated copies, so we only need to consider the register A at each iteration. Thus, we have:

$$A_{i+1} = (A_i \lll 5) + f_i(A_{i-1}, A_{i-2} \ggg 2, A_{i-3} \ggg 2) + (A_{i-4} \ggg 2) + K_i + W_i.$$

4.2   Previous Attacks on SHA-1

Lots of research on SHA-1 has been conducted recently, but the major breakthrough has been published by Wang et al. [22]. They provided the first collision attack against the full SHA-1 algorithm, requiring only 2^{69} message modifications, which is lower than the 2^{80} hash computations expected for an ideal 160-bit hash function. This attack is possible thanks to a non-linear main differential path for SHA-1, given in the original paper. They also use a tool called *message modification technique* that allows to build a message pair of messages conforming to the main differential on the early and middle steps (approximatively step 22), thanks to clever modifications of message bits. Later, an unpublished result [20,21] claimed that using another non-linear main differential with more complex message modification techniques, one can keep the conformance up to step 25 approximatively and thus lower the complexity down to 2^{63} message modifications. The main problem with this approach is that message modifications can be costly during the collision attack and only the ones for the differential path in [22] are known. Note however that some recent work [17] tried to theorize this method.

Very recently, an interesting approach has been published by De Cannière and Rechberger [3] in order to find non-linear main differential paths in an automatic way. By introducing a sharp method to compute the probability of conformance and the number of messages (called nodes) one has to deal with at each step, they can use a heuristic algorithm to converge to a valid non-linear main differential path (prebuild from the Wang *et al.*'s disturbance vector). This algorithm allowed to compute a 2-block collision on a 64-step reduced version of SHA-1 (and more recently on a 70-step reduced version [4]). Note that this automatic tool did not improve the complexity of the previously explained collision attack against full SHA-1, since a non-linear main differential path was already known for that case.

We will show that using the boomerang attack for hash functions, we can improve the collision attacks for SHA-1 of a factor 32. We managed to place five auxiliary differentials maintaining conformance up to step 28 or even further (compared to step 25 approximatively for neutral bits or message modification). Another advantage of our new method is that once an auxiliary differential path is settled, the cost for using it is null, unlike the message modification case which can be quite demanding in terms of complexity [17]. The complex part of the boomerang attack in the explicit conditions approach only takes place during the main differential construction.

4.3 Building Auxiliary Differential Paths

In this section, our goal is to give an insight on how to build auxiliary differential paths for SHA-1. We want those paths to conform to the main differential one as far as possible. Since in the explicit conditions approach the main path is not yet known at this stage, a natural method would be to find auxiliary differentials leading to a collision on a late step. We also want the auxiliary differential paths to be as light as possible. If not so, the number of necessary conditions to have $p_\delta \simeq 1$ would quickly grow and this would be a problem while using the main path automated generator, which needs a lot of degrees of freedom in the message and in the registers.

Building a good auxiliary differential path is very close to building a main differential one. As observed in the latter case, the sparser the better. So in order to find good paths, we will use a well known tool for SHA-1 or SHA-0, introduced in [5]: the local collisions. This technique seems to make the attacker's job much more easier and minimize the number of differences one has to deal with. The idea is to avoid the inserted differences (called perturbations or disturbance vector) to spread among the registers by applying the necessary corrections on the expanded message (the perturbations and the corrections will therefore define the difference in the message). The problem arise that since the message is expanded, we do not have full control over the disturbance vector and thus this vector must respect the expansion as well. This is important when one has to build a main differential path, but here the problem is much more relaxed as we only deal with a few number of steps.

Local collisions. By inserting a difference on W_i^j at step $i+1$, another difference will appear on A_{i+1}^j. Note that a propagation of the difference to other bits of A_{i+1} may occur due to carry effect. To avoid this, we can set $W_i^j = A_{i+1}^j$. Then, at step $i+2$, the difference in A_{i+1}^j needs to be corrected and this can be done by setting $W_{i+1}^{j+5} = \overline{A_{i+1}^j}$. For steps $i+3$ to $i+5$, the behaviour highly depends on the boolean function f_i we are using (and thus the round we are into). Finally, at step $i+6$, we set $W_{i+5}^{j-2} = \overline{A_{i+1}^j}$ to correct the difference in A_{i+1}^j. At this point, we achieve the local collision: no more difference will appear in the next steps. We give in Table 2 all the constraints corresponding to the first round case ($f_i = f_{IF}$). If one respects all those constraints, the local collision occurs with probability 1.

Now that we know how to build local collisions, how do we use them ? We want the number of perturbations inserted in the early steps to be as low as

Table 2. Constraints for a local collision with a perturbation on W_i^j for the first round of SHA-1

step	type	constraints
$i+1$	no carry	$W_i^j = a,\ A_{i+1}^j = a$
$i+2$	correction	$W_{i+1}^{j+5} = \overline{a}$
$i+3$	no correction	$A_{i-1}^{j+2} = A_i^{j+2}$
	correction	$A_{i-1}^{j+2} \neq A_i^{j+2},\ W_{i+2}^j = \overline{a}$
$i+4$	no correction	$A_{i+2}^{j-2} = 0$
	correction	$A_{i+2}^{j-2} = 1,\ W_{i+3}^{j-2} = \overline{a}$
$i+5$	no correction	$A_{i+3}^{j-2} = 1$
	correction	$A_{i+3}^{j-2} = 0,\ W_{i+4}^{j-2} = \overline{a}$
$i+6$	correction	$W_{i+5}^{j-2} = \overline{a}$

possible (at most 5 in practice), in order to minimize the number of constraints on the message and the registers. Moreover, we want the auxiliary path to collide at some middle step k (with $k \geq 25$ in practice). It seems pretty clear that one will achieve this minimization by setting all the corrected perturbations in the 16 first message blocks on the same bit position, as remarked for the main differential path. Our goal is thus to have the first uncorrected perturbation as late as possible. By brute-forcing all the possibilities of this 16-bit mask and all the possibilities of propagation and corrections into the local collisions (corresponding to the f_{IF} case for round 1), we managed to find a lot of candidates (i.e. no difference in registers A_{k-4} to A_k). However, we added a filter: no perturbation should occur from W_{15} to W_{k-1} (note that a perturbation on W_k necessarily exists since we have the first difference on A_{k+1}). Indeed, a corrected perturbation introduced after step 14 would force some constraints on the message and the register outside the early steps where we have degrees of freedom. This would harden the final search of colliding messages. We even sharpen the filter by setting no perturbation from W_{11} to W_{k-1} to avoid problems with wrong bit position corrections due to the rotation in the expansion for the case SHA-1 (if the perturbation would occur on a bit j, some corrections would apply on bit $j+1$ and thus introduce unwanted differences). Finally, our auxiliary differential path will have no difference from register A_{12} to A_k. Note that a general rotation on the bit position does not change the validity of an auxiliary path.

We give in Table 3 the disturbance vector and the differences on the message for an auxiliary differential path with only three perturbations. In this example, the first uncorrected perturbation (underlined) comes on W_{24}^j and thus we get a collision at step 24. Here the three perturbations apply on step 1, 3, 11 and the corresponding constraints to force $p_\delta = 1$ are depicted in Table 4. Each bit a, b, c, d, e, f can take any value, as long as the $\overline{a}$, $\overline{b}$, $\overline{c}$, $\overline{d}$, $\overline{e}$, $\overline{f}$ constraints are fulfilled.

Table 3. Example of an auxiliary differential path, with the perturbation mask and its corresponding message differences for the 32 first steps. The rotation in the expansion is not taken in account.

	W_0 to W_{15}	W_{16} to W_{31}
perturbation mask	1010000000100000	
differences on W^j	1010000000100000	0000000010110110
differences on W^{j+5}	0101000000010000	0000000001011011
differences on W^{j-2}	0001111100000011	0000000000001110

We previously claimed that we were looking for an auxiliary differential path with $k \geq 25$, so why do we presented a $k = 24$ one ? In fact, even if a perturbation appears at step 25, there is a great probability, depending on the main differential path, that our pair remains conformant for some more steps[4]. We experimentally

---

[4] Said in other words, our auxiliary differential path will have a non-zero output difference.

Table 4. Example of an auxiliary differential path in the case $j = 2$: the constraints on the registers and on the message blocks. The MSB's are on the right and "-" stands for no constraint.

```
  i  |                     A_i                      |                     W_i
-----+----------------------------------------------+----------------------------------------------
 -1: | ------------------------------d----          |
 00: | ------------------------------d----          | ------------------------------------a--
 01: | ------------------------------e-a--          | ----------------------------ā-------
 02: | ------------------------------e---1          | ------------------------------b--
 03: | ----------------------------b-0              | --------------------------b̄------ā
 04: | ----------------------------0                | -------------------------------ā
 05: | ----------------------------0                | -------------------------------ā
 06: | ----------------------------                 | -----------------------------b̄
 07: | ----------------------------                 | -----------------------------b̄
 08: | ----------------------------                 | ------------------------------
 09: | --------------------------f----              | ------------------------------
 10: | --------------------------f----              | --------------------------------c--
 11: | --------------------------c--                | -------------------------c̄------
 12: | --------------------------0                  | ------------------------------
 13: | --------------------------0                  | ------------------------------
 14: | ----------------------------                 | ------------------------------c̄
 15: | ----------------------------                 | ------------------------------c̄
```

observed that this greatly depends on the bit position j where we plan to apply our auxiliary path, and the perturbation vector of the main path. For some very few values of j, the auxiliary differential has a small probability to succeed. However, in general, we have a good probability that a first perturbation at step n does not change the main differential conformance of a pair of messages up to step $n + 4$. We just have to choose the j values by avoiding critical positions.

4.4 Placing Auxiliary Differential Paths

We set ourselves in the case of a 2-block collision attack for SHA-1. For more details, we refer to [3]. The first part is thus to find a valid main path for the first block (with no difference on the IV). At this stage, our goal is to get the biggest clique of auxiliary differential paths by placing them in a main one. Since the main differential path automated tool from De Cannière and Rechberger is a heuristic algorithm, placing auxiliary paths in a main one is not a formal science. We tried different techniques but the best one seemed to be to force as much space between the constraints as we could. Note that when placing several auxiliary differential paths, some of them may have constraints in common. Even if not dramatic, we preferred to avoid this situation and strengthen the independence between the auxiliary paths (and thus use them as a clique as for neutral bits). Moreover, some positions are forbidden as the constraints on the message must apply on no-difference bits of the message only (otherwise, one of the message pair would not follow the auxiliary path). Lots of parameters are available when implementing or using the main differential automated tool and they highly influence the number of auxiliary constraints one can force. However, due to space restrictions, we omitted those details here.

We quickly recall in Table 5 the notations used in [3], but we encourage the reader to glance through the original paper. The final main path presented in Tables 8 and 9 contains the constraints of five independent auxiliary paths given in Table 3 and Table 4 at positions $j = \{9, 12, 15, 18, 21\}$.

Note that the auxiliary differential path used here has constraints on the IV ($A_{-1}^{j+2} = A_0^{j+2}$ in Table 4). It expresses the equality between two bits and thus happens with probability $1/2$ for each auxiliary differential. The prepended message computed to get the IV used in Table 8 is given in Table 6.

4.5   Using Auxiliary Differential Paths

Once a differential path is settled, one can easily generate a message instance conformant up to the end of the early steps since at this point the message blocks can be fixed independently. De Cannière and Rechberger use this fast generating technique coupled with a refinement of the differential path. Advanced approaches such as neutral bits or message modifications can decrease the complexity of the final attack, even if their power is reduced for SHA-1 compared to the SHA-0 case[5]. The boomerang attack for hash functions can be viewed as a generalization of those techniques and thus can be used as a neutral bits or message modification tool:

Neutral bit based. The easiest approach is to use a generalization of Biham and Chen neutral bit implementation guidelines together with two levels of message diversification. First, one constructs a base message with a large clique of simultaneously neutral bits which are in addition compatible with the auxiliary differential path. Then, one launches an enumeration that starts from this initial message and applies the neutral bits (using a Gray code encoding for efficiency). This yields many message pairs that follow the main differential path quite far. When the enumeration finds a message conformant up to round 25, a second level of enumeration diversifies this message using the auxiliary paths. The advantage of this technique is that it is quite easy to implement and that the neutral bits and the auxiliary paths can be addressed using very similar treatments. The main drawback is the gap between the range of ordinary neutral bits and the range of the auxiliary paths, which is a bit too wide and thus wastes degree of freedom in the message, compared to the theoretic complexity gain.

Message modification based. From a theoretical point of view, a message modification approach seems better. Indeed, the current best attack is message modification based and using it avoids the initial loss seen with the neutral bit approach. However, in addition to the implementation difficulties, using message

---

[5] some conditions on the message words coming from the late steps have to be fulfilled and in SHA-1 the rotation in the message expansion greatly increase the number of impacted message bits. This harden the neutral bits or message modification work since one has to check that the conditions remain valid after their use.

modification involves a much higher cost per message pair than the neutral bit approach. As a consequence, the apparent theoretical gain is less clear in practice.

Right now, our implementation of these ideas is not fast enough to allow full scale attacks. However, once an initial pair is found, the multiplicative effect works very well. For example, in Table 7 is the first message of a pair conformant until step 29, following the differential path from Table 8. Using the auxiliary differential paths provide 2^5 new conformant messages, the conformance limit is always between step 27 and 29. Note that this group of message words was generated using the neutral bit technique. This has the side effect of slightly changing the main characteristic during the message generation. More precisely, some bits (a, b or c in Table 4) of the auxiliary characteristics are flipped. Bit a is changed for the characteristic in positions 9,12,15; bit b for 12,15,18 and bit c for 15,18. The flipped bits are underlined in the given message. Of course, the slightly modified characteristic is still correct and compatible with the auxiliary ones (the 5 auxiliary differential paths remain valid).

4.6   Complexity Analysis for a Full Collision Attack

The literature has provided two ways of computing the complexity of a 2-block collision attack against SHA-1 : the number of conditions introduced by Wang *et al.* or the number of nodes introduced by C. De Cannière and C. Rechberger. Whatever the original collision attack we are using, our improvement decreases the complexity of a factor 32 since no message modification technique nor neutral bit can keep the conformance later than step 25. Moreover, the probability that a message being valid at step 25 is also valid at step 28 is lower than 2^{-5}.

We do not provide here any main path with auxiliary differentials for the second block since one needs the first block output values. However, experiments showed the same behaviour as for the first block case and the authors believe that the same technique can apply for the second part of the 2-block collision attack.

The reader could argue that we gave a differential path for the first block with a prepended message leading to an IV with chosen properties, and this will not be available for the second block stage of the attack. First, one has to note that the IV defined by the specifications of SHA-1 is strongly structured. Moreover, in a 2-block collision for SHA-1 the first block part costs much less than the second one (about a factor of 8), due to the possible misbehaviour of the final steps for the first block. Thus, by executing several times a first block research, the general complexity is not increased and we have enough degrees of freedom to start properly the second block: assuming the positions where we are placing the auxiliary differentials paths for the second block, the probability of satisfying the 5 constraints is 2^{-5} and 32 trials are required. However, this is not the case here since when reaching the end of the first block, the idea is to look at the available positions for including auxiliary differential. If enough positions are available, we try to construct a compatible main path. Thus, instead of having a single possibility with probability 2^{-5}, we have many. Experimentally, less than 4 tests of prepended messages are needed to apply the boomerang attack with five auxiliary paths.

5 Conclusion

In this paper, we showed that the boomerang attack which was initially devised as a cryptanalytic tool for block ciphers can be adapted to apply on iterated hash functions. Since the attacker model is quite different, due to the absence of keys and the impossibility to use a chosen ciphertext attack, the adaptation is not straightforward. Nonetheless, this new method leads to an improved cryptanalytic technique.

In order to illustrate this technique, we applied it to SHA-1 and obtained a significant improvement for collision attacks on this hash function. We believe that this method would also be powerful against other hash functions. Applying boomerang attack against SHA-0 or MD5 would be an interesting research topic. It may also be worth looking for more general auxiliary differential paths, for example by letting some local collisions slightly behave in a non-linear manner. Another future work could be to find a way to place more auxiliary differential paths in the main differential one, and thus lower the final complexity.

Acknowledgements

The authors would like to thank Christophe De Cannière and Christian Rechberger for their helpful advices when implementing their non-linear differential path automatic search tool.

References

1. Biham, E., Chen, R.: Near-Collisions of SHA-0. In: Franklin, M. (ed.) CRYPTO 2004. LNCS, vol. 3152, pp. 290–305. Springer, Heidelberg (2004)
2. Biham, E., Chen, R., Joux, A., Carribault, P., Lemuet, C., Jalby, W.: Collisions of SHA-0 and Reduced SHA-1. In: Cramer, R.J.F. (ed.) EUROCRYPT 2005. LNCS, vol. 3494, pp. 36–57. Springer, Heidelberg (2005)
3. De Cannière, C., Rechberger, C.: Finding SHA-1 Characteristics: General Results and Applications. In: Lai, X., Chen, K. (eds.) ASIACRYPT 2006. LNCS, vol. 4284, pp. 1–20. Springer, Heidelberg (2006)
4. Rechberger, C., De Cannière, C., Mendel, F.: In: Rump Session of Fast Software Encryption – FSE 2007 (2007)
5. Chabaud, F., Joux, A.: Differential Collisions in SHA-0. In: Krawczyk, H. (ed.) CRYPTO 1998. LNCS, vol. 1462, pp. 56–71. Springer, Heidelberg (1998)
6. Damgård, I.: A Design Principle for Hash Functions. In: Brassard, G. (ed.) CRYPTO 1989. LNCS, vol. 435, pp. 416–427. Springer, Heidelberg (1990)
7. Dobbertin, H.: Cryptanalysis of MD4. In: Gollmann, D. (ed.) Fast Software Encryption. LNCS, vol. 1039, pp. 53–69. Springer, Heidelberg (1996)
8. Joux, A., Peyrin, T.: Message modification, neutral bits and boomerangs. In: Proceedings of NIST 2nd Cryptographic Hash Workshop (2006)
9. Kelsey, J., Kohno, T., Schneier, B.: Amplified Boomerang Attacks Against Reduced-Round MARS and Serpent. In: Schneier, B. (ed.) FSE 2000. LNCS, vol. 1978, pp. 75–93. Springer, Heidelberg (2001)

10. Klima, V.: Tunnels in Hash Functions: MD5 Collisions Within a Minute. ePrint archive (2006), http://eprint.iacr.org/2006/105.pdf
11. Merkle, R.C.: One Way Hash Functions and DES. In: Brassard, G. (ed.) CRYPTO 1989. LNCS, vol. 435, pp. 428–446. Springer, Heidelberg (1990)
12. National Institute of Standards and Technology. FIPS 180: Secure Hash Standard (May 1993) available from http://csrc.nist.gov
13. National Institute of Standards and Technology. FIPS 180-1: Secure Hash Standard (April 1995) available from http://csrc.nist.gov
14. National Institute of Standards and Technology. FIPS 180-2: Secure Hash Standard (August 2002) available from http://csrc.nist.gov
15. Rivest, R.L.: RFC1321: The MD5 Message-Digest Algorithm (April 1992) available from http://www.ietf.org/rfc/rfc1321.txt
16. R.L. Rivest. RFC 1320: The MD4 Message Digest Algorithm (April 1992), http://www.ietf.org/rfc/rfc1320.txt
17. Sugita, M., Kawazoe, M., Imai, H.: Gröbner Basis based Cryptanalysis of SHA-1. In: Fast Software Encryption – FSE'07. LNCS, Springer, Heidelberg (2007), http://eprint.iacr.org/2006/098.pdf (to appear)
18. Wagner, D.: The Boomerang Attack. In: Knudsen, L.R. (ed.) FSE 1999. LNCS, vol. 1636, pp. 156–170. Springer, Heidelberg (1999)
19. Wang, X., Lai, X., Feng, D., Chen, H., Yu, X.: Cryptanalysis of the Hash Functions MD4 and RIPEMD. In: Cramer, R.J.F. (ed.) EUROCRYPT 2005. LNCS, vol. 3494, pp. 1–18. Springer, Heidelberg (2005)
20. Wang, X., Yao, A.C., Yao, F.: Cryptanalysis on SHA-1. In: Proceedings of NIST Cryptographic Hash Workshop (2005)
21. Wang, X., Yin, Y.L., Yu, H.: New Collision Search for SHA-1. In: Rump Session of CRYPTO (2005)
22. Wang, X., Yin, Y.L., Yu, H.: Finding Collisions in the Full SHA-1. In: Shoup, V. (ed.) CRYPTO 2005. LNCS, vol. 3621, pp. 17–36. Springer, Heidelberg (2005)
23. Wang, X., Yu, H.: How to Break MD5 and Other Hash Functions. In: Cramer, R.J.F. (ed.) EUROCRYPT 2005. LNCS, vol. 3494, pp. 19–35. Springer, Heidelberg (2005)
24. Wang, X., Yu, H., Yin, Y.L.: Efficient Collision Search Attacks on SHA-0. In: Shoup, V. (ed.) CRYPTO 2005. LNCS, vol. 3621, pp. 1–16. Springer, Heidelberg (2005)

Appendix

Table 5. Notations used in [3] for a differential path: x represents a bit of the first message and x^* stands for the same bit of the second message

(x, x^*)	$(0,0)$	$(1,0)$	$(0,1)$	$(1,1)$	(x, x^*)	$(0,0)$	$(1,0)$	$(0,1)$	$(1,1)$
?	✓	✓	✓	✓	3	✓	✓	-	-
-	✓	-	-	✓	5	✓	-	✓	-
x	-	✓	✓	-	7	✓	✓	✓	-
0	✓	-	-	-	A	-	✓	-	✓
u	-	✓	-	-	B	✓	✓	-	✓
n	-	-	✓	-	C	-	-	✓	✓
1	-	-	-	✓	D	✓	-	✓	✓
#	-	-	-	-	E	-	✓	✓	✓

Table 6. Message prepended to start with the IV used in Table 8

W_0	0x63e045ce	W_8	0x24b67e5d
W_1	0x362a3ed8	W_9	0x3898e2dd
W_2	0x5c333351	W_{10}	0x18be4543
W_3	0x76481862	W_{11}	0x60746d11
W_4	0x71a360ab	W_{12}	0x4cd56e7c
W_5	0x25e16eb9	W_{13}	0x1589d326
W_6	0x0419a9c2	W_{14}	0x19bab19c
W_7	0x5977272f	W_{15}	0x5fa6c656

Table 7. First message (in binary and hexadecimal) of an example pair following differential path from Table 8, conformant until step 29. Bits underlined are the bits flipped in the the main differential path from Table 8 due to the neutral bits technique.

M_0	11111101100111111111011111111011	0xfd9ff7fb
M_1	01110101000001010011111101110001	0x75053f71
M_2	00011100000111010111001100011111	0x1c1d731f
M_3	00000111001110000000010011111001	0x07380279
M_4	11110101101011101000100000101001	0xf5ae8829
M_5	00110101111110101100101101010011	0x35facb53
M_6	00010000011111001010101100011001	0x107cab19
M_7	10100110111111100110001101101001	0xa6fe6369
M_8	01001000001100111010100101011101	0x4833a95d
M_9	01100000000110110110100111101100	0x601b69ec
M_{10}	10100011010010100100111001100100	0xa34a4e64
M_{11}	01011100100111101011111100100111	0x5c9ebf27
M_{12}	10111011010000110101001001110111	0xbb435277
M_{13}	10100101011101110100110011010100	0xa5774cd4
M_{14}	11111110011110111011010000000000	0xfe7bb400
M_{15}	10110101001110111010110101101011	0xb53bad6b

Table 8. Steps 1 to 39 of the main differential path of the first block. The constraints needed for the first auxiliary differential path (in position j=9) are underlined.

i	A_i	W_i
-4:	0010100101001101110010010100011	
-3:	0000011110000100011001010100010	
-2:	1101100001000010100111110101111	
-1:	0101101111011110110<u>1</u>01111010001	
00:	0100001010110111011110<u>1</u>01100110 11	1uu1110110011111010110--<u>0</u>111111011
01:	n1n01011100101100100<u>1</u>-<u>0</u>100100110	nuu101-10001011--<u>1</u>11111101u1n0n1
02:	1nu11--011111011111101<u>1</u>01<u>1</u>11111u1	--n11-----0-10-111100<u>0</u>1<u>1</u>0n0111uu
03:	nnu00-----0-00-011000011<u>0</u>111110n	x-nn-1--1--01010001001--<u>1</u>u111001
04:	u010u11-0--00010010110-10<u>1</u>0un0u1	uu-u0-------11-0--101100<u>1</u>n1n10nu
05:	1001u00-0--000000000001u<u>0</u>0011010	nn-u0-------11010111--1--<u>1</u>1n100u1
06:	011unnnnnnnnnnnnnnnn1---110n001uu	00n-------1-1--1--0011110<u>0</u>011001
07:	u110-01000000u010110nu111uu1010n	1nu001------1--1-100-1-1<u>0</u>-un-0n-
08:	1111010111111---011unu110-0--nu1	-un0----------11---------u0111nu
09:	-0010---1--1--01-0u-<u>1</u>0nnnnu01010	--u0-------------1--1001-u1--100
10:	--------1--1--0--01-<u>1</u>01nu1111u10	xxu00-----0--1--1--0--<u>1</u>--u----n-
11:	0---------0--1--1--0n-<u>1</u>00nn0u1n0	-xn--1--0--0--1--<u>0</u>---11-0010--x-
12:	0---0-------0--0--0--01-<u>0</u>10n1-nn	x----------------------------u
13:	00----------0--0--0--0010<u>0</u>n0n-00	--10------------------0--1n1----
14:	-0--0-----------------10001u0un-	---1---------1--0--0--1-<u>0</u>00---xn
15:	n----------------------unnn1101	-x-10-------1--0--0--1-<u>0</u>u-n--u-
16:	--1--------------------1--nu001	-n0--------------------1u0-----
17:	n-0--------------------111-0n	xxn-----------------1---1u-x--n-
18:	-11--------------------101-	x-u1------------------0----0--
19:	-----------------------u-	x--------------------11n------
20:	-------------------------	--x--------------------------x
21:	------------------------x	--n--------------------xx-----
22:	-------------------------	x--------------------1-----x
23:	-----------------------x-	-x-------------------x---x-
24:	-----------------------x-	xu-------------------x---xx
25:	------------------------x	-x-------------------x---x-
26:	-------------------------	----------------------------xx
27:	-----------------------x-	-x-------------------x---x-
28:	-----------------------x-	xx-------------------x---xx
29:	------------------------x	xx-------------------x---x-
30:	-------------------------	----------------------------x
31:	-------------------------	-x-------------------------x-
32:	-----------------------x-	xx-------------------x---xx
33:	------------------------x	-x-------------------xx---x-
34:	-------------------------	x--------------------------x
35:	-----------------------x-	-x-------------------x---x-
36:	-----------------------x-	-x-------------------x---x-
37:	-------------------------	-x-------------------------x-
38:	-------------------------	--------------------------x-
39:	-----------------------x-	---------------------x-----
	. . .	. . .

Table 9. Steps 40 to 80 of the main differential path of the first block

i	A_i	W_i
	...	...
40:	--------------------------------	x------------------------------x-
41:	--------------------------------	x--------------------------------
42:	--------------------------------	x------------------------------x-
43:	------------------------------x-	x-----------------------x------
44:	--------------------------------	--------------------------------
45:	------------------------------x-	x-----------------------x------
46:	--------------------------------	x--------------------------------
47:	------------------------------x-	------------------------x------
48:	--------------------------------	x--------------------------------
49:	------------------------------x-	------------------------x------
50:	--------------------------------	x------------------------------x-
51:	--------------------------------	--------------------------------
52:	--------------------------------	x--------------------------------
53:	--------------------------------	x--------------------------------
54:	--------------------------------	--------------------------------
55:	--------------------------------	--------------------------------
56:	--------------------------------	--------------------------------
57:	--------------------------------	--------------------------------
58:	--------------------------------	--------------------------------
59:	--------------------------------	--------------------------------
60:	--------------------------------	--------------------------------
61:	--------------------------------	--------------------------------
62:	--------------------------------	--------------------------------
63:	--------------------------------	--------------------------------
64:	--------------------------------	----------------------------x--
65:	----------------------------x--	-----------------------x------
66:	--------------------------------	----------------------------x--
67:	--------------------------------	----------------------------x--x
68:	----------------------x---	----------------------x------x
69:	--------------------------------	----------------------------x--x
70:	--------------------------------	---------------------------x--x-
71:	------------------------x----	---------------------x------x-
72:	--------------------------------	----------------------------xx-x-
73:	------------------------x---	----------------------x--x--x--
74:	----------------------x-----	----------------------x------xx--
75:	--------------------------------	----------------------------x--xx-
76:	--------------------------------	----------------------------x--x-x-
77:	--------------------x------	--------------------x------x-x-
78:	--------------------------------	-----------------------xx-----
79:	------------------------x-x---	------------------x-xx--x----
80:	----------------------x-------	

Amplifying Collision Resistance:
A Complexity-Theoretic Treatment

Ran Canetti[1,*], Ron Rivest[2], Madhu Sudan[2],
Luca Trevisan[3,**], Salil Vadhan[4,***], and Hoeteck Wee[3,†]

[1] IBM Research
canetti@us.ibm.com
[2] MIT CSAIL
{rivest,madhu}@mit.edu
[3] UC Berkeley
{luca,hoeteck}@cs.berkeley.edu
[4] Harvard University
salil@eecs.harvard.edu

Abstract. We initiate a complexity-theoretic treatment of hardness amplification for collision-resistant hash functions, namely the transformation of weakly collision-resistant hash functions into strongly collision-resistant ones in the standard model of computation. We measure the level of collision resistance by the maximum probability, over the choice of the key, for which an efficient adversary can find a collision. The goal is to obtain constructions with short output, short keys, small loss in adversarial complexity tolerated, and a good trade-off between compression ratio and computational complexity. We provide an analysis of several simple constructions, and show that many of the parameters achieved by our constructions are almost optimal in some sense.

Keywords: collision resistance, hash functions, hardness amplification, combiners.

1 Introduction

Constructing collision-resistant hash functions is a central problem in cryptography, both from the foundational and the practical points of view. The goal is to construct length-decreasing functions for which it is infeasible to find two distinct inputs with the same output. This problem has received much attention over the past two decades. Still, coming up with constructions that are efficient enough to be of use in practice and at the same time enjoy rigorous security guarantees (say, based on the hardness of some well-studied problem) has turned out to be elusive. We also seem unable to construct collision-resistant functions from potentially simpler primitives, c.f. [25]. The

---

* Supported by NSF grants CFF-0635297 and Cybertrust 0430450.
** Supported by NSF grant CCF-0515231.
*** Supported by NSF grants CNS-0430336 and CCF-0133096, and ONR grant N00014-04-1-0478.
† Work done while visiting IBM Research, IPAM and Columbia University, the latter supported by NSF grants CCF-0515231 and CCF-0347839.

A. Menezes (Ed.): CRYPTO 2007, LNCS 4622, pp. 264–283, 2007.

problem is highlighted by the repeated attacks on the popular MD4, MD5 and SHA1 hash functions (refer to [20] and references therein).

Given this state of affairs, it is natural to ask whether one can "bootstrap" collision resistance by constructing "full-fledged" collision-resistant hash functions (CRHF) from "weak" ones. That is, are there general mechanisms for transforming hash functions, for which it is "somewhat easy" (but not completely trivial) to find collisions, into one for which it is infeasible to find collisions? In addition to providing rigorous ways to improve the collision resistance of hash functions, such mechanisms could in themselves suggest methodologies for constructing hash functions "from scratch".

Several works propose design principles for hash functions, e.g. [17,4,14,3]. These mechanisms can indeed be regarded as "hardness amplification" mechanisms for collision-resistant hash functions. However, with the exception of [4], which concentrates on increasing the domain size of the hash function, all the analyses provided for these mechanisms use idealized models of computation, such as modeling the underlying building blocks as random functions. Consequently, we do not currently have constructions that are guaranteed to provide some level of collision resistance in the standard model of computation, under the sole assumption that the underlying building blocks have some weaker collision resistance properties. (Recently, the closely related problem of constructing "combiners" for hash functions has been studied in the standard model [2,19]; we discuss this problem in more detail below.)

This state of the art should be contrasted with the "sister problem" of constructing one-way functions. Here we have a well-established theory of hardness amplification [27] (see also [11]). That is, we have concrete notions of "strength" of one-way functions, and constructions that are guaranteed to provide "strong" one-way functions based on the sole assumption that the underlying building block is a "weak" one-way function. Several lower bounds for "black-box" hardness amplification are also known, e.g. [23,15].

We note that collision resistance often exhibits very different properties than one-wayness. For one, constructing collision-resistant hash functions calls for different design principles (e.g. the proposed expander-based one-way function of [10] is very bad as a collision-resistant function). Furthermore, both practice and theory indicate that collision resistance is considerably harder to achieve than one-wayness, e.g. [6,26,25]. Still, except for some specific points highlighted within, we show that it is possible to translate much of the analysis used in the study of amplification of one-wayness to the setting of collision resistance.

1.1 This Work

We initiate a study of amplification of collision resistance, in a standard reductionist complexity-theoretic framework. That is, we first provide a measure for the "level" of collision resistance of hash functions. We then consider some constructions and quantitatively analyze the amount in which they amplify the collision resistance, along with a number of efficiency parameters (discussed below).

Model for hash functions. Following [4], we model hash functions as a *family* of functions, where a function in the family is specified via a *key*. Security is analyzed for the case where the key is chosen at random (from the space of keys) and made public. We point out several advantages of this approach. Refer to [21] for a more detailed discussion. First, it allows for a natural modeling of the adversary as an algorithm (a circuit) that takes for input a key κ identifying a function h_κ in the family and tries to output a collision $x_0 \neq x_1$ such that $h_\kappa(x_0) = h_\kappa(x_1)$. (Such modeling is not possible for single functions since for any length-reducing function there always exists an adversary that outputs a collision for that function in constant time.) Second, this approach supports a simple and natural quantitative measure for the level of collision resistance: the level of collision resistance is the maximum probability, over the choice of the key, with which an efficient adversary can find a collision. Third, current constructions of hash functions can be naturally regarded as keyed function families. For instance, we may interpret the initialization vector (IV) in SHA0 and SHA1 as a key. Finally, several collision-finding attacks seems to depend on specific values or properties of the key in use and work for some keys but not others. Specific examples include Dobbertin's attack on MD5 [6], time-memory trade-off attacks, and attacks on Gibson's hash function [8]. In particular, it may well be possible that even "broken" functions still have a significant fraction of keys for which attacks are less successful. On the other hand, it may not be sufficient to simply view an IV as a key, because the IV may not be incorporated into the computation in a sufficiently strong way; see the discussion at the end of the introduction.

Parameters. We consider the following parameters for hash functions and hardness amplification. First and foremost is the level of collision resistance. The goal in hardness amplification is to reduce the maximum probability that an efficient adversary can find collisions from $1 - \delta$ to ϵ, where ϵ and δ are typically $o(1)$. Another salient parameter is the output length. Other parameters include the key size, the number of applications of the underlying hash function, the the running time (or, complexity) of the adversaries considered and the "compression ratio" (i.e the ratio of input length to output length). By itself, the compression ratio is less interesting since we may apply a transformation due to Merkle and Damgård [17,4] to increase the compression ratio arbitrarily; this increases the number of applications of the underlying function but maintains the same key size and output length. Our goal is to construct hash functions with a high level of collision resistance, while maintaining short outputs, short keys, and a good trade-off between compression ratio and number of operations.

Constructions. We analyze two construction for hardness amplification. The first is based on simple concatenation (possibly folklore) and the second uses error-correcting codes and was suggested by Knudsen and Preneel [14]. Then, we analyze two additional constructions for reducing the key size and the output length respectively.

Amplification via concatenation. The first construction is simple concatenation: we hash the input using several independently chosen functions and concatenate the hash values. Formally, given a family $\mathcal{H} = \{h_\kappa\}$ of hash functions, and a parameter q, define the family $\mathcal{H}' = \{h'_{\kappa_1,...,\kappa_q}\}$ so that $h'_{\kappa_1,...,\kappa_q}(x) = h_{\kappa_1}(x) \circ ... \circ h_{\kappa_q}(x)$, where $\kappa_1, ..., \kappa_q$ are independently chosen keys in the family $\mathcal{H}$. The analysis is essentially the same as that for classic hardness amplification for one-way functions [27]. The underlying intuition is that finding collisions in $h'_{\kappa_1,...,\kappa_q}$ is hard as long as finding collisions in one of $h_{\kappa_1}, \ldots, h_{\kappa_q}$ is hard. If the initial maximal probability of finding collisions is δ, the maximal probability of finding collisions in the new hash family is $(1 - \delta)^{\Omega(q)} = e^{-\Omega(\delta q)}$. This means that the improvement in the level of collision resistance is exponential in q whereas the output length is linear in q.

Amplification via codes. In the second construction, we first encode the input with an error-correcting code wherein the codeword has length q over some large alphabet. Next, we hash the encoded input using q independently chosen functions (one for each of the q symbols in the codeword) and concatenate the hash values as before. In order to find a collision for this construction, one has to find collisions in *many* of the q underlying hash functions (as opposed to *all* q functions as in the previous construction). This construction was previously analyzed in an idealized setting in [14].

The analysis relies on the idea that finding collisions in the new hash function is hard as long as finding collisions in *several* of the q functions is hard (as opposed to finding collision in just a single function). Indeed, if the initial maximal probability of finding collisions is δ, then we expect that it is hard to find collisions in δq functions. To exploit this, we use a code with minimum distance $(1 - O(\delta))q$, and for such codes, we may achieve a rate of $\Omega(\delta)$. Consequently this construction allows us to hash an input that is longer by a factor of $\Theta(\delta q)$ (compared to the first construction) while still using only q invocations of hash functions from the given family. When compared to amplifying the domain size via the Merkle-Damgård transformation and then applying the first construction, the second construction offers a $\Theta(\delta q)$ factor improvement in the number of hashing operations. The price we pay for this improvement is that for the same δ, ϵ (i.e., for fixed levels of collision resistance in the underlying and target hash functions), the choice of q for the second construction is a constant multiplicative factor larger than that for the first construction.

We remark that this analysis yields also hardness amplification for one-way functions with a logarithmic factor improvement in the security reduction.

Reducing the key size. Next, we demonstrate how to modify both constructions so that the key size increases only by an *additive* logarithmic term (at the price of increasing the output length by a constant multiplicative factor). This is done by choosing the q keys via randomness-efficient sampling using expander graphs. The sampler we require for the concatenation construction is fairly standard (e.g. randomness-efficient samplers were exploited in a similar manner in [5]), whereas the coding-theoretic construction requires a modified analysis of a previous sampler [9].

Reducing the output length. Starting with a family $\mathcal{H}$ of hash functions with output length ℓ_{out} and parameter q, the first two constructions yield a family with output length $q\ell_{out}$. We show that for any Δ, we may in fact reduce the output length to $q \cdot (\ell_{out} - \Delta)$. More generally, we show how to transform any family $\mathcal{H}$ with output length ℓ_{out} into one with output length $\ell_{out} - \Delta$ with a negligible loss in the level of collision resistance. However, the complexity of computing the function increases by a multiplicative factor of 2^Δ, so the construction is only useful for logarithmic values of Δ.

Limitations. We point out some of the limitations of our constructions and try to justify them. A first limitation is that, given a guarantee on the resilience of $\mathcal{H}$ against adversaries of a given size, we can only guarantee resilience of the new hash family $\mathcal{H}'$ against adversaries of much smaller size. A similar limitation is shared by existing hardness amplification results for one-way functions. This may be expected, given that all our constructions, as well as all existing constructions for hardness amplification of one-way functions are "black box". Indeed, evidence that such limitation may be inherent in "black-box constructions" is given in [11, Chapter 2, Ex 16, p. 96]. In addition, our constructions increase both the complexity of the hashing and the output length. To explain the blow-up in these parameters, we provide lower bounds on the number of hashing operations and output length:

- We establish a matching lower bound (up to multiplicative constants) on the number of hashing operations used in our first two constructions. The bound holds for black-box constructions that do not use the input as keys for the underlying hash functions. In particular, the number of hashing operations must have an inverse dependency on δ, the initial maximal probability of finding collisions. The bound is derived from that for hardness amplification for one-way functions in [15].
- Assuming in addition some natural restrictions on the reduction used in the proof of security, we show that the output length of the new hash function is at least $\Omega(\frac{1}{\delta} \cdot \ell_{out})$. Our constructions achieve output length $O(\frac{1}{\delta} \cdot \ell_{out} \cdot \log \frac{1}{\epsilon})$.

While the guarantees provided by our constructions may be too weak to be of real practical significance, this is unfortunately the state of the art for general constructions. Providing better guarantees remains a fascinating open problem.

Combiners. Our results pertaining to the output length (namely the fourth construction and lower bounds thereof) build on the recent work on **black-box combiners** for collision resistance [2,19,12]. We briefly recall the notion and results and explain the connection to hardness amplification.

Black-box combiners for collision resistance. A black-box combiner for collision resistance is a procedure that given t functions $h^1, \ldots, h^t$ with output length ℓ_{out}, computes a single function $\tilde{h}$ with the following property: there is an efficient transformation that given a collision for $\tilde{h}$, outputs collision for each of $h^1, \ldots, h^t$. This guarantees that finding collisions on $\tilde{h}$ is hard as long as finding collisions on *one* of $h^1, \ldots, h^t$ is hard. Concatenating the outputs of $h^1, \ldots, h^t$ on the same input yields a combiner with output length $t \cdot \ell_{out}$. Boneh, Boyen and Pietrzak [2,19] showed that this trivial combiner is essentially optimal by giving a $t \cdot (\ell_{out} - O(\log n))$ lower bound for *deterministic* black-box combiners.

Black-box combiners for collision resistance arise naturally in the context of our work. Indeed, our first hardness amplification construction may be viewed as choosing $\kappa_1, \ldots, \kappa_q$ at random and applying the trivial (deterministic) combiner to $h_{\kappa_1}, \ldots, h_{\kappa_q}$. In addition, since we deal with *families* of functions rather than with single functions, it makes sense in our model to consider also *randomized* combiners (still, for single functions). We can then incorporate any randomness used by the combiner in the key of the new hash family. Two natural questions arise here: Can we beat the [2,19] bound by using randomized combiners? Alternatively, can the bound be improved by removing the additive logarithmic factor?

We answer both questions negatively. We first extend the lower bound of [2,19] to derive a $t \cdot (\ell_{out} - O(\log n))$ lower bound on the output length of randomized black-box combiners. Our lower bound for the output length for hardness amplification builds on this lower bound. We then construct a randomized black-box combiner with output length $t \cdot (\ell_{out} - \Omega(\log n))$. This result is interesting in itself, since it is the first non-trivial combiner that beats concatenation. Furthermore, this combiner underlies our fourth construction mentioned above, which reduces the output length of hash functions. Putting these two results together, we deduce that the optimal randomized black-box combiner has output length $t \cdot (\ell_{out} - \Theta(\log n))$.

Combiners for families of hash functions. So far, we've discussed the relationship between combiners for single functions and hardness amplification for function families. In addition, one may directly study combiners for *families* of functions: Given t families of hash functions with output length ℓ_{out}, construct a single family of hash functions that is collision-resistant as long as one of the t families is collision-resistant. We note that it is possible to construct a combiner having output length $t \cdot (\ell_{out} - O(\log n))$ using our randomized black-box combiner. The concurrent work of Fischlin and Lehmann [7] studies a very similar problem, albeit in an idealized model that only admits generic attacks on the hash functions.

Extensions. Our positive results for hardness amplification of collision resistance may be extended to several other variants of collision resistance. Details of these extensions are deferred to the final version of the paper.

Resistance to correlations. As noted in previous work (e.g. [1]), collision resistance can be regarded as a special case of "resistance to finding correlations." That is, for a given k-ary relation R, say that a family of functions $\mathcal{H}$ is R-**resistant** if it is hard given a random $h \in \mathcal{H}$ to find $x_1, \dots x_k$ such that $R(h(x_1), \dots, h(x_k))$ holds. In this terminology, collision resistance is simply R_{eq}-resistance where $R_{eq}(y_1, y_2)$ iff $y_1 = y_2$. Can R-resistance be amplified for other relations? Can collision resistance be derived from (or imply) R-resistance for other relations R? These are interesting questions.

As a small step in this direction, we consider amplification for the "near collision" relation R_{near}, where $R_{near}(y_1, y_2)$ iff the Hamming distance between y_1 and y_2 is small (see e.g. [16, Sec 9.2.6]). We observe that by encoding the hash value with an error-correcting code, we may transform a standard collision-resistant hash family to a near-collision-resistant hash family. Conversely, given a near-collision-resistant hash family, one can construct a standard collision-resistant hash family with shorter output by "decoding" the hash value to the nearest codeword of a covering code. This yields an amplification theorem for resistance to near-collisions, as a corollary of our amplification theorems for collision resistance.

Target collision resistance. Our results extend also to the related notion of target collision resistance (namely, universal one-way hash functions [18]). Here we may use the same constructions as for collision resistance, except to replace the Merkle-Damgård domain expansion with that of Shoup [24], and the same analysis goes through. We stress that the extension should not be taken for granted, because techniques for collision resistance do not always extend readily to target collision resistance; domain expansion is a good example.

Discussion. We discuss some additional aspects of the analysis in this work. First, we address only collision resistance, which is one out of many desired properties of "cryptographic hash functions". In fact, we do not even address properties such as resistance to finding additional collisions, once a collision is found. Concentrating on plain collision resistance allows for clearer understanding. In fact, constructing hash functions achieving even this specific property seems to be challenging enough, as evidenced by the attacks on MD5 and SHA1.

Another point worth highlighting is that our analysis can be viewed as a demonstration of the benefits in having *families* of hash functions, where there is some assurance that finding collisions in one function in the family does not render other functions in the family completely insecure. This may suggest a methodology for *constructing* practical collision-resistant functions: Design such functions as keyed functions, where the key is intimately incorporated in the evaluation of the function. This might give some hope that finding collisions for one value of the key might not help much in finding collisions for other values of the key. Then, apply a generic amplification mechanism such as the ones studied here to guarantee strong collision resistance *even when a significant fraction of the keys result in weak functions*. We stress that, in order to be of value, the key has to be incorporated in the computation of the function in a strong way. This fact is exemplified (in the negative) by the MD/SHA line of functions: Although these functions are often modeled as families of functions that are keyed via the IV, the actual constructions do not incorporate the IV in the computation in a strong way. And, indeed, the very recent attacks against such functions (e.g. [26]) seem to work equally well for all values of the IV. Similarly suspect are related methods for creating a hash function family from a fixed hash function by treating a portion of the input as key.

Finally, we stress that even though we use asymptotic notation to make our results more readable, they actually provide concrete bounds on the parameters achieved. Moreover, we provide uniform reductions in all of our proofs of security, so even though the positive results are stated for nonuniform adversaries, it is easy to derive an analogue of those results for uniform adversaries.

Organization. We begin with by reviewing quantitative definitions of collision resistance for CRHFs in Section 2. We present all of our constructions for hardness amplification, key size reduction and output length reduction in Section 3, and our lower bounds in Section 4. Given that randomized black-box combiners are a recurring tool in this paper, we define them in Section 2 and present the construction in Section 3 and the matching lower bound in Section 4.

2 Preliminaries

2.1 Quantitative Definitions of Collision Resistance

A family of hash functions is a collection of polynomial-time computable functions $\mathcal{H} = \{\mathcal{H}_n : \{0,1\}^{\ell_{\text{key}}(n)} \times \{0,1\}^{\ell_{\text{in}}(n)} \rightarrow \{0,1\}^{\ell_{\text{out}}(n)}\}$, where n is the security parameter, satisfying $\ell_{\text{out}}(n) < \ell_{\text{in}}(n)$. We refer to $\ell_{\text{in}}, \ell_{\text{out}}, \ell_{\text{key}}$ as the input length, output length and key size of the hash function. We use $h_\kappa : \{0,1\}^{\ell_{\text{in}}(n)} \to \{0,1\}^{\ell_{\text{out}}(n)}$ to denote the function $\mathcal{H}_n(\kappa, \cdot)$ associated with the key $\kappa \in \{0,1\}^{\ell_{\text{key}}(n)}$. We call a pair (x_0, x_1) satisfying $x_0 \neq x_1$ and $h_\kappa(x_0) = h_\kappa(x_1)$ a collision for h_κ.

For any n, we say that $\mathcal{H}_n$ is an (s, ϵ)-CRHF (collision-resistant hash function) if for every nonuniform A of size s,

$$\Pr[\kappa \leftarrow \{0,1\}^{\ell_{\text{key}}(n)}; A(\kappa) \text{ outputs a collision for } h_\kappa] < \epsilon$$

(The quantity ϵ is what we refer to in the introduction as the level of collision resistance.) For notational simplicity, we omit references to n whenever the context is clear (e.g. $\mathcal{H} : \{0,1\}^{\ell_{\text{key}}} \times \{0,1\}^{\ell_{\text{in}}} \to \{0,1\}^{\ell_{\text{out}}}$).

We will also refer to asymptotic notions of CRHFs. As with one-way functions, we want to consider the entire class of nonuniform polynomial-time adversaries (although we do provide uniform reductions in our proofs of security). Formally, we say that $\mathcal{H}$ is a strong CRHF if for every polynomial $p(\cdot)$ and every sufficiently large n, $\mathcal{H}$ is a $(p(n), \frac{1}{p(n)})$-CRHF. Similarly, we say that $\mathcal{H}$ is a weak CRHF if there exists a constant c such that for every polynomial $p(\cdot)$ and every sufficiently large n, $\mathcal{H}$ is a $(p(n), 1 - \frac{1}{n^c})$-CRHF. Standard cryptographic applications of hash functions actually require strong CRHFs, so whenever the strength of the CRHF is not qualified, we will refer to strong CRHFs.

Public-coin vs. secret-coin hash functions. As noted in [13], a distinction needs to be made between public-coin and secret-coin hash functions. In a public-coin hash function, the key corresponds to a uniformly generated random string and the key generation algorithm computes the identity function. In a secret-coin hash function, the distribution of the key may be any samplable distribution. For simplicity and clarity, our definition and exposition refer to public-coin hash functions. It is easy to see that all of our constructions (Constructions 1, 2 and 4) apart from the reduction in key size using randomness-efficient sampling extend to secret-coin hash functions.

2.2 Black-Box Combiners for Collision Resistance

We generalize the notion of black-box combiners from [2,19] so as allow randomized constructions.

Definition 1. *We say that (C, R) is a* randomized black-box (t', t)-combiner for collision resistance *if C, R are deterministic poly-time oracle TMs, and there exists some negligible function $\nu(\cdot)$ such that for all $h^1, \ldots, h^t : \{0,1\}^{\ell_{\text{in}}} \to \{0,1\}^{\ell_{\text{out}}}$:*

CONSTRUCTION. *For every r, $C^{h^1,\ldots,h^t}(r, \cdot)$ computes a function $\tilde{h}_r : \{0,1\}^{\ell_{\text{in}}'} \to \{0,1\}^{\ell_{\text{out}}'}$, where $\ell_{\text{in}}' > \ell_{\text{out}}'$.*

REDUCTION. *With probability $1 - \nu(n)$ over r: if $(\tilde{x}_0, \tilde{x}_1)$ is a collision for $\tilde{h}_r$, then $R^{h^1,\ldots,h^t}(r, \tilde{x}_0, \tilde{x}_1)$ outputs t pairs $(x_0^1, x_1^1), \ldots, (x_0^t, x_1^t)$ such that for at least $t - t' + 1$ values $i \in \{1, \ldots, t\}$, (x_0^i, x_1^i) is a collision for h^i.*

Intuitively, the guarantee is that if it is hard to find collisions on some t' of the functions $h^1, \ldots, h^t$, then with overwhelming probability over r, it is hard to find collisions on $\tilde{h}_r$. Our definition generalizes that in [2,19] in that we provide both C and R with additional "randomness" r, which is interpreted as a key. Specifically, in the previous definitions, C computes a single function, whereas in our definition C computes a family of functions $\{\tilde{h}_r\}$. In our construction, R is deterministic, whereas our lower bound (as with previous work) extends to randomized reductions R.

3 Constructions

The goal of hardness amplification is to deduce the existence of strong CRHFs from weak CRHFs. Fix a security parameter n. The parameters for the new CRHF $\mathcal{H}'$ will be different from those for the starting CRHF $\mathcal{H}$: we use $\ell_{\mathsf{in}}, \ell_{\mathsf{out}}, \ell_{\mathsf{key}}$ to denote the parameters for a $(s, 1 - \delta)$-CRHF that we start with, and $\ell_{\mathsf{in}}', \ell_{\mathsf{out}}', \ell_{\mathsf{key}}'$ to denote the parameters for the (s', ϵ)-CRHF that we are about to construct. Typical values of the parameters are $\delta = \frac{1}{\mathrm{poly}(n)}$ and $\epsilon = \mathrm{neg}(n)$. As outlined in the introduction, we begin two basic constructions for hardness amplification (Sections 3.1 and 3.2) and then show how to reduce the key size (Section 3.3) and output length (Section 3.4). A summary of the parameters is given in Fig 1.

Domain expansion. We compensate the loss in compression ratio in our constructions by first applying Merkle-Damgård domain expansion [4,17], noting that domain expansion for collision resistance preserves the hardness parameter.

Proposition 0 ([4,17]). Fix some security parameter n. Suppose there exists a (s, ϵ)-CRHF $\mathcal{H}_n$ from $\{0, 1\}^{\ell_{\mathsf{key}}} \times \{0, 1\}^{\ell_{\mathsf{in}}}$ to $\{0, 1\}^{\ell_{\mathsf{out}}}$ computable in time T. Then, Construction 0 yields an (s', ϵ)-CRHF $\mathcal{H}'_n$ from $\{0, 1\}^{\ell_{\mathsf{key}}'} \times \{0, 1\}^{\ell_{\mathsf{in}}'}$ to $\{0, 1\}^{\ell_{\mathsf{out}}'}$ with the following parameters:

- $\ell_{\mathsf{out}}' = \ell_{\mathsf{out}}$ and $\ell_{\mathsf{key}}' = \ell_{\mathsf{key}}$
- # hash calls $= \frac{\ell_{\mathsf{in}}' - \ell_{\mathsf{in}}}{\ell_{\mathsf{out}} - \ell_{\mathsf{in}}}$
- security reduction : $s' = s - \ell_{\mathsf{in}}' \cdot T$

3.1 Amplification Via Concatenation

We begin with a description and the analysis of the basic concatenation construction. The analysis we provide is very similar to that for hardness amplification for one-way functions via direct product [27,11]. The presentations is somewhat simpler. We also make a small modification to the analysis that facilitates the analysis of the coding-theoretic construction, discussed in the next section.

Construction 1 (basic). *Pick* $q = \lceil \frac{2}{\delta} \ln \frac{2}{\epsilon} \rceil$ *independent keys* $\kappa_1, \ldots, \kappa_q$. *On input* $x \in \{0, 1\}^{\ell_{\mathsf{in}}}$, *output* $h_{\kappa_1}(x) \circ h_{\kappa_2}(x) \circ \cdots \circ h_{\kappa_q}(x)$

In using the same input to the hash functions under all of the q keys $\kappa_1, \ldots, \kappa_q$, we ensure that a collision x_0, x_1 for the key $(\kappa_1, \ldots, \kappa_q)$ is also a collision for the underlying hash function on each of the keys $\kappa_1, \ldots, \kappa_q$.

Proposition 1 (Construction 1). *Fix some security parameter n. Suppose there exists a $(s, 1 - \delta)$-CRHF $\mathcal{H}_n$ from $\{0,1\}^{\ell_{key}} \times \{0,1\}^{\ell_{in}}$ to $\{0,1\}^{\ell_{out}}$. Then, Construction 1 yields an (s', ϵ)-CRHF $\mathcal{H}'_n$ from $\{0,1\}^{\ell_{key}'} \times \{0,1\}^{\ell_{in}'}$ to $\{0,1\}^{\ell_{out}'}$ with the following parameters:*

- $\ell_{in}' = \ell_{in}$ *and* $\ell_{out}' = \Theta(\frac{\ell_{out}}{\delta} \log \frac{1}{\epsilon})$ *and* $\ell_{key} = \Theta(\frac{\ell_{key}}{\delta} \log \frac{1}{\epsilon})$
- *# hash calls* $= \Theta(\frac{\ell_{in}'}{\delta \ell_{in}} \log \frac{1}{\epsilon})$
- *security reduction :* $s' = s \cdot \Theta(\frac{1}{\epsilon} \log \frac{1}{\epsilon} \log \frac{1}{\delta})^{-1}$

Proof. Suppose A finds collisions on $\mathcal{H}'_n$ with probability at least ϵ, and consider the following algorithm A' for finding collisions on $\mathcal{H}_n$: on input κ,

1. chooses $\kappa_1, \ldots, \kappa_q$ at random, $i \in [q]$ at random, and sets $\kappa_i = \kappa$.
2. runs $A(\kappa_1, \ldots, \kappa_q)$ to obtain x_0, x_1, and outputs x_0, x_1.

To analyze the success probability for A', first fix any set S of keys κ of density $\frac{\delta}{2}$. Intuitively, S represents the set of keys for which it is hard for A' to find a collision.

$$\Pr_{\kappa_1,\ldots,\kappa_q}[A(\kappa_1,\ldots,\kappa_q) \text{ outputs a collision} \bigwedge \text{ at least one of the } \kappa_j\text{'s lies in } S]$$

$$\geq \epsilon - (1 - \tfrac{\delta}{2})^q \geq \tfrac{\epsilon}{2}$$

Hence,

$$\Pr_{\kappa_1,\ldots,\kappa_q,i}[A(\kappa_1,\ldots,\kappa_q) \text{ outputs a collision} \bigwedge \kappa_i \in S] \geq \tfrac{\epsilon}{2q}$$

On the other hand,

$$\Pr_{\kappa_1,\ldots,\kappa_q,i}[A(\kappa_1,\ldots,\kappa_q) \text{ outputs a collision} \bigwedge \kappa_i \in S]$$

$$= \tfrac{\delta}{2} \cdot \Pr_{\kappa \in S}[A'(\kappa) \text{ outputs a collision for } h_\kappa]$$

$$\leq \tfrac{\delta}{2} \cdot \max_{\kappa \in S} \Pr[A'(\kappa) \text{ outputs a collision for } h_\kappa]$$

This implies that for any set S of density $\frac{\delta}{2}$,

$$\max_{\kappa \in S} \Pr[A'(\kappa) \text{ outputs a collision for } h_\kappa] \geq \tfrac{\epsilon}{\delta q}$$

Hence,

$$\Pr_\kappa\left[\Pr[A'(\kappa) \text{ outputs a collision for } h_\kappa] \geq \tfrac{\epsilon}{\delta q}\right] \geq 1 - \tfrac{\delta}{2}$$

By running A' a total of $\frac{\delta q}{\epsilon} \log \frac{1}{\delta} = O(\frac{1}{\epsilon} \log \frac{1}{\epsilon} \log \frac{1}{\delta})$ times, we find collisions on $\mathcal{H}_n$ for a $1 - \frac{\delta}{2}$ fraction of keys with probability $1 - \frac{\delta}{2}$. This means we find collisions on $\mathcal{H}_n$ for a random key with probability at least $1 - \delta$. $\square$

3.2 Amplification Via Codes

Note how the basic construction loses an $O(q)$ factor in the compression ratio because we repeat the same input for each of the q keys. The following work-around was suggested in [14]. We first encode the input x using an error-correcting code C to obtain q symbols $C(x)_1, \ldots, C(x)_q \in \{0,1\}^{\ell_{in}}$, and then we hash each of the q blocks with independently chosen hash functions $h_{\kappa_1}, \ldots, h_{\kappa_q}$ and output the concatenation. Note that the adversary may upon receiving the q keys only produce collisions wherein the codewords disagree only on the "easy" keys. For the analysis to go through, we argue that w.h.p., a $\frac{\delta}{4}$ fraction of the keys (and not just one key) must be "hard". If we pick C to be a code with relative distance $1 - \frac{\delta}{8}$, we are guaranteed there is a $\frac{\delta}{8}$ fraction of positions wherein the codewords disagree and the corresponding keys are "hard".

Construction 2 (coding-theoretic). *Pick* $q = \lceil \frac{16}{\delta} \ln \frac{2}{\epsilon} \rceil$ *independent keys* $\kappa_1, \ldots, \kappa_q$. *Let* $C : \{0,1\}^{\ell_{in}'} \rightarrow (\{0,1\}^{\ell_{in}})^q$ *be an error-correcting code with minimum relative distance* $1 - \frac{\delta}{8}$ *(e.g., the Reed-Solomon code), where* $\ell_{in}' = \Theta(\delta q \ell_{in})$. *On input* $x \in \{0,1\}^{\ell_{in}'}$, *output* $h_{\kappa_1}(C(x)_1) \circ h_{\kappa_2}(C(x)_2) \circ \cdots \circ h_{\kappa_q}(C(x)_q)$.

Proposition 2 (Construction 2). *Fix some security parameter* n. *Suppose there exists a* $(s, 1 - \delta)$-*CRHF* $\mathcal{H}_n$ *from* $\{0,1\}^{\ell_{key}} \times \{0,1\}^{\ell_{in}}$ *to* $\{0,1\}^{\ell_{out}}$. *Then, Construction 2 yields an* (s', ϵ)-*CRHF* $\mathcal{H}_n'$ *from* $\{0,1\}^{\ell_{key}'} \times \{0,1\}^{\ell_{in}'}$ *to* $\{0,1\}^{\ell_{out}'}$ *with the following parameters:*

- $\ell_{in}' = \Theta(\ell_{in} \log \frac{1}{\epsilon})$ *and* $\ell_{out}' = \Theta(\frac{\ell_{out}}{\delta} \log \frac{1}{\epsilon})$ *and* $\ell_{key}' = \Theta(\frac{\ell_{key}}{\delta} \log \frac{1}{\epsilon})$
- *# hash calls* $= \Theta(\frac{1}{\delta} \log \frac{1}{\epsilon})$
- *security reduction* : $s' = s \cdot \Theta(\frac{1}{\epsilon} \log \frac{1}{\delta})^{-1}$

Proof. Suppose A finds collisions on $\mathcal{H}_n'$ with probability at least ϵ, and consider the following algorithm A' for finding collisions on $\mathcal{H}_n$: on input κ,

1. chooses $\kappa_1, \ldots, \kappa_q$ at random, $i \in [q]$ at random, and sets $\kappa_i = \kappa$.
2. runs $A(\kappa_1, \ldots, \kappa_q)$ to obtain x_0, x_1, and outputs $C(x_0)_i, C(x_1)_i$.

To analyze the success probability for A', first fix any set S of keys κ of density $\frac{\delta}{2}$. By a Chernoff bound (the multiplicative variant), we have

$$\Pr_{\kappa_1, \ldots, \kappa_q} [A(\kappa_1, \ldots, \kappa_q) \text{ outputs a collision } (x_0, x_1) \bigwedge \text{at least } \tfrac{\delta}{4} \text{ fraction of } \kappa_j\text{'s lies in } S]$$

$$\geq \epsilon - e^{-\delta q/16} \geq \tfrac{\epsilon}{2}$$

Conditioned on the above event, for a $\frac{\delta}{8}$ fraction of j's in $\{1, 2, \ldots, q\}$, we have $C(x_0)_j \neq C(x_1)_j$ and $\kappa_j \in S$ (since the former occurs for a $1 - \frac{\delta}{8}$ fraction of j's and the latter occurs for a $\frac{\delta}{4}$ fraction of j's). Hence,

$$\Pr_{\kappa_1, \ldots, \kappa_q, i} [A(\kappa_1, \ldots, \kappa_q) \text{ outputs a collision } (x_0, x_1) \wedge \kappa_i \in S \wedge C(x_0)_i \neq C(x_1)_i] \geq \tfrac{\delta \epsilon}{16}$$

On the other hand,

$$\Pr_{\kappa_1,\ldots,\kappa_q,i}[A(\kappa_1,\ldots,\kappa_q) \text{ outputs a collision } (x_0,x_1) \wedge \kappa_i \in S \wedge C(x_0)_i \neq C(x_1)_i]$$

$$= \tfrac{\delta}{2} \cdot \Pr_{\kappa \in S} \Pr[A'(\kappa) \text{ outputs a collision for } h_\kappa]$$

$$\leq \tfrac{\delta}{2} \cdot \max_{\kappa \in S} \Pr[A'(\kappa) \text{ outputs a collision for } h_\kappa]$$

This implies that for any set S of density $\tfrac{\delta}{2}$,

$$\max_{\kappa \in S} \Pr[A'(\kappa) \text{ outputs a collision for } h_\kappa] \geq \tfrac{\epsilon}{8}$$

Hence,

$$\Pr_{\kappa}\left[\Pr[A'(\kappa) \text{ outputs a collision for } h_\kappa] \geq \tfrac{\epsilon}{8}\right] \geq 1 - \tfrac{\delta}{2}$$

Again by running A' a total of $O(\tfrac{1}{\epsilon} \log \tfrac{1}{\delta})$ times, we can find collisions on $\mathcal{H}_n$ with probability $1 - \delta$. □

3.3 Reducing the Key Size

From a theoretical point of view, it is useful to have hash functions with short descriptions (i.e. short keys). Short keys may also be of interest from a practical point of view, although for the most common application of collision-resistant hash functions (digital signatures) the key would be standardized and only distributed once. Starting with a 160-bit key, the above transformations could yield a key that is much longer. Fortunately, there is no inherent cause for this blow-up: we may reduce the key size in each of the above constructions using randomness-efficient sampling [9], namely, we want to sample q keys in $\{0,1\}^{\ell_{\text{key}}}$ using r bits of randomness, where $r \ll q\ell_{\text{key}}$.

To accomplish this, we will use the randomness-efficient hitter in [9, Appendix C], with a slightly different analysis showing that for the parameters we are interested in, the construction satisfies a stronger sampler-like property. The weaker hitter guarantee is sufficient to reduce the key size for Construction 1, whereas the stronger sampler-like property is necessary for Construction 2. For our application, we will also require that that the hitter satisfy a certain reconstructibility property, previously used in [5]. This is used in the security reduction to generate challenges for the adversary breaking $\mathcal{H}'$ given a key for $\mathcal{H}$.

We stress here that for specific concrete parameters, we may use different choices of hitters and samplers for ease of implementation and optimality for those specific parameters.

Lemma 1. *There exists a constant c such that for every $\delta, \epsilon > 0$, there is an efficient randomized procedure $G : \{0,1\}^r \to (\{0,1\}^{\ell_{\text{key}}})^q$ with the following properties:*

— *(sampler) for every subset $S \subseteq \{0,1\}^{\ell_{\text{key}}}$ of density δ, with probability at least $1-\epsilon$, at least $\tfrac{\delta q}{2c}$ of the strings output by G lie in S.*
— *(complexity) the randomness complexity r is $\ell_{\text{key}} + O(\log \tfrac{1}{\epsilon})$ and the sample complexity q is $O(\tfrac{1}{\delta} \log \tfrac{1}{\epsilon})$.*

— *(reconstructible) there exists an efficient algorithm that on input (i, x), outputs a uniformly random element from the set $\{\sigma \mid G(\sigma)_i = x\}$.*

Proof (sketch). The construction (based on that in [9]) proceeds in three stages:

- First, we construct a hitter that generates $\frac{c}{\delta}$ samples in $\{0,1\}^{\ell_{\text{key}}}$ using ℓ_{key} random bits with the following property: for every subset S of $\{0,1\}^{\ell_{\text{key}}}$ with density δ, with probability at least $\frac{2}{3}$, at least one sample lies in S. We may obtain such a hitter using Ramanujan graphs of degree $\frac{c}{\delta}$ and vertex set $\{0,1\}^{\ell_{\text{key}}}$, wherein we pick a random vertex v, and the samples are the indices of the neighbors of v [9].

- Next, we construct a sampler that generates $d = O(\log \frac{1}{\epsilon})$ samples in $\{0,1\}^{\ell_{\text{key}}}$ using $\ell_{\text{key}} + O(d)$ random bits with the following property: for every subset S' of $\{0,1\}^{\ell_{\text{key}}}$ with density $\frac{2}{3}$, with probability at least $1 - \epsilon$, at least $\frac{1}{2}$ of the samples lie in S'. We may obtain such a sampler by taking a random walk of length $d - 1$ on a constant-degree expander with vertex set $\{0,1\}^{\ell_{\text{key}}}$ [9].

- Finally, we compose the sampler and the hitter as follows: we consider a random walk of length $d-1$ on the expander, and use each of the d vertices along the path as random coins for the hitter. Overall, we will run the hitter d times, which generate a total of $q = d \cdot \frac{c}{\delta}$ samples using a total of $\ell_{\text{key}} + O(d)$ random bits. This yields the desired query and randomness complexity.

The sampler guarantee follows fairly readily. Fix S of density δ. Let S' be the set of random coins for the hitter such that at least one sample lies in S, so S' has density at least $\frac{2}{3}$. We know that with probability at least $1 - \epsilon$ (over the random walk), we generate at least $\frac{d}{2}$ samples in S', which in turn yields $\frac{d}{2} = \frac{\delta q}{2c}$ samples that lie in S.

Finally, we check each of the two components in our construction is reconstructible, from which it follows that the combined construction is also reconstructible. For the expander-based hitter, this means that given i, x, we need to compute the vertex v whose i'th neighbor is labeled x. For the expander-based sampler, we need to given i, x, sample a start vertex and a path such that the i'th vertex on the path is labeled x. Indeed, both properties are readily satisfied for standard explicit constructions of constant-degree expanders.   □

The next construction is obtained from Construction 2 by replacing independent sampling of the q keys with randomness-efficient sampling using G, and using a code with slightly different parameters:

Construction 3 (reduced key size). *Run G to obtain q keys $\kappa_1, \ldots, \kappa_q \in \{0,1\}^{\ell_{\text{key}}}$. Let $C : \{0,1\}^{\ell_{\text{in}}'} \rightarrow (\{0,1\}^{\ell_{\text{in}}})^q$ be an error-correcting code with minimum relative distance $1 - \frac{\delta}{4c}$ (e.g., the Reed-Solomon code), where $\ell_{\text{in}}' = \Theta(\delta q \ell_{\text{in}})$. On input $x \in \{0,1\}^{\ell_{\text{in}}'}$, output $h_{\kappa_1}(C(x)_1) \circ h_{\kappa_2}(C(x)_2) \circ \cdots \circ h_{\kappa_q}(C(x)_q)$.*

It is straight-forward to verify that an analogue of Proposition 2 holds for Construction 3 if the CRHF is public-coin, and with essentially the same parameters except that the key size is now reduced to $\ell_{\text{key}} + O(\log \frac{1}{\epsilon})$ (i.e., the randomness complexity of G). We now state our main result for hardness amplification of collision-resistance, which is essentially a restatement of Proposition 2 for independent sampling and for randomness-efficient sampling:

Parameters	Construction 0	Construction 1	Construction 2	Construction 4
input length	$\ell_{in}{}'$	ℓ_{in}	$\Theta(\ell_{in}\log\frac{1}{\epsilon})$	$\ell_{in}-\Delta-\log\ell_{in}$
output length	ℓ_{out}	$\Theta(\frac{\ell_{out}}{\delta}\log\frac{1}{\epsilon})$	$\Theta(\frac{\ell_{out}}{\delta}\log\frac{1}{\epsilon})$	$\ell_{out}-\Delta$
# hash calls	$\frac{\ell_{in}{}'-\ell_{in}}{\ell_{out}-\ell_{in}}$	$\Theta(\frac{1}{\delta}\log\frac{1}{\epsilon})$	$\Theta(\frac{1}{\delta}\log\frac{1}{\epsilon})$	$\Theta(2^{\Delta}\ell_{in})$
key size	ℓ_{key}	$\Theta(\frac{\ell_{key}}{\delta}\log\frac{1}{\epsilon})$	$\Theta(\frac{\ell_{key}}{\delta}\log\frac{1}{\epsilon})$	$\Theta(\ell_{in}{}^2+\Delta)$
(public-coin)	ℓ_{key}	$\ell_{key}+\Theta(\log\frac{1}{\epsilon})$	$\ell_{key}+\Theta(\log\frac{1}{\epsilon})$	$\Theta(\ell_{in}{}^2+\Delta)$

Fig. 1. Summary of parameters for Constructions 0, 1, 2, & 4. In order to compare constructions 1 and 2 on inputs of the same length, we could apply the Merkle-Damgård transformation first, in which case the latter offers a $\Theta(\log\frac{1}{\epsilon})$ factor improvement in the number of hashing operations. For the key size, the second line refers that achieved using Construction 3 for public-coin hash functions.

Theorem 1. *Fix some security parameter* n. *Suppose there exists a* $(s, 1-\delta)$-*CRHF* $\mathcal{H}_n$ *from* $\{0,1\}^{\ell_{key}}\times\{0,1\}^{\ell_{in}}$ *to* $\{0,1\}^{\ell_{out}}$. *Then, there exists an* (s', ϵ)-*CRHF* $\mathcal{H}'_n$ *from* $\{0,1\}^{\ell_{key}}\times\{0,1\}^{\ell_{in}{}'}$ *to* $\{0,1\}^{\ell_{out}{}'}$ *with the following parameters:*

- $\ell_{in}{}' = \Theta(\ell_{in}\log\frac{1}{\epsilon})$ *and* $\ell_{out}{}' = \Theta(\frac{\ell_{out}}{\delta}\log\frac{1}{\epsilon})$ *and* $\ell_{key}{}' = \Theta(\frac{\ell_{key}}{\delta}\log\frac{1}{\epsilon})$
- *# hash calls* = $\Theta(\frac{1}{\delta}\log\frac{1}{\epsilon})$
- *security reduction :* $s' = s\cdot\Theta(\frac{1}{\epsilon}\log\frac{1}{\delta})^{-1}$

Moreover, if the CRHF is public-coin, then we may reduce $\ell_{key}{}'$ *to* $\ell_{key}+\Theta(\log\frac{1}{\epsilon})$.

3.4   Reducing the Output Length

We show that it is possible to reduce the output size of any CRHF by an additive factor of Δ, with a negligible loss in the the probability of finding collisions, but at the price of an exponential (in Δ) multiplicative increase in the complexity of the function, along with a similar decrease in the size of adversaries tolerated. This imposes a limitation of $\Delta = O(\log n)$ for all reasonable settings.

Proposition 3. *Suppose there exists a* (s, ϵ)-*CRHF* $\mathcal{H}$ *from* $\{0,1\}^{\ell_{in}}$ *to* $\{0,1\}^{\ell_{out}}$. *Let* $\Delta = O(\log n)$. *Then, there exists a* $(s - \mathrm{poly}(2^{\Delta}, n), \epsilon + 2^{-\Omega(\ell_{in})})$-*CRHF from* $\{0,1\}^{\ell_{in}-\Delta-\log\ell_{in}-2}$ *to* $\{0,1\}^{\ell_{out}-\Delta}$. *The complexity of the new CRHF is increases by a factor* $\mathrm{poly}(2^{\Delta}, \ell_{in})$.

This result follows the randomized black-box combiner in the following theorem, setting $t' = t = 1$.

Theorem 2. *There is a randomized black-box* (t', t)-*combiner* (C, R) *achieving parameters* $\ell_{in}{}' = \ell_{in} - \Delta - \log\ell_{in} - 2$ *and* $\ell_{out}{}' = (t - t' + 1)\cdot(\ell_{out} - \Delta)$ *for any positive* Δ *such that* $\ell_{in}{}' > \ell_{out}{}' > 0$. *The running times of* C *and* R *are polynomial in* n *and* 2^{Δ} *and the randomness complexity of* C *is* $O(\ell_{in}{}^2 + \Delta)$.

We may in fact use this combiner instead of the trivial combiner for our hardness amplification constructions. However, since we do not optimize on the output length of our hardness amplification within constant multiplicative factors, it does not make sense to try to cut down on the additive terms.

Overview of combiner. We begin with the case $t' = t = 1$ and suppose $h = h^1$ is "highly regular", and we have a partition of $\{0,1\}^{\ell_{in}}$ into $2^{\ell_{in}-\Delta}$ sets $\{S_{\tilde{x}} \mid y \in \{0,1\}^{\ell_{in}-\Delta}\}$ each of size 2^t with the following property: for every $\tilde{x}$, $S_{\tilde{x}}$ contains a unique string x such that $h(x)$ has prefix 0^{Δ}. Then, we define $\tilde{h}(\tilde{x})$ to be the $(\ell_{out} - \Delta)$-bit suffix of $h(x)$. It is easy to see how every collision $(\tilde{x}, \tilde{x}')$ for $\tilde{h}$ yields a collision (x_0, x_1) for h. To arrive at the general construction (which is where randomness plays a role),

- We replace 0^{Δ} with a string $z \in \{0,1\}^{\Delta}$ that is relatively popular in the sense that it occurs in at least an $\Omega(1/2^{\Delta})$ fraction of the images of h. Such a z can be identified by evaluating h on $O(\ell_{in} \cdot 2^{2\Delta})$ random inputs. To bring the randomness complexity down to $O(\ell_{in} + \Delta)$, we choose these inputs using the randomness-efficient Boolean sampler for approximating the mean within an additive error of $\frac{1}{2} \cdot 2^{-\Delta}$ with probability $1 - 2^{-2\ell_{in}}$ in [9].
- We replace the fixed partitioning with a random partitioning induced by a family $\mathcal{G}$ of ℓ_{in}-wise independent functions from $\{0,1\}^{\ell_{in}}$ to $\{0,1\}^{\ell_{in}-\Delta-\log \ell_{in}-2}$. Given $g \in \mathcal{G}$, we take $S_{\tilde{x}} = g^{-1}(\tilde{x})$. This gives us a partition of $\{0,1\}^{\ell_{in}}$ into sets each of size $\tilde{O}(2^{\Delta}\ell_{in})$. With overwhelming probability over g, for every $\tilde{x}$, there exists $x \in S_{\tilde{x}}$ such that $h(x)$ has prefix z (we set x to be the lexicographically first string with this property).

Construction and analysis. We formally state the construction for $t' = t = 1$. For simplicity, we present the construction using independent samples u_i and defer the randomness-efficient version to the full version.

Construction 4. *Let $\mathcal{G} = \{g : \{0,1\}^{\ell_{in}} \to \{0,1\}^{\ell_{in}-\Delta-\log \ell_{in}-2}\}$ be a family of $6\ell_{in}$-wise independent hash functions that such that given y, the set $g^{-1}(y)$ is computable in time $\text{poly}(2^{\Delta}, n)$. (This can be achieved using univariate polynomials of degree $6\ell_{in}$). On input $\tilde{x} \in \{0,1\}^{\ell_{in}-\Delta-\log \ell_{in}-2}$ and randomness $r \in \{0,1\}^{O(\Delta+\ell_{in}^2)}$, we compute $\tilde{h}_r(\tilde{x}) \in \{0,1\}^{\ell_{out}-\Delta}$ as follows:*

1. *Parse r as $g \in \mathcal{G}$ and $u_1, \ldots, u_m \in \{0,1\}^{\ell_{in}}$, where $m = \Theta(2^{2\Delta}\ell_{in})$.*
2. *Let $z \in \{0,1\}^{\Delta}$ be the lexicographically first string that occurs at least a $1/2^{\Delta}$ fraction of times as a prefix among $h(u_1), \ldots, h(u_m)$ (where $h = h^1$);*
3. *Compute $S_{\tilde{x}} = g^{-1}(\tilde{x})$ in order to find a string x in $S_{\tilde{x}}$ such that $h(x)$ has prefix z. Choose the lexicographically first string if there are more than 1; output $0^{\ell_{out}-\Delta}$ if no such string exists or if $|S_{\tilde{x}}| > 8\ell_{in} \cdot 2^{\Delta}$.*
4. *Output the $(\ell_{out} - \Delta)$-bit suffix of $h(x)$.*

For general t', t, we may simply apply the above construction to each of $h^1, \ldots, h^{t-t'+1}$ and concatenate the output; it will be clear from the analysis that we may use the same randomness r for all t functions. Theorem 2 follows readily once we establish the following technical claim for $t' = t = 1$.

Claim. With probability $1 - 2^{-\Omega(\ell_{in})}$ over $r = (g, u_1, \ldots, u_m)$, the following statements hold simultaneously:

- $|\Gamma_z| \geq 2^{\ell_{in} - \Delta - 1}$, where z is as in the construction and $\Gamma_z = \{x \in \{0,1\}^{\ell_{in}} \mid h(x)$ has prefix $z\}$;
- for all $\tilde{x}$, we have $S_{\tilde{x}} \cap \Gamma_z \neq \emptyset$ (where $S_{\tilde{x}} = g^{-1}(\tilde{x})$);
- for all $\tilde{x}$, we have $|S_{\tilde{x}}| \leq 8\ell_{in} \cdot 2^{\Delta}$.

Suppose we have a collision $(\tilde{x}_0, \tilde{x}_1)$ for $\tilde{h}_r$, where the conditions in the technical claim do hold for r. Then, we could in $\text{poly}(2^{\Delta}, \ell_{in})$ time compute $(x_0, x_1) \in S_{\tilde{x}_0} \times S_{\tilde{x}_1}$ such that $h(x_0) = z \circ h_r(\tilde{x}_0)$ and $h(x_1) = z \circ h_r(\tilde{x}_1)$. This implies (x_0, x_1) is a collision for h.

Proof (of claim). By a Chernoff bound, we have that for each Δ-bit prefix w, if w occurs in a p_w fraction of outputs of h as a prefix, then with probability at least $1 - 2^{-2\ell_{in}}$ over the u_i's, w will occur at most a $p_w + \frac{1}{2} \cdot 2^{-\Delta}$ fraction of times (as a prefix) among the $h(u_i)$'s. Taking a union bound over all $2^{\Delta} < 2^{\ell_{in}}$ prefixes, we see that with probability at least $1 - 2^{-\ell_{in}}$, the prefix z must satisfy $p_z \geq \frac{1}{2} \cdot 2^{-\Delta}$ and thus $|\Gamma_z| \geq 2^{\ell_{in} - \Delta - 1}$. We assume in the rest of the proof that this is the case. Then, for each $y \in \{0,1\}^{\ell_{in} - \Delta - \log \ell_{in} - 2}$: $E[|S_{\tilde{x}} \cup \Gamma_z|] = |\Gamma_z| \cdot 2^{-\ell_{in} + t + \log \ell_{in} + 2} \geq 2\ell_{in}$. Applying a tail bound for $6\ell_{in}$-wise independence [22], we obtain:

$$\Pr_g[S_{\tilde{x}} \cap \Gamma_z = \emptyset] \leq 2^{-2\ell_{in}}$$

Taking a union bound over all $y \in \{0,1\}^{\ell_{in} - \Delta - \log \ell_{in} - 2}$, we have:

$$\Pr_g[\exists y : S_{\tilde{x}} \cap \Gamma_z = \emptyset] \leq 2^{-2\ell_{in}} \cdot 2^{\ell_{in} - \Delta - \log \ell_{in} - 2} = 2^{-\Omega(\ell_{in})}$$

Finally, for each y, $E[|S_{\tilde{x}}|] = 4\ell_{in} \cdot 2^{\Delta}$. Again, by using the tail bound for $6\ell_{in}$-wise independence and a union bound, we have $\Pr[\exists y : |S_{\tilde{x}}| > 8\ell_{in} \cdot 2^{\Delta}] < 2^{-\Omega(\ell_{in})}$. $\square$

4 Limitations

We begin by presenting the class of constructions for which we prove lower bounds:

Definition 2. *We say that (C, R) is a* black-box $(1 - \delta, \epsilon)$-amplifier *for collision resistance if $C = (C_{key}, C_{hash})$ is a pair of deterministic (oracle) TMs, and $R = (R_{key}, R_{coll})$ is a pair of randomized (oracle) TMs, and both pairs of TMs run in time $\text{poly}(n, \frac{1}{\delta}, \frac{1}{\epsilon})$. In addition, for all $\mathcal{H} = \{\{0,1\}^{\ell_{key}} \times \{0,1\}^{\ell_{in}} \to \{0,1\}^{\ell_{out}}\}$:*

CONSTRUCTION. *C compute $\mathcal{H}' = \{\{0,1\}^{\ell_{key}'} \times \{0,1\}^{\ell_{in}'} \to \{0,1\}^{\ell_{out}'}\}$ where $\ell_{out}' > \ell_{in}'$ as follows: given a key κ' and a string x, we run $C_{key}(\kappa')$ to obtain $\kappa_1, \ldots, \kappa_q$ and then set $h'_{\kappa'}(x)$ to be $C_{hash}^{h_{\kappa_1}, \ldots, h_{\kappa_q}}(\kappa', x)$.*

REDUCTION. *There exists a constant c such that for every TM A that outputs a collision on $h'_{\kappa'}$ with probability at least ϵ and any subset S of $\{0,1\}^{\ell_{key}}$ of density at least $\delta/2$, there exists $\kappa \in S$ such that*

$$\Pr_{\sigma, R_{coll}}\left[R_{key}(\kappa; \sigma) = \kappa'; R_{coll}^{\mathcal{H}}(i, \sigma, A(\kappa')) \text{ outputs a collision on } h_\kappa\right] > \left(\frac{\delta\epsilon}{n}\right)^c$$

Note that a black-box amplifier should provide an efficient reduction that converts any adversary A that finds collisions in $h'_{\kappa'}$ with probability ϵ into an adversary A' that finds collisions in h_κ with probability $1 - \delta$. Indeed, Definition 2 guarantees that for a $1 - \frac{\delta}{2}$ fraction of keys κ, $R^{A,\mathcal{H}}(\kappa)$ outputs a collision for h_κ with probability $(\frac{\delta\epsilon}{n})^c$. Running R a total of $O((\frac{n}{\delta\epsilon})^c \log \frac{1}{\delta})$ yields the desired reduction. The above reduction is more restrictive than an arbitrary black-box reduction due to the following structural restrictions we place on the construction and the reduction, and this makes our result weaker.

Construction. We do not allow constructions that use the input as a key into the underlying family hash functions. We enforce this constraint by having a key generation algorithm C_{key} select the members $h_{\kappa_1}, \ldots, h_{\kappa_q}$ of the underlying family given only the new key κ', and restrict the actual computation C_{hash} to only query $h_{\kappa_1}, \ldots, h_{\kappa_q}$. We will refer to q as the query complexity of the construction, the idea being that C_{hash} will query each of the functions $h_{\kappa_1}, \ldots, h_{\kappa_q}$ at least once by having C_{key} not generate extraneous keys.

Reduction. The restriction on the reduction states that the reduction only requires a single collision from A' to break $\mathcal{H}$ with noticeable probability. This is true of the reductions used in our constructions and of all known reductions used in hardness amplification for one-way functions (c.f. [15]): all these reductions generate multiple challenges to the adversary and if the adversary successfully answers any of the challenges, the reduction succeeds with high probability.

We present lower bounds for the query complexity of the construction q and the output length ℓ_{out}'.

Theorem 3. *Suppose (C, R) is a black-box $(1 - \delta, \epsilon)$-amplifier for collision resistance with $\epsilon \leq \frac{\delta}{2}$. Then,*

$$q \geq \Omega(\tfrac{1}{\delta} \log \tfrac{1}{\epsilon}) \quad \text{and} \quad \ell_{\text{out}}' \geq \tfrac{1}{\delta} \cdot (\ell_{\text{out}} - O(\log n + \log \tfrac{1}{\epsilon} + \log \tfrac{1}{\delta})) - 2$$

The lower bound for q follows closely the lower bound in [15], by arguing that C_{key} must compute a randomness-efficient hitting sampler, and is omitted due to lack of space. To obtain a lower bound for ℓ_{out}', we begin with an observation of a connection between black-box hardness amplification and randomized black-box combiners. Intuitively, a $(1 - \delta)$-CRHF could comprise $\lfloor \frac{1}{\delta} \rfloor$ functions, of which it is hard to find collisions on just one of them. In this case, the black-box $(1-\delta, \epsilon)$-amplifier acts like a randomized black-box $(1, \lfloor \frac{1}{\delta} \rfloor)$-combiner. To derive a lower bound for the latter, we use the probabilistic argument in Pietrzak's work [19]. We also note that the probabilistic argument is already sufficient to obtain the lower bounds for deterministic black-box combiners, therefore simplifying the lower bounds in [2,19] by eliminating an additional randomization argument therein.

Proof. Set t to be a power of 2 in the interval $[\frac{1}{\delta}, \frac{2}{\delta})$. Pick t random functions $f_1, \ldots, f_t : \{0,1\}^{\ell_{\text{in}}} \to \{0,1\}^{\ell_{\text{out}}}$ and identify $\{0,1\}^{\ell_{\text{key}}}$ with $\{1, 2, \ldots, t\}$ and $\mathcal{H}$ with $\{f_1, \ldots, f_t\}$. Consider the following procedure $\tilde{R}$ for finding collisions in $f_1, \ldots, f_t$ given oracle access to these functions:

— picks $x'_0, x'_1 \in \{0,1\}^{\ell_{in}'}$ and $\kappa' \in \{0,1\}^{\ell_{key}'}$ at random;
— for each $i = 1, 2, \ldots, t$, sample a random σ_i such that $R_{key}(i; \sigma_i) = \kappa'$, and output $R_{coll}^{f_1, \ldots, f_t}(i, \sigma_i, (x'_0, x'_1))$.

We note that for all $f_1, \ldots, f_t$ and for all κ', the function $h'_{\kappa'}$ maps $\{0,1\}^{\ell_{in}'}$ to $\{0,1\}^{\ell_{out}'}$. By the standard lower bound on collision probability or a simple application of Cauchy-Schwartz, we have

$$\Pr_{x'_0, x'_1} [(x'_0, x'_1) \text{ is a collision for } h'_{\kappa'}] \geq 2^{-\ell_{out}'} - 2^{-\ell_{in}'} \geq 2^{-\ell_{out}'-1}$$

Consider a procedure A that outputs collisions on every $h'_{\kappa'}$ by repeatedly choosing (x'_0, x'_1) at random until it finds a collision. By our choice of t, each $\{i\}$ is a subset of $\{0,1\}^{\ell_{key}}$ of density $\frac{1}{t} \geq \delta/2$, for $i = 1, 2, \ldots, t$. The reduction then guarantees that

$$\Pr_{\sigma, R_{coll}} \left[R_{key}(i; \sigma) = \kappa'; R_{coll}^{\mathcal{H}}(i, \sigma, A(\kappa')) \text{ outputs a collision on } f_i \right] > \left(\frac{\delta \epsilon}{n}\right)^c$$

In fact, the above statement is true even if we restrict A to only output collisions for κ' lying in some subset S' of $\{0,1\}^{\ell_{key}'}$ of density ϵ. By a probabilistic argument, this implies that for every subset S' of $\{0,1\}^{\ell_{key}'}$ of density ϵ, there exists $\kappa' \in S'$ such that:

$$\Pr \left[\sigma \leftarrow R_{key}(i; \cdot) = \kappa'; R_{coll}^{\mathcal{H}}(i, \sigma, A(\kappa')) \text{ outputs a collision on } h_\kappa \right] > \left(\frac{\delta \epsilon}{n}\right)^c$$

Call such a κ' i-good. Then, for each i, a $1 - \epsilon$ fraction of κ' is i-good. By a union bound, there exists a $1 - t\epsilon$ fraction of κ' that are i-good, for all $i = 1, 2, \ldots, t$. Hence,

$$\Pr_{\tilde{R}} \left[\tilde{R}^{f_1, \ldots, f_t} \text{ outputs collisions for each of } f_1, \ldots, f_t \right]$$

$$\geq (1 - t\epsilon) \cdot 2^{-\ell_{out}'-1} \cdot \left(\frac{\delta \epsilon}{n}\right)^{ct}$$

Note that the preceding inequality holds for all functions $f_1, \ldots, f_t$ and thus also holds for random functions $f_1, \ldots, f_t$. On the other hand, by the birthday paradox and independence of the t functions, we know that the probability (over random functions) $\tilde{R}$ outputs collisions in each of $f_1, \ldots, f_t$ is at most $\left(\frac{Q^2}{2^{\ell_{out}}}\right)^t$, where $Q = \text{poly}(n, \frac{1}{\delta}, \frac{1}{\epsilon})$ is the query complexity of $\tilde{R}$. Comparing the two bounds and solving for ℓ_{out}' yields the desired bound. $\qquad \square$

The above argument also yields a lower bound on the output length for (t', t)-combiners. The idea is to use R to find $t - t' + 1$ collisions amongst random functions $f_1, \ldots, f_t$ and observe that the probability is bounded by $\binom{t}{t-t'+1} \cdot \left(\frac{Q^2}{2^{\ell_{out}}}\right)^{t-t'+1}$. This establishes the optimality of our construction in Theorem 2 (up to constant factors in the $O(\log n)$ term):

Theorem 4. *Suppose (C, R) is a randomized black-box (t', t)-combiner for CRHFs. Let Q be an upper bound on the query complexity of R. Then,*

$$\ell_{out}' \geq (t - t' + 1)(\ell_{out} - 2\log Q) - t - 1$$

Acknowledgments. We would like to thank Krzysztof Pietrzak for helpful discussions on combiners.

References

1. Anderson, R.: The classification of hash functions. In: Cryptography and Coding '93 (1993)
2. Boneh, D., Boyen, X.: On the impossibility of efficiently combining collision resistant hash functions. In: Dwork, C. (ed.) CRYPTO 2006. LNCS, vol. 4117, Springer, Heidelberg (2006)
3. Coron, J.-S., Dodis, Y., Malinaud, C., Puniya, P.: Merkle-Damgård revisited: How to construct a hash function. In: Dwork, C. (ed.) CRYPTO 2006. LNCS, vol. 4117, Springer, Heidelberg (2006)
4. Damgård, I.: A design principle for hash functions. In: Brassard, G. (ed.) CRYPTO 1989. LNCS, vol. 435, Springer, Heidelberg (1990)
5. De Santis, A., Di Crescenzo, G., Persiano, G.: Randomness-optimal characterization of two NP proof systems. In: Rolim, J.D.P., Vadhan, S.P. (eds.) RANDOM 2002. LNCS, vol. 2483, Springer, Heidelberg (2002)
6. Dobbertin, H.: Cryptanalysis of MD4. In: Fast Software Encryption (1996)
7. Fischlin, M., Lehmann, A.: Security-amplifying combiners for collision-resistant hash functions. In: these proceedings (2007)
8. Gibson, J.K.: Discrete logarithm hash function that is collision free and one way. IEE Proceedings - E 138(6), 407–410 (1991)
9. Goldreich, O.: A sample of samplers - a computational perspective on sampling. ECCC TR97-020 (1997)
10. Goldreich, O.: Candidate one-way functions based on expander graphs. Cryptology ePrint Archive, Report 2000/063 (2000)
11. Goldreich, O.: Foundations of Cryptography: Basic Tools. Cambridge University Press, Cambridge (2001)
12. Herzberg, A.: Tolerant combiners: Resilient cryptographic design. Cryptology ePrint Archive, Report 2002/135 (2002)
13. Hsiao, C.-Y., Reyzin, L.: Finding collisions on a public road, or do secure hash functions need secret coins? In: Franklin, M. (ed.) CRYPTO 2004. LNCS, vol. 3152, Springer, Heidelberg (2004)
14. Knudsen, L.R., Preneel, B.: Construction of secure and fast hash functions using nonbinary error-correcting codes. IEEE Transactions on Information Theory 48(9), 2524–2539 (2002)
15. Lin, H., Trevisan, L., Wee, H.: On hardness amplification of one-way functions. In: Kilian, J. (ed.) TCC 2005. LNCS, vol. 3378, Springer, Heidelberg (2005)
16. Menezes, A.J., van Oorschot, P.C., Vanstone, S.A.: Handbook of Applied Cryptography. CRC Press, Boca Raton, USA (1996)
17. Merkle, R.C.: One way hash functions and DES. In: Brassard, G. (ed.) CRYPTO 1989. LNCS, vol. 435, Springer, Heidelberg (1990)
18. Naor, M., Yung, M.: Universal one-way hash functions and their cryptographic applications. In: Proc. 20th STOC (1989)
19. Pietrzak, K.: Non-trivial black-box combiners for collision-resistant hash-functions don't exist. In: Naor, M. (ed.) EUROCRYPT 2007. LNCS, vol. 4515, pp. 23–33. Springer, Heidelberg (2007
20. Preneel, B.: Hash functions - present state of art. ECrypt Conference on Hash Functions (2005)
21. Rogaway, P.: Formalizing human ignorance: Collision-resistant hashing without the keys. In: Nguyen, P.Q. (ed.) VIETCRYPT 2006. LNCS, vol. 4341, Springer, Heidelberg (2006)
22. Schmidt, J.P., Siegel, A., Srinivasan, A.: Chernoff-Hoeffding bounds for applications with limited independence. SIAM J. Discrete Math 8(2), 223–250 (1995)

23. Shaltiel, R.: Towards proving strong direct product theorems. Computational Complexity 12(1–2), 1–22 (2003)
24. Shoup, V.: A composition theorem for universal one-way hash functions. In: Preneel, B. (ed.) EUROCRYPT 2000. LNCS, vol. 1807, Springer, Heidelberg (2000)
25. Simon, D.R.: Finding collisions on a one-way street: Can secure hash functions be based on general assumptions? In: Nyberg, K. (ed.) EUROCRYPT 1998. LNCS, vol. 1403, Springer, Heidelberg (1998)
26. Wang, X., Yin, Y.L., Yu, H.: Finding collisions in the full SHA-1. In: Shoup, V. (ed.) CRYPTO 2005. LNCS, vol. 3621, Springer, Heidelberg (2005)
27. Yao, A.: Theory and applications of trapdoor functions. In: Proc. 23rd FOCS (1982)

How Many Oblivious Transfers Are Needed for Secure Multiparty Computation?[*]

Danny Harnik[**], Yuval Ishai[***], and Eyal Kushilevitz[†]

Department of Computer Science, Technion, Haifa, Israel
{harnik,yuvali,eyalk}@cs.technion.ac.il

Abstract. Oblivious transfer (OT) is an essential building block for secure multiparty computation when there is no honest majority. In this setting, current protocols for $n \geq 3$ parties require *each pair of parties* to engage in a single OT for *each gate* in the circuit being evaluated. Since implementing OT typically requires expensive public-key operations (alternatively, expensive setup or physical infrastructure), minimizing the number of OTs is a highly desirable goal.

In this work we initiate a study of this problem in both an information-theoretic and a computational setting and obtain the following results.
- If the adversary can corrupt up to $t = (1 - \epsilon)n$ parties, where $\epsilon > 0$ is an arbitrarily small constant, then a total of $O(n)$ OT channels between pairs of parties are necessary and sufficient for general secure computation. Combined with previous protocols for "extending OTs", $O(nk)$ invocations of OT are sufficient for computing arbitrary functions with computational security, where k is a security parameter.
- The above result does not improve over the previous state of the art in the important case where $t = n - 1$, when the number of parties is small, or in the information-theoretic setting. For these cases, we show that an arbitrary function $f : \{0,1\}^n \to \{0,1\}^*$ can be securely computed by a protocol which makes use of a single OT (of strings) between each pair of parties. This result is tight in the sense that at least one OT between each pair of parties is necessary in these cases. A major disadvantage of this protocol is that its communication complexity grows exponentially with n. We present natural classes of functions f for which this exponential overhead can be avoided.

1 Introduction

Secure multiparty computation (MPC) [46,23,7,11] provides a powerful and general tool for distributing computational tasks between mutually distrusting

---

[*] Research supported by grant 1310/06 from the Israel Science Foundation and the Technion VPR fund. Part of this research was done while visiting IPAM.
[**] Supported in part by a fellowship from the Lady Davis Foundation.
[***] Supported by grant 2004361 from the U.S.-Israel Binational Science Foundation.
[†] Supported by grant 2002354 from the U.S.-Israel Binational Science Foundation.

A. Menezes (Ed.): CRYPTO 2007, LNCS 4622, pp. 284–302, 2007.

parties without compromising the privacy of their inputs. We consider the problem of secure computation in the case where a majority of the parties can be corrupted. In this case, secure computation of nontrivial functions implies the existence of *oblivious transfer* (OT) [44,40,17] — a secure two-party protocol which allows a receiver to select one of two strings held by a sender and learn this string (but not the other) without revealing its selection. Moreover, OT can be used as a building block for general MPC protocols that tolerate an arbitrary number of corrupted parties [47,24,22,33,21]. These protocols involve a large number of OT invocations which typically constitute their efficiency bottleneck. Indeed, standard implementations of OT require expensive public-key operations, whereas alternative "information-theoretic" implementations of OT require either a trusted setup [3] or physical infrastructure [13] and may be viewed as being at least as expensive. Thus, minimizing the number of OTs in MPC protocols is a highly desirable goal.

How many OTs are needed to secure the world? In a world consisting of just two parties, this question was essentially answered by Beaver [4] (see also [31]). In a pure information-theoretic setting, ignoring computational efficiency issues, computing a two-argument function whose shorter input has length ℓ generally requires $\Theta(\ell)$ OTs. (Some specific functions require fewer OTs; see [15,5] for a more refined study of the "OT complexity" of information-theoretic secure two-party computation.) Quite remarkably, it is possible to do much better if computational security is required. Assuming the existence of one-way functions, a "seed" of k OTs, where k is a security parameter, can be used for implementing an arbitrary polynomial number of OTs.[1] This implies that k OTs are sufficient for secure two-party computation of arbitrary functions, even ones whose input length is much bigger than k.

Given Beaver's result, it is natural to expect that k OTs would be sufficient for computationally secure MPC protocols involving an arbitrary number of parties. Unfortunately, known protocols are very far from achieving this goal. Beaver's OT extension technique crucially relies on the fact that the number of OTs required by Yao's two-party protocol [46] is equal to the length of the *shorter* input. No similar protocols are known for $n \geq 3$ parties. To make things worse, the number of OT invocations in current protocols (e.g., [22,21]) does not only depend on the length of the inputs but also on the complexity of the function f being computed. Specifically, these protocols require *each pair of parties* to invoke an OT protocol for *each gate* of a circuit computing f.[2] In the computational setting it is possible to apply Beaver's OT extension protocol between each pair of parties, requiring only k OTs for each of the $\binom{n}{2}$ pairs. Thus, the state of the art prior to the current work can be summarized as follows:

---

[1] In contrast, it is not known how to implement even a single OT using a one-way function alone, and the possibility of a black-box construction of this type was ruled out by Impagliazzo and Rudich [30].

[2] Some MPC protocols do not rely on OT but rather on other "public-key" primitives such as threshold homomorphic encryption [19]; however, in these protocols too the number of public-key operations grows linearly with the circuit size of f.

- In the information-theoretic setting, the number of OTs needed by n parties to compute a circuit of size s is $O(n^2 s)$.
- In the computational setting (assuming one-way functions exist) the total number of OTs is $O(n^2 k)$.

The above state of affairs leaves much to be desired and gives rise to several natural questions: Can one reduce the quadratic dependence on the number of parties while maintaining security against a dishonest majority? Can the dependence on the circuit size in the information-theoretic case and the dependence on the security parameter in the computational case be eliminated?

1.1 Our Contribution

We answer the above questions affirmatively, obtaining several upper and lower bounds on the OT complexity of both information-theoretic and computationally secure MPC with no honest majority. Before describing our results, we outline (and justify) some essential details of our model.

Model. We consider a network of n parties that are connected via a synchronous network of secure point-to-point channels (secure channels are necessary in the information-theoretic setting, and can be cheaply implemented in a computational setting via the use of a hybrid encryption). The parties wish to compute a function f, which by default is a polynomial-time computable function taking one input bit from each party and returning an output of an arbitrary length (our results can be generalized to the case where each party has an ℓ-bit input – see below). Our goal is to design an OT-efficient protocol which securely computes f in the presence of a *semi-honest* (aka "honest-but-curious") adversary which may corrupt at most t parties, where the security threshold t satisfies $n/2 < t < n$. In the computational setting, restricting the attention to security in the semi-honest model is justified by the fact that it is possible to use one-way functions (and no additional OTs) for upgrading security to the malicious model [23,21]. Finally, we allow each pair of parties to invoke an ideal OT oracle during the execution of the protocol and count the number of invocations of this oracle. (This model is also referred to as the "OT-hybrid" model.) Using a suitable composition theorem [10,21], each call to the OT oracle can be substituted with an actual secure OT protocol. It is important to stress that our basic OT primitive is *string* OT; that is, the sender's strings are of arbitrary length (yet this length counts towards the communication complexity of our protocols). This is justified by the fact that OT of long strings can be easily reduced to a single invocation of OT of short keys of length k by using symmetric encryption and no public-key operations. Furthermore, most efficient implementations of OT (cf. [38]) directly realize OT of k-bit strings rather than bits.

In the above model, we obtain the following main results.

Number of OT channels. We start by examining the required number of "OT channels" between pairs of parties, that is, the number of distinct pairs that should jointly invoke the OT primitive. We show that if the adversary can corrupt up to $t = (1 - \epsilon)n$ parties, where $\epsilon > 0$ is an arbitrarily small constant, then a total of $O(n)$ OT channels between pairs of parties are sufficient and necessary for general MPC. This is a quadratic improvement over previous protocols, which require OTs between each pair of parties. The $O(n)$ upper bound relies on a technique of Bracha [9] for distributing computations among several committees, a technique for combining oblivious transfers [27], and explicit constructions of dispersers [25,42,26]. Using OT extension protocols [4,31], the $O(n)$ bound implies that $O(nk)$ invocations of OT are sufficient for computing arbitrary functions with computational security[3] when $t = (1 - \epsilon)n$. The lower bound (in a more general form) relies on results from extremal graph theory. We note that the $\Omega(n)$ lower bound holds also if the OT channels are chosen *dynamically*, namely the identity of the pairs of parties which can invoke the OT oracle can be chosen during the execution of the protocol.

Coping with a bigger security threshold. The above results do not improve over the previous state of the art in the important case where $t = n - 1$, when the number of parties is small, or in the information-theoretic setting. For these cases, we show that an arbitrary function $f : \{0, 1\}^n \to \{0, 1\}^*$ can be securely computed by a protocol which makes use of a single OT between each pair of parties. We also show that this result is tight, in the sense that when $t = n - 1$ at least one OT between each pair of parties is necessary. At a high level, the protocol proceeds by $n - 1$ iterations, where in the end of the i-th iteration the first $i + 1$ parties hold additive shares of the truth table of $f(x_1, \ldots, x_i, \cdot, \ldots, \cdot)$, namely f restricted by the inputs of the first i parties. A major disadvantage of this protocol is that its communication complexity grows exponentially with n. We present natural classes of functions f for which this exponential overhead can be avoided. These include sparse polynomials, decision trees, deterministic and nondeterministic finite automata, and CNF and DNF formulas, which capture useful secure computation tasks (cf. [2]). Some of these efficient protocols rely on expander-based constructions of extractors for bit-fixing sources [32].

In the case where each party holds an ℓ-bit input (rather than a 1-bit input) the above upper and lower bounds on the number of OTs grow by a factor of ℓ, whereas the bounds on the number of OT channels remain unchanged.

2   Preliminaries

Throughout the paper we use the following notation: By $P_1, \ldots, P_n$ we denote the n parties, the security threshold t is the maximal number of parties that the adversary can corrupt, and k stands for a security parameter when considering computational or statistical security. When an n-party function f has a single output, we assume by default that the output is given to the first party P_1.

---

[3] This protocol inherits the security and assumptions of the underlying OT extension protocol. In particular, the protocol of [4] can be based on one-way functions but is only proved to be secure against *non-adaptive* adversaries, whereas the protocol of [31] in the random oracle model can be shown to be adaptively secure.

Our model for secure multiparty computation follows standard definitions from the literature [10,21]. The availability of an OT primitive is captured by considering an *OT-hybrid* model, in which each pair of parties can invoke an ideal OT oracle. By a *t-secure protocol* for f we refer by default to a protocol which is *perfectly* secure in the *semi-honest* model against an adversary that may *adaptively* corrupt at most t parties. Perfect security will sometimes be relaxed to statistical or computational security.

3   Counting OT Channels: Upper and Lower Bounds

A closely related question to the number of required OT calls is the number of required OT channels in a network. That is, given a network of n parties, we look at a graph where each node stands for a party and an edge stands for the ability to run an OT between two parties. On each such edge, we assume the ability to execute arbitrarily many OT calls. In addition, there exist private communication channels between *every* pair of parties. The question is how many OT channels are needed in order to simulate a full network of OTs (a network in which every two parties can execute an OT functionality).

More precisely, define the n party OT function f_{OT} as a function that takes inputs from two parties (if more than two parties provide inputs then the function outputs an abort symbol). The first party inputs two string s_0, s_1 and the second inputs a bit c. The output s_c is received by the second party. The question at hand is how many OT channels are required for the network to be able to securely compute the function f_{OT}.

OT Channels: Static vs. Dynamic. When discussing OT channels, special care needs to be taken when modelling the network. The simpler case is when the network is *static*, i.e., the OT channels are set in advance and known to the adversary (this case is suitable for an implementation of OTs based on some physical infrastructure). A stronger model (for the honest parties) allows a *dynamic* network, in which the parties may set the OT channels as part of the protocol (while trying to hide this information from the adversary). Our upper bounds do not take advantage of the dynamic setting and work also in the static setting. We prove our lower bounds initially in the static case and then extend them to the dynamic case.

OT Channels and Counting OTs. Counting OT channels is an interesting question in its own right, and may directly capture the case where OTs are implemented via some physical infrastructure (e.g., noisy point-to-point channels). Moreover, its connection to the number of OT calls needed for secure computation is two fold:

– As means of achieving **upper bounds** on the number of OT calls in a *computational* setting. In the *two-party* computational setting, it is known how to achieve a polynomial number of OT calls at the price of just k calls [4,31] (where k is the security parameter). Therefore, if we only need, say, $O(n)$ OT channels, then we can simulate the whole network at the price of

$O(nk)$ OT calls, which is better than the trivial upper bound of $\binom{n}{2}k$ OT calls.

- As a mechanism for proving **lower bounds** on the number of OT calls (see Theorem 4). Namely, the minimal number of channels needed to securely compute the functionality f_{OT} is in particular a lower bound on the number of OT calls necessary.

We note that the function f_{OT} is just a single example of a function for which the lower bounds hold; a similar lower bound holds for any function that is *complete* for n-party computation, in the sense that it can be used as an oracle for computing arbitrary functions t-securely. A sufficient condition for completeness is that for every pair of inputs, one of the two-party criteria from [34,6,28] is met for some restriction of the other inputs.

3.1   Upper Bounds for $t = (1 - \delta)n$: The Committees Method

We turn to the case that $t = (1-\delta)n$ (we mostly think of δ as a constant fraction, but the discussion is not restricted to this case). Consider the following strategy: from the n parties choose m committees, each of size d, where a party can (and will) participate in several different committees. Assume that each committee has a full network of OT channels between them.

Using each committee we generate a candidate for an OT protocol between party A and party B as follows: The sender and receiver additively share their inputs (s_0, s_1 and c respectively) between all committee members. The committee members now run a secure computation protocol that computes a random additive sharing of s_c between the committee members (this is done using their OT channels and the "GMW protocol" [22]). Now each committee member sends his share of s_c to the receiver B who reconstructs the output. This constitutes a secure OT protocol as long as not all of the committee has been corrupted (if all of the committee is corrupted then there is no security at all). In total, for each of the m committees we have a candidate for an OT protocol, which is secure if not all of the underlying committee is corrupted. It is known how one can combine OT candidates protocols to 1-secure OT protocol as long as a majority of the candidates are secure. This method is called an ($\lceil \frac{m+1}{2} \rceil, m$)-*robust combiner* for OT and its existence was pointed out in [27] (and [37]) based on amplification techniques from [14,45].

Our goal is therefore to solve the following combinatorial problem: find a collection of "small" committees such that every adversary, corrupting at most $t = (1 - \delta)n$ of the parties, covers less than half of the committees. A simple probabilistic argument shows that such collections exist and moreover a random collection whose size depends only on δ (and not on n) is a good solution with high probability.

An Explicit Construction. We next give an explicit choice of committees that satisfies the above requirements. Consider a bipartite graph with m vertices on the left (the committees) and n vertices on the right (the parties). Every committee has d edges connecting it to all the parties it consists of (that is, the graph is d-regular on its left side). The requirement for the committees protocol to be secure is that every set of $m/2$ vertices on the left are connected to more than $(1 - \delta)n$ vertices on the right. This is exactly the setting of a *disperser*[4] with very high min-entropy (min-entropy of $\log(n/2)$ out of the possible $\log n$). There are several explicit constructions that we can use for this task including Goldreich and Wigderson [25], Reingold et al. [42] and Gradwohl et al. [26], all of which have near optimal degree (up to constants with respect to the lower bounds of [41]). Specifically we can work with $d = \lceil \frac{1}{\delta} \rceil$, and $m = n + o(n)$ (or even $m = n - o(n)$ if using the construction of [26]).

Corollary 1. *There exists an explicit construction of a network consisting of $(n + o(n))\binom{\lceil 1/\delta \rceil}{2}$ OT channels such that the network can t-securely compute f_{OT} in the presence of an adversary that corrupts up to $t = (1 - \delta)n$ of the parties. Specifically:*

- *If δ is a constant then the network needs $O(n)$ OT channels.*
- *As long as $\delta \geq \frac{1}{\sqrt{n}}$, the construction requires strictly less than the $\binom{n}{2}$ OT channels of the full network.*

The above corollary can be combined with OT extension protocols [4,31] to yield a $((1-\delta)n)$-secure protocol for an *arbitrary* function f that uses a total of $O(kn)$ OTs. This protocol inherits the security and assumptions of the underlying OT extension protocol (see Footnote 3).

Related works using committees. The idea of virtually performing tasks by committees has been used in distributed computing and cryptography. Originating in the work of Bracha [9] in the context of Byzantine agreement, committees have been used in the same context by [8,12], for MPC [29] and for leader election [39,48,35]. Committees have recently been used by Fitzi et al. [18] to achieve Perfectly Secure Message Transmission (PSMT) in a partial network of secure channels. It should be noted that while the task of PSMT is reminiscent of our question regarding OT channels, there are inherent differences. For example, our committees protocol (above) can effectively achieve an OT call even between two parties that are isolated in the OT graph (not connected by an OT channel to any other party). In PSMT, on the other hand, there is no chance of achieving secure communication with a node that is not connected by any secure channel.[5]

On non-adaptive adversaries. In the case that the adversary is non-adaptive but the network is dynamic, one can do much better. In fact, only a single good committee is needed. Indeed, a randomly chosen committee of size k/δ has probability of $1 - 2^{-\Omega(k)}$ of being good. Note, however, that this simple protocol can be trivially broken by an adaptive adversary who first learns the identity of the committee members and then corrupts all of them.

---

[4] A definition and discussion on dispersers can be found in, e.g. [43].
[5] Recall that in our OT channels model we assume a *full* network of secure channels to be intact.

3.2 Lower Bounds for $t = n - 1$: Full OT Network Is Necessary

In this section we look at the strictest security scenario, where the adversary can corrupt all but one of the parties. We show that given an almost full network of OT channels except for one missing channel, it is impossible to complete the network (i.e., securely compute the function f_{OT}). As a first step we consider a static network with just 3 parties, A, B and C.

Claim 2. *Let A, B and C form a network where C has OT channels with both A and B, but there is no OT channel between A and B. Then, any 2-secure protocol for f_{OT} over this partial network can be used (as a black box) to obtain a two-party OT protocol in the plain model.*

Proof: We transform the given 3-party protocol π_3 into a two-party OT protocol π_2 in two steps: first we eliminate all invocations of the OT oracle, and then we obtain a two-party protocol by letting one of the parties simulate A and the other simulate B and C. These steps are captured by the following two lemmas.

Lemma 1. *Any 3-party protocol π_3 as in Claim 2 can be used (as a black box) to obtain a protocol π_3' over a network with no OT channels, such that π_3' is secure against an adversary that corrupts either $\{A, C\}$ or $\{B, C\}$.*

The protocol π_3' is obtained from π_3 by implementing each OT call via the trivial protocol in which C sends its input to the other OT participant (either A or B). Note that this trivial OT protocol is perfectly secure against an adversary that corrupts either $\{A, C\}$ or $\{B, C\}$, since the input of C is guaranteed to be known to the adversary.

We can now use π_3' to implement OT between A and B in a 3-party network without OT channels. An important observation is that C has no inputs in this protocol.

Lemma 2. *Let Π be a protocol between A, B, C that computes a function for which C has no inputs, and suppose that Π is secure against $\{A, C\}$ and $\{B, C\}$. Then Π is also secure against $\{A\}$ and $\{B\}$.*

The lemma follows simply by observing that when C has no input the view of a corrupted A can be simulated using the simulation of $\{A, C\}$ and likewise for B.

We can now use π_3' to get a two-party OT protocol π_2 by letting one party simulate A and the other party simulate B, C. Due to Lemma 2 we get that the protocol is secure against corruption of either party and hence constitutes an OT protocol in the plain model between two parties. ∎

We Generalize Claim 2 to hold for a *dynamic* network of n parties (rather than a static 3-party network). The proof appears in the full version.

Claim 3. *Any $(n-1)$-secure protocol for f_{OT} over an n-party partial network with at most $\binom{n}{2} - 1$ OT channels can be used (as a black box) to obtain a two-party OT protocol in the plain model.*

As corollaries of the previous lemma, we get the lower bounds that we were seeking for the number of OT invocations:

Theorem 4. *Any n-party protocol Π that $(n-1)$-securely computes f_{OT} using less than $\binom{n}{2}$ OT calls can be used (as a black box) to implement a two-party OT protocol in the plain model. In particular, there is no such Π with perfect or statistical security, and its existence with computational security cannot be based on one-way functions in black-box way.*

3.3   Lower Bounds for Corruption of $t = n - d$ Parties

We show impossibility results for this case that are based on extremal graph theory and give tight bounds (for different ranges of d). The bounds hold also in the dynamic network model.

Theorem 5 (Lower bound for general d). *Consider a network of n parties in the presence of an adversary that can corrupt $t = n - d$ parties.*

1. *Suppose the network (even a dynamic one) has $o(n^2/d)$ OT channels. Then any $(n-d)$-secure protocol for f_{OT} in this network be used (as a black box) to implement a two-party OT protocol in the plain model.*
2. *Suppose d is a constant and the network (even a dynamic one) has less than $(1-c))\binom{n}{2}$ OT channels (for some constant c). Then any $(n-d)$-secure protocol for f_{OT} in this network be used (as a black box) to implement a two-party OT protocol in the plain model.*

Proof: The two claims follow the same principle. The crux is that unless every two sets of parties of size d be connected in the OT graph, then one can build an OT in the plain model. Suppose that there exist two disconnected sets of size at least d then define each of the groups as A and B and the rest of the graph as C. We now reduce this setting to the case of the previous section where we have parties A, B and C such that A and B are not connected and the adversary may corrupt either $\{A, C\}$ or $\{B, C\}$ (since these are sets of size at most $n - d$). By Claim 2, such a setting would allow building a two-party OT protocol in the plain model.

Proving (1) for the static case. Due to the outline above, we examine graphs where every two sets of size d must contain at least one edge between them. This means, in particular, that in the graph of *non-edges* there exists no clique of size $2d$. By Turan's Theorem, such a graph can have at most $(1 - 1/(2d))n^2/2$ non-edges, and hence the OT graph must have at least $\frac{n^2}{4d}$ edges.

Proving (2). The above argument is limited since it only considers the fact that the non-edges graph must not contain a clique of size $2d$. But actually, the graph cannot even contain a bipartite $d \times d$ clique which is a stricter constraint. For such a graph there are rather tight results when d is a constant. Namely, for constant d the Erdös-Stone-Simonovits Theorem [16] states that the non-edges graph must have at most $o(n^2)$ missing edges, which amounts to the OT graph containing $(1 - o(1))\binom{n}{2}$ OT channels. As in Claim 3, even if the layout of OT channels is not known in advance it can simply be guessed. Since the number of missing edges is constant, the probability that a random guess is correct about the eventual network is inverse polynomial, which suffices to extend any static network result (with constant d) to a similar result for dynamic networks.

Sketch of proof of (1) in the dynamic case. We provide an argument that there must be at least $\frac{n^2}{2d}$ edges in the graph (when assuming that every two sets of size d must contain at least one edge between them). This argument provides a slightly better bound than Turan's Theorem (since Turan's theorem discusses anti-cliques rather than bipartite $d \times d$ anti-cliques) but, more crucially, gives us information that is useful for proving the bound in the dynamic case. Specifically, we get a guarantee that if there are less than $\frac{n^2}{2d}$ edges in the graph then in every partition of the graph to d sized sets, at least one of the sets has no neighbor with some other d-sized set (the other set is not necessarily in the partition). This information, gives rise to an efficient procedure that finds (with noticeable probability) sets A, B and C where A and B are of size d and have no connecting edges in the dynamically set network. This, in turn, allows to build a two-party OT in the plain model from a secure protocol for f_{OT} in the dynamic setting. The complete proof appears in the full version.   ∎

4   Upper Bounds for the Case of $t = n - 1$

4.1   The Tables Method

The tables method is a generic secure computation protocol that computes a function by an iterative process on the truth table of the function. The truth table of a function $f : \{0,1\}^n \to \{0,1\}^m$ is simply a $2^n m$ bit long string such that the ith cell (or entry) contains the value $f(x)$ where x is the string representing the integer i. Denote the bits of the string x by $x_1, \ldots, x_n$. The idea is to use the fact that restricting the value of the variable x_1 to be either 0 or 1 amounts to looking either at the first or at the second half of the table. Denote by T^f the truth table of f and by $T^f|_b$ the new table when fixing the first input variable x_1 to b, for $b \in \{0,1\}$. Thus, $T^f|_0$ is simply the first half of the table T^f while the second half is $T^f|_1$.[6] Similarly, denote the table of f after fixing the j most significant bits of x as $T^f|_{x_1,\ldots,x_j}$.

The idea of the protocol is that at the jth iteration the j parties $P_1, \ldots, P_j$ jointly distribute additive shares of the table $T^f|_{x_1,\ldots,x_j}$ between themselves. At the end of the protocol all parties hold a share of the table $T^f|_{x_1,\ldots,x_n}$ which consists simply of the single value $f(x)$. The full protocol TABLES is presented in Figure 1.

---

[6] The first and second halves of a table correspond to fixing of the "most significant" bit, x_1. For every other x_i, the fixing of the ith bit x_i amounts to a different partition of the table into two halves.

TABLES $f(x_1, \ldots, x_n)$

Let $f : \{0,1\}^n \to \{0,1\}^m$ be a function to be computed by n parties $P_1, \ldots, P_n$, each holding a single input bit $x_1, \ldots, x_n$ respectively.

- **Initialization stage:** P_1 computes $T^f|_{x_1}$, i.e. the truth table of f when restricted to his input x_1. Let $S_1^1 = T^f|_{x_1}$.
- **Iteration stage:** The following steps are repeated sequentially for each $j \in [n-1]$. At the beginning of the jth iteration, each of the parties $P_1, \ldots, P_j$ holds a share $S_1^j, \ldots, S_j^j$ (respectively) such that $\bigoplus_{i \in [j]} S_i^j = T^f|_{x_1, \ldots, x_j}$. At the end of the iteration, the table $T^f|_{x_1, \ldots, x_{j+1}}$ is shared among the parties $P_1, \ldots, P_{j+1}$.

  1. For each $i \in [j]$, party P_i chooses a random mask $R_i \in \{0,1\}^{2^{n-j}m}$ and calculates $T_i^0 = S_i^j|_0 \oplus R_i$ and $T_i^1 = S_i^j|_1 \oplus R_i$.
  2. P_{j+1} runs an OT protocol with every P_i such that $i \in [j]$. They run the protocol $OT(T_i^0, T_i^1; x_{j+1})$ with P_i as sender and P_{j+1} as receiver.
  3. For each $i \in [j]$, party P^i sets $S_i^{j+1} = R_i$ while party P_{j+1} sets $S_{j+1}^{j+1} = \bigoplus_{i \in [j]} T_i^{x_{j+1}}$.

- **Output stage:** Each party sends its share S_i^n to P_1 who outputs $\bigoplus_{i \in [n]} S_i^n$.

Fig. 1. The TABLES protocol

Theorem 6. *Protocol TABLES is an $(n-1)$-secure protocol for the function f. The protocol involves a single OT call between each pair of players.*

A proof of Theorem 6 appears in the full version. Note that the protocol can be easily generalized to handle ℓ-bit inputs rather than single bits. In such a case, each party runs ℓ consecutive iterations, one for each input bit. The number of OTs between each pair of players grows to ℓ.

4.2 Applying the Tables Method

The advantage of the tables method is that it requires exactly one OT call between each pair of parties (overall, $\binom{n}{2}$ OT calls) and presents a plausibility result for $(n-1)$-secure computation of any function on n bits, matching the lower bound on the number of OTs for the case of $t = n - 1$ (Theorem 4).

The main problem with the tables method, however, is that the strings sent in the (string) OT are of length $2^n m$. This makes the protocol inefficient except when the input domain of f is of feasible size. In the following we show that for certain classes of functions one can get efficient protocols that still require a minimal number of OTs. For example, we describe how to securely compute any function in NC^0, namely a function in which each bit of the output depends on a constant number of input bits. Note that, under standard cryptographic assumptions, there exist non-trivial cryptographic primitives such as one-way functions and pseudorandom generators in NC^0 [1].

Proposition 7. *For every function $f \in NC^0$ there exists an efficient $(n-1)$-secure computation protocol using just $\binom{n}{2}$ OT calls.*

Proof sketch: For a function $f : \{0,1\}^n \to \{0,1\}^m$ in NC^0 it is guaranteed that each output bit is a function of $c = O(1)$ input bits. We call these the c input bits that *affect* the output bit.

The straightforward protocol runs m separate TABLES protocols, one for each output bit. Since each output bit is affected by only c parties, then each TABLES protocol can be executed only by the c parties that affect this output, using a table of size 2^c which is constant. However, if each protocol is run separately then the number of OT calls would grow to $\binom{c}{2}m$, which may be bigger than $\binom{n}{2}$ when m is large. Using a careful scheduling of the TABLES protocols, all OT calls between each pair of parties can be computed using a *single* OT invocation. (See full version for details.) Thus, the overall number of OTs remains $\binom{n}{2}$, matching the lower bound. ∎

Note that the above mentioned schedule works for every function f (not necessarily in NC^0). The efficiency though is only guaranteed for limited types of functions. More precisely, efficiency is guaranteed for every function where each output bit is affected only by a *logarithmic* number of input variables.

Extension to bounded degree polynomials. A straightforward extension of the above proposition follows from observing that if the output stage is not executed then the above protocol efficiently computes an additive secret-sharing for each output bit. At the same cost, the parties can get an additive secret-sharing of the *sum* of various outputs. This is done simply by each party taking a local sum of the various shares that it holds to create a new share for the sum. This forms an efficient low communication $(n-1)$-secure protocol for all *logarithmic degree polynomials* whose representation as the sum of monomials has only a *polynomially* many terms.

4.3   Oblivious Linear Branching Programs

This section puts forward a generalization of the tables method that extends the class of functions that we can securely compute by an efficient protocol. The class of functions that we deal with is a linear version of oblivious branching programs.

Definition 8 (Oblivious Linear Branching Programs). *A linear branching program LBP on an n-bit input is an ordered set of triples $\{(i_1; M_1^0, M_1^1), \ldots, (i_s; M_s^0, M_s^1)\}$ and an initial vector $v_0 \in \{0,1\}^{w_0}$. Each triple contains an index $i_j \in [n]$ and a pair of boolean $w_{j-1} \times w_j$ matrices M_j^0, M_j^1 (where $w_j \geq 1$). The size of the program is s and its width w is the maximal w_j over all $j \in [s]$. On input $x \in \{0,1\}^n$ the output of the program is $LBP(x) = v_0 M_1^{x_{i_1}} \cdots M_s^{x_{i_s}}$.*

Theorem 9. *There exists an $(n-1)$-secure computation protocol for computing the output of a linear branching program LBP. The protocol makes at most sn OT calls on ℓ-bit strings (where s and ℓ are the size and width of LBP, respectively).*

The protocol runs along the lines of the TABLES protocol (Figure 1), with the main difference being that at step (1), the ith party computes $T_i^0 = S_i^j M_j^0 \oplus R_i$ and $T_i^1 = S_i^j M_j^1 \oplus R_i$ (rather than $T_i^0 = S_i^j|_0 \oplus R_i$ and $T_i^1 = S_i^j|_1 \oplus R_i$). The complete protocol appears in the full version.

4.4   Functions Captured by Linear Branching Programs

As mentioned before, the protocol for linear branching programs is a generalization of the tables method. As such, it captures the same applications as the previous method, but it also captures other functions that could not be efficiently computed in the previous method. We highlight some function classes that can be computed using this methodology:

Tables. The LBP model is indeed a generalization as exemplified by the following presentation: consider the initial vector v_0 that is the truth table of the function f. For each iteration, the two matrices M_j^0 and M_j^1 are simply two projection matrices (and hence also linear operations). M_j^0 leaves only the first half of the coordinates while M_j^1 leaves the second half of the coordinates.

Oblivious branching programs. Similar to linear branching programs, oblivious branching programs inquire a single variable at each layer and move to a new state according to its answer. This can be viewed as a layered graph where each node has two outgoing edges labeled 0 and 1 going to the next level. The width of the program is the maximal number of nodes in a layer and the number of layers is the length of the program. The simulation of such a branching program by an LBP looks at the intermediate states as indicator vectors of length w (all zeros except a single one indicating the current state). The matrix for input 0 has as its ith row, the indicator vector that the ith state should move to in case that the input bit is 0 and likewise for the second matrix. Other models of computation or functions that are captured by their view as a branching program include **decision trees**, **oblivious automata** and *membership in small (polynomial size) set*.

Oblivious counting branching programs. As in the oblivious branching program case, a non-deterministic branching program allows going from one state to a number of states. A *counting branching program* outputs the number of accepting paths that a non-deterministic branching program has. Such non-determinism can easily be incorporated into LBPs by allowing the state vector to vary from an indicator with a single one. The ith row of the matrix will have a 1 for each possible move from the ith state to the next level. If the operations are executed over a large enough field, then the intermediate vector holds in each location the number of paths that lead up to this state. Thus, over a large field this implements a counting branching program. If working over the field $GF(2)$ then this is simply a *parity branching program* that indicates the parity of the number of paths that lead to a state. Unfortunately, the most natural non-deterministic model is not captured by LBPs. This is a non-deterministic procedure that asks whether their exists an accepting path to the program at all (an operation that is no longer a linear one). In Section 5, we present protocols for secure computation for this model.

Sparse Polynomials. LBPs allow for a simple and efficient computation of a monomial over input bits.[7] In addition, an LBP can incorporate in it a number of parallel LBP computations and have the last operation sum their outputs (simply by incorporating this linear operation in the last pair of matrices). Thus LBPs can compute a polynomial as a sum of all of its monomials. For the program to be efficient, the only limitation is that the number of monomials is polynomial. Note that this captures a larger family of functions than in Section 4.2. A closely related question is can one compute a DNF formula using LBPs (DNFs are the OR of monomials rather than their sum). This is a special case of the non-deterministic question addressed in the next section.

5   Secure Computation for Non-deterministic LBP

In this section we suggest a method of securely computing a nondeterministic (or existential) linear branching program. As opposed to counting branching programs that give the sum of the number of solutions (and are easy to compute by LBPs), asking whether or not the exists a solution is a non-linear operation and therefore is not captured by the general framework. A good example is the computation of DNF formulas. Like sparse polynomials these are a polynomial size collection of monomials over n input bits but the question is whether x satisfies at least one of the monomials (rather than their sum). The problem arises from the fact that the OR operation is not a linear one and hence it is not captured by the LBP model. A natural approach is to first compute the sum of the monomials over a large enough field (to avoid a wraparound), and then check whether this sum is zero or not. However, revealing the sum is not a good solution as it leaks more information on the inputs than the desired output (it differentiates, for instance, whether there was a single satisfied monomial or many of them).

We propose a generic method that securely computes the existential analogue of any counting LBP. The method is secure against adversaries that can corrupt up to $t \leq n - \Omega(k)$ where k is the security parameter, and adds a statistical error of at most 2^{-k}. For simplicity we state and prove the theorem formally for limited LBPs where each party has a single input bit and note that as in the previous sections, this protocol may be generalized to more complex LBPs, at the price of additional OT invocations.

Theorem 10. *Let L be a LBP of length n computing a function $f : \{0,1\}^n \to \mathbb{Z}_p$ where p is a $(k/2)$-bit prime and k is the security parameter. Then there exists an efficient n-party statistically t-secure protocol with $t = n - O(k)$ for the predicate g defined as:*

---

[7] For example, an LBP for the AND function over n bits simply uses vectors and matrices of dimension 1 and takes $v_0 = (0)$ and the ith triplet is $(i, 0, 1).$,

$$g(x) = \begin{cases} 0, & f(x) = 0; \\ 1, & otherwise. \end{cases}$$

The protocol requires $4\binom{n}{2}$ *OT calls.*

Proof: The basic idea is to add a randomization stage in each of the iterations of the secure computation protocol for L. This randomization should give an output with the following properties: The output should be uniformly distributed over the domain if $f(x) \neq 0$ but should always be 0 if $f(x) = 0$. Therefore, if the output is not 0 then we know for sure that $f(x) \neq 0$ but learn nothing else about $f(x)$. If the output is 0, then it is most likely that $f(x) = 0$. An error only happens if the uniformly distributed output happened to hit 0 which happens with probability that is inverse of the domain size (we will choose the domain to be of size 2^k).

Cayley expanders and a matrix representation. For the randomization steps we use a constant degree *Cayley expander graph* with a specific structure. A Cayley graph is described over a multiplicative group by a small set of generators $\{G_1, \ldots, G_d\}$. For each element (vertex) v in the group, its neighbors are $\{G_1 \cdot v, \ldots, G_d \cdot v\}$. We can use any expander graph with a constant degree such that its generators can be represented as *affine transformation* over $\mathbb{Z}_p^m$. In particular we can use the expander graph of Margulis [36] and Gaber and Galil [20]. This is an expander over $\mathbb{Z}_N \times \mathbb{Z}_N$ for some integer N, and we take N to be a prime p in the order of $2^{k/2}$. The expander has degree 8 and as we required can be presented by 8 affine transformations.

For simplicity we will describe the construction over $\mathbb{Z}_p^2$ (as in the Margulis graph) although this can be generalized. Suppose that each step is an affine transformation, e.g. a step moves from vertex $v \in \mathbb{Z}_p^2$ to the vector $Av + e$ where $A \in [\mathbb{Z}]_{2,2}$ is a 2×2 matrix and $e \in \mathbb{Z}_p^2$ is a vector. For each such generator we define the corresponding 3×3 matrix $G \in [\mathbb{Z}]_{3,3}$ as:

$$G = \begin{pmatrix} a_{11} & a_{12} & e_1 \\ a_{21} & a_{22} & e_2 \\ 0 & 0 & 1 \end{pmatrix}$$

Notice that multiplying the vector $v = (v_1, v_2, 1)$ by G simply amounts to a step in the expander from vertex (v_1, v_2) with the third coordinate remaining 1. If $A_1, \ldots, A_n$ denotes a series of steps where each $A_i \in \{G_1, \ldots, G_d\}$ and let $v = (0, 0, 1)$ then $A_n A_{n-1} \ldots A_1 v$ stands for a random walk starting at vertex $(0, 0)$ and the first 2 coordinates of the output hold the end vertex of the walk. On the other hand, $A_n A_{n-1} \ldots A_1 \bar{0}$ simply equals $\bar{0}$ (where $\bar{0}$ stands for the vector $(0, 0, 0)$).

The randomization technique. Basically, each party in its turn will contribute a random step A_i in the expander. Our goal is that at the end of the execution the output will be the multiplication $A_n \ldots A_1 v$ where v is the vector $(0, 0, f(x))$. Hence, if $f(x) = 0$ the output will simply be $\bar{0}$. On the other hand if $f(x) \neq 0$ then the output represents the end of a random walk starting at $(0, 0)$. We use the following result of Kamp and Zuckerman [32], which states that an adversary that does not know $\Omega(k)$ of the n expander steps has essentially no knowledge about the outcome of the random walk. The precise statement is that a random walk on a good expander (where each step is represented by a single symbol) is an extractor for a *symbol fixing source*.[8]

Theorem 11 (adapted from Kamp and Zuckerman [32], Theorem 3.1). *Let $a_1, \ldots, a_n$ be a series of steps on an expander of degree d, size d^m and second eigenvalue $\lambda \leq d^{-\alpha}$ and let t be such that $n - t \geq \frac{1}{2\alpha}\left(m + \frac{2}{\log d}\log\frac{1}{2\varepsilon}\right)$. Then conditioned on the view of an adversary that observes at most t elements in the series, the output of the walk is ε-close to uniform.*

In our application the graph has parameters $d = 8$, $\alpha \approx 0.06$ (due to [20]) and the graph is of size 2^k, thus $m = k/3$. When choosing $\varepsilon = 2^{-k}$ the requirement in the Theorem translates to $n - t \geq \Omega(k)$.

Corollary 12. *Let $A_1, \ldots, A_n$ be a sequence of randomly chosen generator matrices for a good expander graph (e.g. the Margulis graph) with vertex set $\mathbb{Z}_p^2$ (for prime p in the order of $2^k/2$). Let $v = (0, 0, c)$ for $c \neq 0$ and denote $u = A_n \ldots A_1 v$. Then conditioned on the view of an adversary that observes at most $t = n - \Omega(k)$ of the sequence the pair (u_1, u_2) is 2^{-k}-close to the uniform distribution on $\mathbb{Z}_p^2$.*

Proof: (of Corollary 12) The corollary follows directly from Theorem 11 in the case that $c = 1$. It is left to show that it also holds for any $c \neq 0$. This can be seen by breaking the vector v into the sum of c vectors of the type $(0, 0, 1)$. For each of the c vectors the random walk gives an almost uniform distribution. When summing up we have a uniform distribution multiplied by c. Since we are working in $\mathbb{Z}_p$ then multiplication amounts to a permutation on the elements of $\mathbb{Z}_p$ and the output remains close to uniform. Note that this is the only place where we require that p is prime. ∎

The actual protocol. The protocol is the same protocol as the general one for computing LBPs only at each iteration, the acting party (that redistributes the shares) chooses a random step in the extractor. The matrix operations of the LBP will always be multiplied from the right, while the random steps of the expander will be multiplied from the left. Technically, the following changes are applied:

- Instead of starting with the vector v_0 of length w_0, the protocol starts with a matrix B_0 of 3 vectors ($B_0 \in [\mathbb{Z}_p]_{3, w_0}$). The first two rows of B_0 are all zero vectors and the third is the vector v_0. Accordingly, the protocol runs throughout with 3 row matrices rather than single row vectors.
- At step (1) of the iteration stage, rather than computing two values, party P_i computes $2d$ values, two for each generator of the expander. They are for each $\tau \in [d]$: $T_i^{0\tau} = G_\tau S_i^j M_j^0 \oplus R_i$ and $T_i^{1\tau} = G_\tau S_i^j M_j^1 \oplus R_i$.

---

[8] A symbol fixing source is a randomness source for which t of n symbols are fixed while the rest are uniformly distributed.

- At step (2) of the iteration, the acting party P_i chooses a random $\tau \in [d]$ and runs a $\binom{2d}{1}$-OT protocol with each party according to his input x_j and τ. Such a protocol requires $\log d + 1$ OT calls.
- At the output step, all parties send the *first two* rows of their shares (but not the third row!) to the designated party P_1. This party calculates the sum and outputs 0 if the sum was $(0,0)$ and 1 otherwise.

The correctness and security of the overall protocol follows since the modified LBP protocol forms a $n - 1$-secure computation for a sharing of the following vector

$$u = A_n \ldots A_1 B_0 M_1^{x_{i_1}} \ldots M_n^{x_{i_n}}$$

In addition, $B_0 M_1^{x_{i_1}} \ldots M_n^{x_{i_n}}$ is simply the vector $(0, 0, f(x))$. Therefore, combined with corollary 12, we get that if P_1 outputs the correct value (up to an error probability of at most $1/p^2 \leq 2^{-k}$). Moreover, if $f(x) \neq 0$ then an adversary that corrupts up to t parties (including P_1) sees a value that is statistically close to uniform, hence leaking no additional information on $f(x)$ or x. This concludes the proof of Theorem 10. ∎

Acknowledgements. We thank Ronen Shaltiel for pointers on constructions of dispersers.

References

1. Applebaum, B., Ishai, Y., Kushilevitz, E.: Cryptography in NC^0 In: 45th FOCS, pp. 166–175 (2004)
2. Barkol, O., Ishai, Y.: Secure computation of constant-depth circuits with applications to database search problems. In: Shoup, V. (ed.) CRYPTO 2005. LNCS, vol. 3621, pp. 395–411. Springer, Heidelberg (2005)
3. Beaver, D.: Precomputing oblivious transfer. In: Coppersmith, D. (ed.) CRYPTO 1995. LNCS, vol. 963, pp. 97–109. Springer, Heidelberg (1995)
4. Beaver, D.: Correlated pseudorandomness and the complexity of private computations. In: 28th STOC, pp. 479–488 (1996)
5. Beimel, A., Malkin, T.: A quantitative approach to reductions in secure computation. In: Naor, M. (ed.) TCC 2004. LNCS, vol. 2951, pp. 238–257. Springer, Heidelberg (2004)
6. Beimel, A., Malkin, T., Micali, S.: The all-or-nothing nature of two-party secure computation. In: Wiener, M.J. (ed.) CRYPTO 1999. LNCS, vol. 1666, pp. 80–97. Springer, Heidelberg (1999)
7. BenOr, M., Goldwasser, S., Wigderson, A.: Completeness theorems for non-cryptographic fault-tolerant distributed computation. In: 20th STOC, pp. 1–10 (1988)
8. Berman, P., Garay, J., Perry, K.: Bit optimal distributed consensus. In: Computer Science Research, pp. 313–332. Plenum Publishing Corporation (1992)
9. Bracha, G.: An $o(\log n)$ expected rounds randomized byzantine generals protocol. Journal of the ACM 34(4), 910–920 (1987)
10. Canetti, R.: Security and composition of multiparty cryptographic protocols. Journal of Cryptology 13(1), 143–202 (2000)

11. Chaum, D., Crépeau, C., Damgård, I.: Multiparty unconditionally secure protocols. In: 20th STOC, pp. 11–19 (1988)
12. Coan, B., Welch, J.: Modular construction of a byzantine agreement protocol with optimal message bit complexity. Information and Computation, 97(1) (1992)
13. Crépeau, C., Kilian, J.: Achieving oblivious transfer using weakened security assumptions. In: 29th FOCS, pp. 42–52 (1988)
14. Damgård, I., Kilian, J., Salvail, L.: On the (im)possibility of basing oblivious transfer and bit commitment on weakened security assumptions. In: Stern, J. (ed.) EUROCRYPT 1999. LNCS, vol. 1592, pp. 56–73. Springer, Heidelberg (1999)
15. Dodis, Y., Micali, S.: Lower bounds for oblivious transfer reductions. In: Stern, J. (ed.) EUROCRYPT 1999. LNCS, vol. 1592, pp. 42–55. Springer, Heidelberg (1999)
16. Erdos, P., Simonovits, M.: A limit theorem in graph theory. Stud. Sci. Math. Hung 1, 51–57 (1966)
17. Even, S., Goldreich, O., Lempel, A.: A randomized protocol for signing contracts. Communications of the ACM 28(6), 637–647 (1985)
18. Fitzi, M., Franklin, M., Garay, J., Vardhan, H.: Towards optimal and efficient perfectly secure message transmission. In: Vadhan, S.P. (ed.) TCC 2007. LNCS, vol. 4392, Springer, Heidelberg (2007)
19. Franklin, M., Haber, S.: Joint encryption and message-efficient secure computation. J. Cryptology 9(4), 217–232 (1996)
20. Gabber, O., Galil, Z.: Explicit constructions of linear-sized superconcentrators. JCSS 22(3), 407–420 (1981)
21. Goldreich, O.: Foundations of Cryptography, vol. 2. Cambridge University Press, Cambridge (2004)
22. Goldreich, O., Micali, S., Wigderson, A.: How to play any mental game - a completeness theorem for protocols with honest majority. In: 19th STOC, pp. 218–229 (1987)
23. Goldreich, O., Micali, S., Wigderson, A.: Proofs that yield nothing but their validity, or all languages in NP have zero-knowledge proof system. Journal of the ACM 38(1), 691–729 (1991)
24. Goldreich, O., Vainish, R.: How to solve any protocol problem - an efficiency improvement. In: Pomerance, C. (ed.) CRYPTO 1987. LNCS, vol. 293, pp. 73–86. Springer, Heidelberg (1988)
25. Goldreich, O., Wigderson, A.: Tiny families of functions with random properties: A quality-size trade-off for hashin. Rand. Structs. and Algs. 11(4), 315–343 (1997)
26. Gradwohl, R., Kindler, G., Reingold, O., Ta-Shma, A.: On the error parameter of dispersers. In: APPROX-RANDOM, pp. 294–305 (2005)
27. Harnik, D., Kilian, J., Naor, M., Reingold, O., Rosen, A.: On tolerant combiners for oblivious transfer and other primitives. In: Cramer, R.J.F. (ed.) EUROCRYPT 2005. LNCS, vol. 3494, pp. 96–113. Springer, Heidelberg (2005)
28. Harnik, D., Naor, M., Reingold, O., Rosen, A.: Completeness in two-party secure computation - a computational view. In: 36th STOC, pp. 252–261 (2004)
29. Hirt, M., Maurer, U.: Player simulation and general adversary structures in perfect multiparty computation. Journal of Cryptology 13(1), 31–60 (2000)
30. Impagliazzo, R., Rudich, S.: Limits on the provable consequences of one-way permutations. In: 21st STOC, pp. 44–61 (1989)
31. Ishai, Y., Kilian, J., Nissim, K., Petrank, E.: Extending oblivious transfers efficiently. In: Boneh, D. (ed.) CRYPTO 2003. LNCS, vol. 2729, pp. 145–161. Springer, Heidelberg (2003)
32. Kamp, J., Zuckerman, D.: Deterministic extractors for bit-fixing sources and exposure-resilient cryptography. In: 44th FOCS, pp. 92–101 (2003)

33. Kilian, J.: Founding cryptography on oblivious transfer. In: 20th STOC, pp. 20–31 (1988)
34. Kilian, J.: A general completeness theorem for two-party games. In: 23rd STOC, pp. 553–560 (1991)
35. King, V., Saia, J., Sanwalani, V., Vee, E.: Towards secure and scalable computation in peer-to-peer networks. In: 47th FOCS, pp. 87–98 (2006)
36. Margulis, G.: Explicit constructions of concentrators. Problemy peredaci informacii 9(4), 71–80 (1973)
37. Meier, R., Przydatek, B., Wullschleger, J.: Robuster combiners for oblivious transfer. In: Vadhan, S.P. (ed.) TCC 2007. LNCS, vol. 4392, Springer, Heidelberg (2007)
38. Naor, M., Pinkas, B.: Efficient oblivious transfer protocols. In: SIAM Symposium on Discrete Algorithms (SODA 2001), pp. 448–457 (2001)
39. Ostrovsky, R., Rajagopalan, S., Vazirani, U.: Simple and efficient leader election in the full information model. In: 26th STOC, pp. 234–242 (1994)
40. Rabin, M.O.: How to exchange secrets by oblivious transfer. TR-81, Harvard (1981)
41. Radhakrishnan, J., Ta-Shma, A.: Bounds for dispersers, extractors, and depth-two superconcentrators. SIAM J. Discrete Math. 13(1), 2–24 (2000)
42. Reingold, O., Vadhan, S., Wigderson, A.: Entropy waves, the zig-zag graph product, and new constant-degree expanders and extractors. ECCC, 8(18) (2001)
43. Shaltiel, R.: Recent developments in explicit constructions of extractors. Bulletin of the EATCS 77, 67–95 (2002)
44. Wiesner, S.: Conjugate coding. SIGACT News 15(1), 78–88 (1983)
45. Wullschleger, J.: Oblivious transfer amplification. In: Naor, M. (ed.) EUROCRYPT 2007, vol. 4515, pp. 555–572. Springer, Heidelberg (2007)
46. Yao, A.C.: Protocols for secure computations. In: 23rd FOCS, pp. 160–164 (1982)
47. Yao, A.C.: How to generate and exchange secrets. In: 27th FOCS, pp. 162–167 (1986)
48. Zuckerman, D.: Randomness-optimal sampling, extractors, and constructive leader election. In: 28th STOC, pp. 286–295 (1996)

Simulatable VRFs with Applications to Multi-theorem NIZK

Melissa Chase and Anna Lysyanskaya

Computer Science Department
Brown University
Providence, RI 02912
{mchase,anna}@cs.brown.edu

Abstract. This paper introduces simulatable verifiable random functions (sVRF). VRFs are similar to pseudorandom functions, except that they are also verifiable: corresponding to each seed SK, there is a public key PK, and for $y = F_{PK}(x)$, it is possible to prove that y is indeed the value of the function seeded by SK. A *simulatable* VRF is a VRF for which this proof can be simulated, so a simulator can pretend that the value of $F_{PK}(x)$ is any y.

Our contributions are as follows. We introduce the notion of sVRF. We give two constructions: one from general assumptions (based on NIZK), but inefficient, just as a proof of concept; the other construction is practical and based on a special assumption about composite-order groups with bilinear maps. We then use an sVRF to get a direct transformation from a single-theorem non-interactive zero-knowledge proof system for a language L to a multi-theorem non-interactive proof system for the same language L.

1 Introduction

It has been more than twenty years since the discovery of zero-knowledge proofs. In that time, they have attracted interest from the theoretical computer science community (leading to the study of interactive proof systems and PCPs), theoretical cryptography community, and, more recently, cryptographic practice.

The proof protocols that have been implemented so far [Bra99, CH02, BCC04], even though zero-knowledge in spirit, are not, strictly speaking, zero-knowledge proofs as we usually define them. Typically, they are honest-verifier interactive zero-knowledge proofs (sometimes, actually, arguments of knowledge) with the interactive step removed using the Fiat-Shamir paradigm [FS87, GK03]. Interaction is an expensive resource, and so using a heuristic such as the Fiat-Shamir transform in order to remove interaction is more attractive than using an interactive proof.

SINGLE-THEOREM NIZK. In contrast to the Fiat-Shamir-based protocols adopted in practice, that do not in fact provide more than just a heuristic security guarantee [GK03], there are also well-known provable techniques for achieving zero-knowledge in non-interactive proofs. Blum et al. [BFM88, DMP88, BDMP91]

A. Menezes (Ed.): CRYPTO 2007, LNCS 4622, pp. 303–322, 2007.

introduced the notion of a *non-interactive zero-knowledge* (NIZK) proof system. In such a proof system, some parameters of the system are set up securely ahead of time. Specifically, a common random string σ is available to all participants. The prover in such a proof system is given an $x \in L$ for some language L and a witness w attesting that $x \in L$. (For example, L can be the language of all pairs (n, e) e is relatively prime to $\phi(n)$. The witness w can be the factorization of n.) The prover computes a proof π, and the proof system is zero-knowledge in the following sense: the simulator can pick its own σ' for which it can find a proof π' for the statement $x \in L$. The values (σ', π') output by the simulator are indistinguishable from (σ, π) that are generated by first picking a random σ and then having the honest prover produce π for $x \in L$ using witness w. Blum et al. also gave several languages L with reasonably efficient NIZK proof systems.

Let us explain the Blum et al. NIZK proof system for the language L in the example above: $L = \{(n, e) \mid \gcd(\phi(n), e) = 1\}$. First, recall that if e is not relatively prime to $\phi(n)$, then the probability that for a random $x \in \mathbb{Z}_n^*$ there exists y such that $y^e = x \bmod n$ is upper-bounded by $1/2$. On the other hand, if e *is* relatively prime to $\phi(n)$, then for all $x \in \mathbb{Z}_n^*$ there exists such a y. So the proof system would go as follows: parse the common random string σ as a sequence $z_1, \ldots, z_\ell$ of elements of $\mathbb{Z}_n^*$, and for each z_i, compute y_i such that $y_i^e = z_i \bmod n$. The proof π consists of the values $y_1, \ldots, y_\ell$. The verifier simply needs to check that each y_i is the eth root of z_i. For any specific instance (n, e), the probability (over the choice of the common random string σ) that a cheating prover can come up with a proof that passes the verification is $2^{-\ell}$. By the union bound, letting $\ell = k(|n| + |e|)$ guarantees that the probability, over the choice of σ, that a cheating prover can find an instance (n, e) and a proof π passing the verification, is negligible in k.

Note that the proof system described above, although expensive, is not prohibitively so. Proof systems of this type have been shown to yield themselves to further optimizations [DCP97]. So why aren't such proofs attractive in practice?

MULTI-THEOREM NIZK. The problem with NIZK as initially defined and explained above was that one proof π completely used up the common random string σ, and so to produce more proofs, fresh common randomness was required. Blum et al. [BDMP91] showed a single-prover multi-theorem NIZK proof system for 3SAT, and since 3SAT is NP-complete, the result followed for any language in NP, assuming quadratic residuocity. Feige, Lapidot and Shamir [FLS99] constructed a multi-prover, multi-theorem NIZK proof system for all NP based on trapdoor permutations. Recently, Groth, Ostrovsky and Sahai [GOS06] gave a multi-theorem NIZK proof system for circuit satisfiability with very compact common parameters and achieving perfect zero-knowledge (with computational soundness), based on the assumption that the Boneh, Goh, Nissim [BGN05] cryptosystem is secure.

In each of the multi-theorem NIZK results mentioned above, to prove that $x \in L$ for a language L in NP, the prover would proceed as follows: first, reduce x to an instance of the right NP-complete problem, also keeping track of the witness w. Then invoke the multi-theorem NIZK proof system constructed for this NP-complete problem. In other words, even if the language L itself had an efficient *single-theorem* NIZK, existing multi-theorem NIZK constructions have no way of exploiting it. The Feige et al. result, which is the most attractive because it is based on general assumptions, is especially bad in this regard: their construction explicitly includes a step that transforms every instance x into a new instance x' via a Cook-Levin reduction. These reductions are what makes NIZK prohibitively expensive to be considered for use in practice.

In this paper, we give a construction for achieving multi-theorem NIZK for any language L based on single-theorem NIZK for L, without having to reduce instances of L to instances of any NP-complete languages. This construction is based on a new building block: a *simulatable verifiable random function* (sVRF).

SIMULATABLE VRFS. Verifiable random functions (VRFs) were introduced by Micali, Rabin, and Vadhan [MRV99]. They are similar to pseudorandom functions [GGM86], except that they are also verifiable. That is to say, associated with a secret seed SK, there is a public key PK, domain D_{PK}, range R_{PK} and a function $F_{PK}(\cdot) : D_{PK} \mapsto R_{PK}$ such that (1) $y = F_{PK}(x)$ is efficiently computable given the corresponding SK; (2) a proof $\pi_{PK}(x)$ that this value y corresponds to the public key PK is also efficiently computable given SK; such a proof can exist only for a unique value y; (3) based purely on PK and oracle calls to $F_{PK}(\cdot)$ and the corresponding proof oracle, no adversary can distinguish the value $F_{PK}(x)$ from a random value without explicitly querying the function on input x. Several constructions of VRFs in the plain model exist [MRV99, Lys02, Dod02, DY05]. In the common-random-string model, Goldwasser and Ostrovsky [GO92] showed that existence of VRFs (with polynomial-size domains; one can also call such VRFs verifiable pseudorandom generators, or VPRGs) is a necessary and sufficient condition for multi-theorem NIZK for all NP. Dwork and Naor [DN00] showed that (approximate) VPRGs in the standard model are necessary and sufficient for zaps (zaps are witness-indistinguishable proof protocols consisting of two rounds; the first round is a message from the verifier to the prover than can be reused for future instances).

We introduce *simulatable* VRFs (sVRFs). In the common parameters model, $F_{PK}(\cdot)$ is a VRF in the sense defined above for all honest settings of the common parameters. However, there is also a way to simulate the common parameters such that, corresponding to a PK, for any $x \in D_{PK}$, $y \in R_{PK}$, it is possible to simulate a proof π that $F_{PK}(x) = y$. The resulting simulation is indistinguishable from the view obtained when the parameters are set up correctly.

USING AN SVRF TO TRANSFORM SINGLE-THEOREM NIZK TO MULTI-THEOREM NIZK. A simulatable VRF with domain of size $\ell(k)$ and binary range allows a prover to come up with a fresh verifiably random string R of appropriate length $\ell(k)$ every time he wants to prove a new theorem. He simply comes up with a new PK for a VRF, and evaluates F_{PK} on input i to obtain the ith bit of R, R_i. The VRF allows him to prove that R was chosen correctly. He can then XOR R with a truly random public string σ_1 to obtain a string σ to be used in a single-theorem NIZK. The resulting construction is zero-knowledge because of the simulatability properties of both the sVRF and the single-theorem NIZK. It is sound because σ_1 is a truly random string, and so it inherits the soundness from the single-theorem NIZK (note that it incurs a penalty in the soundness error). Note that because our sVRF construction is in the public parameters model, the resulting multi-theorem proof system is also in the public parameters model (rather than the common random string model).

Since we give an *efficient* instantiation of sVRFs, our results essentially mean that studying efficient single-theorem NIZK proof systems for languages of interest is a good idea, because our construction gives an *efficient* transformation from such proof systems to multi-theorem ones.

USING AN sVRF INSTEAD OF THE RANDOM ORACLE. An sVRF shares some characteristics with a programmable random oracle: assuming that the parameters of the system were picked by the simulator, the simulator can *program* it to take certain values on certain inputs. One cannot necessarily use it instead of the hash function in constructions where the adversary gets the code for the hash function. But it turns out that it can sometimes replace the random oracle in constructions where the adversary is allowed oracle access to the hash function and requires some means to be sure that the output is correct. For example, using an sVRF instead of H in the RSA-FDH construction [BR93, Cor00] would make the same proof of security hold without the random oracle. Of course, it is not a useful insight: an sVRF is already a signature, so it is silly to use it as a building block in constructing another signature. The reason we think the above observation is worth-while is that it is an example of when using an sVRF instead of an RO gives provable instead of heuristic guarantees.

CONSTRUCTING AN sVRF. Our main result is a direct construction of a simulatable VRF based on the Subgroup Decision assumption (SDA) [BGN05], and an assumption related to the Q-BDHI assumption [BB04b]. Dodis and Yampolskiy [DY05] used the Q-BDHI assumption to extend the Boneh-Boyen short signature scheme [BB04a] and derive a VRF. The Dodis-Yampolskiy VRF is of the form $F_s(x) = e(g,g)^{1/(s+x)}$, where g is a generator of some group G_1 of prime order q, and $e : G_1 \times G_1 \mapsto G_2$ is a bilinear map. The secret key is s while the public key is g^s. The DY proof that $y = F_s(x)$ is the value $\pi = g^{1/(s+x)}$ whose correctness can be verified using the bilinear map.

Our sVRF is quite similar, only it is in a composite-order group with a bilinear map: the order of G_1 is an RSA modulus $n = pq$. This is what makes simulatability possible. In our construction, the public parameters consist of (g, A, D, H), all generators of G_1. As before, the secret key is s, but now the public key is A^s. $F_s(x) = e(H, g)^{1/(s+x)}$, and the proof is a randomized version of the DY proof: $\pi = (\pi_1, \pi_2, \pi_3)$, where $\pi_1 = H^{r/(s+x)}/D^r$, $\pi_2 = g^{1/r}$ and $\pi_3 = A^{(s+x)/r}$. It turns out that, when A generates the entire G_1, there is a unique $y = F_s(x)$ for which a proof exists. However, when A belongs to the order-p subgroup of G_1 (as is going to be the case when the system parameters are picked by the simulator), the verification tests correctness only as far as the order-p subgroup is concerned, and so the order-q component of $F_s(x)$ is unconstrained. The proof of security requires that a strengthening of Q-BDHI hold for the prime-order subgroups of G_1, and that the SDA assumption holds so that A picked by the simulator is indistinguishable from the correct A.

We also give, as proof of concept, a construction under general assumptions, based on multi-theorem NIZK.

ORGANIZATION OF THE REST OF THIS PAPER. In Section 2 we define sVRFs. In Section 3 we give an sVRF construction based on general assumptions as a proof of concept. In Section 4 we give our main result and its proof of security. Finally, in Section 5 we give the transformation from single-theorem NIZK to multi-theorem NIZK using sVRFs.

2   On Defining sVRFs

We begin by adapting the definition of Micali, Rabin and Vadhan [MRV99] in the public parameters model.

Definition 1 (VRF in the public parameters model). *Let Params($\cdot$) be an algorithm generating public parameters* p *on input security parameter* 1^k. *Let $D(\mathrm{p})$ and $R(\mathrm{p})$ be families of efficiently samplable domains for all* $\mathrm{p} \in Params$. *The set of algorithms* (G, Eval, Prove, Verify) *constitutes a verifiable random function (VRF) for parameter model Params, input domain $D(\cdot)$ and output range $R(\cdot)$ if*

Correctness. *Informally, correctness means that the verification algorithm* Verify *will always accept* $(\mathrm{p}, PK, x, y, \pi)$ *when* $y = F_{PK}(x)$, *and* π *is the proof of this fact generated using* Prove. *More formally,* $\forall k, \mathrm{p} \in Params(1^k)$, $x \in D(\mathrm{p})$,

$$\Pr[(PK, SK) \leftarrow G(\mathrm{p}); y = \mathtt{Eval}(\mathrm{p}, SK, x); \pi \leftarrow \mathtt{Prove}(\mathrm{p}, SK, x);$$
$$b \leftarrow \mathtt{Verify}(\mathrm{p}, PK, x, y, \pi) \ : \ b = 1] = 1$$

Pseudorandomness. *Informally, pseudorandomness means that, on input* (p, PK), *even with oracle access to* Eval($\mathrm{p}, SK, \cdot$) *and* Prove($\mathrm{p}, SK, \cdot$), *no adversary can distinguish* $F_{PK}(x)$ *from a random element of $R(\mathrm{p})$ without explicitly querying for it. More formally,* $\forall$ *PPT* $\mathcal{A}$, $\exists$ *negligible* ν *such that*

$$\Pr[\mathrm{p} \leftarrow Params(1^k); (PK, SK) \leftarrow G(\mathrm{p});$$
$$(Q_e, Q_p, x, \mathtt{state}) \leftarrow \mathcal{A}^{\mathtt{Eval}(\mathrm{p}, SK, \cdot), \mathtt{Prove}(\mathrm{p}, SK, \cdot)}(\mathrm{p}, PK);$$
$$y_0 = \mathtt{Eval}(\mathrm{p}, SK, x); y_1 \leftarrow R(\mathrm{p}); b \leftarrow \{0, 1\};$$
$$(Q'_e, Q'_p, b') \leftarrow \mathcal{A}^{\mathtt{Eval}(\mathrm{p}, SK, \cdot), \mathtt{Prove}(\mathrm{p}, SK, \cdot)}(\mathtt{state}, y_b)$$
$$: \ b' = b \wedge x \notin (Q_e \cup Q_p \cup Q'_e \cup Q'_p)] \leq 1/2 + \nu(k)$$

where Q_e and Q_p denote, respectively, the contents of the query tape that records $\mathcal{A}$'s queries to its Eval *and* Prove *oracles in the first query phase, and Q'_e and Q'_p denote the query tapes in the second query phase.*

Verifiability. *For all k, for all $p \in Params(1^k)$, there do not exist $(PK, x, y_1, \pi_1, y_2, \pi_2)$ such that $y_1 \neq y_2$, but $\texttt{Verify}(p, PK, x, y_1, \pi_1) = \texttt{Verify}(p, PK, x, y_2, \pi_2) = ACCEPT$.*

Note that verifiability in the definition above can be relaxed so as to only hold computationally (as opposed to unconditionally).

Simulatability, as defined below, is the novel aspect of sVRFs, setting them apart from VRFs as previously defined. First, we give the definition, and then we discuss variations.

Definition 2 (Simulatable VRF). *Let $(Params, G, \texttt{Eval}, \texttt{Prove}, \texttt{Verify})$ be a VRF (according to Definition 1). They constitute a simulatable VRF if there exist algorithms $(SimParams, SimG, SimProve)$ such that for all PPT $\mathcal{A}$, $\mathcal{A}$'s views in the following two games are indistinguishable:*

Game Real $p \leftarrow Params(1^k)$ *and then $\mathcal{A}(p)$ gets access to the following oracle $\mathcal{R}$: On query NewPK, $\mathcal{R}$ obtains and stores $(PK, SK) \leftarrow G(p)$, and returns PK to $\mathcal{A}$. On query (PK, x), $\mathcal{R}$ verifies that (PK, SK) has been stored for some SK. If not it returns "error". If so, it returns $y = \texttt{Eval}(p, SK, x)$ and $\pi \leftarrow \texttt{Prove}(p, SK, x)$.*

Game Simulated $(p, t) \leftarrow SimParams(1^k)$, *and then $\mathcal{A}(p)$ gets access to the following oracle $\mathcal{S}$: On query NewPK, $\mathcal{S}$ obtains and stores $(PK, SK) \leftarrow SimG(p, t)$, and returns PK to $\mathcal{A}$. On query (PK, x), $\mathcal{S}$ verifies if (PK, SK) has been stored for some SK. If not, it returns "error". If so, $\mathcal{S}$ (1) checks if x has previously been queried, and if so, returns the answer stored; (2) otherwise, $\mathcal{S}$ obtains $y \leftarrow R(p)$ and $\pi \leftarrow SimProve(p, SK, x, y, t)$, and returns and stores (y, π).*

2.1 Simplifying the Definition

The games in the above definition need to store multiple public keys and secret keys, as well as responses to all the queries issued so far, and consistently respond to multiple queries corresponding to all these various keys. It is clear that this level of security is desirable: we want an sVRF to retain its security properties under composition with other instances within the same system. A natural question is whether we can simplify the games by restricting the adversary to just one NewPK query or just one (PK, x) query per PK without weakening the security guarantees. In fact, the four possible combinations of such restrictions yield four distinct security notions, as we show in the full version of this paper.

Although we cannot simplify Definition 2 in this way, we can give a seemingly simpler definition (one that only allows one NewPK query from the adversary) that is strictly stronger than Definition 2 in that it requires that the adversary cannot distinguish the real game from the simulated one, *even with the knowledge of the trapdoor t*.

Definition 3 (Trapdoor-indistinguishable sVRF). *Let $(Params, G, \texttt{Eval}, \texttt{Prove}, \texttt{Verify})$ be a VRF (as in Definition 1). They constitute a trapdoor-indistinguishable (TI) sVRF if there exist algorithms $(SimParams, SimG,$*

SimProve) such that the distribution Params(1^k) is computationally indistinguishable from the distribution SimParams(1^k) and for all PPT $\mathcal{A}$, $\mathcal{A}$'s views in the following two games are indistinguishable:

Game Real Proofs. $(\mathrm{p}, t) \leftarrow SimParams(1^k)$, $(PK, SK) \leftarrow G(\mathrm{p})$ and then $\mathcal{A}(\mathrm{p}, t, PK)$ gets access to the following oracle $\mathcal{R}$: On query x, $\mathcal{R}$ returns $y = \mathtt{Eval}(\mathrm{p}, SK, x)$ and $\pi \leftarrow \mathtt{Prove}(\mathrm{p}, SK, x)$.

Game Simulated Proofs. $(\mathrm{p}, t) \leftarrow SimParams(1^k)$, $(PK, SK) \leftarrow SimG(\mathrm{p}, t)$, and then $\mathcal{A}(\mathrm{p}, t, PK)$ gets access to the following oracle $\mathcal{S}$: On query x, $\mathcal{S}$ (1) checks if x has previously been queried, and if so, returns the answer stored; (2) otherwise, obtains $y \leftarrow R(\mathrm{p})$ and $\pi \leftarrow SimProve(\mathrm{p}, SK, x, y, t)$, and returns and stores (y, π).

By a fairly standard hybrid argument, we have the following lemma (see the full version for the proof):

Lemma 1. *If (Params, G, $\mathtt{Eval}$, $\mathtt{Prove}$, $\mathtt{Verify}$) is a TI-sVRF, it is an sVRF.*

2.2   Weak Trapdoor-Indistinguishable sVRF

We now define a somewhat weaker notion of TI sVRFs, in which a simulator can only give fake proofs for those values of the output range that it has sampled itself in some special way.

Definition 4 (Weak TI-sVRF). *Let (G, $\mathtt{Eval}$, $\mathtt{Prove}$, $\mathtt{Verify}$) be a VRF in the Params(1^k) model with domain $D(\cdot)$ and range $R(\cdot)$. They constitute a weak trapdoor-indistinguishable (TI) sVRF if there exist algorithms (SimParams, SimG, SimProve, SimSample) such that the distribution Params(1^k) is computationally indistinguishable from the distribution SimParams(1^k) and for all PPT $\mathcal{A}$, $\mathcal{A}$'s views in the following two games are indistinguishable:*

Game Real Proofs. $(\mathrm{p}, t) \leftarrow SimParams(1^k)$, $(PK, SK) \leftarrow G(\mathrm{p})$ and then $\mathcal{A}(\mathrm{p}, t, PK)$ gets access to the following oracle: On query x, the oracle returns $y = \mathtt{Eval}(\mathrm{p}, SK, x)$ and $\pi \leftarrow \mathtt{Prove}(\mathrm{p}, SK, x)$.

Game Simulated Proofs. $(\mathrm{p}, t) \leftarrow SimParams(1^k)$, $(PK, SK) \leftarrow SimG(\mathrm{p}, t)$, and then $\mathcal{A}(\mathrm{p}, t, PK)$ gets access to the following oracle: On query x, the oracle (1) checks if x has previously been queried, and if so, returns the answer stored; (2) otherwise, obtains $(y, w) \leftarrow SimSample(\mathrm{p}, t, SK, x)$ and $\pi \leftarrow SimProve(\mathrm{p}, SK, x, y, t, w)$, and returns and stores (y, π).

We now show that a weak TI-sVRF where *SimSample* outputs a uniformly random element of a sufficiently large set can be converted to a TI-sVRF with binary range. Let (G, $\mathtt{Eval}$, $\mathtt{Prove}$, $\mathtt{Verify}$) be a weak TI-sVRF in the *Params* model with domain $D(\mathrm{p})$, and range $R(\mathrm{p}) \subseteq \{0,1\}^{m(k)}$ for some polynomial m for all $\mathrm{p} \in Params(1^k)$. Consider the following algorithms:

***Params****(1^k) Pick $r \leftarrow \{0,1\}^{m(k)}$, $\mathrm{p} \leftarrow Params(1^k)$; return $\mathrm{p}^* = (r, \mathrm{p})$.
G^* On input $\mathrm{p}^* = (r, \mathrm{p})$, output $(PK^*, SK^*) \leftarrow G(\mathrm{p})$.

Eval* and Prove* On input $\mathbf{p}^* = (r, \mathbf{p})$, SK^*, and $x \in D(\mathbf{p})$, compute $y = $ Eval$(\mathbf{p}, SK^*, x)$. Let $y^* = y \cdot r$, where by ".", we denote the inner product, i.e. $y \cdot r = \bigoplus_{i=1}^{|y|} y_i r_i$. Eval* outputs y^*. Prove* picks $\pi \leftarrow$ Prove$(\mathbf{p}, SK^*, x)$ and outputs $\pi^* = (\pi, y)$.

Verify* On input $\mathbf{p}^* = (r, \mathbf{p})$, PK^*, $x \in D(\mathbf{p})$, $y^* \in \{0, 1\}$, $\pi^* = (\pi, y)$: accept iff Verify$(\mathbf{p}, PK, x, y, \pi)$ accepts and $y^* = r \cdot y$.

Lemma 2. *Suppose $(G, \text{Eval}, \text{Prove}, \text{Verify})$ is a weak TI-sVRF with $(SimParams, SimSample, SimG, SimProve)$ as in Definition 4. Let ρ be such that for all $(\mathbf{p}, t) \in SimParams(1^k)$, for all $x \in D(\mathbf{p})$, for all $(SK, PK) \in SimG(\mathbf{p}, t)$, $|SimSample(\mathbf{p}, t, SK, x)| \geq \rho(k)$, and SimSample is a uniform distribution over its support. Let μ be such that for all $\mathbf{p} \in Params(1^k)$, $|D(\mathbf{p})| \leq \mu(k)$. If there exists a negligible function ν such that $\mu(k)\rho(k)^{-\frac{1}{3}} = \nu(k)$ then $(G^*, \text{Eval}^*, \text{Prove}^*, \text{Verify}^*)$ as constructed above are a TI-sVRF in the Params* model with domain $D(\mathbf{p})$, and range $\{0, 1\}$.*

Proof. Correctness, verifiability and pseudorandomness follow easily from the respective properties of the weak TI-sVRF (recall that a weak TI-sVRF is still a VRF – the "weak" part refers to simulatability only). In particular, pseudorandomness follows by standard techniques such as the leftover hash lemma.

We must show TI-simulatability. We first prove a useful claim. Consider specific values $(\mathbf{p}, t) \in SimParams(1^k)$, $(PK, SK) \in SimG(\mathbf{p}, t)$. Since t and SK are fixed, the distributions $R'(x) = SimSample(\mathbf{p}, t, SK, x)$ and $Bad(x) = \{r \in \{0, 1\}^{m(k)} : |\Pr[y \leftarrow R'(x) : y \cdot r = 1] - .5| \geq |R'(x)|^{-\frac{1}{3}}\}$ are well-defined. In English, $Bad(x)$ is the set of those r's for which the random variable $y \cdot r$ (where y is sampled uniformly at random from $R'(x)$, i.e. sampled according to $SimSample(\mathbf{p}, t, SK, x)$) is biased by at least $|R'(x)|^{-\frac{1}{3}}$ from a random bit.

Claim. $\forall x \in D(\mathbf{p})$, $\Pr[r \leftarrow \{0, 1\}^{m(k)} : r \in Bad(x)] \leq |R'(x)|^{-\frac{1}{3}}$.

Proof. (Of claim.) Suppose $x \in D(\mathbf{p})$ is fixed. Let $Weight(r) = \sum_{y \in R'(x)} y \cdot r$. By definition of $Bad(x)$, $r \in Bad(x)$ if and only if $|Weight(r)/|R'(x)| - .5| \geq |R'(x)|^{-\frac{1}{3}}$. It is easy to see that, if the probability is taken over the choice of r, then $Exp[Weight(r)/|R'(x)|] = .5$. On the other hand, for any pair $y_1 \neq y_2 \in R'(x)$, $y_1 \cdot r$ is independent from $y_2 \cdot r$, and so $Weight(r) = \sum_{y \in R'(x)} y \cdot r$ is a sum of pairwise independent random variables. Thus, $Var[Weight(r)] = \sum_{y \in R'(x)} Var[y \cdot r] = |R'(x)|/4$, and $Var[Weight(r)/|R'(x)|] = 1/4|R'(x)|$. Plugging Exp and Var for $Weight(r)/|R'(x)|$ into Chebyshev's inequality, we get $\Pr[|Weight(r)/|R'(x)| - .5| \geq |R'(x)|^{-\frac{1}{3}}] \leq |R'(x)|^{-\frac{1}{3}}$ which completes the proof.

Now we will show that the simulatability property holds. Consider the following algorithms:

SimParams*. On input 1^k, obtain $(\mathbf{p}, t) \leftarrow SimParams(1^k)$, $r \leftarrow \{0, 1\}^{m(k)}$. Output $\mathbf{p}^* = (r, \mathbf{p})$, $t^* = t$.

SimG*. On input $(\mathbf{p}^*, t^*)$, where $\mathbf{p}^* = (r, \mathbf{p})$ obtain $(PK, SK) \leftarrow SimG(\mathbf{p}, t^*)$. Output $PK^* = PK, SK^* = SK$.

SimProve*. On input $(\mathbf{p}^*, SK^*, x, y^*, t^*)$ where $\mathbf{p}^* = (r, \mathbf{p})$, repeat the following up to k times until $y \cdot r = y^*$: $(y, w) \leftarrow SimSample(\mathbf{p}, t, SK, x)$. If after k calls to $SimSample$, $y \cdot r \neq y^*$, output "fail". Else obtain $\pi \leftarrow SimProve(\mathbf{p}, t, SK, x, (y, w))$. Output $\pi^* = (\pi, y)$.

We define two intermediate games in which the adversary is given an oracle that is similar to Game Simulated Proofs from the TI-sVRF definition in that it does not use `Eval` and `Prove`; instead of `Eval`, it uses $SimSample$ (from the weak TI-sVRF definition) to obtain (y, w), and then outputs $y^* = y \cdot r$. The two games generate the proofs in different ways: Game Intermediate Real Proof just uses w and $SimProve$ of the weak TI-sVRF definition to generate π, while Game Intermediate Simulated Proof uses $SimProve^*$ defined above. More precisely:

Game Intermediate Real Proofs. $(\mathbf{p}^*, t^*) \leftarrow SimParams^*(1^k)$, $(PK^*, SK^*) \leftarrow SimG^*(\mathbf{p}^*, t^*)$, and then $\mathcal{A}(\mathbf{p}^*, t^*, PK^*)$ gets access to the following oracle: On query x, the oracle (1) checks if x has previously been queried, and if so, returns the answer stored; (2) otherwise, obtains $(y, w) \leftarrow SimSample(\mathbf{p}, t, SK, x)$, $y^* = y \cdot r$, and $\pi \leftarrow SimProve(\mathbf{p}, SK, x, y, t, w)$, $\pi^* = (\pi, y)$, and returns and stores (y^*, π^*).

Game Intermediate Simulated Proofs. $(\mathbf{p}^*, t^*) \leftarrow SimParams^*(1^k)$, $(PK^*, SK^*) \leftarrow SimG^*(\mathbf{p}^*, t^*)$, and then $\mathcal{A}(\mathbf{p}^*, t^*, PK^*)$ gets access to the following oracle: On query x, the oracle (1) checks if x has previously been queried, and if so, returns the answer stored; (2) otherwise, obtains $(y', w') \leftarrow SimSample(\mathbf{p}, t, SK, x)$, $y^* = y' \cdot r$, and $\pi^* \leftarrow SimProve^*(\mathbf{p}, SK, x, y^*, t)$, and returns and stores (y^*, π^*).

We now argue that these intermediate games are indistinguishable from Game Real Proofs and Game Simulated Proofs as specified by the definition of TI-sVRF, instantiated with $(SimParams, SimG, SimSample, SimProve)$ that follow from simulatability of our weak TI-sVRF, and with $(SimParams^*, SimG^*, SimProve^*)$ defined above. First, it is straightforward to see that an adversary distinguishing between Game Real Proofs and Game Intermediate Real Proofs directly contradicts the simulatability property of weak TI-sVRFs.

The only difference between Game Intermediate Simulated Proofs and Game Simulated Proofs, is the choice of the bit y^*: in the former, it is chosen using $SimSample$, i.e. indistinguishably from the way it is chosen in the real game. In the latter, it is chosen at random. If we condition on the event that for all x, $r \notin Bad(x)$, these two distributions are statistically close.

The only thing left to show is that the two intermediate games defined above are indistinguishable. If we condition on the event that we never fail, then the two games are identical. Note that if for all x, $r \notin Bad(x)$, then the probability that we fail on a particular query is $\leq (1/2 + |R'(x)|^{-\frac{1}{3}})^k$ which is negligible.

Thus we have shown that if the probability that $r \in Bad(x)$ for some x is negligible, then Game Real Proofs is indistinguishable from Game Simulated

Proofs. By the union bound, combined with the claim, $\Pr[r \leftarrow \{0,1\}^{m(k)} : \exists x \in D(\mathbf{p})$ such that $r \in Bad(x)] \leq |D(\mathbf{p})||R'(x)|^{-\frac{1}{3}}$, which is equal to $\nu(k)$ by the premise of the lemma.   □

From Lemmas 1 and 2, we see that from a weak TI-sVRF satisfying the conditions of Lemma 2, we can construct an equally efficient sVRF with range $\{0,1\}$.

Remark. Note that, even though the support of $SimSample(\mathbf{p}, t, SK, x)$ is quite large, the construction above only extracts one bit of randomness from it. Although it can be easily extended to extract a logarithmic number of random bits, there does not seem to be a black-box construction extracting a superlogarithmic number of bits. Suppose ext is a procedure that extracts ℓ bits from y, so $y^* = ext(y)$ is of length ℓ. Then how would $SimProve^*$ work to generate a proof that y^* is correct? It needs to call $SimProve(\mathbf{p}, SK, x, y, t, w)$ for some y such that $y^* = ext(y)$ and w is an appropriate witness. It seems that the only way to obtain such a pair (y, w) is by calling $SimSample(\mathbf{p}, t, SK, x)$; in expectation, 2^ℓ calls to $SimSample$ are needed to obtain an appropriate pair (y, w); if ℓ is superlogarithmic, this is prohibitively inefficient.

3   Construction Based on General Assumptions

In the common-random-string (CRS) model, sVRFs can be constructed from any one-way function and an unconditionally sound multi-theorem non-interactive zero-knowledge proof system (NIZKProve, NIZKVerify) for NP (we review the notion of NIZK in Section 5). Pseudorandom functions (PRFs) can be obtained from one-way functions [HILL99, GGM86] (in the sequel, by $F_s(x)$ we denote a PRF with seed s and input x). In the CRS model, one-way functions also imply unconditionally binding computationally hiding non-interactive commitment [Nao91] (in the sequel, denoted as $Commit(x, q, r)$, where x is the value to which one commits, q is the public parameter, and r is the randomness). We describe the construction below. In the full version, we prove it is an sVRF.

Params. Corresponding to the security parameter k, choose a common random string σ of length $\ell(k)$, where $\ell(k)$ bits suffice for multi-theorem NIZK [BDMP91, FLS99, GOS06]. Choose a random $2k$-bit string q as the public parameter for the Naor commitment scheme. The parameters are $\mathbf{p} = (\sigma, q)$.

Domain and range. The function has domain $D(\mathbf{p}) = \{0,1\}^{p_1(k)}$, and range $R(\mathbf{p}) = \{0,1\}^{p_2(k)}$, where p_1 and p_2 are functions bounded by a polynomial.

G Pick a random seed s for a pseudorandom function $F_s : \{0,1\}^{p_1(k)} \mapsto \{0,1\}^{p_2(k)}$. Let $PK = Commit(s, q, r)$, $SK = (s, r)$, where r is the randomness needed for the commitment.

Eval On input x, output $y = F_s(x)$.

Prove On input x, run NIZKProve using CRS σ to output a NIZK proof π of the following statement: $\exists(s, r) \mid PK = Commit(s, q, r) \wedge y = F_s(x)$.

Verify On input (PK, y, π), verify the proof π using the NIZKVerify algorithm.

4   Efficient Construction

We first present a construction for a weak TI-sVRF with a large output range. As we have shown, this can then be transformed into an sVRF with range $\{0, 1\}$. The security relies on the following assumptions.

Definition 5 $((Q, \nu)$-BDHI [BB04a]). *A family $\mathcal{G}$ of groups satisfies the $(Q(k), \nu(k))$-bilinear Diffie-Hellman inversion assumption if no PPT $\mathcal{A}$, on input (instance, challenge) can distinguish if its challenge is of type 1 or type 2 with advantage asymptotically higher than $\nu(k)$ where instance and challenge are defined as follows: instance $= (G_1, G_2, q, e, g, g^\alpha, g^{\alpha^2}, g^{\alpha^3}, \ldots, g^{\alpha^{Q(k)}})$ where q is a prime of length $poly(k)$, G_1, G_2 are groups of order q returned by $\mathcal{G}(q)$, $e : G_1 \times G_1 \to G_2$ is a bilinear map, $g \leftarrow G_1$, $\alpha \leftarrow \mathbb{Z}_q^*$, challenge of type 1 is $e(g, g)^{\frac{1}{\alpha}}$, while challenge of type 2 is $e(g, g)^R$ for random $R \leftarrow \mathbb{Z}_q^*$.*

Definition 6 $((Q, \nu)$-BDHBI). *An family $\mathcal{G}$ of groups satisfies the $(Q(k), \nu(k))$ bilinear Diffie-Hellman basegroup inversion assumption if no PPT $\mathcal{A}$, on input (instance, challenge) can distinguish if its challenge is of type 1 or type 2 with advantage asymptotically higher than $\nu(k)$, where instance and challenge are defined as follows: instance $= (G_1, G_2, q, e, g, g^\alpha, g^{\alpha^2}, g^{\alpha^3}, \ldots, g^{\alpha^{Q(k)}}, g^\beta)$ where q is a prime of length $poly(k)$, G_1, G_2 are groups of order q returned by $\mathcal{G}(q)$, $e : G_1 \times G_1 \to G_2$ is a bilinear map, $g \leftarrow G_1$, $\alpha \leftarrow \mathbb{Z}_q^*$, $\beta \leftarrow \mathbb{Z}_q^*$, challenge of type 1 is $g^{\frac{1}{\alpha\beta}}$, while challenge of type 2 is g^R for random $R \leftarrow \mathbb{Z}_q^*$.*

The assumption in Definition 6 is a new assumption which can be shown to imply Q-BDHI. We will assume that it holds for the prime order subgroup of composite order bilinear groups that can be efficiently instantiated [BGN05].

Definition 7 (SDA [BGN05]). *A family $\mathcal{G}$ of groups satisfies the subgroup decision assumption if no PPT $\mathcal{A}$, on input (instance, challenge) can distinguish if its challenge is of type 1 or type 2, where instance and challenge are defined as follows: instance $= (G_1, G_2, n, e, h)$ where $n = pq$ is a product of two primes of length $poly(k)$ (for k a sec. param.), G_1, G_2 are groups of order n returned by $\mathcal{G}(q, p)$, $e : G_1 \times G_1 \to G_2$ is a bilinear map, h is a random generator of G_1, challenge of type 1 is g, a random generator of G_1, while challenge of type 2 is g_p, a random order-p element of G_1.*

The weak TI-sVRF construction is as follows:

Params. On input 1^k, choose groups G_1, G_2 of order $n = pq$ for primes p, q, where $|p|$ and $|q|$ are polynomial in k, with bilinear map $e : G_1 \times G_1 \to G_2$. Choose random generators g, H, A, D for G_1. *Params* will output $\mathbf{p} = (G_1, G_2, n, e, g, H, A, D)$.

Domain and range. The input domain $\mathcal{D}(\mathbf{p})$ consists of integers $1 \leq x \leq l(k)$ where $l(k) < 2^{|q|-1}$ (We will later see the connection between $l(k)$ and $Q(k)$ by which our assumption is parameterized.) Note that $\mathcal{D}(\mathbf{p})$ depends only on k, not on $\mathbf{p}$. $R(\mathbf{p}) = G_2$.

G On input p, pick $s \leftarrow \mathbb{Z}_n^*$, output $SK = s$, $PK = A^s$.

Eval On input $(\mathbf{p}, SK, x)$, output $e(H, g)^{\frac{1}{s+x}}$.

Prove On input $(\mathbf{p}, SK, x)$, pick $r \leftarrow \mathbb{Z}_n^*$, and output $\pi = (\pi_1, \pi_2, \pi_3)$, where
$\pi_1 = H^{\frac{r}{s+x}}/D^r$, $\pi_2 = g^{\frac{1}{r}}$, $\pi_3 = A^{\frac{x+s}{r}}$.

Verify On input $(\mathbf{p}, SK, x, y, \pi)$, parse $\pi = (\pi_1, \pi_2, \pi_3)$ and verify that $e(\pi_1, \pi_2)$
$e(D, g) = y$, $e(\pi_3, g) = e(A^x PK, \pi_2)$, $e(\pi_1, \pi_3)e(D, A^x PK) = e(H, A)$.

Theorem 1. $(G, \mathtt{Eval}, \mathtt{Prove}, \mathtt{Verify})$ *as described above constitute a weak TI-sVRF for parameter model Params, input domain $\mathcal{D}$ of size l, and output range G_2 (where G_2 is as output by Params) under the SDA assumption combined with the $(l(k), \nu(k)/l^2(k))$-BDHBI, where ν is an upper bound on the asymptotic advantage that any probabilistic polynomial-time algorithm has in breaking the simulatability game of Definition 4.*

Proof. **Correctness** follows from construction.

Verifiability: Suppose there exists an adversary who, given parameters $\mathbf{p} = (G_1, G_2, n, e, g, H = g^h, A = g^a, D = g^d)$ generated by *Params* can produce $PK, y, y', \pi = (\pi_1, \pi_2, \pi_3), \pi' = (\pi'_1, \pi'_2, \pi'_3)$ such that $\mathtt{Verify}(\mathbf{p}, PK, y, \pi) = \mathtt{Verify}(\mathbf{p}, PK, y', \pi') = 1$. Then we will show that $y = y'$.

Let $\lambda, \mu, \mu', \sigma, \phi, \theta, \sigma', \phi', \theta' \in \mathbb{Z}_n$ be the exponents such that $PK = g^\lambda, y = g^\mu, y' = g^{\mu'}, \pi_1 = g^\sigma, \pi_2 = g^\phi, \pi_3 = g^\theta, \pi'_1 = g^{\sigma'}, \pi'_2 = g^{\phi'}, \pi'_3 = g^{\theta'}$.

If the verifications succeed, then we get that the following equations hold in $\mathbb{Z}_n$: $\sigma\phi + d = \mu$, $\quad \theta = (ax + \lambda)\phi$, $\quad \theta\sigma + d(ax + \lambda) = ha$.

Solving this system of equations gives us: $ha = \mu(ax + \lambda)$. Similarly, if (y', π') satisfy the verification equations, then we know that $ha = \mu'(ax + \lambda)$. H, A are generators for G_1, so $h, a \in \mathbb{Z}_n^*$, and therefore, $\mu'(ax + \lambda) \in \mathbb{Z}_n^*$, and $\mu(ax + \lambda) \in \mathbb{Z}_n^*$. This in turn means that $\mu', \mu, (ax + \lambda) \in \mathbb{Z}_n^*$.

From the solutions to the above equations, we know $\mu(ax + \lambda) = \mu'(ax + \lambda)$. Since $(ax + \lambda) \in \mathbb{Z}_n^*$, we can compute a unique inverse $(ax + \lambda)^{-1}$, and conclude that $\mu = \mu'$, and $y = y'$.

Note that this argument relies crucially on the fact that $h, a \in \mathbb{Z}_n^*$. In our simulation, we will instead choose $a = 0 \bmod q$, which will allow us to avoid this binding property.

Pseudorandomness follows under the Q-BDHI Assumption from pseudorandomnesss of the Dodis-Yampolskiy VRF [DY05].

Simulatability: Consider the following simulator algorithms:

SimParams(1^k). Choose groups G_1, G_2 of order $n = pq$ for prime p, q, where $|p|$ and $|q|$ are polynomial in k, with bilinear map $e : G_1 \times G_1 \rightarrow G_2$. Let G_p be the order p subgroup of G_1, and let G_q be the order q subgroup of G_1. Let $(A, g_p, H_p, D_p) \leftarrow G_p^4$ and $(g_q, H_q, D_q) \leftarrow G_q^3$. Let $g = g_p g_q$, $H = H_p H_q$, and $D = D_p D_q$. Output $\mathbf{p} = (G_1, G_2, n, e, g, H, A, D)$, $t = (g_p, g_q, H_p, H_q, D_p, D_q)$.
This is identical to *Params* except that $A \in G_p$, so that the verification algorithm cannot properly verify the G_q components of y and π.

SimG$(\mathbf{p}, t)$ $(SK, PK) \leftarrow G(\mathbf{p})$.

SimSample. On input (p, t, SK, x), pick $w \leftarrow \mathbb{Z}_q^*$.

Let $y = e(H_p, g_p)^{\frac{1}{s+x}} e(g_q, g_q)^w$. Output (y, w). (Note y's G_p component will be correct, while its G_q component will be random.)

SimProve. On input (p, SK, x, y, t, w), pick $r \leftarrow \mathbb{Z}_n^*$;

let $\pi_1 = (H_p^{\frac{r}{s+x}} / D_p^r)(g_q^{wr} / D_q^r)$, $\pi_2 = g^{\frac{1}{r}}$, $\pi_3 = A^{\frac{x+s}{r}}$. Output $\pi = (\pi_1, \pi_2, \pi_3)$. (Note that π's G_p components are correct, while its G_q components are chosen so as to allow us to fake the proof.)

Lemma 3. *The distribution $Params(1^k)$ is indistinguishable from the distribution $SimParams(1^k)$ by the Subgroup Decision Assumption.*

Proof. The only difference between these two distributions is that in *Params*, A is chosen at random from G_1, and in *SimParams*, A is chosen at random from G_p. Thus, these two distributions are indistinguishable by the Subgroup Decision assumption by a straightforward reduction. $\square$

Lemma 4. *For the algorithms described above, Game Real Proofs and Game Simulated Proofs (as in Definition 4) are indistinguishable with advantage more that $\nu(k)$ by the $(l(k), \nu(k)/l^2(k))$-BDHBI assumption.*

Before we prove this lemma, we will describe and prove an intermediate assumption that follows from the assumptions that we have already made. We state this assumption in terms of any prime order bilinear group. However, we will later assume that this assumption (and the Q-BDHBI assumption) holds over the prime order subgroup of a composite order bilinear group.

Definition 8 $((Q, \nu)$-Intermediate assumption). *A family $\mathcal{G}$ of groups satisfies the $(Q(k), \nu(k))$-intermediate assumption if for all subsets X of $\mathbb{Z}_{2^{a(k)}-1}$ (where $a(k)$ is a polynomial), of size $Q(k) - 1$ for all $x^* \in \mathbb{Z}_{2^{a(k)}-1} \setminus X$, no PPT $\mathcal{A}$, on input (instance, challenge) can distinguish if its challenge is of type 1 or type 2 with advantage asymptotically higher than $\nu(k)$, for instance and challenge defined as follows: instance $= (G_1, G_2, q, e, g, H, D, \{(H^{r_x \frac{1}{s+x}} / D^{r_x}, g^{\frac{1}{r_x}})\}_{\forall x \in X})$ where q is an $a(k)$-bit prime, G_1, G_2 are groups of order q returned by $\mathcal{G}(q)$, $e : G_1 \times G_1 \rightarrow G_2$ is a bilinear map, $(g, H, D) \leftarrow G_1^3$, and $\{r_x\}_{x \in X}$ and s were all picked at random from $\mathbb{Z}_q^*$; challenge of type 1 is $(H^{r^* \frac{1}{s+x^*}} / D^{r^*}, g^{\frac{1}{r^*}})$ where $r^* \leftarrow \mathbb{Z}_q^*$, while challenge of type 2 is (g^{R_1}, g^{R_2}) for R_1 and R_2 random from $\mathbb{Z}_q^*$.*

Lemma 5. *(l, ν)-BDHBI assumption implies (l, ν)-intermediate assumption.*

Proof. Suppose there exists an adversary $\mathcal{A}$ who breaks the intermediate assumption for set X of cardinality $l - 1$, and $x^* \notin X$. Then we show an algorithm $\mathcal{B}$ that can break l-BDHBI Assumption.

Algorithm $\mathcal{B}$ will behave as follows: Receive $G, q, e, g, g^\alpha, \ldots g^{\alpha^l}, g^\beta$, and $Z = g^{\frac{1}{\alpha\beta}}$ or $Z = g^R$ for random $R \in \mathbb{Z}_q^*$.

Choose random values $\Delta_1, \Delta_2 \leftarrow \mathbb{Z}_q^*$. Implicitly, let $\gamma = \gamma(\alpha) = \Delta_1(\alpha - \Delta_2) \prod_{x \in X}(\alpha + (x - x^*))$. Compute $H = g^\gamma$. Note that since this exponent is just an l degree polynomial in α, we can compute this value using $g, \ldots g^{\alpha^l}$. If we implicitly define $s = \alpha - x^*$, we will get $H = g^{\Delta_1(\alpha-\Delta_2)\prod_{x\in X}(s+x)}$. (Note that now we know neither s, nor α explicitly.) Note that because of Δ_1, H is uniformly distributed over G_1, and is independent of g. Now we want to provide D. Implicitly we will define $d = \frac{\gamma-\delta}{\alpha}$, where $\delta = \Delta_1\Delta_2\prod_{x\in X}(x - x^*)$ is the constant term of the polynomial in α (represented by $\gamma(\alpha)$). Note now that δ is a quantity $\mathcal{B}$ can compute, while d is only defined implicitly. Since d is a polynomial expression in α, $D = g^d$ can be expressed as a sum of terms $g, g^\alpha, \ldots, g^{\alpha^{l-1}}$, and computed using the given values. Finally, note that, because of Δ_2, D is uniformly distributed over G_1, and is independent of (g, H).

For all $\hat{x} \in X$: Let $\gamma'(\hat{x}) = \Delta_1(\alpha - \Delta_2)\prod_{x\in X, x\neq\hat{x}}(\alpha + (x - x^*)) = \frac{\gamma}{\hat{x}+s}$. Compute $v = g^{\gamma'(\hat{x})} = g^{\frac{\gamma}{s+\hat{x}}} = H^{\frac{1}{s+\hat{x}}}$. We then choose a random $r_{\hat{x}} \leftarrow \mathbb{Z}_n^*$. We compute and output $(v^{r_{\hat{x}}}/D^{r_{\hat{x}}}, g^{\frac{1}{r_{\hat{x}}}})$.

For x^*: Implicitly define $r^* = \frac{1}{\beta}$. Compute $u_1 = Z^\delta$. If $Z = g^{\frac{1}{\alpha\beta}}$, then this is equal to $g^{\frac{\delta}{\alpha\beta}} = g^{\frac{\gamma}{\alpha\beta} - \frac{\gamma-\delta}{\alpha\beta}} = g^{\frac{\gamma}{\alpha\beta}}/g^{\frac{\gamma-\delta}{\alpha\beta}} = H^{r^*\frac{1}{s+x^*}}/D^{r^*}$. Otherwise, this is equal to g^{R_1} for random R_1. Compute $u_2 = (g^\beta) = (g^{\frac{1}{r^*}})$. Output (u_1, u_2).

Finally, if $\mathcal{A}$ guesses that he received $(H^{r^*\frac{1}{s+x^*}}/D^{r^*}, g^{\frac{1}{r^*}})$, $\mathcal{B}$ guesses that $Z = g^{\frac{1}{\alpha\beta}}$, else that $Z = g_q^R$. If $\mathcal{A}$'s guess is correct, then $\mathcal{B}$'s guess is correct. $\square$

Proof. (of Lemma 4) We first define a series of hybrid games:

Game Hybrid i: Obtain $(p, t) \leftarrow SimParams(1^k)$, and $(PK, SK) \leftarrow$
$SimG(\mathsf{p}, t)$ and then $\mathcal{A}(\mathsf{p}, t, PK)$ gets access to the following oracle: The oracle begins by storing $j = 0$. On query x, the oracle (1) checks if x has previously been queried, and if so, returns the answer stored. Otherwise, (2) if $j < i$ the oracle obtains $(y, w) \leftarrow SimSample(\mathsf{p}, t, SK, x)$ and $\pi \leftarrow SimProve(\mathsf{p}, SK, x, y, w, t)$, returns and stores (y, π), and increments j. (3) Or if $j \geq i$, the oracle computes $y = \mathtt{Eval}(\mathsf{p}, SK, x)$ and $\pi \leftarrow \mathtt{Prove}(\mathsf{p}, SK, x)$, returns and stores (y, π) and increments j.

Note that in this case, $G(\mathsf{p})$ is identical to $SimG(\mathsf{p}, t)$ for all p, t, so Game Hybrid 0 is identical to Game Real Proofs. Game Hybrid Q, where Q is the maximum number of distinct oracle queries (not including repeated queries) that the adversary is allowed to make, is identical to Game Simulated Proofs. Thus, we have only to show the following lemma:

Lemma 6. *Suppose the (l, ν)-BDHBI Assumption holds in one of the two subgroups of a composite bilinear group. Then, when the size of the domain is at most l, no PPT adversary can distinguish Game Hybrid $i-1$ from Game Hybrid i with advantage higher than νl.*

Proof. Suppose there exists an adversary $\mathcal{A}$ who can distinguish Game Hybrid $i - 1$ from Game Hybrid i when the domain $\mathcal{D}$ is of size l. Then we show an algorithm $\mathcal{B}$ that can break the l-intermediate assumption with advantage ϵ.

First we make a guess x^* about which input $\mathcal{A}$ will give in its ith distinct oracle query. Since $|\mathcal{D}| = l$, and all values given to $\mathcal{A}$ will be independent of x^*, we will be correct with probability $1/l$.

Now, we will show an algorithm $\mathcal{B}$, which can, with nonnegligible probability, break the intermediate assumption for set $X = \mathcal{D} \setminus \{x^*\}$ and the x^* chosen above. $\mathcal{B}$ will receive $G, p, q, e, g_p, g_q, H_q, D_q, \{(H_q^{\frac{r_x}{sq+x}}/D_q^{r_x}, g_q^{\frac{1}{r_x}})\}_{\forall x \in X}, (Z_1, Z_2)$ for $g_q, H_q, D_q \leftarrow G_q$, and randomly chosen (but unknown) $\{r_x\}_{x \in X}, s_q \leftarrow \mathbb{Z}_q^*$. Here, either $(Z_1, Z_2) = (H_q^{r^* \frac{1}{sq+x^*}}/D_q^{r^*}, g_q^{\frac{1}{r^*}})$ or $(Z_1, Z_2) = (g_q^{R_1}, g_q^{R_2})$ for random $R_1, R_2 \leftarrow \mathbb{Z}_q^*$.

First, $\mathcal{B}$ prepares the parameters as follows: Choose $H_p, A, D_p \leftarrow G_p$ and compute $g = g_p g_q$, $H = H_p H_q$, $D = D_p D_q$. Set $\mathsf{p} = (G_1, G_2, n, e, g, H, A, D)$. Let $s_p \leftarrow \mathbb{Z}_p^*$, and $PK = A^{s_p}$. Implicitly, set $s \in \mathbb{Z}_n^*$ to the the element such that $s \bmod p = s_p$, and $s \bmod q = s_q$. $\mathcal{B}$ sends p and trapdoor $t = (g_p, g_q, H_p, H_q, D_p, D_q)$ to $\mathcal{A}$.

Now $\mathcal{B}$ must answer $\mathcal{A}$'s queries. We assume (WLOG) that $\mathcal{A}$ does not repeat queries.

When $\mathcal{A}$ sends its jth query, $\hat{x}$, $\mathcal{B}$ proceeds as follows:
If $j < i$: if $\hat{x} = x^*$, then $\mathcal{B}$ has guessed wrong about which value $\mathcal{A}$ will choose in his ith distinct query (if it is used again later, it will be repeated and thus not distinct), so $\mathcal{B}$ aborts. Otherwise, $\mathcal{B}$ chooses a random $w' \in \mathbb{Z}_q^*$. Let $y = e(H_p^{\frac{1}{s_p+\hat{x}}}, g_p)e(H_q, g_q)^{w'}$. Choose a random $r \leftarrow \mathbb{Z}_n^*$. Let $\pi_1 = (H_p^{r \frac{1}{s_p+\hat{x}}}/D_p^r)(H_q^{w'r}/D_q^r)$. Let $\pi_2 = g^{\frac{1}{r}}$ and $\pi_3 = A^{\frac{\hat{x}+s_p}{r}}$. If we implicitly set $w = w' h_q$, (where $H_q = g_q^{h_q}$) then these value will be distributed as in the output of $SimSample$ and $SimProve$. Output $(y, \pi = (\pi_1, \pi_2, \pi_3))$.

If $j = i$: If $\hat{x} \neq x^*$, then $\mathcal{B}$ has guessed wrong, so it aborts. Otherwise, choose random $r_p \leftarrow \mathbb{Z}_p^*$. Implicitly set $r \in \mathbb{Z}_n^*$ to be the element such that $r \bmod q = r^*$ and $r \bmod p = r_p$. Compute $\pi_1 = H_p^{r_p \frac{1}{x^*+s_p}}/D_p^{r_p} Z_1$. Note that, if $Z_1 = H_q^{r^* \frac{1}{sq+x^*}}/D_q^{r^*}$, then this is equal to $H^{\frac{r}{s+x^*}}/D^r$. Otherwise, this is equal to $H_p^{r_p \frac{1}{x^*+s_p}}/D_p^{r_p} g^{R_1}$. Now compute $\pi_2 = g_p^{\frac{1}{r_p}} Z_2$, if $Z_2 = g_q^{\frac{1}{r^*}}$, then this value will be $g^{\frac{1}{r}}$. Otherwise it will be $g_p^{\frac{1}{r_p}} g_q^{R_2}$. Compute $\pi_3 = A^{\frac{s_p+x^*}{r_p}} = A^{\frac{s+x^*}{r}}$. Finally, compute $y = e(\pi_1, \pi_2)e(D, g)$. Output $(y, \pi = (\pi_1, \pi_2, \pi_3))$ to the adversary.

If $j > i$, we know $\hat{x} \neq x^*$, and $\hat{x} \in X$. Let $V_1 = H_q^{r_{\hat{x}} \frac{1}{sq+\hat{x}}}/D_q^{r_{\hat{x}}}$, and $V_2 = g^{\frac{1}{r_{\hat{x}}}}$, as provided in $\mathcal{B}$'s input. $\mathcal{B}$ chooses a random $r_p \leftarrow \mathbb{Z}_p^*$. Implicitly, set $r \in \mathbb{Z}_n^*$ for this query to be the element such that $r \bmod p = r_p$, and $r \bmod q = r_{\hat{x}}$. $\mathcal{B}$ computes $\pi_1 = (H_p^{r_p \frac{1}{s_p+\hat{x}}}/D_p^{r_p})V_1 = H^{r \frac{1}{s+\hat{x}}}/D^r$, $\pi_2 = g_p^{\frac{1}{r_p}} V_2 = g^{\frac{1}{r}}$, and $\pi_3 = A^{\frac{s_p+\hat{x}}{r_p}} = A^{\frac{\hat{x}+s}{r}}$. Finally, $\mathcal{B}$ computes $y = e(\pi_1, \pi_2)e(D, g)$ and outputs $(y, \pi = (\pi_1, \pi_2, \pi_3))$ to $\mathcal{A}$.

Finally, $\mathcal{B}$ gets $\mathcal{A}$'s guess bit b. If $\mathcal{A}$ guesses that this is Game Hybrid $i - 1$, $\mathcal{B}$ guesses that $(Z_1, Z_2) = (H_q^{r^* \frac{1}{sq+x^*}}/D_q^{r^*}, g_q^{\frac{1}{r^*}})$; otherwise $\mathcal{B}$ guesses that $(Z_1, Z_2) = (g_q^{R_1}, g_q^{R_2})$. If $\mathcal{A}$ guesses correctly, $\mathcal{B}$'s guess will also be correct.

$\mathcal{B}$ has a $\frac{1}{l}$ probability of not aborting. Suppose that when $\mathcal{B}$ aborts, it returns a random bit. Then $\mathcal{B}$'s guess is correct with probability $(1 - \frac{1}{l}) * \frac{1}{2} + \frac{1}{l} * (\frac{1}{2} + \epsilon) =$

$\frac{1}{2} + \frac{\epsilon}{l}$, where ϵ is $\mathcal{A}$'s advantage. Thus, if $\mathcal{A}$'s advantage is $\epsilon > \nu l$ then $\mathcal{B}$'s advantage is higher than ν, contradicting the assumption.   □

For the theorem to follow, we observe that the overall reduction from breaking the simulatability game to breaking the BDHBI assumption uses at most $(l + 1)$ hybrids, and so the adversary's advantage ϵ translates into the reduction's advantage ϵ/l^2 in breaking BDHBI.   □

Remark. Since the construction above satisfies the premise of Lemma 2, it can be converted to an sVRF with binary range using the construction in Section 2.2.

5   Multi-theorem NIZK from One-Theorem NIZK Via sVRFs

Here, we omit the definition of single-theorem and multi-theorem NIZK, but refer the reader to Blum et al. [BDMP91] and Feige, Lapidot, Shamir [FLS99]. Instead, we informally sketch this definition:

Algorithms NIZKProve and NIZKVerify. The algorithm NIZKProve takes as input the common random string σ of length $\ell(k)$, and values (x, w), $|x| \leq q(k)$, such that $x \in L$, and w is a witness to this. NIZKProve outputs a *proof* Π. NIZKVerify is the algorithm that takes (σ, x, Π) as input, and outputs *ACCEPT* or *REJECT*.

Perfect completeness. For all $x \in L$, for all witnesses w for x, for all values of the public random string σ, and for all outputs π of NIZKProve(σ, x, w), NIZKVerify$(\sigma, x, \pi) = ACCEPT$.

Soundness $s(k)$. For all adversarial prover algorithms $\mathcal{A}$, for a randomly chosen σ, the probability that $\mathcal{A}$ can produce (x, π) such that $x \notin L$ but NIZKVerify$(\sigma, x, \pi) = ACCEPT$, is $s(k)$.

Single-theorem ZK. There exists an algorithm SimProveOne that, on input 1^k and $x \in L$, $|x| \leq q(k)$, outputs simulated CRS σ^S together with a simulated proof Π^S, such that (σ^S, Π^S) are distributed indistinguishably from (σ, Π) produced by generating a random CRS σ, and obtaining Π by running NIZKProve.

Multi-theorem ZK. There exist algorithms SimCRS and NIZKSimProve, as follows: SimCRS(1^k) outputs (σ, s). For all x, NIZKSimProve(σ, s, x) outputs a *simulated proof* Π^S. Even for a sequence of adversarially and adaptively picked $(x_1, \ldots, x_m)$ (m is polynomial in k), if for all $1 \leq i \leq m$, $x_i \in L$, then the simulated proofs $\Pi_1^S, \ldots, \Pi_m^S$ are distributed indistinguishably from proofs $\Pi_1, \ldots, \Pi_m$ that are computed by running NIZKProve(σ, x_i, w_i), where w_i is some witness that $x_i \in L$.

Suppose that, for a language L, we are given a single-theorem NIZK proof system (ProveOne, VerOne) in the CRS model, with perfect completeness and unconditional soundness error $s(k)$. Let $\ell(k)$ denote the function such that an $\ell(k)$-bit random string serves as the CRS for this proof system. Let $q(k)$ denote the polynomial upper bound on the size of the input x. Suppose also that we are given a simulatable VRF $(G, \text{Eval}, \text{Prove}, \text{Verify})$ in the parameter model *Params*, whose domain is $[1, \ell(k)]$, with range $\{0, 1\}$. Consider the following construction for multi-theorem NIZK in the common reference string model for instances of size k:

Generate common parameters. The algorithm NIZKParams: Obtain $\sigma_1 \leftarrow \{0, 1\}^{\ell(k)}$. Let $\mathbf{p} \leftarrow Params(1^k)$. The values $(\sigma_1, \mathbf{p})$ are the parameters of the system.

Prove. The algorithm NIZKProve: On input instance $x \in L$ with witness w, and common parameters $(\sigma_1, \mathbf{p})$ do: Obtain $(PK, SK) \leftarrow G(1^k, \mathbf{p})$. Let R be the $\ell(k)$-bit string computed as follows: for $1 \leq i \leq \ell(k)$, $R_i = \text{Eval}(\mathbf{p}, SK, i)$, where R_i denotes the ith bit of R. For $1 \leq i \leq \ell(k)$, let $\pi_i \leftarrow \text{Prove}(\mathbf{p}, SK, i)$. Let $\sigma = \sigma_1 \oplus R$. Obtain $\Pi' \leftarrow \text{ProveOne}(\sigma, x, w)$. Output the proof $\Pi = (PK, R, \pi_1, \ldots, p_{\ell(k)}, \Pi')$.

Verify. The algorithm NIZKVerify: On input x and Π, and common parameters $(\sigma_1, \mathbf{p})$, do: (1) for $1 \leq i \leq \ell(i)$, check that $\text{Verify}(\mathbf{p}, PK, i, R_i, \pi_i)$ accepts; (2) let $\sigma = \sigma_1 \oplus R$; check that $\text{VerOne}(\sigma, x, \Pi')$ accepts; if all these checks passed, accept, otherwise, reject.

Theorem 2. *If for a language L, $(\text{ProveOne}, \text{VerOne})$ is a single-theorem NIZK proof system in the $\ell(k)$-bit CRS model for instances of length up to $q(k)$ with perfect completeness and unconditional soundness error $s(k)$, and $(G, \text{Eval}, \text{Prove}, \text{Verify})$ in the parameter model $Params(1^k)$, is a strong simulatable VRF with domain $[1, \ell(k)]$ and range $\{0, 1\}$, then the above construction is a multi-theorem NIZK proof system in the public parameters model that comprises the $\ell(k)$-bit CRS and $Params(1^k)$, with perfect completeness and unconditional soundness error $s(k)2^{u(k)}$, where u denotes the bit length of a PK output by $G(\mathbf{p})$ on input $\mathbf{p} \leftarrow Params(1^k)$.*

Proof. (Sketch) The perfect completeness property follows from the perfect completeness property of the single-theorem NIZK.

Let us show the multi-theorem zero-knowledge property. Recall that, by the definition of (strong) sVRF, we have a simulator consisting of *SimParams*, *SimG* and *SimProve* such that, if (PK, SK) were generated by *SimG*, then for a randomly sampled y from the range of the sVRF, and for any x in the domain, *SimProve* can generate a fake proof that $y = \text{Eval}(SK, x)$. (See Section 2.)

Also recall that by the definition of NIZK, there exists a simulator SimProveOne such that no adversary $\mathcal{A}$ can distinguish between the following two distributions for any $x \in L$ and any witness w for x: (1) choose $\sigma \leftarrow \{0, 1\}^{\ell(k)}$, and let $\Pi \leftarrow \text{ProveOne}(\sigma, x, w)$; give (σ, Π) to $\mathcal{A}$; (2) $(\sigma, \Pi) \leftarrow \text{SimProveOne}(1^k, x)$; give (σ, Π) to $\mathcal{A}$.

Consider the following simulator S for our multi-theorem NIZK construction. The simulator will consist of SimCRS that generates the simulated parameters, and of NIZKSimProve that generates the simulated proof. SimCRS works as follows: generate $(\mathbf{p}, t) \leftarrow SimParams$, and $\sigma_1 \leftarrow \{0, 1\}^{\ell(k)}$; publish

(σ_1, p) as the parameters of the system. NIZKSimProve works like this: generate $(\sigma, \Pi') \leftarrow$ SimProveOne$(1^k, x)$. Then let $R = \sigma \oplus \sigma_1$. Let $(PK, SK) \leftarrow SimG(p, t)$. For $1 \le i \le \ell(k)$, let $\pi_i = SimProve(p, SK, x, R_i, t)$. Output the proof $\Pi = (PK, R, \pi_1, \ldots, p_{\ell(k)}, \Pi')$. In the full version, we show that the view that the adversary obtains in the simulation is indistinguishable from the view obtained when interacting with the prover.

We now show soundness. We are given that, for $\sigma \leftarrow \{0,1\}^{\ell(k)}$, the probability that there exists $x \notin L$ and a proof Π' such that Verify$(\sigma, x, \Pi') = 1$, is $s(k)$.

Consider $p \leftarrow Params$, and $(PK, SK) \leftarrow G(1^k)$. Let R be as defined in NIZKProve: $R_i = $ Eval(SK, i). Note that by the verifiability property of the sVRF, there is a *unique* R for which there exists a proof of correctness $(\pi_1, \ldots, \pi_{\ell(k)})$. The probability, over the choice of σ_1, that there exists $x \notin L$ and a proof Π' such that Verify$(R \oplus \sigma_1, x, \Pi') = 1$ (if such an x exists, we say that PK is *bad* for σ_1), is still $s(k)$, since we first fixed p and PK, and then randomly chose σ_1. By the union bound, since there are $2^{u(k)}$ possible PK's, for every p, the probability that there exists a bad PK for a particular σ_1, is $s(k)2^{u(k)}$.   □

Remark. Note that if an NIZK proof system is in the hidden-random-string (HRS) model (such as those due to Feige, Lapidot and Shamir [FLS99] and Kilian and Petrank [KP98]), then we can take advantage of it as follows: the hidden random string can be obtained the way that σ is currently obtained by the prover in the construction above; only in the construction above, the prover reveals the entire string σ and the proof that each bit of σ is computed correctly; while in the HRS model, the prover only reveals the subset of bits of the hidden random string that he needs to reveal. This observation was inspired by Dwork and Naor's construction of zaps from VRFs and verifiable PRGs [DN00] based on NIZK using HRS model. We give more details on consequences in the HRS model in the full version.

Acknowledgments. We thank Leo Reyzin and Markulf Kohlweiss for helpful discussions. We thank UCLA's Institute for Pure and Applied Mathematics for hosting us while part of this research was carried out. Melissa Chase is supported by NSF grant CNS-0374661 and NSF Graduate Research Fellowship. Anna Lysyanskaya is supported by NSF CAREER grant CNS-0374661 and NSF grant CNS-0627553.

References

[BB04a]   Boneh, D., Boyen, X.: Efficient selective id secure identity based encryption without random oracles. In: Cachin, C., Camenisch, J.L. (eds.) EUROCRYPT 2004. LNCS, vol. 3027, pp. 223–238. Springer, Heidelberg (2004)

[BB04b]   Boneh, D., Boyen, X.: Short signatures without random oracles. In: Cachin, C., Camenisch, J.L. (eds.) EUROCRYPT 2004. LNCS, vol. 3027, pp. 54–73. Springer, Heidelberg (2004)

[BCC04]   Brickell, E., Camenisch, J., Chen, L.: Direct anonymous attestation. In: 11th ACM CCS, pp. 225–234. ACM press, New York (2004)

[BDMP91] Blum, M., De Santis, A., Micali, S.,: Non-interactive zero knowledge. SIAM Journal of Computing 20(6) 1084–1118 (1991)

[BFM88] Blum, M., Feldman, P., Micali, S.: Non-interactive zero-knowledge and its applications (extended abstract). In: 20th Annual ACM STOC, pp. 103–112. ACM Press, New York (1988)

[BGN05] Boneh, D., Goh, E., Nissim, K.: Evaluating 2-DNF formulas on ciphertexts. In: Kilian, J. (ed.) TCC 2005. LNCS, vol. 3378, pp. 325–341. Springer, Heidelberg (2005)

[BR93] Bellare, M., Rogaway, P.: Random oracles are practical: A paradigm for designing efficient protocols. In: 1st ACM CCS, pp. 62–73. ACM press, New York (1993)

[Bra99] Brands, S.: Rethinking Public Key Infrastructure and Digital Certificates— Building in Privacy. PhD thesis, Eindhoven Institute of Technology, Eindhoven, The Netherlands (1999)

[CH02] Camenisch, J., Van Herreweghen, E.: Design and implementation of the idemix anonymous credential system. Technical Report Research Report RZ 3419, IBM Research Division (May 2002)

[Cor00] Coron, J.-S.: On the exact security of full domain hash. In: Bellare, M. (ed.) CRYPTO 2000. LNCS, vol. 1880, pp. 229–235. Springer, Heidelberg (2000)

[DCP97] De Santis, A., Di Crescenzo, G., Persiano, G.: Randomness-efficient non-interactive zero-knowledge (extended abstract). In: Degano, P., Gorrieri, R., Marchetti-Spaccamela, A. (eds.) ICALP 1997. LNCS, vol. 1256, pp. 716–726. Springer, Heidelberg (1997)

[DMP88] De Santis, A., Micali, S., Persiano, G.: Non-interactive zero-knowledge proof systems. In: Pomerance, C. (ed.) CRYPTO 1987. LNCS, vol. 293, pp. 52–72. Springer, Heidelberg (1988)

[DN00] Dwork, C., Naor, M.: Zaps and their applications. In: FOCS, pp. 283–293 (2000)

[Dod02] Dodis, Y.: Efficient construction of (distributed) verifiable random functions. In: Desmedt, Y.G. (ed.) PKC 2003. LNCS, vol. 2567, pp. 1–17. Springer, Heidelberg (2002)

[DY05] Dodis, Y., Yampolskiy, A.: A verifiable random function with short proofs and keys. In: Vaudenay, S. (ed.) PKC 2005. LNCS, vol. 3386, pp. 416–432. Springer, Heidelberg (2005)

[FLS99] Feige, U., Lapidot, D., Shamir, A.: Multiple noninteractive zero knowledge proofs under general assumptions. SIAM Journal on Computing 29(1), 1–28 (1999)

[FS87] Fiat, A., Shamir, A.: How to prove yourself: Practical solutions to identification and signature problems. In: Odlyzko, A.M. (ed.) CRYPTO 1986. LNCS, vol. 263, pp. 186–194. Springer, Heidelberg (1987)

[GGM86] Goldreich, O., Goldwasser, S., Micali, S.: How to construct random functions. Journal of the ACM 33(4), 792–807 (1986)

[GK03] Goldwasser, S., Tauman Kalai , Y.: On the (in)security of the Fiat-Shamir paradigm. In: 44th FOCS, pp. 102–115. IEEE Computer Society Press, Los Alamitos (2003)

[GO92] Goldwasser, S., Ostrovsky, R.: Invariant signatures and non-interactive zero-knowledge proofs are equivalent. In: Brickell, E.F. (ed.) CRYPTO 1992. LNCS, vol. 740, pp. 224–228. Springer, Heidelberg (1993)

[GOS06]   Groth, J., Ostrovsky, R., Sahai, A.: Perfect non-interactive zero knowledge for NP. In: Vaudenay, S. (ed.) EUROCRYPT 2006. LNCS, vol. 4004, pp. 339–358. Springer, Heidelberg (2006)

[HILL99]   Håstad, J., Impagliazzo, R., Levin, L.A., Luby, M.: A pseudorandom generator from any one-way function. SIAM Journal of Computing 28(4), 1364–1396 (1999)

[KP98]   Kilian, J., Petrank, E.: An efficient noninteractive zero-knowledge proof system for NP with general assumptions. Journal of Cryptology 11(1), 1–27 (1998)

[Lys02]   Lysyanskaya, A.: Unique signatures and verifiable random functions from the DH-DDH separation. In: Yung, M. (ed.) CRYPTO 2002. LNCS, vol. 2442, pp. 597–612. Springer, Heidelberg (2002)

[MRV99]   Micali, S., Rabin, M., Vadhan, S.: Verifiable random functions. In: 40th FOCS, pp. 120–130. IEEE Computer Society Press, Los Alamitos (1999)

[Nao91]   Naor, M.: Bit commitment using pseudorandomness. Journal of Cryptology 4(2), 51–158 (1991)

Cryptography in the Multi-string Model

Jens Groth* and Rafail Ostrovsky**

University of California, Los Angeles, CA 90095
{jg,rafail}@cs.ucla.edu

Abstract. The common random string model introduced by Blum, Feldman and Micali permits the construction of cryptographic protocols that are provably impossible to realize in the standard model. We can think of this model as a trusted party generating a random string and giving it to all parties in the protocol. However, the introduction of such a third party should set alarm bells going off: Who is this trusted party? Why should we trust that the string is random? Even if the string is uniformly random, how do we know it does not leak private information to the trusted party? The very point of doing cryptography in the first place is to prevent us from trusting the wrong people with our secrets.

In this paper, we propose the more realistic multi-string model. Instead of having one trusted authority, we have several authorities that generate random strings. We do not trust any single authority; we only assume a majority of them generate the random string honestly. This security model is reasonable, yet at the same time it is very easy to implement. We could for instance imagine random strings being provided on the Internet, and any set of parties that want to execute a protocol just need to agree on which authorities' strings they want to use.

We demonstrate the use of the multi-string model in several fundamental cryptographic tasks. We define multi-string non-interactive zero-knowledge proofs and prove that they exist under general cryptographic assumptions. Our multi-string NIZK proofs have very strong security properties such as simulation-extractability and extraction zero-knowledge, which makes it possible to compose them with arbitrary other protocols and to reuse the random strings. We also build efficient simulation-sound multi-string NIZK proofs for circuit satisfiability based on groups with a bilinear map. The sizes of these proofs match the best constructions in the single common random string model.

We suggest a universally composable commitment scheme in the multi-string model. It has been proven that UC commitment does not exist in the plain model without setup assumptions. Prior to this work, constructions were only known in the common reference string model and the registered public key model. One of the applications of the UC commitment scheme is a coin-flipping protocol in the multi-string model. Armed with the coin-flipping protocol, we can securely realize any multi-party computation protocol.

Keywords: Common random string model, multi-string model, non-interactive zero-knowledge, universally composable commitment, multi-party computation.

---

* Computer Science Department. Work partially done while visiting IPAM and supported in part by NSF ITR/Cybertrust grant No. 0456717 and Cybertrust grant No. 0430254.
** Computer Science Department and Department of Mathematics. Research partially done while visiting IPAM, and supported in part by IBM Faculty Award, Xerox Innovation Group Award, NSF Cybertrust grant no. 0430254, and U.C. MICRO grant.

A. Menezes (Ed.): CRYPTO 2007, LNCS 4622, pp. 323–341, 2007.

1 Introduction

In the common random string model, the parties executing a protocol have access to a uniformly random bit-string. A generalization of this model is the common reference string (CRS) model, where the string may have a non-uniform distribution. Blum, Feldman and Micali [BFM88] introduced the CRS model to construct non-interactive zero-knowledge (NIZK) proofs. Some setup assumption was needed, since only languages in BPP can have non-interactive or two-round NIZK proofs in the plain model [GO94]. There are other examples of protocols that cannot be realized in the standard model but are possible in the CRS model, for instance universally composable (UC) commitment [CF01]. The CRS-model is therefore widely used in cryptographic protocols.

Using the CRS-model creates a problem: Where does the CRS come from? One option is to have a trusted third party that generates the CRS, but this raises a trust issue. It is very possible that the parties cannot find a party that they all trust. Would Apple trust a CRS generated by Microsoft? Would US government agencies be willing to use a CRS generated by their Russian counterparts?

Alternatively, the parties could generate the CRS themselves at the beginning of the protocol. If a majority is honest, they could for instance use multi-party computation to generate a CRS. However, this makes the whole protocol more complicated and requires them to have an initial round of interaction. They could also trust a group of parties to jointly generate a CRS; however, this leaves them with the task of finding a volunteer group of authorities to run a multi-party computation protocol whenever a CRS is needed. There is also no guarantee that different sets of parties can agree on trusting the same group of authorities, so potentially this method will require authorities to participate in many generations of CRS's.

Barak, Canetti, Nielsen and Pass [BCNP04] suggest the registered public key model as a relaxed setup that makes multi-party computation possible. In the registered public key model, parties can only register correctly generated keys. While there is no longer a common reference string in the registered public key model, the underlying problem still persists: who is the trusted party that will check that the parties only register correctly generated public keys?

THE MULTI-STRING MODEL. We propose the multi-string model as a solution to the above mentioned problem. In this model, we have a number of authorities that assist the protocol execution by providing random strings. If a majority of these authorities are honest the protocol will be secure.

There are two reasons that the multi-string model is attractive. First, the authorities play a minimal role in the protocol. They simply publish random strings, they do not need to perform any computation, be aware of each other or any other parties, or have any knowledge about the specifics of the protocol to be executed. This permits easy implementation, the parties wishing to execute a protocol can for instance simply download a set of random strings from agreed upon authorities on the internet. Second, the security of the protocols only needs to rely on a majority of the authorities being honest at the time they created the strings. Even if they are later corrupted, the random strings can still be used. Also, no matter how untrustworthy the other parties in your protocol are, you can trust the protocol if a majority of the authorities is honest. The honesty of a small group of parties that are minimally involved can be magnified and used by a larger set of parties.

The multi-string model is a very reasonable setup assumption. The next question is whether there are interesting protocols that can be securely realized in the multi-string model. We will answer this question affirmatively by constructing non-interactive zero-knowledge proofs, UC commitment and general UC-secure multi-party computation in the multi-string model in the presence of adaptive adversaries.

1.1 Non-interactive Zero-Knowledge

A zero-knowledge proof [GMR89, GMW87] is a two-party protocol, where a prover tries to convince a verifier about the truth of some statement, typically membership of an NP-language. The proof should have the following three properties: completeness, soundness and zero-knowledge. Completeness means that a prover who has an NP-witness can convince the verifier. Soundness means that if the statement is false, then it is impossible to convince the verifier. Zero-knowledge means that the verifier does not learn anything else from the proof than the fact that the statement is true. Interactive zero-knowledge proofs are known to exist in the standard model, however, non-interactive and 2-round zero-knowledge proofs only exist for trivial languages [GO94]. Instead, much research has gone into constructing non-interactive zero knowledge proofs in the CRS-model, see for instance [BFM88, BDMP91, FLS99, Dam92], [DP92, DDP99, DDP02, KP98, Sah01, DDO+02, GOS06b, GOS06a].

MULTI-STRING NIZK. We define the notion of multi-string NIZK proofs in Section 2. In the definitions, we let the adversary see many honestly generated strings and pick the ones it likes. We also allow the adversary to generate some of the strings itself, possibly in a malicious and adaptive manner. Our definition of multi-string NIZK proofs calls for completeness, soundness and zero-knowledge to hold in a threshold manner. If t_c out of n common reference strings are honest, then the prover holding an NP-witness for the truth of the statement should be able to create a convincing proof. If t_s out of n common reference strings are honest, then it should be infeasible to convince the verifier about a false statement. If t_z out of n common reference strings are honestly generated, then it should be possible to simulate the proof without knowing the witness.

It is desirable to minimize t_c, t_s, t_z. As we shall see, $t_c = 0$ is achievable, however, multi-string soundness and multi-string zero-knowledge are complementary in the sense that there is a lower bound $t_s + t_z > n$ for non-trivial languages, see Section 2.

A natural question is under which assumptions we can obtain multi-string NIZK proofs. We prove that if hard on average languages exist in NP then single-string NIZK implies the existence of multi-string NIZK and vice versa.

BEYOND VANILLA MULTI-STRING NIZK. It is undesirable to require a group of authorities to produce random strings for each proof we want to make. We prefer it to be possible to use the same strings over and over again, so each authority has to produce only one single random string. We must therefore consider a setting, where multiple protocols may be running concurrently and may be requiring the use of multi-string NIZK proofs. When the protocol designer has to prove security in such a setting, it may very well be that some of the proofs are simulated, while we still need other proofs to be sound. Moreover, in some cases we may want to extract the witness from a proof. To deal with this realistic setting, where we have both simulations of some proofs and witness extraction of other proofs going on at the same time, we introduce the notions of simulation-extractable multi-string NIZK and extraction zero-knowledge multi-string NIZK.

In simulation-extractable multi-string NIZK, we require that it be possible to extract a witness from the proof if t_s strings are honestly generated, even if the adversary sees simulated proofs for arbitrary other statements. In extraction zero-knowledge, we require that if there are t_z honest strings, then even if the adversary sees extractions of witnesses in some proofs, the other proofs remain zero-knowledge and reveal nothing. We offer a multi-string NIZK proof based on general assumptions, which is both simulation-extractable and extraction zero-knowledge.

MULTI-STRING NIZK PROOFS FROM BILINEAR GROUPS. Recently Groth, Ostrovsky and Sahai [GOS06b, GOS06a] have shown how to construct NIZK proofs from groups with a bilinear map. Their CRS contains a description of a bilinear group and a set of group elements. The group elements can be chosen such that the CRS gives either perfect soundness or perfect zero-knowledge. Soundness strings and simulation strings are computationally indistinguishable, so this gives a NIZK proof in the CRS model.

There is a major technical hurdle to overcome when trying to apply their techniques in the multi-string model: the single-string NIZK proofs rely on the common reference string to contain a description of a bilinear group. In the multi-string model, the authorities generate their random strings completely oblivious of the other authorities. There is therefore no agreement on which bilinear group to use. One might try to let the prover pick the bilinear group, however, this too causes problems since now we need to set up the random strings such that they will work for many choices of bilinear groups.

We resolve these problems by inventing a novel technique to "translate" common reference strings in one group to common reference strings in another group. Each authority picks its own bilinear group and the prover also picks a bilinear group. Using our translation technique, we can translate simulation reference strings chosen by the authorities to simulation reference strings in the prover's bilinear group. Similarly, we can translate soundness reference strings chosen by the authorities to soundness reference strings in the prover's bilinear group.

The resulting multi-string NIZK proofs for circuit satisfiability have size $\mathcal{O}(n + |C|)k$, where n is the number of random strings, $|C|$ is the size of the circuit, and k is the security parameter, i.e., the size of a group element. We will typically have n much smaller than $|C|$, so this matches the best single-string NIZK proofs [GOS06b, GOS06a] that have complexity $\mathcal{O}(|C|k)$.

1.2 Multi-party Computation

Canetti's UC framework [Can01] defines secure execution of a protocol under concurrent execution of arbitrary protocols. Informally a protocol is UC secure if its execution is equivalent to handing protocol input to an honest trusted party that computes everything securely and returns the resulting outputs.

UC COMMITMENT. It is known that in the plain model, any (well-formed) ideal functionality can be securely realized if a majority of the parties are honest. On the other hand, if a majority may be corrupt, there are certain functionalities that are provably impossible to realize. One such example is UC commitment [CF01]. We demonstrate that in the multi-string model UC commitment can be securely realized. The key idea in this construction is to treat each common random string as the key for a commitment scheme. By applying threshold secret-sharing techniques, we can spread the message out on the n commitment scheme in a way such that we can tolerate a minority of fake common reference strings.

MULTI-PARTY COMPUTATION. Canetti, Lindell, Ostrovsky and Sahai [CLOS02] show that any (well-formed) ideal functionality can be securely realized in the CRS-model, even against adversaries that can adaptively corrupt arbitrary parties and where parties are not assumed to be able to securely erase any of their data. However, it was an open question where this CRS should come from, since the parties provably could not compute it themselves.

Armed with our UC commitment it is straightforward to solve this problem. We simply run a coin-flipping protocol to create a CRS. This result points out a nice feature of the multi-string model; it scales extremely well. We just require a majority of the authorities to be honest. Then no matter which group of parties, even if it is a large group of mostly untrustworthy parties, we can magnify the authorities' honesty to enable this entire group to do secure computation.

REMARK. The multi-string model is described in the UC framework as an ideal functionality that provides random strings and allows the adversary to inject a minority of malicious strings as well. This functionality is easy to implement with a set of authorities that just provide random strings. It is important though that these strings are local to the protocol, we do not guarantee security of other protocols that use the same strings. Canetti, Dodis, Pass and Walfish [RCW07] have demonstrated that it is not possible to have a fixed global common random string that is used for multiple and arbitrary different protocol executions and this result extends to the multi-string model.

REMARK. Building on our multi-string NIZK, an alternative proof of our multiparty computation result was shown by [PPS06].

2 Definitions

Let R be an efficiently computable binary relation. For pairs $(x, w) \in R$ we call x the statement and w the witness. Let L be the NP-language consisting of statements in R.

A multi-string proof system for a relation R consists of probabilistic polynomial time algorithms K, P, V, which we will refer to as respectively the key generator, the prover and the verifier. The key generation algorithm can be used to produce common reference strings σ. In the present paper, we can implement our protocols with a key generator that outputs a uniformly random string of polynomial length $\ell(k)$, however, for the sake of generality, we include a key generator in our definitions.

The prover takes as input $(t_c, t_s, t_z, \sigma, x, w)$, where σ is a set of n common reference strings and $(x, w) \in R$, and produces a proof π. The verifier takes as input $(t_c, t_s, t_z, \sigma, x, \pi)$ and outputs 1 if the proof is acceptable and 0 if rejecting the proof. We call (K, P, V) a (t_c, t_s, t_z, n)-NIZK proof system for R if it has the completeness, soundness and zero-knowledge properties described below. We remark that $(1, 1, 1, 1)$-NIZK proof systems correspond closely to the standard notion of NIZK proofs in the CRS-model.

(t_c, t_s, t_z, n)-COMPLETENESS. We will say that (K, P, V) is (t_c, t_s, t_z, n)-complete if the prover can convince the verifier of a true statement, when at least t_c string have been generated honestly. As we shall see later, our protocols will have perfect (t_c, t_s, t_z, n)-completeness for all $0 \leq t_c \leq n$. In other words, even if the adversary chooses all common reference strings itself, we still have probability 1 of outputting an acceptable proof.

Definition 1. (K, P, V) *is* (t_c, t_s, t_z, n)-*complete if for all non-uniform polynomial time adversaries* $\mathcal{A}$ *we have*

$$\Pr\left[S := \emptyset; (\boldsymbol{\sigma}, x, w) \leftarrow \mathcal{A}^K; \pi \leftarrow P(t_c, t_s, t_z, \boldsymbol{\sigma}, x, w) : \right.$$

$$\left. V(t_c, t_s, t_z, \boldsymbol{\sigma}, x, \pi) = 0 \text{ and } (x, w) \in R \text{ and } |\boldsymbol{\sigma} \setminus S| \geq t_c \right] \approx 0,$$

where K *on query* i *outputs* $\sigma_i \leftarrow K(1^k)$ *and sets* $S := S \cup \{\sigma_i\}$.

(t_c, t_s, t_z, n)-SOUNDNESS. The goal of the adversary in the soundness definition is to forge a proof using n common reference strings, even if t_s of them are honestly generated. The adversary gets to see possible choices of correctly generated common reference strings and can adaptively choose n of them, it may also in these n common reference strings include up to $n - t_s$ fake common reference strings chosen by itself.

Definition 2. *We say* (K, P, V) *is* (t_c, t_s, t_z, n)-*sound if for all non-uniform polynomial time adversaries* $\mathcal{A}$ *we have*

$$\Pr\left[S := \emptyset; (\boldsymbol{\sigma}, x, \pi) \leftarrow \mathcal{A}^K : V(t_c, t_s, t_z, \boldsymbol{\sigma}, x, \pi) = 1 \text{ and } x \notin L \text{ and } |\boldsymbol{\sigma} \setminus S| \geq t_s \right] \approx 0,$$

where K *is an oracle that on query* i *outputs* $\sigma_i \leftarrow K(1^k)$ *and sets* $S := S \cup \{\sigma_i\}$.

(t_c, t_s, t_z, n)-ZERO-KNOWLEDGE. We wish to formulate that if t_z common reference strings are correctly generated, then the adversary learns nothing from the proof. As is standard in the zero-knowledge literature, we will say this is the case, when we can simulate the proof given only the statement x. Let therefore S_1 be an algorithm that outputs (σ, τ), respectively a simulation reference string and a simulation trapdoor. Let furthermore, S_2 be an algorithm that takes input $(t_c, t_s, t_z, \boldsymbol{\sigma}, \boldsymbol{\tau}, x, w)$ and simulates a proof π if $\boldsymbol{\tau}$ contains t_z simulation trapdoors for common reference strings in $\boldsymbol{\sigma}$.

We will strengthen the standard definition of zero-knowledge, by splitting the definition into two parts. The first part simply says that the adversary cannot distinguish real common reference strings from simulation reference strings. The second part, says that *even with access to the simulation trapdoors* the adversary cannot distinguish the prover from the simulator on a set of simulated reference strings.

Definition 3. *We say* (K, P, V, S_1, S_2) *is* (t_c, t_s, t_z, n)-*zero-knowledge if we have reference string indistinguishability and simulation indistinguishability as described below.*

REFERENCE STRING INDISTINGUISHABILITY. *For all non-uniform polynomial time adversaries $\mathcal{A}$ we have*

$$\Pr\left[\sigma \leftarrow K(1^k) : \mathcal{A}(\sigma) = 1\right] \approx \Pr\left[(\sigma, \tau) \leftarrow S_1(1^k) : \mathcal{A}(\sigma) = 1\right].$$

(t_c, t_s, t_z, n)-SIMULATION INDISTINGUISHABILITY. *For all non-uniform interactive polynomial time adversaries $\mathcal{A}$ we have*

$$\Pr\left[S := \emptyset; (\boldsymbol{\sigma}, \boldsymbol{\tau}, x, w) \leftarrow \mathcal{A}^{S_1}(1^k); \pi \leftarrow P(t_c, t_s, t_z, \boldsymbol{\sigma}, x, w) : \right.$$
$$\left. \mathcal{A}(\pi) = 1 \text{ and } (x, w) \in R \text{ and } |\boldsymbol{\sigma} \setminus S| \geq t_z\right]$$
$$\approx \Pr\left[S := \emptyset; (\boldsymbol{\sigma}, \boldsymbol{\tau}, x, w) \leftarrow \mathcal{A}^{S_1}(1^k); \pi \leftarrow S_2(t_c, t_s, t_z, \boldsymbol{\sigma}, \boldsymbol{\tau}, x) : \right.$$
$$\left. \mathcal{A}(\pi) = 1 \text{ and } (x, w) \in R \text{ and } |\boldsymbol{\sigma} \setminus S| \geq t_z\right],$$

where S_1 on query i outputs $(\sigma_i, \tau_i) \leftarrow S_1(1^k)$ and sets $S := S \cup \{\sigma_i\}$, and $\boldsymbol{\tau}$ contains t_z simulation trapdoors corresponding to σ_i's in $\boldsymbol{\sigma}$ generated by S_1.

LOWER BOUNDS FOR MULTI-STRING NIZK PROOFS. Soundness and zero-knowledge are complementary. The intuition is that if an adversary controls enough strings to simulate a proof, then he can prove anything and we can no longer have soundness. We capture this formally in the following theorem.

Theorem 1. *If L is a language with a proof system (K, P, V) that has (t_c, t_s, t_z, n)-completeness, soundness and zero-knowledge then $L \in \mathrm{P/poly}$ or $t_s + t_z > n$.*

Proof. Assume we have an (t_c, t_s, t_z, n)-NIZK proof system for R defining L and $t_s + t_z \leq n$. Given an element x, we wish to decide whether $x \in L$ or not. We simulate t_z common reference strings $(\sigma_i, \tau_i) \leftarrow S_1(1^k)$ and generate $n - t_z$ common reference strings $\sigma_j \leftarrow K(1^k)$ setting $\tau_j = \perp$. We then simulate the proof $\pi \leftarrow S_2(\boldsymbol{\sigma}, \boldsymbol{\tau}, x)$. Output $V(\boldsymbol{\sigma}, x, \pi)$.

Let us analyze this algorithm. If $x \in L$, then by (t_c, t_s, t_z, n)-completeness a prover with access to a witness w would output a proof that the verifier accepts if all common reference strings are generated correctly. By reference string indistinguishability, we will therefore also accept the proof when some of the common reference strings are simulated. By (t_c, t_s, t_z, n)-simulation indistinguishability, where we give (x, w) as non-uniform advice to $\mathcal{A}$, we will output 1 with overwhelming probability on $x \in L$.

On the other hand, if $x \notin L$, then by the (t_c, t_s, t_z, n)-soundness we output 0 with overwhelming probability, since $n - t_z \geq t_s$ common reference strings have been generated correctly. This shows that $L \in \mathrm{BPP/poly}$. By [Adl78] we have $\mathrm{P/poly} = \mathrm{BPP/poly}$, which concludes the proof. $\square$

In general, the verifier wishes to minimize t_s to make it more probable that the protocol is sound, and at the same time the prover wishes to minimize t_z to make it more probable that the protocol is zero-knowledge. In many cases, choosing n odd, and setting $t_s = t_z = \frac{n+1}{2}$ will be a reasonable compromise. However, there are also cases where it might be relevant to have an unbalanced setting. Consider the case, where Alice wants to e-mail a NIZK proof to Bob, but does not know Bob's preferences with respect to common reference strings. She may pick a set of common reference strings and make a multi-string proof. Bob did not participate in deciding which common reference strings to use, however, if they came from trustworthy authorities he may be willing to believe that one of the authorities is honest. On the other hand, Alice gets to choose the authorities, so she may be wiling to believe that all of them are honest. The appropriate choice in this situation, is a multi-string proof with $t_s = 1, t_z = n$.

ADVANCED ZERO-KNOWLEDGE PROOFS. Multi-string NIZK proofs can have more advanced properties. In a multi-string proof of knowledge, the reference strings can be generated with an extraction trapdoor. If you hold at least t_s extraction keys, it is possible to extract a witness from the multi-string NIZK proof. We say a multi-string NIZK proof has extraction zero-knowledge, if it is zero-knowledge even to an adversary that can ask for arbitrary extractions of witnesses. This latter notion is similar in nature to CCA-secure encryption.

Another way of strengthening soundness is to say that even after seeing arbitrary simulated proofs, even on false statements, it should not be able to prove another false statement. We call this simulation soundness. This notion can be extended and combined with proofs of knowledge to simulation-extractability, which means that even if we give the adversary access to see simulated multi-string proofs, it cannot produce another proof without us being able to extract a witness from it.

We refer to the full paper [GO07] for formal definitions of multi-string proofs of knowledge, simulation soundness, simulation-sound extractability and extraction zero-knowledge.

3   Multi-string NIZK Proofs Based on General Assumptions

MULTI-STRING NIZK PROOFS. As a warm-up, we will start out with a simple construction of a multi-string NIZK proof that works for $t_c = 0$ and all choices of t_s, t_z, n so $t_s + t_z > n$. This construction is used in the full paper [GO07] to prove the following theorem connecting single-string NIZK proofs and multi-string NIZK proofs.

Theorem 2. *Assuming hard on average languages exist in NP, the existence of NIZK proofs for NP in the common random string model is equivalent to the existence of multi-string NIZK proofs for NP in the common random strings model. The equivalence preserves perfect completeness.*

We use two tools in the construction: a zap $(\ell_{zap}, P_{zap}, V_{zap})$ and a pseudorandom generator PRG. Zaps, introduced by Dwork and Naor [DN02], are two-round public coin witness-indistinguishable proofs, where the verifier's first message is a random string that can be fixed once and for all and be reused in subsequent zaps.

A common random string in our multi-string NIZK proof will consist of a random value r and an initial message σ for the zap. Given a statement $x \in L$, the prover makes n zaps using respectively initial messages $\sigma_1, \ldots, \sigma_n$ for

$$x \in L \quad \text{or} \quad \text{there are } t_z \text{ common reference strings where } r_i \text{ is a pseudorandom value.}$$

In the simulation, we create simulation reference strings as $r := \mathrm{PRG}(\tau)$ enabling the simulator to make zaps without knowing a witness w for $x \in L$.

MULTI-STRING SIMULATION-EXTRACTABLE NIZK PROOF. We will now construct more advanced multi-string NIZK proofs of knowledge that are $(0, t_s, t_z, n)$-simulation-extractable and $(0, t_s, t_z, n)$-extraction zero-knowledge.

To permit the extraction of witnesses, we include a public key for a CCA2-secure cryptosystem in each common reference string. In a proof, the prover will make a (t_s, n)-threshold secret sharing of the witness and encrypt the shares under the n public keys. To extract the witness, we will decrypt t_s of these ciphertexts and combine the shares to get the witness.

To avoid tampering with the proof, we will use a strong one-time signature. The prover generates a key $(vk_{\mathrm{sots}}, sk_{\mathrm{sots}}) \leftarrow K_{\mathrm{sots}}(1^k)$ that he will use to sign the proof. The implication is that the adversary, who sees simulated proofs, must use a different vk_{sots} in his forged proof, because he cannot forge the strong one-time signature.

The common reference string will contain a value, which in a simulation string will be a pseudorandom $2k$-bit value. The prover will prove that he encrypted a (t_s, n)-threshold secret sharing of the witness, or that he knows how to evaluate t_z pseudorandom functions on vk_{sots} using the seeds of the respective common reference strings. On a real common reference string, this seed is not known and therefore he cannot make such a proof. On the other hand, in the simulation the simulator does know these seeds and can therefore simulate without knowing the witness. Simulation soundness follows from the adversary's inability to guess the pseudorandom functions' evaluations on vk_{sots}, even if he knew the evaluations on many other verification keys.

Zero-knowledge under extraction attack follows from the CCA2-security of the cryptosystem. Even after having seen many extractions, the ciphertexts reveal nothing about the witness, or even whether the trapdoor has been used to simulate a proof.

Common reference string/simulation string: $(pk_1, dk_1), (pk_2, dk_2) \leftarrow K_{\mathrm{CCA2}}(1^k)$; $r \leftarrow \{0,1\}^{2k}; \sigma \leftarrow \{0,1\}^{\ell_{\mathrm{zap}}(k)}$. Return $\Sigma := (pk_1, pk_2, r, \sigma)$.

The simulators and extractors S_1, E_1, SE_1 will generate the simulated reference strings in the same way, except for choosing $\tau \leftarrow \{0,1\}^k$ and $r := \mathrm{PRF}_\tau(0)$. We use the simulation trapdoor τ and the extraction key $\xi := dk_1$.

Proof: $P(0, t_s, t_z, (\Sigma_1, \ldots, \Sigma_n), x, w)$ where $(x, w) \in R$ runs as follows: First, generate a key pair for a strong one-time signature scheme $(vk_{\mathrm{sots}}, sk_{\mathrm{sots}}) \leftarrow K_{\mathrm{sots}}(1^k)$. Use (t_s, n)-threshold secret sharing to get shares $w_1, \ldots, w_n$ of w. Encrypt the shares as $c_{1i} := E_{pk_{1i}}(w_i, vk_{\mathrm{sots}}; r_{1i})$. Also encrypt dummy values $c_{2i} \leftarrow E_{pk_{2i}}(0)$. Consider the statement: "All c_{1i} encrypt $(w_i, vk_{\mathrm{sots}})$, where $w_1, \ldots, w_n$ is a (t_s, n)-secret sharing of a witness w so $(x, w) \in R$ or there exist at least t_z seeds τ_i so $r_i = \mathrm{PRF}_{\tau_i}(0)$ and c_{2i} encrypts $\mathrm{PRF}_{\tau_i}(vk_{\mathrm{sots}})$." We can reduce this statement to a polynomial size circuit C and a satisfiability witness W. For all i's we create a zap $\pi_i \leftarrow P_{\mathrm{zap}}(\sigma_i, C, W)$ for C being satisfiable. We sign everything using the one-time signature $sig \leftarrow \mathrm{Sign}_{sk_{\mathrm{sots}}}(vk_{\mathrm{sots}}, x, \Sigma_1, c_{11}, c_{21}, \pi_1, \ldots, \Sigma_n, c_{1n}, c_{2n}, \pi_n)$. The proof is $\Pi := (vk_{\mathrm{sots}}, c_{11}, c_{21}, \pi_1, \ldots, c_{1n}, c_{2n}, \pi_n, sig)$.

Verification: To verify Π on the form described above, verify the strong one-time signature and verify the n zaps $\pi_1, \ldots, \pi_n$.

Extraction: To extract a witness check that the proof is valid. Next, use the first t_s extraction keys in ξ to decrypt the corresponding t_s ciphertexts. We combine the t_s secret shares to recover the witness w.

Simulated proof: To simulate a proof, pick the first t_z simulation trapdoors in τ. These are τ_i so $r_i = \text{PRF}_{\tau_i}(0)$. As in the proof generate $(vk_{\text{sots}}, sk_{\text{sots}}) \leftarrow K_{\text{sots}}(1^k)$. Create t_z pseudorandom values $v_i := \text{PRF}_{\tau_i}(vk_{\text{sots}})$. Encrypt the values as $c_{2i} \leftarrow E_{pk_{2i}}(v_i)$. For the other reference strings, just let $c_{2i} \leftarrow E_{pk_{2i}}(0)$. Let $w_1, \ldots, w_n$ be a (t_s, n)-threshold secret sharing of 0. We encrypt also these values as $c_{1i} \leftarrow E_{pk_{1i}}(w_i, vk_{\text{sots}})$. Let again C be the circuit corresponding to the statement "All c_{1i} encrypt (w_i, vk_{sots}), where $w_1, \ldots, w_n$ is a (t_s, n)-secret sharing of a witness w *or* there exist at least t_z seeds τ_i so $r_i = \text{PRF}_{\tau_i}(0)$ and c_{2i} encrypts $\text{PRF}_{\tau_i}(vk_{\text{sots}})$." From the creation of the ciphertexts c_{2i} we have a witness W for C being satisfiable. Create zaps $\pi_i \leftarrow P_{\text{zap}}(\sigma_i, C, W)$ for C being satisfiable. Finally, make a strong one-time signature on everything $sig \leftarrow \text{Sign}_{sk_{\text{sots}}}(vk_{\text{sots}}, x, \Sigma_1, c_{11}, c_{21}, \pi_1, \ldots, \Sigma_n, c_{1n}, c_{2n}, \pi_n)$. The simulated proof is $\Pi := (vk_{\text{sots}}, c_{11}, c_{21}, \pi_1, \ldots, c_{1n}, c_{2n}, \pi_n, sig)$.

Theorem 3. *The above protocol is a $(0, t_s, t_z, n)$-NIZK proof for all choices of $t_s + t_z > n$. It has $(0, t_s, t_z, n)$-simulation-soundness, $(0, t_s, t_z, n)$-extraction zero-knowledge and statistical $(0, t_s, t_z, n)$-knowledge. It can be securely implemented if enhanced trapdoor permutations exist, and it can be implemented with random strings if dense cryptosystems [DP92] and enhanced trapdoor permutations exist.*

We refer to the full paper [GO07] for the proof.

4   Multi-string NIZK Proofs from Groups with a Bilinear Map

SETUP. We will use bilinear groups generated by $(p, \mathbb{G}, \mathbb{G}_T, e, g) \leftarrow \mathcal{G}(1^k)$ such that:

- p is a k-bit prime.
- $\mathbb{G}, \mathbb{G}_T$ are cyclic groups of order p.
- g is a generator of $\mathbb{G}$.
- $e : \mathbb{G} \times \mathbb{G} \to \mathbb{G}_T$ is a bilinear map such that $e(g, g)$ generates $\mathbb{G}_T$ and for all $a, b \in \mathbb{Z}_p$ we have: $e(g^a, g^b) = e(g, g)^{ab}$.
- Group operations, group membership, and the bilinear map are efficiently computable.
- Given a description $(p, \mathbb{G}, \mathbb{G}_T, e, g)$ it is verifiable that indeed it is a bilinear group and that g generates $\mathbb{G}$.
- There is a decoding algorithm that given a random string of $(n+1)k$ bits interprets it as n random group elements. The decoding algorithm is reversible, such that given n group elements we can pick at random one of the $(n + 1)k$-bit strings that decode to the n group elements.
- The length of the description of $(p, \mathbb{G}, \mathbb{G}_T, e, g)$ is at most $4k$ bits.[1]

---

[1] It is easy to modify the protocol to work whenever the description of the bilinear group is $\mathcal{O}(k)$ bits.

– When working in the random multi-string model, we will assume $\mathcal{G}$ simply outputs a uniformly random $4k$-bit string, from which $(p, \mathbb{G}, \mathbb{G}_T, e, g)$ can be sampled.

We use the decisional linear assumption introduced by Boneh, Boyen and Shacham [BBS04], which says that given group elements (f, g, h, f^r, g^s, h^t) it is hard to tell whether $t = r + s$ or t is random. Throughout the paper, we use bilinear groups $(p, \mathbb{G}, \mathbb{G}_T, e, g) \leftarrow \mathcal{G}(1^k)$ generated such that the DLIN assumption holds for $\mathcal{G}$.

Example. We will offer a class of candidates for DLIN groups as described above. Consider the elliptic curve $y^2 = x^3 + 1 \bmod q$, where $q = 2 \bmod 3$ is a prime. It is straightforward to check that a point (x, y) is on the curve. Furthermore, picking $y \in \mathbb{Z}_q$ at random and computing $x = (y^2 - 1)^{\frac{q+1}{3}} \bmod q$ gives us a random point on the curve. The curve has a total of $q + 1$ points, where we include also the point at infinity. When generating bilinear groups, we will pick p as a k-bit prime. We then let q be the smallest prime[2] so $p|q + 1$ and define $\mathbb{G}$ to be the order p subgroup of the curve. The target group is the order p subgroup of $\mathbb{F}_{q^2}^*$ and the bilinear map is the modified Weyl-pairing [BF03]. Verification of $(p, \mathbb{G}, \mathbb{G}_T, e, g)$ being a group with bilinear maps is straightforward, since it corresponds to checking that p, q are primes so $p|q + 1$ and $q = 2 \bmod 3$ and g is an order p element on the curve. A random point in the group $\mathbb{G}$ can be sampled by picking a random point (x, y) on the curve and raising it to $\frac{q+1}{p}$. Reverse sampling is possible, since multiplying a group element with a random point of order $\frac{q+1}{p}$ gives a random (x, y) on the curve that would generate the group element.

MULTI-STRING NIZK PROOFS FROM DLIN GROUPS. We will construct a $(0, t_s, t_z, n)$-simulation-sound NIZK proof for circuit satisfiability consisting of $\mathcal{O}((n+|C|)k)$ bits, where $|C|$ is the number of gates in the circuit and k is the security parameter. Typically, n is much smaller than $|C|$, so the complexity matches the best known NIZK proofs for circuit satisfiability in the single common reference string model [GOS06b, GOS06a] that have proofs of size $\mathcal{O}(|C|k)$.

One could hope that the construction from Section 3 could be implemented efficiently using groups with a bilinear map. This strategy does not work because each common reference string is generated at random and independently of the others. This means that even if the common reference strings contain descriptions of groups with bilinear maps, most likely they are different and incompatible groups.

In our construction, we let all the common reference strings describe different groups and we also let the prover pick a group with a bilinear map. Our solution to the problem described above, is to translate simulation reference strings created by the authorities into simulation reference strings in the prover's group. This translation will require the use of a pseudorandom generator, which we construct from the DLIN assumption in the full paper [GO07]. This pseudorandom generator is constructed in such a way that there exist efficient simulation-sound NIZK proofs for a value being pseudorandom [Gro06].

Consider a common reference string with group $\mathbb{G}_i$ and the prover's group $\mathbb{G}$. We will let the common reference string contain a random string r_i. The prover will choose

---

[2] In other words, q is the smallest prime in the arithmetic progression $3p - 1, 6p - 1, 9p - 1, \ldots$. Granville and Pomerance [GP90] has conjectured that it requires $\mathcal{O}(k^2)$ steps in this progression to encounter a prime q.

a string s_i. Consider the pair of strings $(r_i \oplus s_i, s_i)$. Since strings can be interpreted as group elements, we have corresponding sets of group elements in respectively $\mathbb{G}_i$ and $\mathbb{G}$. However, since r_i is chosen at random it is unlikely that both $r_i \oplus s_i$ corresponds to a pseudorandom value in $\mathbb{G}_i$ and at the same time s_i corresponds to a pseudorandom value in $\mathbb{G}$. Of course, the prover has some degree of freedom in choosing the group $\mathbb{G}$, but if one is careful and chooses a pseudorandom generator that stretches the input sufficiently then one can use an entropy argument for it being unlikely that both strings are pseudorandom values.

Now we use non-interactive zaps and NIZK proofs to bridge the two groups. The prover will select s_i so $r_i \oplus s_i$ is a pseudorandom value in $\mathbb{G}_i$ specified by the common reference string and give an NIZK proof for this using that common reference string. In his own group, he gets n values $s_1, \ldots, s_n$ and proves that t_z of those are pseudorandom or C is satisfiable. In the simulation, he knows the simulation trapdoors for t_z reference strings and he can therefore simulate NIZK proofs of $r_i \oplus s_i$ being pseudorandom. This means, he can select the corresponding s_i's as pseudorandom values and use this to prove that there are at least t_z pseudorandom values in his own group, so he does not need to know the satisfiability witness w for C being satisfiable to carry out the proof in his own bilinear group.

There is another technical detail to consider. We want the construction to be efficient in n. Therefore, instead of proving directly that there are t_z pseudorandom values or C is satisfiable, we use a homomorphically encrypted counter. In the simulation, we set the counter to 1 for each pseudorandom value and to 0 for the rest of the values in the prover's group. The homomorphic property enables us to multiply these ciphertexts and get an encrypted count of t_z. It is straightforward to prove that the count is t_z or C is satisfiable.

These ideas describe how to get soundness. We can set up the common reference strings such that they enable us to make simulation-sound NIZK proofs in their bilinear groups. With a few extra ideas, we then get a $(0, t_s, t_z, n)$-simulation-sound NIZK proof for circuit satisfiability when $t_s + t_z > n$.

Common reference string/simulation reference string: Generate a DLIN group $(p, \mathbb{G}, \mathbb{G}_T, e, g) \leftarrow \mathcal{G}(1^k)$. Generate a common reference string for a simulation-sound NIZK proof on basis of this group $\Sigma \leftarrow K_{\text{sim-sound}}(p, \mathbb{G}, \mathbb{G}_T, e, g)$ as in [Gro06]. Pick a random string $r \leftarrow \{0,1\}^{61k}$. Output $\Sigma := (p, \mathbb{G}, \mathbb{G}_T, e, g, \sigma, r)$.

Provided one can sample groups from random strings, this can all be set up in the random multi-string model.

When generating a simulation reference string, use the simulator for the simulation-sound NIZK proof to generate $(\sigma, \tau) \leftarrow S_{\text{sim-sound}}(p, \mathbb{G}, \mathbb{G}_T, e, g)$. Output Σ as described above and simulation trapdoor τ.

Proof: Given $t_z, (\Sigma_1, \ldots, \Sigma_n), C, w$ so $C(w) = 1$ do the following. Pick a group $(p, \mathbb{G}, \mathbb{G}_T, e, g) \leftarrow \mathcal{G}(1^k)$. Pick also keys for a strong one-time signature scheme $(vk_{\text{sots}}, sk_{\text{sots}}) \leftarrow K_{\text{sots}}(1^k)$. Encode vk_{sots} as a tuple of $\mathcal{O}(1)$ group elements from $\mathbb{G}$.

For each common reference string Σ_i do the following. Pick a pseudorandom value with 6 key pairs, 6 input pairs and 36 structured elements, as described in the full paper [GO07]. This gives us a total of 60 group elements from $\mathbb{G}_i$. Concatenate the tuple of 60 group elements with vk_{sots} to get $\mathcal{O}(1)$ group elements from $\mathbb{G}_i$. Make a simulation-sound NIZK proof, using σ_i, for these $\mathcal{O}(1)$ group elements being of a form such that the first 60 of them constitute a pseudorandom value. From [Gro06] we know that the size of this proof is $\mathcal{O}(1)$ group elements from $\mathbb{G}_i$. Choose $s_i \in \{0,1\}^{61k}$ to be a random string such that $r_i \oplus s_i$ parses to the 60 elements from the pseudorandom value.

From now on we will work in the group $(p, \mathbb{G}, \mathbb{G}_T, e, g)$ chosen by the prover. Pick $pk := (f, h)$ as two random group elements. This gives us a CPA-secure cryptosystem, encrypting a message $m \in \mathbb{G}$ with randomness $r, s \in \mathbb{Z}_p$ as $E_{pk}(m; r, s)$ $:= (f^r, h^s, g^{r+s}m)$. For each $i = 1, \ldots, n$ we encrypt $1 = g^0$ as $c_i \leftarrow E_{pk}(1)$. Also, we take s_i and parse it as 60 group elements. Call this tuple z_i. Make a non-interactive zap π using the group $(p, \mathbb{G}, \mathbb{G}_T, e, g)$ and combining techniques of [GOS06a] and [Gro06] for the following statement:

$$C \text{ satisfiable} \quad \vee \quad \left(\prod_{i=1}^{n} c_i \text{ encrypts } g^{t_z} \wedge \forall i : c_i \text{ encrypts } g^0 \text{ or } g^1 \right.$$

$$\left. \wedge \forall i : z_i \text{ is a pseudorandom value } \vee c_i \text{ encrypts } g^0 \right).$$

The zap consists of $\mathcal{O}(n + |C|)$ group elements and has perfect soundness.

Sign everything $sig \leftarrow \text{Sign}_{sk_{sots}}(vk_{sots}, C, \Sigma_1, s_1, \pi_1, c_1, \ldots, \Sigma_n, s_n, \pi_n, c_n, p,$ $\mathbb{G}, \mathbb{G}_T, e, g, f, h, \pi)$.

The proof is $\Pi := (vk_{sots}, s_1, \pi_1, c_1, \ldots, s_n, \pi_n, c_n, p, \mathbb{G}, \mathbb{G}_T, e, g, f, h, \pi, sig)$.

Verification: Given common reference strings $\Sigma_1, \ldots, \Sigma_n$, a circuit C and a proof as described above, do the following. For all i check the simulation-sound NIZK proofs π_i for $r_i \oplus s_i$ encoding a pseudorandom structure in $\mathbb{G}_i$ using common reference string σ_i. Verify $(p, \mathbb{G}, \mathbb{G}_T, e, g)$ is a group with a bilinear map. Verify the zap π. Verify the strong one-time signature on everything. Output 1 if all checks are ok.

Simulated proof: We are given reference strings $\Sigma_1, \ldots, \Sigma_n$. t_z of them are simulation strings, where we know the simulation trapdoors τ_i for the simulation-sound NIZK proofs. We wish to simulate a proof for a circuit C being satisfiable.

We start by choosing a group $(p, \mathbb{G}, \mathbb{G}_T, e, g) \leftarrow \mathcal{G}(1^k)$ and public key $f, h \leftarrow \mathbb{G}$. We create ciphertexts $c_i \leftarrow E_{pk}(g^1)$ for the t_z simulation reference strings, where we know the trapdoor τ_i, and set $c_i \leftarrow E_{pk}(g^0)$ for the rest. We also choose a strong one-time signature key pair $(vk_{sots}, sk_{sots}) \leftarrow K_{sots}(1^k)$.

For t_z of the common reference strings, we know the simulation key τ_i. This permits us to choose an arbitrary string s_i and simulate a proof π_i that $r_i \oplus s_i$ encodes a 60 element pseudorandom structure. This means, we are free to choose s_i so it encodes a pseudorandom structure z_i in $\mathbb{G}^{60}$. For the remaining $n - t_z < t_s$ reference strings, we select s_i so $r_i \oplus s_i$ does encode a pseudorandom value in $\mathbb{G}_i$ and carry out a real simulation-sound NIZK proof π_i for it being a pseudorandom value concatenated with vk_{sots}.

For all i we have c_i encrypting g^b, where $b \in \{0, 1\}$. We have $\prod_{i=1}^{n} c_i$ encrypting g^{t_z}. We also have for the t_z simulation strings, where we know τ_i that s_i encodes a pseudorandom structure, whereas for the other common reference strings we have

c_i encrypts g^0. This means we can create the non-interactive zap π without knowing C's satisfiability witness.

Sign everything $sig \leftarrow \text{Sign}_{sk_{\text{sots}}}(vk_{\text{sots}}, C, \Sigma_1, s_1, \pi_1, c_1, \ldots, \Sigma_n, s_n, \pi_n, c_n, p, \mathbb{G}, \mathbb{G}_T, e, g, f, h, \pi)$. The simulated proof is $\Pi := (vk_{\text{sots}}, s_1, \pi_1, c_1, \ldots, s_n, \pi_n, c_n, p, \mathbb{G}, \mathbb{G}_T, e, g, f, h, \pi, sig)$.

Theorem 4. *Assuming we have a DLIN group as described above, then the construction above gives us a $(0, t_s, t_z, n)$-simulation-sound NIZK proof for circuit satisfiability, where the proofs have size $\mathcal{O}((n + |C|)k)$ bits. The proof has statistical $(0, t_s, t_z, n)$-soundness. The scheme can be set up in the random multi-string model if we can sample groups with bilinear maps from random strings.*

The proof can be found in the full paper [GO07].

5   UC Commitment in the Multi-string Model

In the rest of the paper, we will work in Canetti's UC framework. We refer to Canetti [Can01] for a detailed description. Very briefly, the UC framework compares a real world execution of a protocol with an ideal process where the parties have access to an ideal functionality that handles all protocol execution honestly and securely.

IDEAL FUNCTIONALITIES. Let us first formalize the multi-string model in the UC framework. Figure 1 gives an ideal multi-string functionality $\mathcal{F}_{\text{MCRS}}$. We will construct universally composable commitments, see Figure 2, in the multi-string model.

Functionality $\mathcal{F}_{\text{MCRS}}$

Parameterized by polynomial ℓ_{mcrs}, and running with parties $P_1, \ldots, P_N$ and adversary $\mathcal{S}$.

String generation: On input (crs, sid) from $\mathcal{S}$, pick $\sigma \leftarrow \{0, 1\}^{\ell_{\text{mcrs}}(k)}$ and store it. Send $(\text{crs}, sid, \sigma)$ to $\mathcal{S}$.

String selection: On input $(\text{vector}, sid, \sigma_1, \ldots, \sigma_n)$ where $\sigma_1, \ldots, \sigma_n \in \{0, 1\}^{\ell_{\text{mcrs}}(k)}$ from $\mathcal{S}$ check that more than half of the strings $\sigma_1, \ldots, \sigma_n$ match stored strings. In that case output $(\text{vector}, sid, \sigma_1, \ldots, \sigma_n)$ to all parties and halt.

Fig. 1. The ideal multi-string generator

We will assume the parties can broadcast messages, i.e., have access to an ideal broadcast functionality $\mathcal{F}_{\text{BC}}$.

UC COMMITMENT IN THE MULTI-STRING MODEL. We will describe our UC commitment protocol later but first let us offer some intuition. To prove that our UC commitment is secure, we will describe an ideal process adversary $\mathcal{S}$ that interacts with $\mathcal{F}_{\text{COM}}^{1:N}$ and makes a black-box simulation of $\mathcal{A}$ running with $\mathcal{F}_{\text{MCRS}}$ and $P_1, \ldots, P_N$. There are two general types of issues that can come up in the ideal process simulation. First, when $\mathcal{F}_{\text{COM}}^{1:N}$ tells $\mathcal{S}$ that a party has committed to some message, $\mathcal{S}$ does not know

---

Functionality $\mathcal{F}_{\text{COM}}^{1:N}$

Parameterized by polynomial ℓ, and running with parties $P_1, \ldots, P_N$ and adversary $\mathcal{S}$.

Commitment: On input (**commit**, sid, m) from party P_i check that $m \in \{0,1\}^{\ell(k)}$ and in that case store (sid, P_i, m) and send (**commit**, sid, P_i) to all parties and $\mathcal{S}$. Ignore future (**commit**, $sid, \cdot$) inputs from P_i.

Opening: On input (**open**, sid) from P_i check that (sid, P_i, m) has been stored, and in that case send (**open**, sid, P_i, m) to all parties and $\mathcal{S}$.

Fig. 2. The ideal commitment functionality

which message it is, however, $\mathcal{S}$ has to simulate to $\mathcal{A}$ that this party makes a UC commitment. Therefore, we want to be able to make trapdoor commitments and later open them to any value. Second, when a corrupt party controlled by $\mathcal{A}$ sends a UC commitment, then $\mathcal{S}$ needs to input some message to $\mathcal{F}_{\text{COM}}^{1:N}$. In this case, we therefore need to extract the message from the UC commitment.

As a tool to get both the trapdoor/simulation property and at the same time the extractability property, we will use a tag-based simulation-extractable commitment. Informally, a tag-based simulation-extractable commitment scheme, is a non-interactive commitment scheme that takes as input an arbitrary tag, a message and a randomizer (tag, m, r) and outputs a commitment c. The commitment can be opened for tag simply by revealing m, r. The commitment must be a trapdoor commitment: given a simulation trapdoor we can construct commitments for an arbitrary tag that can be opened to any value we desire. At the same time it must be extractable: given an extraction key we can extract the message from any commitment that uses a tag tag that has not been used in a trapdoor commitment. In addition, we will need that the public key for the tag-based simulation-extractable commitment scheme is pseudorandom such that we can set it up in the common random strings model. Tag-based simulation-extractable commitments are formally defined in the full paper [GO07] where we also give a construction.

Our idea in constructing a UC commitment is to use each of the n common random strings output by $\mathcal{F}_{\text{MCRS}}$ as a public key for a tag-based simulation-extractable commitment scheme. This gives us a set of n commitment schemes, of which at least $t = \lceil \frac{n+1}{2} \rceil$ are secure. Without loss of generality, we will from now on assume we have exactly t secure commitment schemes. In the ideal process, the ideal process adversary simulates $\mathcal{F}_{\text{MCRS}}$ and can therefore pick the strings as simulation-extractable public keys where it knows both the simulation trapdoors and the extraction keys.

To commit to a message m, a party makes a (t, n)-threshold secret sharing of it and commits to the n secret share using the n public keys specified by the random strings. When making a trapdoor commitment, $\mathcal{S}$ makes honest commitments to $n - t$ random shares for the adversarial keys, and trapdoor commitments with the t simulation-extractable keys. Since the adversary knows at most $n - t < t$ shares, we can later open the commitment to any message we want by making suitable trapdoor openings of the latter t shares. To extract a message m from a UC commitment made by the adversary, we extract t shares from the simulation-extractable commitments. We can now combine the shares to get the adversarial message.

One remaining issue is when the adversary recycles a commitment or parts of it. This way, we may risk that it uses a trapdoor commitment made by an honest party, in which case we are unable to extract a message. To guard against this problem, we will let the tag for the simulation-extractable commitment scheme contain the identity of the sender P_i, forcing the adversary to use a different tag, which in turn enables us to extract.

Another problem arises when the adversary corrupts a party, which enables it to send messages on behalf of this party. At this point, however, we learn the message so we just need to force it to reuse the same message if it reuses parts of the trapdoor commitment. We therefore introduce a second commitment scheme, which will be a standard trapdoor commitment scheme, and use this trapdoor commitment scheme to commit to the shares of the message. The tag for the simulation-extractable commitment will include this trapdoor commitment. Therefore, if reusing a tag, the adversary must also reuse the same trapdoor commitment given by this tag, which in turn computationally binds him to use the same share as the one the party committed to before being corrupted.

These ideas give us a UC commitment scheme in the multi-string model. As an additional bonus, the protocol is non-interactive except for a little coordination to ensure that everybody received the same commitment.

Commitment: On input (**vector**, sid, $(ck_1, \sigma_1), \ldots, (ck_n, \sigma_n)$) from $\mathcal{F}_{\mathrm{MCRS}}$ and (**commit**, sid, m) from $\mathcal{Z}$, the party P_i does the following. He makes a (t, n)-threshold secret sharing $s_1, \ldots, s_n$ of m. He picks randomizers r_j and makes commitments $c_j := \mathrm{Com}_{ck_j}(s_j; r_j)$. He also picks randomizers R_j and makes tag-based commitments $C_j := \mathrm{Com}_{\sigma_j}((P_i, c_j); s_j; R_j)$. The commitment is $c := (c_1, C_1, \ldots, c_n, C_n)$. He broadcasts (**broadcast**, sid, c).

Receiving commitment: A party on input (**vector**, sid, $(ck_1, \sigma_1), \ldots, (ck_n, \sigma_n)$) from $\mathcal{F}_{\mathrm{MCRS}}$ and (**broadcast**, sid, P_i, c) from $\mathcal{F}_{\mathrm{BC}}$ broadcasts (**broadcast**, sid, P_i, c).

Once it receives similar broadcasts from all parties, all containing the same P_i, c, it outputs (**commit**, sid, P_i) to the environment.

Opening commitment: Party P_i wishing to open the commitment broadcasts (**open**, sid, $s_1, r_1, R_1, \ldots, s_n, r_n, R_n$).

Receiving opening: A party receiving (**open**, sid, P_i, s_1, , r_1, R_1, $\ldots$, s_n, r_n, R_n) from $\mathcal{F}_{\mathrm{BC}}$ to a commitment it earlier received, checks that all commitments are correctly formed $c_j = \mathrm{Com}_{ck_j}(s_j; r_j)$ and $C_j = \mathrm{Com}_{\sigma_j}((P_i, c_j); s_j; r_j)$. It also checks that $s_1, \ldots, s_n$ all are valid shares of a (t, n)-threshold secret sharing of some message m. In that case it outputs (**open**, sid, P_i, m).

Theorem 5. *The protocol securely realizes $\mathcal{F}_{\mathrm{COM}}^{1:N}$ in the $(\mathcal{F}_{\mathrm{BC}}, \mathcal{F}_{\mathrm{MCRS}})$-hybrid model, assuming tag-based simulation-extractable commitment schemes with pseudorandom keys exist in the common random string model.*

See the proof in the full paper [GO07].

6   Multi-party Computation

COIN-FLIPPING. A nice application of UC commitment is coin-flipping. In a coin-flipping protocol the parties generate a series of uniformly random bits. In other words, all the protocols we have in the CRS-model can be securely realized if we can do coin-flipping.

We will now show how to generate a common random string on the fly. The parties will use the following natural coin-flipping protocol.

Commitment: P_i chooses at random $r_i \leftarrow \{0,1\}^{\ell(k)}$. It submits $(\textbf{commit}, sid, r_i)$ to $\mathcal{F}_{\text{COM}}^{1:N}$. $\mathcal{F}_{\text{COM}}^{1:N}$ on this input sends $(\textbf{commit}, sid, P_i)$ to all parties.

Opening: Once P_i sees $(\textbf{commit}, sid, P_j)$ for all j, it sends $(\textbf{open}, sid, r_i)$ to $\mathcal{F}_{\text{COM}}^{1:N}$. $\mathcal{F}_{\text{COM}}^{1:N}$ on this input sends $(\textbf{open}, sid, P_i, r_i)$ to all parties.

Output: Once P_i sees $(\textbf{commit}, sid, P_j, r_j)$ for all j, it outputs $(\textbf{crs}, sid, \oplus_{j=1}^{N} r_j)$ and halts.

Theorem 6. *The protocol securely realizes (perfectly) the ideal common reference string generator* $\mathcal{F}_{\text{CRS}}$ *in the* $\mathcal{F}_{\text{COM}}^{1:N}$*-hybrid model.*

MULTI-PARTY COMPUTATION. Armed with a coin-flipping protocol, we can generate random strings. Canetti, Lindell, Ostrovsky and Sahai [CLOS02] demonstrated that with access to a common random string, it is possible to do any kind of multi-party computation, even if only a minority of the parties is honest. We therefore get the following corollary to Theorems 5 and 6, which we prove in the full paper [GO07].

Theorem 7. *For any well-formed functionality* $\mathcal{F}$ *there is a non-trivial protocol that securely realizes it in the* $(\mathcal{F}_{\text{BC}}, \mathcal{F}_{\text{MCRS}})$*-hybrid model, provided enhanced trapdoor permutations with dense public keys and augmented non-committing encryption exists.*

Acknowledgments

We thank Silvio Micali and Eyal Kushilevitz for an inspiring discussion in February of 2004 that motivated us to explore this setting.

References

[Adl78]   Adleman, L.M.: Two theorems on random polynomial time. In: proceedings of FOCS '78, pp. 75–83 (1978)
[BBS04]   Boneh, D., Boyen, X., Shacham, H.: Short group signatures. In: Franklin, M. (ed.) CRYPTO 2004. LNCS, vol. 3152, pp. 41–55. Springer, Heidelberg (2004)
[BCNP04]   Barak, B., Canetti, R., Nielsen, J.B., Pass, R.: Universally composable protocols with relaxed set-up assumptions. In: proceedings of FOCS '04, pp. 186–195 (2004)
[BDMP91]   Blum, M., De Santis, A., Micali, S., Persiano, G.: Noninteractive zero-knowledge. SIAM Jornal of Computation 20(6), 1084–1118 (1991)
[BF03]   Boneh, D., Franklin, M.K.: Identity-based encryption from the weil pairing. SIAM Journal of Computing 32(3), 586–615 (2003)
[BFM88]   Blum, M., Feldman, P., Micali, S.: Non-interactive zero-knowledge and its applications. In: proceedings of STOC '88, pp. 103–112 (1988)
[Can01]   Canetti, R.: Universally composable security: A new paradigm for cryptographic protocols. In: proceedings of FOCS '01, pp. 136–145 (2001) Full paper available at, http://eprint.iacr.org/2000/067

[CF01]   Canetti, R., Fischlin, M.: Universally composable commitments. In: Kilian, J. (ed.)
            CRYPTO 2001. LNCS, vol. 2139, pp. 19–40. Springer, Heidelberg (2001), Full
            paper available at http://eprint.iacr.org/2001/055
[CLOS02]   Canetti, R., Lindell, Y., Ostrovsky, R., Sahai, A.: Universally composable two-party
            and multi-party secure computation. In: proceedings of STOC '02, pp. 494–503
            (2002), Full paper available at, http://eprint.iacr.org/2002/140
[Dam92]   Damgård, I.: Non-interactive circuit based proofs and non-interactive perfect
            zero-knowledge with proprocessing. In: Rueppel, R.A. (ed.) EUROCRYPT 1992.
            LNCS, vol. 658, pp. 341–355. Springer, Heidelberg (1993)
[DDO+02]   De Santis, A., Di Crescenzo, G., Ostrovsky, R., Persiano, G., Sahai, A.: Robust non-
            interactive zero knowledge. In: Kilian, J. (ed.) CRYPTO 2001. LNCS, vol. 2139,
            pp. 566–598. Springer, Heidelberg (2001)
[DDP99]   De Santis, A., Di Crescenzo, G., Persiano, G.: Non-interactive zero-knowledge: A
            low-randomness characterization of np. In: Wiedermann, J., van Emde Boas, P.,
            Nielsen, M. (eds.) ICALP 1999. LNCS, vol. 1644, pp. 271–280. Springer, Heidel-
            berg (1999)
[DDP02]   De Santis, A., Di Crescenzo, G., Persiano, G.: Randomness-optimal characteriza-
            tion of two np proof systems. In: Rolim, J.D.P., Vadhan, S.P. (eds.) RANDOM
            2002. LNCS, vol. 2483, pp. 179–193. Springer, Heidelberg (2002)
[DN02]   Damgård, I., Nielsen, J.B.: Perfect hiding and perfect binding universally com-
            posable commitment schemes with constant expansion factor. In: Yung, M. (ed.)
            CRYPTO 2002. LNCS, vol. 2442, pp. 581–596. Springer, Heidelberg (2002)
[DP92]   De Santis, A., Persiano, G.: Zero-knowledge proofs of knowledge without interac-
            tion. In: proceedings of FOCS '92, pp. 427–436 (1992)
[FLS99]   Feige, U., Lapidot, D., Shamir, A.: Multiple non-interactive zero knowledge proofs
            under general assumptions. SIAM Journal of Computing 29(1), 1–28 (1999)
[GMR89]   Goldwasser, S., Micali, S., Rackoff, C.: The knowledge complexity of interac-
            tive proofs. SIAM Journal of Computing 18(1), 186–208 (1985) First Published
            at STOC 1985
[GMW87]   Goldreich, O., Micali, S., Wigderson, A.: How to play ANY mental game, or A
            completeness theorem for protocols with honest majority. In: proceedings of STOC
            '87, pp. 218–229 (1987)
[GO94]   Goldreich, O., Oren, Y.: Definitions and properties of zero-knowledge proof sys-
            tems. Journal of Cryptology 7(1), 1–32 (1994)
[GO07]   Groth, J., Ostrovsky, R.: Cryptography in the multi-string model. In: Menezes, A.
            (ed.) CRYPTO 2007. LNCS, vol. 4622, Springer, Heidelberg (2007) Full paper avail-
            able at http://www.cs.ucla.edu/~rafail/PUBLIC/index.html
[GOS06a]   Groth, J., Ostrovsky, R., Sahai, A.: Non-interactive zaps and new techniques for
            nizk. In: Dwork, C. (ed.) CRYPTO 2006. LNCS, vol. 4117, pp. 97–111. Springer,
            Heidelberg (2006)
[GOS06b]   Groth, J., Ostrovsky, R., Sahai, A.: Perfect non-interactive zero-knowledge for
            np. In: Vaudenay, S. (ed.) EUROCRYPT 2006. LNCS, vol. 4004, pp. 339–358.
            Springer, Heidelberg (2006)
[GP90]   Granville, A., Pomerance, C.: On the Least Prime in Certain Arithmetic Progres-
            sions. Journal of the London Mathematical Society s2-41(2), 193–200 (1990)
[Gro06]   Groth, J.: Simulation-sound nizk proofs for a practical language and con-
            stant size group signatures. In: Lai, X., Chen, K. (eds.) ASIACRYPT
            2006. LNCS, vol. 4284, Springer, Heidelberg (2006), Full Paper available at
            http://www.brics.dk/~jg/NIZKGroupSignFull.pdf

[KP98]   Kilian, J., Petrank, E.: An efficient noninteractive zero-knowledge proof system for np with general assumptions. Journal of Cryptology 11(1), 1–27 (1998)

[PPS06]   Padney, O., Prabhakaran, M., Sahai, A.: personal communication (November 2006)

[RCW07]   Canetti, R.P.R., Dodis, Y., Walfish, S.: Universally composable security with pre-existing setup. In: Vadhan, S.P. (ed.) TCC 2007. LNCS, vol. 4392, pp. 61–85. Springer, Heidelberg (2007), http://eprint.iacr.org/2006/432

[Sah01]   Sahai, A.: Non-malleable non-interactive zero-knowledge and adaptive chosen-ciphertext security. In: proceedings of FOCS '01, pp. 543–553 (2001)

Secure Identification and QKD in the Bounded-Quantum-Storage Model

Ivan B. Damgård[1], Serge Fehr[2,*], Louis Salvail[1,**],
and Christian Schaffner[2,***]

[1] BRICS[†], FICS, Aarhus University, Denmark
{ivan,salvail}@brics.dk
[2] CWI[‡] Amsterdam, The Netherlands
{fehr,c.schaffner}@cwi.nl

Abstract. We consider the problem of secure identification: user U proves to server S that he knows an agreed (possibly low-entropy) password w, while giving away as little information on w as possible, namely the adversary can exclude at most one possible password for each execution of the scheme. We propose a solution in the bounded-quantum-storage model, where U and S may exchange qubits, and a dishonest party is assumed to have limited quantum memory. No other restriction is posed upon the adversary. An improved version of the proposed identification scheme is also secure against a man-in-the-middle attack, but requires U and S to additionally share a high-entropy key k. However, security is still guaranteed if one party loses k to the attacker but notices the loss. In both versions of the scheme, the honest participants need no quantum memory, and noise and imperfect quantum sources can be tolerated. The schemes compose sequentially, and w and k can securely be re-used. A small modification to the identification scheme results in a quantum-key-distribution (QKD) scheme, secure in the bounded-quantum-storage model, with the same re-usability properties of the keys, and without assuming authenticated channels. This is in sharp contrast to known QKD schemes (with unbounded adversary) without authenticated channels, where authentication keys must be updated, and unsuccessful executions can cause the parties to run out of keys.

1 Introduction

SECURE IDENTIFICATION. Consider two parties, a *user* U and a *server* S, which share a common secret-key (or password or Personal Identification Number

* Supported by the Dutch Organization for Scientific Research (NWO).
** QUSEP, Quantum Security in Practice, funded by the Danish Natural Science Research Council.
*** Supported by the European project SECOQC.
† Basic Research in Computer Science (www.brics.dk), and Foundations in Cryptography and Security, funded by the Danish Natural Sciences Research Council.
‡ Centrum voor Wiskunde en Informatica, the national research institute for mathematics and computer science in the Netherlands.

A. Menezes (Ed.): CRYPTO 2007, LNCS 4622, pp. 342–359, 2007.

PIN) w. In order to obtain some service from S, U needs to convince S that he is the legitimate user U by "proving" that he knows w. In practice—think of how you prove to the ATM that you know your PIN—such a proof is often done simply by announcing w to S. This indeed guarantees that a dishonest user U* who does not know w cannot identify himself as U, but of course incurs the risk that U might reveal w to a malicious server S* who may now impersonate U. Thus, from a secure identification scheme we also require that a dishonest server S* obtains (essentially) no information on w.

There exist various approaches to obtain secure identification schemes, depending on the setting and the exact security requirements. For instance zero-knowledge proofs (and some weaker versions), as initiated by Feige, Fiat and Shamir[12,11], allow for secure identification. In a more sophisticated model, where we allow the common key w to be of low entropy and additionally consider a man-in-the-middle attack, we can use techniques from password-based key-agreement (like [14,13]) to obtain secure identification schemes. Common to these approaches is that security relies on the assumption that some computational problem (like factoring or computing discrete logs) is hard and that the attacker has limited computing power.

OUR CONTRIBUTION. In this work, we take a new approach: we consider quantum communication, and we develop two identification schemes which are information-theoretically secure under the *sole* assumption that the attacker can only reliably store quantum states of limited size. This model was first considered in [4]. On the other hand, the honest participants only need to send qubits and measure them immediately upon arrival, no quantum storage or quantum computation is required. Furthermore, our identification schemes are robust to both noisy quantum channels and imperfect quantum sources. Our schemes can therefore be implemented in practice using off-the-shelf technology.

The first scheme is secure against dishonest users and servers but not against a man-in-the-middle attack. It allows the common secret-key w to be non-uniform and of low entropy, like a human-memorizable password. Only a user knowing w can succeed in convincing the server. In any execution of this scheme, a dishonest user or server cannot learn more on w than excluding one possibility, which is unavoidable. This is sometimes referred to as *password-based* identification. The second scheme requires in addition to w a uniformly distributed high-entropy common secret-key k, but is additionally secure against a man-in-the-middle attack. Furthermore, security against a dishonest user or server holds as for the first scheme even if the dishonest party knows k (but not w). This implies that k can for instance be stored on a smartcard, and security of the scheme is still guaranteed even if the smartcard gets stolen, assuming that the affected party notices the theft and thus does not engage in the scheme anymore. Both schemes compose sequentially, and w (and k) may be safely re-used super-polynomially many times, even if the identification fails (due to an attack, or due to a technical failure).

A small modification of the second identification scheme results in a quantum-key-distribution (QKD) scheme secure against bounded-quantum-memory adversaries. The advantage of the proposed new QKD scheme is that no authenticated channel is needed and the attacker can *not* force the parties to run out of authentication keys. The honest parties merely need to share a password w and a high-entropy secret-key k, which they can safely re-use (super-polynomially many times), independent of whether QKD succeeds or fails. Furthermore, like for the identification scheme, losing k does not compromise security as long as the loss is noticed by the corresponding party. One may think of this as a quantum version of password-based authenticated key exchange. The properties of our solution are in sharp contrast to all known QKD schemes without authenticated channels (which do not pose any restrictions on the attacker). In these schemes, an attacker can force parties to run out of authentication keys by making the QKD execution fail (e.g. by blocking some messages). Worse, even if the QKD execution fails only due to technical problems, the parties can still run out of authentication keys after a short while, since they cannot exclude that an eavesdropper was in fact present. This problem is an important drawback of QKD implementations, especially of those susceptible to single (or few) point(s) of failure[9].

OTHER APPROACHES. We briefly discuss how our identification schemes compare with other approaches. We have already given some indication on how to construct *computationally* secure identification schemes. This approach typically allows for very practical schemes, but requires some unproven complexity assumption. Another interesting difference between the two approaches: whereas for (known) computationally-secure password-based identification schemes the underlying computational hardness assumption needs to hold indefinitely, the restriction on the attacker's quantum memory in our approach only needs to hold *during* the execution of the identification scheme, actually only at one single point during the execution. In other words, having a super-quantum-storage-device at home in the basement only helps you cheat at the ATM if you can communicate with it on-line quantumly – in contrast to a computational solution, where an off-line super-computer in the basement can make a crucial difference.

Furthermore, obtaining a satisfactory identification scheme requires *some* restriction on the adversary, even in the quantum setting: considering only passive attacks, Lo[15] showed that for an unrestricted adversary, no password-based quantum identification scheme exists. In fact, Lo's impossibility result only applies if the user U is guaranteed not to learn anything about the outcome of the identification procedure. We can argue, however, that a different impossibility result holds even without Lo's restriction: We first show that secure computation of a classical AND gate (in which both players learn the output) can be reduced to a password-based identification scheme. The reduction works as follows. Let w_0, w_0' and w_1 be three distinct elements from $\mathcal{W}$. If Alice has private input $x_A = 0$ then she sets $w_A = w_0$ and if $x_A = 1$ then she sets $w_A = w_1$, and if Bob has private input $x_B = 0$ then he sets $w_B = w_0'$ and if $x_B = 1$ then he sets $w_B = w_1$. Then, Alice and Bob run the identification scheme on inputs

w_A and w_B, and if the identification is rejected, the output is set to 0 while if it is accepted, the output is set to 1. Security of the identification scheme is easily seen to imply security of the AND computation. Now, the secure computation of an AND gate—with statistical security and using quantum communication— can be shown to require a superpolynomial number of rounds if the adversary is unbounded[18]. Therefore, the same must hold for a secure password-based identification scheme.[1]

Another alternative approach is the classical bounded-storage model[17,2,1]. In contrast to our approach, only classical communication is used, and it is assumed that the attacker's *classical* memory is bounded. Unlike in the quantum case where we do not need to require the honest players to have any quantum memory, the classical bounded-storage model requires honest parties to have a certain amount of memory which is related to the allowed memory size of the adversary: if two legitimate users need n bits of memory in an identification protocol meeting our security criterion, then an adversary must be bounded in memory to $O(n^2)$ bits. The reason is that given a secure password-based identification scheme, one can construct (in a black-box manner) a key-distribution scheme that produces a one-bit key on which the adversary has an (average) entropy of $\frac{1}{2}$. On the other hand it is known that in any key-distribution scheme which requires n bits of memory for legitimate players, an adversary with memory $\Omega(n^2)$ can obtain the key except for an arbitrarily small amount of remaining entropy[8]. It follows that password-based identification schemes in the classical bounded-storage model can only be secure against adversaries with memory at most $O(n^2)$. This holds even for identification schemes with only passive security and without security against man-in-the-middle attacks. Roughly, the reduction works as follows. Alice and Bob agree on a public set of two keys $\{w_0, w_1\}$. Alice picks $a \in_R \{0, 1\}$, Bob picks $b \in_R \{0, 1\}$, and they run the identification scheme with keys w_a and w_b respectively. The outcome of the identification is then made public from which Bob determines a. We argue that if the identification fails, i.e. $a \neq b$, then a is a secure bit. Thus, on average, a has entropy (close to) $\frac{1}{2}$ from an eavesdropper's point of view. Consider $w' \notin \{w_0, w_1\}$. By the security property of the identification scheme, Alice and thus also a passive eavesdropper Eve cannot distinguish between Bob having used w_b or w'. Similarly, we can then switch Alice's key w_a to w_{1-a} and Bob's switched key w' to w_{1-b} without changing Eve's view. Thus, Eve cannot distinguish an execution with $a = 0$ from one with $a = 1$ if $a \neq b$.

This limitation of the classical bounded-storage model is in sharp contrast with what we achieve in this paper, the honest players need no quantum memory at all while our identification scheme remains secure against adversaries with quantum memory linear in the total number of qubits sent. The same separation between the two models was shown for OT and bit commitment[4,3].

---

[1] In fact, we believe that the proof from [18] can be extended to cover secure computation of equality of strings, which is equivalent to password-based identification. This would mean that we could prove the impossibility result directly, without the detour via a secure AND computation. Details are omitted due to the space limitation.

Finally, if one settles for the bounded-quantum-storage model, then in principle one could take a generic construction for general two-party secure-function-evaluation (SFE) based on OT together with the OT scheme from [4,3] in order to implement a SFE for string equality and thus password-based identification. However, this approach leads to a highly impractical solution, as the generic construction requires many executions of OT, whereas our solution is comparable with *one* execution of the OT scheme from [4,3]. Furthermore, SFE does not automatically take care of a man-in-the-middle attack, thus additional work would need to be done using this approach.

2 Preliminaries

2.1 Notation and Terminology

QUANTUM STATES. The state of a *qubit* can be described by a vector in the 2-dimensional Hilbert space $\mathbb{C}^2$ in case of a *pure* state, and by a density matrix/operator on $\mathbb{C}^2$ in the general case of a *mixed* state. Similarly, an *n-qubit state* is characterized by a vector in the n-fold tensor product $(\mathbb{C}^2)^{\otimes n}$ in case of a *pure* n-qubit state, and by a density matrix/operator on $(\mathbb{C}^2)^{\otimes n}$ in case of a *mixed* n-qubit state. The pair $\{|0\rangle, |1\rangle\}$ denotes the standard basis, also known as computational or rectilinear or "+"-basis, for $\mathbb{C}^2$. When the context requires, we also write $|0\rangle_+$ and $|1\rangle_+$ instead of $|0\rangle$ respectively $|1\rangle$. The diagonal or "×"-basis is defined as $\{|0\rangle_\times, |1\rangle_\times\}$ where $|0\rangle_\times = (|0\rangle + |1\rangle)/\sqrt{2}$ and $|1\rangle_\times = (|0\rangle - |1\rangle)/\sqrt{2}$. Measuring a qubit in the $+$-basis (resp. ×-basis) means applying the measurement described by projectors $|0\rangle\langle 0|$ and $|1\rangle\langle 1|$ (resp. projectors $|0\rangle_\times\langle 0|_\times$ and $|1\rangle_\times\langle 1|_\times$). The notation generalizes to n-qubit states: For $x = (x_1, \ldots, x_n) \in \{0,1\}^n$ and $\theta = (\theta_1, \ldots, \theta_n) \in \{+, \times\}^n$, we let $|x\rangle_\theta$ be the n-qubit state $|x\rangle_\theta = |x_1\rangle_{\theta_1} \cdots |x_n\rangle_{\theta_n}$; and measuring a n-qubit state in basis $\theta \in \{+, \times\}^n$ means applying the measurement described by projections $|x\rangle_\theta\langle x|_\theta$ with $x \in \{0,1\}^n$.

The behavior of a (mixed) quantum state in a register E is fully described by its density matrix ρ_E. In order to simplify language, we tend to be a bit sloppy and use E as well as ρ_E as "naming" for the quantum state. We often consider cases where a quantum state E may depend on some classical random variable X in that the state is described by the density matrix ρ_E^x if and only if $X = x$. For an observer who has only access to the state E but not to X, the behavior of the state is determined by the density matrix $\rho_E := \sum_x P_X(x)\rho_E^x$, whereas the joint state, consisting of the classical X and the quantum state E, is described by the density matrix $\rho_{XE} := \sum_x P_X(x)|x\rangle\langle x| \otimes \rho_E^x$, where we understand $\{|x\rangle\}_{x \in \mathcal{X}}$ to be the standard (orthonormal) basis of $\mathbb{C}^{|\mathcal{X}|}$. More general, for any event $\mathcal{E}$ (defined by $P_{\mathcal{E}|X}(x) = P[\mathcal{E}|X=x]$ for all x), we write

$$\rho_{XE|\mathcal{E}} := \sum_x P_{X|\mathcal{E}}(x)|x\rangle\langle x| \otimes \rho_E^x \quad \text{and} \quad \rho_{E|\mathcal{E}} := \mathrm{tr}_X(\rho_{XE|\mathcal{E}}) = \sum_x P_{X|\mathcal{E}}(x)\rho_E^x .$$

We also write $\rho_X := \sum_x P_X(x)|x\rangle\langle x|$ for the quantum representation of the classical random variable X (and similarly for $\rho_{X|\mathcal{E}}$). This notation extends naturally to quantum states that depend on several classical random variables. Given X and E as above, by saying that there exists a random variable Y such that ρ_{XYE} satisfies some condition, we mean that ρ_{XE} can be understood as $\rho_{XE} = \operatorname{tr}_Y(\rho_{XYE})$ for some ρ_{XYE} (with classical Y) and that ρ_{XYE} satisfies the required condition.

X is independent of E (in that ρ_E^x does not depend on x) if and only if $\rho_{XE} = \rho_X \otimes \rho_E$, which in particular implies that no information on X can be learned by observing only E. Similarly, X is random and independent of E if and only if $\rho_{XE} = \frac{1}{|\mathcal{X}|}\mathbb{I} \otimes \rho_E$, where $\frac{1}{|\mathcal{X}|}\mathbb{I}$ is the density matrix of the fully mixed state of suitable dimension. Finally, if two states like ρ_{XE} and $\rho_X \otimes \rho_E$ are ε-close in terms of their trace distance $\delta(\rho, \sigma) = \frac{1}{2}\operatorname{tr}(|\rho - \sigma|)$, which we write as $\rho_{XE} \approx_\varepsilon \rho_X \otimes \rho_E$, then the real system ρ_{XE} "behaves" as the ideal system $\rho_X \otimes \rho_E$ except with probability ε in that for any evolution of the system no observer can distinguish the real from the ideal one with advantage greater than ε [20].

We also need to express that a random variable X is (close to) independent of a quantum state E *when given a random variable* Y. This means that when given Y, the state E gives no (or little) additional information on X. Formally, this is expressed by requiring that ρ_{XYE} is of the form (or close to)

$$\rho_{XYE} = \sum_{x,y} P_{XY}(x,y)|x\rangle\langle x| \otimes |y\rangle\langle y| \otimes \rho_E^y \ ,$$

where $\rho_E^y = \sum_x P_{X|Y=y}(x)\rho_E^{x,y}$ for all y. As shorthand for the right-hand side above, we define $\rho_{X\leftrightarrow Y\leftrightarrow E} := \sum_{x,y} P_{XY}(x,y)|x\rangle\langle x| \otimes |y\rangle\langle y| \otimes \rho_E^y$.[2] To further illustrate its meaning, notice that if the Y-register is measured and value y is obtained, the state $\rho_{X\leftrightarrow Y\leftrightarrow E}$ collapses to $\left(\sum_x P_{X|Y=y}(x)|x\rangle\langle x|\right) \otimes \rho_E^y$, so that indeed no further information on x can be obtained from the E-register. This notation naturally extends to $\rho_{X\leftrightarrow Y\leftrightarrow E|\mathcal{E}}$ simply by considering $\rho_{XYE|\mathcal{E}}$.

(CONDITIONAL) SMOOTH MIN-ENTROPY. We briefly recall the notion of (conditional) *smooth* min-entropy[19,21]. For more details, we refer to the aforementioned literature. Let X be a random variable over alphabet $\mathcal{X}$ with distribution P_X. The notion of min-entropy is given by $H_\infty(X) = -\log\left(\max_x P_X(x)\right)$. More general, for any event $\mathcal{E}$, $H_\infty(X\mathcal{E})$ may be defined similarly simply by replacing P_X by $P_{X\mathcal{E}}$. Note that the "distribution" $P_{X\mathcal{E}}$ is not normalized; $H_\infty(X\mathcal{E})$ is still well defined, though. For an arbitrary $\varepsilon \geq 0$, the smooth version $H_\infty^\varepsilon(X)$ is defined as follows. $H_\infty^\varepsilon(X)$ is the *maximum* of the standard min-entropy $H_\infty(X\mathcal{E})$, where the maximum is taken over all events $\mathcal{E}$ with $\Pr(\mathcal{E}) \geq 1 - \varepsilon$. As ε can be interpreted as an error probability, we typically require ε to be negligible in the security parameter n, denoted as $\varepsilon = negl(n)$.

For a pair of random variables X and Y, the *conditional* smooth min-entropy $H_\infty^\varepsilon(X|Y)$ is defined as $H_\infty^\varepsilon(X|Y) = \max_\mathcal{E} \min_y H_\infty(X\mathcal{E}|Y=y)$, where the quantification over $\mathcal{E}$ is over all events $\mathcal{E}$ (defined by $P_{\mathcal{E}|XY}$) with $\Pr(\mathcal{E}) \geq 1 - \varepsilon$. The

---

[2] The notation is inspired by the classical setting where the corresponding independence of X and Z given Y can be expressed by saying that $X \leftrightarrow Y \leftrightarrow Z$ forms a Markov chain.

following lemma shows that for a small ε, smooth min-entropy is essentially as good as ordinary min-entropy; the proof is given in the full version[5].

Lemma 2.1. *If $H_\infty^\varepsilon(X|Y) = r$ then there exists an event $\mathcal{E}'$ such that $P[\mathcal{E}'] \geq 1 - 2\varepsilon$ and $H_\infty(X|\mathcal{E}', Y=y) \geq r - 1$ for every y with $P_{Y\mathcal{E}'}(y) > 0$.*

2.2 Tools

A NEW MIN-ENTROPY-SPLITTING LEMMA. A technical tool, which will come in handy, is the following new entropy-splitting lemma, which may be of independent interest. Informally, it says that if for a list of random variables, every pair has high (smooth) min-entropy, then all of the random variables except one must have high (smooth) min-entropy. The version given here follows immediately from the version given and proven in the full version[5].

Lemma 2.2 (Entropy-Splitting Lemma). *Let $\varepsilon > 0$. Let $X_1, \ldots, X_m$ be a sequence of random variables over $\mathcal{X}_1, \ldots, \mathcal{X}_m$ such that $H_\infty^\varepsilon(X_i X_j) \geq \alpha$ for all $i \neq j$. Then there exists a random variable V over $\{1, \ldots, m\}$ such that for any independent random variable W over $\{1, \ldots, m\}$*

$$H_\infty^{2\sqrt{\varepsilon}}(X_W | VW, V \neq W) \geq \alpha/2 - \log(m) - \log(1/\varepsilon) \ .$$

QUANTUM UNCERTAINTY RELATION. At the very core of our security proofs lies (a special case of) the quantum uncertainty relation from [3], that lower bounds the (smooth) min-entropy of the outcome when measuring an arbitrary n-qubit state in a random basis $\theta \in \{0,1\}^n$.

Theorem 2.3 (Uncertainty Relation[3]). *Let E be an arbitrary fixed n-qubit state. Let Θ be uniformly distributed over $\{+, \times\}^n$ (independent of E), and let $X \in \{0,1\}^n$ be the random variable for the outcome of measuring E in basis Θ. Then, for any $\lambda > 0$, the conditional smooth min-entropy is lower bounded by $H_\infty^\varepsilon(X|\Theta) \geq \left(\frac{1}{2} - \lambda\right)n$ with $\varepsilon = \mathrm{negl}(n)$.*

Thus, ignoring negligibly small "error probabilities" and linear fractions that can be chosen arbitrarily small, the outcome of measuring any n-qubit state in a random basis has $n/2$ bits of min-entropy, given the basis.

PRIVACY AMPLIFICATION. Finally, we recall the quantum-privacy-amplification theorem of Renner and König[20]. We give the simplified version as used in [4]. Recall that a class $\mathcal{F}$ of hash functions from $\mathcal{X}$ to $\mathcal{Y}$ is called (strongly) universal-2 if for any $x \neq x' \in \mathcal{X}$, and for F uniformly distributed over $\mathcal{F}$, the collision probability $P[F(x) = F(x')]$ is upper bounded by $1/|\mathcal{Y}|$, respectively, for the strong notion, the random variables $F(x)$ and $F(x')$ are uniformly and independently distributed over $\mathcal{Y}$.

Theorem 2.4 (Privacy Amplification[20,4]). *Let X be a random variable distributed over $\{0,1\}^n$, and let E be a q-qubit state that may depend on X. Let F be the random and independent choice of a member of a universal-2 class of hash functions $\mathcal{F}$ from $\{0,1\}^n$ into $\{0,1\}^\ell$. Then*

$$\delta\left(\rho_{F(X)FE}, \frac{1}{2^\ell}\mathbb{I} \otimes \rho_{FE}\right) \leq \frac{1}{2} 2^{-\frac{1}{2}\left(H_\infty(X) - q - \ell\right)} \ .$$

3   The Identification Scheme

3.1   The Setting

We assume the honest user U and the honest server S to share some key $w \in \mathcal{W}$. We do not require $\mathcal{W}$ to be very large (i.e. $|\mathcal{W}|$ may not be lower bounded by the security parameter in any way), and w does not necessarily have to be uniformly distributed in $\mathcal{W}$. So, we may think of w as a human-memorizable password or PIN code. The goal of this section is to construct an identification scheme that allows U to "prove" to S that he knows w. The scheme should have the following security properties: a dishonest server S* learns essentially no information on w beyond that he can come up with a guess w' for w and learns whether $w' = w$ or not, and similarly a dishonest user succeeds in convincing the verifier essentially only if he guesses w correctly, and if his guess is incorrect then the only thing he learns is that his guess is incorrect. This in particular implies that as long as there is enough entropy in w, the identification scheme may be safely repeated.

3.2   The Intuition

The scheme we propose is related to the (randomized) 1-2 OT scheme of [3]. In that scheme, Alice sends $|x\rangle_\theta$ to Bob, for random $x \in \{0, 1\}^n$ and $\theta \in \{+, \times\}^n$. Bob then measures everything in basis $+$ or $\times$, depending on his choice bit c, so that he essentially knows half of x (where Alice used the same basis as Bob) and has no information on the other half (where Alice used the other basis), though, at this point, he does not know yet which bits he knows and which ones he does not. Then, Alice sends θ and two hash functions to Bob, and outputs the hash values s_0 and s_1 of the two parts of x, whereas Bob outputs the hash value s_c that he is able to compute from the part of x he knows. It is proven in [3] that no dishonest Alice can learn c, and for any quantum-memory-bounded dishonest Bob, at least one of the two strings s_0 and s_1 is random for Bob.

This scheme can be extended by giving Bob more options for measuring the quantum state. Instead of measuring all qubits in the $+$ or the $\times$ basis, he may measure using m different strings of bases, where any two possible basis-strings have large Hamming distance. Then Alice computes and outputs m hash values, one for each possible basis-string that Bob might have used. She reveals θ and the hash functions to Bob, so he can compute the hash value corresponding to the basis that he has used, and no other hash value. Intuitively, such an extended scheme leads to a randomized 1-m OT.

The scheme can now be transformed into a secure identification scheme as follows, where we assume (wlog) that $\mathcal{W} = \{1, \ldots, m\}$. The user U, acting as Alice, and the server S, acting as Bob, execute the randomized 1-m OT scheme where S "asks" for the string indexed by his key w, such that U obtains random strings $s_1, \ldots, s_m$ and S obtains s_w. Then, to do the actual identification, U sends s_w to S, who accepts if and only if it coincides with his string s_w. Intuitively, such a construction is secure against a dishonest server since unless he asks for the right string (by guessing w correctly) the string U sends him is random and thus gives no information on w. On the other hand, a dishonest user does not know which of the m strings S asked for and wants to see from him. We realize this intuitive idea in the next section. In the actual protocol, U does not have to explicitly compute all the s_i's, and also we only need a single hash function (to compute s_w). We also take care of some subtleties, for instance that the s_i are not necessarily random if Alice (i.e. the user) is dishonest.

3.3 The Basic Scheme

Let $c : \mathcal{W} \to \{+, \times\}^n$ be the encoding function of a binary code of length n with $m = |\mathcal{W}|$ codewords and minimal distance d. c can be chosen such that n is linear in $\log(m)$ or larger, and d is linear in n. Furthermore, let $\mathcal{F}$ and $\mathcal{G}$ be strongly universal-2 classes of hash functions[3] from $\{0,1\}^n$ to $\{0,1\}^\ell$ and from $\mathcal{W}$ to $\{0,1\}^\ell$, respectively, for some parameter ℓ. For $x \in \{0,1\}^n$ and $I \subseteq \{1,\ldots,n\}$, we define $x|_I \in \{0,1\}^n$ to be the restriction of x to the coordinates x_i with $i \in I$. If $|I| < n$ then applying $f \in \mathcal{F}$ to $x|_I$ is to be understood as applying f to $x|_I$ padded with sufficiently many 0's.

Q-ID:
1. U picks $x \in_R \{0,1\}^n$ and $\theta \in_R \{+, \times\}^n$, and sends state $|x\rangle_\theta$ to S.
2. S measures $|x\rangle_\theta$ in basis $c = c(w)$. Let x' be the outcome.
3. U picks $f \in_R \mathcal{F}$ and sends θ and f to S. Both compute $I_w := \{i : \theta_i = c(w)_i\}$.
4. S picks $g \in_R \mathcal{G}$ and sends g to U.
5. U computes and sends $z := f(x|_{I_w}) \oplus g(w)$ to S.
6. S accepts if and only if $z = z'$ where $z' := f(x'|_{I_w}) \oplus g(w)$.

Proposition 3.1 (User security). *Let the initial state of a dishonest server* S^*, *whose quantum memory at step 3 is bounded by q qubits, be independent of the honest user's key W. Then, the joint state $\rho_{W E_{S^*}}$ after the execution of Q-ID is such that there exists a random variable W' that is independent of W and such that*

$$\rho_{W W' E_{S^*} | W' \neq W} \approx_\varepsilon \rho_{W \leftrightarrow W' \leftrightarrow E_{S^*} | W' \neq W} \, ,$$

where $\varepsilon = negl(d - 4\log(m) - 4q - 4\ell)$.

The proposition guarantees that whatever a dishonest S^* does is essentially as good as trying to guess W by some arbitrary (but independent) W' and learning whether the guess was correct or not, but nothing beyond that. Such a property is obviously the best one can hope for, since S^* may always honestly execute the protocol with a guess for W and observe whether he accepts U.

We would like to point out that the security definition used in Proposition 3.1, and in fact any security claim in this paper, guarantees *sequential composability*, as the output state is guaranteed to have the same independency property as is required from the input state (except if the attacker guesses w).

---

[3] Actually, we only need $\mathcal{G}$ to be *strongly* universal-2.

Proof. For readability, we do not keep track of negligibly small error probabilities and of linear fractions that can be chosen arbitrarily small, but (sometimes) merely give some indication of a small error by using the word "essentially". It is straightforward but rather tedious to keep rigorous track of these errors.

We consider and analyze a *purified* version of *Q-ID* where in step 1, instead of sending $|x\rangle_\theta$ to S^* for a random x, U prepares a fully entangled state $2^{-n/2} \sum_x |x\rangle|x\rangle$ and sends the second register to S^* while keeping the first. Then, in step 3 when the memory bound has applied, he measures his register in the random basis $\theta \in_R \{+, \times\}^n$ in order to obtain x. Standard arguments imply that this purified version produces exactly the same common state, consisting of the classical information on U's side and S^*'s quantum state.

Recall that before step 3 is executed, the memory bound applies to S^*, which means that S^* has to measure all but q of the qubits he holds, which consists of his initial state and his part of the EPR pairs. Before doing the measurement, he may append an ancilla register and apply an arbitrary unitary transform. As a result of S^*'s measurement, S^* gets some outcome y, and the common state collapses to a $(n + q)$-qubit state (which depends on y), where the first n qubits are with U and the remaining q with S^*. The following analysis is for a fixed y, and works no matter what y is.

We use upper case letters W, X, Θ, F, G and Z for the random variables that describe the respective values w, x, θ etc. in an execution of the purified version of *Q-ID*. We write $X_j = X|_{I_j}$ for any j, and we let E'_{S^*} be S^*'s q-qubit state at step 3, after the memory bound has applied. Note that W is independent of X, Θ, F, G and E'_{S^*}.

For $1 \leq i \neq j \leq m$, fix the value of X, and correspondingly of X_i and X_j, at the positions where $\mathfrak{c}(i)$ and $\mathfrak{c}(j)$ coincide, and focus on the remaining (at least) d positions. The uncertainty relation (Theorem 2.3) implies that the restriction of X to these positions has essentially $d/2$ bits of min-entropy given Θ. Since every bit in the restricted X appears in one of X_i and X_j, the pair X_i, X_j also has essentially $d/2$ bits of min-entropy given Θ. Lemma 2.2 implies that there exists W' (called V in Lemma 2.2) such that if $W \neq W'$ then X_W has essentially $d/4 - \log(m)$ bits of min-entropy, given W and W' (and Θ). Privacy amplification then guarantees that $F(X_W)$ is ε'-close to random and independent of F, W, W', Θ and E'_{S^*}, conditioned on $W \neq W'$, where $\varepsilon' = \frac{1}{2} \cdot 2^{-\frac{1}{2}(d/4-\log(m)-q-\ell)}$. It follows that $Z = F(X_W) \oplus G(W)$ is ε'-close to random and independent of F, G, W, W', Θ and E'_{S^*}, conditioned on $W \neq W'$. Formally, we want to upper bound

$$\delta(\rho_{WW'E_{\mathsf{S}^*}|W'\neq W}, \; \rho_{W \leftrightarrow W' \leftrightarrow E_{\mathsf{S}^*}|W'\neq W}) \; .$$

Since the output state E_{S^*} is, without loss of generality, obtained by applying some unitary transform to the set of registers $(Z, F, G, W', \Theta, E'_{\mathsf{S}^*})$, the distance above is equal to $\delta(\rho_{WW'(Z,F,G,\Theta,E'_{\mathsf{S}^*})|W'\neq W}, \; \rho_{W\leftrightarrow W'\leftrightarrow(Z,F,G,\Theta,E'_{\mathsf{S}^*})|W'\neq W})$. We then get:

$$\rho_{WW'(Z,F,G,\Theta,E'_{\mathsf{S}^*})|W'\neq W} \approx_{\varepsilon'} \tfrac{1}{2^t}\mathbb{I} \otimes \rho_{WW'(F,G,\Theta,E'_{\mathsf{S}^*})|W'\neq W}$$

$$= \tfrac{1}{2^t}\mathbb{I} \otimes \rho_{W\leftrightarrow W'\leftrightarrow(F,G,\Theta,E'_{\mathsf{S}^*})|W'\neq W} \approx_{\varepsilon'} \rho_{W\leftrightarrow W'\leftrightarrow(Z,F,G,\Theta,E'_{\mathsf{S}^*})|W'\neq W} \; ,$$

where approximations follow from privacy amplification and the exact equality comes from the independency of W, which, when conditioned on $W' \neq W$, translates to independency given W'. The claim follows, with $\varepsilon = 2\varepsilon'$.   □

Proposition 3.2 (Server security). *Let the initial state of an (unbounded) dishonest user* U^* *be independent of the honest server's key* W, *and let* $H_\infty(W) \geq 1$. *Then, there exists* W', *independent of* W, *such that if* $W \neq W'$ *then* S *accepts with probability at most* $m^2/2^{\ell-1}$, *and the common state* $\rho_{W E_{\mathsf{U}^*}}$ *after the execution of* Q-ID *satisfies*

$$\rho_{WW'E_{\mathsf{U}^*}|W'\neq W} \approx_{m^2/2^{\ell-1}} \rho_{W \leftrightarrow W' \leftrightarrow E_{\mathsf{U}^*}|W'\neq W} \ .$$

The formal proof is given in the full version[5]. The idea is the following. We let U^* execute Q-ID with a server that is *unbounded* in quantum memory. Such a server can obviously obtain x and thus compute $s_j = f(x|_{I_j}) \oplus g(j)$ for all j. Note that s_w is the message z that U^* is required to send in the last step. Now, if the s_j's are all distinct, then z uniquely defines w' such that $z = s_{w'}$, and thus S accepts if and only if $w' = w$, and U^* does not learn anything beyond. The strong universal-2 property of g guarantees that the s_j's are all distinct except with probability $m^2/2^\ell$.

We call an identification scheme ε-*secure against impersonation attacks* if user and sender security are satisfied as in Propositions 3.1 and 3.2. The following holds.

Theorem 3.3. *If* $H_\infty(W) \geq 1$, *then the identification scheme* Q-ID *(with suitable choice of parameters) is* ε-*secure against impersonation attacks for any unbounded user and for any server with quantum memory bound* q, *where* $\varepsilon = \mathrm{negl}(n - 33\log(m) - 11q)$.

Proof. We choose $\ell = \frac{1}{8}(d + 4\log(m) - 4q)$. Then user security holds except with an "error" negligible in $d - 4\log(m) - 4q - 4\ell = d/2 - 6\log(m) - 2q$, and thus negligible in $d - 12\log(m) - 4q$. And server security holds except with an "error" negligible in $\ell - 1 - 2\log(m) = \frac{1}{8}(d - 12\log(m) - 4q) - 1$, and thus negligible in $d - 12\log(m) - 4q$. Using a code c which asymptotically meets the Gilbert-Varshamov bound[22], d may be chosen arbitrarily close to $n \cdot h^{-1}(1 - \log(m)/n)$, where h^{-1} is the inverse function of the binary entropy function $h : p \mapsto -(p \cdot \log(p) + (1 - p) \cdot \log(1 - p))$ restricted to $0 < p \leq \frac{1}{2}$. For this d to be larger than $12\log(m)$, clearly n needs to be larger than $24\log(m)$, so that $h^{-1}(1 - \log(m)/n) > h^{-1}(1 - \frac{1}{24})$ which turns out to be larger than $\frac{4}{11}$. The claim follows by normalizing $\frac{4}{11}n - 12\log(m) - 4q$ for n.   □

3.4   An Error-Tolerant Scheme

We now consider an imperfect quantum channel with "error rate" ϕ. The scheme Q-ID is sensitive to such errors in that they cause $x|_{I_w}$ and $x'|_{I_w}$ to be different and thus an honest server S is likely to reject an honest user U. This problem can be overcome by means of error-correcting techniques: U chooses a linear error-correcting code that allows to correct a ϕ-fraction of errors, and then in step 2, in addition to θ and f, U sends a description of the code and the syndrome s of $x|_{I_w}$ to S; this additional information allows S to recover $x|_{I_w}$ from its noisy version $x'|_{I_w}$ by standard techniques. However, this technique introduces a new problem: the syndrome s of $x|_{I_w}$ may give information on w to a dishonest server. Hence, to circumvent this problem, the code chosen by U must have the additional property that for a dishonest user, who has high min-entropy on $x|_{I_w}$, the syndrome s is (close to) independent of w.

This problem has recently been addressed and solved in the classical setting by Dodis and Smith[7]. They present a family of efficiently decodable linear codes allowing to correct a constant fraction of errors, and where the syndrome of a string is close to uniform if the string has enough min-entropy and the code is chosen at random from the family.[4] It remains to verify that their analysis can be translated to our setting where the adversary may have "quantum information".

Lemma 5 of [7] guarantees that for every $0 < \lambda < 1$ and for an infinite number of n''s there exists a δ-biased (as defined in [7]) family $\mathcal{C} = \{C_j\}_{j \in \mathcal{J}}$ of $[n', k', d']_2$-codes with $\delta < 2^{-\lambda n'/2}$, and which allows to efficiently correct a constant fraction of errors. Theorem 3.2 of [10] (which generalizes Lemma 4 in [7] to the quantum setting) guarantees that if a string Y has t bits of min-entropy then for a randomly chosen code $C_j \in \mathcal{C}$, the syndrome of Y is close to random and independent of j and any q-qubit state that may depend on Y, where the closeness is given by $\delta \cdot 2^{(n'+q-t)/2}$. In our application, $Y = X_W$, $n' \approx n/2$ and $t \approx d/4 - \log(m) - \ell$, where the additional loss of ℓ bits of entropy comes from learning the ℓ-bit string z. Choosing $\lambda = 1 - \frac{t}{2n'}$ gives an ensemble of code families that allow to correct a linear number of errors and the syndrome is ε-close to uniform given the quantum state, where $\varepsilon \leq 2^{-n'/2+t/4} \cdot 2^{(n'+q-t)/2} = 2^{-(t-2q)/4}$, which is exponentially small provided that there is a linear gap between t and $2q$. Thus, the syndrome gives essentially no additional information. The error rate ϕ that can be tolerated this way depends in a rather complicated way on λ, but choosing λ larger, for instance $\lambda = 1 - \frac{t+\nu q}{2n'}$ for a constant $\nu > 0$, allows to tolerate a higher error rate but requires q to be a smaller (but still constant) fraction of t.

Another imperfection has to be taken into account in current implementations of the quantum channel: imperfect sources. An imperfect source transmits more than one qubit in the same state with probability η independently each time a new transmission takes place. To deal with imperfect sources, we freely give away (x_i, θ_i) to the adversary when a multi-qubit transmission occurs in position i. It is not difficult to see that parameter ε in Proposition 3.1 then becomes essentially $\varepsilon = negl((1 - \eta)d - 4\log(m) - 4q - 4\ell)$ in this case.

It follows that a quantum channel with error-rate ϕ and multi-pulse rate η, called the (ϕ, η)-weak quantum model in [4], can be tolerated for some small enough (but constant) ϕ and η.

---

[4] As a matter of fact, the error correction in [7] is done by sending the string XOR'ed with a random code word, rather than sending the syndrome, but obviously the latter is equivalent to the first.

4 Defeating Man-in-the-Middle Attacks

4.1 The Approach

In the previous section, we "only" proved security against impersonation attacks, but we did not consider a man-in-the-middle attack, where the attacker sits between an honest user and an honest server and controls their (quantum and classical) communication. And indeed, Q-ID is highly insecure against such an attack: the attacker may measure the first qubit in, say, basis +, and then forward the collapsed qubit (together with the remaining untouched ones) and observe if S accepts the session. If not, then the attacker knows that he introduced an error and hence that the first qubit must have been encoded and measured using the ×-basis, which gives him one bit of information on the key w. The error-tolerant scheme seems to prevent this particular attack, but it is by no means clear that it is secure against *any* man-in-the-middle attack.

To defeat a man-in-the-middle attack that tampers with the quantum communication, we perform a check of correctness on a random subset. The check allows to detect if the attacker tampers too much with the quantum communication, and the scheme can be aborted before sensitive information is leaked to the attacker. In order to protect the classical communication, one might use a standard information-theoretic authentication code. However, the key for such a code can only be securely used a limited number of times. A similar problem occurs in QKD: even though a successful QKD execution produces fresh key material that can be used in the next execution, the attacker can have the parties run out of authentication keys by repeatedly enforcing the executions to fail. In order to overcome this problem, we will use some special authentication scheme allowing to re-use the key under certain circumstances, as discussed in Sect. 4.3.

4.2 The Setting

Similar to before, we assume that the user U and the server S share a not necessarily uniform, low-entropy key w. In order to handle the stronger security requirements of this section, we have to assume that U and S in addition share a uniform high-entropy key k. We require that a man-in-the-middle attacker needs to guess w correctly in order to break the scheme, and if his guess is incorrect then he learns no more information on w besides that his guess is wrong, and he essentially learns no information on k. Furthermore, we require security against impersonation attacks, as defined in the previous section, *even if the dishonest party knows k*. It follows that k can for instance be stored on a smartcard, and security is still guaranteed even if the smartcard gets stolen, assuming that the theft is noticed and the corresponding party does/can not execute the scheme anymore. We would also like to stress that by our security notion, not only w but also k may be safely reused, even if the scheme was under attack.

4.3 An Additional Tool: Extractor MACs

An important tool used in this section is an authentication scheme, i.e., a Message Authentication Code (MAC), that also acts as an extractor, meaning that if there is high min-entropy in the message, then the key-tag pair cannot be distinguished from the key and a random tag. Such a MAC, introduced in [6], is called an extractor MAC, EXTR-MAC for short. For instance $MAC^*_{\alpha,\beta}(x) = [\alpha x] + \beta$, where $\alpha, x \in GF(2^n)$, $\beta \in GF(2^\ell)$ and $[\,.\,]$, denotes truncation to the ℓ first bits, is an EXTR-MAC: impersonation and substitution probability are $1/2^\ell$, and, for an arbitrary message X, a random key $K = (A, B)$ and the corresponding tag $T = [A \cdot X] + B$, the tag-key pair (T, K) is $2^{-(H_2(X)-\ell)/2}$-close to (U, K), where U is the uniform distribution, respectively, ρ_{TKE} is $2^{(H_2(X)-\ell-q)/2}$-close to $\frac{1}{2^\ell}\mathbb{I} \otimes \rho_{KE} = \frac{1}{2^\ell}\mathbb{I} \otimes \rho_K \otimes \rho_E$ if we allow a q-qubit state E that may depend only on X. A useful feature of an EXTR-MAC is that if an adversary gets to see the tag of a message on which he has high min-entropy, then the key for the MAC can be safely re-used (sequentially). Indeed, closeness of the real state, ρ_{TKE}, to the ideal state, $\frac{1}{2^\ell}\mathbb{I} \otimes \rho_{KE} = \frac{1}{2^\ell}\mathbb{I} \otimes \rho_K \otimes \rho_E$, means that no matter how the state evolves, the real state behaves like the ideal one (except with small probability), but of course in the ideal state, K is still "fresh" and can be reused.

4.4 The Scheme

As for Q-ID, let $\mathfrak{c} : W \to \{+, \times\}^n$ be the encoding function of a binary code of length n with $m = |W|$ codewords and minimal distance d. For some parameter ℓ, let $\mathcal{F}$, $\mathcal{G}$ and $\mathcal{H}$ be strongly universal-2 classes of hash functions from $\{0,1\}^n$ to $\{0,1\}^\ell$, W to $\{0,1\}^\ell$, and $\{0,1\}^n$ to $\{0,1\}^{2\ell}$, respectively. Also, let MAC : $\{0,1\}^{2\ell} \times \{0,1\}^* \to \{0,1\}^\ell$ be a standard MAC for a message of arbitrary length L, with a 2ℓ-bit key and an error probability at most $\lceil L/\ell \rceil \cdot 2^{-\ell}$, and let MAC^* : $\mathcal{K} \times \mathcal{M} \to \{0,1\}^\ell$ be an EXTR-MAC with an arbitrary key space $\mathcal{K}$, a (finite) message space $\mathcal{M}$ that will become clear later, and an error probability $2^{-\ell}$. Furthermore, let $\{syn_j\}_{j \in \mathcal{J}}$ be the family of syndrome functions[5] corresponding to a family $\mathcal{C} = \{C_j\}_{j \in \mathcal{J}}$ of linear error correcting codes of size $n' = n/2$, as discussed in Section 3.4: any C_j allows to efficiently correct a δ-fraction of errors for some constant $\delta > 0$, and for a random $j \in \mathcal{J}$, the syndrome of a string with $t = d/4 - \log(m) - 5\ell$ bits of min-entropy is $2^{-(t-2q)/4}$-close to uniform (given j and any q-qubit state). Finally, we let $\ell^* \leq \ell$ be a parameter linear in $n - \ell$, whose exact value will be specified in the proof.

Recall, by the set-up assumption, the user U and the server S share a password $w \in W$ as well as a uniform high-entropy key, which we define to be a random authentication key $k \in \mathcal{K}$. The scheme is given in the box below.

Proposition 4.1 (Security against man-in-the-middle). *Let the initial state of a man-in-the-middle attacker with quantum memory q be independent*

---

[5] We agree on the following convention: for a bit string y of arbitrary length, $syn_j(y)$ is to be understood as $syn_j(y0\cdots0)$ with enough padded zeros if its bit length is smaller than n', and as $(syn_j(y'), y'')$, where y' consist of the first n' and y'' of the remaining bits of y, if its bit length is bigger than n'.

Q-ID$^+$:

1. U picks $x \in_R \{0,1\}^n$ and $\theta \in_R \{+,\times\}^n$, and sends the n-qubit state $|x\rangle_\theta$ to S. Write $I_w := \{i : \theta_i = \mathfrak{c}(w)_i\}$.

2. S picks a random subset $T \subset \{1,\ldots,n\}$ of size ℓ^*, it computes $c = \mathfrak{c}(w)$, replaces every c_i with $i \in T$ by $c_i \in_R \{+,\times\}$ and measures $|x\rangle_\theta$ in basis c. Let x' be the outcome, and let $test' := x'|_T$.

3. U sends θ, $j \in_R \mathcal{J}$, $s := syn_j(x|_{I_w})$, $f \in_R \mathcal{F}$, $h \in_R \mathcal{H}$ and $tag^* := MAC_k^*(\theta, j, s, f, h, x|_{I_w})$ to S.

4. S picks $g \in \mathcal{G}$, and sends T and g to U.

5. U sends $test := x|_T$, $z := f(x|_{I_w}) \oplus g(w)$ and $tag := MAC_{h(x|_{I_w})}(g, T, test, z)$ to S.

6. S recovers $x|_{I_w}$ from $x'|_{I_w}$ with the help of $test$ and s, and it accepts if and only if (1) both MAC's verify correctly, (2) $test$ coincides with $test'$ wherever the bases coincide, and (3) $z = f(x|_{I_w}) \oplus g(w)$.

---

of the keys W and K. Then, there exists W', independent of W, such that the common state ρ_{KWE} after the execution of Q-ID$^+$ satisfies

$$\rho_{KWW'E|W' \neq W} \approx_\varepsilon \rho_K \otimes \rho_{W \leftrightarrow W' \leftrightarrow E|W' \neq W} ,$$

where $\varepsilon = negl(d - 4\log(m) - 8q - 20\ell)$.

Proof. We use capital letters (W, Θ, etc.) for the values (w, θ, etc.) occurring in the scheme whenever we view them as random variables, and we write X_W and X'_W for the random variables taking values $x|_{I_w}$ and $x'|_{I_w}$, respectively. To simplify the argument, we neglect error probabilities that are of order ε, as well as linear fractions that can be chosen arbitrarily small. We merely give indication of a small error by (sometimes) using the word "essentially".

First note that due to the security of the MAC and its key, if the attacker substitutes θ, j, s, f or h in step 3, or if S recovers an incorrect string as $x|_{I_w}$, then S will reject at the end of the protocol. We can define W' (independent of W) as in the proof of Proposition 3.1 such that if $W \neq W'$ then X_W has essentially $d/4 - \log(m)$ bits of min-entropy, given W, W' and Θ. Furthermore, given $TAG^*, F(X_W), H(X_W), TEST$ (as well as K, F, H, T, W, W' and Θ), X_W has still essentially $t = d/4 - \log(m) - 5\ell$ bits of min-entropy, if $W \neq W'$. By the property of the code family $\mathcal{C}$, it follows that if $t > 2q$ with a linear gap then the syndrome $S = syn_J(X_W)$ is essentially random and independent of $J, TAG^*, F(X_W), H(X_W), TEST, K, F, H, T, W, W', \Theta$ and E, conditioned on $W \neq W'$. Furthermore, it follows from the privacy-amplifying property of MAC^* and of f and h that if $d/4 - \log(m) - 5\ell > q$ with a linear gap, then the set of values $(TAG^*, F(X_W), H(X_W))$ is essentially random and independent of $K, F, H, TEST, T, W, W', \Theta$ and E, conditioned on $W \neq W'$. Finally, K is independent of the rest, and E is independent of $K, F, H, TEST, T, W, \Theta$. It follows that $\rho_{KWW'E|W' \neq W} \approx \rho_K \otimes \rho_{W \leftrightarrow W' \leftrightarrow E|W' \neq W}$, *before* he learns S's decision to accept or reject.

It remains to argue that S's decision does not give any additional information on W. We will make a case distinction, which does not depend on w, and we will show for both cases that S's decision to accept or reject is independent of w, which proves the claim. But first, we need the following observation. Recall that outside of the test set T, S measured in the bases dictated by w, but within T in random bases. Let I'_w be the subset of positions $i \in I_w$ with $c_i = \mathfrak{c}(w)_i$ (and thus also $= \theta_i$), and let $T' = T \cap I'_w$. In other words, we remove the positions where S measured in the "wrong" basis. The size of T' is essentially $\ell^*/4$, and given its size, it is a random subset of I'_w of size $|T'|$. It follows from the theory of random sampling, specifically from Lemma 4 of [16], that $\nu(x|_{I'_w}, x'|_{I'_w})$ essentially equals $\nu(x|_{T'}, x'|_{T'})$ (except with probability negligible in the size of T'), where $\nu(\cdot, \cdot)$ denotes the fraction of errors between the two input strings. Due to some technical reason, for the sampling technique to work it is required that $|T'|$ is upper bounded by $\alpha \cdot |I'_w|$, where the constant $\alpha > 0$ depends on the allowed tolerance in estimating the error fraction, and as such on δ, the fraction of errors the code C_j is able to correct. We refer to [16] for more details. Important for us is that ℓ^* can be chosen linear in $n-\ell$. Furthermore, since the set $V = \{i \in T : \theta_i = c_i\}$ of positions where U and S compare x and x' is a superset of T' of essentially twice the size, $\nu(x|_V, x'|_V)$ is essentially lower bounded by $\frac{1}{2}\nu(x|_{T'}, x'|_{T'})$. Putting things together, we get that $\nu(x|_{I'_w}, x'|_{I'_w})$ is essentially upper bounded by $2\nu(x|_V, x'|_V)$. Also note that $\nu(x|_V, x'|_V)$ does not depend on w. We can now do the case distinction: *Case 1:* If $\nu(x|_V, x'|_V) \leq \frac{\delta}{2}$ (minus an arbitrarily small value), then $x|_{I'_w}$ and $x'|_{I'_w}$ differ in at most a δ-fraction of their positions, and thus S correctly recovers $x|_{I_w}$ (using $test = x|_T$ to get $x|_{I_w \setminus I'_w}$ and using s to correct the rest), no matter what w is, and it follows that S's decision only depends on the attacker's behavior, but not on w. *Case 2:* Otherwise, either S cannot correctly recover $x|_{I_w}$ and thus rejects, or it can correctly recover $x|_{I_w}$ and hence can verify tag with the correct key $h(x|_{I_w})$. S is therefore guaranteed to get the correct $test = x|_T$ (or else rejects) and thus rejects as $test$ and $test'$, restricted to V, differ in more than a $\frac{\delta}{2}$-fraction of their positions. Hence, S always rejects in case 2. $\square$

For a dishonest user or server who knows k (but not w), breaking Q-ID$^+$ is equivalent to breaking Q-ID, up to a change in the parameters. Doing the maths on the parameters, it thus follows:

Theorem 4.2. *If $H_\infty(W) \geq 1$, then the identification scheme Q-ID$^+$ is ε-secure against a man-in-the-middle attacker with quantum memory bound q, and, even with a leaked k, Q-ID$^+$ is ε-secure against impersonation attacks for any unbounded user and for any server with quantum memory bound q, where $\varepsilon = negl(n - 100\log(m) - 19q)$.*

It is easy to see that Q-ID$^+$ can tolerate a noisy quantum communication up to any error rate $\phi < \delta$. Similar to the discussion in Section 3.4, tolerating a higher error rate requires the bound on the adversary's quantum memory to be smaller but still linear in the number of qubits transmitted. Imperfect sources can also be addressed in a similar way as for Q-ID. It follows that Q-ID$^+$ can also be shown secure in the (ϕ, η)-weak quantum model provided ϕ and η are small enough constants.

5 Application to QKD

As already pointed out in Section 4.1, current QKD schemes have the shortcoming that if there is no classical channel available that is authenticated by physical means, and thus messages need to be authenticated by an information-theoretic authentication scheme, an attacker can force the parties to run out of authentication keys simply by making an execution (or several executions if the parties share more key material) fail. Even worse, even if there is no attacker, but some execution(s) of the QKD scheme fails due to a technical problem, parties could run out of authentication keys. This shortcoming could make the technology impractical in situations where denial of service attacks or technical interruptions often occur.

The identification scheme $Q\text{-}ID^+$ from the previous section immediately gives a QKD scheme *in the bounded-quantum-storage model* that allows to re-use the authentications key(s). Actually, we can inherit the key-setting from $Q\text{-}ID^+$, where there are two keys, a human-memorizable password and a uniform, high-entropy key, where security is still guaranteed even if the latter gets stolen and the theft is noticed. In order to agree on a secret key sk, the two parties execute $Q\text{-}ID^+$, and extract sk from $x|_{I_w}$ by applying yet another strongly universal-2 function, for instance chosen by U in step 3, where n needs to be increased accordingly to have the additional necessary amount of entropy in $x|_{I_w}$. The analysis of $Q\text{-}ID^+$ immediately implies that if honest S accepts, then he is convinced to share sk with the legitimate U which knows w. In order to convince U, S can then use part of sk to one-time-pad encrypt w, and send it to U. The rest of sk is then a secure secret key, shared between U and S. In order to have a better "key rate", instead of using sk (minus the part used for the one-time-pad encryption) as secret key, one can also run a standard QKD scheme on top of $Q\text{-}ID^+$ and use sk as a one-time authentication key.

References

1. Aumann, Y., Ding, Y.Z., Rabin, M.O.: Everlasting security in the bounded storage model. IEEE Transactions on Information Theory 48(6), 1668–1680 (2002)
2. Cachin, C., Crépeau, C., Marcil, J.: Oblivious transfer with a memory-bounded receiver. In: 39th Annual IEEE Symposium on Foundations of Computer Science (FOCS), pp. 493–502. IEEE Computer Society Press, Los Alamitos (1998)
3. Damgård, I.B., Fehr, S., Renner, R., Salvail, L., Schaffner, C.: A tight high-order entropic quantum uncertainty relation with applications. In: Menezes, A. (ed.) CRYPTO 2007. LNCS, vol. 4622, pp. 360–378. Springer, Heidelberg
4. Damgård, I.B., Fehr, S., Salvail, L., Schaffner, C.: Cryptography in the bounded quantum-storage model. In: 46th Annual IEEE Symposium on Foundations of Computer Science (FOCS), pp. 449–458. IEEE Computer Society Press, Los Alamitos (2005)
5. Damgård, I.B., Fehr, S., Salvail, L., Schaffner, C.: Secure identification and QKD in the bounded-quantum-storage model (2007), available at http://eprint.iacr.org/2007/

6. Dodis, Y., Katz, J., Reyzin, L., Smith, A.: Robust fuzzy extractors and authenticated key agreement from close secrets. In: Dwork, C. (ed.) CRYPTO 2006. LNCS, vol. 4117, pp. 232–250. Springer, Heidelberg (2006)
7. Dodis, Y., Smith, A.: Correcting errors without leaking partial information. In: 37th Annual ACM Symposium on Theory of Computing (STOC), pp. 654–663. ACM Press, New York (2005)
8. Dziembowski, S., Maurer, U.M.: On generating the initial key in the bounded-storage model. In: Cachin, C., Camenisch, J.L. (eds.) EUROCRYPT 2004. LNCS, vol. 3027, pp. 126–137. Springer, Heidelberg (2004)
9. Elliott, C., Pearson, D., Troxel, G.: Quantum cryptography in practice. In: SIG-COMM '03: Proceedings of the 2003 conference on Applications, technologies, architectures, and protocols for computer communications, pp. 227–238 (2003)
10. Fehr, S., Schaffner, C.: Randomness extraction via delta-biased masking in the presence of a quantum attacker (2007), available at http://eprint.iacr.org/2007/
11. Feige, U., Fiat, A., Shamir, A.: Zero knowledge proofs of identity. In: 19th Annual ACM Symposium on Theory of Computing (STOC), pp. 210–217. ACM Press, New York (1987)
12. Fiat, A., Shamir, A.: How to prove yourself: Practical solutions to identification and signature problems. In: Odlyzko, A.M. (ed.) CRYPTO 1986. LNCS, vol. 263, pp. 186–194. Springer, Heidelberg (1987)
13. Gennaro, R., Lindell, Y.: A framework for password-based authenticated key exchange. In: Biham, E. (ed.) EUROCRPYT 2003. LNCS, vol. 2656, pp. 524–543. Springer, Heidelberg (2003)
14. Katz, J., Ostrovsky, R., Yung, M.: Efficient password-authenticated key exchange using human-memorable passwords. In: Pfitzmann, B. (ed.) EUROCRYPT 2001. LNCS, vol. 2045, pp. 473–492. Springer, Heidelberg (2001)
15. Lo, H.-K.: Insecurity of quantum secure computations. Physical Review A 56(2), 1154–1162 (1997)
16. Lo, H.-K., Chau, H.F., Ardehali, M.: Efficient quantum key distribution scheme and a proof of its unconditional security. Journal of Cryptology 18(2), 133–165 (2005)
17. Maurer, U.M.: A provably-secure strongly-randomized cipher. In: Damgård, I.B. (ed.) EUROCRYPT 1990. LNCS, vol. 473, pp. 361–373. Springer, Heidelberg (1991)
18. Nielsen, J.B., Pedersen, T.B., Salvail, L.: Secure two-party quantum computation against semi-honest adversaries. In preparation (2007)
19. Renner, R.: Security of Quantum Key Distribution. PhD thesis, ETH Zürich, (2005), http://arxiv.org/abs/quant-ph/0512258
20. Renner, R., König, R.: Universally composable privacy amplification against quantum adversaries. In: Kilian, J. (ed.) TCC 2005. LNCS, vol. 3378, pp. 407–425. Springer, Heidelberg (2005)
21. Renner, R., Wolf, S.: Simple and tight bounds for information reconciliation and privacy amplification. In: Roy, B. (ed.) ASIACRYPT 2005. LNCS, vol. 3788, pp. 199–216. Springer, Heidelberg (2005)
22. Thommesen, C.: The existence of binary linear concatenated codes with reed - solomon outer codes which asymptotically meet the gilbert- varshamov bound. IEEE Transactions on Information Theory 29(6), 850–853 (1983)

A Tight High-Order Entropic Quantum Uncertainty Relation with Applications

Ivan B. Damgård[1,*], Serge Fehr[2,**], Renato Renner[3,***,†],
Louis Salvail[1,‡], and Christian Schaffner[2,†]

[1] Basic Research in Computer Science (BRICS), funded by the Danish National
Research Foundation, Department of Computer Science, University of Aarhus,
Denmark
{ivan,salvail}@brics.dk
[2] Center for Mathematics and Computer Science (CWI), Amsterdam, Netherlands
{fehr,c.schaffner}@cwi.nl
[3] Cambridge University, UK
r.renner@damtp.cam.ac.uk

Abstract. We derive a new entropic quantum uncertainty relation involving min-entropy. The relation is tight and can be applied in various quantum-cryptographic settings.

Protocols for quantum 1-out-of-2 Oblivious Transfer and quantum Bit Commitment are presented and the uncertainty relation is used to prove the security of these protocols in the bounded-quantum-storage model according to new strong security definitions.

As another application, we consider the realistic setting of Quantum Key Distribution (QKD) against quantum-memory-bounded eavesdroppers. The uncertainty relation allows to prove the security of QKD protocols in this setting while tolerating considerably higher error rates compared to the standard model with unbounded adversaries. For instance, for the six-state protocol with one-way communication, a bit-flip error rate of up to 17% can be tolerated (compared to 13% in the standard model).

Our uncertainty relation also yields a lower bound on the min-entropy key uncertainty against known-plaintext attacks when quantum ciphers are composed. Previously, the key uncertainty of these ciphers was only known with respect to Shannon entropy.

1 Introduction

A problem often encountered in quantum cryptography is the following: through some interaction between the players, a quantum state ρ is generated and then

---

* FICS, Foundations in Cryptography and Security, funded by the Danish Natural Sciences Research Council.

** Supported by the Dutch Organization for Scientific Research (NWO).

*** Supported by HP Labs Bristol.

† Supported by the European project SECOQC.

‡ QUSEP, Quantum Security in Practice, funded by the Danish Natural Science Research Council.

A. Menezes (Ed.): CRYPTO 2007, LNCS 4622, pp. 360–378, 2007.

measured by one of the players (call her Alice in the following). Assuming Alice is honest, we want to know how unpredictable her measurement outcome is to the adversary. Once a lower bound on the adversary's uncertainty about Alice's measurement outcome is established, it is usually easy to prove the desired security property of the protocol. Many existing constructions in quantum cryptography have been proved secure following this paradigm.

Typically, Alice does not make her measurement in a fixed basis, but chooses at random among a set of different bases. These bases are usually chosen to be pairwise *mutually unbiased*, meaning that if ρ is such that the measurement outcome in one basis is fixed then this implies that the uncertainty about the outcome of the measurement in the other basis is maximal. In this way, one hopes to keep the adversary's uncertainty high, even if ρ is (partially) under the adversary's control.

An inequality that lower bounds the adversary's uncertainty in such a scenario is called an *uncertainty relation*. There exist uncertainty relations for different measures of uncertainty, but cryptographic applications typically require the adversary's min-entropy to be bounded from below.

In this paper, we introduce a new general and tight entropic uncertainty relation. Since the relation is expressed in terms of high-order entropy (i.e. min-entropy), it is applicable to a large class of natural protocols in quantum cryptography. In particular, the new relation can be applied in situations where an n-qubit state ρ has each of its qubits measured in a random and independent basis sampled uniformly from a fixed set $\mathcal{B}$ of bases. $\mathcal{B}$ does not necessarily have to be mutually unbiased, but we assume a lower bound h (i.e. an *average entropic uncertainty bound*) on the average Shannon entropy of the distribution P_ϑ, obtained by measuring an arbitrary 1-qubit state in basis $\vartheta \in \mathcal{B}$, meaning that $\frac{1}{|\mathcal{B}|} \sum_\vartheta \mathrm{H}(P_\vartheta) \geq h$.

Uncertainty Relation (informal): Let $\mathcal{B}$ be a set of bases with an average entropic uncertainty bound h as above. Let P_θ denote the probability distribution defined by measuring an arbitrary n-qubit state ρ in basis $\theta \in \mathcal{B}^n$. For a $\theta \in_R \mathcal{B}^n$ chosen uniformly at random, it holds except with negligible probability that

$$\mathrm{H}_\infty(P_\theta) \gtrsim nh \ . \tag{1}$$

Observe that (1) cannot be improved significantly since the min-entropy of a distribution is at most equal to the Shannon entropy. Our uncertainty relation is therefore asymptotically tight when the bound h is tight.

Any lower bound on the Shannon entropy associated to a set of measurements $\mathcal{B}$ can be used in (1). In the special case where the set of bases is $\mathcal{B} = \{+, \times\}$ (i.e. the two BB84 bases), h is known precisely using Maassen and Uffink's entropic relation, see inequality (2) below. We get $h = \frac{1}{2}$ and (1) results in $\mathrm{H}_\infty(P_\theta) \gtrsim \frac{n}{2}$. Uncertainty relations for the BB84 coding scheme [3] are useful since this coding is widely used in quantum cryptography. Its resilience to imperfect quantum channels, sources, and detectors is an important advantage in practice.

We now discuss applications of our high-order uncertainty relation to important scenarios in cryptography: two-party cryptography, quantum key distribution and quantum encryption.

Application I: Two-Party Cryptography in the Bounded-Quantum-Storage Model. Entropic uncertainty relations are powerful tools for the security analysis of cryptographic protocols in the bounded-quantum-storage model. In this model, the adversary is unbounded in every respect, except that at a certain time, his quantum memory is reduced to a certain size (by performing some measurement). In [13], an uncertainty relation involving min-entropy was shown and used in the analysis of protocols for Rabin oblivious transfer (*ROT*) and bit commitment. This uncertainty relation only applies in the case when n qubits are all measured in one out of two mutually unbiased bases.

A major difference between our result (1) and the one from [13] is that while both relations bound the min-entropy conditioned on an event, this event happens in our case with probability essentially 1 (on average) whereas the corresponding event from [13] only happens with probability about $1/2$. In Sect. 4, we prove the following:

1-2 OT in the Bounded-Quantum-Storage Model: *There exists a non-interactive protocol for 1-out-of-2 oblivious transfer (1-2 OT) of ℓ-bit messages, secure against adversaries with quantum memory size at most $n/4 - 2\ell$. Here, n is the number of qubits transmitted in the protocol and ℓ can be a constant fraction of n. Honest players need no quantum memory.*

Since all flavors of *OT* are known to be equivalent under classical information-theoretic reductions, and a *ROT* protocol is already known from [13], the above result may seem insignificant. This is not the case, however, for several reasons: First, although it may in principle be possible to obtain a protocol for *1-2 OT* from the *ROT* protocol of [13] using the standard black-box reduction, the fact that we need to call the *ROT* primitive many times would force the bound on the adversary's memory to be *sublinear* (in the number of transmitted qubits). Second, the techniques used in [13] do not seem applicable to *1-2 OT*, unless via the inefficient generic reduction to *ROT*. And, third, we prove security according to a stronger definition than the one used in [13], namely a quantum version of a recent classical definition for information theoretic *1-2 OT* [10]. The definition ensures that all (dishonest) players' inputs are well defined (and can be extracted when formalized appropriately). In particular, this implies security under sequential composition whereas composability of the protocol from [13] was not proven.

Furthermore, our techniques for *1-2 OT* imply almost directly a non-interactive bit commitment scheme (in the bounded-quantum-storage model) satisfying a composable security definition. As an immediate consequence, we obtain secure *string* commitment schemes. This improves over the bit commitment construction of [13], respectively its analysis, which does *not* guarantee composability and thus does *not* necessarily allow for string commitments. This application can be found in Sect. 5.

Application II: Quantum Key Distribution. We also apply our uncertainty relation to quantum key distribution (QKD) settings. QKD is the art of distributing a secret key between two distant parties, Alice and Bob, using only a completely insecure quantum channel and authentic classical communication. QKD protocols typically provide information-theoretic security, i.e., even an adversary with unlimited resources cannot get any information about the key. A major difficulty when implementing QKD schemes is that they require a low-noise quantum channel. The tolerated noise level depends on the actual protocol and on the desired security of the key. Because the quality of the channel typically decreases with its length, the maximum tolerated noise level is an important parameter limiting the maximum distance between Alice and Bob.

We consider a model in which the adversary has a limited amount of quantum memory to store the information she intercepts during the protocol execution. In this model, we show that the maximum tolerated noise level is larger than in the standard scenario where the adversary has unlimited resources. For *one-way QKD protocols* which are protocols where error-correction is performed non-interactively (i.e., a single classical message is sent from one party to the other), we show the following result:

QKD Against Quantum-Memory-Bounded Eavesdroppers: *Let $\mathcal{B}$ be a set of orthonormal bases of $\mathcal{H}_2$ with average entropic uncertainty bound h. Then, a* one-way *QKD-protocol produces a secure key against eavesdroppers whose quantum-memory size is sublinear in the length of the raw key at a positive rate as long as the bit-flip probability p of the quantum channel fulfills $H_{bin}(p) < h$ where $H_{bin}(\cdot)$ denotes the binary Shannon-entropy function.*

Although this result does not allow us to improve (i.e. compared to unbounded adversaries) the maximum error-rate for the BB84 protocol (the four-state protocol), the six-state protocol can be shown secure against adversaries with memory bound sublinear in the secret-key length as long as the bit-flip error-rate is less than 17%. This improves over the maximal error-rate of 13% for the same protocol against unbounded adversaries. We also show that the generalization of the six-state protocols to more bases (not necessarily mutually unbiased) can be shown secure (against memory-bounded adversaries) for a maximal error-rate up to 20% provided the number of bases is large enough.

The quantum-memory-bounded eavesdropper model studied here is not comparable to other restrictions on adversaries considered in the literature (e.g. *individual attacks*, where the eavesdropper is assumed to apply independent measurements to each qubit sent over the quantum channel [18,23]). In fact, these assumptions are generally artificial and their purpose is to simplify security proofs rather than to relax the conditions on the quality of the communication channel from which secure key can be generated. We believe that the quantum-memory-bounded eavesdropper model is more realistic.

Application III: Key-Uncertainty of Quantum Ciphers. In [15], symmetric quantum ciphers encrypting classical messages with classical secret-keys are considered. It is shown that under known-plaintext attacks, the Shannon uncertainty of the secret-key can be much higher for some quantum ciphers than for any classical one. The Shannon secret-key uncertainty $H(K|C, M)$ of classical ciphers C encrypting messages M of size m with keys K of size $k > m$ is always such that $H(K|C, M) \leq k - m$. In the quantum case, the Shannon secret-key uncertainty is defined as the minimum residual uncertainty about key K given the best measurement (POVM) $P_M(C)$ applied to quantum cipher C given plaintext M. Examples of quantum ciphers are provided with $k = m + 1$ such that $H(K|P_M(C)) = m/2 + 1$ and with $k = 2m$ such that $H(K|P_M(C)) \geq 2m - 1$. All ciphers in [15] have their keys consisting of two parts. The first part chooses one basis out a set $\mathcal{B}$ of bases while the other part is used as a classical one-time-pad. The message is first encrypted with the one-time-pad before being rotated in the basis indicated by the key. In this case, Theorem 4 in [15] states that the Shannon secret-key uncertainty adds up under repetitions with independent and random keys[1]: if $H(K|P_M(C)) \geq h$ then n repetitions with independent keys satisfy $H(K_1, \ldots, K_n | P_{M_1,\ldots,M_n}(C_1, \ldots, C_n)) \geq nh$. Our uncertainty relation allows to obtain a stronger result. The analysis in [15] shows that these quantum ciphers with Shannon secret-key uncertainty h satisfy the condition of our uncertainty relation. As result we obtain a lower bound on the min-entropy key uncertainty given the outcome of any quantum measurement applied to all ciphers and given all plaintexts. When $H(K|P_M(C)) \geq h$ our uncertainty relation tells us that $H_\infty(K_1, \ldots, K_n | P_{M_1,\ldots,M_n}(C_1, \ldots, C_n)) \gtrsim nh$. Notice that unlike the two previous applications, this time the result holds unconditionally. Details of this application will be provided in the full version.

History and Related Work. The history of uncertainty relations starts with Heisenberg who showed that the outcomes of two non-commuting observables A and B applied to any state ρ are not easy to predict simultaneously. However, Heisenberg only speaks about the variance of the measurement results. Because his result had several shortcomings (as pointed out in [19,16]), more general forms of uncertainty relations were proposed by Bialynicki-Birula and Mycielski [7] and by Deutsch [16]. The new relations were called *entropic uncertainty relations*, because they are expressed using Shannon entropy instead of the statistical variance and, hence, are purely information theoretic statements. For instance, Deutsch's uncertainty relation [16] states that $H(P) + H(Q) \geq -2 \log \frac{1+c}{2}$, where P, Q are random variables representing the measurement results and c is the maximum inner product norm between any eigenvectors of A and B. First conjectured by Kraus [21], Maassen and Uffink [24] improved Deutsch's relation to the optimal

$$H(P) + H(Q) \geq -2 \log c . \tag{2}$$

Although a bound on Shannon entropy can be helpful in some cases, it is usually not good enough in cryptographic applications. The main tool to reduce the adversary's information—privacy amplification [5,20,4,27,25]—only works if a bound on the adversary's min-entropy (in fact collision entropy) is known.

---

[1] The proof of Theorem 4 in [15] is incorrect but can easily be fixed without changing the statement.

Unfortunately, knowing the Shannon entropy of a distribution does in general not allow to bound its higher order Rényi entropies.

An entropic uncertainty relation involving Rényi entropy of order 2 (i.e. *collision entropy*) was introduced by Larsen [22,30]. Larsen's relation quantifies precisely the collision entropy for the set $\{A_i\}_{i=1}^{d+1}$ of *all* maximally non-commuting observables, where d is the dimension of the Hilbert space. Its use is therefore restricted to quantum coding schemes that take advantage of *all* $d + 1$ observables, i.e. to schemes that are difficult to implement in practice. Uncertainty relations in terms of Rényi entropy have also been studied in a different context by Bialynicki-Birula [6].

2   Preliminaries

2.1   Notation and Terminology

For any positive integer d, $\mathcal{H}_d$ stands for the complex Hilbert space of dimension d and $\mathcal{P}(\mathcal{H})$ for the set of density operators, i.e., positive semi-definite trace-1 matrices, acting on $\mathcal{H}$. The pair $\{|0\rangle, |1\rangle\}$ denotes the computational or rectilinear or "+" basis for the 2-dimensional Hilbert space $\mathcal{H}_2$. The diagonal or "×" basis is defined as $\{|0\rangle_\times, |1\rangle_\times\}$ where $|0\rangle_\times = (|0\rangle + |1\rangle)/\sqrt{2}$ and $|1\rangle_\times = (|0\rangle - |1\rangle)/\sqrt{2}$. The circular or "⊘" basis consists of vectors $(|0\rangle + i|1\rangle)/\sqrt{2}$ and $(|0\rangle - i|1\rangle)/\sqrt{2}$. Measuring a qubit in the +-basis (resp. ×-basis) means applying the measurement described by projectors $|0\rangle\langle0|$ and $|1\rangle\langle1|$ (resp. projectors $|0\rangle_\times\langle0|_\times$ and $|1\rangle_\times\langle1|_\times$). When the context requires it, we write $|0\rangle_+$ and $|1\rangle_+$ instead of $|0\rangle$ and $|1\rangle$, respectively. If we want to choose the + or ×-basis according to the bit $b \in \{0,1\}$, we write $[+, \times]_b$.

The behavior of a (mixed) quantum state in a register E is fully described by its density matrix ρ_E. We often consider cases where a quantum state may depend on some classical random variable X, in that the state is described by the density matrix ρ_E^x if and only if $X = x$. For an observer who has access to the state but not X, the behavior of the state is determined by the density matrix $\rho_E := \sum_x P_X(x)\rho_E^x$, whereas the joint state, consisting of the classical X and the quantum register E is described by the density matrix $\rho_{XE} := \sum_x P_X(x)|x\rangle\langle x| \otimes \rho_E^x$, where we understand $\{|x\rangle\}_{x\in\mathcal{X}}$ to be the standard (orthonormal) basis of $\mathcal{H}_{|\mathcal{X}|}$. Joint states with such classical and quantum parts are called *cq-states*. We also write $\rho_X := \sum_x P_X(x)|x\rangle\langle x|$ for the quantum representation of the classical random variable X. This notation extends naturally to quantum states that depend on several classical random variables (i.e. to ccq-states, cccq-states etc.). Given a cq-state ρ_{XE} as above, by saying that there exists a random variable Y such that ρ_{XYE} satisfies some condition, we mean that ρ_{XE} can be understood as $\rho_{XE} = \text{tr}_Y(\rho_{XYE})$ for some ccq-state ρ_{XYE} and that ρ_{XYE} satisfies the required condition.[2]

---

[2] The quantum version is similar to the case of distributions of classical variables where given X, the existence of a certain Y is understood that there exists a joint distribution P_{XY} with $\sum_y P_{XY}(\cdot, y) = P_X$.

We would like to point out that $\rho_{XE} = \rho_X \otimes \rho_E$ holds if and only if the quantum part is independent of X (in that $\rho_E^x = \rho_E$ for any x), where the latter in particular implies that no information on X can be learned by observing only ρ_E. Similarly, X is uniformly random and independent of the quantum state in register E if and only if $\rho_{XE} = \frac{1}{|\mathcal{X}|}\mathbb{1} \otimes \rho_E$, where $\frac{1}{|\mathcal{X}|}\mathbb{1}$ is the density matrix of the fully mixed state of suitable dimension. Finally, if two states like ρ_{XE} and $\rho_X \otimes \rho_E$ are ε-close in terms of their trace distance $\delta(\rho, \sigma) = \frac{1}{2}\mathrm{tr}(|\rho - \sigma|)$, which we write as $\rho_{XE} \approx_\varepsilon \rho_X \otimes \rho_E$, then the real system ρ_{XE} "behaves" as the ideal system $\rho_X \otimes \rho_E$ except with probability ε in that for any evolution of the system no observer can distinguish the real from the ideal one with advantage greater than ε [27].

2.2 Smooth Rényi Entropy

We briefly recall the notion of (conditional) *smooth* min-entropy [25,28]. For more details, we refer to the aforementioned literature. Let X be a random variable over alphabet $\mathcal{X}$ with distribution P_X. The standard notion of min-entropy is given by $\mathrm{H}_\infty(X) = -\log\left(\max_x P_X(x)\right)$ and that of max-entropy by $\mathrm{H}_0(X) = \log\left|\{x \in \mathcal{X} : P_X(x) > 0\}\right|$. More general, for any event $\mathcal{E}$ (defined by $P_{\mathcal{E}|X}(x)$ for all $x \in \mathcal{X}$) $\mathrm{H}_\infty(X\mathcal{E})$ may be defined similarly simply by replacing P_X by $P_{X\mathcal{E}}$. Note that the "distribution" $P_{X\mathcal{E}}$ is not normalized; $\mathrm{H}_\infty(X\mathcal{E})$ is still well defined, though. For an arbitrary $\varepsilon \geq 0$, the smooth version $\mathrm{H}_\infty^\varepsilon(X)$ is defined as follows. $\mathrm{H}_\infty^\varepsilon(X)$ is the *maximum* of the standard min-entropy $\mathrm{H}_\infty(X\mathcal{E})$, where the maximum is taken over all events $\mathcal{E}$ with $\Pr(\mathcal{E}) \geq 1 - \varepsilon$. Informally, this can be understood that if $\mathrm{H}_\infty^\varepsilon(X) = r$ then the standard min-entropy of X equals r as well, except with probability ε. As ε can be interpreted as an error probability, we typically require ε to be negligible in the security parameter n.

For random variables X and Y, the *conditional* smooth min-entropy $\mathrm{H}_\infty^\varepsilon(X \mid Y)$ is defined as $\mathrm{H}_\infty^\varepsilon(X \mid Y) = \max_\mathcal{E} \min_y \mathrm{H}_\infty(X\mathcal{E} \mid Y = y)$, where the quantification over $\mathcal{E}$ is over all events $\mathcal{E}$ (defined by $P_{\mathcal{E}|XY}$) with $\Pr(\mathcal{E}) \geq 1 - \varepsilon$. In Sect. 6, we work with smooth min-entropy conditioned on a quantum state. We refer the reader to [25] for the definition of this quantum version. We will make use of the following chain rule for smooth min-entropy [28], which in spirit was already shown in [8].

Lemma 1. $\mathrm{H}_\infty^{\varepsilon+\varepsilon'}(X \mid Y) > \mathrm{H}_\infty^\varepsilon(XY) - \mathrm{H}_0(Y) - \log\left(\frac{1}{\varepsilon'}\right)$ *for all* $\varepsilon, \varepsilon' > 0$.

2.3 Azuma's Inequality

In the following and throughout the paper, the expected value of a real random variable R is denoted by $\mathbb{E}[R]$. Similarly, $\mathbb{E}[R|\mathcal{E}]$ and $\mathbb{E}[R|S]$ denote the conditional expectation of R conditioned on an event $\mathcal{E}$ respectively random variable S.

Definition 1. *A list of real random variables* $R_1, \ldots, R_n$ *is called a* martingale difference sequence *if* $\mathbb{E}[R_i \mid R_1, \ldots, R_{i-1}] = 0$ *with probability 1 for every* $1 \leq i \leq n$, *i.e., if* $\mathbb{E}[R_i \mid R_1 = r_1, \ldots, R_{i-1} = r_{i-1}] = 0$ *for every* $1 \leq i \leq n$ *and* $r_1, \ldots, r_{i-1} \in \mathbb{R}$.

The following lemma follows directly from Azuma's inequality [2,1].

Lemma 2. *Let $R_1, \ldots, R_n$ be a martingale difference sequence such that $|R_i| \leq c$ for every $1 \leq i \leq n$. Then, $\Pr\left[\sum_i R_i \geq \lambda n\right] \leq \exp\left(-\frac{\lambda^2 n}{2c^2}\right)$ for any $\lambda > 0$.*

3   The Uncertainty Relation

We start with a classical tool which itself might be of independent interest.

Theorem 1. *Let $Z_1, \ldots, Z_n$ be n (not necessarily independent) random variables over alphabet $\mathcal{Z}$, and let $h \geq 0$ be such that*

$$\mathrm{H}(Z_i \mid Z_1 = z_1, \ldots, Z_{i-1} = z_{i-1}) \geq h \tag{3}$$

for all $1 \leq i \leq n$ and $z_1, \ldots, z_{i-1} \in \mathcal{Z}$. Then for any $0 < \lambda < \frac{1}{2}$

$$\mathrm{H}_\infty^\varepsilon(Z_1, \ldots, Z_n) \geq (h - 2\lambda)n ,$$

where $\varepsilon = \exp\left(-\frac{\lambda^2 n}{32 \log(|\mathcal{Z}|/\lambda)^2}\right)$.

If the Z_i's are *independent* and have Shannon-entropy at least h, it is known (see [28]) that the smooth min-entropy of $Z_1, \ldots, Z_n$ is at least nh for large enough n. Informally, Theorem 1 guarantees that when the independence -condition is relaxed to a lower bound on the Shannon entropy of Z_i *given any previous history*, then we still have min-entropy of (almost) nh except with negligible probability ε.

Proof (sketch). The idea is to use Lemma 2 for cleverly chosen R_i's. For any i we write $Z^i := (Z_1, \ldots, Z_i)$ (with Z^0 being the "empty symbol"), and similarly for other sequences. We want to show that $\Pr\left[P_{Z^n}(Z^n) \geq 2^{-(h-2\lambda)n}\right] \leq \varepsilon$. By the definition of smooth min-entropy, this then implies the claim. Note that $P_{Z^n}(Z^n) \geq 2^{-(h-2\lambda)n}$ is equivalent to

$$\sum_{i=1}^n \left(\log\left(P_{Z_i \mid Z^{i-1}}(Z_i \mid Z^{i-1})\right) + h \right) \geq 2\lambda n .$$

We set $S_i := \log P_{Z_i \mid Z^{i-1}}(Z_i \mid Z^{i-1})$. For such a sequence of real-valued random variables $S_1, \ldots, S_n$, it is easy to verify that $R_1, \ldots, R_n$ where $R_i := S_i - \mathbb{E}[S_i \mid S^{i-1}]$ forms a martingale difference sequence. If the $|R_i|$ were bounded by c, we could use Lemma 2 to conclude that

$$\Pr\left[\sum_{i=1}^n \left(S_i - \mathbb{E}[S_i \mid S^{i-1}]\right) \geq \lambda n\right] \leq \exp\left(-\frac{\lambda^2 n}{2c^2}\right) .$$

As by assumption $\mathbb{E}[S_i \mid S^{i-1}] \leq -h$, this would give us a bound similar to what we want to show. In order to enforce a bound on $|R_i|$, S_i needs to be truncated whenever $P_{Z_i \mid Z^{i-1}}(Z_i \mid Z^{i-1})$ is smaller than some $\delta > 0$. It is then a subtle and technically involved matter of choosing δ and ε appropriately in order to finish the proof, as shown in the full version of the paper [12].   □

We now state and prove the new entropic uncertainty relation in its most general form. A special case will then be introduced (Corollary 1) and used in the security analysis of all protocols we consider in the following.

Definition 2. *Let $\mathcal{B}$ be a finite set of orthonormal bases in the d-dimensional Hilbert space $\mathcal{H}_d$. We call $h \geq 0$ an* average entropic uncertainty bound *for $\mathcal{B}$ if every state in $\mathcal{H}_d$ satisfies $\frac{1}{|\mathcal{B}|} \sum_{\vartheta \in \mathcal{B}} H(P_\vartheta) \geq h$, where P_ϑ is the distribution obtained by measuring the state in basis ϑ.*

Note that by the convexity of the Shannon entropy H, a lower bound for all *pure* states in $\mathcal{H}_d$ suffices to imply the bound for all (possibly mixed) states.

Theorem 2. *Let $\mathcal{B}$ be a set of orthonormal bases in $\mathcal{H}_d$ with an average entropic uncertainty bound h, and let $\rho \in \mathcal{P}(\mathcal{H}_d^{\otimes n})$ be an arbitrary quantum state. Let $\Theta = (\Theta_1, \ldots, \Theta_n)$ be uniformly distributed over $\mathcal{B}^n$ and let $X = (X_1, \ldots, X_n)$ be the outcome when measuring ρ in basis Θ, distributed over $\{0, \ldots, d-1\}^n$. Then for any $0 < \lambda < \frac{1}{2}$ and $\lambda' > 0$,*

$$H_\infty^{\varepsilon + \varepsilon'}(X \mid \Theta) \geq (h - 2\lambda - \lambda') n$$

with $\varepsilon = \exp\left(-\frac{\lambda^2 n}{32(\log(|\mathcal{B}| \cdot d/\lambda))^2}\right)$ and $\varepsilon' = 2^{-\lambda' n}$.

Proof. Define $Z_i := (X_i, \Theta_i)$ and $Z^i := (Z_1, \ldots, Z_i)$. Let z^{i-1} be arbitrary in $(\{0, \ldots, d-1\} \times \mathcal{B})^{i-1}$. Then

$$H(Z_i \mid Z^{i-1} = z^{i-1}) = H(X_i \mid \Theta_i, Z^{i-1} = z^{i-1}) + H(\Theta_i \mid Z^{i-1} = z^{i-1}) \geq h + \log|\mathcal{B}|,$$

where the inequality follows from the fact that Θ_i is chosen uniformly at random and from the definition of h. Note that h lower bounds the average entropy for any system in $\mathcal{H}_d$, and thus in particular for the i-th subsystem of ρ, with all previous d-dimensional subsystems measured. We use the chain rule for smooth min-entropy (Lemma 1) and Theorem 1 to conclude that,

$$H_\infty^{\varepsilon + \varepsilon'}(X \mid \Theta) > H_\infty^\varepsilon(Z) - H_0(\Theta) - \log\left(\tfrac{1}{\varepsilon'}\right) \geq (h - 2\lambda)n - \lambda' n \ ,$$

for ε and ε' as claimed. $\qquad\square$

For the special case where $\mathcal{B} = \{+, \times\}$ is the set of BB84 bases, we can use the uncertainty relation of Maassen and Uffink [24] (see (2) with $c = 1/\sqrt{2}$), which, using our terminology, states that $\mathcal{B}$ has average entropic uncertainty bound $h = \frac{1}{2}$. Theorem 2 then immediately gives the following corollary.

Corollary 1. *Let $\rho \in \mathcal{P}(\mathcal{H}_2^{\otimes n})$ be an arbitrary n-qubit quantum state. Let Θ be uniformly distributed over $\{+, \times\}^n$, and let X be the outcome when measuring ρ in basis Θ. Then for any $0 < \lambda < \frac{1}{2}$ and $\lambda' > 0$,*

$$H_\infty^{\varepsilon + \varepsilon'}(X \mid \Theta) \geq \left(\tfrac{1}{2} - 2\lambda - \lambda'\right) n$$

where $\varepsilon = \exp\left(-\frac{\lambda^2 n}{32(2 - \log(\lambda))^2}\right)$ and $\varepsilon' = 2^{-\lambda' n}$.

Maassen and Uffink's relation being optimal means there exists a quantum state ρ—namely the product state of eigenstates of the subsystems, e.g. $\rho = |0\rangle\langle 0|^{\otimes n}$—for which $H(X \mid \Theta) = \frac{n}{2}$. On the other hand, we have shown that $(\frac{1}{2} - \lambda)n \leq H_\infty^\varepsilon(X \mid \Theta)$ for $\lambda > 0$ arbitrarily close to 0. For the product state ρ, the X_i's are independent and we know from [28] that in this case $H_\infty^\varepsilon(X \mid \Theta)$ approaches $H(X \mid \Theta) = \frac{n}{2}$. It follows that the relation cannot be significantly improved even when considering Rényi entropy of lower order than min-entropy (but higher than Shannon entropy).

Another tight corollary is obtained if we consider the set of measurements $\mathcal{B} = \{+, \times, \oslash\}$. In [29], Sánchez-Ruiz has shown that for this $\mathcal{B}$ the average entropic uncertainty bound $h = \frac{2}{3}$ is optimal. It implies that $H_\infty^\varepsilon(X \mid \Theta) \approx H(X \mid \Theta) = \frac{2n}{3}$ for negligible ε. In the full version [12], we compute the average uncertainty bound for the set of *all bases* of a d-dimensional Hilbert space.

4 Application: Oblivious Transfer

4.1 Privacy Amplification and a Min-Entropy-Splitting Lemma

Recall, a class $\mathcal{F}$ of hash functions from, say, $\{0,1\}^n$ to $\{0,1\}^\ell$ is called *two-universal* [9,31] if $\Pr[F(x) = F(x')] \leq 1/2^\ell$ for any distinct $x, x' \in \{0,1\}^n$ and for F uniformly distributed over $\mathcal{F}$.

Theorem 3 (Privacy Amplification [27,25]). *Let $\varepsilon \geq 0$. Let ρ_{XUE} be a ccq-state, where X takes values in $\{0,1\}^n$, U in the finite domain $\mathcal{U}$ and register E contains q qubits. Let F be the random and independent choice of a member of a two-universal class of hash functions $\mathcal{F}$ from $\{0,1\}^n$ into $\{0,1\}^\ell$. Then,*

$$\delta\left(\rho_{F(X)FUE}, \tfrac{1}{2^\ell}\mathbb{1} \otimes \rho_{FUE}\right) \leq \frac{1}{2} 2^{-\frac{1}{2}\left(H_\infty^\varepsilon(X|U)-q-\ell\right)} + 2\varepsilon \ . \tag{4}$$

The theorem stated here is slightly different from the version given in [27,25] in that the classical and the quantum parts of the adversary's knowledge are treated differently. A derivation of the above theorem starting from the result in [25] can be found in the full version [12].

A second tool we need is the following Min-Entropy-Splitting Lemma. Note that if the joint entropy of two random variables X_0 and X_1 is large, then one is tempted to conclude that at least one of X_0 and X_1 must still have large entropy, e.g. half of the original entropy. Whereas this is indeed true for Shannon entropy, it is in general not true for min-entropy. The following lemma, though, which appeared in a preliminary version of [33], shows that it *is* true in a randomized sense. For completeness, the proof can be found in the full version [12].

Lemma 3 (Min-Entropy-Splitting Lemma). *Let $\varepsilon \geq 0$, and let X_0, X_1 be random variables (over possibly different alphabets) with $H_\infty^\varepsilon(X_0 X_1) \geq \alpha$. Then, there exists a binary random variable C over $\{0,1\}$ such that $H_\infty^\varepsilon(X_{1-C} C) \geq \alpha/2$.*

The corollary below follows rather straightforwardly by noting that (for normalized as well as non-normalized distributions) $H_\infty(X_0 X_1 \mid Z) \geq \alpha$ holds exactly if $H_\infty(X_0X_1 \mid Z = z) \geq \alpha$ for all z, applying the Min-Entropy-Splitting Lemma, and then using the Chain Rule, Lemma 1.

Corollary 2. *Let $\varepsilon \geq 0$, and let X_0, X_1 and Z be random variables such that $H_\infty^\varepsilon(X_0X_1 \mid Z) \geq \alpha$. Then, there exists a binary random variable C over $\{0,1\}$ such that $H_\infty^{\varepsilon+\varepsilon'}(X_{1-C} \mid ZC) \geq \alpha/2 - 1 - \log(1/\varepsilon')$ for any $\varepsilon' > 0$.*

4.2   The Definition

In *1-2 OT$^\ell$*, the sender Alice sends two ℓ-bit strings S_0, S_1 to the receiver Bob in such a way that Bob can choose which string to receive, but does not learn anything about the other. On the other hand, Alice does not get to know which string Bob has chosen. The common way to build *1-2 OT$^\ell$* is by constructing a protocol for (Sender-)Randomized *1-2 OT$^\ell$*, which then can easily be converted into an ordinary *1-2 OT$^\ell$* (see, e.g., [14]). *Rand 1-2 OT$^\ell$* essentially coincides with ordinary *1-2 OT$^\ell$*, except that the two strings S_0 and S_1 are not *input* by the sender but generated uniformly at random during the protocol and *output* to the sender.

For the formal definition of the security requirements of a quantum protocol for *Rand 1-2 OT$^\ell$*, let us fix the following notation: Let C denote the binary random variable describing receiver R's choice bit, let S_0, S_1 denote the ℓ-bit long random variables describing sender S's output strings, and let Y denote the ℓ-bit long random variable describing R's output string (supposed to be S_C). Furthermore, for a fixed candidate protocol for *Rand 1-2 OT$^\ell$*, and for a fixed input distribution for C, the overall quantum state in case of a dishonest sender $\tilde{\mathsf{S}}$ is given by the ccq-state $\rho_{CY\tilde{\mathsf{S}}}$. Analogously, in the case of a dishonest receiver $\tilde{\mathsf{R}}$, we have the ccq-state $\rho_{S_0S_1\tilde{\mathsf{R}}}$.

Definition 3 (Rand 1-2 OT$^\ell$). *An ε-secure Rand 1-2 OT$^\ell$ is a quantum protocol between S and R, with R having input $C \in \{0,1\}$ while S has no input, such that for any distribution of C, if S and R follow the protocol, then S gets output $S_0, S_1 \in \{0,1\}^\ell$ and R gets $Y = S_C$ except with probability ε, and the following two properties hold:*

ε-Receiver-security: If R is honest, then for any $\tilde{\mathsf{S}}$, there exist random variables S'_0, S'_1 such that $\Pr[Y = S'_C] \geq 1 - \varepsilon$ and $\delta(\rho_{CS'_0S'_1\tilde{\mathsf{S}}}, \rho_C \otimes \rho_{S'_0S'_1\tilde{\mathsf{S}}}) \leq \varepsilon$.
ε-Sender-security: If S is honest, then for any $\tilde{\mathsf{R}}$, there exists a binary random variable D such that $\delta(\rho_{S_{1-D}S_DD\tilde{\mathsf{R}}}, \frac{1}{|2^\ell|}\mathbb{1} \otimes \rho_{S_DD\tilde{\mathsf{R}}}) \leq \varepsilon$.

If any of the above holds for $\varepsilon = 0$, then the corresponding property is said to hold perfectly. If one of the properties only holds with respect to a restricted class $\mathfrak{S}$ of $\tilde{\mathsf{S}}$'s respectively $\mathfrak{R}$ of $\tilde{\mathsf{R}}$'s, then this property is said to hold and the protocol is said to be secure against $\mathfrak{S}$ respectively $\mathfrak{R}$.

Receiver-security, as defined here, implies that whatever a dishonest sender does is as good as the following: generate the ccq-state $\rho_{S'_0S'_1\tilde{\mathsf{S}}}$ independently of C, let R know S'_C, and output $\rho_{\tilde{\mathsf{S}}}$. On the other hand, sender-security implies that whatever a dishonest receiver does is as good as the following: generate the ccq-state $\rho_{S_D D \tilde{R}}$, let S know S_D and an independent uniformly distributed S_{1-D}, and output $\rho_{\tilde{R}}$. In other words, a protocol satisfying Definition 3 is a secure implementation of the natural Rand 1-2 OT$^\ell$ ideal functionality, except that it allows a dishonest sender to influence the distribution of S_0 and S_1, and the dishonest receiver to influence the distribution of the string of his choice. This is in particular good enough for constructing a standard 1-2 OT$^\ell$ in the straightforward way.

We would like to point out the importance of requiring the existence of S'_0 and S'_1 in the formulation of receiver-security in a quantum setting: requiring only that the sender learns no information on C, as is sufficient in the classical setting (see e.g. [10]), does not prevent a dishonest sender from obtaining S_0, S_1 by a suitable measurement *after* the execution of the protocol in such a way that he can choose $S_0 \oplus S_1$ at will, and S_C is the string the receiver has obtained in the protocol. This would for instance make the straightforward construction of a bit commitment[3] based on 1-2 OT insecure.

4.3   The Protocol

We introduce a quantum protocol for Rand 1-2 OT$^\ell$ that will be shown perfectly receiver-secure against any sender and ε-sender-secure against any quantum-memory-bounded receiver for a negligible ε. The first two steps of the protocol are identical to Wiesner's "conjugate coding" protocol [32] from circa 1970 for *"transmitting two messages either but not both of which may be received"*.

The simple protocol is described in Fig. 1, where for $x \in \{0,1\}^n$ and $I \subseteq \{1,\dots,n\}$ we define $x|_I$ to be the restriction of x to the bits x_i with $i \in I$. The sender S sends random BB84 states to the receiver R, who measures all received qubits according to his choice bit C. S then picks randomly two functions from a fixed two-universal class of hash functions $\mathcal{F}$ from $\{0,1\}^n$ to $\{0,1\}^\ell$, where ℓ is to be determined later, and applies them to the bits encoded in the $+$ respectively the bits encoded in $\times$-basis to obtain the output strings S_0 and S_1. Note that we may apply a function $f \in \mathcal{F}$ to a n'-bit string with $n' < n$ by padding it with zeros (which does not decrease its entropy). S announces the encoding bases and the hash functions to the receiver who then can compute S_C. Intuitively, a dishonest receiver who cannot store all the qubits until the right bases are announced, will measure some qubits in the wrong basis and thus cannot learn both strings simultaneously.

We would like to stress that although protocol description and analysis are designed for an ideal setting with perfect noiseless quantum communication and with perfect sources and detectors, all our results can easily be extended to a more realistic noisy setting along the same lines as in [13].

It is clear by the non-interactivity of RAND 1-2 QOT$^\ell$ that a dishonest sender cannot learn anything about the receiver's choice bit. The proof of receiver-security according to Definition 3 can be found in the full version [12]; the idea,

---
[3] The committer sends two random bits of parity equal to the bit he wants to commit to, the verifier chooses to receive at random one of those bits.

RAND 1-2 QOT$^\ell$: Let c be R's choice bit.
1. S picks $x \in_R \{0,1\}^n$ and $\theta \in_R \{+, \times\}^n$, and sends $|x_1\rangle_{\theta_1}, \ldots, |x_n\rangle_{\theta_n}$ to R.
2. R measures all qubits in basis $[+, \times]_c$. Let $x' \in \{0,1\}^n$ be the result.
3. S picks two hash functions $f_0, f_1 \in_R \mathcal{F}$, announces θ and f_0, f_1 to R, and outputs $s_0 := f_0(x|_{I_0})$ and $s_1 := f_1(x|_{I_1})$ where $I_b := \{i : \theta_i = [+, \times]_b\}$.
4. R outputs $s_c = f_c(x'|_{I_c})$.

Fig. 1. Quantum Protocol for *Rand 1-2 OT$^\ell$*

though, simply is to have a dishonest $\tilde{\mathsf{S}}$ execute the protocol with a receiver that has *unbounded quantum memory* and that way can compute S'_0 and S'_1.

Proposition 1. RAND 1-2 QOT$^\ell$ *is perfectly receiver-secure.*

4.4 Security Against Memory-Bounded Dishonest Receivers

We model dishonest receivers in RAND 1-2 QOT$^\ell$ under the assumption that the maximum size of their quantum storage is bounded. Such adversaries are only required to have bounded quantum storage when Step 3 in RAND 1-2 QOT$^\ell$ is reached; before and after that, the adversary can store and carry out arbitrary quantum computations involving any number of qubits. Let $\mathfrak{R}_q$ denote the set of all possible quantum dishonest receivers $\tilde{\mathsf{R}}$ in RAND 1-2 QOT$^\ell$ which have quantum memory of size at most q when Step 3 is reached. We stress once more that apart from the restriction on the size of the quantum memory available to the adversary, no other assumption is made. In particular, the adversary is not assumed to be computationally bounded and the size of his classical memory is not restricted.

Theorem 4. RAND 1-2 QOT$^\ell$ *is ε-secure against $\mathfrak{R}_q$ for a negligible (in n) ε if $n/4 - 2\ell - q \in \Omega(n)$.*

For improved readability, we merely give a sketch of the proof; the formal proof that takes care of all the ε's is given in the full version [12].

Proof (sketch). It remains to show sender-security. Let X be the random variable that describes the sender's choice of x, where we understand the distribution of X to be conditioned on the classical information that $\tilde{\mathsf{R}}$ obtained by measuring all but γn qubits. A standard purification argument, that was also used in [13], shows that the same X can be obtained by measuring a quantum state in basis $\theta \in_R \{+, \times\}^n$, described by the random variable Θ: for each qubit $|x_i\rangle_{\theta_i}$ the sender S is instructed to send to R, S instead prepares an EPR pair $|\Phi\rangle = \frac{1}{\sqrt{2}}(|00\rangle + |11\rangle)$ and sends one part to R while keeping the other, and when Step 3 is reached, S measures her qubits.

The uncertainty relation, Theorem 1, implies that the smooth min-entropy of X given Θ is approximately $n/2$. Let now X_0 and X_1 be the two substrings of X consisting of the bits encoded in the basis $+$ or $\times$, respectively. Then the

Min-Entropy-Splitting Lemma, respectively Corollary 2, implies the existence of a binary D such that X_{1-D} has approximately $n/4$ bits of smooth min-entropy given Θ and D. From the random and independent choice of the hash functions F_0, F_1 and from the Chain Rule, Lemma 1, it follows that X_{1-D} has still about $n/4-\ell$ bits of smooth min-entropy when conditioning on Θ, D, F_D and $F_D(X_D)$. The Privacy Amplification Theorem 3, then guarantees that $S_{1-D} = F_{1-D}(X_{1-D})$ is close to random, given $\Theta, D, F_D, S_D, F_{1-D}$ and $\tilde{R}$'s quantum state of size q, if $n/4 - 2\ell - q$ is positive and linear in n. $\qquad\square$

We note that by adapting recent and more advanced techniques [33] to the quantum case, the security of RAND 1-2 QOT$^\ell$ can be proven against $\mathfrak{R}_q$ if $n/4 - \ell - q \in \Omega(n)$.

5   Application: Quantum Bit Commitment

The binding criterion for classical commitments usually requires that after the committing phase and for any dishonest committer, there exists a bit $d \in \{0,1\}$ that can only be opened with negligible probability. In the quantum world, the binding property cannot be defined the same way. If the commitment is unconditionally concealing, the committer can place himself in superposition of committing to 0 and 1 and only later make a measurement that fixes the choice. For this reason, the previous standard approach (see e.g. [17]) was to use a weaker binding condition only requiring that the probabilities p_0 and p_1 (to successfully open $b = 0$ and $b = 1$ respectively), satisfy $p_0 + p_1 \lesssim 1$. The bit commitment scheme proposed in [13] was shown to be binding in this weak sense.

We first argue that this weak notion is not really satisfactory. For instance, it does not capture the expected behavior of a commitment scheme by allowing a dishonest committer who can open the commitment with probability $1/2$ to any value, and with probability $1/2$ is unable to open it at all (depending on some event occurring during the opening). Another shortcoming of this notion is that committing bit by bit does not yield a secure string commitment—the argument that one is tempted to use requires independence of the p_b's between the different executions, which in general does not hold. We now argue that this notion is *unnecessarily* weak, even when taking into account a committer committing in superposition. We propose the following definition.

Definition 4. *An unconditionally secure commitment scheme is called* binding, *if for every (dishonest) committer there exists a classical binary random variable D whose distribution cannot be influenced by the (dishonest) committer after the commit phase and with the property that the committer's probability to successfully open the commitment to $1 - D$ is negligible.*

Note that this definition still allows a committer to commit to a superposition and otherwise honestly follow the protocol. D is then simply defined to be the outcome when the register that carries the superposition is measured. On the other hand, the definition captures exactly what one expects from a commitment scheme, except that the bit, to which the committer can open the commitment, is not fixed right after the commit phase. However, once committed, the dishonest committer *cannot influence* its distribution anymore, and thus this is not of any help to him, because he can always pretend not to know that bit.

It is also clear that with this stronger notion of the binding property, the obvious reduction from a string to a bit commitment scheme by committing bit-wise can be proven secure: the i-th execution of the bit commitment scheme guarantees a random variable D_i such that the committer cannot open the i-th bit commitment to $1 - D_i$, and thus there exists a random variable S, namely $S = (D_1, D_2, \ldots)$, such that the committer cannot open the list of commitments to any other string than S.

We show in the following that the quantum bit-commitment scheme COMM from [13] fulfills the stronger notion of binding from Definition 4 above. Let $\mathfrak{C}_q$ denote the set of all possible quantum dishonest committers $\tilde{C}$ in COMM which have quantum memory of size at most q at the start of the opening phase. Then the following holds.

Theorem 5. *The quantum bit-commitment scheme* COMM *from [13] is binding according to Definition 4 against* $\mathfrak{C}_q$ *if* $n/4 - q \in \Omega(n)$.

Proof (Sketch). By considering a purified version of the scheme and using the uncertainty relation, one can argue that X has (smooth) min-entropy about $n/2$ given Θ. The Min-Entropy-Splitting Lemma implies that there exists D such that X_{1-D} has smooth min-entropy about $n/4$ given Θ and D. Privacy amplification implies that $F(X_{1-D})$ is close to random given Θ, D, F and $\tilde{C}$'s quantum register of size q, where F is a two-universal one-bit-output hash function, which in particular implies that $\tilde{C}$ cannot guess X_{1-D}. $\qquad\square$

6   Application: Quantum Key Distribution

Let $\mathcal{B}$ be a set of orthonormal bases on a Hilbert space $\mathcal{H}$, and assume that the basis vectors of each basis $\vartheta \in \mathcal{B}$ are parametrized by the elements of some fixed set $\mathcal{X}$. We then consider QKD protocols consisting of the steps described in Fig. 2. Note that the quantum channel is only used in the preparation step. Afterwards, the communication between Alice and Bob is only classical (over an authentic channel).

As shown in [25] (Lemma 6.4.1), the length ℓ of the secret key that can be generated in the privacy amplification step of the protocol described above is given by[4]

$$\ell \approx \mathrm{H}_\infty^\varepsilon(X \mid E) - \mathrm{H}_0(C) \;,$$

where E denotes the (quantum) system containing all the information Eve might have gained during the preparation step of the protocol and where $\mathrm{H}_0(C)$ is the number of error correction bits sent from Alice to Bob. Note that this formula

---

[4] The approximation in this and the following equations holds up to some small additive value which depends logarithmically on the desired security ε of the final key.

One-Way QKD: let $N \in \mathbb{N}$ be arbitrary

1. *Preparation:* For $i = 1 \dots N$, Alice chooses at random a basis $\vartheta_i \in \mathcal{B}$ and a random element $X_i \in \mathcal{X}$. She encodes X_i into the state of a quantum system (e.g., a photon) according to the basis ϑ_i and sends this system to Bob. Bob measures each of the states he receives according to a randomly chosen basis ϑ'_i and stores the outcome Y_i of this measurement.

2. *Sifting:* Alice and Bob publicly announce their choices of bases and keep their data at position i only if $\vartheta_i = \vartheta'_i$. In the following, we denote by X and Y the concatenation of the remaining data X_i and Y_i, respectively. X and Y are sometimes called the *sifted raw key*.

3. *Error correction:* Alice computes some error correction information C depending on X and sends C to Bob. Bob computes a guess $\hat{X}$ for Alice's string X, using C and Y.

4. *Privacy amplification:* Alice chooses at random a function f from a two-universal family of hash functions and announces f to Bob. Alice and Bob then compute the final key by applying f to their strings X and $\hat{X}$, respectively.

Fig. 2. General form for *one-way* QKD protocols

can be seen as a generalization of the well known expression by Csiszár and Körner for classical key agreement [11].

Let us now assume that Eve's system E can be decomposed into a classical part Z and a purely quantum part E'. Then, using the chain rule (Lemma 3.2.9 in [25]), we find

$$\ell \approx \mathrm{H}^\varepsilon_\infty(X \mid ZE') - \mathrm{H}_0(C) \gtrsim \mathrm{H}^\varepsilon_\infty(X \mid Z) - \mathrm{H}_0(E') - \mathrm{H}_0(C) \ .$$

Because, during the preparation step, Eve does not know the encoding bases which are chosen at random from the set $\mathcal{B}$, we can apply our uncertainty relation (Theorem 2) to get a lower bound for the min-entropy of X conditioned on Eve's classical information Z, i.e., $\mathrm{H}^\varepsilon_\infty(X \mid Z) \geq Mh$, where M denotes the length of the sifted raw key X and h is the average entropic uncertainty bound for $\mathcal{B}$. Let q be the bound on the size of Eve's quantum memory E'. Moreover, let e be the average amount of error correction information that Alice has to send to Bob per symbol of the sifted raw key X. Then

$$\ell \gtrsim M(h - e) - q \ .$$

Hence, if the memory bound only grows sublinearly in the length M of the sifted raw key, then the *key rate*, i.e., the number of key bits generated per bit of the sifted raw key, is lower bounded by

$$\mathrm{rate} \geq h - e \ .$$

The Binary-Channel Setting. For a binary channel (where $\mathcal{H}$ has dimension two), the average amount of error correction information e is given by the binary

Shannon entropy[5] $H_{bin}(p) = -(p\log(p) + (1-p)\log(1-p))$, where p is the bit-flip probability of the quantum channel (for classical bits encoded according to some orthonormal basis as described above). The achievable key rate of a QKD protocol using a binary quantum channel is thus given by $rate_{binary} \geq h - H_{bin}(p)$. Summing up, we have derived the following theorem.

Theorem 6. *Let $\mathcal{B}$ be a set of orthonormal bases of $\mathcal{H}_2$ with average entropic uncertainty bound h. Then, a one-way QKD-protocol as in Fig. 2 produces a secure key against eavesdroppers whose quantum-memory size is sublinear in the length of the raw key (i.e., sublinear in the number of qubits sent from Alice to Bob) at a positive rate as long as the bit-flip probability p fulfills $H_{bin}(p) < h$.*

For the BB84 protocol, we have $h = \frac{1}{2}$ and $H_{bin}(p) < \frac{1}{2}$ is satisfied as long as $p \leq 11\%$. This bound coincides with the known bound for security against an unbounded adversary. So, the memory-bound does not give an advantage here.[6]

The situation is different for the six-state protocol where $h = \frac{2}{3}$. In this case, security against memory-bounded adversaries is guaranteed (i.e. $H_{bin}(p) < \frac{2}{3}$) as long as $p \leq 17\%$. If one requires security against an unbounded adversary, the threshold for the same protocol lies below 13%, and even the best known QKD protocol on binary channels with one-way classical post-processing can only tolerate noise up to roughly 14.1% [26]. It has also been shown that, in the unbounded model, no such protocol can tolerate an error rate of more than 16.3%.

The performance of QKD protocols against quantum-memory bounded eavesdroppers can be improved further by making the choice of the encoding bases more random. For example, they might be chosen from the set of all possible orthonormal bases on a two-dimensional Hilbert space. As shown in the full version [12], the average entropic uncertainty bound is then given by $h \approx 0.72$ and $H_{bin}(p) < 0.72$ is satisfied if $p \lesssim 20\%$. For an unbounded adversary, the thresholds are the same as for the six-state protocol.

7   Open Problems

It is interesting to investigate whether the uncertainty relation (Theorem 2) still holds if the measurement bases $(\Theta_1, \ldots, \Theta_n)$ are randomly chosen from a relatively small subset of $\mathcal{B}^n$ (rather than from the entire set $\mathcal{B}^n$). Such an extension would reduce the amount of randomness that is needed in applications. In particular, in the context of QKD with quantum-memory-bounded eavesdroppers, it would allow for more efficient protocols that use a relatively short initial secret key in order to select the bases for the preparation and measurement of the states and, hence, avoid the sifting step.

Another open problem is to consider protocols using higher-dimensional quantum systems. The results mentioned in the previous paragraph show that for

---

[5] This value of e is only achieved if an optimal error-correction scheme is used. In practical implementations, the value of e might be slightly larger.
[6] Note, however, that the analysis given here might not be optimal.

high-dimensional systems, the average entropic uncertainty bound converges to its theoretical maximum. The maximal tolerated channel noise might thus be higher for such protocols (depending on the noise model for higher-dimensional quantum channels).

References

1. Alon, N., Spencer, J.: The Probabilistic Method, 2nd edn. Series in Discrete Mathematics and Optimization. Wiley, Chichester (2000)
2. Azuma, K.: Weighted sums of certain dependent random variables. Tôhoku Mathematical Journal 19, 357–367 (1967)
3. Bennett, C.H., Brassard, G.: Quantum cryptography: Public key distribution and coin tossing. In: Proceedings of IEEE International Conference on Computers, Systems, and Signal Processing, pp. 175–179. IEEE Computer Society Press, Los Alamitos (1984)
4. Bennett, C.H., Brassard, G., Crépeau, C., Maurer, U.M.: Generalized privacy amplification. IEEE Transactions on Information Theory 41, 1915–1923 (1995)
5. Bennett, C.H., Brassard, G., Robert, J.-M.: Privacy amplification by public discussion. SIAM J. Comput. 17(2), 210–229 (1988)
6. Bialynicki-Birula, I.: Formulation of the uncertainty relations in terms of the Rényi entropies. Physical Review A. 74, 52101 (2006)
7. Bialynicki-Birula, I., Mycielski, J.: Uncertainty relations for information entropy. Communications in Mathematical Physics 129(44) (1975)
8. Cachin, C.: Smooth entropy and Rényi entropy. In: Fumy, W. (ed.) EUROCRYPT 1997. LNCS, vol. 1233, pp. 193–208. Springer, Heidelberg (1997)
9. Carter, J.L., Wegman, M.N.: Universal classes of hash functions. In: 9th Annual ACM Symposium on Theory of Computing (STOC), pp. 106–112. ACM Press, New York (1977)
10. Crépeau, C., Savvides, G., Schaffner, C., Wullschleger, J.: Information-theoretic conditions for two-party secure function evaluation. In: Vaudenay, S. (ed.) EUROCRYPT 2006. LNCS, vol. 4004, pp. 538–554. Springer, Heidelberg (2006)
11. Csiszár, I., Körner, J.: Broadcast channels with confidential messages. IEEE Transactions on Information Theory 24(3), 339–348 (1978)
12. Damgård, I.B., Fehr, S., Renner, R., Salvail, L., Schaffner, C.: A tight high-order entropic quantum uncertainty relation with applications (2007), available at http://arxiv.org/abs/quant-ph/0612014
13. Damgård, I.B., Fehr, S., Salvail, L., Schaffner, C.: Cryptography in the bounded quantum-storage model. In: 46th Annual IEEE Symposium on Foundations of Computer Science (FOCS), pp. 449–458. IEEE Computer Society Press, Los Alamitos (2005)
14. Damgård, I.B., Fehr, S., Salvail, L., Schaffner, C.: Oblivious transfer and linear functions. In: Dwork, C. (ed.) CRYPTO 2006. LNCS, vol. 4117, pp. 427–444. Springer, Heidelberg (2006)
15. Damgård, I.B., Pedersen, T.B., Salvail, L.: On the key-uncertainty of quantum ciphers and the computational security of one-way quantum transmission. In: Cachin, C., Camenisch, J.L. (eds.) EUROCRYPT 2004. LNCS, vol. 3027, pp. 91–108. Springer, Heidelberg (2004)
16. Deutsch, D.: Uncertainty in quantum measurements. Physical Review Letters 50(9), 631–633 (1983)

17. Dumais, P., Mayers, D., Salvail, L.: Perfectly concealing quantum bit commitment from any quantum one-way permutation. In: Preneel, B. (ed.) EUROCRYPT 2000. LNCS, vol. 1807, pp. 300–315. Springer, Heidelberg (2000)
18. Fuchs, C.A., Gisin, N., Griffiths, R.B., Niu, C.-S., Peres, A.: Optimal eavesdropping in quantum cryptography. Physical Review A 56, 1163–1172 (1997)
19. Hilgevood, J., Uffink, J.: The mathematical expression of the uncertainty principle. In: Microphysical Reality and Quantum Description, Kluwer Academic Publishers, Dordrecht (1988)
20. Impagliazzo, R., Levin, L.A., Luby, M.: Pseudo-random generation from one-way functions. In: 21st Annual ACM Symposium on Theory of Computing (STOC), pp. 12–24. ACM Press, New York (1989)
21. Kraus, K.: Complementary observables and uncertainty relations. Physical Review D 35(10), 3070–3075 (1987)
22. Larsen, U.: Superspace geometry: the exact uncertainty relationship between complementary aspects. Journal of Physics A: Mathematical and General 23(7), 1041–1061 (1990)
23. Lütkenhaus, N.: Security against individual attacks for realistic quantum key distribution. Physical Review A. 61, 52304 (2000)
24. Maassen, H., Uffink, J.B.M.: Generalized entropic uncertainty relations. Physical Review Letters 60(12), 1103–1106 (1988)
25. Renner, R.: Security of Quantum Key Distribution. PhD thesis, ETH Zürich (2005), http://arxiv.org/abs/quant-ph/0512258
26. Renner, R., Gisin, N., Kraus, B.: An information-theoretic security proof for QKD protocols. Phys. Rev. A. 72(012332) (2005)
27. Renner, R., König, R.: Universally composable privacy amplification against quantum adversaries. In: Kilian, J. (ed.) TCC 2005. LNCS, vol. 3378, pp. 407–425. Springer, Heidelberg (2005)
28. Renner, R., Wolf, S.: Simple and tight bounds for information reconciliation and privacy amplification. In: Roy, B. (ed.) ASIACRYPT 2005. LNCS, vol. 3788, pp. 199–216. Springer, Heidelberg (2005)
29. Sánchez -Ruiz, J.: Entropic uncertainty and certainty relations for complementary observables. Physics Letters A 173(3), 233–239 (1993)
30. Sánchez -Ruiz, J.: Improved bounds in the entropic uncertainty and certainty relations for complementary observables. Physics Letters A 201(2–3), 125–131 (1995)
31. Wegman, M.N., Carter, J.L.: New classes and applications of hash functions. In: 20th Annual IEEE Symposium on Foundations of Computer Science (FOCS), pp. 175–182. IEEE Computer Society Press, Los Alamitos (1979)
32. Wiesner, S.: Conjugate coding. SIGACT News 15(1), 78–88 (1983), original manuscript written circa 1970
33. Wullschleger, J.: Oblivious-Transfer amplification. In: Naor, M. (ed.) EUROCRYPT 2007. LNCS, vol. 4515, pp. 555–572. Springer, Heidelberg (2007)

Finding Small Roots of Bivariate Integer Polynomial Equations: A Direct Approach

Jean-Sébastien Coron

University of Luxembourg

Abstract. Coppersmith described at Eurocrypt 96 an algorithm for finding small roots of bivariate integer polynomial equations, based on lattice reduction. A simpler algorithm was later proposed in [9], but it was asymptotically less efficient than Coppersmith's algorithm. In this paper, we describe an analogous simplification but with the same asymptotic complexity as Coppersmith. We illustrate our new algorithm with the problem of factoring RSA moduli with high-order bits known; in practical experiments our method is several orders of magnitude faster than [9].

Keywords: Coppersmith's theorem, lattice reduction, cryptanalysis.

1 Introduction

At Eurocrypt 96, Coppersmith described how lattice reduction can be used to find small roots of polynomial equations [5,6,7]. Coppersmith's technique has found numerous applications for breaking variants of RSA; for example, cryptanalysis of RSA with $d < N^{.29}$ [3], polynomial-time factorization of $N = p^r q$ for large r [4], and cryptanalysis of RSA with small secret CRT-exponents [18,1]. Coppersmith's technique was also used to obtain an improved security proof for OAEP with small public exponent [23], and to show the deterministic equivalence between recovering the private exponent d and factoring N [10,19].

There are two main theorems from Coppersmith. The first one concerns finding small roots of $p(x) = 0 \mod N$ when the factorization of N is unknown. Coppersmith proved that any root x_0 with $|x_0| < N^{1/\delta}$ can be found in polynomial time, where $\delta = \deg p$. The technique consists in building a lattice that contains the solutions of the modular polynomial equation; all small solutions are shown to belong to an hyperplane of the lattice; an equation of this hyperplane is obtained by considering the last vector of an LLL-reduced basis; this gives a polynomial $h(x)$ such that $h(x_0) = 0$ over the integers, from which one can recover x_0. The method can be extended to handle multivariate modular polynomial equations, but the extension is heuristic only.

Coppersmith's algorithm was further simplified by Howgrave-Graham in [13]. Howgrave-Graham's approach is more direct and consists in building a lattice of polynomials that are multiples of $p(x)$ and N; then by lattice reduction one computes a polynomial with small coefficients such that $h(x_0) = 0 \mod N^k$; if the coefficient of $h(x)$ are sufficiently small then $h(x_0) = 0$ must hold over $\mathbb{Z}$ as

A. Menezes (Ed.): CRYPTO 2007, LNCS 4622, pp. 379–394, 2007.

well, which enables to recover x_0. Howgrave-Graham's approach seems easier to analyze, in particular for the heuristic extension to multivariate modular equations, for which there is much more freedom in selecting the polynomial multiples than for the univariate case. Howgrave-Graham's approach was actually used in all subsequent applications of Coppersmith's technique [1,3,4,18,19,20].

Coppersmith's second theorem concerns finding small roots of bivariate integer polynomial equations $p(x,y) = 0$ over the integers (not modulo N). Coppersmith proved that if $|x_0| < X$ and $|y_0| < Y$ with $XY < W^{2/(3\delta)}$ then such root (x_0, y_0) can be found in polynomial-time, where $W := \max_{ij} |p_{ij}| X^i Y^j$. As for the univariate case, the algorithm consists in building a lattice containing the solutions of the polynomial equation; all small solutions are shown to belong to an hyperplane of the lattice, that is obtained by considering the last vector of an LLL-reduced basis. The equation of the hyperplane gives another polynomial $h(x,y)$ with the same root (x_0, y_0) as $p(x,y)$, which enables to recover (x_0, y_0). There can be improved bounds depending on the shape of the polynomial $p(x,y)$; see [2] for a complete analysis. As for the univariate case, the method extends heuristically to more variables. However, as mentioned in [8], the analysis is more difficult to follow than for the univariate case.

For Coppersmith's second theorem, a simplification was later proposed at Eurocrypt 2004 [9], analogous to Howgrave-Graham's simplification for the univariate case. It consists in generating an arbitrary integer n of appropriate size and constructing a lattice of polynomials that are multiples of $p(x,y)$ and n; then by lattice reduction one computes a polynomial with small coefficients such that $h(x_0, y_0) = 0 \mod n$; if the coefficients of $h(x,y)$ are sufficiently small, then $h(x_0, y_0) = 0$ holds over $\mathbb{Z}$, which enables to recover (x_0, y_0) by taking the resultant of $h(x,y)$ and $p(x,y)$. As for the univariate case, this approach seems easier to implement; it was later used in [11] for partial key exposure attacks on RSA, and in [16] to break one variant of RSA.

However, as opposed to the univariate case, this later simplification is not fully satisfactory because asymptotically its complexity is worse than for Coppersmith's second theorem. Namely, the algorithm in [9] is polynomial time under the stronger condition $XY < W^{2/(3\delta)-\varepsilon}$, for any constant $\varepsilon > 0$; but for $XY < W^{2/(3\delta)}$ the algorithm has exponential-time complexity :

$$\exp\left(\mathcal{O}(\log^{2/3} W)\right),$$

whereas Coppersmith's algorithm is polynomial time.

Therefore in this paper we describe a new algorithm for the bivariate integer case, with a simplification analogous to Howgrave-Graham and [9], but with the same polynomial-time complexity as in Coppersmith's algorithm; namely for $XY < W^{2/(3\delta)}$ our algorithm has complexity

$$\mathcal{O}(\log^{15} W)$$

using LLL [17] and $\mathcal{O}(\log^{11} W)$ using the improved L^2 algorithm [21]. This is done by taking a well chosen integer n (rather than arbitrary) when building the lattice of polynomials; this enables to eliminate most columns of the lattice and then apply LLL on a sub-lattice of smaller dimension. Our new algorithm is easy to implement and performs well in practice. In Section 4 we show the results of practical experiments for the factoring with high-order bits known attack against RSA; we show that the running time is improved by several orders of magnitude compared to [9].

2 Preliminaries

Let $u_1, \ldots, u_\omega \in \mathbb{Z}^n$ be linearly independent vectors with $\omega \leq n$. A lattice L spanned by $\langle u_1, \ldots, u_\omega \rangle$ is the set of all integer linear combinations of $u_1, \ldots, u_\omega$. Such a set of vectors u_i's is called a lattice *basis*. We say that the lattice is full rank if $\omega = n$.

Any two bases of the same lattice L are related by some integral matrix of determinant ± 1. Therefore, all the bases have the same Gramian determinant $\det_{1 \leq i,j \leq \omega} < u_i, u_j >$. One defines the *determinant* of the lattice L as the square root of the Gramian determinant. If the lattice L is full rank, then the determinant of L is equal to the absolute value of the determinant of the $\omega \times \omega$ matrix whose rows are the basis vectors $u_1, \ldots, u_\omega$.

Theorem 1 (LLL). *Let L be a lattice spanned by $(u_1, \ldots, u_\omega) \in \mathbb{Z}^n$, where the Euclidean norm of each vector is bounded by B. The LLL algorithm, given $(u_1, \ldots, u_\omega)$, finds in time $\mathcal{O}(\omega^5 n \log^3 B)$ a vector b_1 such that:*

$$\|b_1\| \leq 2^{(\omega-1)/4} \det(L)^{1/\omega}.$$

In order to obtain a better complexity, one can use an improved version of LLL due to Nguyen and Stehlé, called the L^2 algorithm [21]. The L^2 algorithm achieves the same bound on $\|b_1\|$ but in time $\mathcal{O}(\omega^4 n(\omega + \log B) \log B)$.

In this paper we also consider lattices generated by a set of vectors that are not necessarily linearly independent. Let $u_1, \ldots, u_m \in \mathbb{Z}^n$ with $m \geq n$; the lattice L generated by $\langle u_1, \ldots, u_m \rangle$ consists of all integral linear combinations of $u_1, \ldots, u_m$. A lattice basis for L can be obtained by triangularization of $u_1, \ldots, u_m$; a polynomial-time triangularization algorithm is described in [12]; more details will be given in Section 3.1.

We prove a simple lemma that will be useful when analyzing the determinant of such lattices; it shows that the determinant of a full rank lattice generated by a matrix of row vectors is not modified when performing elementary operations on the *columns* of the matrix :

Lemma 1. *Let M be an integer matrix with m rows and n columns, with $m \geq n$. Let L be the lattice generated by the rows of M. Let M' be a matrix obtained by elementary operations on the columns of M, and let L' be the lattice generated by the rows of M'. Then if L is full rank, L' is full rank with $\det L' = \det L$.*

Proof. See Appendix.

3   Our New Algorithm

We consider a polynomial $p(x, y)$ with coefficients in $\mathbb{Z}$ with maximum degree δ independently in x, y :

$$p(x, y) = \sum_{0 \leq i, j \leq \delta} p_{i,j} x^i y^j.$$

We are looking for an integer pair (x_0, y_0) such that $p(x_0, y_0) = 0$ and $|x_0| < X$ and $|y_0| < Y$. We assume that $p(x, y)$ is irreducible over the integers.

Let k be an integer > 0. We consider the set of polynomials :

$$s_{a,b}(x, y) = x^a \cdot y^b \cdot p(x, y), \quad \text{for } 0 \leq a, b < k \tag{1}$$

$$r_{i,j}(x, y) = x^i \cdot y^j \cdot n, \quad \text{for } 0 \leq i, j < k + \delta \tag{2}$$

where the integer n is generated in the following way.

Let indexes (i_0, j_0) be such that $0 \leq i_0, j_0 \leq \delta$; let S be the matrix of row vectors obtained by taking the coefficients of the polynomials $s_{a,b}(x, y)$ for $0 \leq a, b < k$, but only in the monomials $x^{i_0+i} y^{j_0+j}$ for $0 \leq i, j < k$. There are k^2 such polynomials $s_{a,b}(x, y)$ and k^2 such monomials, so the matrix S is a square matrix of dimension k^2 (see Figure 1 for an illustration); we take :

$$n := |\det S|.$$

We will show in Lemma 3 that for a well chosen (i_0, j_0), the value $|\det S|$ is lower bounded; in particular, this implies that $|\det S| > 0$ and therefore matrix S is invertible.

	x^2y^2	x^2y	xy^2	xy
$s_{1,1}(x, y)$	a	b	c	d
$s_{1,0}(x, y)$		a		c
$s_{0,1}(x, y)$			a	b
$s_{0,0}(x, y)$				a

$S =$

Fig. 1. Matrix S with $p(x, y) = axy + bx + cy + d$, for $k = 2$ and $(i_0, j_0) = (1, 1)$. We get $n = |\det S| = a^4$.

Let $h(x, y)$ be a linear combination of the polynomials $s_{a,b}(x, y)$ and $r_{i,j}(x, y)$. Since we have that $s_{a,b}(x_0, y_0) = 0 \mod n$ for all a, b and $r_{i,j}(x_0, y_0) = 0 \mod n$ for all i, j, we obtain :

$$h(x_0, y_0) = 0 \mod n.$$

The following lemma, due to Howgrave-Graham [13], shows that if the coefficients of polynomial $h(x, y)$ are sufficiently small, then $h(x_0, y_0) = 0$ holds over the integers. For a polynomial $h(x, y) = \sum_{i,j} h_{ij} x^i y^j$, we define $\|h(x, y)\|^2 := \sum_{i,j} |h_{ij}|^2$.

Lemma 2 (Howgrave-Graham). *Let $h(x,y) \in \mathbb{Z}[x,y]$ which is a sum of at most ω monomials. Suppose that $h(x_0, y_0) = 0 \mod n$ where $|x_0| \leq X$ and $|y_0| \leq Y$ and $\|h(xX, yY)\| < n/\sqrt{\omega}$. Then $h(x_0, y_0) = 0$ holds over the integers.*

Proof. We have:

$$
|h(x_0, y_0)| = \left| \sum h_{ij} x_0^i y_0^i \right| = \left| \sum h_{ij} X^i Y^j \left(\frac{x_0}{X}\right)^i \left(\frac{y_0}{Y}\right)^j \right|
$$
$$
\leq \sum \left| h_{ij} X^i Y^j \left(\frac{x_0}{X}\right)^i \left(\frac{y_0}{Y}\right)^j \right| \leq \sum \left| h_{ij} X^i Y^j \right|
$$
$$
\leq \sqrt{\omega} \|h(xX, yY)\| < n
$$

Since $h(x_0, y_0) = 0 \mod n$, this gives $h(x_0, y_0) = 0$. $\square$

We consider the lattice L generated by the row vectors formed with the coefficients of polynomials $s_{a,b}(xX, yY)$ and $r_{i,j}(xX, yY)$. In total, there are $k^2 + (k+\delta)^2$ such polynomials; moreover these polynomials are of maximum degree $\delta + k - 1$ in x, y, so they contain at most $(\delta + k)^2$ coefficients. Let M be the corresponding matrix of row vectors; M is therefore a rectangular matrix with $k^2 + (k+\delta)^2$ rows and $(k+\delta)^2$ columns (see Figure 2 for an illustration). Observe that the rows of M do not form a basis of L (because there are more rows than columns), but L is a full rank lattice of dimension $(k + \delta)^2$ (because the row vectors corresponding to polynomials $r_{i,j}(xX, yY)$ form a full rank lattice).

	x^2y^2	x^2y	xy^2	xy	x^2	y^2	x	y	1
$s_{1,1}(xX, yY)$	aX^2Y^2	bX^2Y	cXY^2	dXY					
$s_{1,0}(xX, yY)$		aX^2Y		cXY	bX^2		dX		
$s_{0,1}(xX, yY)$			aXY^2	bXY		cY^2		dY	
$s_{0,0}(xX, yY)$				aXY			bX	cY	d
$r_{2,2}(xX, yY)$	nX^2Y^2								
$r_{2,1}(xX, yY)$		nX^2Y							
$r_{1,2}(xX, yY)$			nXY^2						
$r_{1,1}(xX, yY)$				nXY					
$r_{2,0}(xX, yY)$					nX^2				
$r_{0,2}(xX, yY)$						nY^2			
$r_{1,0}(xX, yY)$							nX		
$r_{0,1}(xX, yY)$								nY	
$r_{0,0}(xX, yY)$									n

$M =$ (applies to the block of rows above)

Fig. 2. Lattice of polynomials with $p(x,y) = axy + bx + cy + d$, for $k = 2$ and $(i_0, j_0) = (1, 1)$

Let L_2 be the sublattice of L where the coefficients corresponding to all monomials of the form $x^{i_0+i} y^{j_0+j}$ with $0 \leq i, j < k$ are set to zero (those monomials correspond to the matrix left-hand block in Fig. 2). There are k^2 such monomials, so L_2 is a full rank lattice of dimension :

$$\omega = (\delta + k)^2 - k^2 = \delta^2 + 2 \cdot k \cdot \delta. \tag{3}$$

A matrix basis for L_2 can be obtained by first triangularizing M using elementary row operations and then taking the corresponding submatrix (see Fig. 3). A polynomial-time triangularization algorithm is described in [12]; more details will be given in Section 3.1.

	x^2y^2	x^2y	xy^2	xy	x^2	y^2	x	y	1
$s_{1,1}(xX,yY)$	aX^2Y^2	bX^2Y	cXY^2	dXY					
$s_{1,0}(xX,yY)$		aX^2Y		cXY	bX^2	dX			
$s_{0,1}(xX,yY)$			aXY^2	bXY		cY^2		dY	
$s_{0,0}(xX,yY)$				aXY			bX	cY	d
$q_0(xX,yY)$					$*$	$*$	$*$	$*$	$*$
$q_1(xX,yY)$						$*$	$*$	$*$	$*$
$q_2(xX,yY)$							$*$	$*$	$*$
$q_3(xX,yY)$								$*$	$*$
$q_4(xX,yY)$									$*$

Fig. 3. Triangularized lattice of polynomials with $p(x,y) = axy + bx + cy + d$, for $k = 2$ and $(i_0, j_0) = (1,1)$. The 5 polynomials $q_i(xX, yY)$ generate lattice L_2, with coefficients only in the 5 monomials x^2, y^2, x, y and 1. Algorithm LLL is applied on the corresponding 5-dimensional lattice.

We apply the LLL algorithm on lattice L_2. From theorem 1, we obtain a non-zero polynomial $h(x,y)$ that satisfies $h(x_0, y_0) = 0 \mod n$ and :

$$\|h(xX, yY)\| \leq 2^{(\omega-1)/4} \cdot \det(L_2)^{1/\omega}. \tag{4}$$

From lemma 2, this implies that if :

$$2^{(\omega-1)/4} \cdot \det(L_2)^{1/\omega} \leq \frac{n}{\sqrt{\omega}}, \tag{5}$$

then $h(x_0, y_0) = 0$ must hold over the integers.

Now we claim that polynomial $h(x,y)$ cannot be a multiple of $p(x,y)$. Assume the contrary; then the row vector coefficients of $h(x,y)$ is a linear combination of the row vector coefficients of polynomials $s_{a,b}(x,y)$ only. Given that matrix S contains the coefficients of $s_{a,b}(x,y)$ for monomials $x^{i+i_0} y^{j+j_0}$ and given that $h(x,y)$ does not contain such monomials (because $h(x,y)$ lies in L_2), this gives a linear combination of the rows of S equal to zero with non-zero coefficients; a contradiction since matrix S is invertible.

The polynomial $p(x,y)$ being irreducible, this implies that $p(x,y)$ and $h(x,y)$ are algebraically independent with a common root (x_0, y_0); therefore, taking :

$$Q(x) = \text{Resultant}_y(h(x,y), p(x,y))$$

gives a non-zero integer polynomial such that $Q(x_0) = 0$. Using any standard root-finding algorithm, we can recover x_0, and finally y_0 by solving $p(x_0, y) = 0$. This terminates the description of our algorithm.

It remains to compute the determinant of lattice L_2. First we consider the same matrices of row vectors as previously, except that we remove the $X^i Y^j$ powers. Therefore let M' be the same matrix as M, except that we take the coefficients of polynomials $s_{a,b}(x,y)$ and $r_{i,j}(x,y)$, instead of $s_{a,b}(xX,yY)$ and $r_{i,j}(xX,yY)$; matrix M' has $k^2 + (k+\delta)^2$ rows and $(k+\delta)^2$ columns. We put the coefficients corresponding to monomials $x^{i+i_0} y^{j+j_0}$ for $0 \leq i,j < k$ on the left hand block, which has therefore k^2 columns; matrix M' has then the following form :

$$M' = \begin{bmatrix} S & T \\ nI_{k^2} & 0 \\ 0 & nI_w \end{bmatrix}$$

where S is the previously defined square matrix of dimension k^2, while T is a matrix with k^2 rows and $\omega = k^2 + 2k\delta$ columns. Let L' be the lattice generated by the rows of M', and let L'_2 be the sublattice where all coefficients corresponding to monomials $x^{i+i_0} y^{j+j_0}$ for $0 \leq i,j < k$ are set to zero. Note that lattice L' corresponds to lattice L without the $X^i Y^j$ powers, whereas lattice L'_2 corresponds to lattice L_2.

Since $n = |\det S|$, we can find an integer matrix S' satisfying :

$$S' \cdot S = nI_{k^2},$$

namely S' is (up to sign) the adjoint matrix (or comatrix) of S, verifying $S' \cdot S = (\det S)I_{k^2}$. By elementary operations on the rows of M', we can therefore subtract $S' \cdot S$ to the nI_{k^2} block of M'; this gives the following matrix :

$$M'_2 = \begin{bmatrix} I_{k^2} & 0 & 0 \\ -S' & I_{k^2} & 0 \\ 0 & 0 & I_\omega \end{bmatrix} \cdot M' = \begin{bmatrix} S & T \\ 0 & T' \\ 0 & nI_\omega \end{bmatrix}, \tag{6}$$

where $T' = -S' \cdot T$ is a matrix with k^2 rows and ω columns. By elementary operations on the rows of M'_2, we obtain :

$$M'_3 = U \cdot M'_2 = \begin{bmatrix} S & T \\ 0 & T'' \\ & 0 \end{bmatrix},$$

where T'' is a square matrix of dimension ω. We obtain that T'' is a row matrix basis of lattice L'_2, which gives :

$$\det L' = \left| \det \begin{bmatrix} S & T \\ 0 & T'' \end{bmatrix} \right| = |\det S| \cdot |\det T''| = |\det S| \cdot \det L'_2 = n \cdot \det L'_2. \tag{7}$$

We now proceed to compute $\det L'$. The polynomial $p(x,y)$ being irreducible, the gcd of its coefficients is equal to 1. This implies that by elementary operation of the *columns* of M', we can obtain a matrix whose left upper $k^2 \times k^2$ block is the identity matrix and the right upper block is zero. From lemma 1, this does not change the determinant of the generated lattice. Let V be the corresponding unimodular transformation matrix of dimension $(\delta + k)^2$; this gives :

$$M_4' = M' \cdot V = \begin{bmatrix} I_{k^2} & 0 \\ & nV \end{bmatrix}.$$

By elementary row operations on M_4' based on V^{-1} we obtain :

$$M_5' = \begin{bmatrix} I_{k^2} & 0 \\ 0 & V^{-1} \end{bmatrix} \cdot M_4' = \begin{bmatrix} I_{k^2} & 0 \\ & nI_{(\delta+k)^2} \end{bmatrix} = \begin{bmatrix} I_{k^2} & 0 & \\ nI_{k^2} & 0 & \\ 0 & & nI_\omega \end{bmatrix},$$

which again by elementary row operations gives :

$$M_6' = U' \cdot M_5' = \begin{bmatrix} I_{k^2} & 0 \\ 0 & nI_\omega \\ 0 & 0 \end{bmatrix}.$$

Finally this implies :

$$\det L' = \det \begin{bmatrix} I_{k^2} & 0 \\ 0 & nI_\omega \end{bmatrix} = n^\omega \tag{8}$$

Combining equations (7) and (8), we obtain :

$$\det L_2' = n^{\omega-1}.$$

Recall that the columns of L_2' correspond to monomials $x^i y^j$ for $0 \le i,j < \delta+k$, excluding monomials $x^{i_0+i} y^{j_0+j}$ for $0 \le i,j < k$. The columns of lattice L_2 are obtained from the columns of L_2' by multiplication with the corresponding $X^i Y^j$ powers; this gives :

$$\det L_2 = \det L_2' \cdot \frac{\displaystyle\prod_{0 \le i,j < \delta+k} X^i Y^j}{\displaystyle\prod_{0 \le i,j < k} X^{i_0+i} Y^{j_0+j}}$$

$$= n^{\omega-1} \cdot \frac{(XY)^{(\delta+k-1)\cdot(\delta+k)^2/2 - (k-1)\cdot k^2/2}}{(X^{i_0} Y^{j_0})^{k^2}}$$

From inequality (5) we obtain the following condition for Howgrave-Graham's lemma to apply :

$$2^{\omega \cdot (\omega-1)/4} \cdot \frac{(XY)^{(\delta+k-1)\cdot(\delta+k)^2/2 - (k-1)\cdot k^2/2}}{(X^{i_0} Y^{j_0})^{k^2}} \le \frac{n}{\omega^{\omega/2}}. \tag{9}$$

It remains to bound $n = |\det S|$ as a function of the coefficients of $p(x,y)$. Let

$$W = \max_{i,j} |p_{ij}| X^i Y^j$$

The following lemma shows that for the right choice of (i_0, j_0), the determinant of S is bounded in absolute value :

Lemma 3. *Given (u, v) such that $W = |p_{uv}|X^u Y^v$, let indices (i_0, j_0) that maximize the quantity $8^{(i-u)^2 + (j-v)^2} |p_{ij}| X^i Y^j$. Then*

$$\left(\frac{W}{X^{i_0} Y^{j_0}}\right)^{k^2} 2^{-6k^2 \delta^2 - 2k^2} \le |\det S| \le \left(\frac{W}{X^{i_0} Y^{j_0}}\right)^{k^2} \cdot 2^{k^2}. \tag{10}$$

Proof. The proof is very similar to the proof of Lemma 3 in [7]; see Appendix B.

Combining inequalities (9) and (10) with $n = |\det S|$ and $\sqrt{\omega} \le 2^{\omega/2}$, we obtain the sufficient condition :

$$2^{\omega \cdot (\omega - 1)/4} \cdot (XY)^{(\delta + k - 1) \cdot (\delta + k)^2 / 2 - (k-1) \cdot k^2/2} \le W^{k^2} \cdot 2^{-6k^2 \delta^2 - 2k^2} \cdot 2^{-\omega^2/2}.$$

This condition is satisfied if :

$$XY < W^\alpha \cdot 2^{-9\delta},$$

where

$$\alpha = \frac{2k^2}{\delta \cdot (3k^2 + k(3\delta - 2) + \delta^2 - \delta)}.$$

Finally we obtain the sufficient condition :

$$XY < W^{2/(3\delta) - 1/k} \cdot 2^{-9\delta}. \tag{11}$$

The running time is dominated by the time it takes to run LLL on a lattice of dimension $\delta^2 + 2k\delta$, with entries bounded by $\mathcal{O}(W^{k^2})$. Namely, the entries of a matrix basis for L_2 can be reduced modulo $n \cdot X^i Y^j$ on the columns corresponding to monomial $x^i y^j$, because of polynomials $r_{ij}(xX, yY) = n \cdot X^i Y^j x^i y^j$. This implies that we can obtain a matrix basis for L_2 whose entries are bounded by $\mathcal{O}(nX^{\delta+k}Y^{\delta+k})$. From inequality (10) we have $n = \mathcal{O}(W^{k^2})$; using (11) this implies that the matrix entries can be bounded by $\mathcal{O}(W^{k^2})$. From theorem 1 and taking $k > \delta$, the running time is therefore bounded by :

$$\mathcal{O}\left(\delta^6 k^{12} \log^3 W\right)$$

using the LLL algorithm, and $\mathcal{O}\left(\delta^5 k^9 \log^2 W\right)$ using the improved L^2 algorithm. Finally, under the weaker condition

$$XY < W^{2/(3\delta)},$$

one can set $k = \lfloor \log W \rfloor$ and do exhaustive search on the high order $\mathcal{O}(\delta)$ unknown bits of x_0. The running time is then polynomial in 2^δ and $\log W$. Moreover, for a fixed δ, the running time is $\mathcal{O}(\log^{15} W)$ using the LLL algorithm, and $\mathcal{O}(\log^{11} W)$ using the improved L^2 algorithm. Thus we have shown :

Theorem 2 (Coppersmith). *Let $p(x, y)$ be an irreducible polynomial in two variables over $\mathbb{Z}$, of maximum degree δ in each variable separately. Let X and Y be upper bounds on the desired integer solution (x_0, y_0), and let $W = \max_{i,j} |p_{ij}| X^i Y^j$. If $XY < W^{2/(3\delta)}$, then in time polynomial in $(\log W, 2^\delta)$, one can find all integer pairs (x_0, y_0) such that $p(x_0, y_0) = 0$, $|x_0| \le X$, and $|y_0| \le Y$.*

As in [7], there can be improved bounds depending on the shape of the polynomial $p(x, y)$:

Theorem 3 (Coppersmith). *With the hypothesis of Theorem 2, except that $p(x, y)$ has total degree δ, the appropriate bound is :*

$$XY < W^{1/\delta}.$$

Proof. See the full version of this paper.

3.1   Computing a Basis of L_2

In the previous section one needs to compute a basis for lattice L_2, which is then given as input to the LLL algorithm. Such lattice basis can be obtained by triangularization of matrix M; a matrix A is upper triangular if $A_{ij} = 0$ for $i > j$ (as illustrated in Figure 3). A triangularization algorithm is described in [12]; for an $m \times n$ matrix of row vectors, its running time is $\mathcal{O}(n^{3+\varepsilon} m \log^{1+\varepsilon} B)$ for any $\varepsilon > 0$, when the matrix entries are bounded by B in absolute value.

Observe that we don't need to triangularize the full matrix M. Namely from our analysis of the previous section, equation (6) can be used to obtain a set of row vectors that generate L_2; a triangularization algorithm is then applied to derive a lattice basis for L_2. For this we need to compute matrix S' such that $S' \cdot S = (\det S) \cdot I$; we note that this is implemented in Shoup's NTL library [22].

Another possibility is to compute the Hermite Normal form (HNF) of M. An $m \times n$ matrix A of rank n is in HNF if it is upper triangular and $a_{ii} > 0$ for all $1 \le i \le n$ and $0 \le a_{ij} < a_{jj}$ for all $1 \le j \le n$ and $1 \le i < j$. A classical result says that if an $m \times n$ matrix M is of rank n then there exists a $m \times m$ unimodular matrix U such that $U \cdot M$ is in HNF; moreover the HNF is unique. An algorithm for computing the HNF is also described in [12], with the same asymptotic complexity as triangularization. A HNF algorithm is also implemented in Shoup's NTL library [1].

3.2   Difference with the Algorithm in [9]

In [9] a similar lattice L is built but with an integer n which is co-prime with the constant coefficient of $p(x, y)$. This implies that the full lattice L must be considered, whose dimension $d_L = (\delta + k)^2$ grows quadratically with k instead of linearly as in our sub-lattice of dimension $\omega = \delta^2 + 2k\delta$.

With the full lattice L the LLL fudge factor is then $2^{(d_L-1)/4} = 2^{\mathcal{O}(k^2)}$ instead of $2^{(\omega-1)/4} = 2^{\mathcal{O}(k)}$. This translates in the bound for XY into the condition $XY < W^{2/(3\delta)-1/k} \cdot 2^{-\mathcal{O}(k^2+\delta)}$ instead of $XY < W^{2/(3\delta)-1/k} \cdot 2^{-9\delta}$. This implies that in [9], in order to reach the bound $XY < W^{2/(3\delta)}$, one must do exhaustive

---

[1] The LLL algorithms implemented in Shoup's NTL library can in principle receive as input a matrix with $m \ge n$, but for large dimensions we got better results when a lattice basis was provided instead.

search on the high order $\mathcal{O}((\log W)/k + k^2)$ bits of X. The optimum is to take $k := \mathcal{O}(\log^{1/3} W)$; this gives a sub-exponential time complexity :

$$\exp\left(\mathcal{O}(\log^{2/3} W)\right),$$

instead of the polynomial-time complexity as in Coppersmith's algorithm and our new algorithm.

3.3 Extension to More Variables

Our algorithm can be extended to solve integer polynomial equations with more than two variables, but as for Coppersmith's algorithm, the extension is heuristic only.

Let $p(x, y, z)$ be a polynomial in three variables over the integers, of degree δ independently in x, y and z. Let (x_0, y_0, z_0) be an integer root of $p(x, y, z)$, with $|x_0| \leq X$, $|y_0| \leq Y$ and $|z_0| \leq Z$. As for the bivariate case, we can select indices (i_0, j_0, k_0) that maximize the quantity $X^i Y^j Z^k |p_{ijk}|$ and consider the matrix S formed by the coefficients of polynomials $s_{abc}(x, y, z) = x^a y^b z^c \cdot p(x, y, z)$ for $0 \leq a, b, c < m$ for some parameter m, but only in the monomials $x^{i_0+i} y^{j_0+j} z^{k_0+k}$ for $0 \leq i, j, k < m$. Then we take $n := |\det S|$ and define the additional polynomials $r_{ijk}(x, y, z) = x^i y^j z^k n$ for $0 \leq i, j, k < \delta + m$. Then one builds the lattice L formed by all linear combinations of polynomials $s_{abc}(xX, yY, zZ)$ and $r_{ijk}(xX, yY, zZ)$, and consider the sublattice L_2 obtained by setting to 0 the coefficients of monomials corresponding to matrix S. Lattice L_2 has dimension $\omega = (\delta + m)^3 - m^3$ and using the same analysis as in Section 3, one obtains that $\det L_2' = n^{\omega-1}$ where L_2' is the same lattice as L_2 but without the $X^i Y^j Z^k$ powers.

One then applies LLL to sublattice L_2; if the ranges X, Y, Z are small enough, we are guaranteed to find a polynomial $h_1(x, y, z)$ such that $h_1(x_0, y_0, z_0) = 0$ over $\mathbb{Z}$ and $h_1(x, y, z)$ is not a multiple of $p(x, y, z)$, but this is not enough. The second vector produced by LLL gives us a second polynomial $h_2(x, y, z)$ that can satisfy the same property by bounding its norm as in [3]. One can then take the resultant between the three polynomials $p(x, y, z)$, $h_1(x, y, z)$ and $h_2(x, y, z)$ in order to obtain a polynomial $f(x)$ such that $f(x_0) = 0$. But we have no guarantee that the polynomials $h_1(x, y, z)$ and $h_2(x, y, z)$ will be algebraically independent; this makes the method heuristic only.

4 Practical Experiments

As mentioned previously, a direct application of Coppersmith's theorem for the bivariate integer case is to factor $N = pq$ when half of the most significant bits (or least significant bits) of p are known.

Theorem 4 (Coppersmith [7]). *Given $N = pq$ and the high-order $1/4 \log_2 N$ bits of p, one can recover the factorization of N in time polynomial in $\log N$.*

Namely, given the most significant bits of p, one can write :

$$N = (P_0 + x) \cdot (Q_0 + y),$$

where P_0 and Q_0 contain the most significant bits of p and q. This gives a bivariate integer polynomial equation, for which Theorem 2 can be applied directly. One gets $W = P_0 \cdot X \simeq N^{1/2} \cdot X$ which gives $XY < W^{2/3} \simeq N^{1/3} \cdot X^{2/3}$. With $X = Y$ this gives $|x_0| \leq X = N^{1/4}$.

The result of practical experiments are summarized in Table 1, using Shoup's NTL library [22]. For comparison we have implemented our algorithm and the algorithm in [9]. Table 1 shows that our new algorithm is significantly more efficient; for example, for a 1024-bits modulus with $282 = 256 + 26$ bits of p given, our algorithm takes 1 second instead of 13 minutes for the algorithm in [9]; this is due to the fact that LLL is applied on a lattice of smaller dimension.

Table 1. Running times for factoring $N = pq$ given the high-order bits of p, using our algorithm and the algorithm in [9], with Shoup's NTL library on a 1.6 GHz PC under Linux

Parameters			New algorithm		Algorithm in [9]	
N	k	bits of p given	Dimension	LLL	Dimension	LLL
512 bits	4	144 bits	9	<1 s	25	20 s
512 bits	5	141 bits	11	<1 s	36	2 min
1024 bits	5	282 bits	11	1 s	36	13 min
1024 bits	12	266 bits	25	42 s	169	-

The problem of factoring $N = pq$ given the high-order (or low-order) bits of p can also be solved using a simple variant of the one variable modular case, as shown by Howgrave-Graham in [13]. Therefore we have also implemented Howgrave-Graham's algorithm to provide a comparison; experimental results are given in Table 2. We obtain that for the particular case of factoring with high-order bits known, our algorithm and Howgrave-Graham's algorithm have roughly the same running time, and work with the same lattice dimension (but the two lattices are different).

Table 2. Running times for factoring $N = pq$ given the high-order bits of p, using Howgrave-Graham's algorithm with Shoup's NTL library on a 1.6 GHz PC running under Linux.

N	k	bits of p given	Dimension	LLL
512 bits	4	144 bits	9	<1 s
512 bits	5	141 bits	11	<1 s
1024 bits	5	282 bits	11	1 s
1024 bits	12	266 bits	25	37 s

5   Conclusion

We have described a new algorithm for finding small roots of bivariate polynomial equations over the integers, which is simpler than Coppersmith's algorithm but with the same asymptotic complexity. Our simplification is analogous to the simplification brought by Howgrave-Graham for the univariate modular case; it improves on the algorithm in [9] which was not polynomial-time for certain parameters. In practical experiments, our algorithm performs several order of magnitude faster than the algorithm in [9].

References

1. Bleichenbacher, D., May, A.: New Attacks on RSA with Small Secret CRT-Exponents. In: Yung, M., Dodis, Y., Kiayias, A., Malkin, T.G. (eds.) PKC 2006. LNCS, vol. 3958, Springer, Heidelberg (2006)
2. Blomer, J., May, A.: A Tool Kit for Finding Small Roots of Bivariate Polynomials over the Integers. In: Cramer, R.J.F. (ed.) EUROCRYPT 2005. LNCS, vol. 3494, pp. 251–267. Springer, Heidelberg (2005)
3. Boneh, D., Durfee, G.: Crypanalysis of RSA with private key d less than $N^{0.292}$. In: Stern, J. (ed.) EUROCRYPT 1999. LNCS, vol. 1592, Springer, Heidelberg (1999)
4. Boneh, D., Durfee, G., Howgrave-Graham, N.A.: Factoring $n = p^r q$ for large r. In: Wiener, M.J. (ed.) CRYPTO 1999. LNCS, vol. 1666, Springer, Heidelberg (1999)
5. Coppersmith, D.: Finding a Small Root of a Univariate Modular Equation. In: Maurer, U.M. (ed.) EUROCRYPT 1996. LNCS, vol. 1070, Springer, Heidelberg (1996)
6. Coppersmith, D.: Finding a Small Root of a Bivariate Integer Equation; Factoring with High Bits Known. In: Maurer, U.M. (ed.) EUROCRYPT 1996. LNCS, vol. 1070, Springer, Heidelberg (1996)
7. Coppersmith, D.: Small solutions to polynomial equations, and low exponent vulnerabilities. J. of Cryptology 10(4), 233–260 (1997) Revised version of two articles of Eurocrypt '96
8. Coppersmith, D.: Finding small solutions to small degree polynomials. In: Silverman, J.H. (ed.) CaLC 2001. LNCS, vol. 2146, Springer, Heidelberg (2001)
9. Coron, J.S.: Finding Small Roots of Bivariate Polynomial Equations Revisited. In: Cachin, C., Camenisch, J.L. (eds.) EUROCRYPT 2004. LNCS, vol. 3027, Springer, Heidelberg (2004)
10. Coron, J.S., May, A.: Deterministic Polynomial-Time Equivalence of Computing the RSA Secret Key and Factoring. Journal of Cryptology 20(1) (2007)
11. Ernst, M., Jochemsz, E., May, A., de Weger, B.: Partial Key Exposure Attacks on RSA up to Full Size Exponents. In: Cramer, R.J.F. (ed.) EUROCRYPT 2005. LNCS, vol. 3494, pp. 371–386. Springer, Heidelberg (2005)
12. Hafner, J., McCurley, K.: Asymptotically fast triangularization of matrices over rings. SIAM J. Comput. 20, 1068–1083 (1991)
13. Howgrave-Graham, N.A.: Finding small roots of univariate modular equations revisited. In: Darnell, M. (ed.) Cryptography and Coding. LNCS, vol. 1355, pp. 131–142. Springer, Heidelberg (1997)
14. Howgrave-Graham, N.A.: Approximate integer common divisors. In: Silverman, J.H. (ed.) CaLC 2001. LNCS, vol. 2146, Springer, Heidelberg (2001)

15. Howgrave-Graham, N.A.: Computational Mathematics Inspired by RSA. PhD thesis, University of Bath (1998)
16. Jochemz, E., May, A.: A Strategy for Finding Roots of Multivariate Polynomials with New Applications in Attacking RSA Variants. In: Lai, X., Chen, K. (eds.) ASIACRYPT 2006. LNCS, vol. 4284, Springer, Heidelberg (2006)
17. Lenstra, A.K., Lenstra Jr., H.W., Lovász, L.: Factoring polynomials with rational coefficients. Mathematische Ann. 261, 513–534 (1982)
18. May, A.: Cryptanalysis of Unbalanced RSA with Small CRT-Exponent. In: Yung, M. (ed.) CRYPTO 2002. LNCS, vol. 2442, pp. 242–256. Springer, Heidelberg (2002)
19. May, A.: Computing the RSA Secret Key is Deterministic Polynomial Time Equivalent to Factoring. In: Franklin, M. (ed.) CRYPTO 2004. LNCS, vol. 3152, pp. 213–219. Springer, Heidelberg (2004)
20. May, A.: Secret Exponent Attacks on RSA-type Schemes with Moduli $N = p^r q$. In: Bao, F., Deng, R., Zhou, J. (eds.) PKC 2004. LNCS, vol. 2947, pp. 218–230. Springer, Heidelberg (2004)
21. Nguyen, P.Q., Stehlé, D.: Floating-Point LLL Revisited. In: Cramer, R.J.F. (ed.) EUROCRYPT 2005. LNCS, vol. 3494, Springer, Heidelberg (2005)
22. Shoup, V.: Number Theory C++ Library (NTL) version 5.4. Available at, http://www.shoup.net
23. Shoup, V.: OAEP reconsidered. In: Kilian, J. (ed.) CRYPTO 2001. LNCS, vol. 2139, Springer, Heidelberg (2001)

A   Proof of Lemma 1

Let R be a matrix basis of L and let U be the unimodular matrix such that :

$$U \cdot M = \begin{bmatrix} R \\ 0 \end{bmatrix}$$

(a unimodular matrix U satisfies $\det U = \pm 1$). Let V be the unimodular matrix such that $M' = M \cdot V$. Then :

$$U \cdot M \cdot V = U \cdot M' = \begin{bmatrix} R' \\ 0 \end{bmatrix},$$

where $R' = R \cdot V$ is a matrix basis for L'. Then

$$\det L' = |\det R'| = |\det(R \cdot V)| = |\det R| \cdot |\det V| = |\det R| = \det L.$$

B   Proof of Lemma 3

The proof is very similar to the proof of Lemma 3 in [7]. It consists in showing that a matrix related to S is diagonally dominant, which enables to derive a lower bound for its determinant.

Let $W = \max_{i,j} |p_{ij}| X^i Y^j$ and let indices (u, v) such that $W = |p_{uv}| X^u Y^v$. Let indices (i_0, j_0) that maximize the quantity

$$8^{(i-u)^2+(j-v)^2} |p_{ij}| X^i Y^j.$$

The matrix S is obtained by taking the coefficients of the polynomials $x^a y^b p(x, y)$ for $0 \leq a, b < k$, taking only the coefficients of monomials $x^{i_0+i} y^{j_0+j}$ for $0 \leq i, j < k$. We must show :

$$\left(\frac{W}{X^{i_0} Y^{j_0}}\right)^{k^2} 2^{-6k^2 \delta^2 - 2k^2} \leq |\det S| \leq \left(\frac{W}{X^{i_0} Y^{j_0}}\right)^{k^2} 2^{k^2} \qquad (12)$$

We let $\mu(i, j) = ki + j$ be an index function; the matrix element $S_{\mu(a,b),\mu(i,j)}$ is the coefficient of $x^{i_0+i} y^{j_0+j}$ in $x^a y^b p(x, y)$, namely :

$$S_{\mu(a,b),\mu(i,j)} = p_{i_0+i-a, j_0+j-b}$$

We multiply each $\mu(i, j)$ column of S by

$$8^{2(i_0-u)i + 2(j_0-v)j} X^{i_0+i} Y^{j_0+j}$$

and we multiply each $\mu(a, b)$ row by

$$8^{-2(i_0-u)a - 2(j_0-v)b} X^{-a} Y^{-b}$$

to create a new matrix S' whose element is :

$$S'_{\mu(a,b),\mu(i,j)} = p_{i_0+i-a, j_0+j-b} X^{i_0+i-a} Y^{j_0+j-b} 8^{2(i_0-u)(i-a) + 2(j_0-v)(j-b)}$$

and we have :

$$\det S' = \det S \cdot \left(X^{i_0} Y^{j_0}\right)^{k^2} \qquad (13)$$

Now we show that S' is a diagonally dominant matrix. Let denote $\tilde{p}_{ij} = p_{ij} X^i Y^j$; the elements of matrix S' are :

$$S'_{\mu(a,b),\mu(i,j)} = \tilde{p}_{i_0+i-a, j_0+j-b} 8^{2(i_0-u)(i-a) + 2(j_0-v)(j-b)}$$

From maximality of (i_0, j_0) we have :

$$|\tilde{p}_{i_0+i-a, j_0+j-b}| \cdot 8^{(i-a+i_0-u)^2 + (j-b+j_0-v)^2} \leq |\tilde{p}_{i_0 j_0}| 8^{(i_0-u)^2 + (j_0-v)^2}$$

which gives :

$$|\tilde{p}_{i_0+i-a, j_0+j-b}| \cdot 8^{2(i-a)(i_0-u) + 2(j-b)(j_0-v)} \leq |\tilde{p}_{i_0 j_0}| 8^{-(i-a)^2 - (j-b)^2}$$

and then :

$$|S'_{\mu(a,b),\mu(i,j)}| \leq |\tilde{p}_{i_0,j_0}| 8^{-(i-a)^2 - (j-b)^2}$$

Each diagonal element $S'_{\mu(a,b),\mu(a,b)}$ of matrix S' is equal to $\tilde{p}_{i_0,j_0}$, and using :

$$\sum_{(i,j)\neq(a,b)} 8^{-(i-a)^2 - (j-b)^2} \leq \sum_{(i,j)\neq(0,0)} 8^{-i^2 - j^2} \leq -1 + \sum_{(i,j)} 8^{-i^2 - j^2}$$

$$\leq -1 + \left(\sum_i 8^{-i^2}\right)^2 \leq \frac{3}{4}$$

we obtain that the sum of the absolute values of the off-diagonal entries in each $\mu(a,b)$ row is at most $\frac{3}{4}|\tilde{p}_{i_0,j_0}|$. Therefore matrix S' is diagonally dominant and each eigenvalue λ must verify :

$$\frac{1}{4}|\tilde{p}_{i_0,j_0}| \leq |\lambda| \leq \frac{7}{4}|\tilde{p}_{i_0,j_0}|$$

which gives :

$$|\tilde{p}_{i_0,j_0}|^{k^2} 2^{-2k^2} \leq |\det S'| \leq |\tilde{p}_{i_0,j_0}|^{k^2} 2^{k^2} \tag{14}$$

From the optimality of (i_0, j_0), we have :

$$8^{(i_0-u)^2+(j_0-v)^2}|\tilde{p}_{i_0,j_0}| \geq 8^{0+0}|\tilde{p}_{u,v}| = W$$

which gives :

$$8^{-2\delta^2} W \leq |\tilde{p}_{i_0,j_0}| \leq W$$

Combining with (14) we obtain :

$$W^{k^2} 2^{-6k^2\delta^2-2k^2} \leq |\det S'| \leq W^{k^2} \cdot 2^{k^2}$$

and using (13) we obtain (12).

A Polynomial Time Attack on RSA with Private CRT-Exponents Smaller Than $N^{0.073}$

Ellen Jochemsz[1],[*] and Alexander May[2]

[1] Department of Mathematics and Computer Science,
TU Eindhoven, 5600 MB Eindhoven, the Netherlands
e.jochemsz@tue.nl
[2] Faculty of Computer Science
TU Darmstadt, 64289 Darmstadt, Germany
may@informatik.tu-darmstadt.de

Abstract. Wiener's famous attack on RSA with $d < N^{0.25}$ shows that using a small d for an efficient decryption process makes RSA completely insecure. As an alternative, Wiener proposed to use the Chinese Remainder Theorem in the decryption phase, where $d_p = d \mod (p-1)$ and $d_q = d \mod (q-1)$ are chosen significantly smaller than p and q. The parameters d_p, d_q are called private CRT-exponents. Since Wiener's proposal in 1990, it has been a challenging open question whether there exists a polynomial time attack on small private CRT-exponents. In this paper, we give an affirmative answer to this question, and show that a polynomial time attack exists if d_p and d_q are smaller than $N^{0.073}$.

Keywords: RSA, CRT, cryptanalysis, small exponents, Coppersmith's method.

1 Introduction

In the RSA cryptosystem, the public modulus $N = pq$ is a product of two primes of the same bitsize. The public and private exponent e and d satisfy

$$ed = 1 \mod (p-1)(q-1).$$

In many applications of RSA, either e or d is chosen to be small, for efficient modular exponentiation in the encryption/verifying or in the decryption/signing phase. It is well-known that it is dangerous to choose a small private exponent, since Wiener [22] showed that the RSA scheme is insecure if $d < N^{0.25}$, which was extended to $d < N^{0.292}$ by Boneh and Durfee [4].

---

[*] The work described in this paper has been supported in part by the European Commission through the IST Programme under Contract IST-2002-507932 ECRYPT. The information in this document reflects only the author's views, is provided as is and no guarantee or warranty is given that the information is fit for any particular purpose. The user thereof uses the information at its sole risk and liability.

A. Menezes (Ed.): CRYPTO 2007, LNCS 4622, pp. 395–411, 2007.
© International Association for Cryptologic Research 2007

As an alternative approach, Wiener proposed to use the Chinese Remainder Theorem (CRT) for decryption/signing as described by Quisquater and Couvreur in [18], and to use small private CRT-exponents instead of a small private exponent. In that case, the public exponent e and private CRT-exponents d_p and d_q satisfy $ed_p = 1 \mod (p-1)$ and $ed_q = 1 \mod (q-1)$. To obtain a fast decryption/signing phase, d_p and d_q are chosen significantly smaller than p and q. In time-critical applications, for instance for signing procedures on smartcards, this technique is especially useful. Whether there exists a polynomial time attack on this RSA-CRT system with small d_p and d_q has been a challenging open question since Wiener's work (see also the comments in Boneh-Durfee [4], the STORK roadmap [19], and the ECRYPT document on the hardness of the main computational problems in cryptography [9]).

So far, the best attack on this variant is a square-root attack [3] that enables an adversary to factor N in time and space $\tilde{O}(\min\{\sqrt{d_p}, \sqrt{d_q}\})$, which is exponential in the bitsize of d_p and d_q. All other attacks on RSA with small private CRT-exponents can be divided in two categories.

First, there are attacks on the special case where p and q are 'unbalanced' (not of the same bitsize). May [16] described two attacks that work up to a smallest prime factor of $N^{0.382}$. Recently, Bleichenbacher and May [2] improved this to $N^{0.468}$.

Secondly, there are attacks on a special case where not only d_p and d_q, but also e is chosen to be small. Galbraith, Heneghan and McKee [10] and Sun and Wu [20] have made proposals to use RSA-CRT in a way that 'balances' the cost of encryption and decryption by forcing both e and d_p, d_q to be small. In these articles, several attacks are described, after which the authors propose parameters that are not affected by these attacks. Bleichenbacher and May [2] in turn described a new attack on RSA-CRT with balanced exponents, forcing Galbraith, Heneghan, McKee and Sun, Wu to revise their parameter suggestions in [11] and [21], respectively.

However, the attacks in both categories are not applicable in the standard RSA case with small CRT-exponents d_p and d_q, that is, when p and q are balanced and e is full size. In this paper, we describe a way to extend one of the attacks of Bleichenbacher, May [2] such that it also works in the standard RSA-CRT case. This leads to the first polynomial time attack on standard RSA with small private CRT-exponents. More precisely, we present the following result.

Theorem 1 (RSA-CRT with Small d_p, d_q). *Under a well-known heuristic assumption (as described in Section 6), for every $\epsilon > 0$ and sufficiently large n, the following holds:*
Let $N = pq$ be an n-bit RSA modulus, and p, q primes of bitsize $\frac{n}{2}$. Let $e < \phi(N)$, $d_p < p-1$, and $d_q < q-1$ be the public exponent and private CRT-exponents, satisfying $ed_p \equiv 1 \mod (p-1)$ and $ed_q \equiv 1 \mod (q-1)$. Let $bitsize(d_p) \leq \delta n$ and $bitsize(d_q) \leq \delta n$. Then N can be factored in time polynomial in $\log(N)$ provided that

$$\delta < 0.0734 - \epsilon.$$

The rest of the paper is organized as follows. In Section 2, we give a brief introduction to Coppersmith's lattice-based method for finding small roots of polynomials [5]. In Section 3, we recall the Bleichenbacher-May attack [2]. In Section 4, we show how an extension of the attack leads to our new attack on standard RSA-CRT with $\delta < 0.0734 - \epsilon$. Furthermore, we generalize our bound to public exponents e of arbitrary size, and show that this leads to a polynomial time attack on one of the revised parameter choices in [21]. In Section 5, we explain in detail how we use Coppersmith's original method for the implementation of the attack. In Section 6, we discuss the only heuristic part of the attack, namely how to retrieve a common root from a number of polynomials. We conclude in Section 7 by giving experimental data for our attack.

2   Finding Small Roots of Polynomials

Many attacks in RSA cryptanalysis use a similar technique, which originated from Coppersmith's work on finding small roots of polynomials [5]. In essence, the attack starts with a polynomial equation in some of the unknowns of the RSA variant, such as p, q, d, or d_p and d_q in the case of RSA-CRT. An example is the usual RSA equation

$$ed = 1 + k(N + 1 - (p + q)),$$

with the unknowns d, k, p, and q.

Such an equation yields a polynomial f which has a certain root that an attacker wishes to find. In the example, the polynomial

$$f(x_1, x_2, x_3) = ex_1 - 1 - x_2(N + 1 - x_3)$$

has the root $(x_1^{(0)}, x_2^{(0)}, x_3^{(0)}) = (d, k, p + q)$. Finding the root is equivalent to factoring N, since p, q can be computed from $p + q$ using $N = pq$. The goal is to derive a polynomial time attack provided that the size of the root is below a certain bound.

In our new attack on standard RSA-CRT (Section 4), our goal is to find a root of a four-variate polynomial $f(x_1, x_2, x_3, x_4)$. We follow the strategy of Jochemsz and May [13], that we will sketch here.

Let $(x_1^{(0)}, x_2^{(0)}, x_3^{(0)}, x_4^{(0)})$ be a root of the polynomial $f(x_1, x_2, x_3, x_4)$ that is small in the sense that $|x_1^{(0)}| < X_1$, $|x_2^{(0)}| < X_2$, $|x_3^{(0)}| < X_3$, $|x_4^{(0)}| < X_4$, for some known upper bounds X_j, for $j = 1, \ldots, 4$. Moreover, we define W as the maximal absolute coefficient of $f(x_1X_1, x_2X_2, x_3X_3, x_4X_4)$. That is, $W := \|f(x_1X_1, x_2X_2, x_3X_3, x_4X_4)\|_\infty$, where $\|f(x_1, x_2, x_3, x_4)\|_\infty = \max |a_{i_1i_2i_3i_4}|$ for a polynomial $f(x_1, x_2, x_3, x_4) = \sum a_{i_1i_2i_3i_4} x_1^{i_1} x_2^{i_2} x_3^{i_3} x_4^{i_4}$.

A basis B of a lattice L is defined via so-called shift polynomials of the form $x_1^{i_1} x_2^{i_2} x_3^{i_3} x_4^{i_4} f(x_1, x_2, x_3, x_4)$. The choice of the combinations $\{i_1, i_2, i_3, i_4\}$ that are used is described by a set S. The set M then consists of all monomials that appear in the shift polynomials. The choice of S is crucial and depends on the monomials that appear in f. We will give the precise definition of S in Section 4 for our specific polynomial.

Then, LLL-reduction [15] is performed on B to find small vectors in the lattice L. From a result of [13], we know that under the condition

$$X_1^{s_1} X_2^{s_2} X_3^{s_3} X_4^{s_4} < W^s, \text{ for } s_j = \sum_{x_1^{i_1} x_2^{i_2} x_3^{i_3} x_4^{i_4} \in M \backslash S} i_j \text{ and } s = |S|, \qquad (1)$$

the first vectors in the reduced basis are small enough to ensure that we find a list $f_0, \ldots, f_\ell$ of at least three polynomials that all have the root $(x_1^{(0)}, x_2^{(0)}, x_3^{(0)}, x_4^{(0)})$ over the integers. The polynomials $\{f, f_0, \ldots, f_\ell\}$ will reveal their common root $(x_1^{(0)}, x_2^{(0)}, x_3^{(0)}, x_4^{(0)})$ under the assumption that three variables can be eliminated from the polynomial system of equations $\{f = 0, f_0 = 0, \ldots, f_\ell = 0\}$. Resultant computations are often used for this elimination process, but we choose to use Gröbner Bases, as we will explain in Section 6. Experiments must be done to verify that the elimination assumption holds in practice.

3   The Bleichenbacher-May Attack

In [2], Bleichenbacher and May describe two new attacks on RSA-CRT. One of them is meant for the case that both e and d_p and d_q are chosen to be smaller than in standard RSA-CRT. For notation, we use $e = N^\alpha$, $d_p < N^\delta$, and $d_q < N^\delta$ for some $\alpha \in [0, 1]$ and $\delta \in [0, \frac{1}{2}]$. Clearly, if an attack on this so called 'balanced' RSA works in the case $\alpha = 1$, then it threatens the security of standard RSA with small private CRT-exponents.

The attack of Bleichenbacher and May uses a lattice of dimension 3. The attack works whenever $\delta < \min\{\frac{1}{4}, \frac{2}{5} - \frac{2}{5}\alpha\}$, and therefore gives no result in the case $\alpha = 1$. However, we present a generalization of the attack for higher dimensional lattices that is applicable also for $\alpha = 1$. To explain our new attack, we first describe the basics of the BM-attack [2].

Bleichenbacher and May start with the two RSA-CRT equations $ed_p = 1 + k(p-1)$ and $ed_q = 1 + l(q-1)$, and rewrite these as

$$ed_p + k - 1 = kp \quad \text{and} \quad ed_q + l - 1 = lq.$$

Multiplying the two equations yields

$$e^2 d_p d_q + ed_p(l-1) + ed_q(k-1) - (N-1)kl - (k+l-1) = 0.$$

This can be transformed into the linear equation $e^2 x_1 + e x_2 - (N-1)x_3 - x_4 = 0$, if we substitute $x_1 = d_p d_q$, $x_2 = d_p(l-1) + d_q(k-1)$, $x_3 = kl$, $x_4 = k+l-1$.

The given linear equation leads directly to a lattice attack with a lattice of dimension 3. This attack works provided that $\delta < \min\{\frac{1}{4}, \frac{2}{5} - \frac{2}{5}\alpha\}$.

Although linearization of an equation makes the analysis easier and keeps the lattice dimension small, better results can sometimes be obtained by using a non-linear polynomial equation directly. In the next section, we will pursue this approach and use a polynomial with the variables $x_1, \ldots, x_4$ corresponding to d_p, d_q, k, and l, respectively.

4   The New Attack on RSA-CRT

The equation we introduced in the previous section

$$e^2 d_p d_q + e d_p(l-1) + e d_q(k-1) - (N-1)kl - (k+l-1) = 0$$

yields a polynomial $f(x_1, x_2, x_3, x_4) = e^2 x_1 x_2 + e x_1 x_4 - e x_1 + e x_2 x_3 - e x_2 - (N-1)x_3 x_4 - x_3 - x_4 + 1$ with monomials $1, x_1, x_2, x_3, x_4, x_1 x_2, x_1 x_4, x_2 x_3, x_3 x_4$ and a small root

$$(x_1^{(0)}, x_2^{(0)}, x_3^{(0)}, x_4^{(0)}) = (d_p, d_q, k, l), \text{ with } \begin{cases} |x_1^{(0)}| < X_1 = N^\delta, \\ |x_2^{(0)}| < X_2 = N^\delta, \\ |x_3^{(0)}| < X_3 = N^{\alpha+\delta-\frac{1}{2}}, \\ |x_4^{(0)}| < X_4 = N^{\alpha+\delta-\frac{1}{2}}. \end{cases}$$

We will follow the strategy for finding small integer roots of Jochemsz and May [13] as sketched in Section 2, to analyze which attack bound corresponds to this polynomial f.

In the basic strategy of [13], the set S that describes which monomials $x_1^{i_1} x_2^{i_2} x_3^{i_3} x_4^{i_4}$ are used for the shift polynomials, is simply the set that contains all monomials of f^{m-1} for a given integer m. The set M is defined as the set of all monomials that appear in $x_1^{i_1} x_2^{i_2} x_3^{i_3} x_4^{i_4} f(x_1, x_2, x_3, x_4)$, with $x_1^{i_1} x_2^{i_2} x_3^{i_3} x_4^{i_4} \in S$. Since f has a non-zero constant coefficient, all monomials of S are included in M. More precisely, S and M can be described as

$$x_1^{i_1} x_2^{i_2} x_3^{i_3} x_4^{i_4} \in S \Leftrightarrow \begin{cases} i_1 = 0, \ldots, m-1-i_3, \\ i_2 = 0, \ldots, m-1-i_4, \\ i_3 = 0, \ldots, m-1, \\ i_4 = 0, \ldots, m-1, \end{cases}$$

$$x_1^{i_1} x_2^{i_2} x_3^{i_3} x_4^{i_4} \in M \Leftrightarrow \begin{cases} i_1 = 0, \ldots, m-i_3, \\ i_2 = 0, \ldots, m-i_4, \\ i_3 = 0, \ldots, m, \\ i_4 = 0, \ldots, m. \end{cases}$$

However, in [13] it is also advised to explore the possibility of extra shifts of one or more variables. Since X_1 and X_2 are significantly smaller than X_3 and X_4 for $\alpha > \frac{1}{2}$, we find that the attack bound is superior for $\alpha = 1$ if we use extra shifts of x_1 and x_2. Therefore, we take

$$x_1^{i_1} x_2^{i_2} x_3^{i_3} x_4^{i_4} \in S \Leftrightarrow \begin{cases} i_1 = 0, \ldots, m-1-i_3+t, \\ i_2 = 0, \ldots, m-1-i_4+t, \\ i_3 = 0, \ldots, m-1, \\ i_4 = 0, \ldots, m-1, \end{cases}$$

$$x_1^{i_1} x_2^{i_2} x_3^{i_3} x_4^{i_4} \in M \Leftrightarrow \begin{cases} i_1 = 0, \ldots, m-i_3+t, \\ i_2 = 0, \ldots, m-i_4+t, \\ i_3 = 0, \ldots, m, \\ i_4 = 0, \ldots, m, \end{cases}$$

for some t that has to be optimized as a function of m and α.

Our goal is to find at least three polynomials f_0, f_1, f_2 that share the root $(x_1^{(0)}, x_2^{(0)}, x_3^{(0)}, x_4^{(0)})$ over the integers. From Section 2 we know that these polynomials can be computed by lattice reduction techniques as long as

$$X_1^{s_1} X_2^{s_2} X_3^{s_3} X_4^{s_4} < W^s, \text{ for } s_j = \sum_{x_1^{i_1} x_2^{i_2} x_3^{i_3} x_4^{i_4} \in M \backslash S} i_j \text{ and } s = |S|.$$

For a given integer m and $t = \tau m$, our last definition of S and M yields the bound

$$(X_1 X_2)^{(\frac{5}{12} + \frac{5}{3}\tau + \frac{9}{4}\tau^2 + \tau^3) m^4 + o(m^4)} (X_3 X_4)^{(\frac{5}{12} + \frac{5}{3}\tau + \frac{3}{2}\tau^2) m^4 + o(m^4)}$$
$$< W^{(\frac{1}{4} + \tau + \tau^2) m^4 + o(m^4)}.$$

To obtain the asymptotic bound, we let m grow to infinity and let all terms of order $o(m^4)$ contribute to some error term ϵ. If we substitute the values for X_1, X_2, X_3, X_4, W, we obtain

$$\left(\tfrac{5}{12} + \tfrac{5}{3}\tau + \tfrac{9}{4}\tau^2 + \tau^3\right) \cdot 2\delta + \left(\tfrac{5}{12} + \tfrac{5}{3}\tau + \tfrac{3}{2}\tau^2\right) \cdot (2\alpha + 2\delta - 1)$$
$$< \left(\tfrac{1}{4} + \tau + \tau^2\right) \cdot (2\alpha + 2\delta),$$

which leads to

$$\delta < \frac{5 - 4\alpha + 20\tau - 16\alpha\tau + 18\tau^2 - 12\alpha\tau^2}{14 + 56\tau + 66\tau^2 + 24\tau^3} - \epsilon.$$

For $\alpha = 1$, we find an optimal value of $\tau \approx 0.381788$, and we get

$$\delta < 0.0734 - \epsilon.$$

Hence, for a 1024-bit modulus, d_p and d_q are in the attack space if they are less then 75 bits. Analogously, for a 2048-bit modulus, d_p and d_q are in the attack space if they are at most 150 bits.

4.1   Extending the Attack to Other Values of α

In Section 4, we assumed that $x_1^{(0)}, x_2^{(0)}$ are smaller than $x_3^{(0)}, x_4^{(0)}$, t.i. $\alpha \geq \frac{1}{2}$. For $\alpha < \frac{1}{2}$, symmetrically one uses extra x_3 and x_4-shifts instead of extra x_1 and x_2-shifts. Because of the symmetry, one can immediately see that the attack bound is

$$(X_1 X_2)^{(\frac{5}{12} + \frac{5}{3}\tau + \frac{3}{2}\tau^2) m^4 + o(m^4)} (X_3 X_4)^{(\frac{5}{12} + \frac{5}{3}\tau + \frac{9}{4}\tau^2 + \tau^3) m^4 + o(m^4)}$$
$$< W^{(\frac{1}{4} + \tau + \tau^2) m^4 + o(m^4)}.$$

The above bound leads to

$$\delta < \frac{5 - 4\alpha + 20\tau - 16\alpha\tau + 27\tau^2 - 30\alpha\tau^2 + 12\tau^3 - 24\alpha\tau^3}{14 + 56\tau + 66\tau^2 + 24\tau^3} - \epsilon.$$

Note that this bound only holds for $\alpha + \delta > \frac{1}{2}$, since we assume that the values of k and l are unknown to the attacker. Both conditions are only met if $\alpha \geq \frac{1}{6}$.

However, in Section 7.1 we provide experimental evidence that our heuristic attack is successful only when $\alpha \geq \frac{1}{4}$.

In the revised paper by Sun, Hinek, Wu [21], the authors propose as new parameters $\{\alpha = 0.577, \delta = 0.186\}$. For this choice, we find the bound $\delta < 0.192$, which breaks the new proposal in polynomial time.

5   Implementation Using Coppersmith's Original Method

Although we have derived our attack bound from the strategy of Jochemsz, May [13], we deviate from their strategy for the implementation of the attack. Basically, we make use of Coppersmith's original technique [5] instead of Coron's reformulation [6]. This does not change the asymptotic bound of the attack, but it has a major practical advantage. Namely, the lattices used in the attacks are high-dimensional, and Coppersmith's original method requires only the reduction of a lower-dimensional sublattice[1]. Since the LLL-process is the most costly factor in our attack, this leads to a significant improvement in practice. Furthermore, we slightly adapt Coppersmith's original method such that we directly obtain triangular lattice bases, which simplifies the determinant calculations.

So let us first explain how to apply Coppersmith's technique for our attack. We introduce the shift polynomials

$$g_{i_1 i_2 i_3 i_4}(x_1, x_2, x_3, x_4) = x_1^{i_1} x_2^{i_2} x_3^{i_3} x_4^{i_4} f(x_1, x_2, x_3, x_4),$$

for $x_1^{i_1} x_2^{i_2} x_3^{i_3} x_4^{i_4} \in S$ for a set of monomials S, as specified in Section 4. As before, we define the set M as the set of all monomials that appear in the shift polynomials. We use the notation $s = |S|$ for the total number of shifts and $d = |M| - |S|$ for the difference of the number of monomials and the number of shifts. Notice that the maximal coefficient of $f(x_1 X_1, x_2 X_2, x_3 X_3, x_4 X_4)$ is $e^2 X_1 X_2$, and the monomial corresponding to it is $x_1 x_2$. We define S' as the set of monomials $x_1^{i_1+1} x_2^{i_2+1} x_3^{i_3} x_4^{i_4}$, for $x_1^{i_1} x_2^{i_2} x_3^{i_3} x_4^{i_4} \in S$. Naturally, $|S'| = |S| = s$. We now build a $(d + s) \times (d + s)$ matrix B_1 as follows.

The upper left $d \times d$ block is diagonal, where the rows represent the monomials $x_1^{i_1} x_2^{i_2} x_3^{i_3} x_4^{i_4} \in M \backslash S'$. The diagonal entry of the row corresponding to $x_1^{i_1} x_2^{i_2} x_3^{i_3} x_4^{i_4}$ is $(X_1^{i_1} X_2^{i_2} X_3^{i_3} X_4^{i_4})^{-1}$. The lower left $s \times d$ block contains only zeros.

The last s columns of the matrix B_1 represent the shift polynomials $g_{i_1 i_2 i_3 i_4} = x_1^{i_1} x_2^{i_2} x_3^{i_3} x_4^{i_4} f$, for $x_1^{i_1} x_2^{i_2} x_3^{i_3} x_4^{i_4} \in S$. The first d rows correspond to the monomials in $M \backslash S'$, and the last s rows to the monomials of S'. The entry in the column corresponding to $g_{i_1 i_2 i_3 i_4}$ is the coefficient of the monomial in $g_{i_1 i_2 i_3 i_4}$.

This description asks for a simple example. Let us use the set S as described in Section 4 with $m = 1$ and $t = 0$, which results in the lattice basis B_1 given in

---

[1] In these CRYPTO'07 proceedings, a new article by Coron [7] shows how to adapt his method such that it also requires only the reduction of a sublattice instead of the reduction of the full lattice, and hence his new technique could be applied here, too.

Figure 1. We only use the polynomial $f(x_1, x_2, x_3, x_4)$ itself as a shift polynomial. Therefore, $s = 1$ and we have $d+s = 9$ monomials. The rows represent the monomials $1, x_1, x_2, x_3, x_4, x_3x_4, x_2x_3, x_1x_4, x_1x_2$ and the last column corresponds to the coefficients of these monomials in f.

$$
\begin{pmatrix}
1 & 0 & 0 & 0 & 0 & 0 & 0 & 0 & -1 \\
0 & \frac{1}{X_1} & 0 & 0 & 0 & 0 & 0 & 0 & -e \\
0 & 0 & \frac{1}{X_2} & 0 & 0 & 0 & 0 & 0 & -e \\
0 & 0 & 0 & \frac{1}{X_3} & 0 & 0 & 0 & 0 & -1 \\
0 & 0 & 0 & 0 & \frac{1}{X_4} & 0 & 0 & 0 & -1 \\
0 & 0 & 0 & 0 & 0 & \frac{1}{X_3X_4} & 0 & 0 & 1-N \\
0 & 0 & 0 & 0 & 0 & 0 & \frac{1}{X_2X_3} & 0 & e \\
0 & 0 & 0 & 0 & 0 & 0 & 0 & \frac{1}{X_1X_4} & e \\
0 & 0 & 0 & 0 & 0 & 0 & 0 & 0 & e^2
\end{pmatrix}
$$

Fig. 1. Matrix B_1 for the case $m = 1$, $t = 0$

In general, the determinant of the matrix B_1 is

$$
\det(B_1) = \left(\prod_{x_1^{i_1} x_2^{i_2} x_3^{i_3} x_4^{i_4} \in M \backslash S'} (X_1^{i_1} X_2^{i_2} X_3^{i_3} X_4^{i_4})^{-1} \right) \cdot (e^2)^s.
$$

Let
$$
\mathbf{v}(x_1, x_2, x_3, x_4) = (1, x_1, x_2, x_3, x_4, x_3x_4, x_2x_3, x_1x_4, x_1x_2).
$$

Note that in our example,

$$
\mathbf{v}(x_1, x_2, x_3, x_4) \cdot B_1 := (1, \tfrac{x_1}{X_1}, \tfrac{x_2}{X_2}, \tfrac{x_3}{X_3}, \tfrac{x_4}{X_4}, \tfrac{x_3x_4}{X_3X_4}, \tfrac{x_2x_3}{X_2X_3}, \tfrac{x_1x_4}{X_1X_4}, f(x_1, x_2, x_3, x_4)).
$$

So,

$$
\|\mathbf{v}(d_p, d_q, k, l) \cdot B_1\| = \|(1, \tfrac{d_p}{X_1}, \tfrac{d_q}{X_2}, \tfrac{k}{X_3}, \tfrac{l}{X_4}, \tfrac{kl}{X_3X_4}, \tfrac{d_qk}{X_2X_3}, \tfrac{d_pl}{X_1X_4}, 0)\| \leq \sqrt{d}.
$$

Since the X_j upper bound the root, there is always such a vector $\mathbf{v}$ which, if one substitutes the unknowns $\{d_p, d_q, k, l\}$ for the variables $\{x_1, x_2, x_3, x_4\}$, becomes a vector with Euclidean norm smaller than $\sqrt{d}$ after multiplication with the matrix B_1.

Let us perform a unimodular transformation U_1 on B_1 to create a matrix B_2 such that

$$
B_2 = U_1 \cdot B_1 = \left(\begin{array}{c|c} A_{d \times d} & 0_{d \times s} \\ \hline A'_{s \times d} & I_{s \times s} \end{array} \right).
$$

Now if the rows of B_1 form a basis of a lattice L, then the rows of B_2 form a basis of the same lattice. Moreover, the rows of

$$
B_3 = \left(A_{d \times d} \mid 0_{d \times s} \right)
$$

are a basis of the sublattice L_0 of L which has zeros in the last s entries. Notice that $\det(L_0) = \det(L)$. Clearly, $\mathbf{v}(d_p, d_q, k, l) \cdot B_1$ is in the lattice L_0 spanned by the rows of B_3.

Since

$$\mathbf{v}(d_p, d_q, k, l) \cdot B_1 = \mathbf{v}(d_p, d_q, k, l) U_1^{-1} B_2,$$

this means that the last s entries of $\mathbf{v}(d_p, d_q, k, l) U_1^{-1}$ must be zero. We use the notation $\lfloor \mathbf{v} \rfloor_{\text{sh}}$ for the vector $\mathbf{v}$ with its length 'shortened' to its first d entries. Then,

$$\lfloor \mathbf{v}(d_p, d_q, k, l) \cdot B_1 \rfloor_{\text{sh}} = \lfloor \mathbf{v}(d_p, d_q, k, l) U_1^{-1} B_2 \rfloor_{\text{sh}} = \lfloor \mathbf{v}(d_p, d_q, k, l) U_1^{-1} \rfloor_{\text{sh}} A.$$

Next, we reduce A using lattice basis reduction to a basis $B = U_2 A$. It follows that

$$\lfloor \mathbf{v}(d_p, d_q, k, l) \cdot B_1 \rfloor_{\text{sh}} = \lfloor \mathbf{v}(d_p, d_q, k, l) U_1^{-1} \rfloor_{\text{sh}} U_2^{-1} B.$$

We use the notation $\mathbf{v}'(d_p, d_q, k, l)$ for the vector $\lfloor \mathbf{v}(d_p, d_q, k, l) U_1^{-1} \rfloor_{\text{sh}} U_2^{-1}$, and B^* (with row vectors $\mathbf{b}_i^*$) for the basis after applying Gram-Schmidt orthogonalization to B. Now we can make three observations. Firstly, the vector $\mathbf{v}'$ is integral. This is because both matrices U_1 and U_2 have integer entries. Secondly, $\|\mathbf{v}(d_p, d_q, k, l) \cdot B_1\| < \sqrt{d}$. Thirdly, it is known [15] that the Gram-Schmidt orthogonalization of the LLL-reduced basis satisfies

$$\|\mathbf{b}_d^*\| \geq 2^{\frac{-(d-1)}{4}} \det(L)^{\frac{1}{d}}.$$

So, if we combine these three facts, we obtain that

$$\sqrt{d} \geq \|\mathbf{v}(d_p, d_q, k, l) \cdot B_1\| = \| \lfloor \mathbf{v}(d_p, d_q, k, l) \cdot B_1 \rfloor_{\text{sh}} \| = \| \mathbf{v}'(d_p, d_q, k, l) B \|$$

$$\geq | \mathbf{v}'(d_p, d_q, k, l)_d | \cdot \|\mathbf{b}_d^*\| \geq | \mathbf{v}'(d_p, d_q, k, l)_d | \cdot 2^{\frac{-(d-1)}{4}} \det(L)^{\frac{1}{d}}.$$

Since the terms $2^{\frac{-(d-1)}{4}}$ and $\sqrt{d}$ do not depend on N, we let them contribute to an error term ϵ. Thus, whenever

$$\det(L)^{\frac{1}{d}} > 1,$$

we must have $| \mathbf{v}'(d_p, d_q, k, l)_d | = 0$.

Hence, the polynomial $f_0(x_1, x_2, x_3, x_4)$ corresponding to the coefficient vector $\mathbf{v}'(x_1, x_2, x_3, x_4)_d$ contains the root (d_p, d_q, k, l) over the integers.

In Appendix A, we show that the bound $\det(L)^{\frac{1}{d}} > 1$ is equivalent to the bound (1) that was given in Section 2. Moreover, we use a result from Jutla [14] to show that the vectors $\mathbf{v}'(x_1, x_2, x_3, x_4)_{d-\ell}, \ell \geq 2$, yield a list of at least three polynomials $f_0, \ldots, f_\ell$ having the same root (d_p, d_q, k, l). In the next section, we show how to retrieve this root from the polynomials $f, f_0, \ldots, f_\ell$.

The running time of our algorithm is dominated by the time to LLL-reduce the lattice basis A. Taking the algorithm of Nguyen, Stehlé [17] this can be achieved in $\mathcal{O}(d^5(d + \log A_m) \log A_m)$, where $\log A_m$ is the maximal bitsize of an entry in A. Our lattice dimension d depends on ϵ^{-1} only, whereas the bitsize of the entries is bounded by a polynomial in $\log N$. Therefore, the construction of $f_0, \ldots, f_\ell$ can be done in time polynomial in $\log N$.

Moreover, $f_0, \ldots, f_\ell$ have a fixed degree that only depends on ϵ^{-1} and coefficients with bitsize polynomial in $\log N$. This will be important for the analysis in the following section.

6   Extracting the Common Root

Assume that we want to retrieve a common root from four polynomials f, f_0, f_1, f_2. Usually, one uses resultants to eliminate variables one by one until one obtains a univariate polynomial $w_0(x_1)$ that has $x_1^{(0)}$ as a root:

$$r_0(x_1, x_2, x_3) = \mathrm{Res}_{x_4}(f, f_0)$$
$$s_0(x_1, x_2) = \mathrm{Res}_{x_3}(r_0, r_1)$$
$$r_1(x_1, x_2, x_3) = \mathrm{Res}_{x_4}(f, f_1) \qquad\qquad w_0(x_1) = \mathrm{Res}_{x_2}(r_3, r_4)$$
$$s_1(x_1, x_2) = \mathrm{Res}_{x_3}(r_1, r_2)$$
$$r_2(x_1, x_2, x_3) = \mathrm{Res}_{x_4}(f, f_2)$$

However, this method only works if the polynomials are algebraically independent. One cannot easily use more than three candidates f_j, besides repeating the scheme for different combinations. Moreover, the last resultant computation can take a significant amount of time and memory, since the degrees of the resultant polynomials grow fast. We use Gröbner Bases instead of resultant methods to extract the root. For a detailed introduction to Gröbner Bases, we refer to [8].

Suppose we have a set of polynomials $\{f, f_0, \ldots, f_\ell\}$ that have the small root $(x_1^{(0)}, \ldots, x_n^{(0)})$ in common. Then a Gröbner Basis $G := \{g_1, \ldots, g_t\}$ is a set of polynomials that preserves the set of common roots of $\{f, f_0, \ldots, f_\ell\}$. In other words, the variety of the ideal I generated by $\{g_1, \ldots, g_t\}$ is the same as the variety of the ideal generated by $\{f, f_0, \ldots, f_l\}$. The advantage of having a Gröbner Basis is that the g_i can be computed with respect to some ordering that eliminates the variables. Having such an elimination ordering, it is easy to extract the desired root.

In our experiments in Section 7 we usually found much more polynomials $f_0, \ldots, f_\ell$ than the required amount of $\ell = 2$. Therefore, we have two advantages of Gröbner Bases in comparison with resultants. First, in contrast to resultants the computation time of a Gröbner Basis usually benefits from more overdefined systems which lowers the time for extracting the root. Second, we do not have to search over all subsets of three polynomials until we find an algebraically independent one. Instead, we simply put all the polynomials in our Gröbner Basis computation. The elimination of variables can only fail if the variety $\mathbf{V}(I)$ defined by the ideal I which is generated by $\{f, f_0, \ldots, f_\ell\}$ is not zero-dimensional. Therefore, we make the following heuristic assumption for our attack.

Assumption 1: *The variety $\mathbf{V}(I)$ of the ideal I generated by the polynomials in the construction of Section 5 is zero-dimensional.*

Under Assumption 1, the secret root (d_p, d_q, k, l) can be derived in polynomial time, since we run a Gröbner Basis computation on polynomials of a fixed degree.

Recently, Bauer and Joux [1] made some important progress considering the heuristic involved in Coppersmith methods. Their result, for roots of trivariate polynomials, can in theory be extended to more variables. In this way, one could investigate if Assumption 1 can be replaced by a weaker assumption. In this paper, we made no efforts in this direction. Instead we verified the validity of Assumption 1 by experiments.

7 Experiments

In order to test the attack described in this paper for varying bitsizes of e and d_p, d_q we designed a key generation process similar to the one proposed by Galbraith, Heneghan, and McKee [10].

INPUT: Bitsizes n of N, αn of e, δn of d_p, d_q

(1) Choose d_p, d_q of bitsize δn.
(2) Choose k, l of bitsize $(\alpha + \delta - \frac{1}{2})n$ such that $\gcd(d_p, k) = \gcd(d_q, l) = \gcd(k, l) = 1$.
(3) Compute e using Chinese Remaindering such that

$$\left| \begin{array}{l} e = d_p^{-1} \bmod k \\ e = d_q^{-1} \bmod l \end{array} \right| .$$

(4) Compute $e := e + c \cdot kl$ for some c of bitsize $(1 - \alpha - 2\delta)n$.
(5) Compute $p = \frac{ed_p - 1}{k} - 1$ and $q = \frac{ed_q - 1}{l} - 1$. If either p or q is composite, repeat the whole algorithm.

OUTPUT: CRT-RSA-instance (e, N, d_p, d_q, p, q)

Notice that this key generation algorithm works as long as $\alpha + 2\delta \leq 1$. Namely, in Step 3 we compute a public key e of bitsize $(2\alpha + 2\delta - 1)n$, which is extended in Step 4 to bitsize αn. Therefore, we require that $\alpha \geq 2\alpha + 2\delta - 1$.

The above key generation is a slight variation of the GHM algorithm. In [10], the authors choose e, k, l first and afterwards compute d_p, d_q as inverses of e mod k, l, respectively. Then analogously to Step 4 above, they fill up d_q, d_q to the desired bitsize. Thus, their key generation requires that the sizes of d_p, d_q are at least the sizes of k, l. However, this condition is not fulfilled by a large portion of the RSA instances that we can attack. If the conditions of both key generations are fulfilled, one should however prefer the GHM method. It is more efficient, since one can generate p and q separately.

In the following experiments, we applied our key generation algorithm for varying sizes of e and d_p, d_q. The LLL reduction was carried out using a C-implementation of the provable L^2 reduction algorithm due to Nguyen and Stehlé [17]. The timings were performed on a 1GHz PC running Cygwin.

7.1   Experiments for Small e

All experiments in this section were done for 1000-bit N. For every fixed e, we looked for the maximal bitsize for d_p, d_q that gave us enough small vectors for recovering the secrets. In our experiments, we fixed the attack parameter $m = 2$ and tried different values of t.

In the table below, the third column provides the bound of Bleichenbacher-May which can be achieved using a 3-dimensional lattice. The fourth column provides the bound for an attack of Galbraith, Heneghan, and McKee [10], which is closely related to the attack described in this paper (see Appendix B for details on this GHM-attack). The δ-column gives the theoretical upper bound for the chosen parameters m, t and e. The 'asymp'-column gives the asymptotic bound which is reached when the lattice dimension goes to infinity.

e	d_p, d_q	BM[2]	GHM[10]	δ	asymp	lattice parameters	LLL
250 bit	332 bit	0.250	0.333	0.227	0.287	$m = 2, t = 0, \dim = 27$	2 sec
300 bit	299 bit	0.250	0.300	0.209	0.271	$m = 2, t = 0, \dim = 27$	2 sec
400 bit	239 bit	0.240	0.233	0.173	0.243	$m = 2, t = 0, \dim = 27$	2 sec
500 bit	199 bit	0.200	0.167	0.136	0.214	$m = 2, t = 0, \dim = 27$	2 sec
577 bit	168 bit	0.169	0.115	0.108	0.192	$m = 2, t = 0, \dim = 27$	2 sec
700 bit	119 bit	0.120	0.033	0.064	0.157	$m = 2, t = 0, \dim = 27$	2 sec
800 bit	79 bit	0.080	−0.033	0.027	0.128	$m = 2, t = 0, \dim = 27$	2 sec
900 bit	38 bit	0.040	−0.100	−0.009	0.100	$m = 2, t = 0, \dim = 27$	2 sec
900 bit	40 bit	0.040	−0.100	0.013	0.100	$m = 2, t = 1, \dim = 56$	93 sec
925 bit	29 bit	0.030	−0.117	−0.018	0.093	$m = 2, t = 0, \dim = 27$	2 sec
925 bit	31 bit	0.030	−0.117	0.006	0.093	$m = 2, t = 1, \dim = 56$	87 sec
950 bit	19 bit	0.020	−0.133	−0.027	0.087	$m = 2, t = 0, \dim = 27$	2 sec
950 bit	23 bit	0.020	−0.133	−0.001	0.087	$m = 2, t = 1, \dim = 56$	80 sec

In all the above experiments, we were able to recover the factorization of N. Experimentally, we see that our attack is much better than theoretically predicted. The reason is that for these RSA parameter settings, the shortest vectors are linear combinations of certain subsets of the lattice basis. I.e., the shortest vectors belong to some sublattice and the determinant calculation of the full lattice in Section 4 does not accurately capture the optimal choice of basis vectors. However, to identify the optimal sublattice structure for every fixed size e seems to be a difficult task.

Let us first comment on the results for 250-bit and 300-bit e. As can be seen in Appendix B, there exists an attack by Galbraith, Heneghan, and McKee [10] that is closely related to our new attack. Basically, they use a Coppersmith method for finding modular roots, to find the small root (k, l) of a polynomial f_e modulo e. The polynomial f_e is exactly our polynomial f taken modulo e. Since for $\alpha = 0.25, \alpha = 0.3$, the bound of the GHM-attack is superior to our new attack bound, the GHM-attack should be used for these cases instead of the new attack. However, if one uses the new attack, the lattice reduction algorithm chooses certain sublattices that still lead to the GHM-bound. This explains for these small values of α, why the experimental results are better than expected. These were the only instances that we discovered, where Assumption 1 failed. Since the reduced basis vectors corresponded to the underlying structure of the GHM-attack, we were not able to eliminate three variables. However, we always found a polynomial of the form $(k + l - 1)x_3x_4 - kl(x_3 + x_4 - 1)$ in the Gröbner Basis, which directly yields k and l. The knowledge of k is sufficient to factor N in polynomial time, provided that e is large enough: Notice that

$$p = 1 - k^{-1} \mod e.$$

From a theorem of Coppersmith for factoring with high bits known [5], it follows that we can find p in polynomial time whenever $e \geq N^{\frac{1}{4}}$, which is satisfied in our experiments. We also made attacks for the case $e < N^{\frac{1}{4}}$, where we still got the secrets k, l. However, this information seems to be not sufficient for factoring N efficiently. This is consistent with the GHM-attack, where Galbraith, Heneghan, and McKee state that the attack only succeeds if the factorization of N can be extracted in polynomial time from the knowledge of the exposed k, l.

For $\alpha \geq 2/5$, i.e. e of bitsize at least 400, Assumption 1 was always valid. In all experiments, the Gröbner Basis of all polynomials yields the secret parameters (d_p, d_q, k, l) and therefore the factorization of N. The roots were found by using the F4 Gröbner Basis algorithm implemented in Magma V2.11-14. We would like to note that, when we did not include all candidates $f_0, \ldots, f_\ell$ but used only a few, it sometimes happened that we could eliminate two variables only. In that case, we were still able to retrieve the secrets, since the Gröbner Basis, where x_2 and x_4 were eliminated, then contained a polynomial with the terms $(d_p + (k - 1)x_1 - d_px_3)$ and $(d_q + (l - 1)x_1 - d_qx_3)$ in its factorization.

For e of bitsizes 400 up to 800, we actually rediscovered the bound $\frac{2}{5}(1 - \alpha)$ by Bleichenbacher, May experimentally. Again the lattice reduction algorithm chose certain sublattices which in this case lead to the BM-bound. Even a moderate increasement of the lattice dimension did not give us any improvement in this range of e. Although our asymptotical bound always beats the BM-bound, we are not able to see this effect for small e, since going beyond the BM-bound requires high-dimensional lattice bases.

For e larger than 900 bits we can for the first time see the effect of increasing the lattice dimension and we are able to go slightly beyond the BM-bound. This effect intensifies for full size e, where the BM-bound does not give any results at all.

7.2   Experiments for Full Size e

Here we describe the experiments for RSA with a standard key generation for small CRT-exponents, which usually yields full size e. Namely, the parameters

d_p, d_q are chosen for a fixed bitsize and e is the unique integer modulo $\phi(N)$ which is the inverse of d_p, d_q modulo $p - 1$ and $q - 1$, respectively.

N	d_p, d_q	δ	lattice parameters	LLL-time
1000 bit	10 bit	-0.015	$m = 2, t = 1, \dim = 56$	61 sec
1000 bit	13 bit	-0.002	$m = 2, t = 2, \dim = 95$	1129 sec
1000 bit	15 bit	0.002	$m = 3, t = 1, \dim = 115$	13787 sec
2000 bit	20 bit	-0.015	$m = 2, t = 1, \dim = 56$	255 sec
2000 bit	22 bit	-0.002	$m = 2, t = 2, \dim = 95$	1432 sec
2000 bit	32 bit	0.002	$m = 3, t = 1, \dim = 115$	36652 sec
5000 bit	52 bit	-0.015	$m = 2, t = 1, \dim = 56$	1510 sec
5000 bit	70 bit	-0.002	$m = 2, t = 2, \dim = 95$	18032 sec
10000 bit	105 bit	-0.015	$m = 2, t = 1, \dim = 56$	3826 sec
10000 bit	140 bit	-0.002	$m = 2, t = 2, \dim = 95$	57606 sec

Every experiment gave us sufficiently many polynomials with the desired roots over the integers, such that we could recover the factorization. The Gröbner computation never took more than 100 seconds and consumed a maximum of 300 MB.

Notice that for 10000-bit N, we can recover d_p, d_q of bitsize 140, which would not be possible by a square-root attack.

As in the experiments before, the δ-bound is very inaccurate. For lattice dimensions 56 and 95, we should not obtain any results at all, while experimentally we succeeded for d with bitsizes roughly a 0.010-fraction respectively a 0.013-fraction of N. On the other hand, our asymptotical bound states that we could in theory go up to a 0.073-fraction. Unfortunately, we are a tad bit away from the theoretical bound, since currently the best LLL-reductions only allow to reduce lattice bases of moderate size, when the base matrices have large entries. Let us give a numerical example. Theoretically, for $m = 10$ we find an optimal value of $t = 6$ which yields a bound of 0.063. However, this parameter choice results in a lattice dimension of 4200 which is clearly out of practical reach.

Our result guarantees that one can find the factorization of N for a sufficiently large – but fixed – lattice dimension for CRT-exponents d_p, d_q up to a 0.073-fraction. Moreover, it does not rule out that one can go beyond this bound. Even with our approach, the experimental results seem to indicate that an analysis of sublattice structures could lead to a better theoretical bound. We hope that these open problems stimulate further research in the exciting areas of lattice-based cryptanalysis and fast practical lattice reduction algorithms.

Acknowledgements

We thank Antoine Joux and Ralph-Philipp Weinmann for discussions about Gröbner Bases, Maike Ritzenhofen for doing the Gröbner Basis computations in Magma, and Benne de Weger for his helpful comments.

References

1. Bauer, A., Joux, A.: Toward a Rigorous Variation of Coppersmith's Algorithm on Three Variables. In: Naor, M. (ed.) Eurocrypt 2007. LNCS, vol. 4515, pp. 361–378. Springer, Heidelberg (2007)
2. Bleichenbacher, D., May, A.: New Attacks on RSA with Small Secret CRT-Exponents. In: Yung, M., Dodis, Y., Kiayias, A., Malkin, T.G. (eds.) PKC 2006. LNCS, vol. 3958, pp. 1–13. Springer, Heidelberg (2006)
3. Boneh, D.: Twenty Years of Attacks on the RSA Cryptosystem. Notices of the American Mathematical Society 46, 203–213 (1999)
4. Boneh, D., Durfee, G.: Cryptanalysis of RSA with Private Key d Less Than $N^{0.292}$. IEEE Transactions on Information Theory 46, 1339–1349 (2000)
5. Coppersmith, D.: Small Solutions to Polynomial Equations, and Low Exponent RSA Vulnerabilities. Journal of Cryptology 10, 233–260 (1997)
6. Coron, J.-S.: Finding Small Roots of Bivariate Integer Equations Revisited. In: Cachin, C., Camenisch, J.L. (eds.) EUROCRYPT 2004. LNCS, vol. 3027, pp. 492–505. Springer, Heidelberg (2004)
7. Coron, J.-S.: Finding Small Roots of Bivariate Integer Polynomial Equations: a Direct Approach. In: Menezes, A. (ed.) CRYPTO 2007. LNCS, vol. 4622, pp. 379–394. Springer, Heidelberg (2007)
8. Cox, D., Little, J., O'Shea, D.: Ideals, Varieties, and Algorithms, 2nd edn. Springer, Heidelberg (1998)
9. ECRYPT - Hardness of the Main Computational Problems Used in Cryptography, IST-2002-507932, available at http://www.ecrypt.eu.org/documents/D.AZTEC.4-1.1.pdf
10. Galbraith, S., Heneghan, C., McKee, J.: Tunable Balancing of RSA. In: Boyd, C., González Nieto, J.M. (eds.) ACISP 2005. LNCS, vol. 3574, pp. 280–292. Springer, Heidelberg (2005)
11. Galbraith, S., Heneghan, C., McKee, J.: Tunable Balancing of RSA, full version of [10] http://www.isg.rhul.ac.uk/~sdg/full-tunable-rsa.pdf
12. Howgrave-Graham, N.: Finding Small Roots of Univariate Modular Equations Revisited. In: Darnell, M. (ed.) Cryptography and Coding. LNCS, vol. 1355, pp. 131–142. Springer, Heidelberg (1997)
13. Jochemsz, E., May, A.: A Strategy for Finding Roots of Multivariate Polynomials with New Applications in Attacking RSA Variants. In: Lai, X., Chen, K. (eds.) ASIACRYPT 2006. LNCS, vol. 4284, pp. 267–282. Springer, Heidelberg (2006)
14. Jutla, C.S.: On Finding Small Solutions of Modular Multivariate Polynomial Equations. In: Nyberg, K. (ed.) EUROCRYPT 1998. LNCS, vol. 1403, pp. 158–170. Springer, Heidelberg (1998)
15. Lenstra, A., Lenstra Jr., H., Lovász, L.: Factoring Polynomials with Rational Coefficients. Mathematische Ann. 261, 513–534 (1982)
16. May, A.: Cryptanalysis of Unbalanced RSA with Small CRT-Exponent. In: Yung, M. (ed.) CRYPTO 2002. LNCS, vol. 2442, pp. 242–256. Springer, Heidelberg (2002)
17. Nguyen, P., Stehlé, D.: Floating-Point LLL Revisited. In: Cramer, R.J.F. (ed.) EUROCRYPT 2005. LNCS, vol. 3494, pp. 215–233. Springer, Heidelberg (2006)
18. Quisquater, J.J., Couvreur, C.: Fast decipherment algorithm for RSA public-key cryptosystems. Electronic Letters 18, 905–907 (1982)
19. STORK - Strategic Roadmap for Crypto, IST-2002-38273, available at http://www.stork.eu.org/documents/RUB-D6-2_1.pdf
20. Sun, H.-M., Wu, M.-E.: An Approach Towards RSA-CRT with Short Public Exponent IACR eprint, http://eprint.iacr.org/2005/053

21. Sun, H.-M., Hinek, M.J., Wu, M.-E.: On the Design of Rebalanced RSA-CRT, revised version of [20] http://www.cacr.math.uwaterloo.ca/techreports/2005/cacr2005-35.pdf
22. Wiener, M.: Cryptanalysis of Short RSA Secret Exponents. IEEE Transactions on Information Theory 36, 553–558 (1990)

A   Calculating the Bound and Finding More Polynomials

In this appendix, we show that the bound $\det(L)^{\frac{1}{d}} > 1$ of the implementation of our attack using Coppersmith's original method (Section 5) is equivalent to the bound (1) corresponding to an implementation following Coron's method (as used in Section 2). Moreover, we use a result from Jutla [14] to show that the vectors $\mathbf{v}'(x_1, x_2, x_3, x_4)_{d-\ell}$, $\ell \geq 2$, yield a list of at least three polynomials $f_0, \ldots, f_\ell$ having the same root (d_p, d_q, k, l).

One can check that

$$\det(L)^{\frac{1}{d}} = \det(B_1)^{\frac{1}{d}} = \left(\prod_{x_1^{i_1} x_2^{i_2} x_3^{i_3} x_4^{i_4} \in M \backslash S'} (X_1^{i_1} X_2^{i_2} X_3^{i_3} X_4^{i_4})^{-1} \right)^{\frac{1}{d}} \cdot (e^2)^{\frac{s}{d}}.$$

So the bound $\det(L)^{\frac{1}{d}} > 1$ implies that

$$\left(\prod_{x_1^{i_1} x_2^{i_2} x_3^{i_3} x_4^{i_4} \in M \backslash S'} (X_1^{i_1} X_2^{i_2} X_3^{i_3} X_4^{i_4}) \right) < (e^2)^s. \tag{2}$$

Let us substitute e^2 by $\frac{W}{X_1 X_2}$. We observe that the difference between the monomials of $M \backslash S'$ and $M \backslash S$ is s times the monomial $x_1 x_2$. Multiplying both sides by $(X_1 X_2)^s$ yields

$$X_1^{s_1} X_2^{s_2} X_3^{s_3} X_4^{s_4} < W^s, \quad \text{for } s_j = \sum_{x_1^{i_1} x_2^{i_2} x_3^{i_3} x_4^{i_4} \in M \backslash S} i_j \quad \text{and} \quad s = \sum_{x_1^{i_1} x_2^{i_2} x_3^{i_3} x_4^{i_4} \in S} 1.$$

Notice that this condition is equivalent to the condition (1) given in Section 2.

It follows that if this bound holds, then applying Coppersmith's method gives us a polynomial $f_0(x_1, x_2, x_3, x_4)$ from the coefficient vector $\mathbf{v}'(x_1, x_2, x_3, x_4)_d$, such that f_0 has the desired root (d_p, d_q, k, l) over the integers. But in order to extract the root, we have to construct at least two more polynomials which share the same root.

We will prove now that it is always possible to construct any constant number of polynomials with the same common root provided that condition (1) is satisfied, at the cost of a slightly larger error term ϵ in the construction. Therefore, we use a theorem of Jutla [14], which gives us a lower bound for the length of any Gram-Schmidt vector in an LLL-reduced basis. Namely,

$$\|\mathbf{b}_i^*\| \geq 2^{\frac{-(i-1)}{4}} \left(\frac{\det(L)}{b_{\max}^{m-i}} \right)^{\frac{1}{i}} \quad \text{for } i = 1 \ldots d,$$

where $b_{\max}$ is the maximal length of the Gram-Schmidt orthogonalization of the matrix A (the matrix before starting the LLL-reduction process). Following the analysis of [14], it can be checked that in our attack, $b_{\max} = e^2$. Therefore, $\|\mathbf{b}_i^*\| > 1$ reduces to

$$2^{\frac{-(i-1)}{4}} \left(\frac{\left(\prod_{x_1^{i_1} x_2^{i_2} x_3^{i_3} x_4^{i_4} \in M \setminus S'} (X_1^{i_1} X_2^{i_2} X_3^{i_3} X_4^{i_4})^{-1}\right) \cdot (e^2)^s}{(e^2)^{d-i}} \right)^{\frac{1}{i}} > 1.$$

Since $2^{\frac{-(i-1)}{4}}$ does not depend on N, we let it contribute to an error term ϵ. This simplifies our condition to

$$\prod_{x_1^{i_1} x_2^{i_2} x_3^{i_3} x_4^{i_4} \in M \setminus S'} (X_1^{i_1} X_2^{i_2} X_3^{i_3} X_4^{i_4}) < (e^2)^{s-(d-i)},$$

Notice that for $i = d$, we obtain the same bound as in (2). In Section 4, we have seen that $s = m^4(1 + o(1))$. So as long as $d - i = o(m^4)$, the asymptotic bound does not change and we get just another error term that contributes to ϵ. This is clearly satisfied if $d - i = \ell$ for some constant ℓ. Thus, all polynomials $f_0, \ldots, f_\ell$ corresponding to the coefficient vectors $\mathbf{v}'(x_1, x_2, x_3, x_4)_{d-i}$, $i = 0 \ldots \ell$, share the common root (d_p, d_q, k, l), as desired.

B   A Related Attack by Galbraith, Heneghan, and McKee

In Section 7.1 we noted that for very small e, there is an attack by Galbraith, Heneghan, and McKee [10, Section 5.1] that works better than our new attack. In this appendix, we briefly describe this GHM-attack and its relation to our new attack.

Recall that for our new attack, we multiply the equations

$$ed_p + k - 1 = kp \quad \text{and} \quad ed_q + l - 1 = lq$$

to obtain the polynomial

$$f(x_1, x_2, x_3, x_4) = e^2 x_1 x_2 + e x_1 x_4 - e x_1 + e x_2 x_3 - e x_2 - (N-1)x_3 x_4 - x_3 - x_4 + 1$$

with the small root (d_p, d_q, k, l).

In their attack in [10, Section 5.1], Galbraith, Heneghan, and McKee do essentially the same, but modulo e. Hence, the goal of their attack is to find the modular root (k, l) of the polynomial $f_e(x_3, x_4) = (N - 1)x_3 x_4 + x_3 + x_4 - 1$ modulo e. This polynomial f_e, with monomials $1, x_3, x_4, x_3 x_4$ has a well-known [5] bound

$$X_3 X_4 < e^{\frac{2}{3}}.$$

that specifies for which upper bounds X_3, X_4 of x_3, x_4 the root can be found in polynomial time. Substituting $X_3 = X_4 = N^{\alpha + \delta - \frac{1}{2}}$, and $e = N^\alpha$, we find the attack bound

$$\delta < \frac{1}{2} - \frac{2}{3}\alpha.$$

For very small α (for instance $\alpha = 0.25$ and $\delta = 0.3$), this bound is superior to the bound obtained by our new attack, and for these cases, the GHM-attack should be preferred to the new attack.

Invertible Universal Hashing and the TET Encryption Mode

Shai Halevi

IBM T.J. Watson Research Center,
Hawthorne, NY 10532, USA
shaih@alum.mit.edu

Abstract. This work describes a mode of operation, TET, that turns a regular block cipher into a length-preserving enciphering scheme for messages of (almost) arbitrary length. When using an n-bit block cipher, the resulting scheme can handle input of any bit-length between n and 2^n and associated data of arbitrary length.

The mode TET is a concrete instantiation of the generic mode of operation that was proposed by Naor and Reingold, extended to handle tweaks and inputs of arbitrary bit length. The main technical tool is a construction of invertible "universal hashing" on wide blocks, which is as efficient to compute and invert as polynomial-evaluation hash.

1   Introductions

Adding secrecy protection to existing (legacy) protocols and applications raises some unique problems. One of these problems is that existing protocols sometimes require that the encryption be "transparent", and in particular preclude length-expansion. One example is encryption of storage data "at the sector level", where both the higher-level operating system and the lower-level disk expect the data to be stored in blocks of 512 bytes, and so any encryption method would have to accept 512-byte plaintext and produce 512-byte ciphertext.

Clearly, insisting on a length-preserving (and hence deterministic) transformation has many drawbacks. Indeed, even the weakest common notion of security for "general purpose encryption" (i.e., semantic security [GM84]) cannot be achieved by deterministic encryption. Still, there may be cases where length-preservation is a hard requirement (due to technical, economical or even political constrains), and in such cases one may want to use some encryption scheme that gives better protection than no encryption at all. The strongest notion of security for a length-preserving transformation is "strong pseudo-random permutation" (SPRP) as defined by Luby and Rackoff [LR88], and its extension to "tweakable SPRP" by Liskov et al. [LRW02]. A "tweak" is an additional input to the enciphering and deciphering procedures that need not be kept secret. This report uses the terms "tweak" and "associated data" pretty much interchangeably, except that "associated data" hints that it can be of arbitrary length.

Motivated by the application to "sector level encryption", many modes of operation that implement tweakable SPRP on wide blocks were described in the

A. Menezes (Ed.): CRYPTO 2007, LNCS 4622, pp. 412–429, 2007.

literature in the last few years. Currently there are at least eight such proposals, following three different approaches: The "encrypt-mix-encrypt" approach is used for CMC, EME and EME* [HR03, HR04, Hal04], the "hash-ECB-hash" (due to Naor and Reingold [NR97]) is used in PEP [CS06b], and the "hash-CTR-hash" approach is used by XCB [FM04], HCTR [WFW05] and HCH [CS06a] (and some variation of the last approach is used in ABL4 [MV04]). Among these proposals, the "encrypt-mix-encrypt" modes are the most efficient (at least in software), the "hash-CTR-hash" modes are close behind, and PEP and ABL4 are considerably less efficient (more on efficiency in Section 3.5).

This work presents a ninth mode called TET (for linear-Transformation; ECB; linear-Transformation). TET belongs to the "hash-ECB-hash" family, but in terms of efficiency it is similar to the modes of the "hash-CTR-hash" family, thus complementing the current lineup of modes. We also mention that TET may have some practical advantages with respect to intellectual-property concerns, see further discussion in the appendix.

The main technical contribution of this work is a construction of an efficient invertible universal hashing for wide blocks, which is needed in the "hash-ECB-hash" approach. Given the wide range of applications of universal hashing in general, this invertible universal hashing may find applications beyond the TET mode itself. Another small contribution is a slight modification of the OMAC construction for pseudorandom function due to Iwata and Korasawa [IK03]. (In TET we use that pseudorandom function to handle the message-length and the tweak). This construction too can find other applications.

The Naor-Reingold construction and TET. Recall that the Naor-Reingold construction from [NR97] involves a layer of ECB encryption, sandwiched between two layers of universal hashing, as described in Figure 1. The universal hashing layers must be invertible (since they need to be inverted upon decryption), and their job is to ensure that different queries of the attacker will almost never result in "collisions" at the ECB layer. Namely, for any two plaintext (or ciphertext) vectors $p = \langle p_1, \ldots, p_m \rangle$, $q = \langle q_1, \ldots, q_m \rangle$ and two indexes i, j (such that $(p, i) \neq (q, j)$) it should hold with high probability (over the hashing key) that the i'th block of hashing p is different from the j'th block of hashing q.

As mentioned above, the main contribution of this note is a construction of an invertible universal hashing on wide blocks, which is as efficient to compute and invert as polynomial-evaluation hash. In a nutshell, the hashing family works on vectors in $GF(2^n)^m$, and it is keyed by a single random element $\tau \in_R GF(2^n)$, which defines the following $m \times m$ matrix:

$$
A_\tau \stackrel{\text{def}}{=}
\begin{pmatrix}
\tau & \tau^2 & & \tau^m \\
\tau & \tau^2 & & \tau^m \\
& & \ddots & \\
\tau & \tau^2 & & \tau^m
\end{pmatrix}
$$

Set $\sigma \stackrel{\text{def}}{=} 1 + \tau + \tau^2 + \ldots + \tau^m$, we observe that if $\sigma \neq 0$ then the matrix $M_\tau = A_\tau + I$ is invertible and its inverse is $M_\tau^{-1} = I - (A_\tau/\sigma)$. Thus multiplying

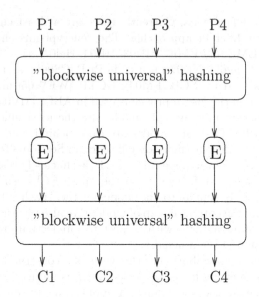

Fig. 1. The Naor-Reingold generic mode: the universal hashing must be invertible, and its job is to prevent collisions in the ECB layer

by M_τ for a random τ (subject to $\sigma \neq 0$) is an invertible universal hashing, and computing and inverting this hash function is about as efficient as computing polynomial evaluation.

The starting point of this work is an implementation of the generic Naor-Reingold mode of operation using the above for the universal hashing layers. We then extend that mode to handle associated data and input of arbitrary length, thus getting the TET mode. Specifically, TET takes a standard cipher with n-bit blocks and turns it into a tweakable enciphering scheme with message space $\mathcal{M} = \{0, 1\}^{n..2^n - 1}$ (i.e., any string of at least n and at most $2^n - 1$ bits) and tweak space $\mathcal{T} = \{0, 1\}^*$. The key for TET consists of two keys of the underlying cipher (roughly one to process the tweak and another to process the data). As we mentioned, TET offers similar performance characteristics to XCB, HCTR and HCH (making it significantly more efficient than PEP and ABL4, and almost as efficient as CMC, and EME/EME*).

A word on notations. Below we use $\oplus$ to denote bit-wise exclusive or (which is the same as addition in $GF(2^n)$), and we use $+/-$ to denote addition/subtraction in other fields or domains (e.g., integer addition). The sum operator $\sum$ is always used to denote finite-field addition.

Organization. Some standard definitions are recalled in Appendix A (which is taken almost verbatim from [HR04, Hal04]). Section 2 describes the hashing scheme that underlies TET, Section 3 describes the TET mode itself, and Section 4 examines the security for this mode. In Appendix B we briefly discuss intellectual-property issues.

2   The Underlying Hashing Scheme

The universality property that is needed for the Naor-Reingold mode of operation is defined next.

Definition 1. *Let $H : \mathcal{K} \times \mathcal{D} \to \mathcal{R}^m$ be a hashing family from some domain $\mathcal{D}$ to m-vectors over the range $\mathcal{R}$, with keys chosen uniformly from $\mathcal{K}$. We denote by $H_k(x)$ the output of H (which is an m-vector over $\mathcal{R}$) on key $k \in \mathcal{K}$ and input $x \in \mathcal{D}$. We also denote by $H_k(x)_i$ the i'th element of that output vector.*

For a real number $\epsilon \in (0, 1)$, we say that $\mathcal{H}$ is "ϵ-blockwise-universal" if for every $x, x' \in \mathcal{D}$ and integers $i, i' \leq m$ such that $(x, i) \neq (x', i')$, it holds that $\Pr_k[H_k(x)_i = H_k(x')_{i'}] \leq \epsilon$, where the probability is taken over the uniform choice of $k \in \mathcal{K}$.

We say that $\mathcal{H}$ is "ϵ-xor-blockwise-universal" if in addition for all fixed $\Delta \in GF(2^n)$ it holds that $\Pr_k[H_k(x)_i \oplus H_k(x')_{i'} = \Delta] \leq \epsilon$.

It was proven in [NR99] that the construction from Figure 1 is a strong PRP on wide blocks provided that the hashing layers are blockwise universal and invertible, and the underlying cipher E is a strong PRP on narrow blocks.

2.1   BPE: A Blockwise Universal Hashing Scheme

To get an invertible blockwise universal hash function, Naor and Reingold proposed in [NR97] to use an unbalanced Feistel network with standard universal hashing. For example, use polynomial-evaluation hash function applied to the first $m - 1$ blocks, xor the result to the last block, and then derive $m - 1$ "pairwise independent" values from the last block and xor them back to the first $m - 1$ blocks. This solution, however, is somewhat unsatisfying in that it entails inherent asymmetry (which is likely to raise problems with implementations). Below we propose a somewhat more elegant blockwise universal hashing based on a simple algebraic trick.

Although for TET we only need a hashing scheme over the field $GF(2^n)$, we describe here the scheme over an arbitrary field. Let $\mathcal{F}$ be a field (with more than $m + 2$ elements) and consider an $m \times m$ matrix over $\mathcal{F}$, $M_\tau \stackrel{\text{def}}{=} A_\tau + I$ for an element $\tau \in \mathcal{F}$, where

$$A_\tau \stackrel{\text{def}}{=} \begin{pmatrix} \tau & \tau^2 & & \tau^m \\ \tau & \tau^2 & & \tau^m \\ & & \ddots & \\ \tau & \tau^2 & & \tau^m \end{pmatrix} \tag{1}$$

It is easy to check that the determinant of M_τ is $\sigma \stackrel{\text{def}}{=} \sum_{i=0}^{m} \tau^i$, and so M_τ is invertible if and only if $\sigma \neq 0$. We observe that when it is invertible, the structure of M_τ^{-1} is very similar to the structure of M_τ itself.

Observation 1. *Let $\mathcal{F}$ be a field and let $\tau \in \mathcal{F}$ be such that $\sigma \overset{\text{def}}{=} \sum_{i=0}^{m} \tau^i \neq 0$, let A_τ be an $m \times m$ matrix with $A_{i,j} = \tau^j$, and let $M_\tau \overset{\text{def}}{=} A_\tau + I$. Then $M_\tau^{-1} = I - (A_\tau / \sigma)$.*

Proof. We first note that $A_\tau^2 = A_\tau(\sigma - 1)$, since for all i, j we have

$$(A_\tau^2)_{i,j} = \sum_{k=1}^{m} \tau^{k+j} = \tau^j \left(\sum_{k=1}^{m} \tau^k \right) = \tau^j \left(\sum_{k=0}^{m} \tau^k - 1 \right) = (A_\tau)_{i,j} \cdot (\sigma - 1)$$

Therefore, assuming $\sigma \neq 0$ we get

$$(A_\tau + I) \cdot (I - \frac{A_\tau}{\sigma}) = A_\tau + I - \frac{A_\tau^2}{\sigma} - \frac{A_\tau}{\sigma} = I + \frac{A_\tau \sigma - A_\tau(\sigma - 1) - A_\tau}{\sigma} = I$$

It follows that computing $\mathbf{y} = M_\tau \mathbf{x}$ and $\mathbf{x} = M_\tau^{-1} \mathbf{y}$ can be done as efficiently as computing polynomial-evaluation hash. Namely, to compute $\mathbf{y} = M_\tau \mathbf{x}$ we first compute $s = \sum_{i=1}^{m} x_i \tau^i$ and set $y_i = x_i + s$, and to invert $\mathbf{x} = M_\tau^{-1} \mathbf{y}$ we re-compute s as $s = \sum_{i=1}^{m} y_i (\tau^i / \sigma)$ and set $x_i = y_i - s$. Moreover, since τ and σ depend only the hashing key, one can speed up the multiplication by τ and τ / σ by pre-computing some tables (cf. [Sho96]).

The blockwise-universal family BPE. Given the observation from above, we define the hashing family BPE (for **B**lockwise **P**olynomial-**E**valuation) and its inverse BPE^{-1} as follows: Let $\mathcal{F}$ be a finite field with $m + 3$ or more elements.

Input: An m-vector of elements from $\mathcal{F}$, $\mathbf{x} = \langle x_1, \dots, x_m \rangle \in \mathcal{F}^m$.
Keys: Two elements $\tau, \beta \in \mathcal{F}$, such that $\sum_{i=0}^{m} \tau^m \neq 0$.
Output: Let α be some fixed primitive element of $\mathcal{F}$, and denote by $\mathbf{b} \overset{\text{def}}{=} \langle \beta, \alpha\beta, \dots, \alpha^{m-1}\beta \rangle$ the m-vector over $\mathcal{F}$ whose i'th entry is $\alpha^{i-1}\beta$. The two hash functions BPE$_{\tau,\beta}(\mathbf{x})$ and BPE$_{\tau,\beta}^{-1}(\mathbf{x})$ are defined as

$$\text{BPE}_{\tau,\beta}(\mathbf{x}) \overset{\text{def}}{=} M_\tau \mathbf{x} + \mathbf{b} \quad \text{and} \quad \text{BPE}_{\tau,\beta}^{-1}(\mathbf{x}) \overset{\text{def}}{=} M_\tau^{-1}(\mathbf{x} - \mathbf{b}) \tag{2}$$

By construction if follows that BPE$_{\tau,\beta}^{-1}(\text{BPE}_{\tau,\beta}(\mathbf{x})) = \mathbf{x}$ for all $\mathbf{x}$ and all τ, β (provided that $\sum_{i=0}^{m} \tau^m \neq 0$). We now prove that these two families (BPE and its inverse) are indeed "blockwise universal".

Claim. Fix a finite field $\mathcal{F}$ and an integer $m \leq |\mathcal{F}| - 3$, and also fix $\mathbf{x}, \mathbf{x}' \in \mathcal{F}^m$ and indexes $i, i' \leq m$ such that $(x, i) \neq (x', i')$, and any $\delta \in \mathcal{F}$.

(i) If $i \neq i'$ then $\Pr_{\tau,\beta}[[\text{BPE}_{\tau,\beta}(\mathbf{x}')]_{i'} - [\text{BPE}_{\tau,\beta}(\mathbf{x}')]_{i'} = \delta] = 1/|\mathcal{F}|$ and similarly $\Pr_{\tau,\beta}\left[[\text{BPE}_{\tau,\beta}^{-1}(\mathbf{x}')]_{i'} - [\text{BPE}_{\tau,\beta}^{-1}(\mathbf{x}')]_{i'} = \delta\right] = 1/|\mathcal{F}|$.

(ii) If $i = i'$ and $\mathbf{x} \neq \mathbf{x}'$ then both $\Pr_{\tau,\beta}[[\text{BPE}_{\tau,\beta}(\mathbf{x}')]_{i'} - [\text{BPE}_{\tau,\beta}(\mathbf{x}')]_{i'} = \delta]$ and $\Pr_{\tau,\beta}\left[[\text{BPE}_{\tau,\beta}^{-1}(\mathbf{x}')]_{i'} - [\text{BPE}_{\tau,\beta}^{-1}(\mathbf{x}')]_{i'} = \delta\right]$ are bounded by $\frac{m}{|\mathcal{F}|-g}$, where $g = GCD(m+1, |\mathcal{F}| - 1)$ if the characteristic of the field $\mathcal{F}$ divides $m + 1$, and $g = GCD(m+1, |\mathcal{F}| - 1) - 1$ otherwise.

Proof. **Case (i),** $i \neq i'$. In this case we have $[\mathrm{BPE}_{\tau,\beta}(\mathbf{x})]_i - [\mathrm{BPE}_{\tau,\beta}(\mathbf{x}')]_{i'} = (\alpha^{i-1} - \alpha^{i'-1})\beta + ((M_\tau \mathbf{x})_i - (M_\tau \mathbf{x}')_{i'})$ which is equal to any fixed δ with probability exactly $1/|\mathcal{F}|$ over the choice of $\beta \in_R \mathcal{F}$ (since α is primitive and so $\alpha^{i-1} \neq \alpha^{i'-1}$). Similarly

$$[\mathrm{BPE}_{\tau,\beta}^{-1}(\mathbf{x})]_i - [\mathrm{BPE}_{\tau,\beta}^{-1}(\mathbf{x}')]_{i'} = \left((I - \frac{A_\tau}{\sigma})(\mathbf{x} - \mathbf{b})\right)_i - \left((I - \frac{A_\tau}{\sigma})(\mathbf{x}' - \mathbf{b})\right)_{i'}$$

$$= ((\frac{A_\tau}{\sigma}\mathbf{b})_i - \mathbf{b}_i) - ((\frac{A_\tau}{\sigma}\mathbf{b})_{i'} - \mathbf{b}_{i'}) + \left((I - \frac{A_\tau}{\sigma})\mathbf{x}\right)_i - \left((I - \frac{A_\tau}{\sigma})\mathbf{x}'\right)_{i'}$$

$$= (\alpha^{i'-1} - \alpha^{i-1})\beta + \left((I - \frac{A_\tau}{\sigma})\mathbf{x}\right)_i - \left((I - \frac{A_\tau}{\sigma})\mathbf{x}'\right)_{i'}$$

where the last equality follows since $(A_\tau \mathbf{b})_i = (A_\tau \mathbf{b})_{i'}$ (because all the rows of A_τ are the same). Again, this sum equals δ with probability exactly 2^{-n}.

Case (ii), $i = i'$ **and** $\mathbf{x} \neq \mathbf{x}'$. In this case we have $[\mathrm{BPE}_{\tau,\beta}(\mathbf{x})]_i - [\mathrm{BPE}_{\tau,\beta}(\mathbf{x}')]_i - \delta = (x_i - x_i' - \delta) + \sum_{j=1}^m (x_j - x_j')\tau^j$, which is zero only when τ is a root of this specific non-zero degree-m polynomial. Similarly for $\mathrm{BPE}_{\tau,\beta}^{-1}$ we have

$$[\mathrm{BPE}_{\tau,\beta}^{-1}(\mathbf{x})]_i - [\mathrm{BPE}_{\tau,\beta}^{-1}(\mathbf{x}')]_i - \delta = \left((I - \frac{A_\tau}{\sigma})(\mathbf{x} - \mathbf{b})\right)_i - \left((I - \frac{A_\tau}{\sigma})(\mathbf{x}' - \mathbf{b})\right)_i - \delta$$

$$= \left((I - \frac{A_\tau}{\sigma})\mathbf{x}\right)_i - \left((I - \frac{A_\tau}{\sigma})\mathbf{x}'\right)_i - \delta = (x_i - x_i' - \delta) + \sum_{j=1}^m \frac{\tau^j}{\sigma}(x_j - x_j')$$

$$\overset{*}{=} \frac{1}{\sigma}\left((x_i - x_i' - \delta)(\sum_{j=0}^m \tau^j) + \sum_{j=1}^m \tau^j(x_j - x_j')\right)$$

$$= \frac{1}{\sigma}\left((x_i - x_i' - \delta) + \sum_{j=1}^m \tau^j((x_j - x_j') + (x_i - x_i' - \delta))\right)$$

where the equality $\overset{*}{=}$ holds since $\sigma = \sum_{i=0}^m \tau^j$. The last expression is zero when τ is a root of the parenthesized polynomial. That polynomial is non-zero since (a) if $x_i - x_i' \neq \delta$ then it has non-zero constant term, and (b) if $x_i - x_i' = \delta$ then there is some index j such that $x_j \neq x_j'$, and thus the coefficient $((x_j - x_j') + (x_i - x_i' - \delta))$ of τ^j is non-zero.

We conclude that for both $\mathrm{BPE}_{\tau,\beta}$ and $\mathrm{BPE}_{\tau,\beta}^{-1}$, a collision in this case implies that τ must be a root of some fixed non-zero degree-m polynomial. Such polynomials have at most m roots, and τ is chosen at random in $\mathrm{GF}(2^n)$ subject to the constraint that $\sigma \neq 0$. Since σ itself is a non-zero degree-m polynomial, then there are at least $2^n - m$ elements $\tau \in \mathrm{GF}(2^n)$ for which $\sigma \neq 0$, and so the collision probability is at most $m/(2^n - m)$.

Moreover, for most values of m we can actually show that there are fewer than m values of τ for which $\sigma = 0$. Specifically, we note that $\sigma = (\tau^{m+1} - 1)/(\tau - 1)$, so $\sigma = 0$ implies that also $\tau^{m+1} - 1 = 0$, which means that τ is an $m + 1$'st root of unity in $\mathcal{F}$. We know that the number of $m + 1$'st roots of unity in $\mathcal{F}$ is exactly $GCD(m + 1, |\mathcal{F}| - 1)$, and one of them is the trivial root $\tau = 1$. The trivial root $\tau = 1$ is also a root of σ if and only if the characteristic of $\mathcal{F}$ divides $m + 1$ (since there are $m + 1$ terms in the sum that defines σ), and all the other $m + 1$'st roots of unity are also root of σ. Hence τ is chosen at random from a set of size $|\mathcal{F}| - g$, where $g = GCD(m + 1, |\mathcal{F}| - 1)$ if the characteristic of $\mathcal{F}$ divides $m + 1$ and $g = GCD(m + 1, 2^n - 1) - 1$ otherwise.

A variant of BPE. It is easy to see that the same claim can be proven also for the variant of BPE that subtracts the vector $\mathbf{b}$ before multiplying by M_τ, namely if we define

$$\widetilde{\mathrm{BPE}}_{\tau,\beta}(\mathbf{x}) \overset{\text{def}}{=} M_\tau(\mathbf{x} - \mathbf{b}) \text{ and } \widetilde{\mathrm{BPE}}_{\tau,\beta}^{-1}(\mathbf{x}) \overset{\text{def}}{=} M_\tau^{-1}\mathbf{x} + \mathbf{b} \qquad (3)$$

then also the hash families $\widetilde{\mathrm{BPE}}$ and $\widetilde{\mathrm{BPE}}^{-1}$ are ϵ-blockwise universal for the same ϵ.

Variable input length. Claim 2.1 refers only to the fixed-input length scenario, where $\mathrm{BPE}_{\tau,\beta}$ is applied always to inputs of the same length. Similar arguments can be used to show universality of BPE, BPE^{-1}, $\widetilde{\mathrm{BPE}}$, and $\widetilde{\mathrm{BPE}}$ also in the variable-input-length scenario, where the same τ and β are used for all the different input lengths.

Claim 2.1 refers only to the fixed-input length scenario, where $\mathrm{BPE}_{\tau,\beta}$ is applied always to inputs of the same length. Similar arguments can show that the four variations BPE, BPE^{-1}, $\widetilde{\mathrm{BPE}}$, and $\widetilde{\mathrm{BPE}}$ are also ϵ-blockwise universal in the variable-input-length scenario, where the same τ and β are used for all the different input lengths.

One complication is that in the variable-input-length scenario, the element $\tau \in \mathcal{F}$ must be chosen such that *for all* m it holds that $1 + \tau + \ldots + \tau^m \neq 0$. This can be achieved by choosing τ as a primitive element in $\mathcal{F}$, which means that it is not an $m + 1$'ts root of unity for any $m < |\mathcal{F}| - 2$, and therefore also not a root of $1 + \tau + \ldots + \tau^m$. As the number of primitive elements in $\mathcal{F}$ is $\phi(|\mathcal{F}| - 1)$ (where ϕ is Euler's totient function), it follows that in this case we choose τ from a set of size exactly $\phi(|\mathcal{F}| - 1)$. Hence the collision probability for any $\mathbf{x}, \mathbf{x}'$ is bounded by $\epsilon = m/\phi(|\mathcal{F}| - 1)$ where m is the length of the longer of $\mathbf{x}, \mathbf{x}'$.

3   The TET Mode of Operation

The BPE hashing scheme immediately implies a mode of operation for implementing a fixed-input-length, non-tweakable enciphering scheme for block-sizes that are a multiple of n bits: namely the Naor-Reingold construction from [NR97] with BPE for the hashing layers (over the field $\mathrm{GF}(2^n)$, where n is the block size of the underlying cipher). In this section I describe how to extend this construction to get a tweakable scheme that supports arbitrary input lengths (and remains secure also when using the same key for different input lengths).

3.1 Tweaks and Variable Input Length

Incorporating a tweak into the basic mode turns out to be almost trivial: Instead of having the element β be part of the key, we derive it from the tweak using the underlying cipher. For example, if we are content with n-bit tweaks then we can just set $\beta \leftarrow E_K(T)$ where k is the cipher key and T is the tweak. Intuitively, this is enough since the multiples of β will be used to mask the input values before they arrive at the ECB layer, so using different pseudo-random values of β for different tweak values means that the ECB layer will be applied on different blocks.

To handle longer tweaks we can replace the simple application of the underlying cipher E with a variable-input-length cipher-based pseudo-random function (e.g., CBC-MAC, PMAC, etc.), using a key which is independent of the cipher key that is used for the ECB layer. In Section 3.3 I describe a particular CBC-MAC-like implementation that suits our needs.

The same fix can be applied also to handle variable input length: namely we derive β from both the tweak and the input length. If we are content with input length of no more than 2^ℓ and tweaks of size $n - \ell$ bits, then we can use $\beta \leftarrow E_K(L, T)$ where T is the tweak value and L is the input length, or else we can use $\beta \leftarrow \mathrm{PRF}_K(L, T)$ for some variable-input-length pseudo-random function. As noted above, using the same hashing key for different input lengths implies that the element τ must satisfy $\sigma_m = 1 \oplus \tau \oplus \ldots \oplus \tau^m \neq 0$ for every possible input length m, and this can be ensured by choosing τ as a random primitive element in $\mathrm{GF}(2^n)$.

3.2 Partial Blocks

It appears harder to extend the mode to handle inputs whose length is not a multiple of n bits. Ideally, we would have liked an elegant way of extending BPE to handle such lengths, and then handle partial blocks in the ECB layer using ciphertext-stealing (cf. [MM82, Fig.2-23]). Unfortunately, I do not know how to extend BPE to handle input length that is not a multiple of n bits while maintaining invertability (except going back to the unbalanced Feistel idea).

Instead, I borrowed a technique that was used also in EME*: When processing an input whose length is not a multiple of n bits, one of the block cipher applications in the ECB layer is replaced with two consecutive applications of the cipher, and the middle value (between the two calls to the underlying cipher) is xor-ed to the partial block. (In addition, the partial block is added to the polynomial-evaluation, so that its value effects all the other blocks.)

In more details, let $\mathbf{x} = \langle x_1, \ldots, x_m \rangle$ be all the full input blocks and let x_{m+1} be a partial block, $\ell = |x_{m+1}|$, $0 < \ell < n$. Instead of just computing $\mathbf{y} = \mathrm{BPE}(\mathbf{x})$, we set the i'th full block to $y_i \leftarrow \mathrm{BPE}(\mathbf{x})_i \oplus (x_{m+1}10..0)$, while leaving x_{m+1} itself unchanged. Then we apply the ECB layer, computing $z_i \leftarrow E_K(y_i)$ for the first $m - 1$ full blocks, and computing $u \leftarrow E_K(y_m)$ and $z_m \leftarrow E_K(u)$ for the last full block. The first bits of u are then xor-ed into the partial block, setting $w_{m+1} = x_{m+1} \oplus u|_{1..\ell}$. Then we do the final BPE layer (adding $(w_{m+1}10..0)$ to each full block), thus getting $w_i \leftarrow \mathrm{BPE}(\mathbf{z})_i \oplus (w_{m+1}10..0)$ and the TET output is the vector $w_1, \ldots, w_m, w_{m+1}$.

3.3   The PRF Function

It is clear that any secure pseudo-random function can be used to derive the element β. We describe now a specific PRF, which is a slight adaptation of the OMAC construction of Iwata and Korasawa [IK03], that seems well suited for our application. The slight modification to OMAC can be thought of as constructing a "tweakable PRF", with an on-line/off-line optimization for the tweak.[1] (In our case, the input-length of TET is the "tweak" for the PRF and the tweak of TET is the input to the PRF.)

We assume that the input length of TET is less than 2^n bits, and we denote by L the input length in bits encoded as an n-bit integer. Also denote the tweak for TET (which is the input to the PRF) by $T = \langle T_1, \ldots, T_{m'} \rangle$ where $|T_1| = \cdots = |T_{m'-1}| = n$ and $1 \leq |T_{m'}| \leq n$.

To compute $\beta \leftarrow \mathrm{PRF}_K(L, T)$ we first compute $X \leftarrow E_K(L)$, then compute β as a CBC-MAC of T, but before the last block-cipher application we xor either the value αX or the value $\alpha^2 X$ (depending on whether the last block is a full block or a partial block). In more details, we set $V_0 = 0$ and then $V_i \leftarrow E_K(V_{i-1} \oplus T_i)$ for $i = 1, \ldots, m' - 1$. Then, if the last block is a full block ($|T_{m'}| = n$) then we set $\beta \leftarrow E_K(\alpha X \oplus V_{m'-1} \oplus T_{m'})$, and if the last block is a partial block ($|T_{m'}| < n$) then we set $\beta \leftarrow E_K(\alpha^2 X \oplus V_{m'-1} \oplus (T_{m'}10..0))$.

Notice that the only difference between this function and the OMAC construction is that OMAC does not have the additional input L and it sets $X \leftarrow E_K(0)$. Proving that this is a secure pseudo-random function is similar to the proof of OMAC [IK03], and is omitted here.

We point out that on one hand, the length L is needed only before processing the last tweak block, so this pseudo-random function is suited for streaming applications where the length of messages is not known in advance.[2] On the other hand, if used with a fixed input length (where L is known ahead of time) then the computation of X can be done off line, in which case we save one block-cipher application during the on-line phase.

3.4   Some Other Details

To get a fully-specified mode of operation one needs to set many other small details. Below I explain my choices for the details that I set, and describe those that are still left unspecified.

---

[1] Formally there is not much difference between a "tweakable" and "non-tweakable" PRF, one can always process the tweak by concatenating it to the input. But here it is convenient to make the distinction since we can offer some tweak-specific performance optimization.

[2] As explained in Section 3.5, TET is not a very good fit for such cases, but this PRF functions can perhaps be used in applications other than TET.

Table 1. Bad τ values for various input lengths, assuming $n = 128$

If the input length is	then these elements are bad values for τ	Bad key probability
512 bytes	$\alpha^{(2^{128}-1)/3}, \alpha^{2\cdot(2^{128}-1)/3}$	2^{-127}
1024 bytes	$\alpha^{i\cdot(2^{128}-1)/5}$ for $i = 1, 2, 3, 4$	2^{-126}
4096 bytes	$\alpha^{i\cdot(2^{128}-1)/257}$ for $i = 1, 2, \ldots, 256$	2^{-120}
65536 bytes	$\alpha^{i\cdot(2^{128}-1)/17}$ for $i = 1, 2, \ldots, 16$	2^{-124}

The element $\alpha \in GF(2^n)$. Recall that BPE uses a fixed primitive element $\alpha \in GF(2^n)$. If the field $GF(2^n)$ is represented with a primitive polynomial, then this fixed element should be set as the polynomial x (or $1/x$), in which case a multiplication by α can be implemented with an n-bit shift and a conditional xor.[3]

The two hashing layers. I chose to use the same hashing keys τ, β for both hashing layers. The security of the mode does not seem to be effected by this. On the other hand, having different keys for the two hashing layers adds a considerable burden to an implementation, especially it if optimizes the GF multiplications by preparing some tables off line.

The hashing key τ. I also chose to derive the hashing key τ from the same cipher key as the hashing key β, rather than being a separate key. (This decision is rather arbitrary, I made it because I could not see any reason to keep τ as a separate key.) Specifically, it can be set as $\tau \leftarrow \mathrm{PRF}_K(0, 0^n) = E_K(\alpha \cdot E_K(0))$. Note that this is not a duplicate of any $\mathrm{PRF}_K(L, T)$, since the input length L is always at least n bits.[4]

Of course, τ must be chosen so that for any message length m it holds that $\sigma_m \neq 0$ (where $\sigma_m = \sum_{i=0}^{m} \tau^m$). Hence if setting $\tau \leftarrow \mathrm{PRF}_K(0, 0)$ results in a bad value for τ then we can keep trying $\mathrm{PRF}_K(0, 1)$, $\mathrm{PRF}_K(0, 2)$, etc. When using TET with fixed input length (containing m complete blocks), we can just include a list of all the "bad τ values" for which $\sigma_m = 0$ with the implementation. This list is fairly easy to construct: Denoting $g = GCD(m+1, 2^n-1)$, when m is even the lists consists of $\alpha^{i\cdot(2^n-1)/g}$ for $i = 1, 2, \ldots, g-1$ (where α is a primitive element). When m is odd it consists of the same elements and also of the element $\alpha^0 = 1$. In Table 1 we list the "bad τ values" for various input lengths assuming $n = 128$.

The approach of having a fixed list of "bad τ values" may not work as well when using TET with variable-input length. One way to handle this case is to insist on τ being a primitive element in $GF(2^n)$, in which case we know that $\sigma_m \neq 0$ for all length m. (We can efficiently test is τ is a primitive element given the prime factorization of 2^n-1). But a better way of handling variable

---

[3] The choice between setting $\alpha = x$ or $\alpha = 1/x$ depends on the endianess of the field representation, and it should be made so that multiplication by α requires left shift and not right shift.

[4] Setting $\tau \leftarrow E_K(0)$ would work just as well in this context, but the effort in proving it is too big for the minuscule saving in running time.

length is to allow different τ's for different input lengths. Specifically, when handling a message of with m full blocks, we try $\mathrm{PRF}_K(0,0)$, $\mathrm{PRF}_K(0,1)$, ... and set τ to the first value for which $\sigma_m \neq 0$. It is not hard to see that this is just as secure as insisting on the same τ for all lengths (since we only use τ to argue about collisions between messages of the same length).

Ordering the blocks for polynomial-evaluation. I chose to order the blocks at the input of BPE in "reverse order", evaluating the polynomial as $\sum_{i=1}^{m} x_i \tau^{m-i+1}$. The reason is to allow processing to start as soon as possible in the case where the input arrives one block at a time. We would like to use Horner's rule when computing BPE($\mathbf{x}$), processing the blocks in sequence as

$$s = (\ldots((x_1\tau \oplus x_2)\tau \oplus x_3)\tau \ldots \oplus x_m)\tau$$

which means that x_1 is multiplied by τ^m, x_2 is multiplied by τ^{m-1}, etc. Similarly when computing BPE$^{-1}(\mathbf{y})$ we would implement the polynomial-evaluation as

$$s = (\ldots((y_1\tau \oplus y_2)\tau \oplus y_3)\tau \ldots \oplus y_m)(\tau/\sigma)$$

which means that y_1 is multiplied by τ^m/σ, y_2 is multiplied by τ^{m-1}/σ, etc.

The hashing direction. For each of the two hashing layers, one can use either of BPE, BPE^{-1}, $\widetilde{\mathrm{BPE}}$, or $\widetilde{\mathrm{BPE}}^{-1}$. For the encryption direction, I chose to use $\widetilde{\mathrm{BPE}}^{-1}$ for the first hashing layer and BPE^{-1} for the second layer. This means that on decryption we use BPE as the first hashing layer and $\widetilde{\mathrm{BPE}}$ for the second layer.

I chose the inverse hash function on encryption and the functions themselves on decryption because inverting the functions may be less efficient than computing them in the forward direction (since one needs to multiply also by τ/σ). In a typical implementation for storage, one would use encryption when writing to storage and decryption when reading back from storage. As most storage is optimized for read (at the expense of the less-frequent write operations), it makes sense to allocate the faster operations for read in this case too.

As for the choice between BPE and $\widetilde{\mathrm{BPE}}$, I chose to add the vector $\mathbf{b}$ in the middle, right before and after the ECB layer. The rationale here is that it is possible to do the computation $\beta \leftarrow \mathrm{PRF}_K(L,T)$ concurrently with the multiplication by M_τ (or its inverse).

Given the choices above, the specification of the TET mode is given in Figure 2. Other details that are not specified here are the choice of the underlying cipher and the block-size n, and the representation of the field $\mathrm{GF}(2^n)$ (including endianess issues).

3.5   Performance of TET

As specified above, the TET mode can be used with variable input length, and in the long version of this note [Hal07] we prove that it is secure when used in

---

function $\mathrm{PRF}_K(L, T_1 \cdots T_{m'})$ // $|L| = |T_1| = \cdots = |T_{m'-1}| = n,\ 1 \le |T_{m'}| \le n$

001   $V_0 \leftarrow 0,\ X \leftarrow E_K(L)$
002   **for** $i \leftarrow 1$ **to** $m' - 1$ **do** $V_i \leftarrow E_K(V_{i-1} \oplus T_i)$
003   **if** $|T_{m'}| = n$ **then return** $E_K(V_{m'-1} \oplus T_{m'} \oplus \alpha X)$
004   **else return** $E_K(V_{m'-1} \oplus T_{m'} \oplus \alpha^2 X)$

Algorithm $\mathrm{TET}_{K_1,K_2}(T; P_1 \cdots P_m P_{m+1})$	**Algorithm** $\mathrm{TET}^{-1}_{K_1,K_2}(T; C_1 \cdots C_m C_{m+1})$
// $|P_1| = \cdots = |P_m| = n,\ 0 \le |P_{m+1}| < n$	// $|C_1| = \cdots = |C_m| = n,\ 0 \le |C_{m+1}| < n$
101 $L \leftarrow mn + |P_{m+1}|$ // input size (bits)	201 $L \leftarrow mn + |C_{m+1}|$ // input size (bits)
102 $i = 0$	202 $i = 0$
103 $\tau \leftarrow \mathrm{PRF}_{K_1}(0, i),\ \sigma \leftarrow 1 \oplus \tau \oplus \ldots \oplus \tau^m$	203 $\tau \leftarrow \mathrm{PRF}_{K_1}(0, i),\ \sigma \leftarrow 1 \oplus \tau \oplus \ldots \oplus \tau^m$
104 **if** $\sigma = 0$ **then** $i \leftarrow i+1$, **goto** 103	204 **if** $\sigma = 0$ **then** $i \leftarrow i+1$, **goto** 203
105 $\beta \leftarrow \mathrm{PRF}_{K_1}(L, T),\ SP \leftarrow 0,\ SC \leftarrow 0$	205 $\beta \leftarrow \mathrm{PRF}_{K_1}(L, T),\ SP \leftarrow 0,\ SC \leftarrow 0$
110 **for** $i \leftarrow 1$ **to** m **do** $SP \leftarrow (SP \oplus P_i) \cdot \tau$	210 **for** $i \leftarrow 1$ **to** m **do** $SC \leftarrow (SC \oplus C_i) \cdot \tau$
111 $SP \leftarrow SP/\sigma$	212 **if** $|C_{m+1}| > 0$ **then**
112 **if** $|P_{m+1}| > 0$ **then**	213 $SC \leftarrow SC \oplus C_{m+1}$ padded with $10..0$
113 $SP \leftarrow SP \oplus P_{m+1}$ padded with $10..0$	
120 **for** $i \leftarrow 1$ **to** m **do**	220 **for** $i \leftarrow 1$ **to** m **do**
121 $PP_i \leftarrow P_i \oplus SP$	221 $CC_i \leftarrow C_i \oplus SC$
122 $PPP_i \leftarrow PP_i \oplus \alpha^{i-1}\beta$	222 $CCC_i \leftarrow CC_i \oplus \alpha^{i-1}\beta$
123 **for** $i \leftarrow 1$ **to** $m - 1$ **do**	223 **for** $i \leftarrow 1$ **to** $m - 1$ **do**
124 $CCC_i \leftarrow E_{K_2}(PPP_i)$	224 $PPP_i \leftarrow E^{-1}_{K_2}(CCC_i)$
125 **if** $|P_{m+1}| > 0$ **then**	225 **if** $|C_{m+1}| > 0$ **then**
126 $MM \leftarrow E_{K_2}(PPP_m)$	226 $MM \leftarrow E^{-1}_{K_2}(CCC_m)$
127 $CCC_m \leftarrow E_{K_2}(MM)$	227 $PPP_m \leftarrow E^{-1}_{K_2}(MM)$
128 $C_{m+1} \leftarrow P_{m+1} \oplus (MM$ truncated$)$	228 $P_{m+1} \leftarrow C_{m+1} \oplus (MM$ truncated$)$
129 **else** $CCC_m \leftarrow E_{K_2}(PPP_m)$	229 **else** $PPP_m \leftarrow E^{-1}_{K_2}(CCC_m)$
130 **for** $i \leftarrow 1$ **to** m **do**	230 **for** $i \leftarrow 1$ **to** m **do**
131 $CC_i \leftarrow CCC_i \oplus \alpha^{i-1}\beta$	231 $PP_i \leftarrow PPP_i \oplus \alpha^{i-1}\beta$
132 $SC \leftarrow (SC \oplus CC_i) \cdot \tau$	232 $SP \leftarrow (SP \oplus PP_i) \cdot \tau$
133 $SC \leftarrow SC/\sigma$	234 **if** $|C_{m+1}| > 0$ **then**
134 **if** $|P_{m+1}| > 0$ **then**	235 $SP \leftarrow SP \oplus P_{m+1}$ padded with $10..0$
135 $SC \leftarrow SC \oplus C_{m+1}$ padded with $10..0$	
140 **for** $i \leftarrow 1$ **to** m **do**	240 **for** $i \leftarrow 1$ **to** m **do**
141 $C_i \leftarrow CC_i \oplus SC$	241 $P_i \leftarrow PP_i \oplus SP$
150 **return** $C_1 \ldots C_m C_{m+1}$	250 **return** $P_1 \ldots P_m P_{m+1}$

Fig. 2. Enciphering and deciphering under TET, with plaintext $P = P_1 \ldots P_m P_{m+1}$, ciphertext $C = C_1 \cdots C_m C_{m+1}$, and tweak T. The element $\alpha \in \mathrm{GF}(2^n)$ is a fixed primitive element.

this manner. However, its efficiency (at least in software) depends crucially on pre-processing that is only possible when used with fixed input length (or at least with a small number of possible lengths). The reason is that on encryption one needs to multiply by τ/σ, which depends on the message length (since $\sigma = \sum_{i=0}^{m} \tau^i$). When used with fixed input length, the value τ/σ can be computed off line, and some tables can be derived to speed up the multiplication by τ/σ. When used with variable input length, however, the value τ/σ must be computed on-line, which at least for software implies a considerable cost. Hence, TET is not very appealing as a variable-input-length mode.

We stress, however, that the motivating application for TET, namely "sector-level encryption", is indeed a fixed-input-length application. Also, there are some limited settings where one can use variable input length without suffering much from the drawback above. For example, a "write once / read many times" application, where the data is encrypted once and then decrypted many times, would only need to worry about computing σ in the initial encryption phase (since σ is not used during decryption). Also, the same value of σ is used for every bit-length from mn to $(m+1)n - 1$, so length variability within this limited range in not effected.[5]

Below we analyze the performance characteristics of TET only for fixed input length. With this assumption, the computation of the PRF function from above takes exactly m' applications of the cipher, where m' is the number of blocks of associated data (full or partial). (This is because the computation of the mask value $X \leftarrow E_K(L)$ can be done off line.) Then we need either m or $m - 1$ GF-multiplies for the polynomial evaluation (depending if we have m or $m - 1$ full blocks), followed by m block-cipher applications for the ECB layer, and again m or $m - 1$ GF multiplies. Altogether, we need $m + m'$ block-cipher applications and either $2m$ or $2m - 2$ GF multiplies. (The shift and xor operations that are also needed are ignored in this description, since they are insignificant in comparison.)

Table 2. Workload for enciphering an m-block input with a 1-block tweak

Mode	CMC	EME*	XCB	HCH	TET
Block-cipher calls	$2m + 1$	$2m + 1 + \lceil m/n \rceil$	$m + 1$	$m + 3$	$m + 1$
GF multiplies	–	–	$2(m + 3)$	$2(m - 1)$	$2m$ or $2(m - 1)$

Table 2 compares the number of block-cipher calls and GF multiplies in CMC, EME*, XCB, HCH, and TET.[6] It is expected that software efficiency will be proportional to these numbers. (As far as I know, the current "common wisdom" is that computing a $GF(2^{128})$ multiplication in software using the approach from [Sho96] with reasonable-size tables, is about as fast as a single application of AES-128.)

As for hardware implementations, all the modes except CMC are parallelizable and pipelinable, so they can be made to run as fast as needed using sufficiently large hardware. Table 3 describes a somewhat speculative efficiency comparison of hypothetical "fully pipelined" implementations of the modes from above (except CMC). In that table I assume (following [YMK05]) that a one-cycle $GF(2^{128})$ multiplication is about three times the size of a module for computing

---

[5] For example, an implementation can handle both 512-byte blocks and 520-byte blocks with a single value of σ (assuming block length of $n = 128$ bits).

[6] The other modes are not included since EME is essentially a special case of EME*, PEP an ABL4 are significantly less efficient than the others, and HCTR is almost identical to HCH.

Table 3. Hardware efficiency: A speculative comparison of pipelined implementations for m-block input and 1-block tweak. Latency is number of cycles until first output block, time is number of cycles until last output block, and size is measured in the equivalent of number of AES-round modules.

Mode	EME*	XCB	HCH	TET
Latency	$m + 30$	$m + 13$	$m + 31$	$2m + 11$
Time	$2m + 10(\lceil m/n \rceil + 2)$	$2m + 27$	$2m + 31$	$2m + 11$
Size	10	13	13	13

the AES round function, and that AES-128 is implemented as 10 such modules. A few other relevant characteristics of these modes are discussed next.

Any input length. All of these modes except CMC support any input length from n bits and up. CMC supports only input length which is a multiple of n bits (but it should be relatively straightforward to extended it using ciphertext-stealing).

Associated data. The modes EME*, XCB and TET support tweaks of arbitrary length. CMC and HCH support only n-bit tweaks (but it is straightforward to extended them to support arbitrary-size tweaks).

Security proofs. The security of XCB was not proved formally, only a sketch was given, and CMC was only proven secure with respect to fixed input-length. The other modes were proven secure with respect to variable input-length. Providing the missing proof seems fairly straightforward to me (but one never knows for sure until the proof is actually written).

Number of keys. Although in principle it is always possible to derive all the needed key material from just one key, different modes are specified with different requirements for key material. In fact using two keys as in TET (one for the ECB layer and one for everything else) offers some unexpected practical advantages over using just one key.

Specifically, implementations sometimes need to be "certified" by standard bodies (such as NIST), and one criterion for certification is that the implementation uses an "approved mode of operation" for encryption. Since standard bodies are slow to approve new modes, it may be beneficial for a TET implementation to claim that it uses the "approved" ECB mode with pre-and post-processing, and moreover the pre- and post-processing is independent of the ECB key. Also, detaching the ECB key from the key that is used elsewhere may make it easier to use a hardware accelerator that supports ECB mode.

4   Security of TET

We relate the security of TET to the security of the underlying primitives from which it is built as follows:

Theorem 1. [TET security] *Fix $n, s \in \mathbb{N}$. Consider an adversary attacking the TET mode with a truly random permutation over $\{0,1\}^n$ in place of the block cipher and a truly random function instead of PRF, such that the total length of all the queries that the attacker makes is at most s blocks altogether.*

The advantage of this attacker in distinguishing TET from a truly random tweakable length-preserving permutation is at most $1.5s^2/\phi(2^n - 1)$ (where ϕ is Euler's totient function). Using the notations from Appendix A, we have

$$\mathbf{Adv}_{\mathrm{TET}}^{\pm \widetilde{\mathrm{prp}}}(s) \leq \frac{3s^2}{2\phi(2^n - 1)}$$

The proof appears in the long version of this note [Hal07]. The intuition is that as long as there are no block collisions in the hash function, then the random permutation in the ECB layer will be applied to new blocks, so it will will output random blocks and the answer that the attacker will see is therefore random.   □

Corollary 1. *With the same setting as in Theorem 1, consider an attacker against TET with a specific cipher E and a specific PRF F, where the attack uses at most total of s' blocks of associated data. Then*

$$\mathbf{Adv}_{\mathrm{TET}[E]}^{\pm \widetilde{\mathrm{prp}}}(t, s, s') \leq \frac{3s^2}{2\phi(2^n - 1)} + 2(\mathbf{Adv}_E^{\pm \mathrm{prp}}(t', s) + \mathbf{Adv}_{\mathrm{PRF}}^{\mathrm{prf}}(t', s'))$$

where $t' = t + O(n(s + s'))$.   □

5   Conclusions

We presented a new method for invertible "blockwise universal" hashing which is about as efficient as polynomial-evaluation hash, and used it in a construction of a tweakable enciphering scheme called TET. This complements the current lineup of tweakable enciphering schemes by providing a scheme in the family of "hash-ECB-hash" which is as efficient as the schemes in the "hash-CTR-hash" family. We also expect that the hashing scheme itself will find other uses beyond TET.

Acknowledgments. I would like to thank the participants of the IEEE SISWG working group for motivating me to write this note. I also thank Doug Whiting and Brian Gladman for some discussions about this mode, and the anonymous CRYPTO reviewers for their comments.

References

[CS06a]   Chakraborty, D., Sarkar, P.: HCH: A new tweakable enciphering scheme using the hash-encrypt-hash approach. In: Barua, R., Lange, T. (eds.) INDOCRYPT 2006. LNCS, vol. 4329, pp. 287–302. Springer, Heidelberg (2006)

[CS06b]   Chakraborty, D., Sarkar, P.: A new mode of encryption providing a tweak-
            able strong pseudo-random permutation. In: Robshaw, M. (ed.) FSE 2006.
            LNCS, vol. 4047, pp. 293–309. Springer, Heidelberg (2006)
[FM04]   Fluhrer, S.R., McGrew, D.A.: The extended codebook (XCB) mode
            of operation. Technical Report, 2007/278, IACR ePrint archive (2004)
            http://eprint.iacr.org/2004/278/
[GM84]   Goldwasser, S., Micali, S.: Probabilistic encryption. Journal of Computer
            and System Sciences 28(2), 270–299 (1984)
[Hal04]   Halevi, S.: EME*: extending EME to handle arbitrary-length messages with
            associated data. In: Canteaut, A., Viswanathan, K. (eds.) INDOCRYPT
            2004. LNCS, vol. 3348, pp. 315–327. Springer, Heidelberg (2004)
[Hal07]   Halevi, S.: Invertible Universal Hashing and the TET Encryption Mode.
            In: Menezes, A. (ed.) CRYPTO 2007. LNCS, vol. 4622, pp. 412–429.
            Springer, Heidelberg (2007) Long version available on-line at, http://
            eprint.iacr.org/2007/014/
[HR03]   Halevi, S., Rogaway, P.: A tweakable enciphering mode. In: Boneh, D. (ed.)
            CRYPTO 2003. LNCS, vol. 2729, pp. 482–499. Springer, Heidelberg (2003)
[HR04]   Halevi, S., Rogaway, P.: A parallelizable enciphering mode. In: Okamoto,
            T. (ed.) CT-RSA 2004. LNCS, vol. 2964, pp. 292–304. Springer, Heidelberg
            (2004)
[IK03]   Iwata, T., Kurosawa, K.: OMAC: One-Key CBC MAC. In: Johansson, T.
            (ed.) FSE 2003. LNCS, vol. 2887, pp. 129–153. Springer, Heidelberg (2003)
[LR88]   Luby, M., Rackoff, C.: How to construct pseudorandom permutations from
            pseudorandom functions. SIAM Journal of Computing 17(2) (1988)
[LRW02]   Liskov, M., Rivest, R.L., Wagner, D.: Tweakable block ciphers. In: Yung,
            M. (ed.) CRYPTO 2002. LNCS, vol. 2442, pp. 31–46. Springer, Heidelberg
            (2002)
[MM82]   Meyr, C.H., Matyas, S.M.: Cryptography: A New Dimension in Computer
            Data Security. John Wiley & Sons, Chichester (1982)
[MV04]   McGrew, D.A., Viega, J.: Arbitrary block length mode. Manuscript
            (2004) Available on-line from http://grouper.ieee.org/groups/1619/
            email/pdf00005.pdf
[NR97]   Naor, M., Reingold, O.: A pseudo-random encryption mode (1997) Manu-
            script available from http://www.wisdom.weizmann.ac.il/~naor/
[NR99]   Naor, M., Reingold, O.: On the construction of pseudo-random permuta-
            tions: Luby-Rackoff revisited. Journal of Cryptology 12(1), 29–66 (1999)
[Sho96]   Shoup, V.: On fast and provably secure message authentication based on
            universal hashing. In: Koblitz, N. (ed.) CRYPTO 1996. LNCS, vol. 1109,
            pp. 74–85. Springer, Heidelberg (1996)
[WFW05]   Wang, P., Feng, D., Wu, W.: HCTR: A variable-input-length enciphering
            mode. In: Feng, D., Lin, D., Yung, M. (eds.) CISC 2005. LNCS, vol. 3822,
            pp. 175–188. Springer, Heidelberg (2005)
[YMK05]   Yang, B., Mishra, S., Karri, R.: A High Speed Architecture for Ga-
            lois/Counter Mode of Operation (GCM). Technical Report, 2005/146,
            IACR ePrint archive (2005), http://eprint.iacr.org/2005/146/

A Preliminaries

A *tweakable enciphering scheme* is a function $\mathbf{E} : \mathcal{K} \times \mathcal{T} \times \mathcal{M} \to \mathcal{M}$ where $\mathcal{M} = \bigcup_{i \in I} \{0,1\}^i$ is the *message space* (for some nonempty index set $I \subseteq \mathbb{N}$)

and $\mathcal{K} \neq \emptyset$ is the *key space* and $\mathcal{T} \neq \emptyset$ is the *tweak space*. We require that for every $K \in \mathcal{K}$ and $T \in \mathcal{T}$ we have that $\mathbf{E}(K, T, \cdot) = \mathbf{E}_K^T(\cdot)$ is a length-preserving permutation on $\mathcal{M}$. The inverse of an enciphering scheme $\mathbf{E}$ is the enciphering scheme $\mathbf{D} = \mathbf{E}^{-1}$ where $X = \mathbf{D}_K^T(Y)$ if and only if $\mathbf{E}_K^T(X) = Y$. A *block cipher* is the special case of a tweakable enciphering scheme where the message space is $\mathcal{M} = \{0, 1\}^n$ (for some $n \geq 1$) and the tweak space is the singleton set containing the empty string. The number n is called the *blocksize*. By $\mathrm{Perm}(n)$ we mean the set of all permutations on $\{0, 1\}^n$. By $\mathrm{Perm}^{\mathcal{T}}(\mathcal{M})$ we mean the set of all functions $\pi : \mathcal{T} \times \mathcal{M} \to \mathcal{M}$ where $\pi(T, \cdot)$ is a length-preserving permutation.

An *adversary* A is a (possibly probabilistic) algorithm with access to some oracles. Oracles are written as superscripts. By convention, the running time of an algorithm includes its description size. The notation $A \Rightarrow 1$ describes the event that the adversary A outputs the bit one.

Security measure. For a tweakable enciphering scheme $\mathbf{E} : \mathcal{K} \times \mathcal{T} \times \mathcal{M} \to \mathcal{M}$ we consider the advantage that the adversary A has in distinguishing $\mathbf{E}$ and its inverse from a random tweakable permutation and its inverse: $\mathbf{Adv}_{\mathbf{E}}^{\pm \widetilde{\mathrm{prp}}}(A) =$

$$\Pr\left[K \xleftarrow{\$} \mathcal{K} : A^{\mathbf{E}_K(\cdot, \cdot)\, \mathbf{E}_K^{-1}(\cdot, \cdot)} \Rightarrow 1\right] - \Pr\left[\pi \xleftarrow{\$} \mathrm{Perm}^{\mathcal{T}}(\mathcal{M}) : A^{\pi(\cdot, \cdot)\, \pi^{-1}(\cdot, \cdot)} \Rightarrow 1\right]$$

The notation shows, in the brackets, an experiment to the left of the colon and an event to the right of the colon. We are looking at the probability of the indicated event after performing the specified experiment. By $X \xleftarrow{\$} \mathcal{X}$ we mean to choose X at random from the finite set $\mathcal{X}$. In writing $\pm\widetilde{\mathrm{prp}}$ the tilde serves as a reminder that the PRP is tweakable and the $\pm$ symbol is a reminder that this is the "strong" (chosen plaintext/ciphertext attack) notion of security. For a block cipher, we omit the tilde.

Without loss of generality we assume that an adversary never repeats an encipher query, never repeats a decipher query, never queries its deciphering oracle with (T, C) if it got C in response to some (T, M) encipher query, and never queries its enciphering oracle with (T, M) if it earlier got M in response to some (T, C) decipher query. We call such queries *pointless* because the adversary "knows" the answer that it should receive.

When $\mathcal{R}$ is a list of resources and $\mathbf{Adv}_{\Pi}^{\mathrm{xxx}}(A)$ has been defined, we write $\mathbf{Adv}_{\Pi}^{\mathrm{xxx}}(\mathcal{R})$ for the maximal value of $\mathbf{Adv}_{\Pi}^{\mathrm{xxx}}(A)$ over all adversaries A that use resources at most $\mathcal{R}$. Resources of interest are the running time t, the number of oracle queries q, and the total number of n-bit blocks in all the queries s. The name of an argument (e.g., t, q, s) will be enough to make clear what resource it refers to.

B   Intellectual-Property Issues

The original motivation for devising the TET mode was to come up with a reasonably efficient mode that is "clearly patent-free". The IEEE security-in-storage working group (SISWG) was working on a standard for length-preserving encryption for storage, and some of the participants expressed the wish to have such a mode. (**Disclaimer:** Not being a patent lawyer, I can only offer my educated guesses for the IP status of the various modes. The assessment below reflects only my opinion about where things stand.)

The modes CMC/EME/EME* from the "encrypt-mix-encrypt" family are all likely to be patent-encumbered, due to US Patent Application 20040131182A1 from the University of California (which as of this writing was not yet issued). Similarly, the XCB mode – which is the first proposed mode in the "hash-CTR-hash" family – is likely to be patent-encumbered due to a US patent application US20070081668A1 from Cisco Systems (also still not issued as of this writing). The status of the other members of the "hash-CTR-hash" family is unclear: they may or may not be covered by the claims of the Cisco patent when it is issued.

This state of affairs left the "hash-ECB-hash" approach as the best candidate for finding patent-free modes: this approach is based on the paper of Naor and Reingold [NR97] that pre-dates all these modes by at least five years, and for which no patent was filed. Specifically for TET, I presented the basic construction from Section 2 in an open meeting of the IEEE SISWG in October of 2002 (as well as in an email message that I sent the SISWG mailing list on October 9, 2002), and I never filed for any patent related to this construction. Thus it seems unlikely to me that there are any patents that cover either the general "hash-ECB-hash" approach or TET in particular.

Reducing Trust in the PKG in Identity Based Cryptosystems

Vipul Goyal

Department of Computer Science
University of California, Los Angeles
vipul@cs.ucla.edu

Abstract. One day, you suddenly find that a private key corresponding to your Identity is up for sale at e-Bay. Since you do not suspect a key compromise, perhaps it must be the PKG who is acting dishonestly and trying to make money by selling your key. How do you find out for sure and even prove it in a court of law?

This paper introduces the concept of Traceable Identity based Encryption which is a new approach to mitigate the (inherent) key escrow problem in identity based encryption schemes. Our main goal is to restrict the ways in which the PKG can misbehave. In our system, if the PKG ever maliciously generates and distributes a decryption key for an Identity, it runs the risk of being caught and prosecuted.

In contrast to other mitigation approaches, our approach does not require multiple key generation authorities.

1   Introduction

The notion of identity based encryption (IBE) was introduced by Shamir [Sha84] as an approach to simplify public key and certificate management in a public key infrastructure (PKI). Although the concept was proposed in 1984 [Sha84], it was only in 2001 that a practical and fully functional IBE scheme was proposed by Boneh and Franklin [BF01]. Their construction used bilinear maps and could be proven secure in the random oracle model. Following that work, a rapid development of identity based PKI has taken place. A series of papers [CHK03, BB04a, BB04b, Wat05, Gen06] striving to achieve stronger notions of security led to efficient IBE schemes in the standard model. There now exist hierarchical IBE schemes [GS02, HL02, BBG05], identity based signatures and authentication schemes [CC03, FS86, FFS88] and a host of other identity based primitive.

In an IBE system, the public key of a user may be an arbitrary string like an e-mail address or other identifier. This eliminates certificates altogether; the sender could just encrypt the message with the identity of the recipient without having to first obtain his public key (and make sure that the obtained public key is the right one). Of course, users are not capable of generating a private key for an identity themselves. For this reason, there is a trusted party called the private key generator (PKG) who does the system setup. To obtain a private key

A. Menezes (Ed.): CRYPTO 2007, LNCS 4622, pp. 430–447, 2007.

for his identity, a user would go to the PKG and prove his identity. The PKG would then generate the appropriate private key and pass it on to the user.

Since the PKG is able to compute the private key corresponding to any identity, it has to be completely trusted. The PKG is free to engage in malicious activities without any risk of being confronted in a court of law. The malicious activities could include: decrypting and reading messages meant for any user, or worse still: generating and distributing private keys for any identity. This, in fact, has been cited as a reason for the slow adoption of IBE despite its nice properties in terms of usability. It has been argued that due to the inherent key escrow problem, the use of IBE is restricted to small and closed groups where a central trusted authority is available [ARP03, LBD+04, Gen03].

One approach to mitigate the key escrow problem problem is to employ multiple PKGs [BF01]. In this approach, the master key for the IBE system is distributed to multiple PKGs; that is, no single PKG has the knowledge of the master key. The private key generation for an identity is done in a threshold manner. This is an attractive solution and successfully avoids placing trust in a single entity by making the system distributed. However, this solution comes at the cost of introducing extra infrastructure and communication. It is burdensome for a user to go to several key authorities, prove his identity to each of them and get a private key component (which has to be done over a secure channel). Further, maintaining multiple independent entities for managing a single PKI might be difficult in a commercial setting (e.g., the PKI has to be jointly managed by several companies).

To the best of our knowledge, without making the PKG distributed, there is no known solution to mitigate the problem of having to place trust in the PKG.

A New Approach. In this paper, we explore a new approach to mitigate the above trust problem. Very informally, the simplest form of our approach is as follows:

- In the IBE scheme, there will be an exponential (or super-polynomial) number of possible decryption keys corresponding to every identity ID.
- Given one decryption key for an identity, it is intractable to find any other.
- A users gets the decryption key corresponding to his identity from the PKG using a secure *key generation protocol.* The protocol allows the user to obtain a single decryption key d_{ID} for his identity *without letting the PKG know which key he obtained.*
- Now if the PKG generates a decryption key d'_{ID} for that identity for malicious usage, with all but negligible probability, it will be different from the key d_{ID} which the user obtained. Hence the key pair $(d_{\mathsf{ID}}, d'_{\mathsf{ID}})$ is a cryptographic proof of malicious behavior of the PKG (since in normal circumstances, only one key per identity should be in circulation).

The PKG can surely decrypt all the user message *passively.* However, the PKG is severely restricted as far as the distribution of the private key d'_{ID} is concerned. The knowledge of the key d'_{ID} enables an entity E to go to the honest user U

(with identity ID and having key d_{ID}) and together with him, sue the PKG by presenting the pair (d'_{ID}, d_{ID}) as a proof of fraud (thus potentially closing down its business or getting some hefty money as a compensation which can be happily shared by E and U). This means that if the PKG ever generates a decryption key for an identity for malicious purposes, it runs the risk that the key could fall into "right hands" which could be fatal.

The above approach can be compared to a regular (i.e., not identity based) PKIs. In a regular PKI, a user will go to a CA and get a certificate binding his identity with his public key. The CA could surely generate one more certificate binding a malicious public key to his identity. However, two certificates corresponding to the same identity constitute a cryptographic proof of fraud. Similarly in our setting, the PKG is free to generate one more decryption key for his identity. However, two decryption keys corresponding to the same identity constitute a proof of fraud. Of course, there are important differences. In a regular PKI, the CA has to actively send the fraudulent certificate to potential encrypters (which is risky for the CA) while in our setting, the PKG could just decrypt the user messages *passively*. However, we believe that the IBE setting is more demanding and ours is nonetheless a step in the right direction.

We call an identity based encryption scheme of the type discussed above as a *traceable identity based encryption* (T-IBE) scheme. This is to reflect the fact that if a malicious decryption key is discovered, it can be traced back either to the corresponding user (if his decryption key is the same as the one found) or to the PKG (if the user has a different decryption key). We formalize this notion later on in the paper. We remark that what we discussed above is a slight simplification of our T-IBE concept. Given a decryption key for an identity, we allow a user to compute certain other decryption keys for the same identity as long as all the decryption keys computable belong to the *same family* (a family can be seen as a subspace of decryption keys). Thus in this case, two decryption keys belonging to different families is a cryptographic proof of malicious behavior of the PKG.

Although the concept of T-IBE is interesting, we do not expect it to be usable on its own. We see this concept more as a stepping stone to achieving what we call a *black-box traceable identity based encryption* discussed later in this section.

Our Constructions. We formalize the notion of traceable identity based encryption and present two construction for it; one very efficient but based on a strong assumption, the other somewhat inefficient but based on the standard decisional BDH assumption.

Our first construction is based on the identity based encryption scheme recently proposed by Gentry [Gen06]. The scheme is the most efficient IBE construction known to date without random oracle. Apart from computational efficiency, it enjoys properties such as short public parameters and a tight security reduction (albeit at the cost of using a strong assumption). Remarkably, we are able to convert Gentry's scheme to a T-IBE scheme without any changes whatsoever to the basic cryptosystem. We are able to construct a secure key generation protocol as per our requirement for the basic cryptosystem and then present new proofs of security to show that the resulting system is a T-IBE system.

Our second construction of traceable identity based encryption is based on the decisional BDH assumption and uses the IBE scheme of Waters [Wat05] and the Fuzzy IBE scheme of Sahai and Waters [SW05] as building blocks. We remark that the construction is not very efficient and requires several pairing operations per decryption.

Black-Box Traceable Identity based Encryption. In traceable identity based encryption, as explained we consider the scenario when a PKG generates and tries to distribute a decryption key corresponding to an identity. T-IBE specifically assumes that the key is a *well-formed* decryption key. However, one can imagine a scenario where the PKG constructs a malformed decryption key which, when used in conjunction with some other decryption process, is still able to decrypt the ciphertexts. In the extreme case, there could a black box (using an unknown key and algorithm) which is able to decrypt the ciphertexts. Given such a box, a third party (such as the court of law), possibly with the cooperation of the PKG and the user, should be able to trace the box back to its source. That is, it should be able to determine whether it was the PKG or the user who was involved in the creation of this black box. We call such a system as *black-box traceable identity based encryption* system. This is a natural extension of the T-IBE concept and is closely related to the concept of black-box traitor tracing in broadcast encryption [CFN94, BSW06]. We leave the construction of a black-box T-IBE scheme as an important open problem.

We stress that black-box T-IBE is really what one would like to use in practice. We intend the current work only to serve as an indication of what might be possible, and as motivation for further work in this direction.

Related Work. To our knowledge, T-IBE is the first approach for any kind of mitigation to the problem of trust in the PKG without using multiple PKGs. On the multiple PKGs side, Boneh and Franklin [BF01] proposed an efficient approach to make the PKG distributed in their scheme using techniques from threshold cryptography. Lee *et al* [LBD+04] proposed a variant of this approach using multiple key privacy agents (KPAs). Also relevant are the new cryptosystems proposed by Gentry [Gen03] and Al-Riyami and Paterson [ARP03]. Although their main motivation was to overcome the key escrow problem, these works are somewhat orthogonal to ours since these cryptosystems are not identity based.

Other Remarks. We only consider identity based *encryption* in this paper. The analogue of T-IBE for identity based signatures appears straightforward to achieve. We also note that it may be possible to profitably combine our approach with the multiple PKG approach and exploit the mutual distrust between the PKGs. For example if two (or more) PKGs collude together to generate a decryption key for an identity, each PKG knows that it has left a cryptographic proof of its fraud with others. We can have a system where the PKG who presents this proof of fraud to the court is not penalized. Now since a PKG has the power to sue another (to close down its business and have one less competitor), this seems to be an effective fraud prevention idea. We do not explore this approach in this paper and defer it to future work.

2   Preliminaries

2.1   Bilinear Maps

We present a few facts related to groups with efficiently computable bilinear maps. Let $\mathbb{G}_1$ and $\mathbb{G}_2$ be two multiplicative cyclic groups of prime order p. Let g be a generator of $\mathbb{G}_1$ and e be a bilinear map, $e : \mathbb{G}_1 \times \mathbb{G}_1 \to \mathbb{G}_2$. The bilinear map e has the following properties:

1. Bilinearity: for all $u, v \in \mathbb{G}_1$ and $a, b \in \mathbb{Z}_p$, we have $e(u^a, v^b) = e(u, v)^{ab}$.
2. Non-degeneracy: $e(g, g) \neq 1$.

We say that $\mathbb{G}_1$ is a bilinear group if the group operation in $\mathbb{G}_1$ and the bilinear map $e : \mathbb{G}_1 \times \mathbb{G}_1 \to \mathbb{G}_2$ are both efficiently computable. Notice that the map e is symmetric since $e(g^a, g^b) = e(g, g)^{ab} = e(g^b, g^a)$.

2.2   Complexity Assumptions

Decisional Bilinear Diffie-Hellman (BDH) Assumption. Let $a, b, c, z \in \mathbb{Z}_p$ be chosen at random and g be a generator of $\mathbb{G}_1$. The decisional BDH assumption [BB04a, SW05] is that no probabilistic polynomial-time algorithm $\mathcal{B}$ can distinguish the tuple $(A = g^a, B = g^b, C = g^c, e(g, g)^{abc})$ from the tuple $(A = g^a, B = g^b, C = g^c, e(g, g)^z)$ with more than a negligible advantage. The advantage of $\mathcal{B}$ is

$$\left| \Pr[\mathcal{B}(A, B, C, e(g, g)^{abc}) = 0] - \Pr[\mathcal{B}(A, B, C, e(g, g)^z) = 0] \right|$$

where the probability is taken over the random choice of the generator g, the random choice of a, b, c, z in $\mathbb{Z}_p$, and the random bits consumed by $\mathcal{B}$.

Decisional Truncated q-ABDHE Assumption. The truncated augmented bilinear Diffie-Hellman exponent assumption (truncated q-ABDHE assumption) was introduced by Gentry [Gen06] and is very closely related to the q-BDHE problem [BBG05] and the q-BDHI problem [BB04a]. Let g be a generator of $\mathbb{G}_1$. The decisional truncated q-ABDHE assumption is: given a vector of $q + 3$ elements

$$(g', g'^{(\alpha^{q+2})}, g, g^\alpha, g^{(\alpha^2)}, \ldots, g^{(\alpha^q)})$$

no PPT algorithm $\mathcal{B}$ can distinguish $e(g, g')^{(\alpha^{q+1})}$ from a random element $Z \in \mathbb{G}_2$ with more than a negligible advantage. The advantage of $\mathcal{B}$ is defined as

$$\left| \Pr[\mathcal{B}(g', g'^{(\alpha^{q+2})}, g, g^{\alpha}, g^{(\alpha^2)}, \ldots, g^{(\alpha^q)}, e(g, g')^{(\alpha^{q+1})}) = 0] \right.$$

$$\left. - \Pr[\mathcal{B}(g', g'^{(\alpha^{q+2})}, g, g^{\alpha}, g^{(\alpha^2)}, \ldots, g^{(\alpha^q)}, Z) = 0] \right|$$

where the probability is taken over the random choice of generator $g, g' \in \mathbb{G}_1$, the random choice of $\alpha \in \mathbb{Z}_p$, the random choice of $Z \in \mathbb{G}_2$, and the random bits consumed by $\mathcal{B}$.

Computational q-BSDH Assumption. The q-Strong Diffie-Hellman assumption (q-SDH assumption) was introduced by Boneh and Boyen [BB04c] for the construction of short signatures where it was also proven to be secure in the generic group model. This assumption was also later used in the construction of short group signatures [BBS04]. Let g be a generator of $\mathbb{G}_1$. The q-SDH assumption is defined in $(\mathbb{G}, \mathbb{G})$ as follows. Given a vector of $q + 1$ elements

$$(g, g^{\alpha}, g^{(\alpha^2)}, \ldots, g^{(\alpha^q)})$$

no PPT algorithm $\mathcal{A}$ can compute a pair $(r, g^{1/(\alpha+r)})$ where $r \in \mathbb{Z}_p$ with more than a negligible advantage. The advantage of $\mathcal{A}$ is defined as

$$\left| \Pr[\mathcal{A}(g, g^{\alpha}, g^{(\alpha^2)}, \ldots, g^{(\alpha^q)}) = (r, g^{1/(\alpha+r)})] \right|$$

The q-BSDH assumption is defined identically to q-SDH except that now $\mathcal{A}$ is challenged to compute $(r, e(g, g)^{1/(\alpha+r)})$. Note that the q-BSDH assumption is already implied by the q-SDH assumption.

2.3   Miscellaneous Primitives

Zero-knowledge Proof of Knowledge of Discrete Log. Informally, a zero-knowledge proof of knowledge (ZK-POK) of discrete log protocol enables a prover to prove to a verifier that it possesses the discrete log r of a given group element R in question. Schnorr [Sch89] constructed an efficient number theoretic protocol to give a ZK-POK of discrete log.

A ZK-POK protocol has two distinct properties: the *zero-knowledge* property and the *proof of knowledge* properties. The former implies the existence of a simulator S which is able to simulate the view of a verifier in the protocol from scratch (i.e., without being given the witness as input). The latter implies the existence of a *knowledge-extractor* Ext which interacts with the prover and extracts the witness using rewinding techniques. For more details on ZK-POK systems, we refer the reader to [BG92].

1-out-of-2 Oblivious Transfer. Informally speaking, a 1-out-of-2 oblivious transfer protocol allows a receiver to choose and receive exactly one of the two string from the sender, such that the other string is computationally hidden from the receiver and the choice of the receiver is computationally hidden from the sender.

Oblivious transfer was first introduced by [Rab81] while the 1-out-of-2 variant was introduced by [EGL85]. Various efficient constructions of 1-out-of-2 oblivious transfer are known based on specific assumptions such factoring or Diffie-Hellman [NP01, Kal05].

3 The Definitions and the Model

A Traceable Identity Based Encryption (T-IBE) scheme consists of five components.

Setup. This is a randomized algorithm that takes no input other than the implicit security parameter. It outputs the public parameters PK and a master key MK.

Key Generation Protocol. This is an interactive protocol between the public parameter generator PKG and the user U. The common input to PKG and U are: the public parameters PK and the identity ID (of U) for which the decryption key has to be generated. The private input to PKG is the master key MK. Additionally, PKG and U may use a sequence of random coin tosses as private inputs. At the end of the protocol, U receives a decryption key d_{ID} as its private output.

Encryption. This is a randomized algorithm that takes as input: a message m, an identity ID, and the public parameters PK. It outputs the ciphertext C.

Decryption. This algorithm takes as input: the ciphertext C that was encrypted under the identity ID, the decryption key d_{ID} for ID and the public parameters PK. It outputs the message m.

Trace. This algorithm associates each decryption key to a family of decryption keys. That is, the algorithm takes as input a well-formed decryption key d_{ID} and outputs a decryption key family number n_F.

Some additional intuition about the relevance of trace algorithm is as follows. In a T-IBE system, there are a super-polynomial number of families of decryption keys. Each decryption key d_{ID} for an identity ID will belong to a unique decryption key family (denoted by the number n_F). Roughly speaking, in the definitions of security we will require that: given a decryption key belonging to a family, it should be intractable to find a decryption key belonging to a different family (although it may be possible to find another decryption key belonging to the same family).

To define security for a traceable identity based encryption system, we first define the following games.

The IND-ID-CPA game. The IND-ID-CPA game for T-IBE is very similar to the IND-ID-CPA for standard IBE [BF01].

Setup. The challenger runs the Setup algorithm of T-IBE and gives the public parameters PK to the adversary.

Phase 1. The adversary runs the Key Generation protocol with the challenger for several adaptively chosen identities $\mathsf{ID}^1, \ldots, \mathsf{ID}^q$ and gets the decryption keys $d_{\mathsf{ID}^1}, \ldots, d_{\mathsf{ID}^q}$.

Challenge. The adversary submits two equal length messages m_0 and m_1 and an identity ID not equal to any of the identities quries in Phase 1. The challenger flips a random coin b and encrypts m_b with ID. The ciphertext C is passed on to the adversary.

Phase 2. This is identical to Phase 1 except that adversary is not allowed to ask for a decryption key for ID.

Guess. The adversary outputs a guess b' of b.

The advantage of an adversary $\mathcal{A}$ in this game is defined as $\Pr[b' = b] - \frac{1}{2}$.

We note that the above game can extended to handle chosen-ciphertext attacks in the natural way by allowing for decryption queries in Phase 1 and Phase 2. We call such a game to be the IND-ID-CCA game.

The FindKey game. The FindKey game for T-IBE is defined as follows.

Setup. The adversary (acting as an adversarial PKG) generates and passes the public parameters PK and an identity ID on to the challenger. The challenger runs a sanity check on PK and aborts if the check fails.

Key Generation. The challenger and the adversary then engage in the key generation protocol to generate a decryption key for the identity ID. The challenger gets the decryption d_{ID} as output and runs a sanity check on it to ensure that it is well-formed. It aborts if the check fails.

Find Key. The adversary outputs a decryption key d'_{ID}. The challenger runs a sanity check on d'_{ID} and aborts if the check fails.

Let SF denote the event that $\mathsf{trace}(d'_{\mathsf{ID}}) = \mathsf{trace}(d_{\mathsf{ID}})$, i.e., d'_{ID} and d_{ID} belong to the same decryption key family. The advantage of an adversary $\mathcal{A}$ in this game is defined as $\Pr[SF]$.

We note that the above game can be extended to include a *decryption phase* where the adversary adaptively queries the challenger with a sequence of ciphertexts $C_1, \ldots, C_m$. The challenger decrypts C_i with its key d_{ID} and sends the resulting message m_i. This phase could potentially help the adversary deduce information about the decryption key family of d_{ID} if it is able to present a *maliciously formed* ciphertext and get the challenger try to decrypt it.

However, if the adversary was somehow restricted to presenting only well-formed ciphertexts, the decrypted message is guaranteed to contain *no information* about the decryption key family (since decryption using every well-formed key would lead to the same message). This can be achieved by adding a ciphertext sanity check phase during decryption. In both of our constructions, we take this route instead of adding a decryption phase to the FindKey game.

The ComputeNewKey game. The ComputeNewKey game for T-IBE is defined as follows.

Setup. The challenger runs the Setup algorithm of T-IBE and gives the public parameters PK to the adversary.

Key Generation. The adversary runs the Key Generation protocol with the challenger for several adaptively chosen identities $ID^1, \ldots, ID^q$ and gets the decryption keys $d_{ID^1}, \ldots, d_{ID^q}$.

New Key Computation. The adversary outputs two decryption keys d_{ID}^1 and d_{ID}^2 for an identity ID. The challenger runs a key sanity check on both of them and aborts if the check fails.

Let DF denote the event that $\mathsf{trace}(d_{ID}^1) \neq \mathsf{trace}(d_{ID}^2)$, i.e., d_{ID}^1 and d_{ID}^2 belong to different decryption key families. The advantage of an adversary $\mathcal{A}$ in this game is defined as $\Pr[DF]$.

We also define a Selective-ID ComputeNewKey game where the adversary has to declare in advance (i.e., before the setup phase) the identity ID for which it will do the new key computation. The advantage of the adversary is similarly defined to be the probability of the event that it is able to output two decryption keys from different decryption key families *for the pre-declared identity* ID. This weakening of the game can be seen as similar to weakening of the IND-ID-CPA game by some previously published papers [CHK03, BB04a, SW05, GPSW06].

Definition 1. *A traceable identity based encryption scheme is* IND-ID-CPA *secure if all polynomial time adversaries have at most a negligible advantage in the* IND-ID-CPA *game, the* FindKey *game and the* ComputeNewKey *game.* IND-ID-CCA *security for T-IBE is defined similarly.*

4   Construction Based on Gentry's Scheme

Our first construction of traceable identity based encryption is based on the identity based encryption scheme recently proposed by Gentry [Gen06]. The scheme is the most efficient IBE construction known to date without random oracle. Apart from computational efficiency, it enjoys properties such as short public parameters and a tight security reduction. This comes at the cost of using a stronger assumption known as the truncated q-ABDHE which is a variant of an assumption called q-BDHE introduced by Boneh, Boyen and Goh [BBG05].

Remarkably, we are able to convert Gentry's scheme to a T-IBE scheme without any changes whatsoever to the basic cryptosystem. We are able to construct a secure key generation protocol as per our requirement for the basic cryptosystem and then present new proofs of security to show the negligible advantage of the adversary in the three games of our T-IBE model. Our proofs are based on the truncated q-ABDHE and the q-BSDH assumption (see Section 2). The end result is a T-IBE scheme which is as efficient as the best known IBE scheme without random oracle. We view this fact as evidence that the additional traceability property does not necessarily come at the cost of a performance penalty.

4.1   The Construction

Let $\mathbb{G}_1$ be a bilinear group of prime order p, and let g be a generator of $\mathbb{G}_1$. In addition, let $e : \mathbb{G}_1 \times \mathbb{G}_1 \rightarrow \mathbb{G}_2$ denote a bilinear map. A security parameter, κ, will determine the size of the groups.

As discussed before, the basic cryptosystem (i.e., Setup, Encryption and Decryption) is identical to Gentry's [Gen06]. For completeness, we describe the whole T-IBE scheme. Parts of this section are taken almost verbatim from [Gen06].

Setup. The PKG picks random generators $g, h_1, h_2, h_3 \in \mathbb{G}_1$ and a random $\alpha \in \mathbb{Z}_p$. It sets $g_1 = g^\alpha$ and then chooses a hash function H from a family of universal one-way hash function. The published public parameters PK and the master key MK are given by

$$\text{PK} = g, g_1, h_1, h_2, h_3, H \qquad \text{MK} = \alpha$$

Key Generation Protocol. This is the protocol through which a user U with an identity ID can securely get a decryption key d_{ID} from PKG . As in [Gen06], PKG aborts if $\text{ID} = \alpha$. The key generation protocol proceeds as follows.

1. The user U selects a random $r \in \mathbb{Z}_p$ and sends $R = h_1^r$ to the PKG .
2. U gives to PKG a zero-knowledge proof of knowledge of the discrete log (as in Section 2) of R with respect to h_1.
3. The PKG now chooses three random numbers $r', r_{\text{ID},2}, r_{\text{ID},3} \in \mathbb{Z}_p$. It then computes the following values

$$(r', h'_{\text{ID},1}), (r_{\text{ID},2}, h_{\text{ID},2}), (r_{\text{ID},3}, h_{\text{ID},3})$$

   where $h'_{\text{ID},1} = (Rg^{-r'})^{1/(\alpha-\text{ID})}$ and $h_{\text{ID},i} = (h_i g^{-r_{\text{ID},i}})^{1/(\alpha-\text{ID})}, i \in \{2,3\}$

   and sends them to the user U.
4. U computes $r_{\text{ID},1} = r'/r$ and $h_{\text{ID},1} = (h'_{\text{ID},1})^{1/r}$. Note that since $h'_{\text{ID},1} = (h_1^r g^{-r'})^{1/(\alpha-\text{ID})}$, $h_{\text{ID},1} = ((h_1^r)^{1/r}(g^{-r'})^{1/r})^{1/(\alpha-\text{ID})} = (h_1 g^{-r_{\text{ID},1}})^{1/(\alpha-\text{ID})}$. It sets the decryption key $d_{\text{ID}} = \{(r_{\text{ID},i}, h_{\text{ID},i}) : i \in \{1,2,3\}\}$.
5. U now runs a key sanity check on d_{ID} as follows. It computes $g^{\alpha-\text{ID}} = g_1/g^{\text{ID}}$ and checks if $e(h_{\text{ID},i}, g^{\alpha-\text{ID}}) \overset{?}{=} e(h_i g^{-r_{\text{ID},i}}, g)$ for $i \in \{1,2,3\}$. U aborts if the check fails for any i.

At the end of this protocol, U has a well-formed decryption key d_{ID} for the identity ID.

Encryption. To encrypt a message $m \in \mathbb{G}_2$ using identity $\text{ID} \in \mathbb{Z}_p$, generate a random $s \in \mathbb{Z}_p$ and compute the ciphertext C as follows

$$C = (g_1^s g^{-s.ID}, \ e(g,g)^s, \ m{\cdot}e(g,h_1)^{-s}, \ e(g,h_2)^s e(g,h_3)^{s\beta})$$

where for $C = (u,v,w,y)$, we set $\beta = H(u,v,w)$.

Decryption. To decrypt a ciphertext $C = (u, v, w, y)$ with identity ID, set $\beta = H(u, v, w)$ and test whether

$$y = e(u, h_{\mathsf{ID},2}h_{\mathsf{ID},3}{}^{\beta})v^{r_{\mathsf{ID},2}+r_{\mathsf{ID},3}\beta}$$

If the above check fails, output $\bot$, else output

$$m = w \cdot e(u, h_{\mathsf{ID},1})v^{r_{\mathsf{ID},1}}$$

For additional intuition about the system and its correctness, we refer the reader to [Gen06]. We also note that the ciphertext sanity check in the decryption algorithm rejects *all* invalid ciphertexts as shown in [Gen06].

Trace. This algorithm takes a well-formed decryption key $d_{\mathsf{ID}} = \{(r_{\mathsf{ID},i}, h_{\mathsf{ID},i}) : i \in \{1, 2, 3\}\}$ and outputs the decryption key family number $n_F = r_{\mathsf{ID},1}$. Hence if $r_{\mathsf{ID},1} = r'_{\mathsf{ID},1}$ for two decryption keys d_{ID} and d'_{ID}, then $\mathsf{trace}(d'_{\mathsf{ID}}) = \mathsf{trace}(d_{\mathsf{ID}})$ (i.e., the two keys belong to the same decryption key family).

4.2   Security Proofs

Theorem 1. *The advantage of an adversary in the* IND-ID-CCA *game is negligible for the above traceable identity based encryption scheme under the decisional truncated q-ABDHE assumption.*

PROOF SKETCH: The proof in our setting very much falls along the lines of the proof of IND-ID-CCA security of Gentry's scheme [Gen06]. Here we just give a sketch highlighting the only difference from the one in [Gen06].

The only difference between [Gen06] and our setting is how a decryption key d_{ID} is issued for an identity ID. In the proof of [Gen06], PKG was free to choose a decryption key d_{ID} on its own and pass it on to the user. PKG in fact chose $r_{\mathsf{ID},i}$ using a specific technique depending upon the truncated q-ABDHE problem instance given. In our setting, however, PKG and the user U engage in a key generation protocol where $r_{\mathsf{ID},1}$ is jointly determined by both of them (via the choice of numbers r and r'). Hence PKG does not have complete control over $r_{\mathsf{ID},1}$.

The above problem can be solved as follows. PKG generates a decryption key $d_{\mathsf{ID}} = \{(r_{\mathsf{ID},i}, h_{\mathsf{ID},i}) : i \in \{1, 2, 3\}\}$ on its own exactly as in [Gen06] and then "forces" the output of U to be the above key during key generation. Recall that during the key generation protocol, U first chooses a random $r \in \mathbb{Z}_p$ and sends $R = h_1^r$ to the PKG . U then gives to PKG a zero-knowledge *proof of knowledge* of the discrete log of R. The proof of knowledge property of the proof system implies the existence of a *knowledge extractor* Ext (see Section 2). Using Ext on U during the proof of knowledge protocol, PKG can extract the discrete log r (by rewinding U during protocol execution) with all but negligible probability. Now PKG sets $r' = rr_{\mathsf{ID},1}$. It then sends $(r', h'_{\mathsf{ID},1} = h_{\mathsf{ID},1}^r), (r_{\mathsf{ID},2}, h_{\mathsf{ID},2}), (r_{\mathsf{ID},3}, h_{\mathsf{ID},3})$ to U.

The user U will compute $r_{\mathsf{ID},1} = r'/r, h_{\mathsf{ID},1} = (h'_{\mathsf{ID},1})^{1/r}$. Hence, PKG has successfully forced the decryption key d_{ID} to be the key chosen by it in advance. ∎

Theorem 2. *Assuming that computing discrete log is hard in $\mathbb{G}_1$, the advantage of an adversary in the* FindKey *game is negligible for the above traceable identity based encryption scheme.*

PROOF: Let there be a PPT algorithm $\mathcal{A}$ that has an advantage ϵ in the FindKey game in the above T-IBE construction. We show how to build a simulator $\mathcal{B}$ that is able to solve discrete log in $\mathbb{G}_1$ with the same advantage ϵ. $\mathcal{B}$ proceeds as follows.

$\mathcal{B}$ runs the algorithm $\mathcal{A}$ and gets the public parameters $\text{PK} = (g, g_1, h_1, h_2, h_3)$ and the identity ID from $\mathcal{A}$. It then invokes the challenger, passes on h_1 to it and gets a challenge $R \in \mathbb{G}_1$. The goal of $\mathcal{B}$ would be to find the discrete log r of R w.r.t. h_1.

$\mathcal{B}$ engages in the key generation protocol with $\mathcal{A}$ to get a decryption key for ID as follows. It sends R to $\mathcal{A}$ and now has to give a *zero-knowledge* proof of knowledge of the discrete log of R. The zero-knowledge property of the proof system implies the existence of a simulator $\mathcal{S}$ which is able to successfully simulate the view of $\mathcal{A}$ in the protocol (by rewinding $\mathcal{A}$), with all but negligible probability. $\mathcal{B}$ uses the simulator $\mathcal{S}$ to simulate the required proof even without of knowledge of r. $\mathcal{B}$ then receives the string $(r', h'_{\text{ID},1}), (r_{\text{ID},2}, h_{\text{ID},2}), (r_{\text{ID},3}, h_{\text{ID},3})$ from $\mathcal{A}$. As before, $\mathcal{B}$ runs a key sanity check by testing if $e(h_{\text{ID},i}, g^{\alpha - \text{ID}}) \overset{?}{=} e(h_i g^{-r_{\text{ID},i}}, g)$ for $i \in \{2, 3\}$. For $i = 1$, $\mathcal{B}$ tests if $e(h'_{\text{ID},i}, g^{\alpha - \text{ID}}) \overset{?}{=} e(Rg^{-r'}, g)$. If any of these tests fail, $\mathcal{B}$ aborts as would an honest user in the key generation protocol.

Now with probability at least ϵ, $\mathcal{A}$ outputs a decryption key (passing the key sanity check and hence well-formed) d'_{ID} such that its decryption key family number n'_F equals the decryption key family number of the key d_{ID}, where d_{ID} is defined (but unknown to $\mathcal{B}$) as $(r'/r, (h'_{\text{ID},1})^{1/r}, (r_{\text{ID},2}, h_{\text{ID},2}), (r_{\text{ID},3}, h_{\text{ID},3}))$. After computing n'_F from d'_{ID} (by running trace on it), $\mathcal{B}$ computes $r = r'/n'_F$. $\mathcal{B}$ outputs r as the discrete log (w.r.t. h_1) of the challenge R and halts.   ∎

Theorem 3. *The advantage of an adversary in the* ComputeNewKey *game is negligible for the above traceable identity based encryption scheme under the computational q-BSDH assumption.*

PROOF: Let there be a PPT algorithm $\mathcal{A}$ that has an advantage ϵ in the ComputeNewKey game in the above T-IBE construction. We show how to build a simulator $\mathcal{B}$ that is able to solve the computational q-BSDH assumption with the same advantage ϵ. $\mathcal{B}$ proceeds as follows.

The functioning of $\mathcal{B}$ in this proof is very similar to that of the simulator in the IND-ID-CPA proof of Gentry's scheme [Gen06]. $\mathcal{B}$ invokes the challenger and gets as input the q-BSDH problem instance $(g, g_1, g_2, \ldots, g_q)$, where $g_i = g^{(\alpha^i)}$.

$\mathcal{B}$ generates random polynomials $f_i(x) \in \mathbb{Z}_p[x]$ of degree q for $i \in \{1, 2, 3\}$. It computes $h_i = g^{f_i(\alpha)}$ using $(g, g_1, g_2, \ldots, g_q)$ and sends the public parameters $\text{PK} = (g, g_1, h_1, h_2, h_3)$ to the algorithm $\mathcal{A}$.

$\mathcal{B}$ now runs the key generation protocol with $\mathcal{A}$ (possibly multiple times) to pass on the decryption keys $d_{\text{ID}^1}, \ldots, d_{\text{ID}^q}$ for the identities $\text{ID}^1, \ldots, \text{ID}^q$ chosen adaptively by $\mathcal{A}$. For an identity ID, $\mathcal{B}$ runs the key generation protocol as follows.

If $\mathsf{ID} = \alpha$, $\mathcal{B}$ uses α to solve the q-BSDH problem immediately. Otherwise, let $F_{\mathsf{ID},i}(x)$ denote the $(q-1)$ degree polynomial $F_{\mathsf{ID},i}(x) = (f_i(x) - f_i(\mathsf{ID}))/(x - \mathsf{ID})$. $\mathcal{B}$ computes the decryption key $d_{\mathsf{ID}} = \{(r_{\mathsf{ID},i}, h_{\mathsf{ID},i}) : i \in \{1,2,3\}\}$ where $r_{\mathsf{ID},i} = f_i(\mathsf{ID})$ and $h_{\mathsf{ID},i} = g^{F_{\mathsf{ID},i}(\alpha)}$. Note that this is a valid private key since $h_{\mathsf{ID},i} = g^{(f_i(\alpha) - f_i(\mathsf{ID}))/(\alpha - \mathsf{ID})} = (h_i g^{-f_i(\mathsf{ID})})^{1/(\alpha - \mathsf{ID})}$. Now $\mathcal{B}$ *forces* the output of $\mathcal{A}$ to be the key d_{ID} during the key generation protocol (see proof of Theorem 1). For more details on why this decryption key appears to be correctly distributed to $\mathcal{A}$, we refer the reader to [Gen06].

Now with probability at least ϵ, $\mathcal{A}$ outputs two decryption keys (passing the key sanity check and hence well-formed) $d_{\mathsf{ID}}^1 = \{(r_{\mathsf{ID},i}^1, h_{\mathsf{ID},i}^1)\}$ and $d_{\mathsf{ID}}^2 = \{(r_{\mathsf{ID},i}^2, h_{\mathsf{ID},i}^2)\}$ for $i \in \{1,2,3\}$ for an identity ID such that $\mathsf{trace}(d_{\mathsf{ID}}^1) \neq \mathsf{trace}(d_{\mathsf{ID}}^2)$. This means that $r_{\mathsf{ID},1}^1 \neq r_{\mathsf{ID},1}^2$. $\mathcal{B}$ then computes

$$(h_{\mathsf{ID},1}^1 / h_{\mathsf{ID},1}^2)^{1/(r_{\mathsf{ID},1}^2 - r_{\mathsf{ID},1}^1)}$$
$$= (h_1 g^{-r_{\mathsf{ID},1}^1} / h_1 g^{-r_{\mathsf{ID},1}^2})^{1/(r_{\mathsf{ID},1}^2 - r_{\mathsf{ID},1}^1)(\alpha - \mathsf{ID})}$$
$$= g^{1/(\alpha - \mathsf{ID})}$$

Finally, $\mathcal{B}$ outputs $-\mathsf{ID}, g^{1/(\alpha - \mathsf{ID})}$ as a solution to the q-BSDH problem instance given and halts. ∎

5   Construction Based on Decisional BDH Assumption

Our second construction of traceable identity based encryption is based on the decisional BDH assumption which is considered to be relatively standard in the groups with bilinear maps. However, the construction is not very efficient and requires several pairing operations per decryption.

We use two cryptosystems as building blocks in this construction: the identity based encryption scheme proposed by Waters [Wat05] and the fuzzy identity based encryption (FIBE) scheme proposed by Sahai and Waters [SW05]. Our first idea is to use an IBE scheme derived from the FIBE construction of Sahai and Waters [SW05]. In FIBE, the encryption is done with a set of attributes which will be defined by the identity in our setting. Additionally, we add a set of *dummy attributes* in the ciphertext. During the key generation protocol, the user gets the set of attributes as defined by his identity as well as a certain subset of the dummy attributes. Very roughly, the subset is such that it can be used to decrypt the ciphertext part encrypted with dummy attributes.

The main properties we need (to add traceability) are derived from the above IBE scheme constructed using FIBE [SW05]. However, as is the case with FIBE, the IBE scheme is only secure in the selective-ID model. To achieve full security, we use the Waters cryptosystem [Wat05] *in parallel* with the FIBE scheme. We remark that Waters cryptosystem is only used to achieve full security and any other fully secure IBE scheme (e.g., [BB04b]) based on a standard assumption could be used. We treat the Waters cryptosystem as a black box as we do not require any specific properties from it. Although we are able to achieve full

security in the IND-ID-CCA game, we do need to use the selective-ID model for
the ComputeNewKey game.

5.1   The Construction

As before, $\mathbb{G}_1$ is a bilinear group of prime order p, and let g be a generator of
$\mathbb{G}_1$. In addition, let $e : \mathbb{G}_1 \times \mathbb{G}_1 \to \mathbb{G}_2$ denote a bilinear map.

We represent the identities as strings of a fixed length ℓ_{ID} (since an identity
ID $\in \mathbb{Z}_p$, ℓ_{ID} is the number of bits required to represent an element in $\mathbb{Z}_p$). Let
ℓ_{sp} be a number which is decided by a statistical security parameter κ_s. Let $\ell =
\ell_{ID} + \ell_{sp}$. We define the following sets: $S = \{1, \ldots, \ell\}$, $S_{ID} = \{1, \ldots, \ell_{ID}\}$, $S_{sp} =
\{\ell_{ID} + 1, \ldots, \ell\}$. We shall denote the ith bit of the identity ID with ID_i. Our
construction follows.

Setup. Run the setup algorithm of the Waters cryptosystem [Wat05] and get
the public parameters PK_w and master key MK_w. Now, for each $i \in S$, choose
two numbers $t_{i,0}$ and $t_{i,1}$ uniformly at random from $\mathbb{Z}_p$ such that all 2ℓ numbers
are different. Also choose a number y uniformly at random in $\mathbb{Z}_p$.

The published public parameters are $PK = (PK_w, PK_{sw})$, where:

$$PK_{sw} = (\{(T_{i,j} = g^{t_{i,j}}) : i \in S, j \in \{0,1\}\}, Y = e(g,g)^y, g)$$

The master key $MK = (MK_w, MK_{sw})$, where:

$$MK_{sw} = (\{(t_{i,j}) : i \in S, j \in \{0,1\}\}, y)$$

Key Generation Protocol. The key generation protocol between PKG and a user
U (with the identity ID) proceeds as follows.

1. U aborts if the published values in the set $\{T_{i,j} : i \in S, j \in \{0,1\}\}$ are not
   all different.
2. PKG generates a decryption key d_w for identity ID using MK_w as per the
   key generation algorithm of the Waters cryptosystem. It sends d_w to U.
3. PKG generates ℓ random numbers $r_1, \ldots, r_\ell$ from $\mathbb{Z}_p$ such that $r_1 + \cdots + r_\ell =
   y$. It computes $R_1 = g^{r_1}, \ldots, R_\ell = g^{r_\ell}$ and sends them to the user.
4. PKG computes the key components $d_{sw,i} = g^{r_i/t_{i,ID_i}}$ for all $i \in S_{ID}$ and
   sends them to U. It also computes key components $d_{sw,i,j} = g^{r_i/t_{i,j}}$ for all
   $i \in S_{sp}, j \in \{0,1\}$ and stores them.
5. PKG and U then engage in ℓ_{sp} executions of a 1-out-of-2 oblivious transfer
   protocol where PKG acts as the sender and U acts as the receiver. In the ith
   execution (where $i \in S_{sp}$), the private input of PKG is the key components
   $d_{sw,i,0}, d_{sw,i,1}$ and the private input of U is a randomly selected bit b_i. The
   private output of U is the key component d_{sw,i,b_i}. For each $i \in S_{sp}$, U now
   runs the following check:

$$e(R_i, g) \stackrel{?}{=} e(T_{i,b_i}, d_{sw,i,b_i})$$

U aborts if any of the above checks fails. This check roughly ensures that for a majority of indices $i \in S_{sp}$, the correct value r_i was used in the creation of $d_{sw,i,0}$ and $d_{sw,i,1}$. Thus for a majority of indices $i \in S_{sp}$, $d_{sw,i,0}$ and $d_{sw,i,1}$ were different from each other.

6. U sets $d_{sw} = (\{d_{sw,i}\}_{i \in S_{ID}}, \{b_i, d_{sw,i,b_i}\}_{i \in S_{sp}})$ and runs a key sanity check on d_{sw} by checking if:

$$Y \stackrel{?}{=} \prod_{i \in S_{ID}} e(T_{i,\mathsf{ID}_i}, d_{sw,i}) \prod_{i \in S_{sp}} e(T_{i,b_i}, d_{sw,i,b_i})$$

U aborts if the above check fails. Finally, U sets its decryption key $d_{\mathsf{ID}} = (d_w, d_{sw})$.

Encryption. To encrypt a message $m \in \mathbb{G}_2$ under an identity ID, break the message into two random shares m_1 and m_2 such that $m_1 \oplus m_2 = m$. Now choose a random value $s \in \mathbb{Z}_p$ and compute the ciphertext $C = (C_w, C_{sw})$. C_w is the encryption of m_1 with ID using the public parameters PK_w as per the encryption algorithm of the Waters cryptosystem and C_{sw} is given by the following tuple.

$$(C' = m_2 Y^s, C'' = g^s, \{(C_i = T_{i,\mathsf{ID}_i}{}^s) : i \in S_{ID}\}, \{(C_{i,j} = T_{i,j}{}^s) : i \in S_{sp}, j \in \{0,1\}\})$$

Decryption. To decrypt the ciphertext $C = (C_w, C_{sw})$ using the decryption key $d_{\mathsf{ID}} = (d_w, d_{sw})$, first run a ciphertext sanity check on C_{sw} by checking if:

$$e(C_i, g) \stackrel{?}{=} e(T_{i,\mathsf{ID}_i}, C''), \quad i \in S_{ID}, \quad \text{and}$$

$$e(C_{i,j}, g) \stackrel{?}{=} e(T_{i,j}, C''), \quad i \in S_{sp}, j \in \{0,1\}$$

If any of the above check fails, output $\bot$. It is easy to see that this check ensures that $\{(C_i = T_{i,\mathsf{ID}_i}{}^s) : i \in S_{ID}\}, \{(C_{i,j} = T_{i,j}{}^s) : i \in S_{sp}, j \in \{0,1\}\}$ where s is the discrete log of C'' w.r.t. g. This ensure that *all* invalid ciphertexts are rejected. In the appendix, we sketch an alternative technique of doing ciphertext sanity check which requires only two pairing operations.

If the ciphertext sanity check succeeds, recover the share m_1 by running the decrypt algorithm of Waters cryptosystem on C_w using d_w. The share m_2 is recovered by the following computations:

$$C' / \prod_{i \in S_{ID}} e(C_i, d_{sw,i}) \prod_{i \in S_{sp}} e(C_{i,b_i}, d_{sw,i,b_i})$$

$$= m_2 e(g,g)^{sy} / \prod_{i \in S_{ID}} e(g^{st_{i,\mathsf{ID}_i}}, g^{r_i/t_{i,\mathsf{ID}_i}}) \prod_{i \in S_{sp}} e(g^{st_{i,b_i}}, g^{r_i/t_{i,b_i}})$$

$$= m_2 e(g,g)^{sy} / \prod_{i \in S} e(g,g)^{sr_i} = m_2$$

Finally, output $m = m_1 \oplus m_2$.

Trace. This algorithm takes a well-formed decryption key $d_{\text{ID}} = (d_w, d_{sw})$ where the component $d_{sw} = (\{d_{sw,i}\}_{i \in S_{ID}}, \{b_i, d_{sw,i,b_i}\}_{i \in S_{sp}})$ and outputs the decryption key family number $n_F = b_{\ell_{ID}+1} \circ b_{\ell_{ID}+2} \circ \ldots \circ b_\ell$, where $\circ$ denotes concatenation.

Security proofs are omitted for the lack of space. They can be found in the full version.

6 Future Work

This work motivates several interesting open problems. The most important one of course is the construction of a *black-box* traceable identity based encryption as discussed in Section 1.

Our second construction based on the decisional BDH assumption is not very efficient and requires several pairing operations per decryption. It is an open problem to design more efficient T-IBE schemes based on an standard assumption. Further, the second construction used the Selective-ID ComputeNewKey game to prove security. It would be interesting to see if this restriction can be removed. Combining the T-IBE approach with the multiple PKG approach also seems to be a promising direction.

Finally, it remains to be seen if the same approach to mitigate the key escrow problem can be profitably used in other related setting like attribute based encryption [SW05, GPSW06].

Acknowledgements

We wish to thank the anonymous reviewers for many useful comments and suggestions. Thanks to Yevgeniy Dodis and Omkant Pandey for helpful discussions. Finally, thanks to Rafail Ostrovsky and Amit Sahai for their feedback and encouragement.

References

[ARP03]   Al-Riyami, S.S., Paterson, K.G.: Certificateless public key cryptography. In: Laih, C.-S. (ed.) ASIACRYPT 2003. LNCS, vol. 2894, pp. 452–473. Springer, Heidelberg (2003)

[BB04a]   Boneh, D., Boyen, X.: Efficient Selective-ID Secure Identity Based Encryption Without Random Oracles. In: Cachin, C., Camenisch, J.L. (eds.) EUROCRYPT 2004. LNCS, vol. 3027, pp. 223–238. Springer, Heidelberg (2004)

[BB04b]   Boneh, D., Boyen, X.: Secure identity based encryption without random oracles. In: Franklin, M. (ed.) CRYPTO 2004. LNCS, vol. 3152, pp. 443–459. Springer, Heidelberg (2004)

[BB04c]   Boneh, D., Boyen, X.: Short signatures without random oracles. In: Cachin, C., Camenisch, J.L. (eds.) EUROCRYPT 2004. LNCS, vol. 3027, pp. 56–73. Springer, Heidelberg (2004)

[BBG05]   Boneh, D., Boyen, X., Goh, E.-J.: Hierarchical identity based encryption
            with constant size ciphertext. In: Cramer [Cra05], pp. 440–456
[BBS04]   Boneh, D., Boyen, X., Shacham, H.: Short Group Signatures. In: Franklin,
            M. (ed.) CRYPTO 2004. LNCS, vol. 3152, Springer, Heidelberg (2004)
[BF01]   Boneh, D., Franklin, M.: Identity Based Encryption from the Weil Pairing.
            In: Kilian, J. (ed.) CRYPTO 2001. LNCS, vol. 2139, pp. 213–229. Springer,
            Heidelberg (2001)
[BG92]   Bellare, M., Goldreich, O.: On defining proofs of knowledge. In: Brickell,
            E.F. (ed.) CRYPTO 1992. LNCS, vol. 740, pp. 390–420. Springer, Heidel-
            berg (1993)
[Bra90]   Brassard, G. (ed.): CRYPTO 1989. LNCS, vol. 435. Springer, Heidelberg
            (1990)
[BSW06]   Boneh, D., Sahai, A., Waters, B.: Fully collusion resistant traitor tracing
            with short ciphertexts and private keys. In: Vaudenay [Vau06] pp. 573–592
[CC03]   Cha, J., Cheon, J.: An identity-based signature from gap diffie-hellman
            groups. In: Desmedt, Y.G. (ed.) PKC 2003. LNCS, vol. 2567, pp. 18–30.
            Springer, Heidelberg (2002)
[CFN94]   Chor, B., Fiat, A., Naor, M.: Tracing traitor. In: Desmedt, Y.G. (ed.)
            CRYPTO 1994. LNCS, vol. 839, pp. 257–270. Springer, Heidelberg (1994)
[CHK03]   Canetti, R., Halevi, S., Katz, J.: A Forward-Secure Public-Key Encryp-
            tion Scheme. In: Biham, E. (ed.) EUROCRPYT 2003. LNCS, vol. 2656,
            Springer, Heidelberg (2003)
[Cra05]   Cramer, R.J.F. (ed.): EUROCRYPT 2005. LNCS, vol. 3494, pp. 22–26.
            Springer, Heidelberg (2005)
[EGL85]   Even, S., Goldreich, O., Lempel, A.: A randomized protocol for signing
            contracts. Commun. ACM 28(6), 637–647 (1985)
[FFS88]   Feige, U., Fiat, A., Shamir, A.: Zero-knowledge proofs of identity. J. Cryp-
            tology 1(2), 77–94 (1988)
[FS86]   Fiat, A., Shamir, A.: How to prove yourself: Practical solutions to identi-
            fication and signature problems. In: Odlyzko, A.M. (ed.) CRYPTO 1986.
            LNCS, vol. 263, pp. 186–194. Springer, Heidelberg (1987)
[Gen03]   Gentry, C.: Certificate-based encryption and the certificate revocation
            problem. In: Biham, E. (ed.) EUROCRPYT 2003. LNCS, vol. 2656, pp.
            272–293. Springer, Heidelberg (2003)
[Gen06]   Gentry, C.: Practical identity-based encryption without random oracles.
            In: Vaudenay [Vau06] pp. 445–464
[GPSW06]   Goyal, V., Pandey, O., Sahai, A., Waters, B.: Attribute-based encryption
            for fine-grained access control of encrypted data. In: Juels, A., Wright,
            R.N., De Capitani di, S. (eds.) ACM Conference on Computer and Com-
            munications Security, pp. 89–98. ACM, New York (2006)
[GS02]   Gentry, C., Silverberg, A.: Hierarchical id-based cryptography. In: Zheng,
            Y. (ed.) ASIACRYPT 2002. LNCS, vol. 2501, pp. 548–566. Springer, Hei-
            delberg (2002)
[HL02]   Horwitz, J., Lynn, B.: Toward hierarchical identity-based encryption. In:
            Knudsen, L.R. (ed.) EUROCRYPT 2002. LNCS, vol. 2332, pp. 466–481.
            Springer, Heidelberg (2002)
[Kal05]   Kalai, Y.T.: Smooth projective hashing and two-message oblivious trans-
            fer. In: Cramer [Cra05], pp. 78–95

[LBD+04] Lee, B., Boyd, C., Dawson, E., Kim, K., Yang, J., Yoo, S.: Secure key issuing in id-based cryptography. In: Hogan, J.M., Montague, P., Purvis, M.K., Steketee, C. (eds.) ACSW Frontiers. CRPIT, vol. 32, pp. 69–74. Australian Computer Society (2004)

[NP01]   Naor, M., Pinkas, B.: Efficient oblivious transfer protocols. In: SODA, pp. 448–457 (2001)

[Rab81]   Rabin, M.O.: How to exchange secrets by oblivious transfer. In: TR-81, Harvard (1981)

[Sch89]   Schnorr, C.-P.: Efficient identification and signatures for smart cards. In: Brassard [Bra90], pp. 239–252

[Sha84]   Shamir, A.: Identity Based Cryptosystems and Signature Schemes. In: Blakely, G.R., Chaum, D. (eds.) CRYPTO 1984. LNCS, vol. 196, pp. 37–53. Springer, Heidelberg (1985)

[SW05]   Sahai, A., Waters, B.: Fuzzy Identity Based Encryption. In: Cramer, R.J.F. (ed.) EUROCRYPT 2005. LNCS, vol. 3494, pp. 457–473. Springer, Heidelberg (2005)

[Vau06]   Vaudenay, S. (ed.): EUROCRYPT 2006. LNCS, vol. 4004. Springer, Heidelberg (2006)

[Wat05]   Waters, B.: Efficient identity-based encryption without random oracles. In: Cramer, [Cra05], pp. 114–127

Appendix

A   Efficient Ciphertext Sanity Check in the Second Construction

To decrypt the ciphertext $C = (C_w, C_{sw})$ using the decryption key $d_{\mathsf{ID}} = (d_w, d_{sw})$, the efficient ciphertext sanity check on C_{sw} is run as follows. First choose $\ell_{ID} + 2\ell_{sp}$ random numbers $s_{i,\mathsf{ID}_i}, i \in S_{ID}$ and $s_{i,j}, i \in S_{sp}, j \in \{0,1\}$. Now check if:

$$e\left(g, \prod_{i \in S_{ID}} C_i^{s_{i,\mathsf{ID}_i}} \prod_{i \in S_{sp}, j \in \{0,1\}} C_{i,j}^{s_{i,j}}\right) \stackrel{?}{=} e\left(C'', \prod_{i \in S_{ID}} T_{i,\mathsf{ID}_i}^{s_{i,\mathsf{ID}_i}} \prod_{i \in S_{sp}, j \in \{0,1\}} T_{i,j}^{s_{i,j}}\right)$$

If the above check fails, output $\perp$. It can be shown that the above check rejects an invalid ciphertext with all but negligible probability (while the previous check was *perfect*).

Pirate Evolution: How to Make the Most of Your Traitor Keys

Aggelos Kiayias* and Serdar Pehlivanoglu*

Computer Science and Engineering, University of Connecticut
Storrs, CT, USA
{aggelos,sep05009}@cse.uconn.edu

Abstract. We introduce a novel attack concept against trace and revoke schemes called pirate evolution. In this setting, the attacker, called an evolving pirate, is handed a number of traitor keys and produces a number of generations of pirate decoders that are successively disabled by the trace and revoke system. A trace and revoke scheme is susceptible to pirate evolution when the number of decoders that the evolving pirate produces exceeds the number of traitor keys that were at his possession. Pirate evolution can threaten trace and revoke schemes even in cases where both the revocation and traceability properties are ideally satisfied: this is because pirate evolution may enable an attacker to "magnify" an initial key-leakage incident and exploit the traitor keys available to him to produce a great number of pirate boxes that will take a long time to disable. Even moderately successful pirate evolution affects the economics of deployment for a trace and revoke system and thus it is important that it is quantified prior to deployment.

In this work, we formalize the concept of pirate evolution and we demonstrate the susceptibility of the trace and revoke schemes of Naor, Naor and Lotspiech (NNL) from Crypto 2001 to an evolving pirate that can produce up to $t \cdot \log N$ generations of pirate decoders given an initial set of t traitor keys. This is particularly important in the context of AACS, the new standard for high definition DVDs (HD-DVD and Blue-Ray) that employ the subset difference method of NNL: for example using our attack strategy, a pirate can potentially produce more than 300 pirate decoder generations by using only 10 traitor keys, i.e., key-leakage incidents in AACS can be substantially magnified.

1 Introduction

A trace and revoke scheme is an encryption scheme that is suitable for digital content distribution to a large set of receivers. In such a scheme, every receiver possesses a decryption key that is capable of inverting the content scrambling mechanism. The defining characteristics of a trace and revoke scheme are the following: (i) *revocation*: the sender can scramble content with a "broadcast pattern" in such a way so that the decryption capability of any subset of the receiver

---

* Research partly supported by NSF CAREER Award CNS-0447808.

A. Menezes (Ed.): CRYPTO 2007, LNCS 4622, pp. 448–465, 2007.

population can be disabled, (ii) *tracing*: given a rogue decryption device (called a pirate decoder) that was produced using the keys of a number of receivers (called traitors) it is possible to render such device useless from future transmissions. This can be done by identifying the traitors and revoking them or in some other fashion (that may not involve the direct identification of any traitor).

Trace and revoke schemes conceptually are a combination of two cryptographic primitives that have been originally suggested and studied independently: broadcast encryption, introduced by Fiat and Naor in [10] and studied further in e.g., [13, 14, 22, 8, 15, 16, 21], and traitor tracing and related codes, introduced by Chor, Fiat and Naor in [6] and studied further in e.g., [33, 23, 20, 2, 28, 29, 30, 17, 18, 19, 31, 34, 5, 27]. The combination of the two primitives appeared first in [24] and explored further in [9]. Trace and revoke schemes for stateless receivers were proposed in [22] and explored further in [15, 16]. The stateless receiver setting is of particular interest since it does not require receivers to maintain state during the life-time of the system; this greatly simplifies the system aspects and deployment management of a trace and revoke scheme.

The security requirements for trace and revoke schemes are relatively well understood when one considers the revocation or tracing components in isolation: the revocation component should be coalition resistant to an adversary that adaptively joins the system, is entirely revoked and subsequently attempts to decrypt a ciphertext. The tracing component should also be coalition resistant: an adversary given a set of keys should be incapable of producing a pirate decoder that cannot have at least one traitor identified. When Naor, Naor and Lotspiech [22] introduced the broadcast encryption framework of subset cover, they made the nice observation that if a broadcast encryption scheme satisfies a property called "bifurcation" then it is possible to construct an efficient tracing procedure that will produce ciphertexts that are unreadable by any given rogue pirate decoder; this satisfies the requirements for a trace and revoke scheme (albeit without identifying traitors directly). They proposed two combinatorial designs (called the complete-subtree and subset-difference method) for broadcast encryption that satisfy bifurcation and thus produced two trace and revoke schemes. The subset-difference scheme is particularly attractive as it enjoys a linear communication overhead during encryption (linear in r the number of revoked users) and it was employed as the basis for the new high definition DVD encryption standard, the AACS [1].

It is common in cryptographic design when a construction combines simultaneously two security functionalities (even when they are well understood in isolation) that the possibility for new forms of attacks springs up. In our case, in a trace and revoke scheme the adversaries that have been considered so far were attacking directly the revocation component (they were revoked and attempted to evade revocation) or the traceability component (they produced a pirate decoder that attempted to evade the tracing algorithm). This raises the question, in a trace and revoke scheme, are these the only relevant attack scenarios?

Pirate Evolution. Pirate evolution is a novel attack concept against a trace and revoke scheme that exploits the properties of the combined functionality of tracing and revocation in such a scheme. In a pirate evolution attack, a pirate obtains a set of traitor keys through a "key-leaking" incident. Using this set of keys the pirate produces an initial pirate decoder. When this pirate decoder is captured and disabled by the transmission system using the tracing mechanism, the pirate "evolves" the first pirate decoder by issuing a second version that succeeds in decrypting ciphertexts that have the broadcast pattern disabling the first decoder. The same step is repeated again and the pirate continues to evolve a new version of the previous decoder whenever the current version of pirate decoder becomes disabled from the system. A pirate that behaves as above will be called an evolving pirate and each version of the pirate decoder will be called a generation (as presumably many copies of the same pirate decoder may be spread by the pirate).

This is a novel attack concept as the adversary here is not trying to evade the revocation or the traceability component. Instead he tries to remain active in the system for as long as possible in spite of the efforts of the administrators of the system. We say that a trace and revoke scheme is immune to pirate evolution if the number of generations that an evolving pirate can produce equals the number of traitor keys that have been corrupted (i.e., the number of traitors). The number of traitors is a natural lower bound to the generations that an evolving pirate can produce: trivially, an evolving pirate can set each version it releases to be equal to the decoder of one of the traitors. Nevertheless, the number of generations that a pirate may produce can be substantially larger depending on the combinatorial properties of the underlying trace and revoke system. We call the maximum number of decoders an evolving pirate can produce, the pirate evolution bound evo of the trace and revoke scheme. Note that this bound will be a function of the number of traitors t as well as of other parameters in the system (such as the number of users).

When evo is larger than t, we say that a trace and revoke scheme is susceptible to pirate evolution. When evo is much larger than t, this means that an initial leaking incident can be "magnified" and be of a scale much larger than what originally expected. Interestingly, a system may satisfy both the tracing and revocation properties in isolation and still be rendered entirely useless if evo is sufficiently large (say super-poly in t). In this case the trace and revoke scheme could be defeated by simply taking too long to catch up with an evolving pirate that could keep exploiting a minor initial key leakage incident to produce a multitude of pirate decoders in succession.

Even when evo is just moderately larger than t, it is an important consideration for a trace and revoke scheme in an actual deployment. The economics of a deployment would be affected and we believe that resilience against pirate evolution attacks should be part of the suggested considerations.

In this work, we introduce and study pirate evolution in the subset cover framework of stateless receivers of [22]. We first formalize the concept of pirate evolution through the means of an attack game played between the evolving pirate and a challenger that verifies certain properties about the pirate decoders produced by the evolving pirate. Next, we prove that it is in fact possible to design trace and revoke schemes that are immune against pirate evolution by presenting a simple design that renders any evolving pirate incapable of producing more pirate decoders than traitors. This result (albeit not very efficient as a trace and revoke scheme) shows that immunity against pirate evolution is attainable in principle. Still, it is interesting to note that immunity may come at a high cost for certain systems and thus it could be desirable in many settings to sacrifice immunity in favor of efficiency if the amount of pirate evolution that is possible is deemed to be within acceptable limits for the economics of a certain system deployment (compare this to the usual example of a bank that allows a few fraudulent transactions if the incurred losses can be factored into the profits).

Next, we focus on the complete-subtree and subset-difference trace and revoke systems of [22]. We demonstrate both these schemes are susceptible to pirate evolution. Each of our pirate evolution attacks requires careful scheduling of how the traitor keys are expended by the pirate; moreover in both cases there is sensitivity to the "geometry" of the leaking incident. For the complete subtree method we present a pirate evolution attack that given t traitor keys, it enables an evolving pirate to produce up to $t \log(N/t)$ generations where N is the total number of users in the system (actually number of leaves in the tree, as the set of currently active users may be much less). For the subset difference method we present a pirate evolution attack that given t traitor keys, it enables an evolving pirate to produce up to $t \log N$ generations of pirate decoders.

In the context of AACS, [1], the new encryption standard for high definition DVDs, where the subset-difference method of [22] is deployed, the pirate evolution attack we present suggests that each single traitor key can be used to yield up to 31 generations of pirate decoders (refer to section 3.3).

2   The Subset-Cover Revocation Framework

The Subset-Cover revocation framework [22] is an abstraction that can be used to formulate a variety of revocation methods. It defines a set of subsets that cover the whole user population and assigns (long-lived) keys to each subset; each user receives a collection of such keys (or derived keys). We denote by N the set of all users where $|N| = N$ and $R \subset N$ the set of users that are to be revoked at a certain instance where $|R| = r$. Note that N is not necessarily the set of currently active users but the number of all users that are anticipated in the lifetime of the system.

The goal of the sender is to transmit a message M to all users such that any $u \in N \setminus R$ can recover the message whereas the revoked users in R can not recover it. Note that the non-recovery property should also extend to any coalition of revoked users. The framework is based on a collection of subsets $\{S_j\}_{j \in \mathcal{J}}$ where $S_j \subseteq N$ such that any subset $S \subseteq N$ can be partitioned into disjoint subsets of $\{S_j\}_{j \in \mathcal{J}}$. Each subset S_j is associated with a *long-lived* key L_j. Users are assumed to be initialized privately with a set of keys such that u has access to L_j if and only if $u \in S_j$. The private data assigned to user u in this intialization step will be denoted by $\mathcal{I}_u$. In particular we define $\mathcal{I}_u = \{j \in \mathcal{J} \mid u \in \mathsf{S}_j\}$ and $\mathcal{K}_u = \{\mathsf{L}_j \mid j \in \mathcal{I}_u\}$. Given a revoked set R, the remaining users $\mathsf{N} \setminus \mathsf{R}$ are partitioned into disjoint $\{\mathsf{S}_{i_1}, \ldots, \mathsf{S}_{i_m}\} \subset \{\mathsf{S}_j\}_{j \in \mathcal{J}}$ so that $\mathsf{N} \setminus \mathsf{R} = \bigcup_{j=1}^{m} \mathsf{S}_{i_j}$. The transmission of the message M is done in a hybrid fashion. First a random session key K is encrypted with all *long-lived* keys L_{i_j} corresponding to the partition, and the message M is encrypted with the session key. Two encryption functions are being used in this framework: (1) $\mathcal{F}_K : \{0,1\}^* \mapsto \{0,1\}^*$ to encrypt the message. (2) $\mathcal{Q}_L : \{0,1\}^l \mapsto \{0,1\}^l$ to encrypt the session key. Each broadacast ciphertext will have the following form:

$$\langle [i_1, i_2, \ldots i_m, \underbrace{\mathcal{Q}_{L_{i_1}}(K), \mathcal{Q}_{L_{i_2}}(K), \ldots \mathcal{Q}_{L_{i_m}}(K)]}_{\text{HEADER}}, \underbrace{\mathcal{F}_K(M)}_{\text{BODY}} \rangle \tag{1}$$

The receiver u decrypts a given ciphertext $\mathcal{C} = \langle [i_1, i_2, \ldots i_m, C_1, C_2, \ldots C_m], M' \rangle$ as follows: (i) Find i_j such that $u \in \mathsf{S}_{i_j}$, if not respond null, (ii) Obtain L_{i_j} from $\mathcal{K}_u$. (iii) Decrypt the session key: $K' = \mathcal{Q}_{L_{i_j}}^{-1}(C_j)$. (iv) Decrypt the message: $M = \mathcal{F}_K^{-1}(M')$. In [22], two methods in the subset cover framework are presented called the Complete Subtree CS and the Subset Difference SD.

Tracing Traitors in the Subset Cover Framework. Beyond revoking sets of users that are not supposed to receive content, trace and revoke schemes are supposed to be able to disable the rogue pirate decoders which are constructed using a set of traitor's keys that are available to the pirate. One way this can be achieved is to identify a traitor given access to a pirate box and then add him to the set of revoked users. Given that the goal of tracing is to disable the pirate box, the NNL tracing algorithm focuses on just this security goal. In the NNL setting, it is sufficient to find a "pattern" which makes the pirate box unable to decrypt.

Regarding the tracing operation, the following assumptions are used for the pirate decoder: (1) the tracing operation is black-box, i.e., it allows the tracer to examine only the outcome of the pirate decoder as an oracle. (2) the pirate decoder is not capable of recording history; (3) the pirate decoder lacks a "locking" mechanism which will prevent the tracer to pose more queries once the box detects that it is under tracing testing. (4) the pirate decoder succeeds in decoding with probability greater than or equal to a threshold q.

Based on the above, the goal of the tracing algorithm is to output *either* a non-empty subset of traitors, *or* a partition of $\mathsf{N} \setminus \mathsf{R} = \bigcup_{j=1}^{m} \mathsf{S}_{i_j}$ for the given revoked users R, such that if this partition is used to distribute content M in the framework as described above it is impossible to be decrypted by the pirate box with sufficiently high probability (larger than the threshold q); at the same time, all good users can still decrypt.

The tracing algorithm can be thought of as a repeated application of the following basic procedure that takes as input a partition: First it is tested whether the box decrypts correctly with the given partition $\bigcup_{j=1}^{m} \mathsf{S}_{i_j}$ (with probability p greater than the threshold). If not, the subset tracing outputs the partition as the output of the tracing algorithm. Otherwise, it outputs one of the subsets containing at least one of the traitors. The tracing algorithm then partitions that subset somehow and inputs the new partition (that is more "refined") to the next iteration of the basic procedure. If the subset resulting by the basic procedure contains only one possible candidate, then we can revoke that user since it is a traitor. Here is how the basic procedure works:

Let p_j be the probability that the box decodes the special tracing ciphertext

$$\langle [i_1, i_2, \ldots i_m, \mathcal{Q}_{L_{i_1}}(R), \mathcal{Q}_{L_{i_2}}(R), \ldots \mathcal{Q}_{L_{i_j}}(R), \mathcal{Q}_{L_{i_{j+1}}}(K), \ldots \mathcal{Q}_{L_{i_m}}(K)], \mathcal{F}_K(M) \rangle$$

where R is a random string of the same length as the key K. Note that $p_0 = p$ and $p_m = 0$, hence there must be some $0 < j \leq m$ for which $|p_{j-1} - p_j| \geq \frac{p}{m}$. Eventually, this leads the existence of a traitor in the subset S_{i_j} under the assumption that it is negligible to break the encryption scheme $\mathcal{Q}$ and the key assignment method.

The above can be turned into a full-fledged tracing algorithm, as long as the Subset-Cover revocation scheme satisfies the "Bifurcation property": any subset S_k can be partitioned into not extremely uneven sets S_{k_1} and S_{k_2}. Both CS and SD methods allow us to partition any subset S_k into two subsets with the Bifurcation property. For the Complete Subset, it is simply taking the subsets rooted at the children of node v_k. For the SD method, given $S_{i,j}$ we take the subsets $S_{i,c}$ and $S_{c,j}$ where v_c is a child of the node v_i and v_j is on the subset rooted at v_c. Formally, we have the following definition for tracing algorithm and encryption procedure after tracing pirate boxes to disable them recovering message:

Definition 1. *For a given set of revoked users* R *and pirate boxes* $B_1, B_2, \cdots, B_s$ *caught by the sender, the encryption function first finds a partition* $\mathcal{S}$ *which renders the pirate boxes useless and outputs the ciphertext. Let* $\mathcal{T}$ *be the tracing function outputting the partition to render the pirate boxes useless, then:* $\mathcal{T}^{B_1, B_2, \cdots, B_s}(\mathsf{R}) = \mathcal{S}$. *Denote the ciphertext created by the encryption scheme interchangeably by following notations:* $\mathcal{C} = \mathcal{E}_{\mathcal{R}}^{B_1, B_2, \cdots, B_s}(M)$ *or* $\mathcal{C} = \mathcal{E}_{\mathcal{S}}(M)$, *where* $\mathcal{E}_k^{-1}(\mathcal{C}) = M$ *for* $k \notin \mathsf{R}$, *and any pirate box* B_i, $0 < i \leq s$, *decrypts the ciphertext with probability less than threshold* q, *i.e.* $Prob[B_i(\mathcal{C}) = M] < q$.

According to the above definition, the sender applies tracing algorithm on the pirate boxes he has access to before broadcasting the message.

3   Pirate Evolution

In this section we introduce the concept of pirate evolution. We present a game based definition that is played with the adversary which is the "evolving pirate." Let t be the number of traitor keys in the hands of the pirate. The traitor keys are made available to the pirate through a key-leaking "incident" $\mathcal{L}$ that somehow chooses a subset of size t from the set $\{\mathcal{I}_1, \ldots, \mathcal{I}_N\}$ (the set of all users' private data assigned by a function $\mathcal{G}$ with a security parameter λ). We permit $\mathcal{L}$ to be also based on the current set of revoked users R. Specifically, if

$\mathsf{T} = \mathcal{L}(\mathcal{I}_1, \mathcal{I}_2, \cdots \mathcal{I}_n, t, \mathsf{R})$ then $|\mathsf{T}| = t$, $\mathsf{T} \subseteq \{\mathcal{I}_u \mid u \in \mathsf{N} \setminus \mathsf{R}\}$. This models the fact that the evolving pirate may be able to select the users that he corrupts. Separating the evolving pirate from the leaking incident is important though as it enables us to describe how a pirate can deal with leaking incidents that are not necessarily the most favorable (the pirate evolution attacks that we will describe in the sequel will operate with any given leaking incident and there will be leaking incidents that are more favorable than others). We note that partial leaking incidents can also be considered within our framework.

Once the leaking incident determines the private user data that will be available to the evolving pirate (i.e., the traitor key material), the evolving pirate $\mathcal{P}$ receives the keys and produces a "master" pirate box $\mathcal{B}$. The pirate is allowed to have oracle access to an oracle $\mathcal{E}_\mathsf{R}(\mathcal{M})$ that returns ciphertexts distributed according to plaintext distribution that is employed by the digital content distribution system (i.e., the access we consider is not adaptive); an adaptive version of the definition (similar to a chosen plaintext attack against symmetric encryption) is also possible.

Given the master pirate box, an iterative process is initiated: the master pirate box spawns successively a sequence of pirate decoders $B_1, B_2, \ldots$ where $B_i = \mathcal{B}(1^{t+\log N}, \ell)$ for $\ell = 1, 2, \ldots$. Note that we loosely think that the master box is simply the compact representation of a vector of pirate boxes; the time complexity allowed for its operation is polynomial in $t + \log N + \log \ell$ (this can be generalized in other contexts if needed — we found it to be sufficient for the evolving pirates strategies we present here). Each pirate box is tested whether it decrypts correctly the plaintexts that are transmitted in the digital content distribution system with success probability at least q. The first pirate box is tested against the "initial" encryption function $\mathcal{E}_\mathsf{R}(\cdot)$, whereas any subsequent box is tested against $\mathcal{E}_\mathsf{R}^{B_1, B_2, \cdots B_{i-1}}(\cdot)$ which is the encryption that corresponds to the conjunctive revocation of the set R and the tracing of all previous pirate boxes. The iteration stops when the master pirate box $\mathcal{B}$ is incapable of producing a pirate decoder with decryption success exceeding the threshold q. Each iteration of the master box corresponds to a "generation" of pirate boxes. The number of successfully decoding pirate generations that the master box can spawn is the output of the game-based definition given below. The trace and revoke scheme is susceptible to pirate evolution if the number of generations returned by the master box is greater than t. Note that the amount of susceptibility varies with the difference between the number of generations and t; the pirate evolution bound evo is the highest number of generations any evolving pirate can produce. Formally, we have the following:

Definition 2. *Consider the game of figure 1 given two probabilistic machines $\mathcal{P}, \mathcal{L}$ and parameters $\mathsf{R} \subseteq \{1, 2, \cdots n\}$, t, $r = |\mathsf{R}|, q$. Let $PE_{\mathcal{P},\mathcal{L}}^{\mathsf{R}}(t)$ be the output of the game. We say that the trace and revoke scheme $\mathsf{TR} = (\mathcal{G}, \mathcal{Q}, \mathcal{F})$ is immune to pirate evolution with respect to key-leaking incident $\mathcal{L}$ if, for any probabilistic polynomial time adversary $\mathcal{P}$, any R and any $t \in \{1, \ldots, |\mathsf{N} - \mathsf{R}|\}$, it holds $PE_{\mathcal{P},\mathcal{L}}^{\mathsf{R}}(t) = t$. We define the pirate evolution bound $\mathsf{evo}[\mathsf{TR}]$ of a trace and revoke scheme TR as the supremum of all $PE_{\mathcal{P},\mathcal{L}}^{\mathsf{R}}(t)$, for any leaking incident $\mathcal{L}$, any set*

$\langle \mathcal{I}_1, \mathcal{I}_2, \cdots \mathcal{I}_N \rangle \leftarrow \mathcal{G}(1^\lambda; \rho; N)$ where $\rho \leftarrow Coins$
$\mathsf{T} \leftarrow \mathcal{L}(\mathcal{I}_1, \mathcal{I}_2, \cdots \mathcal{I}_n, t, \mathsf{R}); \mathsf{K} = \{\mathcal{K}_u \mid u : \mathcal{I}_u \in \mathsf{T}\}$
$\mathcal{B} \leftarrow \mathcal{P}^{\mathcal{E}_\mathsf{R}(\mathcal{M})}(\mathsf{T}, \mathsf{K})$ where $\mathcal{E}_\mathsf{R}(\mathcal{M})$ is an oracle that returns $\mathcal{E}_\mathsf{R}(m)$ with $m \leftarrow \mathcal{M}$
$\ell = 0$
repeat $\ell = \ell + 1$
$\qquad B_\ell \leftarrow \mathcal{B}(1^{t + \log N}, \ell)$
until $Prob[B_\ell(\mathcal{E}_\mathsf{R}^{B_1, B_2, \cdots B_{\ell-1}}(m)) = m] < q$ with $m \leftarrow \mathcal{M}$
output ℓ.

Fig. 1. The attack game played with an evolving pirate

of revoked users R *and any evolving pirate* $\mathcal{P}$; *note that* evo[TR] *is a function of* t *and possibly of other parameters as well. A scheme is susceptible to pirate evolution if its pirate evolution bound satisfies* evo[TR] $> t$.

Note that immunity against pirate evolution attacks is possibly a stringent property; even though we show that it is attainable (cf. the next section) it could be sacrificed in favor of efficiency. Naturally, using a trace and revoke scheme that is susceptible to a pirate evolution that produces many pirate decoders may put the system's managers at a perilous condition once a leaking incident occurs (and as practice has shown leaking incidents are unavoidable). The decision to employ a particular trace-and-revoke scheme in a certain practical setting should be made based on a variety of requirements and constraints and the system's behavior with respect to pirate evolution should be one of the relevant parameters that must be considered in the security analysis.

3.1   A Trace and Revoke Scheme Immune to Pirate-Evolution

In this section we show a simple trace and revoke design that achieves immunity against pirate-evolution. The system simply encrypts the session key with the unique key of each user in the system that is not revoked. This kind of linear length trace and revoke scheme can be formalized in the context of the Subset-Cover framework as follows:

Definition 3. *Let* $|S_j| = 1$ *for all* $j \in \mathcal{J} = \{1, 2, \cdots N\}$, *i.e. the collection is consist of single element sets. Thus, for any user* u, $|\mathcal{K}_u| = |\{\mathsf{L}_u\}| = 1$ *holds. The key assignment* $\mathcal{G}(1^\lambda, N)$ *is done by choosing a random key for each* S_j. *The encryption functions* $\mathcal{Q}$ *and* $\mathcal{F}$ *are encryption functions used in any Subset-Cover framework. We say such trace and revoke scheme* $(\mathcal{G}, \mathcal{Q}, \mathcal{F})$ *is called linear length scheme since the size of cover for non-revoked users in* $\mathsf{N} \setminus \mathsf{R}$ *will be linear in* $|\mathsf{N}| - |\mathsf{R}|$.

The header of a ciphertext $\mathcal{C}$ contains the encryption of session key $\mathcal{Q}_{\mathsf{L}_u}(K)$ if user $u \in \mathsf{N} \setminus \mathsf{R}$. Under the assumption of sufficiently strong $\mathcal{Q}(\cdot)$ no other user will be able recover session key through $\mathcal{Q}_{\mathsf{L}_u}(K)$ and a user u will not be able to recover session key unless the header contains $\mathcal{Q}_{\mathsf{L}_u}(K)$. We show that immunity to pirate evolution is achievable:

Theorem 1. *The trace and revoke scheme as defined in Definition 3 is immune to pirate evolution, i.e. for all polynomial-time adversaries $\mathcal{P}$ and for any key leaking incident $\mathcal{L}$, $PE_{\mathcal{P},\mathcal{L}}^{R}(t) = t$.*

3.2   Pirate Evolution for the Complete Subtree Method

In this section, we demonstrate that the complete subtree method (CS) of [22] is susceptible to pirate evolution. Specifically, we present an evolving pirate that can produce up to $t \log N/t$ pirate boxes, given t traitor keys. Below we present some definitions that will be used throughout this section:

Definition 4. *The partition $S = \mathcal{T}(R)$ is the set of subsets $S_{i_1}, S_{i_2}, \cdots S_{i_m}$ where, $0 < j \leq m$, $i_j \in \mathcal{J}$ corresponds to a node in the full binary tree. Denote the root of subtree containing the users in $S \in \{S_j\}_{j \in \mathcal{J}}$ by $root(S)$, in general we will be using $root()$ as a function outputting root of a given tree. Suppose $T = \mathcal{L}(\mathcal{I}_1, \cdots, \mathcal{I}_n, t, R)$, then we say the Steiner tree $ST(T, S)$ is the minimal subtree of the binary tree rooted at $root(S)$ that connects all the leaves on which the users in $T \cap S$ are placed.*

We denote the unique key of a node v by $L(v)$. It is possible for any $u \in S$ to deduce $L(root(S))$ from its private information $\mathcal{I}_u, \mathcal{K}_u$. A pirate box $Box(L(v))$ is a decoder that uses the key associated to S where $root(S) = v$; it decrypts $\mathcal{E}_S(\cdot)$ iff there exists a $S \in S$ s.t. $root(S) = v$ holds. in other terms, $Box(L(v))$ decrypts $C = \mathcal{E}_R(\cdot)$ iff the header of C contains the encryption $\mathcal{Q}_{L(v)}(K)$.

Figure 2 is an illustration for the partition of non-revoked users and the set of traitors in a broadcasting scheme using the CS method. The description of the evolving pirate relies on a simple observation that is the following lemma:

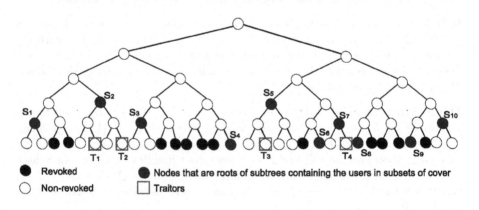

Fig. 2. Complete Subtree with cover and set of traitors

Lemma 1. *For a given set of revoked users R let $\mathcal{T}^{P_i}(R)$ be the partition generated after tracing pirate box $P = Box(L(v))$. Let S be the subset such that $root(S) = v$ holds. Suppose $S = S_L \cup S_R$ where subset S_L (resp. S_R) is left (resp.*

right) part of the subtree rooted at v. It holds that: $S \in \mathcal{T}(R)$ *if and only if* $S_L \in \mathcal{T}^P(R)$ *and* $S_R \in \mathcal{T}^P(R)$.

According to the above lemma, the pirate will be able to produce a new version of pirate box after $P_i = Box(L(root(S)))$ is caught. That is true because after tracing P_i, a traitor u is either in S_L or S_R, and the pirate still will be able to produce a new box by using the key associated to S_L(or S_R depends on which one contains u). The motivation of the evolving pirate is exploiting the above observation to successively generate pirate boxes.

We define the master pirate box $\mathcal{B}$ produced by the adversary $\mathcal{P}^{\mathcal{E}_R(\mathcal{M})}(T, K)$ as producing a vector of pirate boxes. $\mathcal{B}$ constructs the sequence of pirate boxes by walking on the nodes of the forest of Steiner trees $\{ST(T, S) \mid S \in \mathcal{T}(R)\}$. More technically, it recursively runs a procedure called makeboxes on each Tree $= ST(T, S)$ which first creates a pirate box Box by using the unique key assigned to the node $root(\text{Tree})$. It then splits the Tree into two trees. The splitting is needed because tracing Box will result in the partition of the subset S. Thus the splitting procedure is based on the partition of subset S into two equal subsets (in CS tracing works by splitting into the two subtrees rooted at the children of $root(\text{Tree})$). The master box $\mathcal{B}$ then runs makeboxes independently on both of the trees resulted from the partition. Figure 3 is the summary of the evolving pirate strategy. The number of generations that can be produced equals the number of nodes in the forest of Steiner trees $\{ST(T, S) \mid S \in \mathcal{T}(R)\}$.

1.   For each $S \in \mathcal{T}(R)$ run makeboxes$(ST(T, S))$ till the ℓ-th box is produced.
makeboxes(Tree)
1.   Take any user u placed on a leaf of Tree.
2.   Output $Box(L(root(\text{Tree})))$ where $L(root(\text{Tree}))$ is available from $\mathcal{K}_u$
3.   Let ST_L and ST_R be respectively the left and right subtrees of Tree.
4.   run makeboxes(ST_L)
5.   run makeboxes(ST_R)

Fig. 3. The description of master box program $\mathcal{B}(1^{t+\log N}, \ell)$ parameterized by $\mathcal{T}(R), T, \mathcal{K}_u$ for $u \in T$ that is produced by the evolving pirate for the complete subtree method

Theorem 2 is formalizing and proving the correctness of the above procedure, i.e. the next generation should be able to decrypt the message encrypted after tracing has disabled all previous boxes.

Theorem 2. *Let $P_1, P_2, \cdots, P_{v+1}$ be a sequence of pirate boxes constructed by the evolving pirate strategy described in Figure 3. Suppose $\mathcal{C} = \mathcal{E}_R^{P_1, P_2, \cdots, P_v}(M)$, then $Prob[P_{v+1}(\mathcal{C}) = M] \geq q$, provided that v is less than the number of nodes in the forest of trees $\{ST(T, S) \mid S \in \mathcal{T}(R)\}$.*

Leaking Incidents. For the polynomial time adversary $\mathcal{P}$ described in Figure 3, $PE_{\mathcal{P}, \mathcal{L}}^R(t)$ is the number of nodes in the forest of the Steiner trees of the traitors.

Theorem 3 and Theorem 4 give some bounds on this quantity depending on the leaking incident.

Theorem 3. *Let* N *be the set of N users represented by a full binary tree in the Complete Subtree method. For a given* R, *any leaking incident* $\mathcal{L}$ *corrupting t users in a single subset* S $\in \mathcal{T}(R)$ *enables an evolving pirate with respect to* $\mathcal{L}$ *so that* $PE^{R}_{\mathcal{P},\mathcal{L}}(t) \geq 2t - 2 + \log(|S|/t)$.

Theorem 4. *Let* N *be the set of N users represented by a full binary tree. For a given* R, *there exists a leaking incident* $\mathcal{L}$ *corrupting t users in a single subset* S $\in \mathcal{T}(R)$ *so that* $PE^{R}_{\mathcal{P},\mathcal{L}}(t) \geq 2t - 1 + t \log(|S|/t)$.

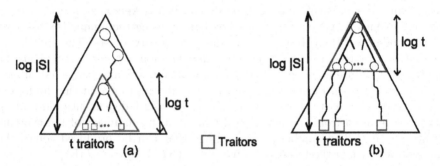

Fig. 4. The illustration of the bounds on the number of pirate boxes produced (a) All of the traitors are descendants of a same node with height $\log t$ (b) None of the paths with length $\log(|S|/t)$ from leaves to the root intersect

Figure 4 shows the cases where it is possible to achieve the bounds given in above two theorems. Figure 4(a) is yielding the bound in Theorem 3 and, Figure 4(b) is yielding the bound in Theorem 4. The maximum number of generations can be achieved following Figure 4(b) in a configuration of the system when there is no revoked user; in this case there is a single element in the partition, namely S containing N users. It follows that the pirate can produce up to $t \log(N/t)$ generations and thus:

Corollary 1. *The pirate evolution bound for the* CS *method satisfies* evo[CS] $\geq$ $t \log(N/t)$.

3.3   Pirate Evolution for the Subset Difference Method

In this section we turn our attention to the Subset Difference (SD) method of [22] that is part of the AACS standard [1]. Compared to the Complete Subtree method, the subsets in the SD method are represented by pairs of nodes. We define the required notations as follows:

Definition 5. *Let* $S_{i,j} \in \{S_{i,j}\}_{(i,j)\in\mathcal{J}}$ *be the set of all leaves in the subtree rooted at* v_i *but not of* v_j. *We define the function set* : $\{(v_i, v_j) \mid (i,j) \in \mathcal{J}\} \rightarrow$

$\{\mathsf{S}_{i,j}\}_{(i,j)\in\mathcal{J}}$ such that the inverse function $set^{-1}()$ maps a subset to its corresponding pair of nodes. Since $\mathsf{S}_{i,j}$ is somehow related to a tree, we still use $root(\mathsf{S}_{i,j})$ to output v_i. Suppose $\mathsf{T} = \mathcal{L}(\mathcal{I}_1, \cdots, t, \mathsf{R})$, then we say the Steiner tree $ST(\mathsf{T}, v_i, v_j)$ is the minimal subtree of the binary tree rooted at v_i, excluding the descendants of v_j, that connects all the leaves in $\mathsf{T} \cap set(v_i, v_j)$ and node v_j. A pirate box $Box(v_i, v_j, \mathcal{K}_u)$ is a decoder that uses the key associated to $set(v_i, v_j)$ as inferred by the private data $\mathcal{K}_u$ assigned to the user u. For simplicity, we also denote the pirate decoder by $Box(v_i, v_j, u)$ (omitting $\mathcal{K}_u$). By definition; $Box(v_i, v_j, u)$ inverts $\mathcal{E}_{\mathcal{S}}(\cdot)$ iff there exists a $\mathsf{S} \in \mathcal{S}$ such that $set(v_i, v_j) = \mathsf{S}$ holds.

The susceptibility of the SD method to pirate evolution relies on the following simple observation regarding the tracing algorithm and the way it operates on a given pirate box:

Lemma 2. *For a given set of revoked users* R *let* $\mathcal{T}^P(\mathsf{R})$ *be the partition generated after tracing pirate box* $P = Box(v_i, v_j, u)$. *Suppose* v_c *is the child of* v_i *that is on the path from* v_i *to* v_j; *note that in this case* $set(v_i, v_j) = set(v_i, v_c) \cup set(v_c, v_j)$. *It holds that:* $set(v_i, v_j) \in \mathcal{T}(\mathsf{R})$ *if and only if* $set(v_i, v_c) \in \mathcal{T}^P(\mathsf{R})$ *and* $set(v_c, v_j) \in \mathcal{T}^P(\mathsf{R})$.

We will exploit the above lemma to successively generate pirate boxes. This is possible because after tracing $P = Box(v_i, v_j, u)$, the traitor u is still in one of the subsets in the partition $\mathcal{T}^P(\mathsf{R})$. We will present an evolving pirate strategy based on the forest of Steiner trees $\{ST(\mathsf{T}, v_i, v_j) \mid set(v_i, v_j) \in \mathcal{T}(\mathsf{R})\}$ by walking on the paths of Steiner trees that will be predefined according to a scheduling of traitors. Unlike our evolving pirate strategy for the CS method, we are focusing on paths instead of nodes because of the inherent structure of the SD method and the way tracing works by merging subsets under a simple condition that is shown in lemma 3. The merging performed by NNL (whose main objective is to curb the ciphertext expansion) is, as we observe, an opportunity for pirate evolution as it leads to the reuse of some nodes in different pairs, i.e. different subsets in the collection $\{\mathsf{S}_{i,j}\}_{(i,j)\in\mathcal{J}}$. In general, the merging will occur whenever it is allowed based on lemma 3 with the following exception: S_1 will not be merged if the partition contains another subset S_2 such that they have resulted from a split-up of a single subset at an earlier iteration of the subset-tracing procedure. The reader may refer to [22].

Lemma 3. *Let* $V_i, v_1, v_2, \cdots v_d, V_j$ *be any sequence of vertices which occur in this order along some root-to-leaf path in the tree corresponding to the subset* $set(V_i, V_j)$, *then* $set(V_i, V_j) = set(V_i, v_1) \cup set(v_1, v_2) \cup \cdots \cup set(v_d, V_j)$.

According to the way tracing works on the SD, whenever the partition contains a series of subsets $\{set(v_i, v_1), set(v_1, v_2), \cdots, set(v_d, v_j)\}$ they can potentially be merged using Lemma 3 into one single subset $set(v_i, v_j)$. To illustrate how evolution for SD method works, we first give an example of a partition of non-revoked users and the set of traitors in Figure 5. Let's focus on the subset rooted at **g** that is magnified in Figure 6(a) and start creating pirate boxes.

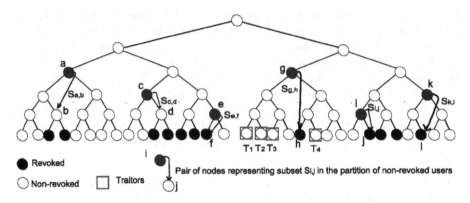

Fig. 5. Subset Difference with cover and set of traitors

Suppose that the evolving pirate uses the keys of the traitor T_4 first; the sequence of pirate boxes created until T_4 is entirely revoked would be $B_1 = Box(1, 5, T_4)$, $B_2 = Box(2, 5, T_4)$ and $B_3 = Box(3, 5, T_4)$. Due to lemma 2 tracing all these boxes would end up with revoking T_4 and $\mathcal{T}^{B_1, B_2, B_3}(\mathsf{R}) = \{set(1, 2), set(2, 3)\}$. Note that in light of lemma 3 the tracing algorithm will merge these two subsets to have the single subset $set(1, 3)$ shown in Figure 6(b). This illustrates the fact that an evolving pirate against the SD method may use the keys of a traitor as many times as the height of the subset it belongs to *without* necessarily restricting the same opportunity for other traitors that are scheduled to be used later. Indeed, we can execute a pirate box construction using the keys of traitor T_3 that would be as many as the height of the tree (compare this to the Complete Subtree method where this is not achievable and using the keys of one traitors strips the opportunity to use such keys for other traitors scheduled later). Proceeding with our example, the master pirate box $\mathcal{B}$ will now be able to create a pirate box $Box(1, 3, T_3)$ (recall that $\mathcal{T}^{B_1, B_2, B_3}(\mathsf{R}) = \{set(1, 3)\}$) followed by another box $Box(1, 2, T_3)$ and so on until T_3 is entirely revoked. Even though we have the opportunity now to make more boxes per traitor compared to the complete subset method, special care is needed to choose the order with which we are expending the traitor keys as we will illustrate below. This is in sharp contrast to the complete subset method where the scheduling of traitors makes no difference in terms of the number of pirate box generations that the master box can spawn. To see the importance of scheduling the traitors appropriately, suppose that we use the traitor T_3 first instead of T_4; then, the sequence of pirate boxes created until T_3 is entirely revoked would be $B_1 = Box(1, 5, T_3)$, $B_2 = Box(1, 2, T_3)$, $B_3 = Box(8, 9, T_3)$ and $B_4 = Box(10, 12, T_3)$ (refer to figure 6(b) for the node numbering). Tracing all these boxes would end up with revoking T_3 and $\mathcal{T}^{B_1, B_2, B_3, B_4}(\mathsf{R}) = \{set(2, 5), set(8, 10), set(10, 11)\}$. Note that this subset collection will also be merged by tracing algorithm, resulting the partition given in figure 6(c). The pirate then will be able to create a pirate box $Box(2, 5, T_4)$ and so on until T_4 is revoked. Observe that now T_4 is isolated in its own subtree,

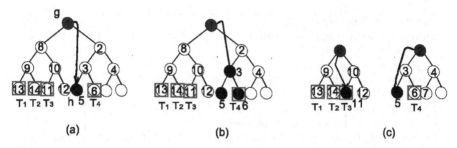

Fig. 6. (a) The subset $S_{1,5}$ of Figure 5 that contains all the traitors. The master pirate box will start producing boxes according to this subset. (b) The partition after pirate evolution using T_4 took place; T_4 is revoked from the system. (c) The partition after pirate evolution using T_3 took place; T_3 is revoked from the system.

and the master pirate box will be able to make fewer boxes using the keys of T_4. Thus, it would have been preferable to start the pirate evolution with traitor T_4.

We observe that the evolving pirate strategy can be based on a representation of the Steiner tree by means of paths hanging off each other hierarchically such that each path stems from an internal node to a traitor placed on a leaf. Each time we choose a traitor, we actually choose a path to walk on to construct pirate boxes. We observe two criteria to maximize the number of pirate decoders. (1) Once a traitor is revoked, we choose a shortest path hanging off the path containing the recently revoked traitor. (2) If there are more than one shortest path, a path with large number of paths hanging off itself would be preferable. Choosing a traitor amounts to choosing a path according to this criteria (in a recursive way). In the next paragraphs we formalize these observations. We introduce a special annotation of the Steiner Tree $ST(\mathsf{T}, u, v)$, where $set(u, v)$ is one of the subsets in the partition, that will enable us to choose the best ordering of the traitors.

Definition 6. *A traitor annotation of a Steiner tree $ST(\mathsf{T}, u, v)$ is the mapping from its nodes to $\mathsf{T} \cup \{\bot\}$ that is defined in Figure 7. We say $ST(\mathsf{T}, u, v)$ is annotated by f. Denote the parent of a node v by $parent(v)$, the sibling by $sibling(v)$, the height by $height(v)$. We define the rank of a traitor s given an annotation f as the number of nodes with 2 children that are annotated by s. We denote the rank of $s \in \mathsf{T}$ by $rank(s)$. Given a Steiner tree ST annotated by f, for any $u \in \mathsf{T}$ the u-path(ST) is the path that is made out of all nodes that are annotated by u. Similarly, we define $\bot$-path(ST) and further we call it as the basic path of the tree ST. We denote u-path(ST) by a vector of nodes, u-path(ST)$=\langle v_1, v_2, \cdots v_s \rangle$ where $v_i = parent(v_{i+1})$ for $0 < i < s$ and $u = f(v_i)$; we also denote v_1 and v_s in this path by $top_f(u)$ and $bottom_f(u)$ respectively.*

Lemma 4. *For a given set of revoked users R and the set of traitors T, let $\mathsf{Tree} = ST(\mathsf{T}, v_i, v_j) \in \{ST(\mathsf{T}, g, r) \mid set(g, r) \in \mathcal{T}(\mathsf{R})\}$ be one of the Steiner trees. Consider the annotation of Tree given in Figure 7. Suppose the shortest path hanging off the $\bot$-path(Tree) is annotated by u. Let $u_1 = top_f(u)$ and $u_s =$*

---

annotation(Tree $ST(\mathsf{T}, i, j)$)

Initially annotate each leaf l with its corresponding traitor $u \in \mathsf{T}$, i.e. $f(l) = u$

$rank(u) = 0$, for each $u \in \mathsf{T}$

$f(j) = \perp$ and $rank(\perp) = 0$.

Annotate each node from bottom to top by following rule:

$$f(parent(v)) = \left\{ \begin{array}{ll} f(v) & sibling(v) \notin \mathsf{Tree} \vee f(v) = \perp \\ f(v) & rank(f(v)) \geq rank(f(sibling(v))) \\ f(sibling(v)) & \text{otherwise} \end{array} \right\}$$

update $rank(f(parent(v))) = rank(f(parent(v))) + 1$ if $sibling(v) \in \mathsf{Tree}$

output f

Fig. 7. Computing the traitor annotation for a given Steiner tree

$bottom_f(u)$ where s is the length of the u-path(Tree). It holds that: (1) There exists a sequence of pirate boxes $B_1, B_2, \cdots B_k$, each using a private key derived from $\mathcal{K}_u$ where $k = height(\mathsf{Tree}) + A_u$ and $A_u \in \{0, 1\}$ such that $A_u = 1$ if and only if $sibling(u_1)$ has a single child in Tree. (2) $\mathcal{T}^{B_1, B_2, \cdots B_k}(\mathsf{R}) = \{\mathcal{T}(\mathsf{R}) \setminus set(v_i, v_j)\} \cup \{set(v_i, parent(u_1)), set(u_1, u_s)\}$.

We next describe our evolving pirate strategy against the SD method. We define the master pirate box $\mathcal{B}$ produced by the adversary $\mathcal{P}^{\mathcal{E}_\mathsf{R}(\mathcal{M})}(\mathsf{T}, \mathsf{K})$ as follows: $\mathcal{B}$ recursively runs a procedure for each subset $\mathsf{S} = set(v_i, v_j) \in \mathcal{T}(\mathsf{R})$ which is called makeboxes, with input the traitor annotated Steiner tree $\mathsf{Tree} = ST(\mathsf{T}, v_i, v_j)$. Observe below that whenever the recursive call is made, the annotation of Tree satisfies that the root is annotated with $\perp$. The basic procedure works as follows:

The root v_i is annotated as $\perp$. Let u-path(Tree)$=\langle u_1, u_2, \cdots u_s \rangle$ be the shortest path hanging off the $\perp$-path(Tree). The master box $\mathcal{B}$ constructs $Box(v_i, v_j, u)$ and more pirate decoders by applying lemma 2. After creating pirate boxes as many as the height of Tree (plus one possibly if $A_u = 1$, cf. lemma 4), the traitor u will be entirely revoked by the system. Lemma 4 tells us that the partition after revoking u will include the subsets $set(v_i, parent(u_1))$ and $set(u_1, u_s)$. We update the path $\langle u_1, u_2, \cdots u_s \rangle$ in $ST(\mathsf{T}, u_1, u_s)$ by annotating it $\perp$ since u is no more in $(set(u_1, u_s))$. The master box $\mathcal{B}$ then runs makeboxes independently on both of the trees $ST(\mathsf{T}, v_i, parent(u_1))$ and $ST(\mathsf{T}, u_1, u_s)$. Refer to figure 8 for the detailed specification of the evolving pirate strategy.

In the following theorem we prove the correctness of the strategy, i.e. that each box will decrypt the ciphertexts that are generated assuming all previous boxes are traced. We also show the maximum number of pirate decoders that can be created.

Theorem 5. Let $P_1, P_2, \cdots, P_{k+1}$ be a sequence of pirate boxes constructed by the pirate evolution strategy described in Figure 8. Suppose $\mathcal{C} = \mathcal{E}_\mathsf{R}^{P_1, P_2, \cdots, P_k}(M)$, then $Prob[P_{k+1}(\mathcal{C}) = M] \geq q$, provided that

$$k < \sum_{set(v_i, v_j) \in \mathcal{T}(\mathsf{R})} \left(rank(\perp) \cdot height(v_i) + \sum_{u \in \mathsf{T} \cap set(v_i, v_j)} C_u + A_u \right)$$

where $C_u = rank(u) \cdot | u\text{-path}(ST(T, v_i, v_j)) |$.

1. For each $S_{i,j} = set(v_i, v_j) \in \mathcal{T}(R)$
2.   Compute $f = \mathtt{annotation}(ST(T, v_i, v_j))$
3.   Run $\mathtt{makeboxes}(ST(T, v_i, v_j), f)$ till the ℓ-th pirate box is produced.
$\mathtt{makeboxes}(\mathsf{Tree}, \text{annotation } f)$
1. Let $\perp$-path in Tree be $\langle k_1, k_2, \cdots k_m \rangle$. Note that $k_1 = v_i$ and $k_m = v_j$.
2. Choose the shortest path hanging off the $\perp$-path, i.e. pick $l = \max(l : sibling(k_l) \in \mathsf{Tree})$ to use the keys of traitor $u = f(sibling(k_l))$; if no such path exists, exit.
3. Denote u-path by $\langle u_1, u_2, \cdots u_s \rangle$
4. Output $Box(k_1, k_m, u), Box(k_2, k_m, u), \cdots Box(k_{l-1}, k_m, u)$
5. Output $Box(k_{l-1}, k_l, u)$ iff $l < m$.
6. Output $Box(u_1, sibling(u_2), u), Box(u_2, sibling(u_3), u), ..Box(u_{s-1}, sibling(u_s), u)$
8. Update $f(u_i) = \perp$, for $0 < i \leq s$
9. $\mathtt{makeboxes}(ST(T, u_1, u_s), f)$
10. $\mathtt{makeboxes}(ST(T, k_1, k_{l-1}), f)$

Fig. 8. The description of master box program $\mathcal{B}(1^{t+\log N}, \ell)$ parameterized by $\mathcal{T}(R)$, T, $\mathcal{K}_u$ for $u \in \mathsf{T}$ that is produced by the evolving pirate for the Subset Difference method

Leaking Incidents. For the evolving pirate $\mathcal{P}$ described in Figure 8, the value of $PE^R_{\mathcal{P},\mathcal{L}}(t)$ follows from theorem 5. Theorem 6 gives some bounds on this quantity depending on the leaking incident for the SD method.

Theorem 6. *Let* N *be the set of* N *users represented by a full binary tree in the* SD *method. For a given* R, *there exists a leaking incident* $\mathcal{L}$ *corrupting* t *users in* $\mathsf{S} \in \mathcal{T}(R)$ *(for simplicity assume* S *is complete subset, and thus* $\log |\mathsf{S}|$ *is an integer), that enables an evolving pirate with respect to* $\mathcal{L}$ *so that*

$$PE^R_{\mathcal{P},\mathcal{L}}(t) = \begin{cases} t \log |\mathsf{S}|, & t \leq \log |\mathsf{S}| + 1 \\ t \log(\frac{|\mathsf{S}|}{2^m}) + 2^m \log(\frac{|\mathsf{S}|}{2^{m-3}}) - \log |\mathsf{S}| - 3, \end{cases}$$

$$t \in \left\{ 2^{m-1} \log(\frac{|\mathsf{S}|}{2^{m-2}}) + 1, \ldots, 2^m \log(\frac{|\mathsf{S}|}{2^{m-1}}) \right\} \ for \ 0 < m < \log |\mathsf{S}|$$

To see the above existence result, consider the following: our goal is to choose t traitors in the set S. Once a leaf is chosen in a complete subtree to place a traitor, we will next choose to place other traitors in the subtrees hanging off the path of that traitor. The first traitor chosen in each hanging subtree, it contributes to the number $PE^R_{\mathcal{P},\mathcal{L}}(t)$ as many pirate generations as the height of the tree (say $h = \log |\mathsf{S}|$). After this first stage of placement, we have h subtrees each containing one traitor and each have different heights. We recursively place the remaining traitors in these subtrees by using the highest possible subtrees first. In short, we can think of the leaking incident of theorem 6 as follows: the first traitor is placed in an arbitrary leaf; then, h stages follow: in stage m (where $m = 0, \ldots, h-1$), the remaining traitors are placed on the subtrees of height $h - m$. At stage $m > 0$ we can place $2^{m-1}(h - m)$ traitors (where we place h traitors at stage 0). By the end of stage m we will have already placed $2^m \log(\frac{|\mathsf{S}|}{2^{m-1}})$ traitors. Note that a traitor placed at stage m will contribute $h - m$ pirate boxes following our evolving pirate strategy. The formula in theorem 6 gives the sum of all pirate generations for each traitor.

The maximum number of generations can be achieved following the leaking incident of Theorem 6 in a configuration of the system when there is no revoked user; in this case there is a single element in the partition, namely S containing N users. The corollary below follows easily from theorem 6.

Corollary 2. *The pirate evolution bound for the* SD *method satisfies* $\mathrm{evo}[\mathrm{SD}] \geq t \log N$ *for* $t \leq \log N$. *It also satisfies that* $\mathrm{evo}[\mathrm{SD}] \geq t \frac{\log N}{2}$ *for* $t \leq \sqrt{N} \cdot \frac{\log N}{2}$.

Relation to the AACS. The AACS standard for Blue-Ray disks and HD-DVDs uses the SD method with $N = 2^{31}$ nodes. It follows that a leaking incident with t traitors enables our evolving pirate strategy to generate up to $31 \cdot t$ generations of pirate boxes in the case that the system has an initial state of ciphertexts with a single element in the partition; note that if the starting configuration of the system has more elements in the partition (e.g., 2^8 elements each corresponding to 2^{23} users) the total number of generations would be $23 \cdot t$ for $t \leq 23$, and so on.

References

1. AACS Specifications (2006), http://www.aacsla.com/specifications/
2. Boneh, D., Franklin, M.: An Efficient Public-Key Traitor Tracing Scheme. In: Wiener, M.J. (ed.) CRYPTO 1999. LNCS, vol. 1666, pp. 338–353. Springer, Heidelberg (1999)
3. Boneh, D., Sahai, A., Waters, B.: Fully Collusion Resistant Traitor Tracing with Short Ciphertexts and Private Keys. In: Vaudenay, S. (ed.) EUROCRYPT 2006. LNCS, vol. 4004, pp. 573–592. Springer, Heidelberg (2006)
4. Boneh, D., Shaw, J.: Collusion-Secure Fingerprinting for Digital Data. IEEE Transactions on Information Theory 44(5), 1897–1905 (1998)
5. Chabanne, H., Hieu Phan, D., Pointcheval, D.: Public Traceability in Traitor Tracing Schemes. In: Cramer, R.J.F. (ed.) EUROCRYPT 2005. LNCS, vol. 3494, pp. 542–558. Springer, Heidelberg (2005)
6. Chor, B., Fiat, A., Naor, M.: Tracing Traitors. In: Desmedt, Y.G. (ed.) CRYPTO 1994. LNCS, vol. 839, pp. 257–270. Springer, Heidelberg (1994)
7. Chor, B., Fiat, A., Naor, M., Pinkas, B.: Tracing Traitors. IEEE Transactions on Information Theory 46(3), 893–910 (2000)
8. Dodis, Y., Fazio, N.: Public Key Broadcast Encryption for Stateless Receivers. In: Feigenbaum, J. (ed.) DRM 2002. LNCS, vol. 2696, pp. 61–80. Springer, Heidelberg (2003)
9. Dodis, Y., Fazio, N., Kiayias, A., Yung, M.: Scalable public-key tracing and revoking. In: Proceedings of the Twenty-Second ACM Symposium on Principles of Distributed Computing (PODC 2003), July 13-16, 2003, Boston, Massachusetts (2003)
10. Fiat, A., Naor, M.: Broadcast Encryption. In: Stinson, D.R. (ed.) CRYPTO 1993. LNCS, vol. 773. Springer, Heidelberg (1994)
11. Fiat, A., Tassa, T.: Dynamic Traitor Tracing. Journal of Cryptology 4(3), 211–223 (2001)
12. Goldreich, O., Goldwasser, S., Micali, S.: How to Construct Random Functions. J. of the ACM 33(4), 792–807 (1986)
13. Gafni, E., Staddon, J., Yin, Y.L.: Efficient Methods for Integrating Traceability and Broadcast Encryption. In: Wiener, M.J. (ed.) CRYPTO 1999. LNCS, vol. 1666, pp. 372–387. Springer, Heidelberg (1999)
14. Garay, J.A., Staddon, J., Wool, A.: Long-Lived Broadcast Encryption. In: Bellare, M. (ed.) CRYPTO 2000. LNCS, vol. 1880, pp. 333–352. Springer, Heidelberg (2000)

15. Halevy, D., Shamir, A.: The LSD Broadcast Encryption Scheme. In: Yung, M. (ed.) CRYPTO 2002. LNCS, vol. 2442, pp. 47–60. Springer, Heidelberg (2002)
16. Jho, N., Hwang, J.Y., Cheon, J.H., Kim, M.H., Lee, D.H., Yoo, E.S.: One-Way Chain Based Broadcast Encryption Schemes. In: Cramer, R.J.F. (ed.) EURO-CRYPT 2005. LNCS, vol. 3494, pp. 559–574. Springer, Heidelberg (2005)
17. Kiayias, A., Yung, M.: Self Protecting Pirates and Black-Box Traitor Tracing. In: Kilian, J. (ed.) CRYPTO 2001. LNCS, vol. 2139, pp. 63–79. Springer, Heidelberg (2001)
18. Kiayias, A., Yung, M.: On Crafty Pirates and Foxy Tracers. In: Sander, T. (ed.) DRM 2001. LNCS, vol. 2320, pp. 22–39. Springer, Heidelberg (2002)
19. Kiayias, A., Yung, M.: Traitor Tracing with Constant Transmission Rate. In: Knudsen, L.R. (ed.) EUROCRYPT 2002. LNCS, vol. 2332, pp. 450–465. Springer, Heidelberg (2002)
20. Kurosawa, K., Desmedt, Y.: Optimum Traitor Tracing and Asymmetric Schemes. In: Nyberg, K. (ed.) EUROCRYPT 1998. LNCS, vol. 1403, pp. 145–157. Springer, Heidelberg (1998)
21. Micciancio, D., Panjwani, S.: Corrupting One vs. Corrupting Many: The Case of Broadcast and Multicast Encryption. In: Bugliesi, M., Preneel, B., Sassone, V., Wegener, I. (eds.) ICALP 2006. LNCS, vol. 4052. Springer, Heidelberg (2006)
22. Naor, D., Naor, M., Lotspiech, J.B.: Revocation and Tracing Schemes for Stateless Receivers. In: Kilian, J. (ed.) CRYPTO 2001. LNCS, vol. 2139, pp. 41–62. Springer, Heidelberg (2001)
23. Naor, M., Pinkas, B.: Threshold Traitor Tracing. In: Krawczyk, H. (ed.) CRYPTO 1998. LNCS, vol. 1462, pp. 502–517. Springer, Heidelberg (1998)
24. Naor, M., Pinkas, B.: Efficient Trace and Revoke Schemes. In: Frankel, Y. (ed.) FC 2000. LNCS, vol. 1962, pp. 1–20. Springer, Heidelberg (2001)
25. Naor, M., Reingold, O.: Number-Theoretic Constructions of Efficient Pseudo-Random Functions. In: FOCS '97, pp. 458–467. IEEE Computer Society, Los Alamitos (1997)
26. Pfitzmann, B.: Trials of Traced Traitors. In: Anderson, R. (ed.) Information Hiding. LNCS, vol. 1174, pp. 49–63. Springer, Heidelberg (1996)
27. Phan, D.H., Safavi-Naini, R., Tonien, D.: Generic Construction of Hybrid Public Key Traitor Tracing with Full- Public-Traceability, pp. 264–275.
28. Safavi-Naini, R., Wang, Y.: Sequential Traitor Tracing. In: Bellare, M. (ed.) CRYPTO 2000. LNCS, vol. 1880, pp. 316–332. Springer, Heidelberg (2000)
29. Safavi-Naini, R., Wang, Y.: Collusion Secure q-ary Fingerprinting for Perceptual Content. In: Sander, T. (ed.) DRM 2001. LNCS, vol. 2320, pp. 57–75. Springer, Heidelberg (2002)
30. Safavi-Naini, R., Wang, Y.: New Results on Frameproof Codes and Traceability Schemes. IEEE Transactions on Information Theory 47(7), 3029–3033 (2001)
31. Safavi-Naini, R., Wang, Y.: Traitor Tracing for Shortened and Corrupted Fingerprints. In: Feigenbaum, J. (ed.) DRM 2002. LNCS, vol. 2696, pp. 81–100. Springer, Heidelberg (2003)
32. Staddon, J.N., Stinson, D.R., Wei, R.: Combinatorial Properties of Frameproof and Traceability Codes. IEEE Transactions on Information Theory 47(3), 1042–1049 (2001)
33. Stinson, D.R., Wei, R.: Combinatorial Properties and Constructions of Traceability Schemes and Frameproof Codes. SIAM Journal on Discrete Math. 11(1), 41–53 (1998)
34. Tardos, G.: Optimal probabilistic fingerprint codes. In: ACM 2003, pp. 116–125 (2003)

A Security Analysis of the NIST SP 800-90 Elliptic Curve Random Number Generator

Daniel R.L. Brown[1] and Kristian Gjøsteen[2]

[1] Certicom Research,
dbrown@certicom.com
[2] Department of Mathematical Sciences, Norwegian
University of Science and Technology, NO-7491 Trondheim, Norway
kristian.gjosteen@math.ntnu.no

Abstract. An elliptic curve random number generator (ECRNG) has been approved in a NIST standard and proposed for ANSI and SECG draft standards. This paper proves that, if three conjectures are true, then the ECRNG is secure. The three conjectures are hardness of the elliptic curve decisional Diffie-Hellman problem and the hardness of two newer problems, the x-logarithm problem and the truncated point problem. The x-logarithm problem is shown to be hard if the decisional Diffie-Hellman problem is hard, although the reduction is not tight. The truncated point problem is shown to be solvable when the minimum amount of bits allowed in NIST standards are truncated, thereby making it insecure for applications such as stream ciphers. Nevertheless, it is argued that for nonce and key generation this distinguishability is harmless.

Keywords: Random Number Generation, Elliptic Curve Cryptography.

1 Introduction

Certain random number generator (RNG) algorithms, such as the Blum-Micali [1] and Kaliski [2] generators, have been proven secure — assuming the conjectured hardness of associated number-theoretic problems. Recently, a new random number generator has been undergoing standardization (see [3,4,5,6]). In this paper, this new generator is called the *Elliptic Curve Random Number Generator (ECRNG)*. Like Kaliski's generator, the ECRNG is based on elliptic curves and is adapted from the Blum-Micali generator. Compared to many other number-theoretic RNGs, the ECRNG is considerably more efficient, because it outputs more bits per number-theoretic operation. The ECRNG is different from the Kaliski RNG in two major respects:

- The ECRNG uses ordinary elliptic curves, not the supersingular elliptic curves of Kaliski RNG. Supersingular curves can make the state transition function a permutation, which is advantageous for making the Blum-Micali proof work, but is disadvantageous because of the Menezes-Okamoto-Vanstone (MOV) attack, which requires a larger curve to make the proof give a useful assurance.[1]

---

[1] This is not to say that MOV attack could be applied against the Kaliski RNG for smaller sized curves.

A. Menezes (Ed.): CRYPTO 2007, LNCS 4622, pp. 466–481, 2007.

For ordinary curves, the state transition function is many-to-one, often two-to-one. This paper adapts the Blum-Micali proof by introducing the x-logarithm problem to overcome the obstacle introduced by state transition function not being a permutation.

- The ECRNG produces at each state update an output with almost as many bits as in the x-coordinate of an elliptic curve point, whereas the Kaliski ECRNG outputs just a single bit. Therefore the ECRNG is considerably more efficient than the Kaliski RNG if operated over the same elliptic curve. The Kaliski RNG outputs a single bit that is a hardcore predicate for the elliptic curve discrete logarithm problem (ECDLP). The ECRNG output function essentially uses a conjectured hardcore function of the ECDLP. The basis of this conjecture is the elliptic curve DDH problem, and the truncated point problem (TPP), defined below.

This paper proves that ECRNG is secure if the following problems are hard:

- The elliptic curve version of the well known *decisional Diffie-Hellman problem* (DDH). This is now widely accepted for certain groups. These groups include most small cofactor order elliptic curve groups defined over finite fields, such as the NIST curves. Boneh [7] gives an excellent survey about the DDH problem.
- The *x-logarithm problem* (XLP): a new problem, which is, given an elliptic curve point, determine whether its discrete logarithm is congruent to the x-coordinate of an elliptic curve point. This problem is discussed further in §5. We provide some evidence that the XLP problem is almost as hard as the DDH problem. The evidence takes the form of loose reduction between a related problem, AXLP (defined below), and the DDH problem.
- The *truncated point problem* (TPP): a new problem, which is, given a bit string of a certain length, determine whether it is obtained by truncating the x-coordinate of a random elliptic curve point. The TPP problem concerns extraction of pseudorandom bits from random elliptic curve points. El Mahassni and Shparlinski [8] give some results about extraction of pseudorandom bits from elliptic curve points. Gürel [9] also gives some results, although with fewer bits extracted than in the ECRNG. We discuss the TPP problem in §7. We find that if too few bits are truncated, then the result is distinguishable from a random bit string. Schoenmakers and Sidorenko [10] independently found a similar result.

Naor and Reingold [11] constructed pseudorandom functions secure as the hardness of the DDH problem, while Gertner and Malkin constructed such a pseudorandom number generator based on the same assumption. Farashahi, Schoenmakers and Sidorenko [12] recently constructed pseudorandom number generators following a modified version of the ECRNG, which are secure as DDH over certain groups.

This paper does not attempt to analyze the various issues surrounding entropy of the secret state of the RNG. Prediction resistance is the ability of RNG to add additional entropy into the secret state to recover completely from a circumstance where an adversary has information about the previous state. Initialization and prediction resistance are general RNG issues, and indeed the standards specifying the ECRNG do not treat the ECRNG especially different from other RNGs with respect initialization and prediction resistance. This paper deliberately restricts itself to ECRNG specific issues.

2   The Elliptic Curve Random Number Generator

Let $\mathbb{F}_q$ be a finite field with q elements. An elliptic curve E over $\mathbb{F}_q$ is defined by a nonsingular cubic polynomial in two variables x and y with coefficients in $\mathbb{F}_q$. This paper considers only cubics in a specially reduced Weierstrass form

$$E(x,y) = y^2 + cxy - (x^3 + ax^{1+c} + b) = 0 \tag{1}$$

where c is 0 if q is odd and 1 if q is odd, since these are most often used in cryptography, and particularly in the ECRNG. We define the rational points of the curve to be

$$E(\mathbb{F}_q) = \{(x,y) \in \mathbb{F}_q^2 : E(x,y) = 0\} \cup \{0\}. \tag{2}$$

An addition law is defined on $E(\mathbb{F}_q)$ using the well-known chord-and-tangent law. For example, $(u,v) + (x,y) = (w,z)$ is computed as follows. Form a line through (u,v) and (x,y), which intersects the curve $E(x,y) = 0$ in three points, namely (u,v), (x,y) and some third point $(w,-z)$, which defines the desired sum by negating the y-coordinate.

In the ECRNG, and in elliptic curve cryptography more generally, one defines some base point P on the curve. One assumes that P has prime order n in the elliptic curve group, so that $nP = 0$. Generally, the number of points in $E(\mathbb{F}_q)$ is hn, where the cofactor h is usually quite small, typically with $h \in \{1, 2, 4\}$. We say that a point Q is *valid* if it is an additive multiple of P. We will generally only consider valid points in this paper, so when we say a random point, we mean a random valid point.

The ECRNG maintains a state, which is an integer $s_i \in [0, \max\{q-1, n-1\}]$. The iteration index i increments upon each output point of the ECRNG. The ECRNG is intended to be initialized by choosing the initial state s_0 uniformly at random from $[0, n-1]$.

For a point $P = (x,y) \in E(\mathbb{F}_q)$, we write $x(P) = \bar{x}$, where $\bar{x} \in \mathbb{Z}$ is obtained by taking the bit representation of the $x \in \mathbb{F}_q$ and considering this is as the bit representation of an integer. When q is prime, we essentially have $\bar{x} = x$, but when q is not a prime, the value of $\bar{x}$ depends on the representation used for the finite field $\mathbb{F}_q$. (We may arbitrarily define $x(0) = 0$, but we will encounter this case negligibly often in our analysis.) Therefore, to fully specify the ECRNG, one needs to define a field representation, because the function $x(\cdot)$ has an important rôle in the ECRNG, as we see below.

The ECRNG has another initialization parameter, which is a point Q. The point should ideally be chosen at random, preferably verifiably at random, such as by deriving it from the output of secure hash function or block cipher.

When the state is s_i, the (raw) output point is defined as

$$R_i = s_i Q. \tag{3}$$

The actual output of the ECRNG applies further processing to R_i. The final output is $r_i = t(x(R_i))$, where t is a function that truncates certain bits from the bit string representation of an elliptic curve point. The purpose of t is to convert the x-coordinate of a pseudorandom EC point to a pseudorandom bit string.

After generating an output point R_i, the state is updated as

$$s_{i+1} = x(s_i P). \tag{4}$$

It is convenient to adopt the following notation. We define the *prestate* at iteration $i + 1$ as $S_{i+1} = s_i P$. Note that $s_{i+1} = x(S_{i+1})$. We may think of the prestate being updated as

$$S_{i+2} = x(S_{i+1})P. \tag{5}$$

The following notation for the ECRNG will be convenient. Let s_0 be the initial state. We define $g_m(Q, s_0) = (R_0, R_1, \ldots, R_m)$ inductively by

$$g_0(Q, s_0) = (s_0 Q) \text{ and } g_m(Q, s_0) = (s_0 Q, g_{m-1}(Q, x(s_0 P))),$$

where we use the convention that a comma indicates concatenation of point sequences. We shall continue to use this convention for the remainder of the paper.

3 Lemmas on Indistinguishability

Random variables X and Y are *computationally indistinguishable* if, given a sample value u that has probability $\frac{1}{2}$ of coming from X and $\frac{1}{2}$ from Y, an adversary cannot distinguish , with a feasible cost of computation and reasonable success rate, whether u comes from X or from Y. Pseudorandomness is indistinguishability from a uniform (equiprobable) distribution. Indistinguishability is a well known notion in cryptology, but for completeness, this section introduces some general notation and lemmas on indistinguishability that are convenient for the proofs of the main theorems.

We write $X \sim Y$ to indicate that random variables X and Y are indistinguishable. Where needed, we write $X \overset{\sigma}{\sim} Y$ to quantify the indistinguishability by some parameters σ, such as success rate or computational cost. We write $X \cong Y$ when random variables X and Y are identically distributed. Obviously, $X \cong Y$ implies $X \sim Y$.

Intuitively, one expects indistinguishability ($\sim$) to be an equivalence relation. Certainly, $\sim$ is reflexive and symmetric, and more interestingly, it is transitive ([13, Ex. 27], for example). This is such a fundamental point it is worth repeating here.

Lemma 1. *If $X \sim Y$ and $Y \sim Z$, then $X \sim Z$.*

A second lemma, which one also intuitively expects, makes proofs cleaner through separating complicated constructions from indistinguishability.

Lemma 2. *If f is an efficiently computable function, and $X \sim Y$, then $f(X) \sim f(Y)$.*

It is worth noting that the converse to this lemma does not necessarily hold: generally, $f(X) \sim f(Y)$ does not imply $X \sim Y$. A constant function f is a trivial counterexample. Nontrivial counterexamples exist too, such as f being a bijection whose inverse is not efficiently computable.

A third lemma, which one again intuitively expects, allows one to analyze distributions by analyzing independent components.

Lemma 3. *If $X \sim Y$ and $W \sim Z$, and X and W are independent variables, as are Y and Z, and X and Z can be efficiently sampled, then $(X, W) \sim (Y, Z)$.*

This lemma also applies under our notational convention that if X and Y are sequences, then (X, Y) is their concatenation.

4   The Decisional Diffie-Hellman Problem

The *decisional Diffie-Hellman problem (DDH)* for a given elliptic curve E and a base point P is to distinguish between a triple $(Q, R, S) = (qP, rP, qrP)$ and a triple $(Q, R, Z) = (qP, rP, zP)$ where q, r, z are integer random variables uniformly distributed in the interval $[0, n-1]$. (Note this q is not to be confused with the field order.) The triple (Q, R, S) is often called a Diffie-Hellman triple. For certain elliptic curves, it is conjectured that the DDH problem is hard.

Conjecture 1. If q, r and z are independent random integers uniformly distributed in $[0, n-1]$, then $(qP, rP, qrP) \sim (qP, rP, zP)$.

This, if true, provides a nontrivial counterexample to the converse to Lemma 2 because random variables $X = (q, r, qr)$ and $Y = (q, r, z)$ are distinguishable, but if one applies the function f defined by $f(x, y, z) = (xP, yP, zP)$, then the conjecture says $f(X) \sim f(Y)$.

The conjectured hardness of the DDH problem, for certain groups, is widely believed among cryptologists. One should be aware that for certain elliptic curves, however, there are efficiently computable so-called pairings that can be used to distinguish Diffie-Hellman triples. Pairings exist for all elliptic curves, but only for a very few are they known to be efficiently computable. For most elliptic curves, one can verify that the known pairings are extremely inefficient and infeasible to use in practice. This has been confirmed for most of the NIST recommended elliptic curves.

5   The x-Logarithm Problem

The *x-Logarithm Problem (XLP)* for elliptic curve $E(\mathbb{F}_q)$ and base point P is to distinguish between dP and xP where: d is an integer chosen uniformly at random in $[0, n-1]$; and $x = x(Z)$ for a point Z chosen uniformly at random in $E(\mathbb{F}_q)$. We conjecture that the x-logarithm problem is hard for most elliptic curves:

Conjecture 2. If d and z are random integers uniformly distributed in the interval $[0, n-1]$, then $dP \sim x(zP)P$.

Now d and $x = x(zP)$ are generally distinguishable. Firstly, known tests on x quickly determine whether there exists a $y \in \mathbb{F}_q$ such that $(x, y) \in E(\mathbb{F}_q)$. Secondly, when the cofactor $h > 1$, we expect to have $x > n$ for at least about half of the x-coordinates x of random points, whereas for d, we always have $d < n$. Therefore, this conjecture, if true, gives another counterexample to the converse of Lemma 2.

An intuitive reason for the plausibility of the XLP conjecture is that given public key dP, one expects that nothing substantial is leaked about the private key d. This intuition derives from the conjectured hardness of the elliptic curve discrete logarithm problem (ECDLP). However, a formal argument that the ability to determine whether $dP = x(zP)P$ for some z, implies an ability to find d is not known to the authors. In fact, conceivably, the ECDLP could be hard, even though certain information about the discrete logarithm, such as whether it is congruent to an x-coordinate, is easily discernible.

Certain bits in the binary representation of the d have been shown by Kaliski [2] to be as hard to find as the whole of d. A bit of information with such a property is known as a *hardcore predicate* for the ECDLP. Kaliski's proof that certain bits of the binary representation of the discrete logarithm are hardcore predicate works with a reduction, as follows. Given $Q = dP$, determine from Q a bit of information about d. Then transform Q to some $Q' = d'P$ in such a way that there is an known relation between d and d', and d' has one less bit of freedom than d. Next determine a bit of information about d', then d'' and so on. The transformation is such that all bits of information learnt are independent and can be easily be reconstituted to learn d in its entirety.

What would be ideal to make Conjecture 2 into a theorem, would be another transformation with comparable properties to Kaliski's for determining the discrete logarithm d using an oracle for solving XLP. Instead we have the following result, Theorem 1, which is not ideal in that it

1. is not a tight reduction,
2. concerns a harder variant of the XLP.

The *Arbitrary-base x-Logarithm Problem (AXLP)* for elliptic curve $E(\mathbb{F}_q)$ is to distinguish between (P, dP) and (P, xP) where: d is an integer chosen uniformly at random in $[0, n - 1]$; and $x = x(Z)$ for points P and Z chosen uniformly at random in $E(\mathbb{F}_q)$. The distinction between AXLP and XLP is that in XLP, the adversary needs only to succeed for fixed base P, whereas in AXLP, the adversary must succeed for any base P.

Note that for r chosen uniformly at random from $[0, n - 1]$, we have that $(P, dP) \cong (rP, rdP)$ and $(P, xP) \cong (rP, rxP)$. This means that any adversary A can be replaced by an equally effective adversary A' that first randomizes its input by multiplying both points by a random integer r. Let

$$f_i = \Pr[A(P, iP) = 1] \text{ and } f = \Pr[A(P, Q) = 1] = \frac{1}{n}\sum_{i=0}^{n-1} f_i \qquad (6)$$

where the probabilities are taken over random P and Q.

Lemma 4. *For any adversary A against AXLP, there exists an adversary B against DDH with advantage*

$$\epsilon = \frac{2}{n} \sum_{i=0}^{n-1} (f_i - f)^2. \tag{7}$$

Proof. The idea is that for a DDH triple (Q, R, S), $\log_P Q = \log_R S$. If we run A with the pairs (P, Q) and (R, S) as input, the output should be correlated. But if (Q, R, S) is a random triple, the output should be uncorrelated.

Let B be the algorithm that on input of (Q, R, S) samples u and v uniformly at random from $0, 1, ..., n - 1$, then runs A on (uP, uR) and (vQ, vS). If the outputs are equal, B outputs 1, otherwise 0.

We compute the advantage of B in distinguishing DDH tuples from random tuples as $\epsilon = |\delta_0 - \delta_1|$, where δ_0 is the probability that B outputs 1 on input of a DDH triple, and δ_1 is the probability that B outputs 1 on input of a random triple.

For a DDH triple $(Q, R, S) = (qP, rP, qrP)$, the two runs of A will have identical the same input distributions. That is, we can deal with each possible logarithm $r = i$ separately and get

$$
\begin{aligned}
\delta_0 &= \Pr[B(Q, R, S) = 1] \\
&= \Pr[A(uP, ruP) = A(vQ, rvQ) = 1] + \Pr[A(uP, ruP) = A(vQ, rvQ) = 0] \\
&= \sum_i (\Pr[A(P, iP) = 1]^2 + \Pr[A(P, iP) = 0]^2)/n \\
&= \sum_i (f_i^2 + (1 - f_i)^2)/n \\
&= \sum_i (1 + 2f_i^2 - 2f_i)/n.
\end{aligned}
\tag{8}
$$

For a random triple, the two runs of A will be independent, and we get

$$
\begin{aligned}
\delta_1 &= \Pr[A(P, jP) = 1 \mid j \text{ random}]^2 + \Pr[A(P, jP) = 0 \mid j \text{ random}]^2 \\
&= f^2 + (1 - f)^2 = 1 + 2f^2 - 2f \\
&= \sum_i (1 + 2f^2 - 2f_i)/n.
\end{aligned}
\tag{9}
$$

Summing up, we get

$$
\begin{aligned}
\epsilon &= \left| \sum_i (1 + 2f_i^2 - 2f_i - (1 + 2f^2 - 2f_i))/n \right| \\
&= (2/n) \left| \sum_i (f_i^2 - f^2) \right| = (2/n) \sum_i (f_i - f)^2
\end{aligned}
\tag{10}
$$

which completes the proof. $\qquad\square$

We now need to show that this adversary against DDH has a significant advantage if the AXLP adversary has a significant advantage.

Let a_i be the probability that the x-coordinate of a random point Z of order n is i modulo n, that is

$$a_i = \Pr[x(Z) \equiv i \pmod{n} \mid Z \text{ random of order } n]. \tag{11}$$

The signed advantage of an adversary A against AXLP is

$$
\begin{aligned}
\epsilon' &= \Pr[A(P, iP) = 1 \mid i = x(Z), Z \text{ random of order } n] - \\
&\quad \Pr[A(P, iP) = 1 \mid i \text{ random}] \\
&= \sum_i \Pr[A(P, iP) = 1](a_i - 1/n) \tag{12} \\
&= \sum_i f_i(a_i - 1/n).
\end{aligned}
$$

Since $\sum_i (a_i - 1/n) = 0$, we can write

$$\epsilon' = \sum_i (f_i - f)(a_i - 1/n). \tag{13}$$

Next, let t be the maximal number of points on a curve with the same x-coordinate modulo n. (That is, t is at most twice the cofactor.) Then $|a_i - 1/n| \leq (t-1)/n$, or alternatively

$$|a_i - 1/n| \frac{n}{t-1} \leq 1. \tag{14}$$

Lemma 5. *For any adversary A against AXLP, we have*

$$\frac{2}{n} \sum_i (f_i - f)^2 \geq \frac{2}{t-1}(\epsilon')^2, \tag{15}$$

where f_i, f, t and ϵ' are as defined above.

Proof. We multiply each term in the sum with $((a_i - 1/n)\, n/(t-1))^2 \leq 1$, getting

$$\frac{2}{n} \sum_i (f_i - f)^2 \geq \frac{2}{(t-1)^2} \frac{1}{n} \sum_i (f_i - f)^2 (a_i - 1/n)^2 n^2 = \mu. \tag{16}$$

Using the fact that the average sum of squares is greater than the square of the average (Jensen's inequality applied to $z \mapsto z^2$), we get that

$$\mu \geq \frac{2}{(t-1)^2} \left(\sum_i \frac{1}{n}(f_i - f)(a_i - 1/n)n \right)^2 = \frac{2}{(t-1)^2}(\epsilon')^2, \tag{17}$$

which concludes the proof. $\qquad\square$

Theorem 1. *If the DDH problem is hard and the cofactor is small, then AXLP is hard.*

Proof. If there exists an adversary A against AXLP with advantage $|\epsilon'|$, then by Lemma 4 and 5 there exists an adversary against DDH with advantage

$$\epsilon \geq \frac{2}{(t-1)^2}(\epsilon')^2. \tag{18}$$

If $|\epsilon'|$ is non-negligible and the cofactor (and hence t) is small, then ϵ is non-negligible. □

This reduction is not tight. If $|\epsilon'| \approx \frac{1}{2}$ and the cofactor is 1, then $\epsilon \approx \frac{1}{2}$. On the the other hand, if $|\epsilon'| \approx 2^{-40}$, then $\epsilon \approx 2^{-80}$. It may be the case that a feasible distinguisher for DDH exists with this advantage, in which case an efficient distinguisher for AXLP with advantage 2^{-40} could not be ruled out.

In practice, the bound given by t is not sharp. For the cofactor two NIST curves, one can prove that for almost all $d \in [0, n-1]$, at most one of d and $d+n$ can be the x-coordinate of a valid point. This is because a valid x-coordinate has a certain trace, and in trinomial or pentanomial basis representation, the trace depends on a few bits of the x-coordinate including the least significant bit. From this we see that $t = 2$ would work. For cofactor four curves, we only have a heuristic estimate for t, which presumably could be confirmed (or denied) by further analysis.

The advantage of B in the proof above can possibly be increased by also comparing A's run on (T, V) and (U, W) and other such pairings, although one does not expect a significant increase compared to increased runtime.

Because AXLP appears to be a hard problem, it seems quite reasonable to conjecture that XLP is also a hard problem. It should be noted that the hardness of XLP suggests that the output of the ECRNG, before truncation, is suitable to use for the generation of an ECC private key. One would conjecture that it would therefore be just as sensible to concatenate enough output of the properly truncated ECRNG, and use these as a ECC private key, since presumably truncation and concatenation should not decrease the security.

6   Security of the Raw ECRNG Outputs Points

The traditional notion of security for an RNG is that its output is indistinguishable from random. We prove below in Theorem 2 that the raw output points of the ECRNG are indistinguishable from random points, that they are pseudorandom as points. Furthermore, we prove that the points are forward secure, as defined below.

The following proof is not substantially different than the proof for the Blum-Micali generator [1] or the Kaliski generator [2]. However, unlike these generators, which used a hardcore bit of the discrete logarithm, the ECRNG uses a hardcore function — as suggested, for example, in [14] — which yields greater efficiency provide that one accepts hardness of the corresponding problem, DDH, to ensure the function is hardcore. A second reason for providing the proof anew here is that the state update transition function is not a permutation. This issue is addressed via recourse to the hardness of the x-logarithm problem. Roughly speaking, hardness of XLP ensures that the state transition function is indistinguishable from a permutation.

In cryptology, *forward secrecy* refers to the following property: present secrets remain secret into the future, even from an adversary who acquires all future secrets. So, in forward secrecy, the secrecy of present secrets extends forward into the future indefinitely and without depending on protection of some future secrets. Many key agreement schemes, and even some digital signature schemes, claim forward secrecy. When implementing these schemes, one likely needs to ensure that any RNGs used have forward secrecy too. In [5,3], *forward secrecy* has been renamed[2] *backtracking resistance* to convey the notion that an adversary cannot use future secret to backtrack to present secrets.

To model forward secrecy, we let adversary see the latest prestate, but still it cannot distinguish previous output points from random points.

Theorem 2. *If the DDH and XLP problems are hard, and $Q, Z_0, \ldots, Z_m, Z_{m+1}$ are independent and uniformly distributed random points, and s_0 is a random integer uniformly distributed in $[0, n-1]$, and $g_m(Q, s_0) = (R_0, \ldots, R_m)$, with the next prestate of the ECRNG being S_{m+1}, then*

$$(Q, R_0, \ldots, R_m, S_{m+1}) \sim (Q, Z_0, \ldots, Z_m, Z_{m+1}). \tag{19}$$

Proof. The case of $m = 0$ is to show $(Q, R_0, S_1) \sim (Q, Z_0, Z_1)$. This follows directly from hardness of the DDH problem. Assume by induction that

$$(Q, R_0, \ldots, R_{m-1}, S_m) \sim (Q, Z_0, \ldots, Z_{m-1}, Z_m). \tag{20}$$

The current outputs and prestate are given by

$$(Q, R_0, \ldots, R_{m-1}, R_m, S_{m+1}) = (Q, R_0, \ldots, R_{m-1}, x(S_m)Q, x(S_m)P) \tag{21}$$

Combining (20) and (21), and applying Lemma 2, we get

$$(Q, R_0, \ldots, R_{m-1}, R_m, S_{m+1}) \sim (Q, Z_0, \ldots, Z_{m-1}, x(Z_m)Q, x(Z_m)P). \tag{22}$$

Hardness of XLP gives $x(Z_m)P \sim Z_{m+1}$. Writing $Q = qP$, Lemma 2 gives

$$(Q, Z_0, \ldots, Z_{m-1}, x(Z_m)Q, x(Z_m)P) \sim (qP, Z_0, \ldots, Z_{m-1}, qZ_{m+1}, Z_{m+1}). \tag{23}$$

Hardness of DDH gives $(qP, qZ_{m+1}, Z_{m+1}) \sim (Q, Z_m, Z_{m+1})$ where $Q = qP$, so Lemma 3 gives

$$(qP, Z_0, \ldots, Z_{m-1}, qZ_{m+1}, Z_{m+1}) \sim (Q, Z_0, \ldots, Z_{m-1}, Z_m, Z_{m+1}). \tag{24}$$

Lemma 1 on transitivity connects (22) to (23) to (24) to complete the inductive step, getting us our desired result.   □

---

[2] Breaking precedent not only with wider usage in cryptology but also with other ANSI standards such as X9.42 and X9.62, which use forward secrecy.

This proof makes essential use of Q being random. The reason for this is more than just to make the proof work. If Q is not random, then it may be the case the adversary knows a d such that $dQ = P$. Then $dR_i = dS_{i+1}$, so that such a distinguisher could immediately recover the secret prestates from the output. Once the distinguisher gets the prestates, it can easily distinguish the output from random. Therefore, it is generally preferable for Q to be chosen randomly, relative to P.

Although Theorem 2 says that hardness of the DDH problem is one of the *sufficient* conditions for indistinguishability of the ECRNG output points, it is not at all clear whether or not hardness of the DDH problem is a *necessary* condition. It is clear that hardness of the *computational* Diffie-Hellman problem (CDH) is a necessary condition in that S_{i+1} is the Diffie-Hellman product of P and R_i to the base Q.

Hardness of XLP, however, is necessary for indistinguishability of the raw output points. Output $R_1 = s_1 Q = x(S_1)Q$, so distinguishing it from random Z_1 is essentially XLP. Distinguishing output $R_j = x(S_{j-1})Q$ from random is almost XLP except that S_{j-1} is not necessarily a random point. However, if distinguishing algorithm A is an XLP solver, then one heuristically expects that A could distinguish R_j from a random point. Algorithm A would only fail if the S_{j-1} were distributed with a bias such that A reports that $x(S_{j-1})Q$ was not of the form $x(Z)P$ for some valid point Z. Therefore one cannot hope to strengthen Theorem 2 by replacing the hardness of XLP with a weaker yet still natural assumption. One could improve the result, however, by proving that XLP is as hard as some other problems, such as DDH or ECDLP.

7   Truncated Point Problem and Security of the Full ECRNG

For appropriate choice of truncation function $t(\cdot)$, we conjecture the following.

Conjecture 3. Let R be a random point and b a random bit string of length matching the output length of $t(\cdot)$. Then $t(x(R)) \sim b$.

We call the problem of distinguishing between $t(x(R))$ and b, the *Truncated Point Problem* (TPP). This paper does not substantially address this conjecture, but rather uses it to prove something about the final output of the ECRNG, rather than just its raw output points.

Theorem 3. *If the DDH, XLP and TPP problems are hard, then the ECRNG has forward secrecy.*

Proof. Apply Theorem 2 to get that the raw outputs are indistinguishable. By the assumed hardness of the TPP problem, each truncated point is indistinguishable from random bit strings. Apply the lemmas as necessary and get that the ECRNG output bit strings are indistinguishable from random bit strings, even from an adversary that gets to see the latest prestate. $\square$

Although hardness of XLP is necessary for the raw output points to be pseudo-random, it does not seem necessary for the full ECRNG output bit strings to be pseudorandom. Likewise, hardness of CDH may not be necessary for security of the full ECRNG, even if it is necessary for the indistinguishability of the raw output points. Truncation of the raw output points may yield bit strings are that unusable even by an XLP distinguisher or a CDH solver to distinguish the ECRNG outputs.

We note that it is straightforward to generalize the construction to any group G with suitable maps $x : G \to \mathbb{Z}_n$ and $t : G \to \{0,1\}^l$. If the corresponding DDH, XLP and TPP problems are hard, the generator will have forward secrecy. If the map x is a permutation, the corresponding XLP will be trivially hard.

The proposed truncation function drops some number of the leftmost bits of the bit representation of the x-coordinate. The number of bits dropped is at least $13 + \log_2(h)$, where h is the cofactor. The number of bits dropped must also be such that resulting length is a multiple of eight. Current (draft) standards allow any number of bits to be dropped that meets these conditions.

We consider if $B \sim t(x(R))$ where R is a random point and B is a random bit string whose output length l matches that of the truncation function. Let k be the number of bits truncated from $x(R)$, which has length $m = k + l$.

It is well-known that the advantage of any distinguisher is bounded above by the statistical distance between the distributions B and $t(x(R))$, and that the optimal distinguisher has advantage equal to the statistical distance. The statistical distance Δ between B and $t(x(R))$ is by definition

$$\Delta = \Delta(B, t(x(R))) = \sum_b |\Pr[t(x(R)) = b] - \Pr[B = b]|. \qquad (25)$$

The easier probability to compute is $\Pr[B = b] = 2^{-l}$ because all 2^l bit strings b are equally likely. The other probability has theoretical formula given by $\Pr[t(x(R))] = \frac{n(b)}{n}$, where $n(b)$ is the number of valid points R such that $t(x(R)) = b$. Note the $n(b)$ is always even, if we ignore the negligibly frequent case $R = 0$, because $x(R) = x(-R)$. Also, as k bits of the x-coordinate are truncated, we have $0 \leq n(b)/2 \leq 2^k$. Let B_i be the number of b such that $n(b) = 2i$. Then

$$\Delta = \sum_{i=0}^{2^k} B_i \left| \frac{2i}{n} - 2^{-l} \right|. \qquad (26)$$

Now we make some heuristic assumptions. Assume that the set X of valid x-coordinates is a random subset of bit strings of length $k + l$, such that each bit string belongs to X with probability $1/(2h)$, where h is the cofactor. Consider cofactor $h = 1$. Our first heuristic assumption implies B_i has a binomial distribution, so its approximate expectation is:

$$E(B_i) \approx 2^{l-2^k} \binom{2^k}{i}, \qquad (27)$$

where E is not to be confused with the elliptic curve equation. This distribution is because there are 2^k bit strings of length $k + l$ that truncate to a given bit string b of length l, and each of these completions of b has probability $\frac{1}{2}$ of belonging to X. Typically a few B_i may veer off considerably from the expected value. Nevertheless, by linearity of expectation, we can substitute these expected values into (25), getting expected statistical distance:

$$E(\Delta) \approx \sum_{i=0}^{2^k} 2^{l-2^k} \binom{2^k}{i} \left| \frac{2i}{n} - 2^{-l} \right|. \tag{28}$$

Take the approximation $n \approx 2^{l+k}$, to get a second heuristic assumption that $\frac{2i}{n} \approx \frac{2i}{2^{k+l}}$. Pulling a common factor through the sum gives

$$E(\Delta) \approx 2^{-2^k-k+1} \sum_{i=0}^{2^k} \binom{2^k}{i} \left| i - 2^{k-1} \right|. \tag{29}$$

The terms with $0 \leq i \leq 2^{k-1}$ are identical to those with $2^k \geq i \geq 2^{k-1}$, and the term with $i = 2^{k-1}$ is zero, so we can eliminate the absolute value signs, getting

$$E(\Delta) \approx 2^{-2^k-k+2} \sum_{i=0}^{2^{k-1}} \binom{2^k}{i} \left(2^{k-1} - i \right). \tag{30}$$

Using the general identity $i\binom{j}{i} = j\binom{j-1}{i-1}$, with a convention that $\binom{j-1}{-1} = 0$, gives

$$E(\Delta) \approx 2^{-2^k-k+2} \sum_{i=0}^{2^{k-1}} \left(2^{k-1}\binom{2^k}{i} - 2^k\binom{2^k-1}{i-1} \right). \tag{31}$$

Pulling out common factor 2^{k-1} from the sum and the general identity $\binom{j}{i} = \binom{j-1}{i-1} + \binom{j-1}{i}$ gives

$$E(\Delta) \approx 2^{-2^k+1} \sum_{i=0}^{2^{k-1}} \left(\binom{2^k-1}{i} - \binom{2^k-1}{i-1} \right). \tag{32}$$

This summation telescopes, giving

$$E(\Delta) \approx 2^{-2^k+1} \binom{2^k-1}{2^{k-1}} \tag{33}$$

For large even J, Stirling's approximation gives a third heuristic assumption that $\binom{J}{J/2} \approx \frac{2^J}{\sqrt{2\pi J}}$, and clearly $\binom{J-1}{J/2} = \frac{1}{2}\binom{J}{J/2}$. Applying this to (33) with $J = 2^k$ gives

$$E(\Delta) \approx \frac{1}{\sqrt{2\pi 2^k}} \tag{34}$$

as the heuristic value for the statistical distance. For $k = 16$, the heuristic for the expected statistical distance is about $\frac{1}{641}$. Although this is just a heuristic, it may be prudent to pay the price of using a larger k when a high degree of indistinguishability is desired. If only unpredictability is desired, say if the ECRNG is used for nonces or keys, but not as one-time pads, then this may not be so critical.

An optimal distinguisher can be constructed by computing $n(b)$, which has a computational cost of about 2^k times the cost of validating a potential x-coordinate. If $\frac{n(b)}{n} > 2^{-l}$ report $t(x(R))$, otherwise report B. Provided that k is small enough, then one heuristically expects that an efficient distinguisher exists with advantage of about that given by (34).

Rather than making all these highly heuristic assumptions, one can also gather some empirical data by sampling random B to infer an approximate distribution for $n(B)$. The inferred distribution can be used as estimates for the quantities B_i and thus the statistical distance Δ. An accurate inference requires a large random sampling. We have carried out quite extensive experiments, and they confirm the heuristic estimates.

One possibility for repairing the ECRNG is to use standard techniques for entropy extraction to process the raw outputs. In order to increase the asymptotic output rate, one may want to extract output from tuples of points instead of single points. This introduces a higher startup cost and requires a buffer to hold the points, but this can be tolerated in many practical applications.

Acknowledgments

The first author thanks the ANSI X9F1 working group for introducing him to the ECRNG, and Certicom colleagues for valuable discussions, especially Matt Campagna for a careful reading. The second author thanks John Kelsey for introducing him to the ECRNG. We would also like to thank the anonymous referees for helpful comments.

References

1. Blum, M., Micali, S.: How to generate cryptographically strong sequences of pseudo-random bits. SIAM Journal on Computing 13, 850–864 (1984)
2. Kaliski, B.S.: A pseudo-random bit generator based on elliptic logarithms. In: Odlyzko, A.M. (ed.) CRYPTO 1986. LNCS, vol. 263, pp. 84–103. Springer, Heidelberg (1987)
3. Barker, E., Kelsey, J.: Recommendation for Random Number Generation Using Deterministic Random Bit Generators. National Institute of Standards and Technology (2006), http://csrc.nist.gov/CryptoToolkit/RNG/SP800-90_June2006.pdf
4. Johnson, D.B.: X9.82 part 3 number theoretic DRBGs. Presentation at NIST RNG Workshop (2004), http://csrc.nist.gov/CryptoToolkit/RNG/Workshop/NumberTheoreticDRBG.pdf
5. Barker, E.: ANSI X9.82: Part 3 —2006, Random Number Generation, Part 3: Deterministic Random Bit Generators. American National Standards Institute (2006), Draft. http://www.x9.org/

6. Standards for Efficient Cryptography Group: SEC 1: Elliptic Curve for Cryptography. Draft 1.7 edn. (2006), http://www.secg.org/
7. Boneh, D.: The decision Diffie-Hellman problem. In: Buhler, J.P. (ed.) Algorithmic Number Theory. LNCS, vol. 1423, pp. 48–63. Springer, Heidelberg (1998), http://crypto.stanford.edu/~dabo/abstracts/DDH.html
8. Mahassni, E.E., Shparlinksi, I.: On the uniformity of distribution of congruential generators over elliptic curves. In: International Conference on Sequences and Their Applications, SETA '01, pp. 257–264. Springer, Heidelberg (2002)
9. Gürel, N.: Extracting bits from coordinates of a point of an elliptic curve. ePrint 2005/324, IACR (2005), http://eprint.iacr.org/
10. Schoenmakers, B., Sidorenko, A.: Cryptanalysis of the dual elliptic curve pseudorandom generator. ePrint 2006/190, IACR (2006), http://eprint.iacr.org/
11. Naor, M., Reingold, O.: Number-theoretic constructions of efficient pseudo-random functions. In: FOCS '97, pp. 458–467. IEEE Computer Society Press, Los Alamitos (1997), http://www.wisdom.weizmann.ac.il/~reingold/publications/GDH.PS
12. Farashahi, R.R., Schoenmakers, B., Sidorenko, A.: Efficient pseudorandom generators based on the DDH assumption. ePrint 2006/321, IACR (2006), http://eprint.iacr.org/
13. Luby, M.: Pseudorandomness and Cryptographic Applications. Princeton University Press, Princeton, NJ (1996)
14. Goldreich, O.: Foundations of Cryptography. Cambridge University Press, Cambridge (2001)
15. Smart, N.P.: A note on the x-coordinate of points on an elliptic curve in characteristic two. Information Processing Letters 80(5), 261–263 (2001)

A   Unpredictability of the Next State from the Current Output

Unpredictability is a much weaker property than indistinguishability, but is also much more important. If the ECRNG outputs are used as cryptographic keys, very little harm may come from them being distinguishable. If they are predictable, however, then all may be lost. Indistinguishability implies unpredictably, so in fact, we have already proven unpredictability.

The theorem below, however, proves a little bit of unpredictability under weaker, arguably more accepted, conjectures, such as hardness of CDH instead of the hardness of the DDH problem.

Theorem 4. *If CDH and XLP are hard, and q and s_0 are independent random integers uniformly distributed in $[0, n-1]$, and $g_m(qP, s_0) = (R_0, \ldots, R_m)$ and $Q = qP$, then an adversary who gets to see only Q and R_m cannot compute the next prestate S_{m+1}.*

Proof. Clearly $S_1 \sim Z$ where $Z = zP$ and z is a random integer uniformly distributed in $[0, n-1]$. Indeed, $s_0 \cong z$, so $S_1 = s_0 P \cong zP = Z$. Assume by induction that $S_{j-1} \sim Z$. Now $S_j = x(S_{j-1})P \sim x(Z)P \sim Z$, with the second indistinguishability flowing from the hardness of XLP. Therefore $S_{m+1} \sim Z$. Since q is independent of z, we have $(Q, S_{m+1}) \sim (Q, Z)$. Now $(Q, R_m) =$

$(Q, qS_{m+1}) \sim (Q, qZ) \sim (Q, Z)$, with the second indistinguishability flowing from Z being able to absorb q by independence.

Suppose adversary A takes (Q, R_m) and outputs $S_{m+1} = q^{-1}R_m$. Then adversary can also take (Q, Z) and output $U = q^{-1}Z$, because otherwise A could distinguish (Q, R_m) from (Q, Z). Let $(X, Y) = (xP, yP)$ with x, y independent random integers uniformly distributed in $[0, n-1]$. We will use A to compute xyP. Pick a random integer u with the same distribution. Let $U = uP$. Apply A to (X, U) to get $V = x^{-1}U = x^{-1}uP$. Let $W = u^{-1}V = x^{-1}P = wP$. Apply A to (W, Y) to get $w^{-1}Y = (x^{-1})^{-1}Y = xY = xyP$, as desired. Because we assumed that CDH is hard, adversary A cannot find xyP, so we get a contradiction. $\square$

The simple proof techniques above do not seem to rule out an adversary who could use two output points to find the next state, or one output point to find the next output point. The obstacle in the first case seems to be that output points obey a relationship that needs to be simulated if we wish to solve CDH. The obstacle in the second case is that the next output can be thought of as a one-way function of the Diffie-Hellman product of public values, and we seem to need to invert it to solve CDH.

B   A Caution About the Truncated Point Problem for Binary Curves

It should be noted that for the NIST recommended curves defined over the binary field $\mathbb{F}_{2^{409}}$, valid elliptic curve points have a fixed rightmost bit in their canonical representation. Therefore, for the curves B-409 and K-409, the truncation function should also drop the rightmost bit. The explanation for this phenomenon (see also [15]) is that one of the conditions for a point to have the correct order can be characterized by the trace of the x-coordinate have a fixed value. The trace depends on the field representation. For trinomial and pentanomial field representations, the trace simplifies to a sum of just a few of the bits, the *trace bits*, in the representation. In all fields, the rightmost bit is a trace bit. For the 409-bit field, the rightmost bit is the only trace bit. For the other four NIST recommended binary fields, there is at least one trace bit among the leftmost truncated bits. Consequently, the constant trace condition does not leak any information after truncation in these cases.

A Generalization of DDH with Applications to Protocol Analysis and Computational Soundness

Emmanuel Bresson[1], Yassine Lakhnech[2], Laurent Mazaré[3],
and Bogdan Warinschi[4,*]

[1] DCSSI Crypto Lab,
emmanuel@bresson.org
[2] VERIMAG Grenoble,
yassine.lakhnech@imag.fr
[3] Amadeus SAS,
laurent.mazare@m4x.org
[4] University of Bristol,
bogdan@cs.bris.ac.uk

Abstract. In this paper we identify the (P, Q)-DDH assumption, as an extreme, powerful generalization of the Decisional Diffie-Hellman (DDH) assumption: virtually all previously proposed generalizations of DDH are instances of the (P, Q)-DDH problem. We prove that our generalization is no harder than DDH through a concrete reduction that we show to be rather tight in most practical cases. One important consequence of our result is that it yields significantly simpler security proofs for protocols that use extensions of DDH. We exemplify in the case of several group-key exchange protocols (among others we give an elementary, direct proof for the Burmester-Desmedt protocol). Finally, we use our generalization of DDH to extend the celebrated computational soundness result of Abadi and Rogaway [1] so that it can also handle exponentiation and Diffie-Hellman-like keys. The extension that we propose crucially relies on our generalization and seems hard to achieve through other means.

Keywords: Diffie-Hellman Assumptions, Protocol Security, Provable Security, Computational Soundness.

1 Introduction

The Decisional Diffie-Hellman (DDH) assumption postulates that, even if given g^x and g^y, it is difficult for any feasible computation to distinguish between g^{xy} and g^r, when x, y and r are selected at random. The simplicity of the statement and several other nice properties (for example random self-reducibility)

---

* The work described in this paper has been supported in part by the European Commission through the IST Programme under Contract IST-2002-507932 ECRYPT. The information in this document reflects only the author's views, is provided as is and no guarantee or warranty is given that the information is fit for any particular purpose. The user thereof uses the information at its sole risk and liability.

A. Menezes (Ed.): CRYPTO 2007, LNCS 4622, pp. 482–499, 2007.

make the DDH assumption a powerful building block for cryptographic primitives and protocols. Examples of its use include provably secure public-key cryptosystems [13,11], pseudo-random functions [19,9], and pseudo-random generators [4]. The assumption has been particularly successful in the design of efficient and provably secure protocols for key-exchange: two parties can exchange a key by sending to each other g^x and g^y (for randomly chosen x and y). Pseudorandomness of the established common key g^{xy} is ensured by the DDH assumption.

Several generalizations of the DDH assumption naturally appear in the context of extending the above scenario from the two-party case to group key-exchange protocols. Perhaps the best known such generalization is the Group Decisional Diffie-Hellman assumption proposed by Steiner et al. [23] and refined by Bresson et al. [7]. Here, the assumption is that given all values $g^{\prod_i x_i}$, for up to $n-1$ exponents, it is hard to distinguish $g^{x_1 \cdots x_n}$ from a random power g^r. The assumption is sufficient to prove secure a protocol where users that privately select powers $x_1, x_2, \ldots x_n$ agree on a common shared key $g^{x_1 \cdots x_n}$. Such generalizations serve two goals. On the one hand they provide simple solutions to the problem that inspired them. More importantly, whenever such general assumptions can be reduced to a more standard assumption (as is the case of many generalizations of DDH), security proofs for protocols can be made more modular: First, prove once and for all the equivalence between the general and the basic assumption. Then, use the more general assumption as a more convenient basic building block for protocols. In this paper we investigate the limits of extending the DDH assumption. Our results are as follows.

GENERALIZATION OF DDH. Our generalization of DDH is as follows:

- The adversary receives elements of the form $g^{p(x_1,x_2,\ldots,x_n)}$; here p ranges over a fixed set of polynomials P. This setting generalizes significantly all of the previous work where only monomials were allowed in the exponents (*i.e.* the adversary was given only elements $g^{\prod_I x_i}$ for some subset I).
- The adversary receives several challenges of the form $g^{q(x_1,x_2,\ldots,x_n)}$; here q ranges over a fixed set of polynomials Q and the adversary has to determine if he is confronted with these challenges or random group elements. The adversary can see the challenges at any moment (*i.e.*, not necessarily at the end).

We call the problem associated to polynomial sets P and Q, the (P,Q)-DDH problem. In spite of its generality we show that the (P,Q)-DDH assumption reduces to the basic DDH assumption under several mild restrictions on the polynomials in P and Q. In particular, polynomials in Q have to be linearly independent from those in P since otherwise the problem becomes trivial. In general, the loss of security in the reduction that we provide from (P,Q)-DDH to DDH may be exponential. This is to be expected, and perhaps unavoidable, due to the general setting in which we work. Fortunately, we identify several situations where the security loss stays within practical bounds and note that all practical scenarios that we are aware of are instances of these situations. Furthermore, we show that the quality of the reduction can be often improved by using the random self reducibility property of DDH. We prove the equivalence with DDH via a hybrid argument which generalizes those used previously for other generalizations of DDH. We give the formal description of the (P, Q)-DDH problem and clarify its relation to basic DDH in Section 2.

APPLICATIONS TO PROTOCOL SECURITY. Next, we demonstrate the versatility of the (P, Q)-DDH assumption through several examples:

- we show that the multi-decisional Diffie-Hellman [6] and the Group Decisional Diffie-Hellman assumptions [7] are instances of the (P, Q)-DDH assumption for appropriately chosen P and Q. Interestingly, for the latter assumption our main theorem yields a better reduction to DDH than in previous works.
- we use the (P, Q)-DDH assumption to provide proofs for some DDH-based key-exchange protocols in the presence of *passive* adversaries. In particular, we supply a simple security proof for the Burmester-Desmedt protocol, and exemplify the use of our assumption for a simple protocol that we introduce.

Our examples show that the (P, Q)-DDH problem is an extremely convenient tool for proving the security of protocols in the presence of passive adversaries. In combination with generic results that map such protocols to protocols secure against active adversaries, our simple proofs form the basis of a powerful two-step methodology for the design of provably secure protocols. 1) Prove the protocol secure against passive adversaries using our flexible assumption 2) map the protocol to one secure against active adversaries using special purpose compilers such as the one developed by Katz and Yung for the case of group-key exchange protocols [14]. We develop the ideas sketched above in Section 3.

APPLICATION TO COMPUTATIONAL SOUNDNESS. Our final application is in the context of computational soundness framework. The general goal of this research direction is to allow symbolic, and thus mechanical reasoning about protocols at an abstract, symbolic level, in such a way that symbolically derived results imply security in the standard cryptographic sense. This would permit to prove the cryptographic security of protocols, but it would avoid the standard hand-made, error-prone cryptographic proofs through the use of automated tools.

In all of the prior work in this direction, the translation of results from the symbolic world to the cryptographic world is done using so-called "soundness theorems". Notice that these theorems have to deal with *all arbitrary uses* of the primitives in *all possible protocols*! This explains perhaps why exponentiation and Diffie-Hellman like keys are conspicuously missing from all existing computational soundness results: one needs to identify precisely, and in a generic way which of all possible uses of exponentiation are secure and which not. The main result of this paper accomplishes precisely that.

Based on our result we incorporate Diffie-Hellman keys in the framework proposed by Abadi and Rogaway [1]. We extend appropriately the symbolic language introduced in [1] and show that it is possible to use the resulting language to symbolically prove indistinguishability of cryptographic distributions. In particular, this result yields a mechanical way of proving security of key-exchange protocols (in the presence of passive adversaries, with no corruption). The symbolic language and the soundness theorem are in Section 4.

Related Work. A generalization of Diffie-Hellman to more general polynomials expressions was investigated by Kiltz in 2001 [15], where a (single) challenge of the form $g^{P(a,b)}$, with the adversary seeing g^a and g^b, is considered. We enlarge the setting in two distinct directions: first we allow many variables instead of just two (and thus, allow the adversary to "see" many polynomials in the exponent), second we allow multiple challenges. Moreover, we provide direct and concrete applications of our main results to the analysis of cryptographic protocols. We note that the work in [15] also studies the case of computational problems in generic groups [21]. Here we concentrate on the decisional case only, and use the standard cryptographic model. Essentially, all previous generalizations of DDH are particular case of our framework. This thus include the so-called "group Diffie-Hellman" assumptions [7], in which the challenge is $g^{x_1 \cdots x_n}$, but also the so-called "parallel Diffie-Hellman" assumption [6], in which the adversary sees $(g^{x_1}, \ldots, g^{x_n})$ and must distinguish tuples of the form $(g^{rx_1}, \ldots, g^{rx_n})$ from random ones $(g^{y_1}, \ldots, g^{y_n})$. Perhaps the closest assumption to the one that we study here is the General Diffie-Hellman Exponent (GDHE) introduced by Boneh et al. in the full version of [5]. We remark that GDHE has been designed to handle bilinear pairings, it has been designed with a single challenge, and its hardness has only been studied in the generic group model. Finally we notice that Square Exponent [16,10,22] and Inverse Exponent can [20] can be seen as instances of our setting.

2 A Generalization of the Decisional Diffie-Hellman Problem

2.1 The DDH Problem

A group family $\mathbb{G}$ is a set of finite cyclic groups $\mathbb{G} = \{G_\lambda\}$ where λ ranges over an infinite index set. We assume in the following that there exists a polynomial-time (in the bit-length of λ) algorithm that given λ and two elements in G_λ outputs their product. (We adopt the multiplicative notation for groups).

Let η be the security parameter. An Instance Generator IG for $\mathbb{G}$ is a probabilistic polynomial-time (in η) algorithm that outputs some index λ and a generator g of G_λ; therefore, IG induces a distribution on set of indexes λ. The Decisional Diffie-Hellman assumption states that for every probabilistic polynomial-time algorithm $\mathcal{A}$, every constant α and all sufficiently large η's, we have:

$$\left| \Pr\left[\mathcal{A}(\lambda, g, g^a, g^b, g^{ab}) = 1\right] - \Pr\left[\mathcal{A}(\lambda, g, g^a, g^b, g^c) = 1\right] \right| < \frac{1}{\eta^\alpha},$$

where the probabilities are taken over the random bits of $\mathcal{A}$, the choice of $\langle \lambda, g \rangle$ according to the distribution $IG(1^\eta)$ and the choice of a, b and c uniformly at random in $[1, |G_\lambda|]$.

In the remaining of the paper we will need to deal with concrete security results. We define the advantage of *any* algorithm $\mathcal{A}$ as the difference of probabilities above. We say that the DDH problem is (ϵ, t)-hard on $\mathbb{G}$ if the advantage of any algorithm running in time t is upper-bounded by ϵ. The (asymptotic) DDH assumption states it is the case for t polynomial and ϵ negligible (in η).

2.2 The (P,Q)-DDH Problem

Here we introduce formally our generalization of the Decisional Diffie-Hellman problem. As discussed in the introduction we generalize the DDH problem in two crucial directions. First, the group elements that the adversary sees are powers of g that are polynomials (instead of monomials as in the original problem and prior generalizations). Second the adversary is confronted with multiple challenges simultaneously. That is, his goal is to distinguish a list of values obtained by raising g to various polynomials from a list of random powers of g.

Let P and Q be two sets of polynomials in $\mathbb{Z}_q[X_1, X_2, \ldots, X_n]$. We assume that these sets are ordered, and write $p_1, p_2, \ldots$ and $q_1, q_2, \ldots$ for their elements, respectively. Informally, the (P, Q)-DDH-problem asks an adversary to distinguish the distributions:

$$\left(\{ g^{p_i(x_1, x_2, \ldots, x_n)} \}_{p_i \in P}, \{ g^{q_j(x_1, x_2, \ldots, x_n)} \}_{q_j \in Q} \right), \text{with } x_i \xleftarrow{\$} \mathbb{Z}_q \qquad (1)$$

$$\text{and } \left(\{ g^{p_i(x_1, x_2, \ldots, x_n)} \}_{p_i \in P}, \{ g^{r_j} \}_{j \in [|Q|]} \right), \text{with } x_i \xleftarrow{\$} \mathbb{Z}_q, r_j \xleftarrow{\$} \mathbb{Z}_q \qquad (2)$$

Notice that our generalization is quite powerful. All previous generalizations of the DDH problem can be seen as instances of the (P, Q)-DDH problem for suitably chosen P and Q. For example:

- For sets $P = \{X_1, X_2\}$ and $Q = \{X_1 X_2\}$, the associated (P, Q)-DDH is the standard DDH problem.
- For sets $P = \{ \prod_{i \in E} X_i \mid E \subsetneq [1, n] \}$ and $Q = \{X_1 X_2 \cdots X_n\}$ the associated (P, Q)-DDH problem corresponds to the group decisional Diffie-Hellman problem.
- For sets $P = \{X_1, X_2, \ldots, X_n\}$ and $Q = \{X_1 X_{n+1}, X_2 X_{n+1}, \ldots, X_n X_{n+1}\}$ the associated (P, Q)-DDH problem is the parallel Diffie-Hellman problem (see for instance [6]).

We call a pair of sets of polynomials (P, Q) a *challenge*. Our formalization of the (P, Q)-DDH problem departs from the more established formulations where an adversary is explicitly given as input samples from either distribution (1) or distribution (2) and has to decide which is the case. However here the size of sets P and Q may be exponential (for instance for the GDH problem the set P contains exponentially many polynomials), and yet we are typically interested in polynomial-time adversaries who may not have the time to read all the inputs. Therefore we provide the adversary with access to the two distributions via oracles.

Definition 1 ((P,Q)-DDH). *Let q be a prime number. Let $\mathbb{G}$ be a group of order q, g a generator of $\mathbb{G}$, and $P, Q \subseteq \mathbb{Z}_q[X_1, X_2, \ldots, X_n]$ two sets of polynomials. We define the oracles* $\mathsf{Real}_{(P,Q)}$ *and* $\mathsf{Fake}_{(P,Q)}$ *as follows. Both oracles first select uniformly at random* $x_i \xleftarrow{\$} \mathbb{Z}_q$, *for $i \in [n]$. Then they answer two types of* *queries. On input* $(\texttt{info}, i)$ *for* $1 \leq i \leq |P|$, *both* $\mathsf{Real}_{(P,Q)}$ *and* $\mathsf{Fake}_{(P,Q)}$ *answer with* $g^{p_i(x_1,x_2,...,x_n)}$. *On each new input* $(\texttt{chall}, j)$ *for some* $1 \leq j \leq |Q|$, *oracle* $\mathsf{Real}_{(P,Q)}$ *answers with* $g^{q_j(x_1,x_2,...,x_n)}$ *whereas oracle* $\mathsf{Fake}_{(P,Q)}$ *selects* $r_j \xleftarrow{\$} \mathbb{Z}_q$ *and answers with* g^{r_j}. *The adversary can intertwine* $\texttt{info}$ *and* $\texttt{chall}$ *queries. His goal is to distinguish between these two oracles.*

We define the advantage of an adversary $\mathcal{A}$ *to solve the* (P,Q)-DDH *problem by:*

$$\mathbf{Adv}_{\mathcal{A}}^{(P,Q)\text{-DDH}} = \left| \Pr\left[\mathcal{A}^{\mathsf{Real}_{(P,Q)}}(g) = 1\right] - \Pr\left[\mathcal{A}^{\mathsf{Fake}_{(P,Q)}}(g) = 1\right] \right|$$

where the probabilities are over the coins of the adversary and those used by the oracles. We say that the (P,Q)-DDH *problem is* (ϵ, t)-*hard in* $\mathbb{G}$, *if for any* $\mathcal{A}$ *running within time* t, $\mathbf{Adv}_{\mathcal{A}}^{(P,Q)\text{-DDH}} \leq \epsilon$.

2.3 Our Main Result: DDH Implies (P,Q)-DDH

Before giving our main theorem, we introduce some necessary notions and notations. For a polynomial p we write $mon(p)$ for the set of monomials occurring in p and write $var(p)$ for the set of variables that occur in p. The notation is naturally extended to sets of polynomials[1]. For a monomial m we denote by $ord(m)$ the order of m (i.e., the sum of the powers of its variables). We say p is *power-free* if any $X_i \in var(p)$ appears at power at most 1 (our results hold only for such polynomials). We write $\mathsf{PF}(\mathbb{Z}_q[X_1, X_2, \ldots, X_n])$ for the set of *power-free polynomials* with variables $\{X_1, \ldots, X_n\}$ and coefficients in $\mathbb{Z}_q$. Finally, we write $\mathsf{Span}(P)$ for the vector space over $\mathbb{Z}_q$ generated by P.

For some choice of the (P, Q) challenge, the (P, Q)-DDH problem is trivial (think of the case when $P = \{x_1, x_2\}$ and $Q = \{x_1 + x_2\}$). We therefore restrict the class of challenges only to the interesting cases where the polynomials in Q are linearly independent from those in P. Our main technical result will state that for all non-trivial challenges solving the (P, Q)-DDH problem reduces to solving DDH.

Definition 2 (Non-trivial challenge). *We say that challenge* (P, Q) *is non-trivial if* $\mathsf{Span}(P) \cap \mathsf{Span}(Q) = \{0\}$ *and polynomials in* Q *are linearly independent.*

First we identify a syntactic condition on the sets P and Q which ensures that the adversary has 0 advantage in breaking the (P, Q)-DDH problem. Our condition enforces that for these challenges, which we call *impossible challenges* the distribution of the $g^q(x_1, x_2, \ldots, x_n)$ (for all polynomials $q \in Q$) is statistically independent from the joint distribution $(g^p)_{p \in P}$. The definition is somewhat technical, and uses the graph $G_{(P,Q)}$ whose vertexes are $mon(P \cup Q)$, and in which there is an edge between monomials m_1 and m_2 if there exists $p \in P$ such that m_1, m_2 are in $mon(p)$. We denote by $mon_P^+(Q)$ the set of monomials reachable in this graph from $mon(Q)$ (that is, the strongly connected components of $G_{(P,Q)}$ containing $mon(Q)$). This set, informally, is the smallest superset

---

[1] For example, for set $P = \{X_1X_3 + X_1X_4, X_2 + X_1X_4\}$ it holds that $var(P) = \{X_1, X_2, X_3, X_4\}$, $mon(P) = \{X_2, X_1X_3, X_1X_4\}$.

of $mon(Q)$ that is stable through linear combinations with any polynomials of P containing a monomial of $mon_P^+(Q)$.

Definition 3 (Impossible Challenge). *We say that a non-trivial challenge (P, Q) is impossible if the two following conditions hold:*

1. $\forall m \in mon_P^+(Q), ord(m) = 1$: *all monomials in $mon_P^+(Q)$ are variables,*
2. $\forall m \in mon_P^+(Q), \forall m' \in mon(P)\backslash mon_P^+(Q), m \notin var(m')$: *any monomial that occurs in P but not in $mon_P^+(Q)$ cannot contain an element of $mon_P^+(Q)$ as a variable.*

The first requirement asks that all polynomials in Q are actually sums of variables. The second requirement asks that all polynomials in P either do not use any variable linked to Q (i.e. from $mon_P^+(Q)$) or are sums of variables. The next lemma formally captures that for all challenges that satisfy these two requirements no adversary can win the associated (P, Q)-DDH problem.

Lemma 4. *If (P, Q) is an impossible challenge then $\mathbf{Adv}_{\mathcal{A}}^{(P,Q)\text{-DDH}} = 0$ for all adversaries $\mathcal{A}$.*

STRATEGIES. The proof of our main theorem is based on a hybrid argument: it uses a sequence of transformations from a non-trivial challenge (P, Q) into an impossible challenge, such that if an adversary succeeds in the original challenge with significantly better probability than in the transformed challenge, then DDH is easy. In our formalization we use power-free polynomials with 2^α variables, that is polynomials in $\mathsf{PF}(\mathbb{Z}_q[X_1, X_2, \ldots, X_{2^\alpha}])$, for some natural number α. It is convenient to identify the index of variables with subsets of $[\alpha]$, and by a slight abuse of notation we identify X_i and $X_{\{i\}}$ (for each $i \in [\alpha]$). Thus, we regard $\mathbb{Z}_q[X_1, X_2, \ldots, X_\alpha]$ as $\mathbb{Z}_q[X_{\{1\}}, X_{\{2\}}, \ldots, X_{\{\alpha\}}]$.

Given a non-trivial challenge (P, Q) with $P, Q \subseteq \mathsf{PF}(\mathbb{Z}_q[X_1, \ldots, X_\alpha])$ we show how to build a sequence of challenges $(P_0, Q_0), (P_1, Q_1), \ldots, (P_l, Q_l)$, with $P_i, Q_i \in \mathsf{PF}(\mathbb{Z}_q[X_1, X_2, \ldots, X_{2^\alpha}])$ such that:

(i). $(P, Q) = (P_0, Q_0)$
(ii). for each adversary $\mathcal{A}$ against the (P_i, Q_i)-DDH there exists an adversary $\mathcal{B}$ against DDH such that:

$$\mathbf{Adv}_{\mathcal{A}}^{(P_i,Q_i)\text{-DDH}} \leq 2.\mathbf{Adv}_{\mathcal{B}}^{\text{DDH}} + \mathbf{Adv}_{\mathcal{A}}^{(P_{i+1},Q_{i+1})\text{-DDH}}$$

(iii). (P_l, Q_l) is an impossible challenge, so $\mathbf{Adv}_{\mathcal{A}}^{(P_l,Q_l)\text{-DDH}} = 0$

Our main result follows by finding an appropriate bound on the length l of the sequence.

One possible way to construct a sequence as above is as follows. Set (P_0, Q_0) to be (P, Q). To obtain (P_{i+1}, Q_{i+1}) out of (P_i, Q_i) we select a pair of variables X_u and X_v that occur together in some monomial in $mon(P \cup Q)$, and *merge* them into a new variable $X_{u \cup v}$. More precisely, in each monomial $m \in mon(P \cup Q)$ where both X_u and X_v occur, we remove these two variables and replace them with $X_{u \cup v}$. (Recall that variables are indexed by subsets of $[\alpha]$.) We call one such transformation a DDH *step*. The procedure ends when we obtain an impossible challenge (P_l, Q_l) (condition (iii) above). We call a sequence of DDH reductions as above a *strategy*, and we represent strategies as lists of pairs of variables $(X_{u_1}, X_{v_1}), ..., (X_{u_l}, X_{v_l})$, with $u_i, v_i \subseteq [\alpha]$ for all i. The *length* of a strategy is the length of the associated list. A strategy σ is *successful* for challenge (P, Q), if the result of applying σ to (P, Q) is an impossible challenge.

Example 5. Take $P = \{X_1, X_2, X_3\}$ and $Q = \{X_1 X_2 X_3\}$. A successful strategy for (P, Q) is $(X_1, X_2), (X_{1,2}, X_3)$. That is, in the first step we replace $X_1 X_2$ by $X_{1,2}$, and obtain $P_1 = P$ and $Q_1 = \{X_{1,2} X_3\}$. In the second step we replace $X_{1,2} X_3$ by $X_{1,2,3}$. The resulting challenge $(P, \{X_{1,2,3}\})$ is impossible.

The following lemma shows the obtained strategies satisfy condition (ii) above.

Lemma 6. *Let (P', Q') be a challenge obtained from challenge (P, Q) by a DDH step. Then for any adversary $\mathcal{A}$ there exists an adversary $\mathcal{B}$ such that:*

$$\mathbf{Adv}_{\mathcal{A}}^{(P,Q)\text{-DDH}} = 2.\mathbf{Adv}_{\mathcal{B}}^{\mathsf{DDH}} + \mathbf{Adv}_{\mathcal{A}}^{(P',Q')\text{-DDH}}$$

Moreover, if $t_{\mathcal{A}}$ is the execution time of $\mathcal{A}$, $N_{\mathcal{A}}$ is a bound on the number of oracle queries made by $\mathcal{A}$, then the execution time $t_{\mathcal{B}}$ of $\mathcal{B}$ is bounded by $t_{\mathcal{A}} + N_{\mathcal{A}} t_{(P,Q)}$, where $t_{(P,Q)}$ is (a bound on) the execution time of the oracle related to challenge (P, Q). If (P, Q) is a non-trivial challenge then (P', Q') is also a non-trivial challenge.

The previous two lemmas yield the following concrete security relation between DDH and (P, Q)-DDH.

Proposition 7. *Let $P, Q \in \mathsf{PF}(\mathbb{Z}_q[X_1, X_2, \ldots, X_\alpha])$ form a non-trivial challenge. If (P, Q) has a successful strategy of length n and if the DDH problem is (ϵ, t)-hard, then the (P, Q)-DDH is (ϵ', t')-hard, for $\epsilon' = 2n \cdot \epsilon$ and $t' + Nt_{(P,Q)} = t$ where N is a bound on the number of oracle queries and $t_{(P,Q)}$ a bound on the execution time of the oracle for challenge (P, Q).*

GENERIC STRATEGIES. We now exhibit a class of strategies, that we call *generic strategies* that are successful for arbitrary challenges (P, Q). Recall that there are two conditions for a challenge (P, Q) to be impossible: all the monomials of $mon_P^+(Q)$ must be variables and these variables must not occur in any other monomial of P. The idea behind generic strategies is rather simple. First, we change monomials in $mon_P^+(Q)$ into monomials of order 1 by successively merging variables. This leads to an intermediate challenge (P', Q') for which all monomials of $mon_{P'}^+(Q')$ are variables. Next, we deal with the fact that some variables in $mon_{P'}^+(Q')$ may occur elsewhere in P'. Then, for any variable x in $mon_{P'}^+(Q')$, if x appears in a monomial m of P' whose order is greater than 2, then m is transformed using a DDH step so that x does not appear anymore in m. After applying these two steps, we obtain an impossible challenge.

Example 8. Consider the challenge (P, Q) where P has as single element the polynomial $p = X_1 X_2 X_3 X_4$ and Q has as single element the polynomial $q =$

$X_1 X_2$. The first step transforms q into a variable by using the DDH step (X_1, X_2). The resulting challenge is $(P', Q') = (\{X_{1,2}X_3X_4\}, \{X_{1,2}\})$. Notice that $X_{1,2}$ appears in P', so we apply the DDH step $(X_{1,2}, X_3)$ and obtain the challenge $(\{X_{1,2,3}X_4\}, \{X_{1,2}\})$. This challenge is impossible therefore we found a successful strategy whose length is 2.

Next, we provide a bound on the length of generic strategies, which in turn gives an upper bound on the length of successful strategies. Let (P, Q) be an arbitrary challenge. First, we define the *order* of Q within P which we denote by $ord_P^+(Q)$. This quantity is defined by $ord_P^+(Q) = \sum_{m \in mon_P^+(Q)} \left(ord(m) - 1 \right)$.

The set $\mathsf{nm}(P, Q)$ of non-maximal elements of $mon_P^+(Q)$ is the set of monomials m which appear in $mon_P^+(Q)$ such that there exists a monomial m' that verifies the following two requirements:

1. m is a strict sub-monomial of m': all the variables of m appear in m' and m is different from m'.
2. m' is in $mon(P)$ but is not in $mon_P^+(Q)$.

Example 9. We still consider the challenge (P, Q) where P contains one element $p = X_1 X_2 X_3 X_4$ and Q has one element $q = X_1 X_2$. Then $mon_P^+(Q)$ contains only q. Moreover q is not maximal because $p = qX_3 X_4$ hence the set of non-maximal elements $\mathsf{nm}(P, Q)$ is also equal to $\{q\}$.

We are able to show that for any non-trivial challenge there exist strategies whose length can be upper-bounded.

Proposition 10 (Bounded strategies). *For any non-trivial challenge (P, Q), there exists a successful strategy of length:*

$$ord_P^+(Q) + \left(2^{|\mathsf{nm}(P,Q)|} - 1 \right) . \left(\alpha + ord_P^+(Q) \right)$$

Combined with Proposition 7, we obtain our main theorem:

Theorem 11 (Relating (P, Q)-DDH to DDH). *Let (P, Q) be a non-trivial challenge on variables X_1 to X_α. If the DDH problem is (ϵ, t)-hard, then (P, Q)-DDH is (ϵ', t')-hard, for*

$$\epsilon' = 2\epsilon \left(ord_P^+(Q) + \left(2^{|\mathsf{nm}(P,Q)|} - 1 \right) . \left(\alpha + ord_P^+(Q) \right) \right)$$

and $t' + N t_{(P,Q)} = t$ where N is a bound on the number of oracle queries.

Several remarks are in order. We restrict challenges to sets of power-free polynomials. Extending our result beyond this class, would require dealing with group elements of the form g^{X^2}. This seems to be a difficult problem since, for instance, the indistinguishability of (g^x, g^{x^2}) and (g^x, g^r) under the DDH assumption is an open problem [2]. On the other hand, we can easily lift the requirement that polynomials in Q are linearly independent, and modifying appropriately the behavior of the $\mathsf{Fake}_{(P,Q)}$ oracle. We choose to use the current formulation for simplicity.

The formulation of our theorem implies that in the worst case, the loss of security in our reduction may be exponential. We note however that in most, if not all, practical cases $\mathsf{nm}(P, Q)$ is empty, and in those cases the loss in security is only linear. Moreover, notice that in the case when P and Q contain only monomials the hypothesis of the theorem implies that $mon_P^+(Q) = mon(Q) = Q$ and $ord_P^+(Q) = \sum_{m \in Q} (ord(m) - 1)$. In the next section we consider a few examples where our theorem gives linear security reductions in several interesting applications. However for some applications (like the Burmester-Desmedt protocol), better reductions can be found using the random self-reducibility property of DDH.

RANDOM SELF-REDUCIBILITY. As said above, the DDH problems has the nice property to be *Random Self-Reducible* (RSR for short). Roughly, this property means that an efficient algorithm for the average case implies an efficient algorithm for the worst case. In the case of DDH, when randomizing an instance, one gets instances, which (1) are uniformly distributed, (2) have all the same solution as the original instance. Thus, being able to solve a single random instance implies that we can solve any instance. As an illustration, let $(X, Y, Z) = (g^x, g^y, [g^{xy}|g^z])$ be an instance of DDH (the notation $Z = [A|B]$ means that the problem is to decide whether Z equals A or B). It is easy to see that for α and β chosen at random, $(X^\alpha, Y^\beta, Z^{\alpha\beta})$ is a new, random instance with the same (decision) solution than the original one.

Here we use RSR as introduced in lemma 5.2 of [3]: from $(g^x, g^y, [g^{xy}|g^z])$ we generate two new instances of DDH: $(g^{\alpha x}, g^y, [g^{\alpha xy}|g^{\alpha z}])$, where α is randomly sampled in $\mathbb{Z}_q$ and $(g^{\alpha x}, g^{\beta y}, [g^{\alpha\beta xy}|g^{\alpha\beta z}])$ where α and β are sampled in $\mathbb{Z}_q$. Using this, we are able to lower the bound given in proposition 7 by giving a finer definition of the weight of a sequence. The idea is that multiple steps can be combined in a single step using RSR. A strategy $(X_{u_1}, X_{v_1}), \ldots, (X_{u_k}, X_{v_k})$ is said to be *randomly self-reducible* (RSR) for a challenge (P, Q) if:

- For step i, X_{u_i} and X_{v_i} have not been introduced in previous steps: for any $j < i$, u_i and v_i are different from $u_j \cup v_j$.
- For step i, X_{v_i} has to be fresh, i.e. this variable was never used in previous steps: for any $j < i$, v_i is different from u_j and v_j.
- Let X_u and X_v be two distinct variables from the strategy, if the product $X_u.X_v$ occurs in P or Q, then there exists a step i such that $u = u_i$ and $v = v_i$ (or $u = v_i$ and $v = u_i$).

Then the weight of such a sequence is 1 as it only counts as a single step and we can extend the result of lemma 6. The idea is that all the kind of "independence" of variables captured by the above conditions allows us to use a single DDH challenge to deal with all the steps (X_{u_i}, X_{v_i}) at once. Formally, we have the following:

Lemma 12. *Let (P', Q') be a challenge obtained from challenge (P, Q) by a RSR strategy. Then for any adversary $\mathcal{A}$ there exists an adversary $\mathcal{B}$ such that:*

$$\mathbf{Adv}_{\mathcal{A}}^{(P,Q)\text{-DDH}} = 2.\mathbf{Adv}_{\mathcal{B}}^{\text{DDH}} + \mathbf{Adv}_{\mathcal{A}}^{(P',Q')\text{-DDH}}$$

Moreover, if $t_{\mathcal{A}}$ is the execution time of $\mathcal{A}$, $N_{\mathcal{A}}$ is a bound on the number of oracle queries made by $\mathcal{A}$, then the execution time $t_{\mathcal{B}}$ of $\mathcal{B}$ is bounded by $t_{\mathcal{A}} + N_{\mathcal{A}} t_{(P,Q)}$, where $t_{(P,Q)}$ is (a bound on) the execution time of the oracle related to challenge (P,Q).

We exemplify the use of RSR strategies in obtaining better reductions in Section 3 for the case of the Burmester-Desmedt protocol.

3   Applications: Simple Proofs for Diffie-Hellman-Based Protocols

In this section, we show the applicability of our main theorem in a few different contexts. First we apply it to reprove equivalence between the Group DDH problem and basic DDH. Our result yield a tighter security reduction than previous result. As explained in the introduction, our theorem can be used to easily obtain relation between the hardness of DDH and various of its extensions. To illustrate the simplicity associated to using the (P, Q)-DDH assumption we show how to use it to link the reverse DDH assumption (which we introduce) and basic DDH. Finally, we demonstrate that our theorem yields simpler proofs of security for group key-exchange protocols in the presence of passive adversaries, and we show how to obtain a proof for the Burmester-Desmedt protocol.

Throughout this section, we work in a group in which the DDH problem is (ϵ, t)-hard and work with polynomials with α variables $X_1, \ldots, X_\alpha$ (we assume α to be equal to the security parameter.)

GDDH. The Group Decisional Diffie-Hellman (GDDH) problem [23] can be formalized with the challenge (P, Q):

- $P = \{\prod_{i \in E} X_i \mid E \subsetneq [1, \alpha]\}$, that is, P contains all the monomials of order up to $\alpha - 1$.
- $Q = \{\prod_{1 \leq i \leq \alpha} X_i\}$.

Clearly, $\mathsf{Span}(P) \cap \mathsf{Span}(Q) = \{0\}$ and therefore we can apply Theorem 11. Notice that sets P and Q contain only monomials and since $X_1 X_2 \cdots X_\alpha$ is trivially maximal in P, it follows that the (P, Q)-DDH problem is (ϵ', t')-hard, with $\epsilon' = 2(\alpha - 1)\epsilon$ and $t' = t - N t_{(P,Q)} \geq t - t' t_{(P,Q)}$. Thus t' is greater than $t/(1 + t_{(P,Q)})$. Moreover, when calling the oracle, the worst case consists in generating all the X_i and multiplying them, which can be done time polynomial in α. Our results contrasts with that of [7] where the reduction is linear but requires an exponential time in α.

REVERSE GDDH. We illustrate the use of non-maximal elements through an example that we call the Reverse GDDH problem. This problem is given by the challenge (P, Q):

- $P = \{\prod_{i \in E} X_i \mid E \subseteq [1, \alpha] \wedge E \neq \{1\}\}$, that is, P contains all the possible monomials except X_1.
- $Q = \{X_1\}$.

Since X_1 is not maximal in P we have that $|\mathsf{nm}(P, Q)| = 1$. By Theorem 11 we obtain that the loss of security is $\epsilon' = 2\alpha\epsilon$, which is linear in the security parameter.

THE BURMESTER-DESMEDT PROTOCOL. Introduced in [8] and later analyzed in [14], this protocol is a two-round key exchange protocol between α parties. In the first round, each user U_i samples a random X_i and broadcasts g^{X_i}. In the second round, U_i broadcasts $g^{X_i X_{i+1} - X_{i-1} X_i}$ (with the convention that $X_0 = X_\alpha$ and $X_{\alpha+1} = X_1$). The common secret is $g^{X_1 X_2 + \cdots + X_\alpha X_1}$.

Recall that in the passive setting, security of such a group key-exchange protocol is roughly modeled as follows. First the (passive) adversary observes bit-strings for the different messages exchanged by the participants (using so-called Execute queries). At some point the adversary decides to challenge the shared secret by trying to distinguish that secret from a random element (the so-called Test query). The adversary is allowed to intertwine his queries. The model, actually corresponds to the (P, Q)-DDH assumption, where the polynomials that correspond to the messages sent by parties are placed in P and the polynomial that corresponds to the shared secret is in Q. Therefore the (P, Q)-DDH assumption that corresponds to polynomials:

- $P = \{X_i \mid 1 \leq i \leq \alpha\} \cup \{X_i X_{i+1} - X_{i-1} X_i \mid 1 \leq i \leq \alpha\}$ corresponds to the broadcast messages.
- $Q = \{\sum_{i=1}^{\alpha} X_i X_{i+1}\}$ corresponds to the shared secret.

is equivalent to the security of the Burmester Desmedt protocol against passive adversaries.

It is easy to check that $\mathsf{Span}(P) \cap \mathsf{Span}(Q) = \{0\}$ (see for instance [14]). Here again Q has only one element and this element is maximal in P. We get $ord_P^+(Q) = \alpha$ and after applying Theorem 11, $\epsilon' = 2\alpha\epsilon$, that is we obtain a linear reduction.

The reduction factor obtained through the use of Theorem 11 is based on generic strategies and is not optimal. Next we show that it is possible to use RSR strategies to obtain better reduction factors (essentially matching the ones that appear in [14]). For simplicity, we assume that α is a multiple of 3. The assumption does not change the asymptotic factors obtained through the reduction bellow. We proceed in two steps: First, we apply the RSR strategy:

$$(X_1, X_2)(X_4, X_5) \ldots (X_{3i+1}, X_{3i+2}) \ldots$$

Let (P', Q') be the resulting challenge. Finally, by applying the following RSR strategy

$$(X_2, X_3)(X_3, X_4)(X_5, X_6)(X_6, X_7) \ldots (X_{3i+2}, X_{3i+3}), (X_{3i+3}, X_{3i+4}) \ldots$$

we obtain an impossible challenge. Using Lemma 12 twice, we get that for any adversary $\mathcal{A}$ against (P, Q)-DDH there exists an adversary $\mathcal{B}$ (of similar time complexity) against DDH such that $\mathbf{Adv}_{\mathcal{A}}^{(P,Q)\text{-DDH}} = 4\mathbf{Adv}_{\mathcal{B}}^{\text{DDH}}$.

CENTRALIZED DIFFIE-HELLMAN. We introduce a toy group key exchange protocol in order to illustrate how our results can be used to easily prove such new protocols. This key distribution protocol considers $\alpha - 2$ users $U_1, \ldots, U_{\alpha-2}$ and a server S. Each user U_i randomly samples a group element X_i, while the server S samples two group elements $X_{\alpha-1}, X_\alpha$. Then each user U_i sends g^{X_i} to S and receives $g^{X_\alpha + X_i X_{\alpha-1}}$. The server also broadcasts $g^{X_{\alpha-1}}$. The shared secret is g^{X_α}. The security of the shared key is captured by the challenge (P, Q), where:

- $P = \{X_i \mid 1 \le i \le \alpha - 1\} \cup \{X_\alpha + X_i X_{\alpha-1} \mid 1 \le i \le \alpha - 2\}$ corresponds to the broadcast messages.
- $Q = \{X_\alpha\}$ corresponds to the shared secret.

Each monomial $X_i X_{\alpha-1}$ appears only once, thus $\mathsf{Span}(P) \cap \mathsf{Span}(Q) = \{0\}$. The set Q has only one element and this element which is maximal in P. Thus $mon(Q) = \{X_\alpha\}$ and $mon_P^+(Q) = \{X_\alpha, X_1 X_{\alpha-1}, \ldots, X_{\alpha-2} X_{\alpha-1}\}$ from which it follows that $ord_P^+(Q)$ is $\alpha - 2$. The loss of security in the reduction is thus only linear.

4   A Symbolic Logic for Diffie-Hellman Exponentials and Encryption

In this section we give a symbolic language for representing messages formed by using nonces, symmetric encryption and exponentiation. In some sense, the language that we give in this section is a formal "notation" for distributions. This notation has the crucial property that it can be used to *automatically* reason about the indistinguishability of distributions that arise in cryptographic protocols, without resorting to reduction proofs. For example, using this language, one can define and reason about the security of keys in multicast protocols (see for example [17]) in a way that is meaningful to standard cryptographic models. The main ingredient that enables for such results is a soundness theorem which explains how results at the abstract level of the notation that we introduce map to results about the indistinguishability of distributions.

SYNTAX. First we make precise the set of symbolic messages that we consider. Let **Keys**, **Nonce** and **Exponents** be three countable disjoint sets of symbols for keys, random nonces, and exponents. We let **Poly** be the set of power-free polynomials with variables in **Exponents** and coefficients in $\mathbb{Z}_q$. The set **Msg** of message expressions is defined by the following grammar:

$$\mathbf{Msg} ::= \mathbf{Keys} \mid \mathbf{g}^{\mathbf{Poly}} \mid \mathbf{Nonce} \mid (\mathbf{Msg}, \mathbf{Msg}) \mid \{\mathbf{Msg}\}_{\mathbf{Keys}} \mid \{\mathbf{Msg}\}_{h(\mathbf{g}^{\mathbf{Poly}})}$$

Equality for expressions is defined modulo polynomial equality. For example, let p and q be two polynomials from **Poly** such that $p = q$ (for classical polynomial equality, e.g. $p = X_1 + X_2 + X_1$ and $q = 2X_1 + X_2$), then $\mathbf{g}^p = \mathbf{g}^q$.

COMPUTATIONAL INTERPRETATION. One should think of the elements of **Msg** as symbolic representation for (ensembles of) distributions . For instance, elements of **Keys** represent (the distributions of) cryptographic keys obtained by running the key generation algorithm of some (fixed) encryption scheme. A term like $\mathbf{g}^X$ represents the distribution of g^x when exponent x is chosen at random, and $h(\mathbf{g}^{X_1 X_2})$ represents the distribution of keys obtained by applying a hash function to $g^{x_1 x_2}$ for random x_1 and x_2. A slightly more complex example is the expression: $(g^x, g^y, \{K\}_{h(g^{xy})})$ that represents the distribution of a conversation between two parties that first exchange a Diffie-Hellman key, and then use this key to encrypt a symmetric key.

Let us precise how symbolic expressions are mapped to distributions. Consider a symmetric encryption scheme $\Pi = (\mathcal{KG}, \mathcal{E}, \mathcal{D})$, a family of groups $\mathbb{G} = (\mathbb{G}_\eta)_{\eta \in \mathbb{N}}$ which come with a publicly known generator $\mathfrak{g}$ for each security parameter, and an efficiently computable function $h : \mathbb{G}_\eta \to \{0,1\}^\eta$ to derive cryptographic keys out of exponentials.

We associate to each expression $E \in \mathbf{Msg}$ and security parameter $\eta \in \mathbb{N}$ a distribution $\widehat{E}$ (to avoid cluttered notation we omit to show the dependency on $\Pi, \mathbb{G}$ and η.) We define this distribution as the output of the following randomized algorithm: For each key symbol K that occurs in E we generate a value $\widehat{K} \xleftarrow{\$} \mathcal{KG}(\eta)$; for each variable $X_i \in \mathbf{Exponents}$ we select $\widehat{X} \xleftarrow{\$} \{1, \ldots, |\mathbb{G}_\eta|\}$; for every nonce $N \in \mathbf{Nonce}$ we select $\widehat{N} \xleftarrow{\$} \{0,1\}^\eta$. The output $\widehat{E}$ is computed inductively on the structure of E: $\widehat{(E_1, E_2)} = \widehat{E_1}.\widehat{E_2}$, $\widehat{\mathfrak{g}^{p(X_1, \ldots, X_n)}} = \mathfrak{g}^{p(\widehat{X_1}, \ldots, \widehat{X_n})}$, $\widehat{\{E\}_K} = \mathcal{E}(\widehat{E}, \widehat{K})$, and $\widehat{\{E\}_{h(\mathfrak{g}^p)}} = \mathcal{E}(\widehat{E}, h(\widehat{\mathfrak{g}^p}))$.

THE SYMBOLIC ADVERSARY. Now we explain how one can reason symbolically about secrecy of message in expressions. Security of encryption in symbolic messages is captured by an axiomatically defined deduction relation $\vdash$. The $\vdash$ relation defines precisely when an expression $E \in \mathbf{Msg}$ can be deduced from a finite set of expressions $S \subseteq \mathbf{Msg}$ (written $S \vdash E$) by a passive eavesdropper. The deduction relation $\vdash$ is an extension of the standard Dolev-Yao inference system [12] and is given by the following rules:

$$\frac{E \in S}{S \vdash E} \qquad \frac{S \vdash (E_1, E_2)}{S \vdash E_1} \qquad \frac{S \vdash (E_1, E_2)}{S \vdash E_2} \qquad \frac{S \vdash \{E\}_K \quad E \vdash k}{S \vdash E}$$

We only consider deduction rule in this axiomatisation. Indeed the $\vdash$ relation is only used to check that a key or an exponentiation can be deduced, thus composition rules are useless.

To the standard Dolev-Yao rules that capture security of encryption, we add several rules for dealing with exponentials, and keys derived from exponentials:

$$\frac{}{E \vdash \mathfrak{g}^1} \qquad \frac{E \vdash \mathfrak{g}^p \quad E \vdash \mathfrak{g}^q}{E \vdash \mathfrak{g}^{\lambda p + q}} \lambda \in \mathbb{Z}_q \qquad \frac{E \vdash \{m\}_{h(\mathfrak{g}^p)} \quad E \vdash \mathfrak{g}^p}{E \vdash m}$$

The first rule says that the adversary knows the generator $\mathfrak{g}$ of the group; the second says that the adversary can multiply group elements that it knows, and raise group elements that it knows to arbitrary powers in $\mathbb{Z}_q$. The last rule allows the adversary to decrypt a ciphertext under a key derived from an exponential, provided that the adversary can compute that exponential.

SYMBOLIC EQUIVALENCE OF EXPRESSIONS. In a symbolic expression, the information revealed via $\vdash$ can be characterized using *patterns* [1,17]. Intuitively, the pattern of expression $E \in \mathbf{Msg}$ is obtained by replacing all its unrecoverable subexpressions (those sub-expressions that occur encrypted under keys that the adversary cannot derive from E) by the symbol $\square$ (undecryptable). For an expression $E \in \mathbf{Msg}$ its pattern is formally defined by the following inductive rules:

$$pattern\big((E_1, E_2)\big) = \big(pattern(E_1), pattern(E_2)\big)$$

$$\begin{aligned}
pattern\big(\{E'\}_K\big) &= \{pattern(E')\}_K && \text{if } E \vdash K \\
pattern\big(\{E'\}_K\big) &= \{\square\}_K && \text{if } E \not\vdash K \\
pattern\big(\{E'\}_{h(\mathbf{g}^p)}\big) &= \{pattern(E')\}_{h(\mathbf{g}^p)} && \text{if } E \vdash \mathbf{g}^p \\
pattern\big(\{E'\}_{h(\mathbf{g}^p)}\big) &= \{\square\}_{h(\mathbf{g}^p)} && \text{if } E \not\vdash \mathbf{g}^p \\
pattern(E') &= E' && \text{if } E' \in \mathbf{Nonce} \cup \mathbf{Keys} \cup \mathbf{g}^{\mathbf{Poly}}
\end{aligned}$$

Two expressions $E_1, E_2 \in \mathbf{Msg}$ are deemed symbolically equivalent if they have the same pattern (an adversary can gather the same information out of both expressions): $E_1 \equiv E_2$ if and only if $pattern(E_1) = pattern(E_2)$.

We would like to claim that equivalent expressions have associated indistinguishable distributions. However, the equivalence defined above is too stringent: For example, expressions $(K_1, \{K_1\}_{K_2})$ and $(K_2, \{K_2\}_{K_3})$ are different, although they clearly have equal distributions. The solution is to relax the equivalence by allowing renaming of key and nonce symbols (and even renaming of polynomials). The above expressions become equivalent by renaming (in the first expression) K_1 and K_2 to K_2 and K_3, respectively.

Renaming the polynomials that occur in exponentials is more subtle. Notice that we would like to identify the expressions $E_1 = (\mathbf{g}^{X_1}, \mathbf{g}^{X_2}, \mathbf{g}^{X_1 X_2})$ and $E_2 = (\mathbf{g}^{X_1}, \mathbf{g}^{X_2}, \mathbf{g}^{X_3})$ by renaming the polynomial $X_1 X_2$ to the polynomial X_3 (this models the DDH assumption). However not all renamings of polynomials should be considered valid: the expression E_1 and $E_3 = (\mathbf{g}^{X_1}, \mathbf{g}^{X_2}, \mathbf{g}^{X_1 + X_2})$ (which are distinguishable since the linear dependency that the adversary can observe in the second expression is absent in the first expression) should not be made indistinguishable by mapping for example $X_1 X_2$ to $X_1 + X_2$. Based on the intuition that underlies our main theorem, we only consider *linear dependence preserving injective* renamings of polynomials (which are renamings that preserve all linear dependencies in the original expression).

Definition 13 (Linear dependance preserving renamings). *Let E be an expression and $\sigma : poly(E) \to \mathbf{Poly}$ be an injective renaming of the polynomials in E. Then σ is said to be linear dependence preserving (ldp) if:*

$$\forall p_1, p_2, \ldots, p_n \in poly(E), \ \forall a_1, \ldots, a_n, b \in \mathbb{Z}, \ \sum_{i=1}^{n} a_i.p_i = b \Leftrightarrow \sum_{i=1}^{n} a_i.p_i \sigma = b$$

For the expression E_1 given above, it can be verified that σ defined by $\sigma(X_1) = X_1, \sigma(X_2) = X_2$ and $\sigma(X_1 X_2) = X_3$ is ldp whereas if we set $\sigma(X_1 X_2) = X_1 + X_2$, σ the resulting renaming is not.

We say that two expressions E_1 and E_2 are *equivalent up to renaming*, and we write $E_1 \cong E_2$ if there exists a renaming σ that is injective on the sets of nonces, keys, and injective and dependence preserving on the set of polynomials, such that $\sigma(m) \equiv n$.

SOUNDNESS THEOREM. We are now ready to state our soundness theorem. Similarly to the original paper of Abadi and Rogaway [1], we implement encryption using a scheme that besides being IND-CPA secure also hides the length of the plaintext. We write IND-CPA* for the resulting security notion. We emphasize that we use the additional requirement only for simplicity – this requirement can be easily lifted by refining the pattern definition as in [18,17]. The implementation that we consider uses a family of groups where the DDH problem is (asymptotically) hard. Finally, we require that the key derivation function h is such that $\mathcal{KG}(\eta)$ and $h(\mathfrak{g}^r)$ output equal distributions when r is selected at random. The soundness result holds for acyclic expressions, that is expressions where encryption cycles do not occur.

Definition 14 (Acyclic expression). *An expression E is acyclic if the two following conditions are satisfied:*

1. *If p is a polynomial such that $h(\mathfrak{g}^p)$ occurs as an encryption key in E, then p is not a linear combination of the polynomials that occur in E (and are different from p).*
2. *There exists an order $\prec$ between keys and polynomials from $poly(E)$: if u appears encrypted using v or $\mathbf{g}^v$ then $u \prec v$. This order must not have any cycles.*

The first condition is intended to avoid encryptions in which the plaintext and the encryption key are linearly dependent, as for example in $\{\mathbf{g}^{X_1}, \mathbf{g}^{X_1+X_2}\}_{h(\mathbf{g}^{X_2})}$. It can be easily shown that the occurrence of such a ciphertext can reveal the encryption key without contradicting IND-CPA-security of the encryption scheme.

The next theorem establishes the main result of this section: the distributions of equivalent expressions are computationally indistinguishable.

Theorem 15 (Symbolic equivalence implies indistinguishability). *Let E_1 and E_2 be two acyclic expressions, such that $E_1 \cong E_2$. Let Π be a symmetric encryption scheme that is IND-CPA* secure and $\mathbb{G}$ be a group such that the DDH assumption holds, then $\widehat{E_1} \approx \widehat{E_2}$.*

To appreciate the power that the above soundness theorem provides, consider the expression:

$$E(F) = \left(\mathbf{g}^{X_1}, \mathbf{g}^{X_2}, \mathbf{g}^{X_3}, \mathbf{g}^{X_1 X_2}, \mathbf{g}^{X_1 X_3}, \mathbf{g}^{X_2 X_3}, \{K\}_{h(\mathbf{g}^{X_1 X_2 X_3})}, \{F\}_K\right)$$

where F is some arbitrary expression. Expression E represents the transcript of the executions of the following (toy) protocol: three parties with secret keys X_1, X_2 and X_3 first agree on a common secret key $h(\mathbf{g}^{X_1 X_2 X_3})$ (by broadcasting the first 6 messages in the expression). Then, one of the parties generates a new key K which it broadcasts to the other parties encrypted under $h(\mathbf{g}^{X_1 X_2 X_3})$. Finally, one of the parties, sends some secret expression F encrypted under K. To argue about the security of the secret expression (against a passive adversary) it is sufficient to show that the distributions associated to the expressions $E(F)$ and $E(0)$ are indistinguishable.

Although conceptually simple, a full cryptographic proof would require several reductions (to DDH and security of encryption), and most likely would involve at least one hybrid argument (for proving the security of encrypting K under $h(g^{X_1 X_2 X_3})$). The tedious details of such a proof can be entirely avoided by using our soundness theorem: it is straightforward to verify that $E(F) \cong E(0)$, and this procedure can be automated. Since $E(F)$ is acyclic, the desired result follows immediately by Theorem 15.

5   Conclusion

In this paper we propose a significant generalization of the DDH problem. We show that in most of the important cases our generalization is not harder than the classical two-parties DDH. As applications, we demonstrate that our generalization enables simple and tight security proofs for several existing key exchange protocols. Moreover, the generalization is instrumental in obtaining a computational soundness theorem that deals with exponentiation and Diffie-Hellman-like keys. We leave as an interesting open problem the question of how to extend this last result to the case of active adversaries.

Acknowledgments. We would like to thank Mihir Bellare, Mathieu Baudet and anonymous reviewers for useful comments and suggestions. Some of the research was carried out while the fourth author was with LORIA, INRIA Lorraine in the CASSIS group. He was supported by ACI Jeunes Chercheurs JC9005 and ARA SSIA Formacrypt.

References

1. Abadi, M., Rogaway, P.: Reconciling two views of cryptography (the computational soundness of formal encryption). In: IFIP TCS2000, pp. 3–22 (2000)
2. Bao, F., Deng, R., Zhu, H.: Variations of Diffie-Hellman problem. In: Qing, S., Gollmann, D., Zhou, J. (eds.) ICICS 2003. LNCS, vol. 2836, pp. 301–312. Springer, Heidelberg (2003)
3. Bellare, M., Boldyreva, A., Micali, S.: Public-key encryption in a multi-user setting: Security proofs and improvements. In: Preneel, B. (ed.) EUROCRYPT 2000. LNCS, vol. 1807, pp. 259–274. Springer, Heidelberg (2000)
4. Blum, M., Micali, S.: How to generate cryptographically strong sequences of pseudo-random bits. SIAM J. of Computing 13, 850–864 (1984)
5. Boneh, D., Boyen, X., Goh, E.-J.: Hierarchical identity based encryption with constant size ciphertext. In: Cramer, R.J.F. (ed.) EUROCRYPT 2005. LNCS, vol. 3494, pp. 440–456. Springer, Heidelberg (2005)
6. Bresson, E., Chevassut, O., Pointcheval, D.: Group Diffie-Hellman key exchange secure against dictionary attacks. In: Zheng, Y. (ed.) ASIACRYPT 2002. LNCS, vol. 2501, pp. 497–514. Springer, Heidelberg (2002)
7. Bresson, E., Chevassut, O., Pointcheval, D.: The group Diffie-Hellman problems. In: Nyberg, K., Heys, H.M. (eds.) SAC 2002. LNCS, vol. 2595, pp. 325–338. Springer, Heidelberg (2003)
8. Burmester, M., Desmedt, Y.: A secure and efficient conference key distribution system (extended abstract). In: De Santis, A. (ed.) EUROCRYPT 1994. LNCS, vol. 950, pp. 275–286. Springer, Heidelberg (1995)
9. Canetti, R.: Towards realizing random oracles: Hash functions that hide all partial information. In: Kaliski Jr., B.S. (ed.) CRYPTO 1997. LNCS, vol. 1294, pp. 455–469. Springer, Heidelberg (1997)

10. Coppersmith, D., Shparlinski, I.: On polynomial approximation of the discrete logarithm and the Diffie-Hellman mapping. J. of Cryptology 13(2), 339–360 (2000)
11. Cramer, R., Shoup, V.: A practical public key cryptosystem provably secure against adaptive chosen ciphertext attack. In: Krawczyk, H. (ed.) CRYPTO 1998. LNCS, vol. 1462, pp. 13–25. Springer, Heidelberg (1998)
12. Dolev, D., Yao, A.: On the security of public key protocols. IEEE IT 29(12), 198–208 (1983)
13. ElGamal, T.: A public key cryptosystem and a signature scheme based on discrete logarithm. IEEE IT 31(4), 469–472 (1985)
14. Katz, J., Yung, M.: Scalable protocols for authenticated group key exchange. In: Boneh, D. (ed.) CRYPTO 2003. LNCS, vol. 2729, pp. 110–125. Springer, Heidelberg (2003)
15. Kiltz, E.: A tool box of cryptographic functions related to the Diffie-Hellman function. In: Indocrypt '01, pp. 339–350 (2001)
16. Maurer, U., Wolf, S.: Diffie-Hellman oracles. In: Koblitz, N. (ed.) CRYPTO 1996. LNCS, vol. 1109, pp. 268–282. Springer, Heidelberg (1996)
17. Micciancio, D., Panjwani, S.: Adaptive security of symbolic encryption. In: Kilian, J. (ed.) TCC 2005. LNCS, vol. 3378, pp. 245–263. Springer, Heidelberg (2005)
18. Micciancio, D., Warinschi, B.: Completeness theorems for the Abadi-Rogaway logic of encrypted expressions. J. of Computer Security. Preliminary version in WITS 2002 (2004)
19. Naor, M., Reingold, O.: Number-theoretic constructions of efficient pseudo-random functions. In: FOCS '97, pp. 458–467 (1997)
20. Sadeghi, A.-R., Steiner, M.: Assumptions related to discrete logarithms: Why subtleties make a real difference. In: Pfitzmann, B. (ed.) EUROCRYPT 2001. LNCS, vol. 2045, pp. 244–261. Springer, Heidelberg (2001)
21. Shoup, V.: Lower bounds for discrete logarithms and related problems. In: Fumy, W. (ed.) EUROCRYPT 1997. LNCS, vol. 1233, pp. 256–266. Springer, Heidelberg (1997)
22. Shparlinski, I.: Security of most significant bits of g^{x^2}. IPL 83(2), 109–113 (2002)
23. Steiner, M., Tsudik, G., Waidner, M.: Diffie-Hellman key distribution extended to group communication. In: ACM CCS 96, pp. 31–37. ACM Press, New York (1996)

Chernoff-Type Direct Product Theorems

Russell Impagliazzo[1,*], Ragesh Jaiswal[1,**], and Valentine Kabanets[2]

[1] University of California San Diego, USA
{russell,rjaiswal}@cs.ucsd.edu
[2] Simon Fraser University, Canada
kabanets@cs.sfu.ca

Abstract. Consider a challenge-response protocol where the probability of a correct response is at least α for a legitimate user, and at most $\beta < \alpha$ for an attacker. One example is a CAPTCHA challenge, where a human should have a significantly higher chance of answering a single challenge (e.g., uncovering a distorted letter) than an attacker. Another example would be an argument system without perfect completeness. A natural approach to boost the gap between legitimate users and attackers would be to issue many challenges, and accept if the response is correct for more than a threshold fraction, for the threshold chosen between α and β. We give the first proof that parallel repetition with thresholds improves the security of such protocols. We do this with a very general result about an attacker's ability to solve a large fraction of many independent instances of a hard problem, showing a Chernoff-like convergence of the fraction solved incorrectly to the probability of failure for a single instance.

1 Introduction

Cryptographic protocols use gaps between the informational and computational abilities of legitimate users and attackers to distinguish the two. Thus, the greater the gap between the ability of legitimate users to solve a type of problem and that of attackers, the more useful the problem is. Ideally, a problem should be reliably easy for legitimate users (in that the chance of failure for legitimate users should be negligible) but reliably hard for attackers (in that the chance of the attacker's success is negligible).

Direct product theorems point out ways to make problems reliably hard for attackers. The idea is that if an attacker has some chance of failing on a single challenge, the chance of solving multiple independent challenges should drop exponentially. Examples of such theorems in cryptography include Yao's theorem that weak one-way functions imply strong one-way functions ([23]) and results of [4,6], showing similar drops even when an attacker cannot know for certain whether a response to a challenge is correct. Direct product theorems are also important in average-case complexity, circuit complexity, and derandomization.

* Research partially supported by NSF Awards CCR-0313241 and CCR-0515332. Views expressed are not endorsed by the NSF.
** Research partially supported by NSF Awards CCR-0313241, CCR-0515332, CCF-0634909 and CNS-0524765. Views expressed are not endorsed by the NSF.

While intuitive, such results are frequently non-trivial. One reason for this is that there are other circumstances where the intuition is incorrect, and many instances are not proportionally harder. Examples of circumstances where direct products fail are parallel repetition for multiple round protocols and for non-verifiable puzzles ([4,6,19]).

While standard direct product theorems are powerful, they can only be used to amplify the gap between legitimate users and attackers if legitimate users are successful with high probability. The legitimate user's chance of solving k independent challenges also drops exponentially, so unless the probability of failure isn't much more than $1/k$ to start, both legitimate users and attackers will almost certainly fail to solve all of the problems.

For example, a CAPTCHA protocol is meant to distinguish between humans and programs, usually using a visual challenge based on distorted text with extraneous lines ([2]). While there seems to be a large gap between the abilities of typical humans and the best current vision algorithms to solve these challenges, algorithms can solve a non-negligible fraction of the puzzles, and many humans (including us) fail a non-negligible fraction of the puzzles. [2] prove that sequential repetition of the protocol increases this gap, and refer to [4] for the "more complicated" case of parallel repetition. Indeed, the results of [4] (and improved by [6]) do apply to parallel repetition of CAPTCHA protocols. However, for the reason above, these results only show that the probability of algorithmic success decreases with repetitions, not that the gap improves.

An obvious, intuitive solution to this problem is to make many independent challenges, but accept if the solver is successful on a larger fraction than expected for an attacker. Here, we prove that, for a large variety of problems, this approach indeed amplifies the gap between legitimate users and attackers. The kind of problems we consider are the *weakly verifiable puzzles* of [6], which include challenge-response protocols such as CAPTCHA as a special case. The puzzles are *weakly verifiable* in the sense that, while the generator of the puzzle can verify a solution, the attacker (who is just given the puzzle, not the way it was generated) cannot necessarily verify whether a proposed solution is acceptable. For P a weakly verifiable puzzle, we denote by $P^{k,T}$ the puzzle that asks k independent challenges from P and accepts if at least $(k - T)$ of the responses are correct solutions to P.

Theorem 1 (Main Theorem). *Let P be a weakly verifiable puzzle so that any solver running in time $t(n)$ has probability at least δ of failure (for sufficiently large n). Let k, $\gamma > 0, T = (1 - \gamma)\delta k$, and $\epsilon > 2 \cdot e^{-\frac{\gamma^2 \delta^2 k}{64}}$, be given parameters (as functions of n). Then no solver running in time $t'(n) = t(n)poly(\epsilon, 1/n, 1/k)$ time can solve $P^{k,T}$ with probability greater than ϵ, for some polynomial poly, for sufficiently large n.*

We call this a Chernoff-type direct product theorem, since it shows that the "tail bound" on the number of correctly solved puzzles drops exponentially in the region beyond its expectation.

Standard Chernoff bounds show that, if the legitimate user can solve the problem with probability of failure less than say $(1-2\gamma)\delta$, then they will succeed in $P^{k,T}$ with all but exponentially small probability. Thus, the above Chernoff-type direct product theorem indeed shows how to amplify any gap between legitimate users and attackers.

1.1 Weakly Verifiable Puzzles

Our result holds for the notion of weakly verifiable puzzles defined by [6].

A *weakly verifiable puzzle* has two components: First, a distribution ensemble $D = D_1, ..., D_n, ...$ on pairs (x, α), where x is called the puzzle and α the check string. n is called the security parameter. Secondly, a polynomial-time computable relation $R((x, \alpha), y)$ where y is a string of a fixed polynomially-related length.

The puzzle is thought of as defining a type of challenge x, with y being the solver's response. However, the correctness of the response is not easily verified (and may not be well-defined) given just x. On the other hand, the party generating the puzzle x also knows α, so can verify correctness.

In [6], the distribution D is restricted to being polynomially-sampleable. In this case, without loss of generality, we can assume that α is the n bit random tape used to generate the puzzle and check string (if not, we can redefine R as R' which first generate the check string from the random tape, then verifies R. Thus, to simplify the notation in our proofs, we usually assume α is a uniformly generated n bit string, and that x is a function of α. A version of our result also holds when D is not polynomial time sampleable, but only for non-uniform adversaries (since many samples from D are required as advice.)

Some examples of how weakly verifiable puzzles arise in different settings include:

1. Consider a challenge-response protocol where a prover is trying to get a verifier to accept them as legitimate (e.g., a CAPTCHA protocol, where the prover is trying to convince the verifier to accept them as human.) We assume that the verifier is polynomial time with no secret inputs, (although an honest prover may have secret inputs.) Let α be the random bits used by the verifier. In the first round, the verifier sends a challenge $x = g(\alpha)$, and the prover sends a response y. The verifier then decides whether to accept by some polynomial time algorithm, $R(\alpha, y)$. Our results are interesting if there is some chance that the honest prover will be rejected, such as an honest human user failing a CAPTCHA challenge based on visual distortion.

2. Consider a secret-agreement protocol with a passive eavesdropper. Let r_A be the random tape used by one party, and r_B that by the other party. Then the conversation C is a function of both r_A, r_B, as is the message m agreed upon. The eavesdropper succeeds if she computes m given C. Then consider $\alpha = (r_A, r_B)$, $x = C$, and $R(C, (r_A, r_B), y)$ if y is the message agreed upon by the two parties using r_A and r_B. Note that there may be some tapes where the parties fail to agree, and thus has no success. Our result shows

that, if the parties agree more probably than the eavesdropper can guess the secret, then running the protocol several times, they will almost certainly have more shared secrets than the eavesdropper can guess. Note that, unlike for challenge-response protocols, here there is no restriction on the amount of interaction between the legitimate parties (as long as the eavesdropper is passive).

3. Let f be a (weak) one-way function, and b a (partially-hidden) bit for f, in the sense that it is sometimes hard to always predict b from $x = f(z)$. Since f may not be one-to-one, b may be hard to predict for either information-theoretic or computational reasons. Here, we let $\alpha = z$, $x = f(\alpha)$, and $R(x, \alpha, b')$ if $b' = b(\alpha)$. Our results say that no adversary given an n tuple of $x_i = f(z_i)$ can produce a string closer in relative Hamming distance to $b(x_1)...b(x_n)$ than the hardness of prediction.

4. In the non-uniform setting, our results apply to any function. If f is a function (possibly non-Boolean, or even multi-valued, as long as it takes on at most a polynomial number of values), we can define α to be (the set of all elements in) $f(x)$. Then $y \in f(x)$ if and only if $y \in \alpha$, so this is testable in polynomial-time given α. This distribution isn't necessarily polynomial-time sampleable, so our results would only apply for non-uniform adversaries (e.g., Boolean circuits.)

Note that in some examples, success may be ill-defined, in that x may not uniquely determine α, and so it may not be information-theoretically possible to know whether $R((x, \alpha), y)$ given only x.

1.2   Related Work

The notion of a Direct Product Theorem, in which solving multiple instances of a problem simultaneously is proven harder than a single instance, was introduced by Yao in [23]. Due to its wide applicability in cryptography and computational complexity, a number of different versions and proofs of such theorems can be found in the literature. [8] contains a good compilation of such results. In this paper, we use some of the proof techniques (namely the *trust halving strategy*) introduced by Impagliazzo and Wigderson in [12]. Such techniques were also used to show a version of the Direct Product Theorem under a more general cryptographic setting by Bellare, Impagliazzo and Naor in [4]. The idea was to show that the soundness error decreases exponentially with parallel repetition in any 3-round challenge-response protocol. This paper also showed that such error amplification might not be possible for a general (> 3)-round protocol. Pietrzak and Wikstrom in [19] extend this negative result. On the positive side, Canetti, Halevi and Steiner in [6] used ideas from [4] to define a general class of *weakly verifiable puzzles* for which they show parallel repetition amplifies hardness, also giving a quantitative improvement over [4]. More recently, Pass and Venkita-subramaniam [18] show similar positive results for constant round public coin protocols. Note that all the results mentioned above, consider parallel repetition without threshold i.e. they consider the hardness of answering all the instances of the parallel repetition question simultaneously.

In this paper, we use the Sampling Lemma (Lemma 1) from [11] in an essential manner. The proof of this Lemma uses ideas from Raz's parallel repetition paper [20].

1.3 Techniques

Our main lemma shows how to use a breaking strategy that solves the threshold puzzle with probability ϵ as a subroutine in an algorithm that solves a single puzzle with probability greater than $(1 - \delta)$. This algorithm is a version of the trust-reducing strategies from [12,4]. In a trust-reducing strategy, the real puzzle is hidden among $(k-1)$ randomly generated puzzles, and the number of mistakes the subroutine makes on the random puzzles is used to compute a probability that the algorithm accepts the answer for the real puzzle.

However, we need to deviate substantially from the analysis in the previous papers. The previous work considered the performance of the strategy on a set H of "hard" instances, and showed that if $|H| \geq \delta 2^n$, then the strategy worked almost certainly on random elements of H. (Thus the fraction of puzzles where the strategy fails with reasonable probability would be at most δ.) In contrast, in the threshold scheme, it suffices for the adversary to answer correctly only on the instances outside a "really hard" set H' of size $(1 - \gamma)\delta$, and make an error on the instances in H'. Since H' could be a large $(1 - \gamma)$ fraction of H, the conditional probability of success on H of any strategy is at most γ.

In order to get around this obstacle, we need a more global way of analyzing the trust-reducing strategy. Our main tool for doing this is a sampling lemma (from [11]) that can be used to show that the strategy has approximately the same success probability on almost all instances. This allows us us to analyze the strategy on random instances, and infer similar success for almost all instances.

2 Preliminaries

Definition 1. *For any distribution $\mathcal{D}$, $x \leftarrow \mathcal{D}$ denotes sampling an element from the distribution $\mathcal{D}$, and $\mathcal{D}(x)$ denotes the probability of sampling the element x.*

Definition 2. *Given two distributions $\mathcal{D}_1$ and $\mathcal{D}_2$ over $\{0,1\}^n$, the statistical distance $Dist(\mathcal{D}_1, \mathcal{D}_2)$ between them is defined as*

$$Dist(\mathcal{D}_1, \mathcal{D}_2) = \frac{1}{2} \sum_{x \in \{0,1\}^n} |\mathbf{Pr}[\mathcal{D}_1(x)] - \mathbf{Pr}[\mathcal{D}_2(x)]|$$

Let $\mathcal{U}$ be the uniform distribution on $\{0,1\}^n$. Consider the following distribution over $\{0,1\}^n$. Pick an m tuple of n-bit string $(x_1, \ldots, x_m)$ uniformly at random and output x_i for a randomly chosen $i \in [m]$. The distribution is equivalent to $\mathcal{U}$ if the tuple is randomly chosen from $\{0,1\}^{nm}$. The next lemma shows that the distribution is close to uniform even when the tuple is chosen randomly from a subset $G \subseteq \{0,1\}^{nm}$ of size $\epsilon 2^{nm}$.

Lemma 1 (Sampling Lemma). *Let $G \subseteq \{0,1\}^{mn}$ be any subset of size $\epsilon 2^{mn}$. Let $\mathcal{U}$ be a uniform distribution on the set $\{0,1\}^n$, and let $\mathcal{D}$ be the distribution defined as follows: pick a tuple $(x_1, \ldots, x_m)$ of n-bit strings uniformly from the set G, pick an index i uniformly from $[m]$, and output x_i. Then the statistical distance between the distributions $\mathcal{U}$ and $\mathcal{D}$ is less than $0.6\sqrt{\frac{\log 1/\epsilon}{m}}$.*

See [11] for the proof of this lemma. The following corollary will be used in the proof of our main result.

Corollary 1. *Let $\mathcal{G}$ be a distribution over $\{0,1\}^{nm}$ (which can be viewed as m-tuples of n-bit strings) such that for any $\bar{x} \in \{0,1\}^{nm}$, $\mathcal{G}(\bar{x}) \leq \frac{1}{\epsilon \, 2^{nm}}$. Let $\mathcal{U}$ be a uniform distribution over $\{0,1\}^n$, and let $\mathcal{D}$ be the distribution defined as follows: pick a tuple $(x_1, \ldots, x_m) \leftarrow \mathcal{G}$, pick an index i uniformly from $[m]$, and output x_i. Then the statistical distance between the distributions $\mathcal{U}$ and $\mathcal{D}$ is less than $0.6\sqrt{\frac{\log 1/\epsilon}{m}}$.*

Proof. We can represent the distribution $\mathcal{G}$ as a convex combination of uniform distributions on subsets of size at least $\epsilon 2^{nm}$. We can then apply the Sampling Lemma to each term in the combination to obtain the corollary. $\qquad\square$

3 Proof of the Main Theorem

The proof is by contradiction. Given a solver $\bar{C}$ that solves the weakly verifiable puzzle $P^{k,T}$ with probability at least ϵ, we give a solver $\mathcal{C}$ which solves the puzzle P with probability at least $(1-\delta)$. The probability of success is over the internal randomness of the solver and uniformly chosen $\alpha \in \{0,1\}^n$.

Let G be the subset of $\bar{\alpha} = (\alpha_1, \ldots, \alpha_k) \in (\{0,1\}^n)^k$ where[1]

$$|\{i : \neg R((x_i, \alpha_i), \bar{C}(x_1, \ldots, x_k)_i)\}| \leq (1-\gamma)\delta k$$

So G denotes the "good" subset of $\bar{\alpha}$'s for the solver $\bar{C}$ where we have minimum guarantee of ϵ. In order to illustrate the ideas of the proof, we first prove the Theorem assuming access to an oracle $\mathcal{O}^G$ deciding the membership of a given tuple $(\alpha_1, \ldots, \alpha_k)$ in the "good" set G. We then drop this assumption by, essentially, imitating the behavior of the oracle.

3.1 Assuming Oracle $\mathcal{O}^G$ Exists

In this subsection, in order to illustrate the ideas of the proof in a simplified setting, we temporarily assume that there is an oracle $\mathcal{O}^G$ which tells if a given tuple $(\alpha_1, \ldots, \alpha_k)$ belongs to the "good" set G. This subsection is to develop the reader's intuition, and is not strictly required for the proof of the real case. For this reason, we will slur over some calculations. In the rest of the section, we show how to drop this assumption by approximating oracle $\mathcal{O}^G$ in a computable

---

[1] For a string α_i, we implicitly denote the puzzle by x_i.

way. The rest of the section is self-contained, so this subsection, while helpful, may be skipped by the reader. Note that this oracle is unrealistic in many ways, one of which is that it's answer depends on α, when in the real case the Solver will only have x, not α.

Consider the randomized Solver $\mathcal{C}$ defined in Figure 1 which is allowed a special output $\perp$ which is considered as an incorrect answer in the analysis.

INPUT: $z = (x, \alpha)$
OUTPUT: y
ORACLE ACCESS: Solver $\bar{C}$ and $\mathcal{O}^G$
PARAMETERS: $\epsilon \geq 2 \cdot e^{-\frac{\gamma^2 \delta^2 k}{64}}$, $timeout = \frac{4n}{\epsilon}$.

1. Repeat lines 2-6 for at most $timeout$ times:
2.   Choose $i \in [k]$ uniformly at random.
3.   Choose $\alpha_1, \ldots, \alpha_{k-1} \in \{0, 1\}^n$ uniformly at random.
4.   Let $\bar{\alpha} \leftarrow (\alpha_1, \ldots, \alpha_{i-1}, \alpha, \alpha_i, \ldots, \alpha_{k-1})$.
5.   **If** $\mathcal{O}^G(\bar{\alpha}) = 1$
6.   then **output** $y = \bar{C}(\bar{x})_i$, //*where the elements of $\bar{x}$ are the puzzles*
                                    // *generated from check strings $\bar{\alpha}$.*
7. **output** $\perp$

Solver 1. Randomized Solver $\mathcal{C}$ given $\bar{C}$ and $\mathcal{O}^G$ as oracle

We want to analyze the success probability of solver $\mathcal{C}$ on a given input $z = (x, \alpha)$. To this end, we need to argue that (1) the probability of the timeout (i.e., of outputting $\perp$ in line 7) is small, and (2) conditioned on the output being different from $\perp$, it is a correct output with high probability (greater than $1 - \delta$).

We will focus on analyzing the conditional success probability in (2), i.e., $\mathbf{Pr}_{i, \alpha_1, \ldots, \alpha_{k-1}}[\mathcal{C}(\alpha) \text{ is correct} \mid \text{output} \neq \perp]$, for a given input $z = (x, \alpha)$ to $\mathcal{C}$. Observing that $\mathcal{C}$ outputs something other than $\perp$ exactly when the tuple $\bar{\alpha}$ built in line 4 is in the set G, we can rewrite this conditional probability as $\mathbf{Pr}_{i \in [k], \bar{\alpha} = (\alpha_1, \ldots, \alpha_k) \in G}[\mathcal{C}(\alpha_i) \text{ is correct} \mid \alpha_i = \alpha]$, where i is chosen uniformly from $[k]$, and $\bar{\alpha}$ uniformly from G.

Let $\mathcal{D}(\alpha) = \mathbf{Pr}_{i \in [k], \bar{\alpha} \in G}[\alpha_i = \alpha]$, and let $\mathcal{U}$ be the uniform distribution on αs. Using our Sampling Lemma, we will argue that the distributions $\mathcal{D}$ and $\mathcal{U}$ are statistically close to each other. Using this closeness, we can finish the analysis of the success probability of solver $\mathcal{C}$ as follows. The conditional success probability of $\mathcal{C}$ for a random input α is $\sum_\alpha \mathbf{Pr}_{i, \bar{\alpha} \in G}[\mathcal{C}(\alpha_i) \text{ is correct} \mid \alpha_i = \alpha] * \mathcal{U}(\alpha)$, which is approximately equal to $\sum_\alpha \mathbf{Pr}_{i, \bar{\alpha} \in G}[\mathcal{C}(\alpha_i) \text{ is correct} \mid \alpha_i = \alpha] * \mathcal{D}(\alpha)$. The latter expression is exactly $\mathbf{Pr}_{i, \bar{\alpha} \in G}[\mathcal{C}(\alpha_i) \text{ is correct}]$, which is at least $1 - (1 - \gamma)\delta = 1 - \delta + \gamma\delta$, by the definition of G. We will show that the statistical distance between $\mathcal{D}$ and $\mathcal{U}$ and the probability of the timeout of $\mathcal{C}$ are less than $\gamma\delta$, which would imply that $\mathcal{C}$ succeeds on more than $1 - \delta$ fraction of inputs.

To demonstrate the structure of the analysis in the general case, we recast the arguments above as follows. We introduce a certain random experiment $\mathcal{E}$

(see Figure 1), which corresponds to the inner loop of the algorithm $\mathcal{C}$. We then relate the success probability of $\mathcal{C}$ to that of $\mathcal{E}$.

> Experiment $\mathcal{E}$
>
> $(\alpha_1, \ldots, \alpha_k) \xleftarrow{\$} G$ // Let $x_1, \ldots, x_k$ be the corresponding puzzles
>
> $i \xleftarrow{\$} [k]$
>
> **output** $(\alpha_i, \bar{C}(x_1, \ldots, x_k)_i)$

Fig. 1. Experiment $\mathcal{E}$

We say that experiment $\mathcal{E}$ *succeeds* if it outputs a pair (α, y) such that $R((x, \alpha), y)$. Since for each k-tuple in G, $\bar{C}$ outputs a correct answer for at least $1 - (1 - \gamma)\delta$ fraction of the elements, the probability of success of this experiment is clearly $\geq 1 - (1 - \gamma)\delta$.

Let $\mathcal{D}$ be the probability distribution on the first elements of outputs of $\mathcal{E}$, i.e., $\mathcal{D}(\alpha)$ is the probability that $\mathcal{E}$ outputs a pair (α, y). Let R_α represent the probability that it outputs such a pair with $R((x, \alpha), y)$, and W_α the probability that it outputs such a pair with $\neg R((x, \alpha), y)$. So, $\mathcal{D}(\alpha) = R_\alpha + W_\alpha$. Clearly, we have that $1 - (1 - \gamma)\delta \leq \mathbf{Pr}[\mathcal{E} \text{ succeeds}] = \sum_{\alpha \in \{0,1\}^n} R_\alpha$.

Since $\mathcal{D}$ is sampled by picking a random element of a set G of tuples of size at least $\epsilon 2^{nk}$, from the sampling lemma it is within $0.6\sqrt{\log(1/\epsilon)/k} \leq \gamma\delta/8$ statistical distance of the uniform distribution. In particular, for $H = \{\alpha | \mathcal{D}(\alpha) \leq (1/2)2^{-n}\}$, $|H| \leq (\gamma\delta/4)2^n$.

Let p_α be the probability that a random $\bar{\alpha}$ containing α is in G. Then the expectation of p_α for random α is at least ϵ, and $\mathcal{D}(\alpha) = p_\alpha / \sum_{\alpha'} p_{\alpha'} = 2^{-n}(p_\alpha / \mathbf{Exp}[p_{\alpha'}])$. So all elements not in H have $p_\alpha \geq \epsilon/2$. For each such element, the probability that we get a timeout in $\mathcal{C}$ is at most $(1 - p_\alpha)^{timeout} \leq e^{-n}$.

Given that $\mathcal{C}$ on α does not time out, the probability of it succeeding is $R_\alpha / \mathcal{D}(\alpha)$. Thus, the overall probability of success is at least $(\sum_\alpha \mathcal{U}(\alpha) R_\alpha / \mathcal{D}(\alpha)) - \mathbf{Pr}[\mathcal{C} \text{ times out}]$. We get $\sum_\alpha \mathcal{U}(\alpha) R_\alpha / \mathcal{D}(\alpha) = \sum_\alpha (\mathcal{U}(\alpha) - \mathcal{D}(\alpha)) R_\alpha / \mathcal{D}(\alpha) + \sum_\alpha R_\alpha \geq \sum_\alpha (-1)|\mathcal{U}(\alpha) - \mathcal{D}(\alpha)| + (1 - (1 - \gamma)\delta) \geq 1 - (1 - \gamma)\delta - Dist(\mathcal{D}, \mathcal{U}) \geq 1 - (1 - 3/4\gamma)\delta$.

The probability of time-out can be bounded by the probability that $\alpha \in H$ plus the probability of time-out given that $\alpha \notin H$. As previously mentioned, this is at most $\delta\gamma/4 + e^{-n}$, giving a total success probability at least $1 - (1 - \gamma/2)\delta - e^{-n} > 1 - \delta$, as desired.

3.2 Removing the Oracle $\mathcal{O}^G$

We will use the same set of ideas as in the previous subsection while removing the dependency on the simplifying assumptions. The most important assumption made was the existence of the oracle $\mathcal{O}^G$ which helped us to determine if a tuple $(\alpha_1, \ldots, \alpha_k) \in G$. The second assumption that we made was that the randomized solver $\mathcal{C}$ receives as input the pair $z = (x, \alpha)$ which is not a valid assumption since α is supposed to be hidden from the solver.

We get around these assumptions, in some sense, by imitating the combined behavior of the randomized solver and the oracle $\mathcal{O}^G$ of the previous subsection. Below we define a new randomized solver which is only given x as an input, with α being hidden from the solver.

INPUT: x // *corresponding to α*
OUTPUT: y
ORACLE ACCESS: Solver $\bar{C}$
PARAMETERS: $\epsilon \geq 2 \cdot e^{-\frac{\gamma^2 \delta^2 k}{64}}$, *timeout* $= \frac{4n}{\epsilon}$, $t_0 = (1 - \gamma)\delta k$, $\rho = 1 - \frac{\gamma \delta}{16}$.

1. Repeat lines 2-10 for at most *timeout* times:

> 2.   // *Subroutine TRS (Trust Reducing Strategy)*
> 3.   Choose $i \in [k]$ uniformly at random.
> 4.   Choose $\alpha_1, \ldots, \alpha_{k-1} \in \{0,1\}^n$ uniformly at random.
> 5.   Let $\bar{x} \leftarrow (x_1, \ldots, x_{i-1}, x, x_i, \ldots, x_{k-1})$.
> 6.   Let $l = \{j : \neg R((x_j, \alpha_j), \bar{C}(x_1, \ldots, x_{k-1})_j), j \neq i\}$
> 7.   **If** $|l| > t_0$
> 8.   **output** $\bar{C}(x_1, \ldots, x_{k-1})_i$ with probability $\rho^{|l|-t_0}$
> 9.   **else**
> 10.   **output** $\bar{C}(x_1, \ldots, x_{k-1})_i$ with probability 1

11. **output** $\perp$

Solver 2. Randomized Solver $\mathcal{C}$ given $\bar{C}$ as oracle

---

To be able to analyze the above solver we abstract out a single execution of the loop $2 - 10$ (the subroutine TRS) and design an experiment $\mathcal{E}_3$ which has similar behavior. To further simplify the analysis we design a simpler experiment $\mathcal{E}_2$ such that (1) analyzing $\mathcal{E}_2$ is easy and (2) $\mathcal{E}_2$ is not too much different from $\mathcal{E}_3$ so that we can easily draw comparisons between them. The description of Experiments $\mathcal{E}_2$ and $\mathcal{E}_3$ is given in Figure 2.

Definition 3. *Experiments $\mathcal{E}_2$ and $\mathcal{E}_3$ are said to* succeed *if they output a correct pair (i.e. a pair (α, y) such that $R((x, \alpha), y)$). The success probability is defined as the probability that a correct pair is produced conditioned on the experiment producing a pair.*

Proof outline. We observe that the success probability of $\mathcal{C}$ on a given input x corresponding to a hidden string α is exactly the success probability of experiment $\mathcal{E}_3$ *conditioned* on the event that $\mathcal{E}_3$ produces a pair $(\alpha, \cdot)$. For a random input x corresponding to a uniformly random string α, the success probability of $\mathcal{C}$ is then $\sum_\alpha \mathbf{Pr}[\mathcal{E}_3$ succeeds $\mid \mathcal{E}_3$ outputs $(\alpha, \cdot)] * \mathcal{U}(\alpha)$, where $\mathcal{U}$ denotes the uniform distribution. On the other hand, the success probability of $\mathcal{E}_3$ can be written as $\sum_\alpha \mathbf{Pr}[\mathcal{E}_3$ succeeds $\mid \mathcal{E}_3$ outputs $(\alpha, \cdot)] * \mathcal{D}_3(\alpha)$, where $\mathcal{D}_3(\alpha)$ is the probability that experiment $\mathcal{E}_3$ produces a pair $(\alpha, \cdot)$ conditioned on $\mathcal{E}_3$ producing some pair (i.e., conditioned on the output of $\mathcal{E}_3$ being different from $\perp$). We

Experiment $\mathcal{E}_2$

$(\alpha_1, \ldots, \alpha_k) \xleftarrow{\$} (\{0,1\}^n)^k$

$i \xleftarrow{\$} [k]$

$J \leftarrow \{j | \neg R((x_j, \alpha_j), \bar{C}(x_1, \ldots, x_k))_j\}$

if $|J| > (1 - \gamma)\delta k$

$\quad t = |J| - (1 - \gamma)\delta k$

else

$\quad t = 0$

output $(\alpha_i, \bar{C}(x_1, \ldots, x_k)_i)$

$\quad$ with probability ρ^t and $\perp$

$\quad$ with probability $(1 - \rho^t)$

Experiment $\mathcal{E}_3$

$(\alpha_1, \ldots, \alpha_k) \xleftarrow{\$} (\{0,1\}^n)^k$

$i \xleftarrow{\$} [k]$

$J \leftarrow \{j | \neg R((x_j, \alpha_j), \bar{C}(x_1, \ldots, x_k))_j\}$

if $|J| > (1 - \gamma)\delta k$

$\quad t = |J| - (1 - \gamma)\delta k$

else

$\quad$ **output** $(\alpha_i, \bar{C}(x_1, \ldots, x_k)_i)$

$\quad\quad$ with probability 1

if $i \in J$

$\quad$ **output** $(\alpha_i, \bar{C}(x_1, \ldots, x_k)_i)$

$\quad\quad$ with probability ρ^{t-1} and $\perp$

$\quad\quad$ with probability $(1 - \rho^{t-1})$

else

$\quad$ **output** $(\alpha_i, \bar{C}(x_1, \ldots, x_k)_i)$

$\quad\quad$ with probability ρ^t and $\perp$

$\quad\quad$ with probability $(1 - \rho^t)$

Fig. 2. Experiments $\mathcal{E}_2$ and $\mathcal{E}_3$

then argue that the distributions $\mathcal{U}$ and $\mathcal{D}_3$ are statistically close, and hence the success probability of C can be lowerbounded with that of $\mathcal{E}_3$. Finally, we lowerbound the success probability of $\mathcal{E}_3$, getting the result for C.

In reality, the success probability of experiment $\mathcal{E}_2$ is easier to analyze than $\mathcal{E}_3$. So we actually show that the conditional success probability of $\mathcal{E}_3$ can be lowerbounded by that of $\mathcal{E}_2$, and then argue that $\mathcal{U}$ is statistically close to $\mathcal{D}_2$, where $\mathcal{D}_2$ is defined for $\mathcal{E}_2$ in the same way as $\mathcal{D}_3$ was defined for $\mathcal{E}_3$ above.

Next we give the details of the proof. We start by analyzing $\mathcal{E}_2$.

Analyzing $\mathcal{E}_2$. Let us partition the k-tuples $(\{0,1\}^n)^k$ into the following subsets:

$$G_0 = G = \{(\alpha_1, \ldots, \alpha_k) \in \{0,1\}^{nk} : |\{j : \neg R((x_j, \alpha_j), \bar{C}(x_1, \ldots, x_k)_j)\}| \leq (1 - \gamma)\delta k\}$$

$$G_1 = \{(\alpha_1, \ldots, \alpha_k) \in \{0,1\}^{nk} : |\{j : \neg R((x_j, \alpha_j), \bar{C}(x_1, \ldots, x_k)_j)\}| = (1 - \gamma)\delta k + 1\}$$

$$\vdots$$

$$G_{k(1-(1-\gamma)\delta)} = \{(\alpha_1, \ldots, \alpha_k) \in \{0,1\}^{nk} : |\{j : \neg R((x_j, \alpha_j), \bar{C}(x_1, \ldots, x_k)_j)\}| = k\}$$

Definition 4. *We let $S_{\not\perp}$ denote the general event that the experiment produces a pair (i.e. does not produce $\perp$) and S_c denote the event that the experiment produces a correct output.*

Claim 2. $\mathbf{Pr}[\mathcal{E}_2 \text{ succeeds}] \geq \left(1 - \frac{t_0 + \gamma\delta k/2}{k}\right)\left(1 - \frac{\rho^{\gamma\delta k/2}}{\epsilon}\right)$, *where* $t_0 = (1-\gamma)\delta k$.

Proof. Let $t_0 = (1 - \gamma)\delta k$ and $\Delta = \gamma\delta k$. Let $\bar{A}$ denote the random tuple chosen in the first step of experiment $\mathcal{E}_2$. Recalling that the success probability of $\mathcal{E}_2$ is defined as the probability of producing a correct pair conditioned on producing a pair as output, we get

$$
\begin{aligned}
\mathbf{Pr}[\mathcal{E}_2 \text{ succeeds}] &= \mathbf{Pr}[S_c | S_{\not{L}}] \\
&= \mathbf{Pr}[S_c, S_{\not{L}}]/\mathbf{Pr}[S_{\not{L}}] \\
&= \sum_{\bar{\alpha} \in \{0,1\}^{nk}} \mathbf{Pr}[S_c, S_{\not{L}}, \bar{A} = \bar{\alpha}]/\mathbf{Pr}[S_{\not{L}}] \\
&= \sum_{\bar{\alpha} \in \{0,1\}^{nk}} \mathbf{Pr}[S_c | S_{\not{L}}, \bar{A} = \bar{\alpha}] \cdot \mathbf{Pr}[S_{\not{L}}, \bar{A} = \bar{\alpha}]/\mathbf{Pr}[S_{\not{L}}].
\end{aligned}
$$

We will split the set of $\bar{\alpha}$'s into the following three sets:

$$
G = G_0, \qquad I = G_1 \cup \ldots \cup G_{\Delta/2}, \qquad B = \{0,1\}^{nk} - G - I,
$$

which stand for "good", "intermediate" and "bad", respectively. We note that $\mathcal{E}_2$ performs *well* on tuples in the good subset, *reasonably well* on tuples in the intermediate subset and *poorly* on the tuples in the bad subset. The intuitive idea is that we counter the poor effect of the bad subset of tuples by exponentially weighing down their contribution in the overall probability of success of $\mathcal{E}_2$.

We have

$$
\begin{aligned}
\mathbf{Pr}[\mathcal{E}_2 \text{ succeeds}] &\geq \sum_{\bar{\alpha} \in G \cup I} \mathbf{Pr}[S_c | S_{\not{L}}, \bar{A} = \bar{\alpha}] \cdot \mathbf{Pr}[S_{\not{L}}, \bar{A} = \bar{\alpha}]/\mathbf{Pr}[S_{\not{L}}] \\
&\geq \sum_{\bar{\alpha} \in G \cup I} \left(1 - \frac{t_0 + \Delta/2}{k}\right) \cdot \mathbf{Pr}[S_{\not{L}}, \bar{A} = \bar{\alpha}]/\mathbf{Pr}[S_{\not{L}}] \\
&\geq \left(1 - \frac{t_0 + \Delta/2}{k}\right) \left(\frac{\sum_{\bar{\alpha} \in G \cup I} \mathbf{Pr}[S_{\not{L}}, \bar{A} = \bar{\alpha}]}{\mathbf{Pr}[S_{\not{L}}]}\right) \\
&= \left(1 - \frac{t_0 + \Delta/2}{k}\right) \cdot \frac{\mathbf{Pr}[S_{\not{L}}] - \mathbf{Pr}[S_{\not{L}}, \bar{A} \in B]}{\mathbf{Pr}[S_{\not{L}}]} \\
&= \left(1 - \frac{t_0 + \Delta/2}{k}\right) \cdot \left(1 - \frac{\mathbf{Pr}[S_{\not{L}}, \bar{A} \in B]}{\mathbf{Pr}[S_{\not{L}}]}\right).
\end{aligned}
$$

Observe that $\mathbf{Pr}[S_{\not{L}}, \bar{A} \in B] \leq \mathbf{Pr}[S_{\not{L}} | \bar{A} \in B] * \mathbf{Pr}[\bar{A} \in B] \leq \rho^{\Delta/2} \cdot 1 = \rho^{\Delta/2}$, and $\mathbf{Pr}[S_{\not{L}}] \geq \mathbf{Pr}[S_{\not{L}}, \bar{A} \in G] \geq \epsilon$. Thus, we get that $1 - \mathbf{Pr}[S_{\not{L}}, \bar{A} \in B]/\mathbf{Pr}[S_{\not{L}}] \geq 1 - \rho^{\Delta/2}/\epsilon$, and the claim follows. $\square$

Let A be the random variable denoting the first element of the pair produced by $\mathcal{E}_2$ conditioned on $\mathcal{E}_2$ producing a pair. We now write down the success probability of $\mathcal{E}_2$ in terms of the conditional probability that $\mathcal{E}_2$ produces a correct pair given that it produces a pair $(\alpha, .)$ for a fixed $\alpha \in \{0,1\}^n$.

$$\mathbf{Pr}[\mathcal{E}_2 \ succeeds] = \mathbf{Pr}[S_c | S_{\not{L}}]$$

$$= \sum_{\alpha \in \{0,1\}^n} \mathbf{Pr}[\mathcal{E}_2 \ succeeds \ on \ A | A = \alpha, S_{\not{L}}] \cdot \mathbf{Pr}[A = \alpha | S_{\not{L}}]$$

$$= \sum_{\alpha \in \{0,1\}^n} \mathbf{Pr}[\mathcal{E}_2 \ succeeds \ on \ A | A = \alpha, S_{\not{L}}] \cdot \mathcal{D}_2(\alpha) \quad (1)$$

where $\mathcal{D}_2$ is a distribution defined as $\mathcal{D}_2(\alpha) = \mathbf{Pr}[A = \alpha | S_{\not{L}}]$.

Note the similarity between the distribution $\mathcal{D}_2$ and distribution $\mathcal{D}$ of the previous section. $\mathcal{D}$ was sampled by producing a randomly chosen element from a randomly chosen tuple in G. Here we allow tuples to be chosen from any G_i but we weigh down the contribution of the tuple by a factor of ρ^i. In other words, $\mathcal{D}_2$ can be sampled in the following manner: Pick a random tuple $\bar{\alpha} \in \{0,1\}^{nk}$, let $\bar{\alpha} \in G_i$, output a randomly chosen element of the tuple with probability ρ^i.

Comparing $\mathcal{D}_2$ and $\mathcal{U}$. We will show that $\mathcal{D}_2$ is statistically close to the uniform distribution $\mathcal{U}$.

Claim 3. $Dist(\mathcal{D}_2, \mathcal{U}) < 0.6\sqrt{\frac{\log 1/\epsilon}{k}}$.

Proof. To sample from $\mathcal{D}_2$, pick $(\alpha_1, \ldots, \alpha_k) \leftarrow \mathcal{G}$, pick a random $i \in [k]$ and output α_i, where $\mathcal{G}$ is a distribution on k-tuples such that $\mathcal{G}(\bar{\alpha})$ is the conditional probability that $\mathcal{E}_2$ outputs the randomly chosen element from $\bar{\alpha}$ given that $\mathcal{E}_2$ produces a pair. More specifically, given $\bar{\alpha} \in G_i$,

$$\mathcal{G}(\bar{\alpha}) = \frac{\rho^i}{|G_0| + \rho |G_1| + \ldots + \rho^{k(1-(1-\gamma)\delta)} |G_{k(1-(1-\gamma)\delta)}|} \leq \frac{1}{|G_0|} \leq \frac{1}{\epsilon \cdot 2^{nk}}.$$

By Corollary 1, we get the conclusion of the Claim. □

Comparing $\mathcal{E}_3$ and $\mathcal{E}_2$. To continue, we need the following definitions.

Definition 5. *Given $\alpha \in \{0,1\}^n$ and a k-tuple $(\alpha_1, \ldots, \alpha_k) \in (\{0,1\}^n)^k$, let $h(\alpha, (\alpha_1, \ldots, \alpha_k)) = \{i : \alpha_i = \alpha\}$. Given a k-tuple $(\alpha_1, \ldots, \alpha_k) \in (\{0,1\}^n)^k$ and solver $\bar{C}$, let $l(\alpha_1, \ldots, \alpha_k) = \{i : \neg R((x_i, \alpha_i), \bar{C}(x_1, \ldots, x_k)_i)\}$*

In other words, for a given element and tuple, h denotes the subset of indices where the element is present, and, for a given tuple, l denotes the subset of indices where $\bar{C}$ is incorrect. Consider the following two quantities:

$$X_\alpha = \underbrace{\sum_{\bar{\alpha} \in G} |h(\alpha, \bar{\alpha}) \cap l(\bar{\alpha})|}_{M_\alpha} + \underbrace{\sum_{\bar{\alpha} \in \{0,1\}^{nk} - G} |h(\alpha, \bar{\alpha}) \cap l(\bar{\alpha})| \cdot \rho^{|l(\bar{\alpha})| - t_0}}_{N_\alpha}$$

$$Y_\alpha = \sum_{\bar{\alpha} \in G} |h(\alpha, \bar{\alpha}) - h(\alpha, \bar{\alpha}) \cap l(\bar{\alpha})| + \sum_{\bar{\alpha} \in \{0,1\}^{nk} - G} |h(\alpha, \bar{\alpha}) - h(\alpha, \bar{\alpha}) \cap l(\bar{\alpha})| \cdot \rho^{|l(\bar{\alpha})| - t_0}$$

It is easy to see that

$$\mathbf{Pr}[\mathcal{E}_2 \text{ succeeds on } A \mid A = \alpha, S_{\cancel{\perp}}] = \frac{Y_\alpha}{X_\alpha + Y_\alpha} = \frac{Y_\alpha}{M_\alpha + N_\alpha + Y_\alpha}. \tag{2}$$

Experiment $\mathcal{E}_3$ is mostly the same as $\mathcal{E}_2$, except when, for a randomly chosen tuple $\bar{\alpha} \in \{0,1\}^{nk} - G$ (line 1), the randomly chosen index i (line 2) lands in the subset $l(\bar{\alpha})$ of indices on which $\bar{C}$ is incorrect. Here $\mathcal{E}_2$ only outputs the pair with probability $\rho^{|l(\bar{\alpha})|-t_0}$ (instead of $\rho^{|l(\bar{\alpha})|-t_0-1}$ as in $\mathcal{E}_3$). Thus we have

$$\mathbf{Pr}[\mathcal{E}_3 \text{ succeeds on } A \mid A = \alpha, S_{\cancel{\perp}}] = \frac{Y_\alpha}{M_\alpha + N_\alpha/\rho + Y_\alpha}. \tag{3}$$

Finally, using (2) and (3), we get:

$$\frac{\mathbf{Pr}[\mathcal{E}_2 \text{ succeeds on } A \mid A = \alpha, S_{\cancel{\perp}}]}{\mathbf{Pr}[\mathcal{E}_3 \text{ succeeds on } A \mid A = \alpha, S_{\cancel{\perp}}]} = \frac{M_\alpha + N_\alpha/\rho + Y_\alpha}{M_\alpha + N_\alpha + Y_\alpha}$$

$$\leq \frac{M_\alpha + N_\alpha + Y_\alpha}{\rho \cdot (M_\alpha + N_\alpha + Y_\alpha)}$$

$$= 1/\rho. \tag{4}$$

Analyzing $\mathcal{C}$. We first note that the subset H of α's for which the above solver does not produce an answer (or produces $\perp$) is small. Consider the following two claims:

Claim 4. *Let $H \subseteq \{0,1\}^n$ be such that, for every $\alpha \in H$, TRS produces an answer with probability $< \epsilon/4$. Then $|H| < \frac{\gamma\delta}{4} \cdot 2^n$.*

Proof. For the sake of contradiction, assume that $|H| \geq \frac{\gamma\delta}{4} \cdot 2^n$. For a randomly chosen tuple $\bar{\alpha} = (\alpha_1, \ldots, \alpha_k)$, the expected number of α_i's from H is $\gamma\delta k/4$. By Chernoff bounds, all but $e^{-\frac{\gamma\delta k}{64}}$ fraction of tuples $\bar{\alpha}$ will contain at least $\gamma\delta k/8$ elements from H.

For a random $\alpha \in H$, consider the distribution on tuples $\bar{\alpha}$ induced by lines 3–5 of Solver 2. That is, $\bar{\alpha}$ is sampled by picking independently uniformly at random $\alpha \in H$, location $i \in [k]$, and $\alpha_1 \ldots \alpha_{k-1} \in \{0,1\}^n$, and producing $\bar{\alpha} = (\alpha_1, \ldots, \alpha_{i-1}, \alpha, \alpha_i, \ldots, \alpha_{k-1})$. Observe that every tuple $\bar{\alpha}'$ containing exactly s elements from H will be assigned by this distribution probability exactly $\frac{4s}{\gamma\delta k}$ times the probability of $\bar{\alpha}'$ under the uniform distribution. So the probability of sampling tuples in G which have more than $\frac{\gamma\delta k}{8}$ elements from H is at least $\frac{\epsilon - e^{-\frac{\gamma\delta k}{64}}}{2} \geq \frac{\epsilon}{4}$, since $\epsilon = 2 \cdot e^{-\frac{\gamma^2\delta^2 k}{64}}$. This means that for a random $\alpha \in H$, a single iteration of the subroutine TRS of Solver 2 will produce a definite answer with probability at least $\epsilon/4$ (note that TRS always produces an answer when $\bar{\alpha}' \in G$). By averaging, there exists a particular $\alpha_0 \in H$ for which TRS succeeds in producing an answer with probability at least $\epsilon/4$. □

Claim 5. *For every $\alpha \in \{0,1\}^n - H$, $\mathbf{Pr}[\mathcal{C}(x) \neq \perp] > 1 - e^{-n}$.*

Proof. From the previous claim we know that for any $\alpha \in \{0, 1\}^n - H$, the subroutine TRS produces an answer with probability at least $\epsilon/4$. So, the probability that Solver 2 fails to produce a definite answer on this input α within *timeout* iterations is at most $(1 - \epsilon/4)^{\frac{4n}{\epsilon}} \leq e^{-n}$. □

The similarity between solver $\mathcal{C}$ and Experiment $\mathcal{E}_3$ yields the following useful fact:

$$\mathbf{Pr}[R((x, \alpha), \mathcal{C}(x)) | \mathcal{C}(x) \neq \perp] = \mathbf{Pr}[\mathcal{E}_3 \text{ succeeds on } A \mid A = \alpha, S_{\not{L}}]. \quad (5)$$

We now analyze the success probability of the solver $\mathcal{C}$. The probability is over uniformly random $\alpha \in \{0, 1\}^n$ and its internal randomness.

$$\mathbf{Pr}[\mathcal{C} \text{ succeeds}] = \frac{1}{2^n} \sum_{\alpha \in \{0,1\}^n} \mathbf{Pr}[R((x, \alpha), \mathcal{C}(x)) \wedge \mathcal{C}(x) \neq \perp]$$

$$= \sum_{\alpha \in \{0,1\}^n} \mathbf{Pr}[R((x, \alpha), \mathcal{C}(x)) \mid \mathcal{C}(x) \neq \perp] * \mathbf{Pr}[\mathcal{C}(x) \neq \perp] * \mathcal{U}(\alpha)$$

$$= \sum_{\alpha \in \{0,1\}^n} \mathbf{Pr}[\mathcal{E}_3 \text{ succeeds on } A \mid A = \alpha, S_{\not{L}}] * \mathbf{Pr}[\mathcal{C}(x) \neq \perp] * \mathcal{U}(\alpha)$$

$$\text{(from (5))} \quad (6)$$

Let $H \subseteq \{0, 1\}^n$ be the set from Claim 4. Let $\bar{H}$ be the complement of H in the set $\{0, 1\}^n$. By Claim 5 and Eq. (4), we get that for every $\alpha \in \bar{H}$,

$$\mathbf{Pr}[\mathcal{E}_3 \text{ succeeds on } A \mid A = \alpha, S_{\not{L}}] * \mathbf{Pr}[\mathcal{C}(x) \neq \perp] * \mathcal{U}(\alpha) \geq$$
$$(1 - e^{-n}) \, \rho \, \mathbf{Pr}[\mathcal{E}_2 \text{ succeeds on } A \mid A = \alpha, S_{\not{L}}] * \mathcal{U}(\alpha). \quad (7)$$

Comparing $\mathcal{C}$ and $\mathcal{E}_2$. We can now compare the success probabilities of Experiment $\mathcal{E}_2$ and the solver $\mathcal{C}$.

Claim 6. $\mathbf{Pr}[\mathcal{C} \text{ succeeds}] \geq \mathbf{Pr}[\mathcal{E}_2 \text{ succeeds}] - \left(Dist(\mathcal{U}, \mathcal{D}_2) + (1 - \rho) + \rho \cdot e^{-n} + \frac{\gamma\delta}{4}\right)$

Proof. Using the lower bound from (7), we can lower-bound $\mathbf{Pr}[\mathcal{C} \text{ succeeds}]$ as follows:

$$\mathbf{Pr}[\mathcal{C} \text{ succeeds}] \geq \rho(1 - e^{-n}) \sum_{\alpha \in \bar{H}} \mathbf{Pr}[\mathcal{E}_2 \text{ succeeds on } A \mid A = \alpha, S_{\not{L}}] * \mathcal{U}(\alpha).$$

Next, observe that

$$\sum_{\alpha \in \bar{H}} \mathbf{Pr}[\mathcal{E}_2 \text{ succeeds on } A \mid A = \alpha, S_{\not{L}}] * \mathcal{U}(\alpha) \geq$$
$$\sum_{\alpha \in \{0,1\}^n} \mathbf{Pr}[\mathcal{E}_2 \text{ succeeds on } A \mid A = \alpha, S_{\not{L}}] * \mathcal{U}(\alpha) - \gamma\delta/4,$$

since $\sum_{\alpha \in H} \mathcal{U}(\alpha) < \gamma\delta/4$. Expressing $\mathcal{U}(\alpha)$ as $(\mathcal{U}(\alpha) - \mathcal{D}_2(\alpha)) + \mathcal{D}_2(\alpha)$, we can rewrite

$$\sum_{\alpha \in \{0,1\}^n} \mathbf{Pr}[\mathcal{E}_2 \text{ succeeds on } A \mid A = \alpha, S_{\not L}] * \mathcal{U}(\alpha)$$

as

$$\sum_{\alpha \in \{0,1\}^n} \mathbf{Pr}[\mathcal{E}_2 \text{ succeeds on } A \mid A = \alpha, S_{\not L}] * \mathcal{D}_2(\alpha) +$$

$$\sum_{\alpha \in \{0,1\}^n} \mathbf{Pr}[\mathcal{E}_2 \text{ succeeds on } A \mid A = \alpha, S_{\not L}] * (\mathcal{U}(\alpha) - \mathcal{D}_2(\alpha)).$$

The first summand is exactly $\mathbf{Pr}[\mathcal{E}_2 \text{ succeeds}]$. The second summand can be lower-bounded by restricting the summation to those $\alpha \in \{0,1\}^n$ where $\mathcal{U}(\alpha) < \mathcal{D}_2(\alpha)$, and observing that the resulting expression is at least $-Dist(\mathcal{U}, \mathcal{D}_2)$.

Putting it all together, we get that

$$\mathbf{Pr}[\mathcal{C} \text{ succeeds}] \geq \rho(1 - e^{-n})(\mathbf{Pr}[\mathcal{E}_2 \text{ succeeds}] - Dist(\mathcal{U}, \mathcal{D}_2) - \gamma\delta/4).$$

Rearranging the terms on the right-hand side yields the claim.   □

The previous claim and Claim 2 yield the following final result.

Claim 7. $\mathbf{Pr}[\mathcal{C} \text{ succeeds}] \geq (1 - \delta) + \frac{\gamma\delta}{32}$.

Proof. Indeed, we have

$$\mathbf{Pr}[\mathcal{C} \text{ succeeds}] \geq \left(1 - \frac{t_0 + \Delta/2}{k}\right)\left(1 - \frac{\rho^{\Delta/2}}{\epsilon}\right) - \left(Dist(\mathcal{U}, \mathcal{D}_2) + (1 - \rho) + \rho \cdot e^{-n} + \frac{\gamma\delta}{4}\right). \quad (8)$$

For $\rho = 1 - \frac{\gamma\delta}{16}, \epsilon = 2e^{-\frac{\gamma^2\delta^2 k}{64}}, \Delta = \gamma\delta k$ and $t_0 = (1 - \gamma)\delta k$, we get $1 - (t_0 + \Delta/2)/k = 1 - \delta + \gamma\delta/2$ and $1 - \rho^{\Delta/2}/\epsilon \geq 1 - e^{-\gamma^2\delta^2 k/64}/2$. By Claim 3, we have that $Dist(\mathcal{U}, \mathcal{D}_2) \leq \gamma\delta/8$. So we can lowerbound the right-hand side of Eq. (8) by

$$1 - \delta - (1 - \delta)e^{-\gamma^2\delta^2 k/64}/2 + (1 - e^{-\gamma^2\delta^2 k/64}/2)\gamma\delta/2 - (\gamma\delta/8 + \gamma\delta/16 + \epsilon^{-n} + \gamma\delta/4),$$

which is at least $1 - \delta + \gamma\delta/32$, for sufficiently large n.   □

4   Open Problems

While the results here are fairly general, there are some obvious possible extensions. First, can similar results be proved for other domains, such as public-coin protocols ([18]). Also, our bounds on the adversary's success probability, although asymptotically exponentially small, are quite weak when applied to concrete problems such as actual CAPTCHA protocols with reasonable numbers of repetitions. Can the bounds be improved quantitatively, analogously to how [6] improved the bounds from [4]? Finally, we would like to find more applications of our results, to such problems as making strong secret agreement protocols from weak ones ([9]).

References

1. Aaronson, S.: Limitations of quantum advice and one-way communication. In: Proceedings of the Nineteenth Annual IEEE Conference on Computational Complexity, pp. 320–332. IEEE Computer Society Press, Los Alamitos (2004)
2. von Ahn, L., Blum, M., Hopper, N.J., Langford, J.: CAPTCHA: Using hard AI problems for security. In: Biham, E. (ed.) Advances in Cryptology – EUROCRPYT 2003. LNCS, vol. 2656, pp. 294–311. Springer, Heidelberg (2003)
3. Babai, L., Fortnow, L., Nisan, N., Wigderson, A.: BPP has subexponential time simulations unless EXPTIME has publishable proofs. Computational Complexity 3, 307–318 (1993)
4. Bellare, M., Impagliazzo, R., Naor, M.: Does parallel repetition lower the error in computationally sound protocols? In: Proceedings of the Thirty-Eighth Annual IEEE Symposium on Foundations of Computer Science, pp. 374–383. IEEE Computer Society Press, Los Alamitos (1997)
5. Chernoff, H.: A measure of asymptotic efficiency for tests of a hypothesis based on the sum of observations. Annals of Mathematical Statistics 23, 493–509 (1952)
6. Canetti, R., Halevi, S., Steiner, M.: Hardness amplification of weakly verifiable puzzles. In: Kilian, J. (ed.) TCC 2005. LNCS, vol. 3378, pp. 17–33. Springer, Heidelberg (2005)
7. Gal, A., Halevi, S., Lipton, R., Petrank, E.: Computing from partial solutions. In: Proceedings of the Fourteenth Annual IEEE Conference on Computational Complexity, pp. 34–45. IEEE Computer Society Press, Los Alamitos (1999)
8. Goldreich, O., Nisan, N., Wigderson, A.: On Yao's XOR-Lemma. Electronic Colloquium on Computational Complexity (TR95-050) (1995)
9. Holenstein, T.: Key agreement from weak bit agreement. In: Proceedings of the 37th ACM Symposium on Theory of Computing, pp. 664–673. ACM Press, New York (2005)
10. Impagliazzo, R.: Hard-core distributions for somewhat hard problems. In: Proceedings of the Thirty-Sixth Annual IEEE Symposium on Foundations of Computer Science, pp. 538–545. IEEE Computer Society Press, Los Alamitos (1995)
11. Impagliazzo, R., Jaiswal, R., Kabanets, V.: Approximately list-decoding direct product codes and uniform hardness amplification. In: Proceedings of the Forty-Seventh Annual IEEE Symposium on Foundations of Computer Science (FOCS06), pp. 187–196. IEEE Computer Society Press, Los Alamitos (2006)
12. Impagliazzo, R., Wigderson, A.: P=BPP if E requires exponential circuits: Derandomizing the XOR Lemma. In: Proceedings of the Twenty-Ninth Annual ACM Symposium on Theory of Computing, pp. 220–229. ACM Press, New York (1997)
13. Klivans, A.R.: On the derandomization of constant depth circuits. In: Goemans, M.X., Jansen, K., Rolim, J.D.P., Trevisan, L. (eds.) RANDOM 2001 and APPROX 2001. LNCS, vol. 2129. Springer, Heidelberg (2001)
14. Klauck, H., Spalek, R., de Wolf, R.: Quantum and classical strong direct product theorems and optimal time-space tradeoffs. In: Proceedings of the Forty-Fifth Annual IEEE Symposium on Foundations of Computer Science, pp. 12–21. IEEE Computer Society Press, Los Alamitos (2004)
15. Levin, L.A.: One-way functions and pseudorandom generators. Combinatorica 7(4), 357–363 (1987)
16. Nisan, N., Rudich, S., Saks, M.: Products and help bits in decision trees. In: Proceedings of the Thirty-Fifth Annual IEEE Symposium on Foundations of Computer Science, pp. 318–329. IEEE Computer Society Press, Los Alamitos (1994)

17. Parnafes, I., Raz, R., Wigderson, A.: Direct product results and the GCD problem, in old and new communication models. In: Proceedings of the Twenty-Ninth Annual ACM Symposium on Theory of Computing, pp. 363–372. ACM Press, New York (1997)
18. Pass, R., Venkitasubramaniam, M.: An efficient parallel repetition theorem for arthur-merlin games. In: STOC'07 (to appear)
19. Pietrzak, K., Wikstrom, D.: Parallel repetition of computationally sound protocols revisited. In: TCC'07, 2007 (to appear)
20. Raz, R.: A parallel repetition theorem. SIAM Journal on Computing 27(3), 763–803 (1998)
21. Shaltiel, R.: Towards proving strong direct product theorems. In: Proceedings of the Sixteenth Annual IEEE Conference on Computational Complexity, pp. 107–119. IEEE Computer Society Press, Los Alamitos (2001)
22. Trevisan, L.: List-decoding using the XOR lemma. In: Proceedings of the Forty-Fourth Annual IEEE Symposium on Foundations of Computer Science, pp. 126–135. IEEE Computer Society Press, Los Alamitos (2003)
23. Yao, A.C.: Theory and applications of trapdoor functions. In: Proceedings of the Twenty-Third Annual IEEE Symposium on Foundations of Computer Science, pp. 80–91. IEEE Computer Society Press, Los Alamitos (1982)

Rerandomizable RCCA Encryption

Manoj Prabhakaran and Mike Rosulek

Department of Computer Science, University of Illinois, Urbana-Champaign
{mmp,rosulek}@uiuc.edu

Abstract. We give the first perfectly rerandomizable, Replayable-CCA (RCCA) secure encryption scheme, positively answering an open problem of Canetti et al. (*CRYPTO* 2003). Our encryption scheme, which we call the *Double-strand Cramer-Shoup scheme*, is a non-trivial extension of the popular Cramer-Shoup encryption. Its security is based on the standard DDH assumption. To justify our definitions, we define a powerful "Replayable Message Posting" functionality in the Universally Composable (UC) framework, and show that any encryption scheme that satisfies our definitions of rerandomizability and RCCA security is a UC-secure implementation of this functionality. Finally, we enhance the notion of rerandomizable RCCA security by adding a receiver-anonymity (or key-privacy) requirement, and show that it results in a correspondingly enhanced UC functionality. We leave open the problem of constructing a scheme achieving this enhancement.

1 Introduction

Non-malleability and rerandomizability are opposing requirements to place on an encryption scheme. Non-malleability insists that an adversary should not be able to use one ciphertext to produce another one which decrypts to a related value. Rerandomizability on the other hand requires that anyone can alter a ciphertext into another ciphertext in an unlinkable way, such that both will decrypt to the same value. Achieving this delicate tradeoff was proposed as an open problem by Canetti et al. [7].

We present the first (perfectly) rerandomizable, RCCA-secure public-key encryption scheme. Because our scheme is a non-trivial variant of the Cramer-Shoup scheme, we call it the *Double-strand Cramer-Shoup* encryption. Like the original Cramer-Shoup scheme, the security of our scheme is based on the Decisional Diffie Hellman (DDH) assumption. Additionally, our method of using ciphertext components from two related groups may be of independent interest.

Going further, we give a combined security definition in the Universally-Composable (UC) security framework by defining a "Replayable Message Posting" functionality $\mathcal{F}_{\mathrm{RMP}}$. As a justification of the original definitions of rerandomizability and RCCA security, we show that any scheme which satisfies these definitions is also a *UC-secure realization* of the functionality $\mathcal{F}_{\mathrm{RMP}}$. (Here we restrict ourselves to static adversaries, as opposed to adversaries who corrupt the parties adaptively.) As an additional contribution on the definitional front,

A. Menezes (Ed.): CRYPTO 2007, LNCS 4622, pp. 517–534, 2007.

in Sect. 7.1, we introduce a notion of receiver anonymity for RCCA encryptions, and a corresponding UC functionality.

$\mathcal{F}_{\mathrm{RMP}}$ is perhaps the most sophisticated functionality that has been UC-securely realized in the standard model, i.e., without super-polynomial simulation, global setups, or an honest majority assumption.

Once we achieve this UC-secure functionality, simple modifications can be made to add extra functionality to our scheme, such as authentication and replay-testability (the ability for a ciphertext's recipient to check whether it was obtained via rerandomization of another ciphertext, or was encrypted independently).

Related work. Replayable-CCA security was proposed by Canetti et al. [7] as a relaxation of standard CCA security. They also raised the question of whether a scheme could be simultaneously *rerandomizable* and RCCA secure. Gröth [18] presented a rerandomizable scheme that achieved a weaker form of RCCA security, and another with full RCCA security in the *generic groups* model. Our work improves on [18], in that our scheme is more efficient, and we achieve full RCCA security in a standard model.

Rerandomizable encryption schemes also appear using the term *universal re-encryption* schemes (*universal* refers to the fact that the rerandomization/re-encryption routine does not require the public key), introduced by Golle et al. [17]. Their CPA-secure construction is based on El Gamal, and our construction can be viewed as a non-trivial extension of their approach, applied to the Cramer-Shoup construction.

The notion of receiver-anonymity (key-privacy) that we consider in Sect. 7.1 is an extension to the RCCA setting, of a notion due to Bellare et al. [3] (who introduced it for the simpler CPA and CCA settings).

As mentioned before, our encryption scheme is based on the Cramer-Shoup scheme [9,10], which in turn is modeled after El Gamal encryption [14]. The security of these schemes and our own is based on the DDH assumption (see, e.g. [4]). Cramer and Shoup [10] later showed a wide range of encryption schemes based on various assumptions which provide CCA security, under a framework subsuming their original scheme [9]. We believe that much of their generalization can be adapted to our current work as well, though we do not investigate this in detail here (see the remark in the concluding section).

Shoup [26] and An et al. [1] introduced a variant of RCCA security, called *benignly malleable*, or gCCA2, security. It is similar to RCCA security, but uses an arbitrary equivalence relation over ciphertexts to define the notion of replaying. However, these definitions preclude rerandomizability by requiring that the equivalence relation be efficiently computable *publicly*. A simple extension of our scheme achieves a modified definition of RCCA security, where the replay-equivalence relation is computable only by the ciphertext's designated recipient. Such a functionality also precludes perfect rerandomization, though our modification does achieve a computational relaxation of the rerandomization requirement.

Motivating applications. Golle et al. [17] propose a CPA-secure rerandomizable encryption scheme for use in mixnets [8] with applications to RFID tag anonymization. Implementing a re-encryption mixnet using a rerandomizable encryption scheme provides a significant simplification over previous implementations, which require distributed key management. Golle et al. call such networks *universal mixnets.* Some attempts have been made to strengthen their scheme against a naïve chosen-ciphertext attack, including by Klonowski et al. [19], who augment the scheme with a rerandomizable RSA signature. However, these modifications still do not prevent all practical chosen-ciphertext attacks, as demonstrated by Danezis [11].

We anticipate that by achieving full RCCA security, our construction will be an important step towards universal mixnets that do not suffer from active chosen-ciphertext attacks. However, mix-net applications tend to also require a "receiver-anonymity" property (see Sect. 7.1) from the underlying encryption scheme. In fact, the utility of rerandomizable RCCA encryption is greatly enhanced by this anonymity property. We do not have a scheme which achieves this. However, our current result is motivated in part by the power of such a scheme. We illustrate its potential with another example application (adapted from a private communication [20]). Consider a (peer-to-peer) network routing scenario, with the following requirements: (1) each packet should carry a *path object* which encodes its entire path to the destination; (2) each node in the network should not get any information from a path object other than the length of the path and the next hop in the path; and (3) there should be a mechanism to broadcast link-failure information so that any node holding a path object can check if the failed link occurs in that path, without gaining any additional information. This problem is somewhat similar to "Onion Routing" [5,12,16,21]. However, adding requirement (3) makes the above problem fundamentally different. Using an *anonymous,* rerandomizable, RCCA-secure encryption scheme one can achieve this selective revealing property as well as anonymity. We defer a more formal treatment of this scenario to future work.

Due to lack of space, we have omitted many details in this paper. We refer the readers to the online version for a detailed presentation [24].

2   Definitions

We call a function ν *negligible in* n if it asymptotically approaches zero faster than any inverse polynomial in n; that is, $\nu(n) = n^{-\omega(1)}$. We call a function *noticeable* if it is non-negligible. A probability is *overwhelming* if it is negligibly close to 1 (negligible in an implicit security parameter). In all the encryption schemes we consider, the security parameter is the number of bits needed to represent an element from the underlying cyclic group.

2.1   Encryption and Security Definitions

In this section we give the syntax of a perfectly rerandomizable encryption scheme, and then state our security requirements, which are formulated as indistinguishability experiments. Later, we justify these indistinguishability-based definitions by showing that any scheme which satisfies them is a secure realization of a powerful functionality in the UC security model, which we define in Sect. 5.

Syntax and correctness of a perfectly rerandomizable encryption scheme. A perfectly rerandomizable encryption scheme consists of four polynomial-time algorithms (polynomial in the implicit security parameter):

1. KeyGen: a randomized algorithm which outputs a *public key PK* and a corresponding *private key SK*.
2. Enc: a randomized encryption algorithm which takes a *plaintext* (from a plaintext space) and a public key, and outputs a *ciphertext*.
3. Rerand: a randomized algorithm which takes a ciphertext and outputs another ciphertext.
4. Dec: a deterministic decryption algorithm which takes a private key and a ciphertext, and outputs either a plaintext or an error indicator $\perp$.

We emphasize that the Rerand procedure takes only a ciphertext as input, and in particular, no public key.

We require the scheme to satisfy the following correctness properties for all key pairs $(PK, SK) \leftarrow$ KeyGen:

- For every plaintext msg and every (honestly generated) ciphertext $\zeta \leftarrow$ $\mathsf{Enc}_{PK}(\mathsf{msg})$, we must have $\mathsf{Dec}_{SK}(\zeta) = \mathsf{msg}$.
- For every independently chosen $(PK', SK') \leftarrow$ KeyGen, the sets of honestly generated ciphertexts under PK and PK' are disjoint, with overwhelming probability over the randomness of KeyGen.
- For every plaintext msg and every (honestly generated) ciphertext $\zeta \leftarrow$ $\mathsf{Enc}_{PK}(\mathsf{msg})$, the distribution of $\mathsf{Rerand}(\zeta)$ is identical to that of $\mathsf{Enc}_{PK}(\mathsf{msg})$.
- For every (purported) ciphertext ζ and every $\zeta' \leftarrow \mathsf{Rerand}(\zeta)$, we must have $\mathsf{Dec}_{SK}(\zeta') = \mathsf{Dec}_{SK}(\zeta)$.

In other words, decryption is the inverse of encryption, and ciphertexts can be labeled "honestly generated" for at most one honestly generated key pair. We require that rerandomizing an honestly generated ciphertext induces the same distribution as an independent encryption of the same message, while the only guarantee for an adversarially generated ciphertext is that rerandomization preserves the value of its decryption (under all private keys).

Perfect vs. computational rerandomization. For simplicity, we only consider statistically perfect rerandomization. However, for most purposes (including our UC functionality), a computational relaxation suffices. Computational rerandomization can be formulated as an indistinguishability experiment against an adversary; given two ciphertexts (of a chosen plaintext), no adversary can have a significant advantage in determining whether they are independent encryptions or if one is a rerandomization of the other. As in our other security experiments, the adversary is given access to a decryption oracle.

Replayable-CCA (RCCA) security. We use the definition from Canetti et al. [7]. An encryption scheme is said to be *RCCA secure* if the advantage of any PPT adversary $\mathcal{A}$ in the following experiment is negligible:

1. **Setup:** Pick $(PK, SK) \leftarrow$ KeyGen. $\mathcal{A}$ is given PK.
2. **Phase I:** $\mathcal{A}$ gets access to the decryption oracle $\mathsf{Dec}_{SK}(\cdot)$.
3. **Challenge:** $\mathcal{A}$ outputs a pair of plaintexts $(\mathsf{msg}_0, \mathsf{msg}_1)$. Pick $b \leftarrow \{0,1\}$ and let $\zeta^* \leftarrow \mathsf{Enc}_{PK}(\mathsf{msg}_b)$. $\mathcal{A}$ is given ζ^*.
4. **Phase II:** $\mathcal{A}$ gets access to a *guarded decryption oracle* $\mathsf{GDec}_{SK}^{(\mathsf{msg}_0, \mathsf{msg}_1)}$ which on input ζ, first checks if $\mathsf{Dec}_{SK}(\zeta) \in \{\mathsf{msg}_0, \mathsf{msg}_1\}$. If so, it returns replay; otherwise it returns $\mathsf{Dec}_{SK}(\zeta)$.
5. **Guess:** $\mathcal{A}$ outputs a bit $b' \in \{0,1\}$. The *advantage* of $\mathcal{A}$ in this experiment is $\Pr[b' = b] - \frac{1}{2}$.

Tightness of decryption. An encryption scheme is said to have *tight decryption* if the success probability of any PPT adversary $\mathcal{A}$ in the following experiment is negligible:

1. Pick $(PK, SK) \leftarrow$ KeyGen and give PK to $\mathcal{A}$.
2. $\mathcal{A}$ gets access to the decryption oracle $\mathsf{Dec}_{SK}(\cdot)$.
3. $\mathcal{A}$ outputs a ciphertext ζ. $\mathcal{A}$ is said to *succeed* if $\mathsf{Dec}_{SK}(\zeta) = \mathsf{msg} \neq \bot$ for some msg, yet ζ is *not* in the range of $\mathsf{Enc}_{PK}(\mathsf{msg})$.

Observe that when combined with correctness property (2), this implies that an adversary cannot generate a ciphertext which successfully decrypts under more than one honestly generated key. Such a property is useful in achieving a more robust definition of our UC functionality $\mathcal{F}_{\mathrm{RMP}}$ in Sect. 5 (without it, a slightly weaker yet still meaningful definition is achievable).

2.2 Decisional Diffie-Hellman (DDH) Assumption

Let $\mathbb{G}$ be a (multiplicative) cyclic group of prime order p. The *Decisional Diffie-Hellman (DDH) assumption in* $\mathbb{G}$ is that the following two distributions are computationally indistinguishable:

$$\{(g, g^a, g^b, g^{ab})\}_{g \leftarrow \mathbb{G}; a,b \leftarrow \mathbb{Z}_p} \quad \text{and} \quad \{(g, g^a, g^b, g^c)\}_{g \leftarrow \mathbb{G}; a,b,c \leftarrow \mathbb{Z}_p}.$$

Here, $x \leftarrow X$ denotes that x is drawn uniformly at random from a set X.

Cunningham chains. Our construction requires two (multiplicative) cyclic groups with a specific relationship: $\mathbb{G}$ of prime[1] order p, and $\widehat{\mathbb{G}}$ of prime order q, where $\widehat{\mathbb{G}}$ is a subgroup of $\mathbb{Z}_p^*$. We require the DDH assumption to hold in both groups (with respect to the same security parameter).

As a concrete example, the DDH assumption is believed to hold in $\mathbb{QR}_p^*$, the group of quadratic residues modulo p, where p and $\frac{p-1}{2}$ are prime (i.e, p is a *safe*

---

[1] It is likely that our security analysis can be extended to groups of orders with large prime factors, as is done in [10]. For simplicity, we do not consider this here.

prime). Given a sequence of primes $(q, 2q+1, 4q+3)$, the two groups $\widehat{\mathbb{G}} = \mathbb{QR}^*_{2q+1}$ and $\mathbb{G} = \mathbb{QR}^*_{4q+3}$ satisfy the needs of our construction. A sequence of primes of this form is called a *Cunningham chain* (of the first kind) of length 3 (see [2]). Such Cunningham chains are known to exist having q as large as 20,000 bits. It is conjectured that there are infinitely many such chains.

3   Motivating the Double-Strand Construction

Conceptually, the crucial enabling idea in our construction is that of using two "strands" of ciphertexts which can be recombined with each other for rerandomization without changing the encrypted value. To motivate this idea, we sketch the rerandomizable scheme of Golle et al. [17], which is based on the El Gamal scheme and secure against chosen plaintext attacks.

Recall that in an El Gamal encryption scheme over a group $\mathbb{G}$ of order p, the private key is $a \in \mathbb{Z}_p$ and the corresponding public key is $A = g^a$. A message $\mu \in \mathbb{G}$ is encrypted into the pair $(g^v, \mu A^v)$ for a random $v \in \mathbb{Z}_p$.

To encrypt a message $\mu \in \mathbb{G}$ in a "Double-strand El Gamal" scheme, we generate two (independent) El Gamal ciphertexts: one of μ (say, C_0) and one of the identity element in $\mathbb{G}$ (say, C_1). Such a double-strand ciphertext (C_0, C_1) can be rerandomized by computing $(C'_0, C'_1) = (C_0 C_1^r, C_1^s)$ for random $r, s \leftarrow \mathbb{Z}_p$ (where the operations on C_0 and C_1 are component-wise).

Our construction adapts this paradigm of rerandomization for Cramer-Shoup ciphertexts, and when chosen ciphertext attacks are considered. The main technical difficulty is in ensuring that the prescribed rerandomization procedure is the *only* way in which "strands" can be used to generate a valid ciphertexts.

Cramer-Shoup encryption. The Cramer-Shoup scheme [9] uses a group $\mathbb{G}$ of prime order p in which the DDH assumption is believed to hold. The private key is $b_1, b_2, c_1, c_2, d_1, d_2 \in \mathbb{Z}_p$ and the public key is $g_1, g_2 \in \mathbb{G}$, $B = \prod_{i=1}^{2} g_i^{b_i}$, $C = \prod_{i=1}^{2} g_i^{c_i}$, and $D = \prod_{i=1}^{2} g_i^{d_i}$.

To encrypt a message msg, first pick $x \in \mathbb{Z}_p$. and for $i = 1, 2$ let and $X_i = g_i^x$. Encode msg into an element μ in $\mathbb{G}$. The ciphertext is $(X_1, X_2, \mu B^x, (CD^m)^x)$ where $m = \mathsf{H}(X_1, X_2, \mu B^x)$ and H is a collision-resistant hash function.

In our scheme the ciphertext will contain two "strands," each one similar to a Cramer-Shoup ciphertext, allowing rerandomization as in the example above. However, instead of pairs we require 5-tuples of g_i, b_i, c_i, d_i (i.e., for $i = 1, \ldots, 5$). To allow for rerandomization, we use a direct encoding of the message for the exponent m (instead of a hash of part of the ciphertext). Finally, we thwart attacks which splice together strands from different encryptions by correlating the two strands with shared random masks.

Our security analysis is more complicated than the ones in [3,9,18]. However all these analyses as well as the current one follow the basic idea that if an encryption were to be carried out using the secret key in a "bad" way, the result will remain indistinguishable from an actual encryption (by the DDH assumption), but will also become statistically independent of the message and the public key.

4   Our Construction

In this section we describe our main construction, the *Double-strand Cramer-Shoup* (DSCS) encryption scheme. First, we introduce a simpler encryption scheme that is used as a component of the main scheme.

4.1   Double-Strand Malleable Encryption Scheme

We now define a rerandomizable encryption scheme which we call the "Double-strand malleable encryption" (DSME). As its name suggests, it is malleable, so it does not achieve our notions of RCCA security. However, it introduces the double-strand paradigm for rerandomization which we will use in our main construction. We will also use our DSME scheme as a component in our main construction, where its malleability will actually be a vital feature.

System parameters. A cyclic multiplicative group $\widehat{\mathbb{G}}$ of prime order q. $\widehat{\mathbb{G}}$ also acts as the message space for this scheme.

Key generation. Pick random generators $\widehat{g}_1, \widehat{g}_2, \widehat{g}_3$ from $\widehat{\mathbb{G}}$, and random $\mathbf{a} = (a_1, a_2, a_3)$ from $(\mathbb{Z}_q)^3$. The private key is $\mathbf{a}$. The public key consists of $\widehat{g}_1, \widehat{g}_2, \widehat{g}_3$, and $A = \prod_{j=1}^{3} \widehat{g}_j^{a_j}$.

Encryption: $\mathsf{MEnc}_{MPK}(u \in \widehat{\mathbb{G}})$:

- Pick random $v, w \in \mathbb{Z}_q$. For $j = 1, 2, 3$: let $V_j = \widehat{g}_j^v$ and $W_j = \widehat{g}_j^w$.
- Output $(\mathbf{V}, uA^v, \mathbf{W}, A^w)$, where $\mathbf{V} = (V_1, V_2, V_3)$ and $\mathbf{W} = (W_1, W_2, W_3)$.

Decryption: $\mathsf{MDec}_{MSK}(U = (\mathbf{V}, A_V, \mathbf{W}, A_W))$:

- **Check ciphertext integrity:** Check if $A_W \overset{?}{=} \prod_{j=1}^{3} W_j^{a_j}$. If not, output $\perp$.
- **Derive plaintext:** Output $A_V / \prod_{j=1}^{3} V_j^{a_j}$.

Rerandomization: $\mathsf{MRerand}(U = (\mathbf{V}, A_V, \mathbf{W}, A_W))$: The only randomness used in MEnc is the choice of v and w in $\widehat{\mathbb{G}}$. We can rerandomize both of these quantities by choosing random $s, t \in \mathbb{Z}_q$ and outputting the following ciphertext:

$$U' = (\mathbf{V}\mathbf{W}^s, A_V \cdot A_W^s, \mathbf{W}^t, A_W^t).$$

Here $\mathbf{V}\mathbf{W}^s$ and $\mathbf{W}^t$ denote component-wise operations. It is not hard to see that if U is in the range of $\mathsf{MEnc}_{MPK}(u)$ (with random choices v and w), then U' is in the range of $\mathsf{MEnc}_{MPK}(u)$ with corresponding random choices $v' = v + sw$ and $w' = tw$.

Homomorphic operation (multiplication by known value): Let $u' \in \widehat{\mathbb{G}}$ and let $U = (\mathbf{V}, A_V, \mathbf{W}, A_W)$ be a DSME ciphertext. We define the following operation:

$$u' \otimes U \stackrel{\text{def}}{=} (\mathbf{V}, u' \cdot A_V, \mathbf{W}, A_W).$$

It is not hard to see that for all private keys MSK, if $\mathsf{MDec}_{MSK}(U) \neq \perp$ then $\mathsf{MDec}_{MSK}(u' \otimes U) = u' \cdot \mathsf{MDec}_{MSK}(U)$, and if $\mathsf{MDec}_{MSK}(U) = \perp$ then $\mathsf{MDec}_{MSK}(U') = \perp$ as well.

Observe that this scheme is malleable under more than just multiplication by a known quantity. For instance, given $r \in \mathbb{Z}_q$ and an encryption of u, one can derive an encryption of u^r. As it turns out, the way we use DSME in the main construction ensures that we achieve our final security despite such additional malleabilities.

4.2 Double-Strand Cramer-Shoup Encryption Scheme

Now we give our main construction: a rerandomizable, RCCA-secure encryption scheme called the "Double-strand Cramer-Shoup" (DSCS) scheme. At the high level, it has two Cramer-Shoup encryption strands, one carrying the message, and the other to help rerandomize it. But unlike in the Cramer-Shoup scheme, we need to allow rerandomization, and so we do not use a prefix of the ciphertext itself in ensuring consistency; instead we use a direct encoding of the plaintext.

Further, we must prevent the possibility of mixing together strands from two *different* encryptions of the same message (say, in the manner in which rerandomizability allows two strands to be mixed together) to obtain a ciphertext which successfully decrypts, which would yield a successful adversarial strategy in our security experiments. For this, we correlate the two strands of a ciphertext with shared random masks. These masks are random exponents which are separately encrypted using the malleable DSME scheme described above (so that they may be hidden from everyone but the designated recipient, but also be rerandomized via the DSME scheme's homomorphic operation).

Finally, we must restrict the ways in which a ciphertext's two strands can be recombined, so that essentially the only way in which the two strands can be used to generate a ciphertext that decrypts successfully is to combine the two strands in the manner prescribed in the Rerand algorithm. To accomplish this, we perturb the exponents of the message-carrying strand by an additional (fixed) vector. Intuitively, this additive perturbance must remain present in the message-carrying strand of a ciphertext, which restricts the ways in which that strand can be combined with things. As a side-effect, our construction requires longer strands (i.e., more components) than in the original Cramer-Shoup scheme.

System parameters. A cyclic multiplicative group $\mathbb{G}$ of prime order p. A space of messages. An injective encoding $\mathsf{encode}_{\mathbb{G}}$ of messages into $\mathbb{G}$. An injective mapping $\mathsf{encode}_{\mathbb{Z}_p}$ of messages into $\mathbb{Z}_p$ (or into $\mathbb{Z}_p^*$, without any significant difference). These functions should be efficiently computable in both directions.

We also require a secure DSME scheme over a group $\widehat{\mathbb{G}}$ of prime order q, where $\widehat{\mathbb{G}}$ is also a subgroup of $\mathbb{Z}_p^*$. This relationship is crucial, as the homomorphic operation $\otimes$ of the DSME scheme must coincide with multiplication in the exponent in $\mathbb{G}$.

Finally, we require a fixed vector $\mathbf{z} = (z_1, \ldots, z_5) \in (\mathbb{Z}_p)^5$ with a certain degenerate property. For our purposes, $\mathbf{z} = (0, 0, 0, 1, 1)$ is sufficient.

Key generation. Generate 5 keypairs for the DSME scheme in $\widehat{\mathbb{G}}$. Call them $A_i, \mathbf{a}_i$ for $i = 1, \ldots, 5$.

Pick random generators $g_1, \ldots, g_5 \in \mathbb{G}$, and random $\mathbf{b} = (b_1, \ldots, b_5), \mathbf{c} = (c_1, \ldots, c_5), \mathbf{d} = (d_1, \ldots, d_5)$ from $(\mathbb{Z}_p)^5$. The private key consists of $\mathbf{b}, \mathbf{c}, \mathbf{d}$ and the 5 private keys for the DSME scheme. The public key consists of $(g_1, \ldots g_5)$, the 5 public keys for the DSME scheme, and the following values:

$$B = \prod_{i=1}^{5} g_i^{b_i}, \qquad C = \prod_{i=1}^{5} g_i^{c_i}, \qquad D = \prod_{i=1}^{5} g_i^{d_i}.$$

Encryption: $\mathsf{Enc}_{PK}(\mathsf{msg})$:

- Pick random $x, y \in \mathbb{Z}_p^*$ and random $u_1, \ldots, u_5 \in \widehat{\mathbb{G}}$.
- For $i = 1, \ldots, 5$: let $X_i = g_i^{(x+z_i)u_i}$; $Y_i = g_i^{y u_i}$; and $U_i = \mathsf{MEnc}_{A_i}(u_i)$.
- Let $\mu = \mathsf{encode}_{\mathbb{G}}(\mathsf{msg})$, and $m = \mathsf{encode}_{\mathbb{Z}_p}(\mathsf{msg})$.
- Output:

$$(\mathbf{X}, \mu B^x, (CD^m)^x, \mathbf{Y}, B^y, (CD^m)^y, \mathbf{U}),$$

where $\mathbf{U} = (U_1, \ldots, U_5), \mathbf{X} = (X_1, \ldots, X_5), \mathbf{Y} = (Y_1, \ldots, Y_5)$.

Decryption: $\mathsf{Dec}_{SK}(\zeta = (\mathbf{X}, B_X, P_X, \mathbf{Y}, B_Y, P_Y, \mathbf{U}))$:

- **Decrypt U_i's:** For $i = 1, \ldots, 5$: set $u_i = \mathsf{MDec}_{\mathbf{a}_i}(U_i)$. If any $u_i = \perp$, immediately output $\perp$.
- **Strip u_i's and z_i's:** For $i = 1, \ldots, 5$: set $\overline{X}_i = X_i^{1/u_i} g_i^{-z_i}$ and $\overline{Y}_i = Y_i^{1/u_i}$.
- **Derive purported plaintext:** Set $\mu = B_X / \prod_{i=1}^{5} \overline{X}_i^{b_i}$, $\mathsf{msg} = \mathsf{encode}_{\mathbb{G}}^{-1}(\mu)$, and $m = \mathsf{encode}_{\mathbb{Z}_p}(\mathsf{msg})$.
- **Check ciphertext integrity:** Check the following conditions:

$$B_Y \overset{?}{=} \prod_{i=1}^{5} \overline{Y}_i^{b_i}; \qquad P_X \overset{?}{=} \prod_{i=1}^{5} \overline{X}_i^{c_i + d_i m}; \qquad P_Y \overset{?}{=} \prod_{i=1}^{5} \overline{Y}_i^{c_i + d_i m}.$$

If any checks fail, output $\perp$. Otherwise output msg.

Rerandomization: $\mathsf{Rerand}(\zeta = (\mathbf{X}, B_X, P_X, \mathbf{Y}, B_Y, P_Y, \mathbf{U}))$: The only randomness used in Enc is the choice of x, y, $\mathbf{u} = (u_1, \ldots, u_5)$, and the randomness used in each instance of MEnc. We can rerandomize each of these quantities by choosing random $r_1, \ldots, r_5 \in \widehat{\mathbb{G}}$, random $s, t \in \mathbb{Z}_p^*$, and constructing a ciphertext which corresponds to an encryption of the same message, with corresponding random choices $u_i' = u_i r_i$, $x' = x + ys$, and $y' = yt$:

- For $i = 1, \ldots, 5$, set $U_i' = \mathsf{MRerand}(r_i \otimes U_i)$; $X_i' = (X_i Y_i^s)^{r_i}$; and $Y_i' = Y_i^{r_i t}$.
- $B_X' = B_X B_Y^s$ and $P_X' = P_X P_Y^s$.
- $B_Y' = B_Y^t$ and $P_Y' = P_Y^t$.

The rerandomized ciphertext is $\zeta' = (\mathbf{X}', B_X', P_X', \mathbf{Y}', B_Y', P_Y', \mathbf{U}')$.

4.3 Complexity

The complexities of the DSCS scheme are summarized in Table 1 and Table 2.[2]
Clearly our scheme is much less efficient than the Cramer-Shoup encryption

Table 1. Number of elements

	$\widehat{\mathbb{G}}$	$\mathbb{Z}_q$	$\mathbb{G}$	$\mathbb{Z}_p$
Public key	20	-	8	-
Private key	-	15	-	15
Ciphertext	40	-	14	-

Table 2. Group operations performed

	exp.		mult.		inv.	
	$\widehat{\mathbb{G}}$	$\mathbb{G}$	$\mathbb{Z}_p^*$	$\mathbb{G}$	$\mathbb{Z}_p^*$	$\mathbb{G}$
Enc	40	16	15	3	0	0
Dec (worst case)	30	35	40	22	10	1
Rerand	36	19	40	7	0	0

scheme. On the other hand, it is much more efficient than the only previously proposed rerandomizable (weak) RCCA-secure scheme [18], which used $O(k)$ group elements to encode a k-bit message (or in other words, to be able to use the group itself as the message space, it used $O(\log p)$ group elements). In fact, if we restrict ourselves to weak RCCA security (as defined in [18]) and a computational version of rerandomizability, our construction can be simplified to have only 10 group elements (we omit the details of that construction in this paper).

Rerandomizable RCCA security is a significantly harder problem by our current state of knowledge. Despite the inefficiency, we believe that by providing the first complete solution (i.e., not in the generic group model) we have not only solved the problem from a theoretical perspective, but also opened up the possibility of efficient and practical constructions.

5   Replayable Message Posting

We define the "Replayable Message Posting" functionality $\mathcal{F}_{\mathrm{RMP}}$ in the Universally Composable (UC) security framework [6,22], also variously known as environmental security [15,25] and network-aware security [23] framework.

This functionality concisely presents the security achieved by a rerandomizable, RCCA-secure encryption scheme. The functionality allows parties to publicly post messages which are represented by abstract *handles*, arbitrary strings provided by the adversary. The adversary is not told the actual message (unless, of course, the recipient is corrupted by the adversary). Only the designated receiver is allowed to obtain the corresponding message from the functionality.

Additionally, $\mathcal{F}_{\mathrm{RMP}}$ provides a reposting functionality: any party can "repost" (i.e., make a copy of) any existing handle. Requesting a repost does not reveal the message. To the other parties (including the adversary and the original message's recipient), the repost appears exactly like a normal message posting; i.e, the

---

[2] Multiplication and inversion operations in $\mathbb{Z}_p^*$ include operations in the subgroup $\widehat{\mathbb{G}}$. We assume that for $\widehat{g}_i$ elements of the public key, $\widehat{g}_i^{-1}$ can be precomputed.

functionality's external behavior is no different for a repost versus a normal post.

A similar functionality $\mathcal{F}_{\text{RPKE}}$ was defined by Canetti et. al [7] to capture (not necessarily rerandomizable) RCCA security. $\mathcal{F}_{\text{RPKE}}$ is itself a modification of the $\mathcal{F}_{\text{PKE}}$ functionality of [6], which modeled CCA security. Both of these functionalities similarly represent messages via abstract handles. However, the most important distinction between these two functionalities is that $\mathcal{F}_{\text{RMP}}$ provides the ability to repost handles as a *feature*; thus, it does not include the notion of "decrypting" handles which are not previously known to the functionality.

We now formally define the behavior of $\mathcal{F}_{\text{RMP}}$. It accepts the following four kinds of requests from parties:

Registration: On receiving a message register from a party sender, the functionality sends (ID-REQ, sender) to the adversary, and expects in response an identifier string id.[3] If the string received in response has been already used, ignore the request. Otherwise respond to sender with the string id, and also send a message (ID-ANNOUNCE, id) to all other parties.

Additionally, we reserve a special identifier $id_\perp$ for the adversary. The adversary need not explicitly register to use this identifier, nor is it announced to the other parties. We also insist that only corrupted parties are allowed to post messages for $id_\perp$ (though honest parties may repost the resulting handles).[4]

Message posting: On receiving a request (post, id, msg) from a party sender, the functionality behaves as follows:[5] If id is not registered, ignore the request.

If id is registered to an uncorrupted party, send (HANDLE-REQ, sender, id) to the adversary; otherwise send (HANDLE-REQ, sender, id, msg) to the adversary. In both cases, expect a string handle in return. If handle has been previously used, ignore this request. Otherwise, record (handle, sender, id, msg) internally and publish (HANDLE-ANNOUNCE, handle, id) to all registered parties.

Note that if the recipient of a message is corrupted, it is reasonable for the functionality to reveal msg to the adversary when requesting the handle.

Message reposting: On receiving a message (repost, handle) from a party sender, the functionality behaves as follows: If handle is not recorded internally, ignore the request.

Otherwise, suppose (handle, sender', id, msg) is recorded internally. If id is registered to an uncorrupted party, send (HANDLE-REQ, sender, id) to the adversary;

---

[3] This can be modified to have the functionality itself pick an id from a predetermined distribution specified as part of the functionality. In this case the functionality will also provide the adversary with some auxiliary information about id (e.g., the randomness used in sampling id). For simplicity we do not use such a stronger formulation.

[4] $id_\perp$ models the fact that an adversary may generate key pairs without announcing them, and broadcast encryptions under those keys.

[5] We assume that msg is from a predetermined message space, with size superpolynomial in the security parameter; otherwise the request is ignored.

otherwise send (HANDLE-REQ, sender, id, msg) to the adversary. In both cases, expect a string handle' in return. If handle' has been previously used, ignore this request. Otherwise, record (handle', sender, id, msg) internally and publish (HANDLE-ANNOUNCE, handle', id) to all registered parties.

As above, if the message's recipient is corrupted, the functionality can legitimately reveal msg to the adversary when requesting the handle.

Message reading: On receiving a message (get, handle) from a party, if a record (handle, sender, id, msg) is recorded internally, and id is registered to this party, then return (id, msg) to it. Otherwise ignore this request.

6   Results

We present two main results below. The first is that the DSCS encryption scheme presented in Sect. 4 achieves the security definitions defined in Sect. 2.1. The second result is that any construction which meets these guarantees is a secure realization of the $\mathcal{F}_{\mathrm{RMP}}$ functionality defined in Sect. 5. For the complete proofs of these results, we refer the reader to the full version of this paper [24].

Theorem 1. *The DSCS scheme (Sect. 4) is a perfectly rerandomizable encryption scheme which satisfies the definitions of RCCA security and tight decryption under the DDH assumption in* $\mathbb{G}$ *and* $\widehat{\mathbb{G}}$.

PROOF OVERVIEW: Here we sketch an outline of the proof of RCCA security.

It is convenient to formulate our proof in terms of alternate encryption and decryption procedures. We remark that this outline is similar to that used in previous proofs related to the Cramer-Shoup construction [3,9,10,13]. However, the implementation is significantly more involved in our case.

Alternate encryption. First, we would like to argue that the ciphertexts hide the message and the public key used in the encryption. For this we describe an alternate encryption procedure AltEnc. AltEnc actually uses the *private key* to generate ciphertexts. In short, instead of using $\{X_i = g_i^x\}$ and $\{Y_i = g_i^y\}$, AltEnc picks random group elements for these ciphertext components, then uses the private key to generate the other components according to the quantities which are computed by Dec.

When AltEnc is used to generate the challenge ciphertext in the RCCA security experiment, it follows from the DDH assumption in $\mathbb{G}$ and $\widehat{\mathbb{G}}$ that for any adversary the experiment's outcome does not change significantly. Additionally, the ciphertexts produced by AltEnc are information-theoretically independent of the message.

Alternate decryption. An adversary may be able to get information about the message used in the encryption not only from the challenge ciphertext, but also from the answers to the decryption queries that it makes. Indeed, since the decryption oracle uses the private key there is a danger that information about the private key is leaked, especially when the oracle answers maliciously crafted ciphertexts. To show that our scheme does leak information in this way, we describe an alternate decryption procedure AltDec to be used in the security experiments, which can be implemented using only the public key(s) and challenge ciphertext (quantities which are already known to the adversary). AltDec will be computationally unbounded, but since it is accessed as an oracle, this does not affect the analysis. More importantly, its functionality is statistically indistinguishable from the honest decryption procedure (even when the adversary is given a ciphertext generated by AltEnc).

By computing discrete logarithms of some components of its input and comparing with the public key and challenge ciphertext, the alternate decryption procedure can check whether its input is "looks like" an honest encryption or a rerandomization of the challenge ciphertext, and give the correct response in these cases. To establish the correctness of this approach, we show that ciphertexts which are rejected by AltDec would be rejected by the normal decryption algorithm with overwhelming probability as well. The $\mathbf{u}$ and $\mathbf{z}$ components of our construction are vital in preventing all other ways of combining the challenge strands and the public key. This is the most delicate part of our proof.

We conclude that with these two modifications – alternate challenge ciphertext and the alternate decryption procedure – the adversary's view in the RCCA security experiment is independent of the secret bit b, and so the adversary's advantage is zero. Furthermore, the outcome of this modified experiment is only negligibly different from the outcome of the original experiment, so the security claim follow.   ◁

Theorem 2. *Every rerandomizable encryption scheme which is RCCA-secure, and has tight decryption*[6] *is a secure realization of* $\mathcal{F}_{\mathrm{RMP}}$ *in the standard UC model.*

PROOF OVERVIEW: For simplicity we consider all communications to use a broadcast channel. The scheme yields a protocol for $\mathcal{F}_{\mathrm{RMP}}$ in the following natural way: public keys correspond to identifiers and ciphertexts correspond to handles. To register oneself, one generates a key pair and broadcasts the public key. To post a message to a party, one simply encrypts the message under his public key. To repost a handle, one simply applies the rerandomization procedure. To retrieve a message in a handle, one simply decrypts it using one's private key.

To prove the security of this protocol, we demonstrate a simulator $\mathcal{S}$ for each adversary $\mathcal{A}$. The simulator $\mathcal{S}$ internally runs $\mathcal{A}$ and behaves as follows:

When $\mathcal{F}_{\mathrm{RMP}}$ receives a registration request from an honest party, it requests an identifier from $\mathcal{S}$. $\mathcal{S}$ generates a key pair, sends the public key as the identifier string, and simulates to $\mathcal{A}$ that an honest party broadcasted this public key.

---

[6] By relaxing the requirement that the scheme have tight decryption (and also the correctness requirement that ciphertexts are not honest ciphertexts for more than one honest key pair), we can still realize a weaker variant of $\mathcal{F}_{\mathrm{RMP}}$. In this variant, handles may be re-used, and the adversary is notified any time an honest party reposts any handle which the adversary posted/reposted. We omit the details here.

When $\mathcal{F}_{\text{RMP}}$ receives a post request addressed to an honest party (or a repost request for such a handle), it requests a new handle from $\mathcal{S}$, without revealing the message. The ith time this happens, $\mathcal{S}$ generates the handle handle_i^H by picking a random message msg_i^H and encrypting it under the given identity (say, public key PK_i). In its simulation, $\mathcal{S}$ uses this ciphertext as the one broadcast by the sender. The correctness of this simulated ciphertext follows from the scheme's RCCA security property.

When $\mathcal{A}$ broadcasts a public key, $\mathcal{S}$ registers it as an identifier in $\mathcal{F}_{\text{RMP}}$. When $\mathcal{A}$ broadcasts a ciphertext ζ, $\mathcal{S}$ behaves as follows:

1. If for some i, $\text{Dec}_{SK_i}(\zeta) = \text{msg}_i^H$ (the ith random message chosen to simulate a ciphertext between honest parties), then $\mathcal{S}$ instructs $\mathcal{F}_{\text{RMP}}$ to repost handle_i^H.
2. If $\text{Dec}_{SK}(\zeta) = \text{msg} \neq \bot$ for any of the private keys SK that it picked while registering honest parties, then $\mathcal{S}$ instructs $\mathcal{F}_{\text{RMP}}$ to post msg, addressed to the corresponding honest party.
3. Otherwise, ζ does not successfully decrypt under any of these private keys. The jth time this happens, $\mathcal{S}$ picks a random message msg_j^A and instructs $\mathcal{F}_{\text{RMP}}$ to post msg_j^A to $\text{id}_\bot$. It also remembers $\text{handle}_j^A = \zeta$.

In all the above cases, $\mathcal{S}$ sends ζ to $\mathcal{F}_{\text{RMP}}$ as the handle for this new message. Tight decryption ensures that at most one of the above decryptions succeeds. Further, if one does succeed, the ciphertext must be in the support of honest encryptions, so the perfect rerandomization condition holds for it.

When $\mathcal{F}_{\text{RMP}}$ receives a post request addressed to a corrupted party (or a repost request for such a handle), it sends the corresponding message msg and identifier id to $\mathcal{S}$, and requests a new handle.

1. If $\text{id} = \text{id}_\bot$ and $\text{msg} = \text{msg}_j^A$ (the jth random message chosen to simulate an adversarial ciphertext to $\mathcal{F}_{\text{RMP}}$), then $\mathcal{S}$ generates a handle by rerandomizing the corresponding handle_j^A.
2. Otherwise, $\mathcal{S}$ generates a handle by encrypting the message under the appropriate public key.

In its simulation, $\mathcal{S}$ uses this ciphertext as the one broadcast by the sender. $\lhd$

7   Extensions

Once our construction is made available as a UC-secure realization of $\mathcal{F}_{\text{RMP}}$, it is easier to extend in a modular fashion. We first describe a few useful extensions which are easily achieved, and then discuss extending the notion of receiver-anonymity to rerandomizable RCCA encryption schemes.

Replay-test. In some applications, it is convenient if the recipient of a ciphertext is able to check whether it is a rerandomization of another ciphertext, or an independent encryption of the same plaintext. We call such a feature a *replay-test* feature. A replay-test precludes having perfect or even statistical rerandomization, and so we must settle for a computational definition of rerandomization.

Redefining RCCA security and rerandomizability for schemes which include a replay-test feature is a non-trivial extension of our current definitions. In particular, note that in a chosen ciphertext attack, the adversary should be allowed to access a replay-test oracle as well as a decryption oracle, while responses from the decryption oracle should be guarded based on the replay-test instead of a check of the plaintext.

However, instead of modifying our security definitions based on standalone experiments, we can directly formulate a new UC functionality. The functionality is identical to $\mathcal{F}_{RMP}$, but it provides an additional **test** command: a party can give two handles, and if it is the designated receiver of both the handles, then the functionality tells it whether the two handles were derived as reposts of the same original post. To do this, the functionality maintains some extra book-keeping internally. This functionality can be easily achieved starting from $\mathcal{F}_{RMP}$: each message is posted with a random nonce appended. To implement **test**, the receiver retrieves the messages of the two handles and compares their nonces.

Authentication. As should be intuitive, authentication can be achieved by signing the messages using a public-key signature scheme, before posting them. In terms of the functionality, a separate **register** feature is provided which allows *senders* to register themselves (this corresponds to publishing the signature verification key). Then the functionality's **get** command is augmented to provide not only the message in the handle, but also who originally posted the handle. The identifiers for receiving messages and sending messages are separate, but they can be tied together (by signing the encryption scheme's public key and publishing it), so that only the signature verification keys need to be considered as identifiers in the system.

Variable-length plaintexts. In our presentation of our encryption scheme, there is a hard limit on the message length, because the message must be encoded as an element in a group of fixed size. However, $\mathcal{F}_{RMP}$ can easily be extended to allow messages of variable lengths: for this the longer message is split into smaller pieces; a serial number and a common random nonce are appended to each piece; the first piece also carries the total number of pieces. Then each piece is posted using the fixed-length $\mathcal{F}_{RMP}$ functionality. The decryption procedure performs the obvious simple integrity checks on a set of ciphertexts and discards them if they are not all consistent and complete. Note that the resulting modification to the $\mathcal{F}_{RMP}$ functionality tells the adversary the length of the message (i.e., number of pieces) while posting or reposting a handle. It is straight-forward to construct a simulator (on receiving a handle and a length, the simulator creates the appropriate number of handles and reports to the adversary; when the adversary reposts handles, the simulator will not make a repost to the functionality unless all handles it generated for one variable-length handle are reposted together).We note that these extensions can be applied one after the other.

7.1 Anonymity

Bellare et al. [3] introduced the notion of receiver-anonymity (or key-privacy) for CPA and CCA secure encryption schemes, which we extend to the RCCA setting. In a system with multiple users, rerandomizability of ciphertexts, without receiver-anonymity, may not be satisfactory. For instance, while rerandomizability allows unlinkability of multiple encryptions in terms of their contents, without anonymity they could all be linked as going to the same receiver.

RCCA receiver-anonymity. An encryption scheme is said to be *RCCA receiver-anonymous* if the advantage of any PPT adversary $\mathcal{A}$ in the following experiment is negligible:

1. **Setup:** Pick $(PK_0, SK_0) \leftarrow$ KeyGen and $(PK_1, SK_1) \leftarrow$ KeyGen. $\mathcal{A}$ is given (PK_0, PK_1)
2. **Phase I:** $\mathcal{A}$ gets access to the decryption oracles $\text{Dec}_{SK_0}(\cdot)$ and $\text{Dec}_{SK_1}(\cdot)$.
3. **Challenge:** $\mathcal{A}$ outputs a plaintext msg. Pick $b \leftarrow \{0,1\}$ and let $\zeta^* \leftarrow \text{Enc}_{PK_b}(\text{msg})$. $\mathcal{A}$ is given ζ^*.
4. **Phase II:** $\mathcal{A}$ gets access to a *guarded decryption oracle* $\text{GDec}^{\text{msg}}_{SK_0, SK_1}(\cdot)$, which on input ζ, first checks if $\text{msg} \in \{\text{Dec}_{SK_0}(\zeta), \text{Dec}_{SK_1}(\zeta)\}$. If so, it returns replay; otherwise it returns the pair $(\text{Dec}_{SK_0}(\zeta), \text{Dec}_{SK_1}(\zeta))$.
5. **Guess:** $\mathcal{A}$ outputs a bit $b' \in \{0,1\}$. The *advantage* of $\mathcal{A}$ in this experiment is $\Pr[b' = b] - \frac{1}{2}$.

Our scheme does not achieve this definition of anonymity. We leave it as an interesting open problem.

Modifications to $\mathcal{F}_{\text{RMP}}$. If a rerandomizable RCCA-secure encryption scheme additionally meets this definition of RCCA anonymity, the scheme can be used to implement an "anonymous" variant of $\mathcal{F}_{\text{RMP}}$. In this variant, the functionality does not reveal the handle's recipient in a HANDLE-ANNOUNCE broadcast, nor in the HANDLE-REQ messages it sends to the adversary (unless the handle's recipient is corrupted).

   In proving this, compared to the proof of Theorem 2, the only change in the simulator for this modified functionality is that it uses a "dummy" public key to generate all simulated ciphertexts addressed to honest recipients, instead of using the correct key.

8   Conclusions and Future Directions

This work leads to several interesting questions. First, can the efficiency of our scheme be improved? Public-key encryption schemes like Cramer-Shoup are much less efficient than private-key schemes. To exploit the best of both worlds, one can use a hybrid encryption scheme which uses a public-key encryption scheme to share a private key, and then encrypt the actual voluminous data with the private-key encryption. It is interesting to consider whether such a hybrid scheme can be devised in a rerandomizable manner. To achieve perfect rerandomization, the public-key scheme would have to be malleable (to rerandomize the private key), and the private-key scheme should allow reencryption which changes the key accordingly.

Second, can rerandomizable RCCA-secure schemes be constructed which are also RCCA receiver-anonymous? As mentioned earlier, many applications require not only rerandomizability of the ciphertexts, but also receiver anonymity.

Third, can CCA-like tradeoffs be defined for encryption schemes with more sophisticated homomorphic features? In this work, we give a tight tradeoff, allowing malleability via the identity function in the strongest manner, while prohibiting all other kinds of malleability. How can one define (and achieve) a similar tradeoff for, say, malleability via multiplication?

Finally, we based our schemes on the DDH assumption. However, as mentioned before, it is likely that the extensions of Cramer and Shoup [10] can be adapted for our problem too. But we point out that our requirements on the "Universal Hash Proofs" would be more demanding than what they require. In particular, when using the double-strand approach, we seem to require 5-universality instead of 2-universality, corresponding to our use of five bases $g_1, \ldots, g_5$ instead of just two.

Acknowledgments

We would like to acknowledge useful discussions with Rui Xue about the work in [10]. We also thank Ran Canetti, Michael Loui, and the anonymous referees for their helpful feedback on earlier drafts of this manuscript.

References

1. An, J.H., Dodis, Y., Rabin, T.: On the security of joint signature and encryption. In: Knudsen, L.R. (ed.) EUROCRYPT 2002. LNCS, vol. 2332, pp. 83–107. Springer, Heidelberg (2002)
2. Andersen, J.K., Weisstein, E.W.: Cunningham chain. From MathWorld–A Wolfram Web Resource (2005), http://mathworld.wolfram.com/CunninghamChain.html
3. Bellare, M., Boldyreva, A., Desai, A., Pointcheval, D.: Key-privacy in public-key encryption. In: Boyd, C. (ed.) ASIACRYPT 2001. LNCS, vol. 2248, pp. 566–582. Springer, Heidelberg (2001)
4. Boneh, D.: The decision diffie-hellman problem. In: Buhler, J.P. (ed.) Algorithmic Number Theory. LNCS, vol. 1423, pp. 48–63. Springer, Heidelberg (1998)
5. Camenisch, J., Lysyanskaya, A.: A formal treatment of onion routing. In: Shoup, V. (ed.) CRYPTO 2005. LNCS, vol. 3621, pp. 169–187. Springer, Heidelberg (2005)
6. Canetti, R.: Universally composable security: A new paradigm for cryptographic protocols. Cryptology ePrint Archive, Report 2000/067 (2005)
7. Canetti, R., Krawczyk, H., Nielsen, J.B.: Relaxing chosen-ciphertext security. In: Boneh, D. (ed.) CRYPTO 2003. LNCS, vol. 2729, pp. 565–582. Springer, Heidelberg (2003)
8. Chaum, D.: Untraceable electronic mail, return addresses, and digital pseudonyms. Commun. ACM 4(2) (February 1981)

9. Cramer, R., Shoup, V.: A practical public key cryptosystem provably secure against adaptive chosen ciphertext attack. In: Krawczyk, H. (ed.) CRYPTO 1998. LNCS, vol. 1462. Springer, Heidelberg (1998)
10. Cramer, R., Shoup, V.: Universal hash proofs and a paradigm for adaptive chosen ciphertext secure public-key encryption. In: Knudsen, L.R. (ed.) EUROCRYPT 2002. LNCS, vol. 2332, pp. 45–64. Springer, Heidelberg (2002)
11. Danezis, G.: Breaking four mix-related schemes based on universal re-encryption. In: Proceedings of Information Security Conference 2006. Springer, Heidelberg (2006)
12. Dingledine, R., Mathewson, N., Syverson, P.F.: Tor: The second-generation onion router. In: USENIX Security Symposium, pp. 303–320. USENIX (2004)
13. Elkind, E., Sahai, A.: A unified methodology for constructing public-key encryption schemes secure against adaptive chosen-ciphertext attack. Cryptology ePrint Archive, Report 2002/042 (2002), http://eprint.iacr.org/
14. Gamal, T.E.: A public key cryptosystem and a signature scheme based on discrete logarithms. In: Blakely, G.R., Chaum, D. (eds.) CRYPTO 1984. LNCS, vol. 196, pp. 10–18. Springer, Heidelberg (1985)
15. Goldreich, O.: Foundations of Cryptography: Basic Applications. Cambridge University Press, Cambridge (2004)
16. Goldschlag, D.M., Reed, M.G., Syverson, P.F.: Onion routing. Commun. ACM 42(2), 39–41 (1999)
17. Golle, P., Jakobsson, M., Juels, A., Syverson, P.: Universal re-encryption for mixnets. In: Proceedings of the 2004 RSA Conference, Cryptographer's track, San Francisco, USA (February 2004)
18. Gröth, J.: Rerandomizable and replayable adaptive chosen ciphertext attack secure cryptosystems. In: Naor, M. (ed.) TCC 2004. LNCS, vol. 2951, pp. 152–170. Springer, Heidelberg (2004)
19. Klonowski, M., Kutylowski, M., Lauks, A., Zagórski, F.: Universal re-encryption of signatures and controlling anonymous information flow. In: WARTACRYPT '04 Conference on Cryptology. Bedlewo/Poznan (2006)
20. Lad, M.: Personal communication (2005)
21. The onion routing program. A program sponsored by the Office of Naval Research, DARPA and the Naval Research Laboratory, http://www.onion-router.net/
22. Pfitzmann, B., Waidner, M.: Composition and integrity preservation of secure reactive systems. In: ACM Conference on Computer and Communications Security, pp. 245–254. ACM Press, New York (2000)
23. Prabhakaran, M.: New Notions of Security. PhD thesis, Department of Computer Science, Princeton University (2005)
24. Prabhakaran, M., Rosulek, M.: Anonymous rerandomizable rcca encryption. Cryptology ePrint Archive, Report 2007/119, (2007), http://eprint.iacr.org/
25. Prabhakaran, M., Sahai, A.: New notions of security: achieving universal composability without trusted setup. In: STOC, pp. 242–251. ACM Press, New York (2004)
26. Shoup, V.: A proposal for an iso standard for public key encryption. Cryptology ePrint Archive, Report 2001/112, (2001), http://eprint.iacr.org/

Deterministic and Efficiently Searchable Encryption

Mihir Bellare[1], Alexandra Boldyreva[2], and Adam O'Neill[2]

[1] Dept. of Computer Science & Engineering, University of California at San Diego,
9500 Gilman Drive, La Jolla, CA 92093, USA
mihir@cs.ucsd.edu
http://www-cse.ucsd.edu/users/mihir
[2] College of Computing, Georgia Institute of Technology,
266 Ferst Drive, Atlanta, GA 30332, USA
{aboldyre,amoneill}@cc.gatech.edu
http://www.cc.gatech.edu/{∼ aboldyre, amoneill}

Abstract. We present as-strong-as-possible definitions of privacy, and constructions achieving them, for public-key encryption schemes where the encryption algorithm is *deterministic*. We obtain as a consequence database encryption methods that permit fast (i.e. sub-linear, and in fact logarithmic, time) search while provably providing privacy that is as strong as possible subject to this fast search constraint. One of our constructs, called RSA-DOAEP, has the added feature of being length preserving, so that it is the first example of a public-key cipher. We generalize this to obtain a notion of efficiently-searchable encryption schemes which permit more flexible privacy to search-time trade-offs via a technique called bucketization. Our results answer much-asked questions in the database community and provide foundations for work done there.

1 Introduction

The classical notions of privacy for public-key encryption schemes, mainly indistinguishability or semantic security under chosen-plaintext or chosen-ciphertext attack [34,43,47,28,10], can only be met when the encryption algorithm is randomized. This paper treats the case where the encryption algorithm is deterministic. We begin by discussing the motivating application.

FAST SEARCH. Remote data storage in the form of outsourced databases is of increasing interest [49]. Data will be stored in encrypted form. (The database service provider is not trusted.) We are interested in a public key setting, where anyone can add to the database encrypted data which a distinguished "receiver" can retrieve and decrypt. The encryption scheme must permit search (by the receiver) for data retrieval. Public-key encryption with keyword search (PEKS) [15,1,17] is a solution that provably provides strong privacy but search takes time linear in the size of the database. Given that databases can be terabytes in size, this is prohibitive. The practical community indicates that they want search on encrypted data to be as efficient as on unencrypted data, where a

A. Menezes (Ed.): CRYPTO 2007, LNCS 4622, pp. 535–552, 2007.

record containing a given field value can be a retrieved in time logarithmic in the size of the database. (For example, via appropriate tree-based data structures.) Deterministic encryption allows just this. The encrypted fields can be stored in the data structure, and one can find a target ciphertext in time logarithmic in the size of the database. The question is what security one can expect. To answer this, we need a definition of privacy for deterministic encryption.

A DEFINITION. One possibility is to just ask for one-wayness, but we would like to protect partial information about the plaintext to the maximum extent possible. To gauge what this could be, we note two inherent limitations of deterministic encryption. First, no privacy is possible if the plaintext is known to come from a small space. Indeed, knowing that c is the encryption under public key pk of a plaintext x from a set X, the adversary can compute the encryption c_x of x under pk for all $x \in X$, and return as the decryption of c the x satisfying $c_x = c$. We address this by only requiring privacy when the plaintext is drawn from a space of large min-entropy. Second, and more subtle, is that the ciphertext itself is partial information about the plaintext. We address this by only requiring non-leakage of partial information when the plaintext and partial information do not depend on the public key. This is reasonable because in real life public keys are hidden in our software and data does not depend on them. While certainly weaker than the classical notions met by randomized schemes, our resulting notion of privacy for deterministic encryption, which we call PRIV, is still quite strong. The next question is how to achieve this new notion.

CONSTRUCTIONS. Our first construction is generic and natural: Deterministically encrypt plaintext x by applying the encryption algorithm of a randomized scheme but using as coins a hash of (the public key and) x. We show that this "Encrypt-with-Hash" deterministic encryption scheme is PRIV secure in the random oracle (RO) model of [12] assuming the starting randomized scheme is IND-CPA secure. Our second construction is an extension of RSA-OAEP [13,31]. The padding transform is deterministic but uses three Feistel rounds rather than the two of OAEP. RSA-DOAEP is proven PRIV secure in the RO model assuming RSA is one-way. This construction has the attractive feature of being length-preserving. (The length of the ciphertext equals the length of the plaintext.) This is important when bandwidth is expensive —senders in the database setting could be power-constrained devices— and for securing legacy code.

HISTORICAL CONTEXT. Diffie and Hellman [26] suggested that one encrypt plaintext x by applying to it an injective trapdoor function. A deterministic encryption scheme is just a family of injective trapdoor functions, so our definition is an answer to the question of how much privacy Diffie-Hellman encryption can provide. (We clarify that not all trapdoor functions meet our definition. For example, plain RSA does not.)

In the symmetric setting, deterministic encryption is captured by ciphers including block ciphers. So far there has been no public key analog. Deterministic encryption meeting our definition provides one, and in particular RSA-DOAEP is the first length-preserving public-key cipher.

EFFICIENTLY SEARCHABLE ENCRYPTION. We introduce the notion of *efficiently searchable encryption* (ESE) schemes. These are schemes permitting fast (i.e. logarithmic time) search. Encryption may be randomized, but there is a deterministic, collision-resistant function of the plaintext that can also be computed from the ciphertext and serves as a tag, permitting the usual (fast) comparison-based search. Deterministic encryption schemes are a special case and the notion of security remains the same. (Our PRIV definition does not actually require encryption to be deterministic.) The benefit of the generalization is to permit schemes with more flexible privacy to search-time trade-offs. Specifically, we analyze a scheme from the database literature that we call "Hash-and-Encrypt." It encrypts the plaintext with a randomized scheme but also includes in the ciphertext a deterministic, collision-resistant hash of the plaintext. (This is an ESE scheme with the hash playing the role of the tag, and so permits fast search.) We prove that this scheme is PRIV secure in the RO model when the underlying encryption scheme is IND-CPA. With this scheme, loss of privacy due to lack of entropy in the message space can be compensated for by increasing the probability δ of hash collisions. (This can be done, for example, by truncating the output of the hash function.) The trade-off is that the receiver gets "false positives" in response to a search query and must spend the time to sift through them to obtain the true answer. This technique is known as *bucketization* in the database literature, but its security was not previously rigorously analyzed.

DISCUSSION. Our schemes only provide privacy for plaintexts that have high min-entropy. (This is inherent in being deterministic or efficiently searchable, not a weakness of our particular constructs.) We do not claim database fields being encrypted have high min-entropy. They might or they might not. The point is that practitioners have indicated that they will not sacrifice search time for privacy. Our claim is to provide the best possible privacy subject to allowing fast search. In some cases, this may very well mean no privacy. But we also comment that bucketization can increase privacy (at the cost of extra processing by the receiver) when the database fields being encrypted do not have high min-entropy.

EXTENSIONS. Our basic PRIV definition, and the above-mentioned results, are all for the CPA (chosen-plaintext attack) case. The definition easily extends to the CCA (chosen-ciphertext attack) case, and we call the resulting notion PRIV-CCA. Our Encrypt-with-Hash deterministic encryption scheme is not just PRIV, but in fact PRIV-CCA, in the RO model even if the underlying randomized encryption scheme is only IND-CPA, as long as the latter has the extra property that no ciphertext is too likely. In Section 6 we detail this and also discuss how RSA-DOAEP and Encrypt-and-Hash fare under CCA.

OPEN. All our constructs are in the RO model. An important open question is to construct ESE or deterministic encryption schemes meeting our definition in the standard model. We note that in the past also we have seen new notions first emerge only with RO constructions achieving them, but later standard model constructs have been found. This happened for example for IBE [14,52] and PEKS [17]. Note that the results of [32] rule out a standard model black-box reduction from deterministic public-key encryption to ordinary public-key encryption, but the former could still be built under other assumptions.

RELATED WORK. In the symmetric setting, deterministic encryption is both easier to define and to achieve than in the asymmetric setting. Consider the experiment that picks a random challenge bit b and key K and provides the adversary with a left-or-right oracle that, given plaintexts x_0, x_1 returns the encryption of x_b under K. Security asks that the adversary find it hard to guess b as long as its queries $(x_0^1, x_1^1), \ldots, (x_0^q, x_1^q)$ satisfy the condition that $x_0^1, \ldots, x_0^q$ are all distinct and also $x_1^1, \ldots, x_1^q$ are all distinct. To the best of our knowledge, this definition of privacy for deterministic symmetric encryption first appeared in [11]. However, it is met by a PRP and in this sense deterministic symmetric encryption goes back to [42].

Previous searchable encryption schemes provably meeting well-defined notions of privacy include [15,35,1,6,16,17] in the public-key setting and [50,33,23] in the symmetric setting. However, all these require linear-time search, meaning the entire database must be scanned to answer each query. In the symmetric setting, further assumptions such as the data being known in advance, and then having the user (who is both the "sender" and "reciever" in this setting) pre-compute a specialized index for the server, has been shown to permit efficiency comparable to ours without sacrificing security [24]. Follow-on work to ours [4] treats ESE (as we mean it here) in the symmetric setting, providing the symmetric analog of what we do in our current paper.

Sub-linear time searchable encryption has been much targeted by the database security community [45,3,36,25,38,39,41,37,19,51]. However, they mostly employ weak, non-standard or non-existing primitives and lack definitions or proofs of security. As a notable exception, Kantarcioglu and Clifton [40] recently called for a new direction of research on secure database servers aiming instead for "efficient encrypted database and query processing with *provable* security properties." They also propose a new cryptographic definition that ensures schemes reveal only the number of records accessed on each query, though a scheme meeting the definition requires tamper-resistant trusted hardware on the server.

Definitions that, like ours, restrict security to high min-entropy plaintexts have appeared before, specifically in the contexts of perfectly one-way probabilistic hash functions (POWHFs) [20,21] and information-theoretically secure one-time symmetric encryption [48,27]. The first however cannot be met by deterministic schemes, and neither handle the public-key related subtleties we mentioned above. (Namely that we must limit security to plaintexts not depending on the public key.) Also our definition considers the encryption of multiple related messages while those of [20,21] consider only independent messages.

USE FOR OTHER APPLICATIONS. We note that one can also use our definitions to analyze other systems-security applications. In particular, a notion of "convergent encryption" is proposed in [2,29] for the problem of eliminating wasted space in an encrypted file system by combining duplicate files across multiple users. Despite pinpointing the correct intuition for security of their scheme, they are only able to formally show (for lack of a suitable security definition) that it achieves the very weak security notion of one-wayness. One can use our definitions to show that their scheme achieves much stronger security.

2   Notation and Conventions

Unless otherwise indicated, an algorithm may be randomized. An adversary is either an algorithm or a tuple of algorithms. In the latter case, we say it is polynomial time if each constituent algorithm is polynomial time. The abbreviation "PT" stands for "polynomial time" and "PTA" for polynomial time algorithm or adversary. If x is a string then $|x|$ denotes its length in bits. By $x_1\| \cdots \|x_n$ we denote an encoding of $x_1, \ldots, x_n$ from which $x_1, \ldots, x_n$ are uniquely recoverable. Vectors are denoted in boldface, for example $\mathbf{x}$. If $\mathbf{x}$ is a vector then $|\mathbf{x}|$ denotes the number of components of $\mathbf{x}$ and $\mathbf{x}[i]$ denotes its ith component ($1 \le i \le |\mathbf{x}|$).

3   Deterministic Encryption and Its Security

ASYMMETRIC ENCRYPTION. An (asymmetric) encryption scheme $\Pi = (\mathcal{K}, \mathcal{E}, \mathcal{D})$ consists of three PTAs. The key-generation algorithm $\mathcal{K}$ takes input the unary encoding 1^k of the security parameter k to return a public key pk and matching secret key sk. The encryption algorithm $\mathcal{E}$ takes pk and a plaintext x to return a ciphertext. The deterministic decryption algorithm $\mathcal{D}$ takes sk and a ciphertext c to return a plaintext. We require that $\mathcal{D}(sk, c) = x$ for all for all k and all $x \in \mathrm{PtSp}(k)$, where the probability is over the experiment

$$(pk, sk) \xleftarrow{\$} \mathcal{K}(1^k) \; ; \; c \xleftarrow{\$} \mathcal{E}(pk, x)$$

and PtSp is a plaintext space associated to Π. Unless otherwise indicated, we assume $\mathrm{PtSp}(k) = \{0, 1\}^*$ for all k. We extend $\mathcal{E}$ to vectors via

> **Algorithm** $\mathcal{E}(pk, \mathbf{x})$
> For $i = 1, \ldots, |\mathbf{x}|$ do $\mathbf{y}[i] \xleftarrow{\$} \mathcal{E}(pk, \mathbf{x}[i])$
> Return $\mathbf{y}$

We say that Π is deterministic if $\mathcal{E}$ is deterministic. Although this is an important case of interest below, this is not assumed by the definition, which also applies when $\mathcal{E}$ is randomized.

PRIVACY ADVERSARIES. A privacy adversary $A = (A_{\mathrm{m}}, A_{\mathrm{g}})$ is a pair of PTAs. We clarify that $A_{\mathrm{m}}, A_{\mathrm{g}}$ share neither coins nor state. A_{m} takes input 1^k but *not* the public key, and returns a plaintext vector $\mathbf{x}$ together with some side information t. A_{g} takes $1^k, pk$ and an encryption $\mathbf{x}$ and tries to compute t.

The adversary must also obey the following rules. First, there must exist functions $v(\cdot), n(\cdot)$ such that $|\mathbf{x}| = v(k)$ and $|\mathbf{x}[i]\| = n(k)$ for k, all $(\mathbf{x}, t)$ output by $A_{\mathrm{m}}(1^k)$, and all $1 \le i \le v(k)$. Second, all plaintext vectors must have the same *equality pattern*, meaning for all $1 \le i, j \le v(k)$ there is a symbol $\diamond \in \{=, \ne\}$

such that $\mathbf{x}[i] \diamond \mathbf{x}[j]$ for all $(\mathbf{x}, t)$ output by $A_m(1^k)$. We say that A has *min-entropy* $\mu(\cdot)$ if

$$\Pr\left[\, \mathbf{x}[i] = x \; : \; (\mathbf{x}, t) \stackrel{\$}{\leftarrow} A_m(1^k) \right] \leq 2^{-\mu(k)}$$

for all $1 \leq i \leq v(k)$ and all $x \in \{0,1\}^*$. We say that A has *high* min-entropy if $\mu(k) \in \omega(\log(k))$.

The definition below is for chosen-plaintext attacks (CPA). In Section 6 we extend the definition to take chosen-ciphertext attacks (CCA) into account.

THE DEFINITION. Let $\Pi = (\mathcal{K}, \mathcal{E}, \mathcal{D})$ be an encryption scheme and A be a privacy adversary as above. We associate to A, Π the following:

Experiment $\mathbf{Exp}_{\Pi,A}^{\mathrm{priv}\text{-}0}(k)$	Experiment $\mathbf{Exp}_{\Pi,A}^{\mathrm{priv}\text{-}1}(k)$
$(pk, sk) \stackrel{\$}{\leftarrow} \mathcal{K}(1^k)$	$(pk, sk) \stackrel{\$}{\leftarrow} \mathcal{K}(1^k)$
$(\mathbf{x}_1, t_1) \stackrel{\$}{\leftarrow} A_m(1^k)$	$(\mathbf{x}_0, t_0) \stackrel{\$}{\leftarrow} A_m(1^k) \; ; \; (\mathbf{x}_1, t_1) \stackrel{\$}{\leftarrow} A_m(1^k)$
$\mathbf{c} \stackrel{\$}{\leftarrow} \mathcal{E}(pk, \mathbf{x}_1)$	$\mathbf{c} \stackrel{\$}{\leftarrow} \mathcal{E}(pk, \mathbf{x}_0)$
$g \stackrel{\$}{\leftarrow} A_g(1^k, pk, \mathbf{c})$	$g \stackrel{\$}{\leftarrow} A_g(1^k, pk, \mathbf{c})$
If $g = t_1$ then return 1	If $g = t_1$ then return 1
Else return 0	Else return 0

The *advantage* of a privacy adversary A against Π is

$$\mathbf{Adv}_{\Pi,A}^{\mathrm{priv}}(k) = \Pr\left[\, \mathbf{Exp}_{\Pi,A}^{\mathrm{priv}\text{-}0}(k) = 1 \right] - \Pr\left[\, \mathbf{Exp}_{\Pi,A}^{\mathrm{priv}\text{-}1}(k) = 1 \right].$$

We say that Π is PRIV secure if $\mathbf{Adv}_{\Pi,A}^{\mathrm{priv}}(\cdot)$ is negligible for every PTA A with high min-entropy.

As usual, in the random oracle (RO) model [12], all algorithms and adversaries are given access to the RO(s). In particular, both A_m and A_g get this access. Let us now discuss some noteworthy aspects of the new definition.

ACCESS TO THE PUBLIC KEY. If A_m were given pk, the definition would be unachievable for deterministic Π. Indeed, $A_m(1^k)$ could output $(\mathbf{x}, t)$ where $\mathbf{x}[1]$ is chosen at random from $\{0,1\}^k$, $|\mathbf{x}| = 1$, and $t = \mathcal{E}(pk, \mathbf{x})$. Then $A_g(1^k, pk, c)$ could return c, and A would have min-entropy 2^{-k} but

$$\mathbf{Adv}_{\Pi,A}^{\mathrm{priv}}(k) \geq 1 - 2^{-k}.$$

Intuitively, the ciphertext is non-trivial information about the plaintext, showing that any deterministic scheme leaks information about the plaintext that depends on the public key. Our definition asks that information unrelated to the public key not leak. Note that this also means that we provide security only for messages unrelated to the public key, which is acceptable in practice, because normal data is unlikely to depend on any public key. In real life, public keys are abstractions hidden in our software, not strings we look at.

VECTORS OF MESSAGES. The classical definitions explicitly only model the encryption of a single plaintext, but a simple hybrid argument from [8] shows that security when multiple plaintexts are encrypted follows. This hybrid argument fails in our setting. One can give examples showing that the version of our definition in which $|\mathbf{x}|$ is restricted to be 1 does not imply the stated version. (See the full paper [9] for details.) This is why we have explicitly considered the encryption of multiple messages.

THE HIGH MIN-ENTROPY REQUIREMENT. In the absence of the high-entropy restriction on A, it is clear that the definition would be unachievable for deterministic Π. To see this, consider $A_m(1^k)$ that outputs $(0,0)$ with probability $1/2$ and $(1,1)$ with probability $1/2$. Then $A_g(1^k, pk, c)$ could return 0 if $\mathcal{E}(pk, 0) = c$ and 1 otherwise, giving A an advantage of $1/2$. This reflects the fact that trial encryption of candidate messages is always a possible attack when encryption is deterministic.

SECURITY FOR MULTIPLE USERS. The classical notions of privacy, as well as ours, only model a single user (SU) setting, where there is just one receiver and thus just one public key. An extension of the classical notions to cover multiple users, each with their own public key, is made in [8,7], and these works go on to show that SU security implies multi-user (MU) security in this case. We leave it open to appropriately extend our definition to the MU setting and then answer the following questions: does SU security imply MU security, and do our schemes achieve MU security? But we conjecture that the answer to the first question is "no" while the answer to the second is "yes."

4   Secure Deterministic Encryption Schemes

We propose two constructions of deterministic schemes that we prove secure under our definition.

4.1   Encrypt-with-Hash

We first propose a generic deterministic encryption scheme that replaces the coins used by a standard encryption scheme with the hash of the message. More formally, let $\mathsf{AE} = (\mathcal{K}, \mathcal{E}, \mathcal{D})$ be any public-key encryption scheme. Say that $\mathcal{E}(pk, x)$ draws its coins from a set $\mathsf{Coins}_{pk}(|x|)$. We write $\mathcal{E}(pk, x; R)$ for the output of $\mathcal{E}$ on inputs pk, x and coins R. Let $H \colon \{0,1\}^* \times \{0,1\}^* \to \{0,1\}^*$ be a hash function with the property that $H(pk, x) \in \mathsf{Coins}_{pk}(|x|)$ for all pk and all $x \in \{0,1\}^*$. The RO-model "*Encrypt-with-Hash*" deterministic encryption scheme $\mathsf{EwH} = (\mathcal{DK}, \mathcal{DE}, \mathcal{DD})$ is defined via

Alg $\mathcal{DK}(1^k)$	**Alg** $\mathcal{DE}^H(pk, x)$	**Alg** $\mathcal{DD}^H((sk, pk), y)$
$(pk, sk) \xleftarrow{\$} \mathcal{K}(1^k)$	$R \leftarrow H(pk, x)$	$x \leftarrow \mathcal{D}(sk, y)$
Return $(pk, (sk, pk))$	$y \leftarrow \mathcal{E}(pk, x; R)$	$R \leftarrow H(pk, x)$
	Return y	If $\mathcal{E}(pk, x; R) = y$ then
		Return x
		Else Return $\bot$

The *max public-key probability* mpk$(\cdot)$ of AE is defined as follows: for every k we let mpk(k) be the maximum taken over all $w \in \{0,1\}^*$ of the quantity

$$\Pr\left[\,(pk, sk) \xleftarrow{\$} \mathcal{K}\ :\ pk = w\,\right].$$

The following then implies that the construction achieves PRIV-security if the starting encryption scheme is IND-CPA.

Theorem 1. *Suppose there is a privacy adversary* $A = (A_m, A_g)$ *against* EwH *with min-entropy* μ, *which outputs vectors of size* v *and makes at most* q_h *queries to its hash oracle. Then there exists an IND-CPA adversary* B *against* AE *such that*

$$\mathbf{Adv}^{\mathrm{priv}}_{\mathsf{EwH},A} \leq \mathbf{Adv}^{\mathrm{ind\text{-}cpa}}_{\mathsf{AE},B} + \frac{2q_h v}{2^\mu} + 2q_h\mathsf{mpk}\,, \tag{1}$$

where mpk *is the max public-key probability of* AE. *Furthermore,* B *makes* v *queries to its LR-oracle and its running-time is at most that of* A *plus* $O(1)$.

The proof is in [9].

We stress that mpk$(\cdot)$ is negligible for any IND-CPA scheme, so its being small here is *not* an extra assumption. The reason we make the term explicit is that for most schemes it is easy to analyze directly and is unconditionally exponentially-small in the security parameter, which provides more precise security guarantees. For example, in ElGamal [30], the public key contains a value g^x, where x is a random exponent in the secret key. In this case, the max public-key probability is $1/|G|$, where $|G|$ is the order of the corresponding group. Also note that in the theorem (and in the rest of the paper), we use the definition of IND-CPA (or -CCA) that allows an adversary to make as many queries as it likes to its LR-oracle. This is known to be equivalent (with loss in security by a factor less than or equal to the total number of LR-queries made) to allowing only one such query [8]. We also clarify that (1) is a relationship between functions of k, so we are saying it holds for all $k \in \mathbb{N}$. For simplicity of notation we omit k here and further in the paper.

4.2 RSA-DOAEP, a Length-Preserving Deterministic Scheme

It is sometimes important to minimize the number of bits transmitted over the network. We devise an efficient deterministic encryption scheme that is optimal in this regard, namely is length-preserving. (That is, the length of the ciphertext equals the length of the plaintext.) Length-preserving schemes can also be needed for securing legacy code. Ours is the first such construction shown secure under a definition of security substantially stronger than one-wayness, and in particular is the first construction of an asymmetric cipher.

THE SCHEME. The construction is based on RSA-OAEP [13,31]. But in place of the randomness in this scheme we use part of the message, and we add an extra round to the underlying transform. Formally, our scheme is parameterized by integers $k_0, k_1 > 0$ satisfying $n > 2k_0$ and $n \geq k_1$. The plaintext space $\mathrm{PtSp}(k)$ consists of all strings of length at least $\max(k_1, 2k_0 + 1)$. We assume here for simplicity that all messages to encrypt have a fixed length $n = n(k)$. Let $\mathcal{F}$ be an RSA trapdoor-permutation generator with modulus length $|N| = k_1$. The key-generation algorithm of the associated RO-model deterministic encryption scheme RSA-DOAEP ("D" for "deterministic") on input 1^k runs $\mathcal{F}$ on the same input and returns (N, e) as the public key and (N, d) as the secret key. Let $s[i \ldots j]$ denote bits i through j of a string s, for $1 \le i \le j \le |s|$. The encryption and decryption algorithms have oracle access to functions $H_1, H_2 \colon \{0,1\}^* \times \{0,1\}^* \to \{0,1\}^{k_0}$ and $R \colon \{0,1\}^* \times \{0,1\}^* \to \{0,1\}^{n-k_0}$, and are defined as follows:

Algorithm $\mathcal{E}^{H_1, H_2, R}((N, e), x)$	Algorithm $\mathcal{D}^{H_1, H_2, R}((N, d), y)$
$x_l \leftarrow x[1 \ldots k_0]$	$x_1 \leftarrow y[1 \ldots n - k_1]$
$x_r \leftarrow x[k_0 + 1 \ldots n]$	$y_1 \leftarrow y[n - k_1 + 1 \ldots n]$
$s_0 \leftarrow H_1((N, e), x_r) \oplus x_l$	$x \leftarrow x_1 \| (y_1^d \bmod N)$
$t_0 \leftarrow R((N, e), s_0) \oplus x_r$	$s_1 \leftarrow x[1 \ldots k_0]$
$s_1 \leftarrow H_2((N, e), t_0) \oplus s_0$	$t_0 \leftarrow x[k_0 + 1 \ldots n]$
$x_1 \leftarrow (s_1 \| t_0)[1 \ldots n - k_1]$	$s_0 \leftarrow H_2((N, e), t_0) \oplus s_1$
$x_2 \leftarrow (s_1 \| t_0)[n - k_1 + 1 \ldots n]$	$x_r \leftarrow R((N, e), s_0) \oplus t_0$
$y \leftarrow x_1 \| (x_2^e \bmod N)$	$x_l \leftarrow H_1((N, e), x_r) \oplus s_0$
Return y	Return $x_l \| x_r$

SECURITY. The following implies that the construction achieves PRIV-security if RSA is one-way.

Theorem 2. *Suppose there exists a privacy adversary $A = (A_\mathrm{m}, A_\mathrm{g})$ against RSA-DOAEP with min-entropy μ that makes at most q_{h_i} queries to oracle H_i for $i \in \{1, 2\}$ and q_r to oracle R, and outputs vectors of size v with components of length n. Let mpk be the max public-key probability of RSA-DOAEP. We consider two cases:*

- *Case 1: $n < k_0 + k_1$. Then there is an inverter I against $\mathcal{F}$ such that*

$$\mathbf{Adv}^{\mathrm{priv}}_{\text{RSA-DOAEP}, A} \le q_{\mathrm{h}_2} v \cdot \sqrt{\mathbf{Adv}^{\mathrm{owf}}_{\mathcal{F}, I} + 2^{4k_0 - 2k_1 + 10}}$$
$$- 2^{2k_0 - k_1 + 5} + \frac{2q_\mathrm{r} v}{2^{k_0}} + \frac{2q_{\mathrm{h}_1} q_\mathrm{r} v}{2^\mu} + 2(q_{\mathrm{h}_1} + q_{\mathrm{h}_2} + q_\mathrm{r}) \mathsf{mpk} \,.$$

  Furthermore the running-time of I is at most that of A plus $O(q_{\mathrm{h}_2} \log(q_{\mathrm{h}_2}) + k_1^3)$.

- *Case 2: $n \ge k_0 + k_1$. Then there is an inverter I against RSA $\mathcal{F}$ such that*

$$\mathbf{Adv}^{\mathrm{priv}}_{\text{RSA-DOAEP}, A} \le v \cdot \mathbf{Adv}^{\mathrm{owf}}_{\mathcal{F}, I}$$
$$+ \frac{2q_\mathrm{r} v}{2^{k_0}} + \frac{2q_{\mathrm{h}_1} q_\mathrm{r} v}{2^\mu} + 2(q_{\mathrm{h}_1} + q_{\mathrm{h}_2} + q_\mathrm{r}) \mathsf{mpk} \,.$$

  Furthermore, the running-time of I is at most that of A plus $O(q_{\mathrm{h}_2} \log(q_{\mathrm{h}_2}))$.

The proof is in [9].

In practice, we will have, e.g. $k_1 = 1024$, and then one can set the parameter k_0 to, say, 160 bits to effectively maximize security regardless of which case of the theorem applies (i.e. independent of the length of the particular plaintext to encrypt). Thus, typically, letting n be the length to whose restriction the message space gives the smallest adversarial min-entropy, the relation between $n - 160$ and 1024 then determines which case of the theorem applies. We note that the weaker security guarantee in Case 1 is analogous to the state-of-the-art for RSA-OAEP itself [31,46].

ENCRYPTING LONG MESSAGES. Typically, to encrypt long messages efficiently using an asymmetric scheme in practice, one employs hybrid encryption. This methodology in particular applies to the Encrypt-with-Hash construction, in which the starting scheme can be a hybrid one. However, with RSA-DOAEP, we do not provide an explicit way to securely utilize hybrid encryption while keeping encryption deterministic, and, in any case, if using some form of hybrid encryption, RSA-DOAEP would no longer be length-preserving (since an encrypted symmetric key would need to be included with the ciphertext). We would therefore like to stress that one can efficiently encrypt long messages using RSA-DOAEP *without* making use of hybrid encryption. Intuitively, this is possible because, somewhat similarly to the randomized case [18], the underlying Feistel-network in the scheme acts as a kind of "all-or-nothing transform" (AONT), such that unless an adversary with large min-entropy inverts the RSA image in a ciphertext then it cannot recover any information about a (long) message, for the practical parameter settings given above.

5   Efficiently Searchable Encryption (ESE)

We now turn to the aforementioned application of outsourced databases, where data is sent to a remote server. The database server is untrusted. The data in each field in the database is encrypted separately under the public key of a receiver, who needs to be able to query the server to retrieve the encrypted records containing particular data. Since databases are often large, a linear scan by the server on each query is prohibitive. Deterministic encryption provides a possible solution to the problem. A query, i.e. a ciphertext, specifies the exact piece of data the server needs to locate, so the server can answer it just like for unencrypted data, and hence search-time stays sub-linear (or logarithmic) in the database size. In general though, encryption permitting efficient search does not to be deterministic *per se*. Accordingly, we first define a new primitive that we call *efficiently searchable encryption (ESE)*, which more generally permits this "efficient searchability."

THE NEW PRIMITIVE. The basic idea is to associate a "tag" to a plaintext, which can be computed both by the client to form a particular query *and* by the server from a ciphertext that encrypts it, so that it can index the data appropriately in standard (e.g. tree-based) data structures and search according to the tags. These functionalities are captured, respectively, by the functions F, G below.

Let $\mathsf{AE} = (\mathcal{K}, \mathcal{E}, \mathcal{D})$ be a public-key encryption scheme with associated plaintext space $\mathrm{PtSp}(k)$. We say AE is a δ-*efficiently searchable encryption (-ESE)* scheme where $\delta(\cdot) < 1$ if there exist PTAs F, G such that for every k we have

1. **Perfect Consistency:** For every $x_1 \in \mathrm{PtSp}(k)$, the probability that $F(pk, x_1) = G(pk, c)$ equals one, where the probability is over

$$(pk, sk) \xleftarrow{\$} \mathcal{K}(1^k) \; ; \; c \xleftarrow{\$} \mathcal{E}(pk, x_1) \, .$$

2. **Computational Soundness:** For every PTA $\mathcal{M}$ that on input 1^k outputs a pair of distinct messages in $\mathrm{PtSp}(k)$, the probability that $F(pk, x_0) = G(pk, \mathcal{E}(pk, x_1))$ is at most $\delta(k)$, where the probability is over

$$(pk, sk) \xleftarrow{\$} \mathcal{K}(1^k) \; ; \; (x_0, x_1) \xleftarrow{\$} \mathcal{M}(1^k) \, .$$

We refer to the output of F, G as the *tag* of the message or a corresponding ciphertext.

Above, consistency ensures that the server can locate at least the desired ciphertexts on a query, because their tags and those of the plaintexts used in forming the query will be the same. Soundness limits the number of false-positives that are located as well, by bounding the number of other plaintexts that may have the same tag, and precludes degeneracy, where the whole database is returned on every query. With flexible trade-offs are desirable here, δ may be quite large. This is why $\mathcal{M}$ is not given input pk; if it were, the soundness condition would not make sense for large δ, since $\mathcal{M}$ could compute tags of messages itself and then output two of the many it finds to agree on their tags. As in our definition of privacy, one way to view this restriction is as saying that, in practice, the data is not picked as depending on any public key.

SECURITY OF ESE. The rule that a privacy adversary output vectors with the same equality pattern has a natural interpretation in the context of ESE. Intuitively, this means that, in the outsourced database application, all the server should learn about the data is which records contain the same attribute values/keywords and how many times each one occurs (called the *occurrence profile/distribution* of the data).

As shown in the full version [9], any deterministic encryption scheme is 0-efficiently searchable under our definition. We will see how under our PRIV definition, relaxing the soundness of a different ESE scheme (i.e. increasing δ) via "bucketization" (cf. [44,22]), with each plaintext being randomly assigned a tag and some number of them corresponding to each tag (i.e. each "bucket"), though requiring the receiver do more work to filter out false-positive results can mitigate the power of a dictionary attack by the server when min-entropy of the data is low. While one might want to try to use such bucketization to hide the occurrence profile of the data as well, or else to have the bucket distribution depend on that of the input, as we explain in the full version [9], neither of these are likely to be possible in practice. So we stick to the PRIV definition in analyzing ESE.

We next analyze a simple probabilistic ESE construction occurring in the database literature.

5.1 Encrypt-and-Hash ESE

This scheme represents an approach suggested in the database literature, in which the tag of a message is its hash. Let $\mathsf{AE} = (\mathcal{K}, \mathcal{E}, \mathcal{D})$ be any pubic-key encryption scheme and $H \colon \{0,1\}^* \times \{0,1\}^* \to \{0,1\}^{l_h}$ for some $l_h > 0$ be a hash function. The RO-model *"Encrypt-and-Hash"* encryption scheme $\mathsf{EaH} = (\mathcal{HK}, \mathcal{HE}, \mathcal{HD})$ is defined via

Alg $\mathcal{HK}(1^k)$	**Alg** $\mathcal{HE}^H(pk, x)$	**Alg** $\mathcal{HD}^H((sk, pk), y\|h)$
$(pk, sk) \xleftarrow{\$} \mathcal{K}(1^k)$	$h \leftarrow H(pk, x)$	$x \leftarrow \mathcal{D}(sk, y)$
Return $(pk, (sk, pk))$	$y \leftarrow \mathcal{E}(pk, x)$	$h' \leftarrow H(pk, x)$
	Return $y\|h$	If $h' = h$ then
		Return x
		Else Return $\perp$

Then EaH is efficiently searchable under our definition. (See the full version [9] for details.) The following implies the construction is PRIV secure if the underlying encryption scheme is IND-CPA, independent of l_h, the proof of which appears in [9].

Theorem 3. *Suppose there is a privacy adversary* $A = (A_m, A_g)$ *against* EaH *that outputs vectors of size* v *and makes at most* q_h *queries to its hash oracle. Then there exists an IND-CPA adversary* B *against* AE *such that*

$$\mathbf{Adv}^{\mathrm{priv}}_{\mathsf{EaH}, A} \leq \mathbf{Adv}^{\mathrm{ind\text{-}cpa}}_{\mathsf{AE}, B} + \frac{2q_h v}{2^{\mu}} + 2q_h \mathsf{mpk} \,,$$

where mpk *is the max public-key probability of* AE. *Furthermore,* B *makes* v *queries to its LR-oracle and its running-time is at most that of* A *plus* $O(1)$.

The above tells us that the construction achieves security when min-entropy of the data is high enough to preclude a dictionary attack by the adversary against the scheme. What about when min-entropy of the data is not high? In this case, the construct allows for bucketization as previously described. To obtain a γ-ESE scheme (assuming that γ is power of two), one can simply set l_h to be $\log \gamma$. The particular RO chosen for an instance of the scheme then determines the plaintext-to-tag mapping. Intuitively, if the number of plaintexts corresponding to any given tag is not too low, the scheme can still provide reasonable security because the adversary will not be able to distinguish ciphertexts of plaintexts with equal tags. The following captures the security gain from using this technique in a quantitative way. The parameter j below represents a lower bound on the minimum bucket-size (i.e. the minimum number of plaintexts associated to any given tag) according to the choice of the RO, which we hope to be as large as possible with a given hash-length l_h for security.

Theorem 4. *Suppose there is a privacy adversary* $A = (A_m, A_g)$ *against* EaH *with min-entropy* μ, *which outputs vectors of length* v *and makes at most* q_h *queries to* H. *Then there exists an IND-CPA adversary* B *against* AE *such that*

$$\mathbf{Adv}^{\mathrm{priv}}_{\mathsf{EaH}, A} \leq \mathbf{Adv}^{\mathrm{ind\text{-}cpa}}_{\mathsf{AE}, B} + \frac{2q_h v}{2^{\mu} j} + 2q_h \mathsf{mpk} \,,$$

for any $0 \leq q_{\mathrm{h}}v \leq 2^{\mu}$ *(larger* $q_{\mathrm{h}}v$ *cannot increase A's advantage) and any* $j > 0$ *with probability at least* $1 - 1/(\exp(2^{\mu} - 2^{l_{\mathrm{h}}}(l_{\mathrm{h}} + (j-1)\ln l_{\mathrm{h}})))$ *over the choice of H. Furthermore, B makes* v *queries to its LR-oracle and its running-time is at most that of A plus* $O(1)$.

The proof is in [9].

Thus when j above is such that $j \ll (2^{\mu - l_{\mathrm{h}}} - l_{\mathrm{h}})/\ln(l_{\mathrm{h}}) + 1$, the given bound on $\mathbf{Adv}^{\mathrm{priv}}_{\mathrm{EaH},A}$ holds with probability extremely close to one. This means that the analysis is only meaningful when μ is large enough relative to l_{h} (say by at least a few bits) for the right-hand side of this inequality to be sigificantly greater than 1, reflecting the fact that, if μ and l_{h} are the same size, bucketization is unlikely to have much effect on security because the minimum-bucket size is likely to be very small (again, with probability taken over the choice of the RO in the scheme). Precise bounds can be obtained for a specific application by plugging in the appropriate values and checking at what value a greater-or-equal minimum bucket-size j becomes overwhelmingly likely, in which case one can use the bound with such a j. We provide an example in the full paper [9]. Note that the trade-off as the hash length is decreased is query-efficiency. On each query, all records whose specified attributes values are in the same buckets as those of the desired result are returned to the user, who can complete the query itself by filtering out false-positives as needed.

We remark that one cannot use a POWHF [20,21] to compute the tags in place of the RO in the construction, because POWHFs are randomized and this will violate the consistency requirement of ESE.

6 CCA and Other Extensions

Our definition, and so far our security proofs, are for the CPA case. Here we discuss extensions to the CCA case and then other extensions such as to hash functions rather than encryption schemes.

PRIV-CCA. Extend $\mathbf{Exp}^{\mathrm{priv}\text{-}b}_{\Pi,A}(k)$ to give A_{g} oracle access to $\mathcal{D}(sk, \cdot)$, for $b \in \{0, 1\}$, which it can query on any string not appearing as a component of $\mathbf{c}$. Note that A_{m} does *not* get this decryption oracle. Let

$$\mathbf{Adv}^{\mathrm{priv\text{-}cca}}_{\Pi,A}(k) = \Pr\left[\mathbf{Exp}^{\mathrm{priv}\text{-}0}_{\Pi,A}(k) = 1\right] - \Pr\left[\mathbf{Exp}^{\mathrm{priv}\text{-}1}_{\Pi,A}(k) = 1\right].$$

We say that Π is *PRIV-CCA secure* if $\mathbf{Adv}^{\mathrm{priv\text{-}cca}}_{\Pi,A}(\cdot)$ is negligible for every PTA A with high min-entropy.

ENCRYPT-WITH-HASH. Deterministic encryption scheme EwH is PRIV-CCA secure even if the starting encryption scheme is only IND-*CPA* but meets an extra condition, namely that no ciphertext occurs with too high a probability. More precisely, the *max-ciphertext probability* $\mathsf{mc}(\cdot)$ of $\mathsf{AE} = (\mathcal{K}, \mathcal{E}, \mathcal{D})$ is defined as follows: we let $\mathsf{mc}(k)$ be the maximum taken over all $x \in \mathrm{PtSp}(k)$ of the quantity

$$\Pr\left[(pk, sk) \xleftarrow{\$} \mathcal{K} ; c_1, c_2 \xleftarrow{\$} \mathcal{E}(pk, x) : c_1 = c_2\right].$$

Then Theorem 1 extends as follows.

Theorem 5. *Suppose there is a PRIV-CCA adversary $A = (A_m, A_g)$ against* EwH *with min-entropy μ, which outputs vectors of size v with components of length n and makes at most q_h queries to its hash oracle and at most q_d queries to its decryption oracle. Let* mpk *and* mc *be max public-key and max-ciphertext probabilities of* AE, *respectively. Then there exists an IND-CPA adversary B against* AE *such that*

$$\mathbf{Adv}^{\mathrm{priv}}_{\mathsf{EwH},A} \leq \mathbf{Adv}^{\mathrm{ind\text{-}cpa}}_{\mathsf{AE},B} + \frac{2q_h v}{2^\mu} + 2q_h\mathsf{mpk} + 2q_d\mathsf{mc}.$$

Furthermore, B makes v queries to its LR-oracle and its running-time is at most that of A plus $O(q_h(T_\mathcal{E} + q_d))$, where $T_\mathcal{E}$ is the maximum time for one computation of $\mathcal{E}$ on a message of length n.

The proof is given in [9].

The requirement that $\mathsf{mc}(\cdot)$ be small is quite mild. Most practical encryption schemes have negligible max-ciphertext probability. Furthermore, any IND-CPA scheme can be easily modified to have low max-ciphertext probability if does not already. In the full paper [9], we detail all this and also show by example that not all IND-CPA schemes have low max-ciphertext probability.

RSA-DOAEP. Although RSA-DOAEP as a stand-alone is demonstrably PRIV-CCA insecure, when properly combined in an "encrypt-then-sign" fashion with a secure digital signature scheme it achieves CCA security in the natural "outsider security" model analogous to that in [5]. This may come at no additional cost, for example in the database-security application we discussed, which also requires authenticity anyway.

EXTENSIONS TO OTHER PRIMITIVES. It is straightforward to adapt our PRIV definition to a more general primitive that we call a (public-key) *hiding scheme*, which we define as a pair HIDE = (Kg, F) of algorithms, where Kg outputs a key K and F takes K and an input x to return an output we call the ciphertext. Note that every public-key encryption scheme $(\mathcal{K}, \mathcal{E}, \mathcal{D})$ has an associated hiding scheme where Kg runs $\mathcal{K}$ to receive output (pk, sk) and then returns pk, and $\mathsf{F}(K, \cdot)$ is the same as $\mathcal{E}(pk, \cdot)$. In general, though, a hiding scheme is not required to be invertible, covering for example the case of hash functions.

ENCRYPT-AND-HASH. In contrast to Encrypt-with-Hash, PRIV-CCA security of ESE scheme EaH requires IND-CCA security of the starting encryption scheme AE in general, in which case the analogous statements to Theorem 3 and Theorem 4 holds when considering PRIV-CCA adversaries against EaH. These are stated and proven in [9].

In fact, the basic construction generalizes to using any deterministic hiding scheme HIDE = (Kg, F) as defined above in place of the RO, where we replace a query $H(pk, x)$ in the scheme by $\mathsf{F}(K, (pk, x))$. Theorem 3 then generalizes as follows.

Theorem 6. *Suppose there is a privacy adversary $A = (A_m, A_g)$ against* EaH *that outputs vectors of size v. Let* mpk *be the max public-key probability of* AE.

Then there exists an IND-CPA adversary B against AE *and a privacy adversary B' against* HIDE *such that*

$$\mathbf{Adv}^{priv}_{\mathsf{EaH},A} \leq \mathbf{Adv}^{ind\text{-}cpa}_{\mathsf{AE},B} + \mathbf{Adv}^{priv}_{\mathsf{HIDE},B'}.$$

Furthermore, B makes v queries to its LR-oracle, B' outputs vectors of length v with components of length n, and the running-times of B, B' are at most that of A plus O(1).

Again, the proof is in [9]. Note that in the RO model it is easy to construct a PRIV secure deterministic hiding scheme (Kg, F), simply by setting Kg to output nothing and F on input x to return $H(x)$, where H is a RO. In this case, we recover the original construction.

Acknowledgments

We would like to thank Brian Cooper and Cynthia Dwork for helpful discussions, and Alex Dent and Ulrich Kühn for feedback on an early draft of this paper. Thanks also to Diana Smetters and Dan Wallach for pointing us to the work of [2,29], and to Nisheeth Vishnoi for help in formulating and proving Theorem 4. Finally, we thank the anonymous reviewers of Crypto 2007 for their comments and suggestions. Mihir Bellare was supported by NSF grants CNS-0524765, CNS-0627779, and a gift from Intel Corporation. Alexandra Boldyreva was supported in part by NSF CAREER award 0545659. Adam O'Neill was supported in part by the above-mentioned grant of the second author.

References

1. Abdalla, M., Bellare, M., Catalano, D., Kiltz, E., Kohno, T., Lange, T., Malone-Lee, J., Neven, G., Paillier, P., Shi, H.: Searchable encryption revisited: Consistency properties, relation to anonymous IBE, and extensions. In: Shoup, V. (ed.) CRYPTO 2005. LNCS, vol. 3621. Springer, Heidelberg (2005)
2. Adya, A., Bolosky, W.J., Castro, M., Cermak, G., Chaiken, R., Douceur, J.R., Howell, J., Lorch, J.R., Theimer, M., Wattenhofer, R.: FARSITE: Federated, available, and reliable storage for an incompletely trusted environment. In: Symposium on Operating System Design and Implementation (OSDI '02). Springer, Heidelberg (2002)
3. Agrawal, R., Kiernan, J., Srikant, R., Xu, Y.: Order preserving encryption for numeric data. In: SIGMOD '04. ACM Press, New York (2004)
4. Amanatidis, G., Boldyreva, A., O'Neill, A.: New security models and provably-secure schemes for basic query support in outsourced databases. In: Working Conference on Data and Applications Security (DBSec '07). LNCS. Springer, Heidelberg (2007)
5. An, J.-H., Dodis, Y., Rabin, T.: On the security of joint signature and encryption. In: Knudsen, L.R. (ed.) EUROCRYPT 2002. LNCS, vol. 2332. Springer, Heidelberg (2002)
6. Baek, J., Safavi-Naini, R., Susilo, W.: Public key encryption with keyword search revisited. Cryptology ePrint Archive, Report 2005/151 (2005)

7. Baudron, O., Pointcheval, D., Stern, J.: Extended notions of security for multicast public key cryptosystems. In: Welzl, E., Montanari, U., Rolim, J.D.P. (eds.) ICALP 2000. LNCS, vol. 1853. Springer, Heidelberg (2000)

8. Bellare, M., Boldyreva, A., Micali, S.: Public-key encryption in a multi-user setting: Security proofs and improvements. In: Preneel, B. (ed.) EUROCRYPT 2000. LNCS, vol. 1807. Springer, Heidelberg (2000)

9. Bellare, M., Boldyreva, A., O'Neill, A.: Deterministic and efficiently searchable encryption. Full Version of this paper (2007), http://www.cc.gatech.edu/~aboldyre/publications.html

10. Bellare, M., Desai, A., Jokipii, E., Rogaway, P.: A concrete security treatment of symmetric encryption. In: FOCS '97, pp. 394–403 (1997)

11. Bellare, M., Kohno, T., Namprempre, C.: Authenticated encryption in SSH: provably fixing the SSH binary packet protocol. In: Conference on Computer and Communications Security (CCS '02). ACM Press, New York (2002)

12. Bellare, M., Rogaway, P.: Random oracles are practical: A paradigm for designing efficient protocols. In: Conference on Computer and Communications Security (CCS '93). ACM Press, New York (1993)

13. Bellare, M., Rogaway, P.: Optimal asymmetric encryption – how to encrypt with RSA. In: De Santis, A. (ed.) EUROCRYPT 1994. LNCS, vol. 950. Springer, Heidelberg (1995)

14. Boneh, D., Boyen, X.: Secure identity based encryption without random oracles. In: Franklin, M. (ed.) Crypto '04, LNCS, vol. 3027. Springer, Heidelberg (2004)

15. Boneh, D., Di Crescenzo, G., Ostrovsky, R., Persiano, G.: Public key encryption with keyword search. In: Cachin, C., Camenisch, J.L. (eds.) EUROCRYPT 2004. LNCS, vol. 3027. Springer, Heidelberg (2004)

16. Boneh, D., Waters, B.: Conjunctive, subset, and range queries on encrypted data (2007)

17. Boyen, X., Waters, B.: Anonymous hierarchical identity-based encryption (without random oracles). In: Dwork, C. (ed.) CRYPTO 2006. LNCS, vol. 4117. Springer, Heidelberg (2006)

18. Boyko, V.: On the security properties of OAEP as an all-or-nothing transform. In: Wiener, M.J. (ed.) CRYPTO 1999. LNCS, vol. 1666, Springer, Heidelberg (1999)

19. Brinkman, R., Feng, L., Doumen, J.M., Hartel, P.H., Jonker, W.: Efficient tree search in encrypted data. Technical Report TR-CTIT-04-15, Enschede (March 2004)

20. Canetti, R.: Towards realizing random oracles: Hash functions that hide all partial information. In: Kaliski Jr., B.S. (ed.) CRYPTO 1997. LNCS, vol. 1294. Springer, Heidelberg (1997)

21. Canetti, R., Micciancio, D., Reingold, O.: Perfectly one-way probabilistic hash functions. In: STOC '98. ACM Press, New York (1998)

22. Ceselli, A., Damiani, E., De Capitani di Vimercati, S., Jajodia, S., Paraboschi, S., Samarati, P.: Modeling and assessing inference exposure in encrypted databases. ACM Trans. Inf. Syst. Secur. 8(1), 119–152 (2005)

23. Chang, Y.-C., Mitzenmacher, M.: Privacy preserving keyword searches on remote encrypted data. In: Ioannidis, J., Keromytis, A.D., Yung, M. (eds.) ACNS 2005. LNCS, vol. 3531. Springer, Heidelberg (2005)

24. Curtmola, R., Garay, J.A., Kamara, S., Ostrovsky, R.: Searchable symmetric encryption: Improved definitions and efficient constructions. In: Juels, A., Wright, R.N., De Capitani di Vimercati, S. (eds.) Conference on Computer and Communications Security (CCS '06). ACM Press, New York (2006)

25. Damiani, E., De Capitani Vimercati, S., Jajodia, S., Paraboschi, S., Samarati, P.: Balancing confidentiality and efficiency in untrusted relational DBMSs. In: Jajodia, S., Atluri, V., Jaeger, T. (eds.) Conference on Computer and Communications Security (CCS '03). ACM Press, New York (2003)
26. Diffie, W., Hellman, M.: New directions in cryptography. IEEE Transactions on Information Theory 22(6), 1130–1140 (1976)
27. Dodis, Y., Smith, A.: Entropic security and the encryption of high entropy messages. In: Kilian, J. (ed.) TCC 2005. LNCS, vol. 3378. Springer, Heidelberg (2005)
28. Dolev, D., Dwork, C., Naor, M.: Non-malleable cryptography. SIAM Journal on Computing 30(2) (2000)
29. Douceur, J.R., Adya, A., Bolosky, W.J., Simon, D., Theimer, M.: Reclaiming space from duplicate files in a serverless distributed file system. In: Conference on Distributed Computing Systems (ICDCS'02) (2002)
30. ElGamal, T.: A public key cryptosystem and signature scheme based on discrete logarithms. IEEE Transactions on Information Theory 31 (1985)
31. Fujisaki, E., Okamoto, T., Pointcheval, D., Stern, J.: RSA-OAEP is secure under the RSA assumption. In: Kilian, J. (ed.) CRYPTO 2001. LNCS, vol. 2139. Springer, Heidelberg (2001)
32. Gertner, Y., Malkin, T., Reingold, O.: On the impossibility of basing trapdoor functions on trapdoor predicates. In: FOCS '01. IEEE Computer Society Press, Los Alamitos (2001)
33. Goh, E.-J.: Secure indexes. Cryptology ePrint Archive, Report, 2003/216 (2003), http://eprint.iacr.org/2003/216/
34. Goldwasser, S., Micali, S.: Probabilistic encryption. Journal of Computer and System Sciences 28(2) (1984)
35. Golle, P., Staddon, J., Waters, B.: Secure conjunctive keyword search over encrypted data. In: Jakobsson, M., Yung, M., Zhou, J. (eds.) ACNS 2004. LNCS, vol. 3089. Springer, Heidelberg (2004)
36. Hacigümüs, H., Iyer, B., Li, C., Mehrotra, S.: Executing SQL over encrypted data in the database-service-provider model. In: Conference on Management of data (SIGMOD '02). ACM Press, New York (2002)
37. Hacigümüs, H., Iyer, B.R., Mehrotra, S.: Efficient execution of aggregation queries over encrypted relational databases. In: Lee, Y., Li, J., Whang, K.-Y., Lee, D. (eds.) DASFAA 2004. LNCS, vol. 2973. Springer, Heidelberg (2004)
38. Hore, B., Mehrotra, S., Tsudik, G.: A privacy-preserving index for range queries. In: Nascimento, M.A., Özsu, M.T., Kossmann, D., Miller, R.J., Blakeley, J.A., Schiefer, K.B. (eds.) VLDB '04. Morgan Kaufmann, San Francisco (2004)
39. Iyer, B.R., Mehrotra, S., Mykletun, E., Tsudik, G., Wu, Y.: A framework for efficient storage security in RDBMS. In: Bertino, E., Christodoulakis, S., Plexousakis, D., Christophides, V., Koubarakis, M., Böhm, K., Ferrari, E. (eds.) EDBT 2004. LNCS, vol. 2992. Springer, Heidelberg (2004)
40. Kantracioglu, M., Clifton, C.: Security issues in querying encrypted data. In: Jajodia, S., Wijesekera, D. (eds.) Data and Applications Security XIX. LNCS, vol. 3654, pp. 325–337. Springer, Heidelberg (2005)
41. Li, J., Omiecinski, E.: Efficiency and security trade-off in supporting range queries on encrypted databases. In: Jajodia, S., Wijesekera, D. (eds.) Data and Applications Security XIX. LNCS, vol. 3654, pp. 69–83. Springer, Heidelberg (2005)
42. Luby, M., Rackoff, C.: How to construct pseudorandom permutations from pseudorandom functions. SIAM J. Comput. 17(2) (1988)
43. Micali, S., Rackoff, C., Sloan, B.: The notion of security for probabilistic cryptosystems. SIAM Journal on Computing 17(2), 412–426 (1988)

44. Mykletun, E., Tsudik, G.: Aggregation queries in the database-as-a-service model. In: Damiani, E., Liu, P. (eds.) Data and Applications Security XX. LNCS, vol. 4127, pp. 89–103. Springer, Heidelberg (2006)
45. Özsoyoglu, G., Singer, D.A., Chung, S.S.: Anti-tamper databases: Querying encrypted databases. In: Working Conference on Data and Applications Security (DBSec '03). LNCS, Springer, Heidelberg (2003)
46. Pointcheval, D.: How to encrypt properly with RSA. RSA Laboratories' Crypto-Bytes, 5(1) (Winter/Spring 2002)
47. Rackoff, C., Simon, D.: Non-interactive zero-knowledge proof of knowledge and chosen ciphertext attack. In: Feigenbaum, J. (ed.) CRYPTO 1991. LNCS, vol. 576. Springer, Heidelberg (1992)
48. Russell, A., Wang, H.: How to fool an unbounded adversary with a short key. IEEE Transactions on Information Theory 52(3), 1130–1140 (2006)
49. Arsenal Digital Solutions. Top 10 reasons to outsource remote data protection, http://www.arsenaldigital.com/services/remote_data_protection.htm
50. Song, D.X., Wagner, D., Perrig, A.: Practical techniques for searches on encrypted data. In: Symposium on Security and Privacy. IEEE Press, New York (2000)
51. Wang, H., Lakshmanan, L.V.S.: Efficient secure query evaluation over encrypted XML databases. In: VLDB '06. VLDB Endowment (2006)
52. Waters, B.: Efficient identity-based encryption without random oracles. In: Cramer, R.J.F. (ed.) EUROCRYPT 2005. LNCS, vol. 3494. Springer, Heidelberg (2005)

Secure Hybrid Encryption from Weakened Key Encapsulation

Dennis Hofheinz and Eike Kiltz[*]

Cryptology and Information Security Research Theme
CWI Amsterdam
The Netherlands
{hofheinz,kiltz}@cwi.nl

Abstract. We put forward a new paradigm for building hybrid encryption schemes from *constrained chosen-ciphertext secure* (CCCA) key-encapsulation mechanisms (KEMs) plus authenticated symmetric encryption. Constrained chosen-ciphertext security is a new security notion for KEMs that we propose. It has less demanding security requirements than standard CCCA security (since it requires the adversary to have a certain plaintext-knowledge when making a decapsulation query) yet we can prove that it is CCCA sufficient for secure hybrid encryption.

Our notion is not only useful to express the Kurosawa-Desmedt public-key encryption scheme and its generalizations to hash-proof systems in an abstract KEM/DEM security framework. It also has a very constructive appeal, which we demonstrate with a new encryption scheme whose security relies on a class of intractability assumptions that we show (in the generic group model) strictly weaker than the Decision Diffie-Hellman (DDH) assumption. This appears to be the first practical public-key encryption scheme in the literature from an algebraic assumption strictly weaker than DDH.

1 Introduction

One of the main fields of interest in cryptography is the design and analysis of encryption schemes in the public-key setting (PKE schemes) that are secure against a very strong type of attacks — indistinguishability against chosen-ciphertext attacks (IND-CCA) [24]. In this work, we are interested in *practical schemes* with proofs of security under *reasonable security assumptions* (without relying on heuristics such as the random oracle model) and in *general methods* for constructing such schemes.

The first practical IND-CCA secure PKE scheme without random oracles was proposed in a seminal paper by Cramer and Shoup [11, 13]. Their construction was later generalized to hash proof systems [12]. In [30, 13] Cramer and Shoup also give a hybrid variant that encrypts messages of arbitrary length. The idea

---

[*] Supported by the research program Sentinels (http://www.sentinels.nl). Sentinels is being financed by Technology Foundation STW, the Netherlands Organization for Scientific Research (NWO), and the Dutch Ministry of Economic Affairs.

A. Menezes (Ed.): CRYPTO 2007, LNCS 4622, pp. 553–571, 2007.

is to conceptually separate the key-encapsulation (KEM) part from the symmetric (DEM) part. Generally, this hybrid approach greatly improved practicality of encryption schemes. A folklore composition theorem (formalized in [13]) shows that if both KEM and DEM are CCA-secure then the hybrid encryption is CCA-secure. Common wisdom was that this sufficient condition was also necessary. However, at CRYPTO 2004, Kurosawa and Desmedt challenged this common wisdom by presenting a hybrid encryption scheme that demonstrates that a weaker security condition on the KEM may suffice for full CCA-secure hybrid encryption. Compared to the original Cramer-Shoup scheme, the scheme by Kurosawa and Desmedt improved efficiency and ciphertext expansion by replacing some of its algebraic components with *information theoretically* secure symmetric primitives. More recently, the KEM part of their scheme was indeed shown to be not CCA secure [15].

One natural open problem from [21] is if there exists a weaker yet natural security condition on the KEM such that, in combination with sufficiently strong symmetric encryption, chosen-ciphertext secure hybrid encryption can be guaranteed.

Extending the work of Cramer and Shoup [12], it was demonstrated in [21, 2, 14] that a variant of hash-proof systems (HPS) can be combined with symmetric encryption and a message authentication code (MAC) to obtain hybrid encryption. If the hash-proof system is *universal₂*, then the encryption scheme is chosen-ciphertext secure. However, the Kurosawa-Desmedt hybrid scheme could not be rigorously explained in this general HPS framework since the underlying hash-proof system is not universal$_2$. (Roughly, this is since universal$_2$ is a *statistical* property whereas the Kurosawa-Desmedt system contains a *computational* component, namely a target collision resistant (TCR) hash function.) In [21] (and [12]) only less efficient "hash-free variants" of their schemes could be explained through hash proof systems; security of all efficient TCR-based schemes had to be proved separately.

Surprisingly, almost all *practical* standard-model encryption schemes [11, 13, 21, 2, 10, 9, 19, 20] are based on the difficulty of Decision Diffie-Hellman (DDH) or stronger assumptions. This is contrasted by the existence of many natural groups in which the DDH assumption is known to be wrong; examples include pairing-groups and certain non prime-order groups like $\mathbb{Z}_p^*$. This often overlooked fact may turn into a serious problem in case DDH turns out to be wrong in all cryptographically interesting groups. In particular, [16] give evidence that groups with easy DDH problem, but hard computational Diffie-Hellman problem exist. [16] interpret this as an argument to rely on weaker assumptions than DDH.

1.1   Our Contributions

A NEW KEM/DEM COMPOSITION THEOREM. We put forward the security notion of *indistinguishability against constrained chosen-ciphertext attacks* (IND-CCCA) for KEMs which is stronger than IND-CPA (CPA stands for chosen-plaintext attacks) yet strictly weaker than IND-CCA. Intuitively, CCCA is separated from CCA security by only allowing an adversary to make a decapsulation query if it has sufficient "implicit knowledge" about the plaintext key to be decapsulated (hence the name "constrained chosen-ciphertext security").[1]

As our main technical contribution we formalize the above notion and prove a composition theorem that shows that *any* IND-CCCA secure KEM combined with *any* authenticated (symmetric) encryption scheme yields IND-CCA secure hybrid encryption. This gives a positive answer to the open question from [21] mentioned before. Authenticated encryption is a quite general symmetric primitive and examples include "encrypt-then-mac" schemes (based on computationally secure primitives), and also more efficient single-pass schemes (see, e.g., [25]).

Constrained chosen-ciphertext secure KEMs formalize a new design paradigm for efficient hybrid encryption. To guarantee chosen-ciphertext security for hybrid encryption schemes it is sufficient to verify a natural security condition on the key encapsulation part. We assess the constructive appeal of this framework by demonstrating that the original Kurosawa-Desmedt scheme [21], along with its variants [2,23] and all hash-proof systems based schemes [12,21], can be thoroughly explained through it. We furthermore present a new IND-CCCA secure KEM from the DDH assumption and show how to build a class of practical KEMs from progressively weaker assumptions than DDH.

CONSTRAINED CHOSEN-CIPHERTEXT SECURE KEM FROM DDH. We propose a new KEM which is IND-CCCA secure under the DDH assumption. Although it relies on different proof techniques (it is not based on hash proof systems), syntactically it is reminiscent to the one by Kurosawa and Desmedt and can in fact be viewed as its *dual* (in the sense that certain parts from the ciphertext and the symmetric key are swapped in our scheme).

CONSTRAINED CHOSEN-CIPHERTEXT SECURE KEM FROM n-LINEAR. Building on [8,18] we introduce a new class of purely algebraic intractability assumptions, the n-Linear assumptions, where $n \geq 1$ is a parameter. They are such that the DDH assumption equals the 1-Linear assumption, the Linear assumption [8] equals the 2-Linear assumption, and the n-Linear assumptions become *strictly weaker* as the parameter n grows. More precisely, 1-Linear = DDH, and n-Linear implies $n + 1$-Linear, but (in the generic group model [29]) $n + 1$-Linear is still hard relative to an n-Linear oracle. In fact, for $n \geq 2$ the n-Linear assumption does not seem to be invalid in any obvious sense even in the groups from [16], in which the DDH problem is easy, and the computational Diffie-Hellman problem is supposedly hard. We generalize the KD scheme and its dual to a class of parametrized KEMs and prove their IND-CCCA security assuming n-Linear. These appear to be the first practical encryption schemes in the literature from a purely algebraic assumption which is strictly weaker than DDH.

COMPUTATIONAL HASH-PROOF SYSTEMS. We propose a purely computational variant of hash-proof systems. Generalizing [12,21], we prove that computational

---

[1] This is reminiscent to the notion of "plaintext awareness" for public-key encryption [5] where it is infeasible for an adversary to come up with a valid ciphertext without being aware of the corresponding plaintext. Our definition is weaker in the sense that it only requires the adversary to have *implicit knowledge* on the plaintext.

hash-proof systems directly imply IND-CCCA secure KEMs. Hence, in combination with authenticated encryption, they yield efficient IND-CCA secure hybrid encryption. The Kurosawa-Desmedt scheme fits this framework, i.e. the underlying HPS is computational. This gives the first full explanation of the Kurosawa-Desmedt scheme in terms of HPS. As a generalization we provide computational hash-proof systems from the n-Linear assumptions hence explaining IND-CCCA security of our class of KEMs from the n-Linear assumptions.

1.2 Discussion and Related Work

In [1] (which is the full version of [2]), Abe et al. address the question from [21] about the existence of a natural weaker security condition for KEMs. They propose the notion of *LCCA secure KEMs with respect to the predicate* $\mathcal{P}^{\mathrm{mac}}$ and prove it sufficient to obtain, in combination with a MAC, IND-CCA secure tag-KEMs (and hence IND-CCA secure hybrid encryption). Though syntactically similar to ours, their notion *mingles* security of the KEM with the MAC part of the symmetric encryption scheme. The conceptual difference in our notion is that we give a general security definition for KEMs that is *completely independent* of any particular symmetric primitive. We think that this is more natural and more closely follows the spirit of the KEM/DEM approach [13], where (for good reason) KEM and DEM are viewed as independent components.

Independent from this work Shacham [28] also proposes a family of hybrid encryption schemes from the n-Linear assumptions. His schemes can be viewed as a (slightly less efficient) Cramer-Shoup variant of our schemes from Section 4.2.

The 2-Linear assumption was introduced by Boneh, Boyen, and Shacham [8] and was later used in gap-groups to build an IND-CCA secure KEM [19]. For $n > 2$, Kiltz [18] introduced the class of *gap n-Linear* assumptions and (generalizing [19]) built a class of IND-CCA secure KEMs from it. Compared to n-Linear, in the latter gap-assumptions an adversary gets access to a DDH oracle which makes (for example) the gap 2-Linear assumption incomparable to DDH. In contrast, our motivation is to build schemes from an assumption weaker than DDH.

2 Hybrid Encryption from Constrained CCA Secure KEMs

2.1 Key Encapsulation Mechanisms

A *key-encapsulation mechanism* $\mathcal{KEM} = (\mathsf{KEM.Kg}, \mathsf{KEM.Enc}, \mathsf{KEM.Dec})$ with key-space $\mathcal{K}(k)$ consists of three polynomial-time algorithms (PTAs). Via $(pk, sk) \xleftarrow{\$} \mathsf{KEM.Kg}(1^k)$ the randomized key-generation algorithm produces public/secret keys for security parameter $k \in \mathbb{N}$; via $(K, C) \xleftarrow{\$} \mathsf{KEM.Enc}(pk)$ the randomized encapsulation algorithm creates an uniformly distributed symmetric key $K \in \mathcal{K}(k)$ together with a ciphertext C; via $K \leftarrow \mathsf{KEM.Dec}(sk, C)$ the possessor of secret key sk decrypts ciphertext C to get back a key K which is an element in $\mathcal{K}$ or a special rejection symbol $\perp$. For consistency, we require that for all $k \in \mathbb{N}$, and all $(K, C) \overset{\$}{\leftarrow}$ KEM.Enc(pk) we have $\Pr[\text{KEM.Dec}(sk, C) = K] = 1$, where the probability is taken over the choice of $(pk, sk) \overset{\$}{\leftarrow}$ KEM.Kg(1^k), and the coins of all the algorithms in the expression above. Here we only consider only KEMs that produce perfectly uniformly distributed keys (i.e., we require that for all public keys pk that can be output by KEM.Kg, the first component of KEM.Enc(pk) has uniform distribution).[2]

CONSTRAINED CHOSEN-CIPHERTEXT SECURITY. The common requirement for a KEM is indistinguishability against chosen-ciphertext attacks (IND-CCA) [13] where an adversary is allowed to adaptively query a decapsulation oracle with ciphertexts to obtain the corresponding session key. We relax this notion to indistinguishability against *constrained chosen-ciphertext attacks* (IND-CCCA). Intuitively, we only allow the adversary to make a decapsulation query if it already has some "a priori knowledge" about the decapsulated key. This partial knowledge about the key is modeled implicitly by letting the adversary additionally provide an efficiently computable Boolean predicate pred : $\mathcal{K} \rightarrow \{0, 1\}$. If $\text{pred}(K) = 1$ then the decapsulated key K is returned, and $\perp$ otherwise. The amount of uncertainty the adversary has about the session key (denoted as *plaintext uncertainty* uncert$_{\mathcal{A}}$) is measured by the fraction of keys the predicate evaluates to 1. We require this fraction to be negligible for every query, i.e. the adversary has to have a high a priori knowledge about the decapsulated key when making a decapsulation query. More formally, for an adversary $\mathcal{A}$ we define the advantage function

$$\mathbf{Adv}^{\text{ccca}}_{\mathcal{KEM}, \mathcal{A}}(k) = \left| \Pr[\mathbf{Exp}^{\text{ccca-1}}_{\mathcal{KEM}, \mathcal{A}}(k) = 1] - \Pr[\mathbf{Exp}^{\text{ccca-0}}_{\mathcal{KEM}, \mathcal{A}}(k) = 1] \right|$$

where, for $b \in \{0, 1\}$, $\mathbf{Exp}^{\text{ccca-b}}_{\mathcal{KEM}, \mathcal{A}}$ is defined by the following experiment.

Experiment $\mathbf{Exp}^{\text{ccca-b}}_{\mathcal{KEM}, \mathcal{A}}(k)$

$(pk, sk) \overset{\$}{\leftarrow}$ KEM.Kg(1^k)

$K_0^* \overset{\$}{\leftarrow} \mathcal{K}(k)$; $(K_1^*, C^*) \overset{\$}{\leftarrow}$ KEM.Enc(pk)

$b' \overset{\$}{\leftarrow} \mathcal{A}^{\text{DEC}(\cdot, \cdot)}(pk, K_b^*, C^*)$

Return b'

$\text{DEC}(\text{pred}_i, C_i)$

$K \leftarrow$ KEM.Dec(sk, C_i)

If $K = \perp$ or $\text{pred}_i(K) = 0$ then $\perp$

Else return $K \in \mathcal{K}$

with the restriction that $\mathcal{A}$ is only allowed to query $\text{DEC}(\text{pred}_i, C_i)$ on predicates pred_i that are provided as PTA[3] and on ciphertexts C_i different from the challenge ciphertext C^*.

For an adversary $\mathcal{A}$, let $t_{\mathcal{A}}$ denote the number of computational steps $\mathcal{A}$ runs (that includes the maximal time to *evaluate* each pred_i once), and let $Q_{\mathcal{A}}$

---

[2] This requirement is met by all popular KEMs and makes our reduction in Theorem 1 tighter. However, we can show Theorem 1 also without this assumption, and derive that the keys are computationally close to uniform from our upcoming KEM security assumption. This comes at the price of a less tight security reduction in Theorem 1.

[3] Technically, we charge the time required to evaluate each pred_i to $\mathcal{A}$'s runtime and require that $\mathcal{A}$ be polynomial-time.

be the number of decapsulation queries $\mathcal{A}$ makes to its decapsulation oracle. For simplicity and without losing on generality, we consider only adversaries for which $t_{\mathcal{A}}$ and $Q_{\mathcal{A}}$ are independent of the environment that $\mathcal{A}$ runs in. To adversary $\mathcal{A}$ in the above experiment we also associate $\mathcal{A}$'s (implicit) plaintext uncertainty $\mathrm{uncert}_{\mathcal{A}}(k)$ when making decapsulation queries, measured by

$$\mathrm{uncert}_{\mathcal{A}}(k) \; = \; \frac{1}{Q} \sum_{1 \leq i \leq Q} \Pr_{K \in \mathcal{K}}[\mathrm{pred}_i(K) = 1] \, ,$$

where $\mathrm{pred}_i : \mathbb{G} \to \{0, 1\}$ is the predicate $\mathcal{A}$ submits in the ith decapsulation query. Let, for integers k, t, Q and $0 \leq \mu \leq 1$,

$$\mathbf{Adv}^{\mathrm{ccca}}_{\mathcal{KEM}, t, Q, \mu}(k) \; = \; \max_{\mathcal{A}} \mathbf{Adv}^{\mathrm{ccca}}_{\mathcal{KEM}, \mathcal{A}}(k),$$

where the maximum is over all $\mathcal{A}$ with $t_{\mathcal{A}} \leq t$, $Q_{\mathcal{A}} \leq Q$, and $\mathrm{uncert}_{\mathcal{A}}(k) \leq \mu$.

A key encapsulation mechanism $\mathcal{KEM}$ is said to be *indistinguishable against constrained chosen ciphertext attacks* (IND-CCCA) if for all PTA adversaries $\mathcal{A}$ with negligible $\mathrm{uncert}_{\mathcal{A}}(k)$ (in any environment), the advantage $\mathbf{Adv}^{\mathrm{ccca}}_{\mathcal{KEM}, \mathcal{A}}(k)$ is a negligible function in k.

It is worth pointing out that by making different restrictions on $\mathrm{uncert}(k)$ our notion of CCCA security leads to an interesting continuum between CPA and CCA security. With the restriction $\mathrm{uncert}(k) = 0$ then CCCA = CPA; with the trivial restriction $\mathrm{uncert}(k) \leq 1$ (which makes is possible to always use the constant predicate $\mathrm{pred}(\cdot) := 1$) then CCCA = CCA. Here, we require a negligible $\mathrm{uncert}(k)$, which syntactically makes IND-CCCA more similar to IND-CPA than to IND-CCA security. Yet, since it in principle allows decryption queries, IND-CCCA is substantially stronger than IND-CPA, and — as we will show — is a good base for hybrid IND-CCA security.

2.2 Authenticated Encryption

An authenticated symmetric encryption (AE) scheme $\mathcal{AE} = (\mathsf{AE.Enc}, \mathsf{AE.Dec})$ is specified by its encryption algorithm $\mathsf{AE.Enc}$ (encrypting $M \in MsgSp(k)$ with keys $K \in \mathcal{K}(k)$) and decryption algorithm $\mathsf{AE.Dec}$ (returning $M \in MsgSp(k)$ or $\perp$). Here we restrict ourselves to deterministic PTAs $\mathsf{AE.Enc}$ and $\mathsf{AE.Dec}$. The AE scheme needs to provide privacy (indistinguishability against one-time attacks) and authenticity (ciphertext authenticity against one-time attacks). This is *simultaneously* captured (similar to the more-time attack case [26]) by defining the ae-ot-advantage of an adversary $\mathcal{B}_{ae}$ as $\mathbf{Adv}^{\mathrm{ae\text{-}ot}}_{\mathcal{AE}, \mathcal{B}_{ae}}(k) =$

$$2 \left| \Pr[K \xleftarrow{\$} \mathcal{K}(k) \, ; \, b \xleftarrow{\$} \{0, 1\} \, ; \, b' \xleftarrow{\$} \mathcal{B}^{\mathrm{LoR}_b(\cdot, \cdot), \mathrm{DoR}_b(\cdot)}_{ae}(1^k) \, : \, b = b'] - 1 \right| \, .$$

Here, $\mathrm{LoR}_b(M_0, M_1)$ returns $\psi \leftarrow \mathsf{AE.Enc}(K, M_b)$, and $\mathcal{B}_{ae}$ is allowed only one query to this left-or-right encryption oracle (one-time attack), with a pair of equal-length messages. Furthermore, the decrypt-or-reject oracle $\mathrm{DoR}_1(\psi)$ returns $M \leftarrow \mathsf{AE.Dec}(K, \psi)$ and $\mathrm{DoR}_0(\psi)$ always returns $\perp$ (reject), $\mathcal{B}_{ae}$ is allowed only one query to this decrypt-or-reject oracle which must be different from the output of the left-or-right oracle.

We say that $\mathcal{AE}$ is a *one-time secure authenticated encryption scheme* (AE-OT secure) if the advantage function $\mathbf{Adv}^{ae\text{-}ot}_{\mathcal{AE},\mathcal{B}_{ae}}(k)$ is negligible for all PTA $\mathcal{B}_{ae}$. Again, for integers k, t, $\mathbf{Adv}^{ae\text{-}ot}_{\mathcal{AE},t}(k) = \max_{\mathcal{B}_{ae}} \mathbf{Adv}^{ae\text{-}ot}_{\mathcal{AE},\mathcal{B}_{ae}}(k)$, where the maximum is over all $\mathcal{B}_{ae}$ that fulfill $t_{\mathcal{B}_{ae}} \leq t$.

2.3   Hybrid Encryption

Let $\mathcal{KEM} = (\mathsf{KEM.Kg}, \mathsf{KEM.Enc}, \mathsf{KEM.Dec})$ be a KEM and let $\mathcal{AE} = (\mathsf{AE.Enc}, \mathsf{AE.Dec})$ be an authenticated encryption scheme. We assume that the two schemes are compatible in the sense that for all security parameters k, we have that the KEM's and the AE's key-space are equal. Then we can consider a hybrid public key encryption scheme (whose syntax and security definition is standard and can be looked up in the full version) that encrypts arbitrary messages $M \in MsgSp$. The construction of $\mathcal{PKE} = (\mathsf{PKE.kg}, \mathsf{PKE.Enc}, \mathsf{PKE.Dec})$ is as follows.

$\mathsf{PKE.kg}(1^k)$	$\mathsf{PKE.Enc}(pk, M)$	$\mathsf{PKE.Dec}(sk, C_{pke} = (C, \psi))$
$(pk, sk) \xleftarrow{\$} \mathsf{KEM.Kg}(1^k)$	$(K, C) \xleftarrow{\$} \mathsf{KEM.Enc}(pk)$	$K \leftarrow \mathsf{KEM.Dec}(sk, C)$
Return (pk, sk)	$\psi \leftarrow \mathsf{AE.Enc}(K, M)$	$M \leftarrow \mathsf{AE.Dec}(K, \psi)$
	Return $C_{pke} = (C, \psi)$	Return M or $\bot$

Here $\mathsf{PKE.Dec}$ returns $\bot$ if either $\mathsf{KEM.Dec}$ or $\mathsf{AE.Dec}$ returns $\bot$.

Theorem 1. *Assume $\mathcal{KEM}$ is secure in the sense of* IND-CCCA *and $\mathcal{AE}$ is secure in the sense of* AE-OT. *Then $\mathcal{PKE}$ is secure in the sense of* IND-CCA. *In particular,*

$$\mathbf{Adv}^{cca}_{\mathcal{PKE},t,Q}(k) \leq \mathbf{Adv}^{ccca}_{\mathcal{KEM},t,Q,Q \cdot \mathbf{Adv}^{ae\text{-}ot}_{\mathcal{AE},t}(k)}(k) + (Q+1)\mathbf{Adv}^{ae\text{-}ot}_{\mathcal{AE},t}(k) + \frac{Q}{|\mathcal{K}|} .$$

Proof. Let $\mathcal{A}$ be an adversary on the IND-CCA security of the hybrid scheme. We will consider a sequence of games, Game 1, Game 2, ..., each game involving $\mathcal{A}$. Let X_i be the event that in Game i, it holds that $b = b'$, i.e., that the adversary succeeds. We will make use of the following simple "Difference Lemma" [13].

Lemma 1. *Let X_1, X_2, B be events, and suppose that $X_1 \wedge \neg B \Leftrightarrow X_2 \wedge \neg B$. Then $|\Pr[X_1] - \Pr[X_2]| \leq \Pr[B]$.*

Game 1. The original PKE IND-CCA game, i.e. we have

$$|\Pr[X_1] - 1/2| = \mathbf{Adv}^{cca}_{\mathcal{PKE},\mathcal{A}}(k) .$$

Game 2. Let $C^*_{pke} = (C^*, \psi^*)$ be the challenge ciphertext in the PKE IND-CCA game. In this game the decryption oracle in the first phase rejects all ciphertexts of the form $C_{pke} = (C^*, *)$. The view of adversary $\mathcal{A}$ is identical

in Games 1 and 2 until a decryption query $(C^*, *)$ is made in the first phase of the IND-CCA experiment (so *before* $\mathcal{A}$ gets to see C^*).
Since the key K encapsulated in C^* is uniformly distributed and independent of $\mathcal{A}$'s view in the first phase, we have

$$|\Pr[X_2] - \Pr[X_1]| \leq \frac{Q}{|\mathcal{K}|} .$$

Note that each ciphertext uniquely determines a key.

Game 3. Replace the symmetric key K^* used to create the PKE challenge ciphertext with a random key K^*, uniformly independently chosen from $\mathcal{K}$. The proof of the following lemma is postponed until later.

Lemma 2. $|\Pr[X_3] - \Pr[X_2]| \leq \mathbf{Adv}_{\mathcal{KEM}, t, Q, Q \cdot \mathbf{Adv}_{\mathcal{AE}, t}^{ae\text{-}ot}(k)}^{ccca}(k).$

Game 4. Reject all ciphertexts C_{pke} of the form $(C^*, *)$. Since ψ^* was generated using a random key $K^* \in \mathcal{K}$ that only leaks to $\mathcal{A}$ through ψ^*, authenticity of $\mathcal{AE}$ implies

$$|\Pr[X_4] - \Pr[X_3]| \leq Q_{\mathcal{A}} \cdot \mathbf{Adv}_{\mathcal{AE}, \mathcal{B}_{ae}}^{ae\text{-}ot}(k)$$

for a suitable adversary $\mathcal{B}_{ae}$ that simulates Game 3, using the LoR_b with two identical messages to obtain the AE part of the challenge ciphertext. $\mathcal{B}_{ae}$ simply uniformly picks one AE part of a decryption query of the form (C^*, ψ) to submit to the decrypt-or-reject oracle $\mathrm{DoR}_1(\cdot)$.
Finally, Game 4 models one-time security of the AE scheme, and we have

$$|\Pr[X_4] - 1/2| \leq \mathbf{Adv}_{\mathcal{AE}, t}^{ae\text{-}ot}(k) .$$

Collecting the probabilities proves the theorem. It leaves to prove Lemma 2.

Proof (Lemma 2). We show that there exists an adversary $\mathcal{B}_{kem}$ against the IND-CCCA security of $\mathcal{KEM}$ with $t_{\mathcal{B}_{kem}} = t_{\mathcal{A}}$, $Q_{\mathcal{B}_{kem}} = Q_{\mathcal{A}}$, and an adversary $\mathcal{B}_{ae}$ against $\mathcal{AE}$ with $t_{\mathcal{B}_{ae}} = t_{\mathcal{A}}$, such that

$$\mathrm{uncert}_{\mathcal{B}_{kem}}(k) \leq Q_{\mathcal{A}} \cdot \mathbf{Adv}_{\mathcal{AE}, \mathcal{B}_{ae}}^{ae\text{-}ot}(k) \tag{1}$$

$$\Pr[X_2] = \Pr[\mathbf{Exp}_{\mathcal{KEM}, \mathcal{B}_{kem}}^{ccca\text{-}1}(k) = 1] \tag{2}$$

$$\Pr[X_3] = \Pr[\mathbf{Exp}_{\mathcal{KEM}, \mathcal{B}_{kem}}^{ccca\text{-}0}(k) = 1] . \tag{3}$$

The adversary $\mathcal{B}_{kem}$ against the CCCA security of $\mathcal{KEM}$ is defined as follows. $\mathcal{B}_{kem}$ inputs (pk, K_b^*, C^*) for an unknown bit b. First, $\mathcal{B}_{kem}$ runs $\mathcal{A}_1$ on input pk. For the ith decryption query (C_i, ψ_i) made by adversary $\mathcal{A}$, adversary $\mathcal{B}_{kem}$ defines the function $\mathrm{pred}_i : \mathcal{K} \rightarrow \{0, 1\}$ as

$$\mathrm{pred}_i(K) := \begin{cases} 0 & : & \text{if AE.Dec}(K, \psi_i) \text{ returns } \perp \\ 1 & : & \text{otherwise} \end{cases}$$

Note that the symmetric ciphertext ψ_i is hard-coded into $\mathrm{pred}_i(\cdot)$ which is clearly efficiently computable. $\mathcal{B}_{kem}$ queries (pred_i, C_i) to its own oracle $\mathrm{DEC}(\cdot, \cdot)$ and receives the following answer. If KEM.Dec(sk, C_i) returns a key $K_i \in \mathcal{K}$ with AE.Dec(K_i, ψ_i) returns $\perp$ then $\mathrm{DEC}(\mathrm{pred}_i, C_i)$ returns the key K_i. Otherwise (if KEM.Dec(sk, C_i) returns $\perp$ or if AE.Dec(K_i, ψ_i) returns $\perp$), $\mathrm{DEC}(\mathrm{pred}_i, C_i)$ returns $\perp$. Note that by the syntax of $\mathcal{AE}$ this perfectly simulates $\mathcal{A}$'s decryption queries.

For $\mathcal{A}$'s encryption challenge for two messages $M_0, M_1, \mathcal{B}_{kem}$ uses its own input (K_b^*, C^*) together with a random bit δ to create a challenge ciphertext $C_{pke}^* = (C^*, \psi^* \leftarrow$ AE.Enc(K^*, M_δ)) of message M_δ. Adversary $\mathcal{B}_{kem}$ runs $\mathcal{A}_2(C_{pke}^*, St_1)$ and inputs a guess bit δ' for δ. Finally, $\mathcal{B}_{kem}$ concludes its game with outputting $b' = 1$ if $\delta = \delta'$ and $b' = 0$, otherwise. This completes the description of $\mathcal{B}_{kem}$.

Adversary $\mathcal{B}_{kem}$ always perfectly simulates $\mathcal{A}$'s decapsulation queries. In case $b = 1$, $\mathcal{B}_{kem}$ uses the real key K_1^* for $\mathcal{A}$'s simulation which implies Equation (2). In case $b = 0$, $\mathcal{B}_{kem}$ uses a random key K_0^* for $\mathcal{A}$'s simulation which implies Equation (3). The complexity bounds for $\mathcal{B}_{kem}$ are clear from the construction, and it is left to show that $\mathrm{uncert}_{\mathcal{B}_{kem}}(k) \leq Q \cdot \mathbf{Adv}_{\mathcal{AE}, \mathcal{B}_{ae}}^{ae\text{-}ot}(k)$ for a suitable $\mathcal{B}_{ae}$.

To this end we build an adversary $\mathcal{B}_{ae}$ against the AE security of $\mathcal{AE}$ as follows. $\mathcal{B}_{ae}$ inputs 1^k and, using its own pair of KEM keys $(pk, sk) \xleftarrow{\$}$ KEM.Kg(1^k), emulates the same simulation for $\mathcal{A}$ as $\mathcal{B}_{kem}$ did above (using sk to answer its own $\mathrm{DEC}(\cdot, \cdot)$ queries). It additionally picks a random index $j^* \in \{1, \ldots, Q\}$. On $\mathcal{A}$'s j^* decryption query (C_{j^*}, ψ_{j^*}), $\mathcal{B}_{ae}$ submits ψ_{j^*} to its own decryption-or-reject oracle $\mathrm{DoR}_b(\cdot)$, and outputs $b' = 0$ iff $\mathrm{DoR}_b(\cdot)$ rejects with $\perp$.

Now $\mathcal{B}_{ae}$ will always output $b' = 0$ if $b = 0$ by definition of DoR_0. In case $b = 1$, $\mathcal{B}_{ae}$ will output $b' = 1$ iff the ciphertext ψ_{j^*} is valid in the sense AE.Dec(K', ψ_{j^*}) $\neq \perp$ for an independent, uniformly (by the AE experiment) chosen key K'. So adversary $\mathcal{B}_{ae}$'s advantage is as follows.

$$\mathbf{Adv}_{\mathcal{AE}, \mathcal{B}_{ae}}^{ae\text{-}ot}(k) = \Pr[K' \xleftarrow{\$} \mathcal{K} : \text{AE.Dec}(K', \psi_{j^*}) \neq \perp]$$

The above equals $\Pr[K' \xleftarrow{\$} \mathcal{K} : \mathrm{pred}_{j^*}(K') = 1]$, where $\mathrm{pred}_{j^*}(\cdot) = $ AE.Dec($\cdot, \psi_{j^*}$) is the predicate $\mathcal{B}_{kem}$ submits to oracle DEC as the j^*th query. For a uniformly chosen $j^* \in \{1, \ldots, Q\}$, the above equals $\mathrm{uncert}_{\mathcal{B}_{ae}}(k)$. Consequently, $\mathbf{Adv}_{\mathcal{AE}, \mathcal{B}_{ae}}^{ae\text{-}ot}(k) \geq \frac{1}{Q} \cdot \mathrm{uncert}_{\mathcal{B}_{ae}}(k)$. $\square$

3   Efficient Key Encapsulation from DDH

3.1   Building Blocks

We describe the building blocks used and assumptions made about them.

GROUP SCHEMES. A group scheme $\mathcal{GS}$ [13] specifies a sequence $(\mathcal{GR}_k)_{k \in \mathbb{N}}$ of group descriptions. For every value of a security parameter $k \in \mathbb{N}$, $\mathcal{GR}_k$ specifies the four tuple $\mathcal{GR}_k = (\hat{\mathbb{G}}_k, \mathbb{G}_k, p_k, g_k)$ (for notational convenience we sometimes drop the index k). $\mathcal{GR}_k = (\hat{\mathbb{G}}, \mathbb{G}, p, g)$ specifies a finite abelian group $\hat{\mathbb{G}}$, along with a prime-order subgroup $\mathbb{G}$, a generator g of $\mathbb{G}$, and the order p of $\mathbb{G}$. We denote the identity element of $\mathbb{G}$ as $1_{\mathbb{G}} \in \mathbb{G}$. We assume the existence of an efficient sampling algorithm $x \xleftarrow{\$} \mathbb{G}$ and an efficient membership algorithm that test if a given element $x \in \hat{\mathbb{G}}$ is contained in the subgroup $\mathbb{G}$.

We further assume the DDH problem is hard in $\mathcal{GS}$, captured by defining the ddh-advantage of an adversary $\mathcal{B}_{\mathrm{ddh}}$ as

$$\mathbf{Adv}^{\mathrm{ddh}}_{\mathcal{GS},\mathcal{B}_{\mathrm{ddh}}}(k) = |\Pr[\mathcal{B}_{\mathrm{ddh}}(g,h,g^a,h^a) = 1] - \Pr[\mathcal{B}_{\mathrm{ddh}}(g,h,g^a,K) = 1]|,$$

where $g, h, K \xleftarrow{\$} \mathbb{G}$ and $a \leftarrow \mathbb{Z}_p^*$.

AUTHENTICATED ENCRYPTION. We need an abstract notion of algebraic authenticated encryption where the keyspace consists of $\mathbb{G}$, secure in the sense of OT-AE. In the full version we recall (following the encrypt-then-mac approach [4,13]) how to build such algebraic AE satisfying all required functionality and security from the following basic primitives:

- A (computationally secure) one-time symmetric encryption scheme with binary k-bit keys (such as AES or padding with a PRNG)
- A (computationally secure) MAC (existentially unforgeable) with k-bit keys
- A (computationally secure) key-derivation function (pseudorandom).

We remark that for our purposes it is also possible to use a more efficient single-pass authenticated encryption scheme (see, e.g., [25]). In both cases the ciphertext expansion (i.e., ciphertext size minus plaintext size) of the AE scheme is only k (security parameter) bits which is optimal with respect to our security notion.

TARGET COLLISION RESISTANT HASHING. $\mathcal{TCR} = (\mathsf{TCR}_k)_{k \in \mathbb{N}}$ is a family of keyed hash functions $\mathsf{TCR}^s_k : \mathbb{G} \to \mathbb{Z}_p$ for each k-bit key s. It is assumed to be target collision resistant (TCR) [13], which is captured by defining the tcr-advantage of an adversary $\mathcal{B}_{\mathrm{tcr}}$ as $\mathbf{Adv}^{\mathrm{tcr}}_{\mathsf{TCR},\mathcal{B}_{\mathrm{tcr}}}(k) =$

$$\Pr[\mathsf{TCR}^s(c^*) = \mathsf{TCR}^s(c) \wedge c \neq c^* \; : \; s \xleftarrow{\$} \{0,1\}^k \; ; \; c^* \xleftarrow{\$} \mathbb{G} \; ; \; c \xleftarrow{\$} \mathcal{B}_{\mathrm{tcr}}(s,c^*)].$$

Note TCR is a weaker requirement than collision-resistance, so that, in particular, any practical collision-resistant function can be used. Also note that our notion of TCR is related to the stronger notion of universal one-way hashing [22], where in the security experiment of the latter the target value c^* is chosen by the adversary (but before seeing the hash key s).

Commonly [13, 21] this function is implemented using a dedicated cryptographic hash function like MD5 or SHA, which we assume to be target collision resistant. Since $|\mathbb{G}| = |\mathbb{Z}_p| = p$ we can alternatively also use a fixed (non-keyed) bijective encoding function $\mathsf{INJ} : \mathbb{G} \to \mathbb{Z}_p$. In that case we have a perfectly collision resistant hash function, i.e. $\mathbf{Adv}^{\mathrm{tcr}}_{\mathsf{INJ},\mathcal{B}_{\mathrm{tcr}}}(k) = 0$. In the full version, we show how to build such bijective encodings for a number of concrete group schemes.

3.2   The Key-Encapsulation Mechanism

Let $\mathcal{GS}$ be a group scheme where $\mathcal{GR}_k$ specifies $(\hat{\mathbb{G}}, \mathbb{G}, g, p)$ and let TCR : $\mathbb{G} \to \mathbb{Z}_p$ be a target collision resistant hash function (for simplicity we assume TCR to be non-keyed). We build a key encapsulation mechanism $\mathcal{KEM} = (\mathsf{KEM.kg}, \mathsf{KEM.Enc}, \mathsf{KEM.Dec})$ with $\mathcal{K} = \mathbb{G}$ as follows.

KEM.Kg(1^k)	KEM.Enc(pk)	KEM.Dec(sk, C)
$x, y, \omega \xleftarrow{\$} \mathbb{Z}_p^*$	$r \xleftarrow{\$} \mathbb{Z}_p^* \; ; \; c \leftarrow g^r$	Parse C as $(c, \pi) \in \hat{\mathbb{G}}^2$
$u \leftarrow g^x \; ; \; v \leftarrow g^y \; ; \; h \leftarrow g^\omega$	$t \leftarrow \mathsf{TCR}(c) \; ; \; \pi \leftarrow (u^t v)^r$	if $c \notin \mathbb{G}$ return $\perp$
$pk \leftarrow (u, v, h) \in \mathbb{G}^3$	$C \leftarrow (c, \pi) \in \mathbb{G}^2$	$t \leftarrow \mathsf{TCR}(c)$
$sk \leftarrow (x, y, \omega) \in (\mathbb{Z}_p)^3$	$K \leftarrow h^r \in \mathbb{G}$	if $c^{xt+y} \neq \pi$ return $\perp$
Return (sk, pk)	Return (C, K)	Return $K \leftarrow c^\omega$

We stress that decryption never explicitly checks if $\pi \in \mathbb{G}$; this check happens implicitly when $c \in \mathbb{G}$ and $c^{xt+y} = \pi$ is checked. A correctly generated ciphertext has the form $C = (c, \pi) \in \mathbb{G} \times \mathbb{G}$, where $c = g^r$ and $\pi = (u^t v)^r = (g^{xt+y})^r = c^{xt+y}$. Hence decapsulation will not reject and compute the key $K = c^\omega = h^r$, as in encapsulation.

Encryption takes four standard exponentiations plus one application of TCR, where the generation of π can also be carried out as one single multi-exponentiation [6]. Decryption takes two exponentiations plus one application of TCR, where the two exponentiations can also be viewed as one sequential exponentiation [6] (which is as efficient as a multi-exponentiation) to simultaneously compute c^{xt+y} and c^ω. The proof of the the following theorem is given in the full version.

Theorem 2. *Let $\mathcal{GS}$ be a group scheme where the DDH problem is hard and assume $\mathcal{TCR}$ is target collision resistant. Then $\mathcal{KEM}$ is secure in the sense of IND-CCCA. In particular,*

$$\mathbf{Adv}^{\mathrm{ccca}}_{\mathcal{KEM}, t, Q, \mathrm{uncert}(k)}(k) \leq \mathbf{Adv}^{\mathrm{ddh}}_{\mathcal{GS}, t}(k) + \mathbf{Adv}^{\mathrm{tcr}}_{\mathcal{TCR}, t}(k) + \mathrm{uncert}(k) + \frac{Q}{p} \; .$$

3.3   Comparison with Cramer-Shoup and Kurosawa-Desmedt

The following table summarizes the key-encapsulation part of the Cramer-Shoup encryption scheme [13], the Kurosawa-Desmedt scheme [21], and ours.

Scheme	Ciphertext	Encapsulated Key
Cramer-Shoup	$g^r, \hat{g}^r, (u^t v)^r$	h^r
Kurosawa-Desmedt	$g^r, \hat{g}^r$	$(u^t v)^r$
Ours	$g^r, (u^t v)^r$	h^r

Here $\hat{g}$ is another element from the public-key. Compared to the Cramer-Shoup scheme, the Kurosawa-Desmedt scheme leaves out the value h^r and defines $(u^t v)^r$ as the session key. Our results shows that it is also possible to leave out the element $\hat{g}^r$ from the ciphertext and that $\pi = (u^t v)^r$ is sufficient to authenticate $c = g^r$. Hence, our scheme can be viewed as the *dual* of (the KEM part of) the Kurosawa-Desmedt scheme [21].

From a technical point of view, our scheme mixes Cramer-Shoup like techniques [12] to obtain a form of "plaintext awareness" for inconsistent ciphertexts with an "algebraic trick" from the Boneh-Boyen identity-based encryption scheme [7] to decrypt consistent ciphertexts.Compared to Cramer-Shoup based proofs [11, 13, 21, 2] the most important technical difference, caused by the mentioned ability to decrypt consistent ciphertexts without knowing the full secret key, is that during our simulation the challenge ciphertexts is never made inconsistent. Intuitively this is the reason why we manage to maintain a consistent simulation using less redundancy in the secret key. This demonstrates that IND-CCCA security can be obtained with constructions that differ from hash proof systems.

On the other hand, the security proofs of all known schemes based on IBE techniques [10, 9, 19, 20, 18] inherently rely on some sort of external consistency check for the ciphertexts. This can be seen as the main reason why security of the IBE-based PKE schemes could only be proved in pairing groups (or relative to a gap-assumption), where the pairing was necessary for helping the proof identifying inconsistent ciphertexts. In our setting, the consistency check is done implicitly, using information-theoretic arguments borrowed from hash proof systems.

3.4   Efficiency

We compare our new DDH-based scheme's efficiency with the one of Kurosawa and Desmedt (in its more efficient "explicit-rejection" variant from [23]). Most importantly, the number of exponentiations for encryption and decryption are equal in both schemes. Although our security result is much more general (our KEM can be combined with any authenticated encryption scheme) this is not an exclusive advantage of our scheme. In fact we can derive the same result for the KD scheme from a more general theorem that we will prove in Section 5. (A similar result about combining the Kurosawa-Desmedt scheme with authenticated encryption was already obtained in [3] in the context of statefull encryption.)

However, there is one crucial difference in case one needs a scheme that is provably secure *solely* on the DDH assumption. Note that security (of the KD scheme and ours) relies on the DDH assumption *and* the assumption that $\mathcal{TCR}$ is target collision resistant. So as long as one does not want to sacrifice *provable* security by implementing the TCR function with a dedicated hash function like SHA-x or MD5 (what potentially renders the whole scheme insecure given the recent progress in attacking certain hash functions), one must either resort to inefficient generic constructions of TCR functions [22, 27], or one can use the "hash-free technique" described in [13]. With this latter technique, one can get rid of the TCR function completely; however, this comes at the cost of additional elements in the public and the secret key, and additional exponentiations during encryption. This overhead is linear in the number of elements that would have been hashed with the TCR. In the Kurosawa-Desmedt scheme, TCR acts on two group elements whereas in our scheme only on one. Hence the hash-free variant of our scheme is more efficient.

More importantly, since in our scheme a TCR is employed which maps *one* group element to integers modulo the group-order this can also be a bijection. In many concrete groups, e.g., when using the subgroup of quadratic residues modulo a safe prime or certain elliptic curves, this bijection can be trivially implemented at zero cost [13, 9], without any additional computational assumption, and without sacrificing provable security. See the full version for more details. In terms of efficiency we view this as the main benefit of our scheme.

4 Key Encapsulation from n-Linear

4.1 Linear Assumptions

Let $n = n(k)$ be a polynomial in k. Generalizing [8,18] we introduce the class of n-Linear assumptions which can be seen as a natural generalization of the DDH assumption and the Linear assumption.

Let $\mathcal{GS}$ be a group scheme. We define the n-lin-advantage of an adversary $\mathcal{B}_{n\text{-lin}}$ as

$$\mathbf{Adv}^{n\text{-lin}}_{\mathcal{GS},\mathcal{B}_{n\text{-lin}}}(k) = \Big| \Pr[\mathcal{B}_{n\text{-lin}}(g_1,\ldots,g_n,g_1^{r_1},\ldots,g_n^{r_n},h,h^{r_1+\ldots+r_n}) = 1]$$
$$- \Pr[\mathcal{B}_{n\text{-lin}}(g_1,\ldots,g_n,g_1^{r_1},\ldots,g_n^{r_n},h,K) = 1]\Big|,$$

where $g_1,\ldots,g_n,h,K \xleftarrow{\$} \mathbb{G}$ and all $r_i \leftarrow \mathbb{Z}_p^*$. We say that the *n-Linear Decisional Diffie-Hellman (n-Linear) assumption relative to group scheme $\mathcal{GS}$ holds if* $\mathbf{Adv}^{n\text{-lin}}_{\mathcal{GS},\mathcal{B}_{n\text{-lin}}}$ is a negligible function in k for all polynomial-time adversaries $\mathcal{B}_{n\text{-lin}}$.

The n-Linear assumptions form a strict hierarchy of security assumptions with 1-Linear = DDH, 2-Linear=Linear [8] and, the larger the n, the weaker the n-Linear assumption. More precisely, for any $n \geq 1$ we have that n-Linear implies $n+1$-Linear. On the other hand (extending the case of $n = 1$ [8]) we can show that in the generic group model [29], the $n+1$-Linear assumption holds, even relative to an n-Linear oracle.

Lemma 3. DDH = *1-Linear* $\overset{\Leftarrow}{\Rightarrow}$ *2-Linear* $\overset{\Leftarrow}{\Rightarrow}$ *3-Linear* $\overset{\Leftarrow}{\Rightarrow}$ ...

4.2 The Key-Encapsulation Mechanism

Let $\mathcal{GS}$ be a group scheme where $\mathcal{GR}_k$ specifies $(\hat{\mathbb{G}}, \mathbb{G}, g, p)$ and let TCR : $\mathbb{G}^{n+1} \to \mathbb{Z}_p$ be a target collision resistant hash function. Generalizing the Kurosawa-Desmedt KEM, for a parameter $n = n(k) \geq 1$, we build $\mathcal{KEM} = $ (KEM.Kg, KEM.Enc, KEM.Dec) as follows.

Key generation KEM.Kg(1^k) generates random group elements $g_1,\ldots,g_n,h \in \mathbb{G}$. Furthermore, it defines $u_j = g_j^{x_j}h^z$ and $v_j = g_j^{y_j}h^{z'}$ for random $z,z' \in \mathbb{Z}_p$ and $x_j,y_j \in \mathbb{Z}_p$ $(j \in \{1,\ldots,n\})$. The public key contains the elements h, $(g_j,u_j)_{1 \leq i \leq n}$, and the secret key contains all corresponding indices.

KEM.Enc(pk)	KEM.Dec(sk,C)
$\forall j \in \{1,\ldots,n\}$: $r_j \xleftarrow{\$} \mathbb{Z}_p^*$; $c_j \leftarrow g_j^{r_j}$	$\forall j \in \{1,\ldots,n\}$: check if $c_j \in \mathbb{G}$
$d \leftarrow h^{r_1+\ldots+r_n}$; $t \leftarrow$ TCR$(c_1,\ldots,c_n,d)$	Check if $d \in \mathbb{G}$
$C \leftarrow (c_1,\ldots,c_n,d)$; $K = \prod_{i=1}^{n}(u_i^t v_i)^{r_i}$	$t \leftarrow$ TCR$(c_1,\ldots,c_n,d)$
Return (C,K)	Return $K \leftarrow d^{zt+z'} \cdot \prod_{j=1}^{n} c_j^{x_j t + y_j}$

Ciphertexts contain $n+1$ group elements, public/secret keys $2n+1$ elements. The scheme instantiated with $n = 1$ precisely reproduces the KEM part of the Kurosawa-Desmedt encryption scheme [21]. Security of the schemes can be explained using the more general framework of computational hash-proof systems. This will be done in Section 5.

Theorem 3. *Let $\mathcal{GS}$ be a group scheme where the n-Linear problem is hard, assume $\mathcal{TCR}$ is target collision resistant. Then $\mathcal{KEM}$ is secure in the sense of IND-CCCA.*

We remark that it is also possible to give the scheme in its explicit-rejection variant [13]. Furthermore, in the full version we also provide a class of alternative schemes generalizing our dual KD scheme from Section 3 to the n-Linear assumption.

5   Key Encapsulation from Hash Proof Systems

In [12], Cramer and Shoup showed that their original scheme in [13] was a special instance of a generic framework based on hash proof systems (HPS). Following [12] we recall the basic ideas of hash proof systems and show (generalizing [21]) how to build IND-CCCA secure key encapsulation based on a computational variant of hash proof systems. Here we use a slightly different notation for HPS that better reflects our primary application of hash-proof systems to key-encapsulation mechanisms.

5.1   Hash Proof Systems

Let $\mathcal{C}, \mathcal{K}$ be sets and $\mathcal{V} \subset \mathcal{C}$ a language. Let $\mathsf{D}_{sk} : \mathcal{C} \to \mathcal{K}$ be a hash function indexed with $sk \in \mathcal{S}$, where $\mathcal{S}$ is a set. A hash function D_{sk} is *projective* if there exists a projection $\mu : \mathcal{S} \to \mathcal{P}$ such that $\mu(sk) \in \mathcal{P}$ defines the action of D_{sk} over the subset $\mathcal{V}$. That is, for every $C \in \mathcal{V}$, the value $K = \mathsf{D}_{sk}(C)$ is uniquely determined by $\mu(sk)$ and C. In contrast, nothing is guaranteed for $C \in \mathcal{C} \setminus \mathcal{V}$, and it may not be possible to compute $\mathsf{D}_{sk}(C)$ from $\mu(sk)$ and C. A *strongly universal$_2$* projective hash function has the additional property that for $C \in \mathcal{C} \setminus \mathcal{V}$, the projection key $\mu(sk)$ actually says nothing about the value of $K = \mathsf{D}_{sk}(C)$, even given an instance (C^*, K^*) such that $C^* \in \mathcal{C} \setminus \mathcal{V}$ and $K^* = \mathsf{D}_{sk}(C)$. More precisely, for all $pk \in \mathcal{P}$, C, all $C^* \in \mathcal{C} \setminus \mathcal{V}$ with $C \neq C^*$, all $K, K^* \in \mathcal{K}$,

$$\Pr_{\substack{sk \in \mathcal{S} \\ \mathsf{D}_{sk}(C^*)=K^* \\ \mu(sk)=pk}} [\mathsf{D}_{sk}(C) = K] = 1/|\mathcal{K}|. \tag{4}$$

A hash proof system $\mathcal{HPS} = (\mathsf{HPS.param}, \mathsf{HPS.pub}, \mathsf{HPS.priv})$ consists of three algorithms where the randomized algorithm $\mathsf{HPS.param}(1^k)$ generates instances of $params = (group, \mathcal{C}, \mathcal{V}, \mathcal{P}, \mathcal{S}, \mathsf{D}_{(.)} : \mathcal{C} \to \mathcal{K}, \mu : \mathcal{S} \to \mathcal{P})$, where $group$ may contain some additional structural parameters. The deterministic public evaluation algorithm $\mathsf{HPS.pub}$ inputs the projection key $pk = \mu(sk)$, $C \in \mathcal{V}$ and a witness

w of the fact that $C \in \mathcal{V}$ and returns $K = \mathsf{D}_{sk}(C)$. The deterministic private evaluation algorithm inputs $sk \in \mathcal{S}$ and returns $\mathsf{D}_{sk}(C)$, without knowing a witness. We further assume there are efficient algorithms given for sampling $sk \in \mathcal{S}$ and sampling $C \in \mathcal{V}$ uniformly together with a witness w.

As computational problem we require that the *subset membership problem* is hard in $\mathcal{HPS}$ which means that the two elements C and C' are computationally indistinguishable, for random $C \in \mathcal{V}$ and random $C' \in \mathcal{C} \setminus \mathcal{V}$. This is captured by defining the advantage function $\mathbf{Adv}^{\mathrm{sm}}_{\mathcal{HPS},\mathcal{A}}(k)$ of an adversary $\mathcal{A}$ as

$$\mathbf{Adv}^{\mathrm{sm}}_{\mathcal{HPS},\mathcal{A}}(k) := \big| \Pr[C_1 \xleftarrow{\$} \mathcal{C} ; b' \xleftarrow{\$} \mathcal{A}(\mathcal{C},\mathcal{V},C_1) : b' = 1]$$

$$- \Pr[C_0 \xleftarrow{\$} \mathcal{C} \setminus \mathcal{V} ; b' \xleftarrow{\$} \mathcal{A}(\mathcal{C},\mathcal{V},C_0) : b' = 1]\big| .$$

5.2 Key Encapsulation from HPS

Using the above notion of a hash proof system, Kurosawa and Desmedt [21] proposed a hybrid encryption scheme which improved the schemes from [12]. The key-encapsulation part of it is as follows. The system parameters of the scheme consist of $params \xleftarrow{\$} \mathsf{HPS.param}(1^k)$.

KEM.Kg(k). Choose random $sk \xleftarrow{\$} \mathcal{S}$ and define $pk = \mu(sk) \in \mathcal{P}$. Return (pk, sk).

KEM.Enc(pk). Pick $C \xleftarrow{\$} \mathcal{V}$ together with its witness ω that $C \in \mathcal{V}$. The session key $K = \mathsf{D}_{sk}(C) \in \mathcal{K}$ is computed as $K \xleftarrow{\$} \mathsf{HPS.pub}(pk, C, \omega)$. Return (K, C).

KEM.Dec(sk, C). Reconstruct the key $K = \mathsf{D}_{sk}(C)$ as $K \leftarrow \mathsf{HPS.priv}(sk, C)$ and return K.

We can prove the following theorem that is a slight generalization of [21].

Theorem 4. *If $\mathcal{HPS}$ is strongly universal$_2$ and the subset membership problem is hard in $\mathcal{HPS}$ then $\mathcal{KEM}$ is secure in the sense of* IND-CCCA.

Unfortunately, the original KEM part of the Kurosawa Desmedt DDH-based hybrid encryption scheme [21] cannot be explained using this framework and hence needed a separate proof of security. This is since the underlying DDH-based hash proof system involves a target collision resistant hash function TCR which is a "computational primitive" whereas the strongly universal$_2$ property from Equation (4) is a *statistical property* which is in particularly not fulfilled by the DDH-based HPS from [12] used in [21]. In fact, the most efficient HPS-based schemes that are known involve computation of a TCR function and hence all need a separate proof of security. We note that this problem is inherited from the original HPS approach [13].

We overcome this problem we defining the weaker notion of *computational hash proof systems*.

5.3 Computational Hash Proof Systems

We now define a weaker computational variant of strongly universal$_2$ hashing. For an adversary $\mathcal{B}$ we define the advantage function $\mathbf{Adv}^{\mathrm{cu2}}_{\mathcal{HPS},\mathcal{B}}(k) = |\Pr[\mathbf{Exp}^{\mathrm{cu2-1}}_{\mathcal{HPS},\mathcal{B}}(k) = 1] - \Pr[\mathbf{Exp}^{\mathrm{cu2-0}}_{\mathcal{HPS},\mathcal{B}}(k) = 1]|$ where, for $b \in \{0,1\}$, $\mathbf{Exp}^{\mathrm{cu2-b}}_{\mathcal{HPS},\mathcal{B}}$ is defined by the following experiment.

Experiment $\mathbf{Exp}^{\mathrm{cu2-b}}_{\mathcal{HPS},\mathcal{B}}(k)$

$params \xleftarrow{\$} \mathsf{HPS.param}(1^k)\,;\ sk \xleftarrow{\$} \mathcal{S}\,;\ pk \leftarrow \mu(sk)$
$C^* \xleftarrow{\$} \mathcal{C} \setminus \mathcal{V}\,;\ K^* \leftarrow \mathsf{D}_{sk}(C^*)\,;\ (C, St) \xleftarrow{\$} \mathcal{B}_1^{\mathrm{EVALD}(\cdot)}(pk, C^*, K^*)$
$K_0 \xleftarrow{\$} \mathcal{K}\,;\ K_1 \leftarrow \mathsf{D}_{sk}(C)\,;\ b' \xleftarrow{\$} \mathcal{B}_2(St, K_b)$
Return b'

where the evaluation oracle $\mathrm{EVALD}(C)$ returns $K = \mathsf{D}_{sk}(C)$ if $C \in \mathcal{V}$ and $\perp$, otherwise. We also restrict to adversaries that only return ciphertexts $C \neq C^*$ and that ensure $C \in \mathcal{C} \setminus \mathcal{V}$. This is without losing generality, since $\mathcal{B}_1$ can check $C \in \mathcal{V}$ with its oracle EVALD. A hash proof system $\mathcal{HPS}$ is said to be *computationally universal$_2$* (CU$_2$) if for all polynomial-time adversaries $\mathcal{B}$ that satisfy these requirements, the advantage function $\mathbf{Adv}^{\mathrm{cu2}}_{\mathcal{HPS},\mathcal{B}}(k)$ is a negligible function in k.

The following theorem strengthens Theorem 4. A proof will be given in the full version.

Theorem 5. *If $\mathcal{HPS}$ is computationally universal$_2$ and the subset membership problem is hard then $\mathcal{KEM}$ from Section 5.2 is IND-CCCA secure. In particular,*

$$\mathbf{Adv}^{\mathrm{ccca}}_{\mathcal{KEM},t,Q,\mathrm{uncert}_{\mathcal{A}}(k)}(k) \leq \mathbf{Adv}^{\mathrm{sm}}_{\mathcal{HPS},t}(k) + (Q+1)\cdot(\mathrm{uncert}_{\mathcal{A}}(k) + \mathbf{Adv}^{\mathrm{cu2}}_{\mathcal{HPS},t}(k))\,.$$

5.4 A Computational HPS from n-Linear

Let $\mathcal{GS}$ be a group scheme where $\mathcal{GR}_k$ specifies $(\hat{\mathbb{G}}, \mathbb{G}, g, p)$. Let $group = (\mathcal{GR}, g_1, \ldots, g_n, h)$, where $g_1, \ldots, g_n, h$ are independent generators of $\mathbb{G}$. Define $\mathcal{C} = \mathbb{G}^{n+1}$ and $\mathcal{V} = \{(g_1^{r_1}, \ldots, g_n^{r_n}, h^{r_1 + \cdots + r_n}) \subset \mathbb{G}^{n+1} : r_1, \ldots, r_n \in \mathbb{Z}_p\}$ The values $(r_1, \ldots, r_n) \in \mathbb{Z}_p^n$ are a witness of $C \in \mathcal{V}$. Let $\mathsf{TCR} : \mathbb{G}^{n+1} \to \mathbb{Z}_p$ be a target collision resistant hash function. Let $\mathcal{S} = \mathbb{Z}_p^{2n+2}$, $\mathcal{P} = \mathbb{G}^{2n}$, and $\mathcal{K} = \mathbb{G}$. For $sk = (x_1, y_1, \ldots, x_n, y_n, z, z') \in \mathbb{Z}^{2n+2}$, define $\mu(sk) = (u_1, \ldots, u_n, v_1, \ldots, v_n)$, where, for $1 \leq i \leq n$, $u_i = g_i^{x_i} h^z$ and $v_i = g_i^{y_i} h^{z'}$. This defines the output of $\mathsf{HPS.param}(1^k)$. For $C = (c_1, \ldots, c_n, d) \in \mathcal{C}$ define

$$\mathsf{D}_{sk}(C) := d^{zt+z'} \cdot \prod_{i=1}^{n} c_i^{x_i t + y_i}, \quad \text{where } t = \mathsf{TCR}(c_1, \ldots, c_n)\,. \tag{5}$$

This defines $\mathsf{HPS.priv}(sk, C)$. Given $pk = \mu(sk)$, $C \in \mathcal{V}$ and a witness $w = (r_1, \ldots, r_n) \in (\mathbb{Z}_p)^n$ such that $C = (c_1, \ldots, c_n, d) = (g_1^{r_1}, \ldots, g_n^{r_n}, h^{r_1 + \cdots + r_n})$ public evaluation $\mathsf{HPS.pub}(pk, C, w)$ computes $K = \mathsf{D}_{sk}(C)$ as

$$K = \prod_{i=1}^{n} (u_i^t v_i)^{r_i}\,.$$

Correctness follows by Equation (5) and the definition of μ. This completes the description of $\mathcal{HPS}$. Clearly, under the n-Linear assumption, the subset membership problem is hard in $\mathcal{HPS}$.

Obviously, the above defined HPS is not strongly universal$_2$ in the sense of Equation (4). But it is still computationally universal$_2$.

Lemma 4. *The n-Linear based HPS is computationally universal$_2$.*

Together with Theorem 5 this proves Theorem 3. For the case $n = 1$ this also gives an alternative security proof for the Kurosawa-Desmedt scheme [21].

Proof. Consider an adversary $\mathcal{B}$ in the CU$_2$ experiment such that $\mathcal{B}_1$ outputs a ciphertext $C \in \mathcal{C} \setminus \mathcal{V}$ and let $K \leftarrow D_{sk}(C)$. Let COL be the event that $C \neq C^*$ but $\mathsf{TCR}(C) = \mathsf{TCR}(C^*)$. We claim that for the following adversary $\mathcal{B}_{\mathrm{tcr}}$ we have $\mathbf{Adv}^{\mathrm{tcr}}_{\mathsf{TCR}, \mathcal{B}_{\mathrm{tcr}}}(k) = \Pr[\text{COL}]$. Adversary $\mathcal{B}_{\mathrm{tcr}}$ inputs (s, C^*) and generates a random instance of *params* with known indices α_i such that $h = g^{\alpha_i}$. Furthermore, $\mathcal{B}_{\mathrm{tcr}}$ picks a random $sk \in \mathcal{S}$ and runs $\mathcal{B}_1$ on $pk = \mu(sk)$, a random $C^* \in \mathcal{C} \setminus \mathcal{V}$, and $K^* = D_{sk}(C^*)$. To answer a query to the evaluation oracle EVALD$(\cdot)$, $\mathcal{B}_{\mathrm{tcr}}$ fist verifies $C = (c_1, \dots, c_n, d) \in \mathcal{V}$ by checking if $\prod c_i^{\alpha_i} = d$. If not, return $\perp$. Otherwise it returns $K = D_{sk}(C)$. If for a decapsulation query C event COL happens, $\mathcal{B}_{\mathrm{tcr}}$ returns C to its TCR experiment and terminates.

Now we claim that conditioned under $\neg$COL, the key $K = D_{sk}(C)$ is a uniform element in $\mathcal{K}$ independent of the adversary's view. This implies that not even a *computationally unbounded* $\mathcal{B}_2$ could succeed in the second stage. Hence, $\mathbf{Adv}^{\mathrm{cu2}}_{\mathcal{HPS}, \mathcal{B}}(k) \leq \mathbf{Adv}^{\mathrm{tcr}}_{\mathsf{TCR}, \mathcal{B}_{\mathrm{tcr}}}(k)$, which proves the lemma.

Let $\log(\cdot) = \log_g(\cdot)$. Consider the view of $\mathcal{B}_2$ consisting of the random variables (pk, C^*, K^*, C), where $sk = (x_1, y_1, \dots, x_n, y_n, z, z') \xleftarrow{\$} \mathbb{Z}^{2n+2}$, $pk = \mu(sk) = (u_1, \dots, u_n, v_1, \dots, v_n)$, $C^* = (c_1^*, \dots, c_n^*, d^*) = (g_1^{r_1^*}, \dots, g_n^{r_n^*}, h^{r^*})$ with $\sum r_i^* \neq r^*$ since $C^* \in \mathcal{C} \setminus \mathcal{V}$, $K^* = D_{sk}(C^*)$, and $C = (c_1, \dots, c_n, d) = (g_1^{r_1}, \dots, g_n^{r_n}, h^r)$ ($\sum r_i \neq r$ since $C \in \mathcal{C} \setminus \mathcal{V}$). From the system parameters $g_1, \dots, g_n, h$, adversary $\mathcal{B}_2$ learns $\omega = \log h$, $\omega_i = \log g_i$, and from pk

$$\text{for } 1 \leq i \leq n : \log u_i = \omega_i x_i + \omega z, \quad \log v_i = \omega_i y_i + \omega z' . \tag{6}$$

From C^* the adversary learns $r_i^* = \log_{g_i} c_i^*$, $r^* = \log_h d^*$, and from K^* (by Equation (5)) the value

$$\log K^* = \sum \omega_i r_i^* (x_i t^* + y_i) + \omega(z t^* + z') , \tag{7}$$

and $t^* = \mathsf{TCR}(c_1^*, \dots, c_n^*, d^*)$. Furthermore, from C, $\mathcal{B}_2$ learns $r_i = \log_{g_i} c_i$ and $r = \log_h d$. Let $K = D_{sk}(C)$. Our claim is that

$$\log K = \sum \omega_i r_i (x_i t + y_i) + \omega(z t + z') , \tag{8}$$

with $t = \mathsf{TCR}(C) \neq t^*$, is a uniform and independent element in $\mathbb{Z}_p$. Consider the set of linear equations over the hidden values $x_1, \dots, x_n, y_1, \dots, y_n, z, z'$ defined by Equations (6), (7), and (8), defined by the matrix $M \in \mathbb{Z}_p^{n+2 \times n+2}$,

$$M = \begin{pmatrix} & x_1 & \cdots & x_n & y_1 & \cdots & y_n & z & z' \\ & \omega_1 & & & & & & \omega & \\ & & \ddots & & & 0 & & \vdots & 0 \\ & & & \omega_n & & & & \omega & \\ & & & & \omega_1 & & & \omega & \\ & & 0 & & & \ddots & & 0 & \vdots \\ & & & & & & \omega_n & \omega & \\ & \omega_1 r_1^* t^* & \cdots & \omega_n r_n^* t^* & \omega_1 r_1^* & \cdots & \omega_n r_n^* & \omega t^* r^* & \omega r^* \\ & \omega_1 r_1 t & \cdots & \omega_n r_n t & \omega_1 r_1 & \cdots & \omega_n r_n & \omega t r & \omega r \end{pmatrix}$$

Since $\det(M) = \omega^2 \prod \omega_i (t - t^*)(\sum_{i=1}^n r_i - r)(\sum_{i=1}^n r_i^* - r^*) \neq 0$, Equation (8) is linearly independent of (6) and (7).

We note that (generalizing [12]) we can also give a computationally universal$_2$ hash-proof system based on Paillier's decision composite residue (DCR) assumption.

References

1. Abe, M., Gennaro, R., Kurosawa, K.: Tag-KEM/DEM: A new framework for hybrid encryption. Cryptology ePrint Archive, Report 2005/027 (2005), http://eprint.iacr.org/

2. Abe, M., Gennaro, R., Kurosawa, K., Shoup, V.: Tag-KEM/DEM: A new framework for hybrid encryption and a new analysis of Kurosawa-Desmedt KEM. In: Cramer, R.J.F. (ed.) EUROCRYPT 2005. LNCS, vol. 3494, pp. 128–146. Springer, Heidelberg (2005)

3. Bellare, M., Kohno, T., Shoup, V.: Stateful public-key cryptosystems: How to encrypt with one 160-bit exponentiation. In: ACM CCS 2006, pp. 380–389. ACM Press, New York (2006)

4. Bellare, M., Namprempre, C.: Authenticated encryption: Relations among notions and analysis of the generic composition paradigm. In: Okamoto, T. (ed.) ASIACRYPT 2000. LNCS, vol. 1976, pp. 531–545. Springer, Heidelberg (2000)

5. Bellare, M., Rogaway, P.: Random oracles are practical: A paradigm for designing efficient protocols. In: ACM CCS 1993, pp. 62–73. ACM Press, New York (1993)

6. Bernstein, D.J.: Pippenger's exponentiation algorithm (2001), Available from http://cr.yp.to/papers.html#pippenger

7. Boneh, D., Boyen, X.: Efficient selective-ID secure identity based encryption without random oracles. In: Cachin, C., Camenisch, J.L. (eds.) EUROCRYPT 2004. LNCS, vol. 3027, pp. 223–238. Springer, Heidelberg (2004)

8. Boneh, D., Boyen, X., Shacham, H.: Short group signatures. In: Franklin, M. (ed.) CRYPTO 2004. LNCS, vol. 3152, pp. 41–55. Springer, Heidelberg (2004)

9. Boyen, X., Mei, Q., Waters, B.: Direct chosen ciphertext security from identity-based techniques. In: ACM CCS 2005, pp. 320–329. ACM Press, New York (2005)

10. Canetti, R., Halevi, S., Katz, J.: Chosen-ciphertext security from identity-based encryption. In: Cachin, C., Camenisch, J.L. (eds.) EUROCRYPT 2004. LNCS, vol. 3027, pp. 207–222. Springer, Heidelberg (2004)

11. Cramer, R., Shoup, V.: A practical public key cryptosystem provably secure against adaptive chosen ciphertext attack. In: Krawczyk, H. (ed.) CRYPTO 1998. LNCS, vol. 1462, pp. 13–25. Springer, Heidelberg (1998)

12. Cramer, R., Shoup, V.: Universal hash proofs and a paradigm for adaptive chosen ciphertext secure public-key encryption. In: Knudsen, L.R. (ed.) EUROCRYPT 2002. LNCS, vol. 2332, pp. 45–64. Springer, Heidelberg (2002)

13. Cramer, R., Shoup, V.: Design and analysis of practical public-key encryption schemes secure against adaptive chosen ciphertext attack. SIAM Journal on Computing 33(1), 167–226 (2003)

14. Gennaro, R., Shoup, V.: A note on an encryption scheme of Kurosawa and Desmedt. Cryptology ePrint Archive, Report 2004/194 (2004)

15. Hofheinz, D., Herranz, J., Kiltz, E.: The Kurosawa-Desmedt key encapsulation is not chosen-ciphertext secure. Cryptology ePrint Archive, Report 2006/207 (2006)

16. Joux, A., Nguyen, K.: Separating decision Diffie-Hellman from computational Diffie-Hellman in cryptographic groups. Journal of Cryptology 16(4), 239–247 (2003)

17. Katz, J., Yung, M.: Unforgeable encryption and chosen ciphertext secure modes of operation. In: Schneier, B. (ed.) FSE 2000. LNCS, vol. 1978, pp. 284–299. Springer, Heidelberg (2001)

18. Kiltz, E.: Chosen-ciphertext secure key-encapsulation based on Gap Hashed Diffie-Hellman. In: PKC 2007. LNCS, vol. 4450, pp. 282–297 (2007)

19. Kiltz, E.: Chosen-ciphertext security from tag-based encryption. In: Halevi, S., Rabin, T. (eds.) TCC 2006. LNCS, vol. 3876, pp. 581–600. Springer, Heidelberg (2006)

20. Kiltz, E.: On the limitations of the spread of an IBE-to-PKE transformation. In: Yung, M., Dodis, Y., Kiayias, A., Malkin, T.G. (eds.) PKC 2006. LNCS, vol. 3958, pp. 274–289. Springer, Heidelberg (2006)

21. Kurosawa, K., Desmedt, Y.: A new paradigm of hybrid encryption scheme. In: Franklin, M. (ed.) CRYPTO 2004. LNCS, vol. 3152, pp. 426–442. Springer, Heidelberg (2004)

22. Naor, M., Yung, M.: Universal one-way hash functions and their cryptographic applications. In: 21st ACM STOC, pp. 33–43. ACM Press, New York (1989)

23. Phong, L.T., Ogata, W.: On a variation of Kurosawa-Desmedt encryption scheme. Cryptology ePrint Archive, Report 2006/031 (2006)

24. Rackoff, C., Simon, D.R.: Non-interactive zero-knowledge proof of knowledge and chosen ciphertext attack. In: Feigenbaum, J. (ed.) CRYPTO 1991. LNCS, vol. 576, pp. 433–444. Springer, Heidelberg (1992)

25. Rogaway, P., Bellare, M., Black, J., Krovetz, T.: OCB: A block-cipher mode of operation for efficient authenticated encryption. In: ACM CCS 2001, pp. 196–205. ACM Press, New York (2001)

26. Rogaway, P., Shrimpton, T.: A provable-security treatment of the key-wrap problem. In: Vaudenay, S. (ed.) EUROCRYPT 2006. LNCS, vol. 4004, pp. 373–390. Springer, Heidelberg (2006)

27. Rompel, J.: One-way functions are necessary and sufficient for secure signatures. In: 22nd ACM STOC, pp. 387–394. ACM Press, New York (1990)

28. Shacham, H.: A Cramer-Shoup encryption scheme from the Linear Assumption and from progressively weaker linear variants. Cryptology ePrint Archive, Report 2007/074 (2007)

29. Shoup, V.: Lower bounds for discrete logarithms and related problems. In: Fumy, W. (ed.) EUROCRYPT 1997. LNCS, vol. 1233, pp. 256–266. Springer, Heidelberg (1997)

30. Shoup, V.: Using hash functions as a hedge against chosen ciphertext attack. In: Preneel, B. (ed.) EUROCRYPT 2000. LNCS, vol. 1807. Springer, Heidelberg (2000)

Scalable and Unconditionally Secure Multiparty Computation

Ivan Damgård and Jesper Buus Nielsen*

Dept. of Computer Science, BRICS, Aarhus University

Abstract. We present a multiparty computation protocol that is unconditionally secure against adaptive and active adversaries, with communication complexity $\mathcal{O}(\mathcal{C}n)k + \mathcal{O}(Dn^2)k + \text{poly}(n\kappa)$, where $\mathcal{C}$ is the number of gates in the circuit, n is the number of parties, k is the bit-length of the elements of the field over which the computation is carried out, D is the multiplicative depth of the circuit, and κ is the security parameter. The corruption threshold is $t < n/3$. For passive security the corruption threshold is $t < n/2$ and the communication complexity is $\mathcal{O}(n\mathcal{C})k$. These are the first unconditionally secure protocols where the part of the communication complexity that depends on the circuit size is linear in n. We also present a protocol with threshold $t < n/2$ and complexity $\mathcal{O}(\mathcal{C}n)k + \text{poly}(n\kappa)$ based on a complexity assumption which, however, only has to hold *during* the execution of the protocol – that is, the protocol has so called everlasting security.

1 Introduction

In secure multiparty computation a set of n parties, $\mathcal{P} = \{P_1, \ldots, P_n\}$, want to compute a function of some secret inputs held locally by some of the parties. The desired functionality is typically specified by a function $f : (\{0,1\}^*)^n \to (\{0,1\}^*)^n$. Party P_i has input $x_i \in \{0,1\}^*$ and output $y_i \in \{0,1\}^*$, where $(y_1, \ldots, y_n) = f(x_1, \ldots, x_n)$. During the protocol a subset $C \subset \mathcal{P}$ of the parties can be *corrupted*. Security means that all parties receive correct outputs and that the messages seen by the corrupted parties $P_i \in C$ during the protocol contain no information about the inputs and outputs of the honest parties $(\mathcal{P} \setminus C)$, other than what can be computed efficiently from the inputs and outputs of the corrupted parties. *Passive security* means that the above condition holds when all parties follow the protocol. *Active security* means that the above condition holds even when the corrupted parties in C might deviate from the protocol in an arbitrary coordinated manner. When a protocol is secure against all subsets of size at most t we say that it is secure against an adversary with *corruption threshold t*.

In the *cryptographic model* each pair of parties is assumed to share an authenticated channel and the corrupted parties are assumed to be poly-time bounded, to allow the use of cryptography. In the *information theoretic model* it is assumed

---

* Funded by the Danish Agency for Science, Technology and Innovation.

A. Menezes (Ed.): CRYPTO 2007, LNCS 4622, pp. 572–590, 2007.

that each pair of parties share a perfectly secure channel and that the parties have an authenticated broadcast channel, and it is assumed that the corrupted parties are computationally unbounded. We talk about *cryptographic security* versus *unconditional security*.

The MPC problem dates back to Yao [Yao82]. The first generic solutions with cryptographic security were presented in [GMW87, CDG87], and later generic solutions with unconditional security were presented in [BGW88, CCD88, RB89, Bea89]. In both cases, security against an active adversary with threshold $t < n/2$ is obtained, and $t < n/2$ is known to be optimal.

Thresh.	Adv.	Communication	Reference
$t < n/2$	passive	$\mathcal{O}(\mathcal{C}n^2)k$	[BGW88]
$t < n/2$	active	$\mathcal{O}(\mathcal{C}n^5)k + \text{poly}(n\kappa)$	[CDD+99]
$t < n/3$	active	$\mathcal{O}(\mathcal{C}n^2)k + \text{poly}(n\kappa)$	[HM01]
$t < n/2$	active	$\mathcal{O}(\mathcal{C}n^2)k + \text{poly}(n\kappa)$	[BH06]
$t < n/2$	passive	$\mathcal{O}(\mathcal{C}n)k$	this paper
$t < n/3$	active	$\mathcal{O}(\mathcal{C}n)k + \mathcal{O}(Dn^2)k + \text{poly}(n\kappa)$	this paper
$t < n/2$	active, limited during protocol	$\mathcal{O}(\mathcal{C}n)k + \text{poly}(n\kappa)$	this paper

Fig. 1. Comparison of some unconditionally secure MPC protocols

The communication complexity of a MPC protocol is taken to be the total number of bits sent and received by the honest parties in the protocol. Over the years a lot of research have been focused on bringing down the communication complexity of active secure MPC [GV87, BB89, BMR90, BFKR90, Bea91, GRR98] [CDM00, CDD00, HMP00, HM01, HN05, DI06, HN06]. Building on a lot of previous work, it was recently shown in [DI06, HN06]that there exist cryptographic secure MPC protocols with communication complexity $\mathcal{O}(\mathcal{C}n)k + \text{poly}(n\kappa)$, where $\mathcal{C}$ is the size of a Boolean circuit computing f, k is the bit length of elements from the field (or ring) over which the computation takes place (at least $k = \log_2(n)$) and $\text{poly}(n\kappa)$ is some complexity independent of the circuit size, typically covering the cost of some fixed number of broadcasts or a setup phase. All earlier protocols had a first term of at least $\mathcal{O}(\mathcal{C}n^2)k$, meaning that the communication complexity depending on $\mathcal{C}$ was quadratic in the number of parties. Having linear communication complexity is interesting as it means that the work done by a single party is independent of the number of parties participating in the computation, giving a fully scalable protocol. Until the work presented in this paper it was, however, not known whether there existed an *unconditionally* secure protocol with linear communication complexity. In fact, this was an open problem even for passive secure protocols. We show that indeed it is possible to construct an unconditional secure protocol where the part of the communication complexity which depends on the circuit size is linear in n. For active security, however, we get a quadratic dependency on the multiplicative depth of the circuit, denoted by D. Our results are compared to some previous unconditionally secure protocol in Fig. 1.

Note that our active secure protocol is more efficient than even the previous most efficient passive secure protocol, as clearly $D \leq C$. Note also that our active secure protocol obtains threshold $t < n/3$. When no broadcast channel is given, this is optimal. If a broadcast channel is given, however, it is possible to get active security for $t < n/2$ [RB89]. By using an unconditionally hiding commitment scheme to commit parties to their shares, it is possible to obtain a protocol with resilience $t < n/2$ and with communication complexity $\mathcal{O}(\mathcal{C}n)\kappa + \mathrm{poly}(n\kappa)$. This protocol is not unconditional secure, but at least has everlasting security in the sense that if the computational assumptions are not broken *during the run of the protocol*, then the protocol is secure against a later computationally unbounded adversary.

In comparison to the earlier work achieving linear complexity, we emphasize that achieving unconditional security is not only of theoretical interest: using "information theoretic" techniques for multiparty computation is *computationally* much more efficient than using homomorphic public-key encryption as in [HN06], for instance. We can therefore transplant our protocols to the computational setting, using cryptography only as a transport mechanism to implement the channels, and obtain a protocol that is computationally more efficient than the one from [HN06], but incomparable to what can be obtained from [DI06]: we get a better security threshold but a non-constant number of rounds.

On the technical side, we make use of a technique presented in [HN06] that was used there to get computational security. It is based on the fact that the standard method for secure multiplication based on homomorphic encryption uses only public, broadcasted messages. The idea is now to select a "king", to whom everyone sends what they would otherwise broadcast. The king does whatever computation is required and returns results to the parties. This can save communication, provided we can verify the king's work in a way that is cheap when amortized over many instances.

This idea is not immediately applicable to unconditionally secure protocols, because the standard method for multiplication in this scenario involves private communication where players send different messages to different players. This can of course not all be sent to the king without violating privacy. We therefore introduce a new multiplication protocol, involving only public communication that is compatible with the "king-paradigm" (the main idea is used in the TRIPLES protocol in Fig. 4). This, together with adaptation of known techniques, is sufficient for passive security.

For active security, we further have to solve the problem that each player frequently has to make a secret shared value public, based on shares he has received in private. Here, we need to handle errors introduced by corrupt players. Standard VSS-based techniques would be quadratic in n, but we show how to use error correction based on Van der Monde matrices to correct the errors, while keeping the overall complexity linear in n (this idea is used in the OPENROBUST protocol in Fig. 9).

2   Preliminaries

Model. We use $\mathcal{P} = \{P_1, \ldots, P_n\}$ to denote a set of n parties which are to do the secure evaluation, and we assume that each pair of parties share a perfectly secure channel. Furthermore, we assume that the parties have access to an authenticated broadcast channel. We allow some subset of corrupted parties $C \subset \mathcal{P}$, of size at most t, to behave in some arbitrary coordinated manner. We call $H = \mathcal{P} \setminus C$ the honest parties. We consider secure circuit evaluation. All input gates are labeled by a party from $\mathcal{P}$. That party provides a secret input for the gate, and the goal of the secure circuit evaluation is to make the output of the circuit public to each party in $\mathcal{P}$. The protocol takes an extra input $\kappa \in \mathbb{N}$, the security parameter, which is given to all parties. The protocol should run in time $\text{poly}(\kappa)$ and the "insecurity" of the protocol should be bounded by $\text{poly}(\kappa)2^{-\kappa}$.

The Ground Field and the Extension Field. For the rest of the paper we fix a finite field $\mathbb{F}$ over which most of our computations will be done. We call $\mathbb{F}$ the ground field. We let $k = \log_2(|\mathbb{F}|)$ denote the bit-length of elements from $\mathbb{F}$. We also fix an extension field $\mathbb{G} \supset \mathbb{F}$ to be the smallest extension for which $|\mathbb{G}| \geq 2^\kappa$. Since $|\mathbb{G}| < 2^{2\kappa}$, a field element from $\mathbb{G}$ can be written down using $\mathcal{O}(\kappa)$ bits. We call $\mathbb{G}$ the extension field.

Van der Monde Matrices. We write a matrix $M \in \mathbb{F}^{(r,c)}$ with r rows and c columns as $M = \{m_{i,j}\}_{i=1,\ldots,r}^{j=1,\ldots,c}$. For $C \subseteq \{1, \ldots, c\}$ we let $M^C = \{m_{i,j}\}_{i=1,\ldots,r}^{j \in C}$ denote the matrix consisting of the columns from M indexed by $j \in C$. We use $M^\top$ to denote the transpose of a matrix, and for $R \subseteq \{1, \ldots, r\}$ we let $M_R = ((M^\top)^R)^\top$. For distinct elements $\alpha_1, \ldots, \alpha_r \in \mathbb{F}$ we use $\text{Van}^{(r,c)}(\alpha_1, \ldots, \alpha_r) \in \mathbb{F}^{(r,c)}$ to denote the Van der Monde matrix $\{\alpha_i^j\}_{i=1,\ldots,r}^{j=0,\ldots,c-1}$, and we use $\text{Van}^{(r,c)} \in \mathbb{F}^{(r,c)}$ to denote *some* Van der Monde matrix of the form $\text{Van}^{(r,c)}(\alpha_1, \ldots, \alpha_r)$ when the elements $\alpha_1, \ldots, \alpha_r$ are inconsequential. It is a well-known fact that all $\text{Van}^{(c,c)}$ are invertible. Consider then $V = \text{Van}^{(r,c)}$ with $r > c$, and let $R \subset \{1, \ldots, r\}$ with $|R| = c$. Since V_R is a Van der Monde matrix $\text{Van}^{(c,c)}$ it follows that V_R is invertible. So, any c rows of a Van der Monde matrix form an invertible matrix. In the following we say that Van der Monde matrices are super-invertible.

Secret Sharing. For the rest of the paper a subset $I \subseteq \mathbb{F}^*$ of $|\mathcal{P}|$ non-zero elements is chosen. Each party in $\mathcal{P}$ is then assigned a unique element i from I, and we index the parties in $\mathcal{P}$ by P_i for $i \in I$. For notational convenience we will assume that $I = \{1, \ldots, |\mathcal{P}|\}$, which is possible when $\mathbb{F}$ has characteristic at least n. All our techniques apply also to the case where $\mathbb{F}$ has smaller characteristic, as long as $|\mathbb{F}| > n$.

By a d-polynomial we mean a polynomial $f(X) \in \mathbb{F}[X]$ of degree *at most d*. To share a value $x \in \mathbb{F}$ with degree d, a uniformly random d-polynomial $f(X) \in \mathbb{F}[X]$ with $f(0) = x$ is chosen, and P_i is given the share $x_i = f(i)$. This is the same as letting $x_0 = x$, choosing $x_1, \ldots, x_d \in \mathbb{F}$ uniformly at random and letting $(y_1, \ldots, y_n) = M^{(d)}(x_0, x_1, \ldots, x_d)$, where $M^{(d)} = \text{Van}^{(n,d+1)}(1, \ldots, n)$.

In the following we call a vector $\boldsymbol{y} = (y_1, \ldots, y_n) \in \mathbb{F}^n$ a d-sharing (of x) if there exists $(x_1, \ldots, x_d) \in \mathbb{F}^d$ such that $M^{(d)}(x, x_1, \ldots, x_d)$ and $\boldsymbol{y}$ agree on the share of all honest parties; I.e., $\boldsymbol{y}_H = M_H^{(d)}(x, x_1, \ldots, x_d)$. In the following we typically use $[x]$ to denote a d-sharing of x with $d = t$, and we always use $\langle x \rangle$ to denote a d-sharing of x with $d = 2t$.

Whenever we talk about a sharing $[x]$ we implicitly assume that party P_i is holding x_i such that $[x] = (x_1, \ldots, x_n)$. We introduce a shorthand for specifying computations on these local shares. If sharings $[x^{(1)}], \ldots, [x^{(\ell)}]$ have been dealt, then each P_i is holding a share $x_i^{(l)}$ of each $[x^{(l)}]$. Consider any function $f : \mathbb{F}^\ell \to \mathbb{F}^m$. By $([y^{(1)}], \ldots, [y^{(m)}]) = f([x^{(1)}], \ldots, [x^{(\ell)}])$, we mean that each P_i computes $(y_i^{(1)}, \ldots, y_i^{(m)}) = f(x_i^{(1)}, \ldots, x_i^{(\ell)})$, defining sharings $[y^{(k)}] = (y_1^{(k)}, \ldots, y_n^{(k)})$ for $k = 1, \ldots, m$. It is well-known that if f is an affine function and each $[x^{(l)}]$ is a consistent d-sharing, then the $[y^{(k)}]$ are consistent d-sharings of $(y^{(1)}, \ldots, y^{(m)}) = f(x^{(1)}, \ldots, x^{(\ell)})$. Furthermore, if $[x_1]$ and $[x_2]$ are consistent d-sharings, then $[y] = [x_1][x_2]$ is a consistent $2d$-sharing of $y = x_1 x_2$. When $d = t$ we therefore use the notation $\langle y \rangle = [x_1][x_2]$.

Error Correction. It is well-known that Van der Monde matrices can be used for error correction. Let $M = \text{Van}^{(r,c)}$ be any Van der Monde matrix with $r > c$. Let $\boldsymbol{x} \in \mathbb{F}^c$ and let $\boldsymbol{y} = M\boldsymbol{x}$. Consider any $R \subset \{1, \ldots, r\}$ with $|R| = c$. By Van der Monde matrices being super-invertible it follows that M_R is invertible. Since $\boldsymbol{y}_R = M_R\boldsymbol{x}$ it follows that $\boldsymbol{x} = M_R^{-1}\boldsymbol{y}_R$, so that $\boldsymbol{x}$ can be computed from any r entries of $\boldsymbol{y}$. It follows that if $\boldsymbol{x}^{(1)} \neq \boldsymbol{x}^{(2)}$ and $\boldsymbol{y}^{(1)} = M\boldsymbol{x}^{(1)}$ and $\boldsymbol{y}^{(2)} = M\boldsymbol{x}^{(2)}$, then $\text{ham}(\boldsymbol{y}^{(1)}, \boldsymbol{y}^{(2)}) \geq r-c+1$, where ham denotes the Hamming distance. Assume now that $r \geq c + 2t$ for some positive integer t, such that $\text{ham}(\boldsymbol{y}^{(1)}, \boldsymbol{y}^{(2)}) \geq 2t + 1$. This allows to correct up to t errors, as described now. Let $\boldsymbol{y} = M\boldsymbol{x}$ and let $\boldsymbol{y}'$ be any vector with $\text{ham}(\boldsymbol{y}, \boldsymbol{y}') \leq t$. It follows that $\text{ham}(\boldsymbol{y}', M\boldsymbol{x}') \geq t + 1$ for all $\boldsymbol{x}' \neq \boldsymbol{x}$. Therefore $\boldsymbol{y}$ can be computed uniquely from $\boldsymbol{y}'$ as the vector $\boldsymbol{y}$ with $\text{ham}(\boldsymbol{y}, \boldsymbol{y}') \leq t$. Then $\boldsymbol{x}$ can be compute from $\boldsymbol{y}$ as $\boldsymbol{x} = M_R^{-1}\boldsymbol{y}_R$ with e.g. $R = \{1, \ldots, c\}$. The Berlekamp-Welch algorithm allows to compute $\boldsymbol{y}$ from $\boldsymbol{y}'$ at a price in the order of performing Gaussian elimination on a matrix from $\mathbb{F}^{(r,r)}$.

Randomness Extraction. We also use Van der Monde matrices for randomness extraction. Let $M = \text{Van}^{(r,c)\top}$ be the transpose of any Van der Monde matrix with $r > c$. We use the computation $(y_1, \ldots, y_c) = M(x_1, \ldots, x_r)$ to extract randomness from $(x_1, \ldots, x_r)$. Assume that $(x_1, \ldots, x_r)$ is generated as follows: First $R \subset \{1, \ldots, r\}$ with $|R| = c$ is picked and a uniformly random $x_i \in_R \mathbb{F}$ is generated for $i \in R$. Then for $j \in T$, with $T = \{1, \ldots, r\} \setminus R$, the values x_j are generated with an arbitrary distribution independent of $\{x_i\}_{i \in R}$. For any such distribution of $(x_1, \ldots, x_r)$ the vector $(y_1, \ldots, y_c)$ is uniformly random in $\mathbb{F}^c$. To see this, note that $(y_1, \ldots, y_c) = M(x_1, \ldots, x_r)$ can be written as $(y_1, \ldots, y_c) = M^R(x_1, \ldots, x_r)_R + M^T(x_1, \ldots, x_r)_T$. Since $M^R = \text{Van}_R^{(r,c)\top}$ we have that M^R is invertible. By definition of the input distribution, the vector $(x_1, \ldots, x_r)_R$ is uniformly random in $\mathbb{F}^c$. Therefore $M^R(x_1, \ldots, x_r)_R$ is uniformly random in $\mathbb{F}^c$. Since $(x_1, \ldots, x_r)_T$ was sampled independent of $(x_1, \ldots, x_r)_R$, it follows that $M^R(x_1, \ldots, x_r)_R + M^T(x_1, \ldots, x_r)_T$ is uniformly random in $\mathbb{F}^c$, as desired.

3 Private, $t < n/2$

We first present a passive secure circuit-evaluation protocol. Later we show how to add robustness. Throughout this section we assume that there are at most $t = \lfloor(n-1)/2\rfloor$ corrupted parties. To prepare for adding robustness, some parts of the passive protocol are slightly more involved than necessary. The extra complications come from the fact that we want to place all dealings of sharings in a preprocessing phase, where the inputs have not been used yet. This will later allow a particularly efficient way of detecting cheating parties, and will furthermore give a circuit-evaluation phase which consists of only opening sharings, limiting the types of errors that can be encountered after the parties entered their inputs into the computation.

3.1 Random Double Sharings

We first present a protocol, DOUBLE-RANDOM(ℓ), which allows the parties to generate sharings $[r_1], \ldots, [r_\ell]$ and $\langle R_1 \rangle, \ldots, \langle R_\ell \rangle$, where each $[r_l]$ is a uniformly random t-sharing of a uniformly random value $r_l \in \mathbb{F}$ and each $\langle R_l \rangle$ is a uniformly random $2t$-sharing of $R_l = r_l$. We consider the case $\ell = n - t$. For larger ℓ, the protocol is simply run in parallel a number of times. As part of the protocol the parties use a fixed matrix $M = \mathrm{Van}^{(n,n-t)^\top}$ for randomness extraction.

---

1. Each $P_i \in \mathcal{P}$: Pick a uniformly random value $s^{(i)} \in_R \mathbb{F}$ and deal a t-sharing $[s^{(i)}]$ and a $2t$-sharing $\langle s^{(i)} \rangle$.
2. Compute
$$([r_1], \ldots, [r_{n-t}]) = M([s^{(1)}], \ldots, [s^{(n)}])$$
$$(\langle R_1 \rangle, \ldots, \langle R_{n-t} \rangle) = M(\langle s^{(1)} \rangle, \ldots, \langle s^{(n)} \rangle) \, ,$$
and output $(([r_1], \langle R_1 \rangle), \ldots, ([r_{n-t}], \langle R_{n-t} \rangle))$.

Fig. 2. Double-Random$(n - t)$

The protocol is given in Fig. 2. Assume that t parties are corrupted, leaving exactly $m = n - t$ honest parties. The m sharings $[s^{(i)}]$ dealt by the honest parties are independent uniformly random sharings of independent, uniformly random values unknown by the corrupted parties. The matrix M being a super-invertible matrix with m rows then implies that the sharings $([r_1], \ldots, [r_m])$ are independent uniformly random t-sharings of uniformly random values unknown by the corrupted parties. In the same way $(\langle R_1 \rangle, \ldots, \langle R_m \rangle)$ are seen to be independent, uniformly random $2t$-sharings of uniformly random elements unknown by the corrupted parties, and $R_l = r_l$. We also use a protocol RANDOM(ℓ) which runs as DOUBLE-RANDOM(ℓ) except that the $2t$-sharings $\langle R \rangle$ are not generated.

Each of the $2n$ dealings communicate $\mathcal{O}(n)$ field elements from $\mathbb{F}$, giving a total communication complexity of $\mathcal{O}(n^2 k)$. Since $n - t = \Theta(n)$ pairs are generated, the communication complexity per generated pair is $\mathcal{O}(nk)$. A general number ℓ of pairs can thus be generated with communication complexity $\mathcal{O}(n\ell k + n^2 k)$.

3.2 Opening Sharings

The next protocol, OPEN$(d, [x])$, is used for reconstructing a d-sharing efficiently. For this purpose a designated party $P_{\text{king}} \in \mathcal{P}$ will do the reconstruction and send the result to the rest of the parties.

---

1. Each $P_i \in \mathcal{P}$: Let $P_{\text{king}} \in \mathcal{P}$ be some agreed-upon party and send the share x_i of $[x]$ to P_{king}.
2. P_{king}: Compute a d-polynomial $f(X) \in \mathbb{F}[X]$ with $f(i) = x_i$ for all $P_i \in \mathcal{P}$, and send $x = f(0)$ to all parties.
3. Each $P_i \in \mathcal{P}$: Output x.

Fig. 3. Open$(d, [x])$

It is clear that if $[x]$ is a d-sharing of x and there are no active corruptions, then all honest parties output x. The communication complexity is $2(n-1) = \mathcal{O}(n)$ field elements from $\mathbb{F}$.

3.3 Multiplication Triples

We then present a protocol, TRIPLES(ℓ) which allows the parties to generate ℓ multiplication triples, which are just triples $([a], [b], [c])$ of uniformly random t-sharings with $c = ab$.

---

1. All parties: Run RANDOM(2ℓ) and DOUBLE-RANDOM(ℓ) and group the outputs in ℓ triples $([a], [b], ([r], \langle R \rangle))$. For each triple in parallel, proceed a follows:
   (a) All parties: Compute $\langle D \rangle = [a][b] + \langle R \rangle$.
   (b) All parties: Run $D \leftarrow$ OPEN$(2t, \langle D \rangle)$.
   (c) All parties: Compute $[c] = D - [r]$, and output $([a], [b], [c])$.

Fig. 4. Triples(ℓ)

The sharings $[a]$ and $[b]$ are t-sharings, so $[a][b]$ is a $2t$-sharing of ab. Therefore $\langle D \rangle = [a][b] + \langle R \rangle$ is a uniformly random $2t$-sharing of $D = ab + R$. The revealing of D leaks no information on a or b, as R is uniformly random. Therefore the protocol is private. Then $[c] = D - [r]$ is computed. Since $[r]$ is a t-sharing, $[c]$ will be a t-sharing, of $D - r = ab + R - r = ab$. The communication complexity of generating ℓ triples is seen to be $\mathcal{O}(n\ell k + n^2 k)$.

3.4 Circuit Evaluation

We are then ready to present the circuit-evaluation protocol. The circuit Circ $= \{G_{gid}\}$ consists of gates G_{gid} of the following forms.

input: $G_{gid} = (gid, \mathtt{inp}, P_j)$, where $P_j \in \mathcal{P}$ provides a secret input $x_{gid} \in \mathbb{F}$.

random input: $G_{gid} = (gid, \mathtt{ran})$, where $x_{gid} \in_R \mathbb{F}$ is chosen as a secret, uniformly random element.

affine: $G_{gid} = (gid, \mathtt{aff}, a_0, gid_1, a_1, \ldots, gid_\ell, a_\ell)$, where $a_0, a_1, \ldots, a_\ell \in \mathbb{F}$ and $x_{gid} = a_0 + \sum_{l=1}^{\ell} a_l x_{gid_l}$.

multiplication: $G_{gid} = (gid, \mathtt{mul}, gid_1, gid_2)$, where $x_{gid} = x_{gid_1} x_{gid_2}$.

output: $G_{gid} = (\mathtt{out}, gid_1)$, where all parties are to learn x_{gid_1}.[1]

What it means to securely evaluate Circ can easily be phrased in the UC framework[Can01], and our implementation is UC secure. We will however not prove this with full simulation proofs in the following, as the security of our protocols follow using standard proof techniques.

Preprocessing Phase. First comes a preprocessing phase, where a number of sharings are generated for some of the gates in Circ. The details are given in Fig. 5. The communication complexity is $\mathcal{O}(n\ell k + n^2 k)$, where ℓ is the number of random gates plus the number of input gates plus the number of output gates.

All gates are handled in parallel by all parties running the following:

random: Let r be the number of random gates in Circ, run RANDOM(r) and associate one t-sharing $[x_{gid}]$ to each $(gid, \mathtt{ran}) \in$ Circ.

input: Let i be the number of input gates in Circ, run RANDOM(i) and associate one t-sharing $[r_{gid}]$ to each $(gid, \mathtt{inp}, P_j) \in$ Circ. Then send all shares of $[r_{gid}]$ to P_j to let P_j compute r_{gid}.

multiplication: Let m be the number of multiplication gates in Circ, run TRIPLES(m) and associate one multiplication triple $([a_{gid}], [b_{gid}], [c_{gid}])$ to each $(gid, \mathtt{mul}, gid_1, gid_2) \in$ Circ.

Fig. 5. Preprocess(Circ)

Evaluation Phase. Then comes an evaluation phase. During the evaluation phase a t-sharing $[x_{gid}]$ is computed for each gate gid, and we say that gid has been computed when this happens. Note that the random gates are computed already in the preprocessing. A non-output gate a said to be **ready** when all its input gates have been computed. An output gate is said to be **ready** when in addition all input gates and random gates in the circuit have been computed.[2] The evaluation proceeds in rounds, where in each round all ready gates are computed in parallel. When several sharings are opened in a round, they are opened in parallel, using one execution of OPEN. The individual gates are handled as detailed in Fig. 6. Note that the evaluation phase consists essentially only of opening sharings and taking affine combinations.

---

[1] Private outputs can be implemented using a standard masking technique.

[2] This definition will ensure that all inputs have been provided before any outputs are revealed.

The evaluation proceeds in rounds, where in each round all ready gates are computed in parallel, as follows:

input: For $(gid, \mathtt{inp}, P_j) \in \text{Circ}$:
   1. P_j: Retrieve the input $x_{gid} \in \mathbb{F}$ and send $\delta_{gid} = x_{gid} + r_{gid}$ to all parties.
   2. All parties: Compute $[x_{gid}] = \delta_{gid} - [r_{gid}]$.
affine: For $(gid, \mathtt{aff}, a_0, gid_1, a_1, \ldots, gid_\ell, a_\ell) \in \text{Circ}$: All parties compute $[x_{gid}] = a_0 + \sum_{l=1}^{\ell} a_l [x_{gid_l}]$.
multiplication: For $(gid, \mathtt{mul}, gid_1, gid_2) \in \text{Circ}$ all parties proceed as follows:
   1. Compute $[\alpha_{gid}] = [x_{gid_1}] + [a_{gid}]$ and $[\beta_{gid}] = [x_{gid_2}] + [b_{gid}]$.
   2. Run $\alpha_{gid} \leftarrow \text{OPEN}([\alpha_{gid}])$ and $\beta_{gid} \leftarrow \text{OPEN}([\beta_{gid}])$.
   3. Let $[x_{gid}] = \alpha_{gid}\beta_{gid} - \alpha_{gid}[b_{gid}] - \beta_{gid}[a_{gid}] + [c_{gid}]$.
output: For $(\mathtt{out}, gid) \in \text{Circ}$: Run $x_{gid} \leftarrow \text{OPEN}([x_{gid}])$.

Fig. 6. Eval(Circ)

The correctness of the protocol is straight-forward except for Step 3 in **multiplication**. The correctness of that step follows from [BB89] which introduced this preprocessed multiplication protocol. The privacy of the protocol follows from the fact that r_{gid} in the input protocol and a_{gid} and b_{gid} in the multiplication protocol are uniformly random elements from $\mathbb{F}$, in the view of the corrupted parties. Therefore $\delta_{gid} = x_{gid} + r_{gid}$ and $\alpha_{gid} = x_{gid_1} + a_{gid}$ and $\beta_{gid} = x_{gid_2} + b_{gid}$ leak no information on x_{gid} respectively x_{gid_1} and x_{gid_2}. Therefore the values of all gates are hidden, except for the output gates, whose values are allowed (required) to leak.

The communication complexity, including preprocessing, is seen to be $\mathcal{O}(n\mathcal{C}k + n^2 k)$, where $\mathcal{C} = |\text{Circ}|$ is the number of gates in the circuit.

4   Robust, $t < n/4$

By now we have a circuit-evaluation protocol which is private and correct as long a no party deviates from the protocol. In this section we add mechanisms to ensure robustness. Throughout this section we assume that there are at most $t = \lfloor (n-1)/4 \rfloor$ corrupted parties. In the following section we then extend the solution to handle $t < n/3$.

4.1   Error Points

In the passive secure protocol there are several points where a party could deviate from the protocol to make it give a wrong output. We comment on two of these here and sketch how they are dealt with. More details follow later. A party which was asked to perform a d-sharing could distribute values which are not d-consistent. We are going to detect a cheater by asking the parties to open a random polynomial combination of all sharings they have dealt. Also, P_{king}

could fail to send the right value in OPEN (Fig. 3). We are going to use error correction to make sure this does not matter, by opening a Van der Monde code of the sharings to be opened and then correcting the t mistakes that the corrupted parties might have introduced.

4.2   Coin-Flip

In the protocol opening a polynomial combination of sharings we need a random value $x \in \mathbb{G}$ from the extension field. Therefore we need a protocol, FLIP(), for flipping a random value from $\mathbb{G}$. The standard protocol does this using a VSS protocol as subprotocol: Each $P_i \in \mathcal{P}$: Pick uniformly random $x_i \in_R \mathbb{G}$ and deal a VSS of x_i among the parties in $\mathcal{P}$. All parties: Reconstruct each x_i and let $x = \sum_{P_i \in \mathcal{P}} x_i$. Any of the known VSS's will do, e.g., [BGW88], since we only call FLIP a small number of times, and so its precise complexity is not important.

4.3   Dispute Control

In the following we use the technique of dispute control[BH06]. We keep a dispute set Disputes, initially empty, consisting of sets $\{P_i, P_j\}$ with $P_i, P_j \in \mathcal{P}$. If $\{P_i, P_j\} \in$ Disputes, then we write $P_i \leftrightarrow P_j$. If during a protocol a dispute arises between P_i and P_j, then $\{P_i, P_j\}$ is added to Disputes. This is done in such a way that: (1) All parties in $\mathcal{P}$ agree on the value of Disputes. (2) If $P_i \leftrightarrow P_j$, then P_i is corrupted or P_j is corrupted. For a given dispute set Disputes and $P_i \in \mathcal{P}$ we let Disputes$_i$ be the set of $P_j \in \mathcal{P}$ for which $P_i \leftrightarrow P_j$, and we let Agree$_i = \mathcal{P} \setminus$ Disputes$_i$.

All sub-protocols will use the same dispute set Disputes. We say that a sub-protocol has dispute control if (1) It can never halt due to a dispute between P_i and P_j if $\{P_i, P_j\}$ is already in Disputes. (2) If it does not generate a dispute, then it terminates with the correct result. (3) If it generates a dispute, then it is secure to rerun the sub-protocol (with the now larger dispute set).

We also keep a set Corrupt $\subset \mathcal{P}$. If during the run of some protocol a party P_i is detected to deviate from the protocol, then P_i is added to Corrupt. This is done in such a way that: (1) All parties in $\mathcal{P}$ agree on the value of Corrupt. (2) If $P_i \in$ Corrupt, then P_i is corrupted.

We enforce that when it happens for the first time that $|\,\text{Disputes}_i\,| > t$, where t is the bound on the number of corrupted parties, then P_i is added to Corrupt. It is easy to see that if $|\,\text{Disputes}_i\,| > t$, then indeed P_i is corrupted.

Secret Sharing with Dispute Control. We use a technique from [BH06] to perform secret sharing with dispute control. When a party P_i is to deal a d-sharing $[x]$ then P_i uses a random d-polynomial where $f(0) = x$ and $f(j) = 0$ for $P_j \in$ Disputes$_i$. Since all sharings will have $d \geq t$ and $P_i \in$ Corrupt if $|\,\text{Disputes}_i\,| > t$, this type of dealing is possible for all $P_i \notin$ Corrupt. The advantage is that all parties will agree on what P_i sent to all $P_j \in$ Disputes$_i$. This can then be exploited to ensure that P_i will never get a new dispute with some $P_j \in$ Disputes$_i$.

4.4 Dealing Consistent Sharings

The robust protocol for sharing values will run the private protocol for dealing sharings followed by a check that the generated sharings are consistent. In Fig. 7 we consider the case where P_i shares ℓ values $(y_1, \ldots, y_\ell)$.

---

1. If $P_i \in$ Corrupt, then output $([y_1], \ldots, [y_\ell]) = ([0], \ldots, [0])$, where $[0] = (0, \ldots, 0)$ is the dummy sharing of 0. Otherwise, proceed as below.
2. P_i: Deal d-sharings $[y_1], \ldots, [y_\ell]$ over $\mathbb{F}$ among the parties in $\mathcal{P}$ along with a d-sharing $[r]$ over $\mathbb{G}$, where $r \in_R \mathbb{G}$ is a uniformly random element from the extension field. By definition all parties in Disputes$_i$ get 0-shares.
3. All parties in $\mathcal{P}$: Run $x \leftarrow$ FLIP(), and compute $[y] = [r] + \sum_{l=1}^{\ell} x^l [y_l]$ in $\mathbb{G}$.
4. All parties in Agree$_i$: Broadcast the share of $[y]$. All parties in Disputes$_i$ are defined to broadcast 0.
5. P_i: In parallel with the above step, broadcast all shares of $[y]$.
6. All parties in $\mathcal{P}$: If the sharing $[y]$ broadcast by P_i is not a d-sharing with all parties in Disputes$_i$ having a 0-share, then output $([y_1], \ldots, [y_\ell]) = ([0], \ldots, [0])$. Otherwise, if the shares broadcast by the other parties are identical to those broadcast by P_i, then output $([y_1], \ldots, [y_\ell])$. Otherwise, let Disputes$' =$ Disputes$\cup\{(P_i, P_j)\}$ for each $P_j \in$ Agree$_i$ broadcasting a share different from that broadcast by P_i.

Fig. 7. Share(P_i, Disputes)

We first argue that if any of the sharings dealt by P_i are not d-sharings, then a dispute will be generated, except with probability poly$(\kappa)2^{-\kappa}$. Namely, let $f_0(X) \in \mathbb{G}[X]$ be the lowest degree polynomial consistent with the honest shares of $[r]$, and for $i = 1, \ldots, \ell$ let $f_l(X) \in \mathbb{G}[X]$ be the lowest degree polynomial consistent with the honest shares of $[y_l]$. It can be seen that $f_l(X)$ is also the lowest degree polynomial $f_l^{(\mathbb{F})}(X) \in \mathbb{F}[X]$ consistent with the honest shares of $[y_l]$.[3] It follows that if the sharings dealt by P_i are not all d-consistent, then one of the polynomials $f_l(X)$ has degree larger than d. Let m be such that $f_m(X)$ has maximal degree among $f_0(X), \ldots, f_\ell(X)$, let d_m be the degree of $f_m(X)$ and write each $f_l(X)$ as $f_l(X) = \alpha_l x^{d_m} + f_l'(X)$, where $f_l'(X)$ has degree lower than d_m. By definition $\alpha_m \neq 0$. Therefore $g(Y) = \sum_{i=0}^{\ell} \alpha_l Y^l$ is a non-zero polynomial over $\mathbb{G}$ with degree at most ℓ, and since x is uniformly random in $\mathbb{G}$, it follows that $g(x) = 0$ with probability at most $\ell/|\mathbb{G}| = $ poly$(\kappa)2^{-\kappa}$. So, we can assume that $g(x) \neq 0$. This implies that $f(X) = \sum_{l=0}^{\ell} x^l f_l(X)$ has degree $d_m > d$. Note that $f(X)$ is consistent with the honest shares of $[y] = [r] + \sum_{l=1}^{\ell} x^l [y_l]$, and let $g(X) \in \mathbb{G}[X]$ be the lowest degree polynomial which is consistent with the honest shares of $[y]$. Let $h(X) = f(X) - g(X)$. Clearly $h(i) = 0$ for all honest

---

[3] The polynomial $f_l(X)$ can be computed from the indexes $i \in \mathbb{F}$ of the honest parties P_i and the shares $x_l^{(i)} \in \mathbb{F}$ of the honest parties P_i using Lagrange interpolation, which is linear. Therefore the coefficients of $f_l(X)$ ends up in $\mathbb{F}$, even when the interpolation is done over the extension $\mathbb{G}$.

parties P_i. Since $h(i)$ has degree at most $d_m < |H|$, where H is the set of honest parties, and $h(i) = 0$ for $i \in H$ it follows that $h(X)$ is the zero-polynomial. So, $g(X) = f(X)$ and $[y]$ thus has degree d_m. Therefore the honest shares of $[y]$ are not on a d-polynomial, and thus some dispute will be generated. It follows that when no dispute is generated, then all the sharings dealt by P_i are d-consistent, except with probability $\text{poly}(\kappa)2^{-\kappa}$. This in particular applies to the sharings $[y_1], \ldots, [y_\ell]$.

As for the privacy, note that when P_i is honest, $\sum_{l=1}^{\ell} x^l[y_l]$ is a d-sharing over $\mathbb{G}$ and $[r]$ is an independent uniformly random d-sharings over $\mathbb{G}$ of a uniformly random $r \in_R \mathbb{G}$. Therefore $[y]$ is a uniformly random d-sharing over $\mathbb{G}$ and leaks no information to the corrupted parties when reconstructed.

If the protocol SHARE(P_i, Disputes) fails, it is rerun using the new larger Disputes$'$. Since Disputes$_i$ grows by at least 1 in each failed attempt, at most t failed attempts will occur. So, if $\lceil \ell/t \rceil$ values are shared in each attempt, the total number of attempts needed to share ℓ values will be $2t$. Since each attempt has communication complexity $\mathcal{O}(\lceil \ell/t \rceil nk) + \text{poly}(n\kappa)$ the total complexity is $\mathcal{O}(\ell nk) + \text{poly}(n\kappa)$, where $\text{poly}(n\kappa)$ covers the cost of the n broadcasts and the run of FLIP() in each attempt. The round complexity is $\mathcal{O}(t)$.

Dealing Inter-Consistent Sharings. The above procedure allows a party P_i to deal a number of consistent d-sharings. This can easily be extend to allow a party P_i to deal consistent t-sharings $[y_1], \ldots, [y_\ell]$ and $2t$-sharings $\langle Y_1 \rangle, \ldots, \langle Y_\ell \rangle$ with $Y_l = y_l$. The check uses a random t-sharing $[r]$ and a random $2t$-sharing $\langle R \rangle$ with $R = r$, and then $[y] = [r] + \sum_{l=1}^{\ell} x^l[y_l]$ and $\langle Y \rangle = \langle R \rangle + \sum_{l=1}^{\ell} x^l[Y_l]$ are opened as above. In addition to the sharings being t-consistent (respectively $2t$-consistent), it is checked that $Y = y$. Note that if $R = r$ and $X_l = x_l$, then indeed $Y = y$. On the other hand, if $R \neq r$ or some $X_l \neq x_l$, then $Y - y = (R-r) + \sum_{l=1}^{\ell} x^l(Y_l - y_l)$ is different from 0 except with probability $\ell/|\mathbb{G}|$, giving a soundness error of $\text{poly}(\kappa)2^{-\kappa}$.

4.5   Random Double Sharings

Recall that the purpose of this protocol is to generate a set of random values that are unknown to all parties and are both t- and $2t$-shared. The robust protocol for this is derived directly from the passive secure protocol, and we also denote it by DOUBLE-RANDOM(ℓ). The only difference between the two is that the above robust procedure for dealing inter-consistent sharings is used as subprotocol. To generate ℓ double sharings, first each party deals $\lceil \ell/(n-t) \rceil$ random pairs using the procedure for dealing inter-consistent sharings. Then the first pair from each party is used to compute $n-t$ pairs as in the passive secure DOUBLE-RANDOM(ℓ), using a matrix $\text{Van}^{(n,n-t)^\top}$. At the same time the second pairs from each party is used to compute $n-t$ more pairs, and so on. This yields a total of $(n-t)\lceil \ell/(n-t) \rceil \geq \ell$ pairs. The communication complexity is $\mathcal{O}(\ell nk) + \text{poly}(n\kappa)$.

4.6   Opening Sharings

We describe how sharings are opened. We assume that the sharings to be opened are consistent d-sharings of the same degree $d \leq 2t$. Reconstruction of a single sharing happens by sending all shares to some king, which then reconstructs. Since the sharing is d-consistent, the king receives at least $n - t > 3t = d + t$ correct sharings and at most t incorrect sharings. Therefore the king can always compute the d-polynomial $f(X)$ of the sharing using Berlekamp-Welch. The details are given in Fig. 8.

---

1. The parties agree on a consistent d-sharings $[x]$ with $d \leq 2t$.
2. Each $P_i \in \mathcal{P}$: Send the share x_i of $[x]$ to P_{king}.
3. P_{king}: Run Berlekamp-Welch on the received shares to get x, and send x to all parties.

Fig. 8. Open$(P_{\text{king}}, d, [x])$

The protocol in Fig. 8 has the obvious flaw that P_{king} could send an incorrect value. This is handled by expanding $n - (2t + 1)$ sharings to n sharings using a linear error correcting code tolerating t errors. Then each P_i opens one sharing, and the possible t mistakes are removed by error correction. The details are given in Fig. 9.

---

1. The parties agree on consistent d-sharings $[x_1], \ldots, [x_\ell]$ with $\ell = n - (2t + 1)$ and $d \leq 2t$.
2. All parties: Compute $([y^{(1)}], \ldots, [y^{(n)}]) = M([x_1], \ldots, [x_\ell])$, where $M = \text{Van}^{(n,\ell)}$.
3. All parties: For each $P_i \in \mathcal{P}$ in parallel, run $y^{(i)} \leftarrow \text{OPEN}(P_i, d, [y^{(i)}])$.
4. All parties: Run Berlekamp-Welch on the values $(y^{(1)}, \ldots, y^{(n)})$ to get $(x_1, \ldots, x_\ell)$.

Fig. 9. OpenRobust$(d, [x_1], \ldots, [x_\ell])$

The communication complexity of opening $\ell = n - (2t + 1)$ values is $\mathcal{O}(n^2 k)$, giving an amortized communication complexity of $\mathcal{O}(nk)$ per reconstruction. An arbitrary ℓ sharings can thus be reconstructed with communication complexity $\mathcal{O}(\ell n k + n^2 k)$.

4.7   Circuit Evaluation

We now have robust protocols DOUBLE-RANDOM (and thus RANDOM) and OPENROBUST. This allows to implement a robust version of TRIPLES exactly as in Fig. 4. Then a robust preprocessing can be run exactly as in Fig. 5. This in turn almost allows to run a robust circuit-evaluation as in Fig. 6. Note in particular that since all sharings computed during the circuit evaluation are linear combinations of sharings constructed in the preprocessing, all sharings will be consistent t-sharings. Therefore Berlekamp-Welch can continuously be used to compute the openings of such sharings. Indeed, the only additional complication in running EVAL(Circ) is in Step 2 in **input**, where it must be ensured that P_j sends the same δ_{gid} to all parties. This is handled by distributing all δ_{gid} using n parallel broadcasts (each P_j broadcast all its δ_{gid} in one message).

Since an ℓ-bit message can be broadcast with communication complexity $\mathcal{O}(\ell) + \text{poly}(n\kappa)$ (see [FH06]), the communication complexity of handling the input gates will be $\mathcal{O}(\ell_i nk) + \text{poly}(n\kappa)$, where ℓ_i is the number of input gates. The communication complexity of handling the remaining gates is seen to be $\mathcal{O}(n\mathcal{C}k + (D+1)n^2k) + \text{poly}(n\kappa)$, where $\mathcal{C} = |\text{Circ}|$ and D is the multiplicative depth of the circuit. The term $(D+1)n^2k$ comes from the fact that OPENROBUST is run for each layer of multiplication gates and run once to handle the output gates in parallel. If we sum the above with the communication complexity of the preprocessing phase we get a communication complexity of $\mathcal{O}(n\mathcal{C}k + (D+1)n^2k) + \text{poly}(n\kappa)$. The round complexity is $\mathcal{O}(t + D + 1)$, where t comes from running the robust sharing protocol.

5 Robust, $t < n/3$

We sketch how the case $t < n/3$ is handled. Assume first that a preprocessing phase has been run where the "usual" consistent t-sharings have been associated to each gate. Since $t < n/3$ it follows that there are at least $2t + 1$ honest shares in each sharing. Therefore Berlekamp-Welch can be used to reconstruct these t-sharings. Since the circuit-evaluation phase consists only of opening t-sharings, it can thus be run exactly as in the case $t < n/4$.

The main problem is therefore to establish the preprocessed t-sharings. The protocol SHARE can be run as for $t < n/4$ as can then DOUBLE-RANDOM. Going over TRIPLES it thus follows that the only problematic step is $D \leftarrow$ OPEN$(2t, \langle D \rangle)$, where a sharing of degree $2t$ is opened. Since there are only $2t + 1$ honest parties, Berlekamp-Welch cannot be used to compute D in Step 3 in Fig. 8. An honest party P_{king} can therefore find itself not being able to contribute correctly to the reconstruction. This is handled by ensuring that when this happens, then P_{king} can complain and the parties together identify a new dispute.

In more detail, when P_{king} is to reconstruct some $2t$-sharing $\langle y \rangle$ as part of the run of OPEN in TRIPLES (Fig. 4), then P_{king} collects shares y_j from each $P_j \in \text{Agree}_{\text{king}}$.[4] If these shares are on some $2t$-polynomial $D(X) \in \mathbb{F}[X]$, then P_{king} sends $D(0)$ to all parties. Otherwise some $P_j \in \text{Agree}_{\text{king}}$ sent an incorrect y_j. This is used to find a new dispute, as detailed in Fig. 10.

---

[4] P_{king} can safely ignore D_j from $P_j \in \text{Disputes}_{\text{king}}$ as P_{king} knows that these P_j are corrupted.

0. Assume that P_{king} was reconstructing some $2t$-sharing $\langle y \rangle$ as part of the opening in TRIPLES (Fig. 4), and assume that the shares y_j for $P_j \in \text{Agree}_{\text{king}}$ are $2t$-inconsistent.

1. Each $P_j \in$ Honest: Broadcast all the shares sent and received during RANDOM and DOUBLE-RANDOM.[a] At the same time P_{king} broadcasts the $2t$-inconsistent shares $\langle y \rangle$.

2. All parties: If some $P_j \in$ Honest claims to have sent sharings which do not have the correct degree, then add P_j to Corrupt and terminate. Otherwise, if some P_i and P_j disagree on a share $R_j^{(i)}$ sent from P_i to P_j, then add the dispute $P_i \leftrightarrow P_j$.[b] Otherwise, proceed as below.

3. All parties: Compute from the broadcast sharings the $2t$-sharing $\langle y' \rangle$ that P_{king} was reconstructing. Since $\langle y' \rangle$ is $2t$-consistent and $\langle y \rangle$ is $2t$-inconsistent on $\text{Agree}_{\text{king}}$, there exists $P_i \in \text{Agree}_{\text{king}}$ where $y_i' \neq y_i$. For each such P_i, add the dispute $P_{\text{king}} \leftrightarrow P_i$.[c]

---

[a] This is secure as the secret inputs did not enter any computation yet in the preprocessing phase.

[b] Note that the dispute will be new, as $P_i \leftrightarrow P_j$ implies that $R_j^{(i)}$ is defined to be 0, and thus no dispute can arise.

[c] The dispute is new as $P_i \in \text{Agree}_{\text{king}}$. Note that we cannot add P_i to Corrupt, as P_{king} could be lying about y_i.

Fig. 10. Detect-Dispute

The protocol in Fig. 10 can be greatly optimized, using a more involved protocol avoiding many of the broadcasts. However, already the simple solution has communication complexity $\text{poly}(n\kappa)$. Since the protocol always leads to a new dispute and at most $\mathcal{O}(t^2)$ disputes are generated in total, it follows that the protocol contributes with a term $\text{poly}(n\kappa)$ to the overall communication complexity. This gives us a robust protocol with communication complexity $\mathcal{O}(n\mathcal{C}k + (D+1)n^2k) + \text{poly}(n\kappa)$ for the case $t < n/3$.

6   Robust, $t < n/2$

It is yet an open problem to get an unconditionally secure protocol with linear communication complexity for the case $t < n/2$. One can however construct a protocol withstanding a computationally bounded adversary *during* the protocol and a computationally unbounded adversary *after* the protocol execution. Due to space limitations we can only sketch the involved ideas.

Transferable Secret Sharing. To allow some P_{king} to reconstruct a sharing and, verifiably, transfer the result to the other parties, each t-sharing $[x]$ is augmented by a Pedersen commitment $C = \text{commit}(x; r)$, known by all parties, and the randomness r is shared as $[r]$. We write $[\![x]\!] = (C, [x], [r])$. In the preprocessing, each P_j generates many random t-sharings $[\![x]\!]$ and corresponding, normal

$2t$-sharings $\langle x \rangle$, where $P_i \in \text{Disputes}_j$ get $x_i = r_i = 0$. By committing using a group of prime order q and secret sharing in $\text{GF}(q)$, the sharings $[\![x]\!] = (C, [x], [r])$ are linear modulo q, and the usual protocols can be used for checking consistency of sharings and for combining random sharings using a Van der Monde matrix (Sec. 4.4 and 4.5). When a consistency check fails, a new dispute is identified using a technique reminiscent of that used when $t < n/3$, but slightly more involved. Then a new attempt is made at generating random shared values. The total overhead of identifying disputes is kept down by generating the random shared values in phases, each containing a limited number of sharings.

Multiplication Triples. After enough pairs $([\![x_l]\!], \langle x_l \rangle)$ have been generated, a multiplication protocol to be described below is run to generate multiplication triples. Generation of triples are done in n sequential phases, where for efficiency checks for correct behavior are done simultaneously for all triples in a phase.

The input to the generation of a multiplication triple is sharings $[\![a]\!]$, $[\![b]\!]$, $[\![r]\!]$, $\langle r \rangle$, $[\![\tilde{b}]\!]$, $[\![\tilde{r}]\!]$, $\langle \tilde{r} \rangle$. The parties compute their shares in $[a][b] + \langle r \rangle$, $[a][\tilde{b}] + \langle \tilde{r} \rangle$, send these to a selected party P_{king}, who reconstructs values D, respectively $\tilde{D}$, and sends these values to all players. Players now compute $[\![c]\!] = D - [\![r]\!]$ and $[\![\tilde{c}]\!] = \tilde{D} - [\![\tilde{r}]\!]$. For simplicity we assume that $2t + 1 = n$, in which case the shares received by P_{king} will always be $2t$-consistent. Hence even an honest P_{king} might be tricked into reconstructing wrong D and $\tilde{D}$, this problems is handled below. Also, a dishonest P_{king} may distribute inconsistent values for $D, \tilde{D}$. We therefore run the share consistency check from Sec. 4.4 over all $[\![c]\!], [\![\tilde{c}]\!]$ in the current phase and disqualify P_{king} if it fails. Now the (supposed) multiplication triples $([\![a]\!], [\![b]\!], [\![c]\!])$ and $([\![a]\!], [\![\tilde{b}]\!], [\![\tilde{c}]\!])$ are checked. First a uniformly random value $X \in_R \text{GF}(q)$ is flipped. Then it is checked that $([\![a]\!], [\![\tilde{b}]\!] + X[\![b]\!], [\![\tilde{c}]\!] + X[\![c]\!])$ is a multiplication triple: Compute $b = \text{OPEN}([\![\tilde{b}]\!] + X[\![b]\!])$, compute $d = \text{OPEN}([\![a]\!]b - ([\![\tilde{c}]\!] + X[\![c]\!]))$, and check that $d = 0$. If $d = 0$, then $([\![a]\!], [\![b]\!], [\![c]\!])$ is taken to be the generated triple. Here OPEN refers to the reconstruction procedure described below, which either disqualifies P_{king} or lets at least one honest player compute d plus a proof that it is correct. He can therefore alert all parties if $d \neq 0$.

For efficiency the same X is used for all checks done in the same phase. If $d \neq 0$, then all messages sent and received during the multiplication and the generating of the sharings involved in the multiplication are broadcast and some new dispute is found. Since $\mathcal{O}(1)$ sharings are involved in multiplication, each being a linear combination of at most $\mathcal{O}(n)$ sharings, at most $\mathcal{O}(n)$ sharings are broadcast to find a dispute. Since at most t^2 disputes are generated, the total overhead is thus independent of the circuit size.

Reconstructing. As usual, the evaluation phase proceeds by computing affine combinations of the t-sharings $[\![x]\!]$ generated in the preprocessing and by opening such sharings. All that we need is thus a protocol for opening such t-sharings. From a beginning some fixed reconstructor $P_{\text{king}} \in \text{Honest}$ is chosen. In a reconstruction of $[\![x]\!] = (C, [x], [r])$, each P_i sends (x_i, r_i) to P_{king}, who computes (x, r)

and sends (x, r) to all parties. The receivers can check that $C = \text{commit}(x; r)$. If the shares received by P_{king} are not t-consistent, then P_{king} complains and a special protocol FIND-CORRUPT-SHARE is run to remove some corrupted party from Honest. The details of how FIND-CORRUPT-SHARE identifies a new corrupted party from the incorrect shares is rather involved, and the details will be given in the full version due to space limitations. Most important is that the communication complexity of FIND-CORRUPT-SHARE is bounded by $\mathcal{O}(|\text{Circ}|k) + \text{poly}(n\kappa)$. Since each run of FIND-CORRUPT-SHARE removes one new party from Honest, it is run at most $t = \mathcal{O}(n)$ times, giving a total overhead of at most $\mathcal{O}(|\text{Circ}|nk) + \text{poly}(n\kappa)$. The procedure FIND-CORRUPT-SHARE will be run until P_{king} sees that the shares (x_i, r_i) from parties in Honest are t-consistent. At this point P_{king} can then interpolate (x, r) and send it to all parties. The above procedure always allows an honest P_{king} to compute (x, r) and send it to all parties, maybe after some runs of FIND-CORRUPT-SHARE. A party P_i not receiving (x, r) such that $C = \text{commit}(x; r)$ therefore knows that P_{king} is corrupt. If P_i does not receive an opening it therefore signs the message "P_i disputes P_{king}" and sends it to all parties. All parties receiving "P_i disputes P_{king}" signed by P_i adds the dispute $P_i \leftrightarrow P_{\text{king}}$ and sends the signed "P_i disputes P_{king}" to all parties. Any party which at some point sees that P_{king} has more than t disputes stops the execution. Then P_{king} is removed from Honest, and some fresh $P_{\text{king}} \in$ Honest is chosen to be responsible for reconstructing the sharings. Then the computation is restarted. When a new reconstructor P_{king} is elected, all gates that the previous reconstructor have (should have) handled might have to be reopened by the new P_{king}. To keep the cost of this down, each reconstructor will handle only $\mathcal{O}(|\text{Circ}|/n)$ gates before a new reconstructor is picked.

Distributing Results. At the end of the evaluation, each sharing $[\![y]\!]$ associated to an output gate has been opened by some P_{king} which was in Honest at the end of his reign. This means that P_{king} was, at the end of his reign, disputed by at most t parties, which in turn implies that at least one of the $t + 1$ *honest* parties holds the opening of the $[\![y]\!]$ handled by P_{king}. But it is not guaranteed that *all* honest parties hold the opening. Therefore, each $P_i \in$ Honest is made responsible for $O/|\text{Honest}|$ of the O output sharings $[\![y]\!]$. All parties holding an opening of $[\![y]\!]$ for which P_i is responsible sends the opening to P_i; At least the one honest party will do so, letting P_i learn all openings. Then each P_i sends the opening of each $[\![y]\!]$ for which it is responsible to all parties. The total communication for this is $\mathcal{O}(Onk)$. Since there is honest majority in Honest, all parties now hold the opening of more than half the outputs, and they know which ones are correct. To ensure that this is enough, a Van der Monde error correction code is applied to the outputs before distribution. This is incorporated into the circuit, as a final layer of affine combinations. The only cost of this is doubling the number of output gates in the circuit. Note that the code only has to correct for erasures and hence can work for $t < n/2$.

References

[BB89]   Bar-Ilan, J., Beaver, D.: Non-cryptographic fault-tolerant computing in constant number of rounds of interaction. In: PODC'89, pp. 201–209 (1989)

[Bea89]   Beaver, D.: Multiparty protocols tolerating half faulty processors. In: Brassard, G. (ed.) CRYPTO 1989. LNCS, vol. 435, pp. 560–572. Springer, Heidelberg (1990)

[Bea91]   Beaver, D.: Efficient multiparty protocols using circuit randomization. In: Feigenbaum, J. (ed.) CRYPTO 1991. LNCS, vol. 576, pp. 420–432. Springer, Heidelberg (1992)

[BFKR90]   Beaver, D., Feigenbaum, J., Kilian, J., Rogaway, P.: Security with low communication overhead (extended abstract). In: Menezes, A.J., Vanstone, S.A. (eds.) CRYPTO 1990. LNCS, vol. 537, pp. 62–76. Springer, Heidelberg (1991)

[BGW88]   Ben-Or, M., Goldwasser, S., Wigderson, A.: Completeness theorems for non-cryptographic fault-tolerant distributed computation (extended abstract). In: 20th STOC, pp. 1–10 (1988)

[BH06]   Beerliova-Trubiniova, Z., Hirt, M.: Efficient multi-party computation with dispute control. In: Halevi, S., Rabin, T. (eds.) TCC 2006. LNCS, vol. 3876, pp. 305–328. Springer, Heidelberg (2006)

[BMR90]   Beaver, D., Micali, S., Rogaway, P.: The round complexity of secure protocols (extended abstract). In: 22nd STOC, pp. 503–513 (1990)

[Can01]   Canetti, R.: Universally composable security: A new paradigm for cryptographic protocols. In: 42nd FOCS, pp. 136–145 (2001)

[CCD88]   Chaum, D., Crépeau, C., Damgård, I.: Multiparty unconditionally secure protocols (extended abstract). In: 20th STOC, pp. 11–19 (1988)

[CDD+99]   Cramer, R., Damgård, I., Dziembowski, S., Hirt, M., Rabin, T.: Efficient multiparty computations secure against an adaptive adversary. In: Stern, J. (ed.) EUROCRYPT 1999. LNCS, vol. 1592, pp. 311–326. Springer, Heidelberg (1999)

[CDD00]   Cramer, R., Damgård, I., Dziembowski, S.: On the complexity of verifiable secret sharing and multiparty computation. In: 32nd STOC, pp. 325–334 (2000)

[CDG87]   Chaum, D., Damgård, I., van de Graaf, J.: Multiparty computations ensuring privacy of each party's input and correctness of the result. In: Pomerance, C. (ed.) CRYPTO 1987. LNCS, vol. 293, pp. 87–119. Springer, Heidelberg (1988)

[CDM00]   Cramer, R., Damgård, I., Maurer, U.: General secure multi-party computation from any linear secret-sharing scheme. In: Preneel, B. (ed.) EUROCRYPT 2000. LNCS, vol. 1807, pp. 316–334. Springer, Heidelberg (2000)

[DI06]   Damgård, I., Ishai, Y.: Scalable secure multiparty computation. In: Dwork, C. (ed.) CRYPTO 2006. LNCS, vol. 4117, pp. 501–520. Springer, Heidelberg (2006)

[FH06]   Fitzi, M., Hirt, M.: Optimally efficient multi-valued byzantine agreement. In: PODC 2006 (2006)

[GMW87]   Goldreich, O., Micali, S., Wigderson, A.: How to play any mental game or a completeness theorem for protocols with honest majority. In: 19th STOC (1987)

[GRR98]   Gennaro, R., Rabin, M., Rabin, T.: Simplified VSS and fast-track multiparty computations with applications to threshold cryptography. In: PODC'98 (1998)

[GV87]   Goldreich, O., Vainish, R.: How to solve any protocol problem - an efficiency improvement. In: Pomerance, C. (ed.) CRYPTO 1987. LNCS, vol. 293, pp. 73–86. Springer, Heidelberg (1988)

[HM01]   Hirt, M., Maurer, U.: Robustness for free in unconditional multi-party computation. In: Kilian, J. (ed.) CRYPTO 2001. LNCS, vol. 2139, pp. 101–118. Springer, Heidelberg (2001)

[HMP00]   Hirt, M., Maurer, U.M., Przydatek, B.: Efficient secure multi-party computation. In: Okamoto, T. (ed.) ASIACRYPT 2000. LNCS, vol. 1976, pp. 143–161. Springer, Heidelberg (2000)

[HN05]   Hirt, M., Nielsen, J.B.: Upper bounds on the communication complexity of optimally resilient cryptographic multiparty computation. In: Roy, B. (ed.) ASIACRYPT 2005. LNCS, vol. 3788, pp. 79–99. Springer, Heidelberg (2005)

[HN06]   Hirt, M., Nielsen, J.B.: Robust multiparty computation with linear communication complexity. In: Dwork, C. (ed.) CRYPTO 2006. LNCS, vol. 4117, pp. 463–482. Springer, Heidelberg (2006)

[RB89]   Rabin, T., Ben-Or, M.: Verifiable secret sharing and multiparty protocols with honest majority. In: 21th STOC, pp. 73–85.

[Yao82]   Yao, A.C.-C.: Protocols for secure computations (extended abstract). In: 23rd FOCS.

On Secure Multi-party Computation in Black-Box Groups

Yvo Desmedt[1,*], Josef Pieprzyk[2], Ron Steinfeld[2], and Huaxiong Wang[2,3]

[1] Dept. of Computer Science, University College London, UK
[2] Centre for Advanced Computing – Algorithms and Cryptography (ACAC)
Dept. of Computing, Macquarie University, North Ryde, Australia
[3] Division of Math. Sci., Nanyang Technological University, Singapore
{josef,rons,hwang}@comp.mq.edu.au, hxwang@ntu.edu.sg

Abstract. We study the natural problem of secure n-party computation (in the passive, computationally unbounded attack model) of the n-product function $f_G(x_1, \ldots, x_n) = x_1 \cdot x_2 \cdots x_n$ in an arbitrary finite group $(G, \cdot)$, where the input of party P_i is $x_i \in G$ for $i = 1, \ldots, n$. For flexibility, we are interested in protocols for f_G which require only *black-box* access to the group G (i.e. the only computations performed by players in the protocol are a group operation, a group inverse, or sampling a uniformly random group element).

Our results are as follows. First, on the negative side, we show that if $(G, \cdot)$ is non-abelian and $n \geq 4$, then no $\lceil n/2 \rceil$-private protocol for computing f_G exists. Second, on the positive side, we initiate an approach for construction of black-box protocols for f_G based on k-of-k threshold secret sharing schemes, which are efficiently implementable over any black-box group G. We reduce the problem of constructing such protocols to a combinatorial colouring problem in planar graphs. We then give two constructions for such graph colourings. Our first colouring construction gives a protocol with optimal collusion resistance $t < n/2$, but has exponential communication complexity $O(n \binom{2t+1}{t}^2)$ group elements (this construction easily extends to general adversary structures). Our second probabilistic colouring construction gives a protocol with (close to optimal) collusion resistance $t < n/\mu$ for a graph-related constant $\mu \leq 2.948$, and has efficient communication complexity $O(nt^2)$ group elements. Furthermore, we believe that our results can be improved by further study of the associated combinatorial problems.

Keywords: Multi-Party Computation, Non-Abelian Group, Black-Box, Planar Graph, Graph Colouring.

1 Introduction

Background. Groups form a natural mathematical structure for cryptography. In particular, the most popular public-key encryption schemes today (RSA [17]

---

* A part of this research was funded by NSF ANI-0087641, EPSRC EP/C538285/1. Yvo Desmedt is BT Chair of Information Security.

A. Menezes (Ed.): CRYPTO 2007, LNCS 4622, pp. 591–612, 2007.

and Diffie-Hellman/ElGamal [8,9]) both operate in abelian groups. However, the discovery of efficient quantum algorithms for breaking these cryptosystems [19] gives increased importance to the construction of alternative cryptosystems in non-abelian groups (such as [13,15]), where quantum algorithms seem to be much less effective.

Motivated by such emerging cryptographic applications of non-abelian groups, we study the natural problem of secure n-party computation (in the passive, computationally unbounded attack model) of the n-product function $f_G(x_1, \ldots, x_n)$ $= x_1 \cdot x_2 \cdots x_n$ in an arbitrary finite group $(G, \cdot)$, where the input of party P_i is $x_i \in G$ for $i = 1, \ldots, n$. For flexibility, we are interested in protocols for f_G which require only *black-box* access to the group G (i.e. the only computations performed by players in the protocol are a group operation $(x, y) \to x \cdot y$, a group inverse $x \to x^{-1}$, or sampling a random group element $x \in_R G$). It is well known that when $(G, \cdot)$ is abelian, a straightforward 2-round black-box protocol exists for f_G which is t-private (secure against t parties) for any $t < n$ and has communication complexity $O(n^2)$ group elements. However, to our knowledge, when $(G, \cdot)$ is non-abelian, no constructions of black-box protocols for f_G have been designed until now. Consequently, to construct a t-private protocol for f_G in some non-abelian group G one currently has to resort to adopting existing non black-box methods, which may lead to efficiency problems (see 'Related Work').

Our Results. Our results are as follows. First, on the negative side, we show that if $(G, \cdot)$ is non-abelian and $n \geq 4$, then no $\lceil n/2 \rceil$-private protocol for computing f_G exists. Second, on the positive side, we initiate an approach for construction of black-box protocols for f_G based only on k-of-k threshold secret sharing schemes (whereas previous non black-box protocols rely on Shamir's t-of-n threshold secret sharing scheme over a ring). We reduce the problem of constructing such protocols to a combinatorial colouring problem in planar graphs. We then give two constructions for such graph colourings. Our first colouring construction gives a protocol with optimal collusion resistance $t < n/2$, but has exponential communication complexity $O(n\binom{2t+1}{t}^2)$ group elements (this construction also easily generalises to general Q^2 adversary structures $\mathcal{A}$ as defined in [11], giving communication complexity $O(n|\mathcal{A}|^2)$ group elements). Our second probabilistic colouring construction gives a protocol with (close to optimal) collusion resistance $t < n/\mu$ for a graph-related constant $\mu \leq 2.948$, and has efficient communication complexity $O(nt^2)$ group elements. Furthermore, we believe that our results can be improved by further study of the associated combinatorial problems. We note that our protocols easily and naturally generalize to other arbitrary functions defined over the group G.

Related Work. There are two known non black-box methods for constructing a t-private protocol for the n-product function f_G for any $t < n/2$. They are both based on Shamir's t-of-n threshold secret sharing scheme [18] and its generalizations.

The first method [3,4,10] requires representing f_G as a boolean circuit, and uses Shamir's secret sharing scheme over the field $GF(p)$ for a prime $p > 2t + 1$. This protocol has total communication complexity $O(t^2 \log t \cdot N_{AND}(f_G))$ bits, where $N_{AND}(f_G)$ denotes the number of AND gates in the boolean AND/NOT circuit for computing f_G. Thus this protocol is efficient only for very small groups G, for which $N_{AND}(f_G)$ is manageable.

The second method [5] (see also [2] for earlier work) requires representing f_G as an arithmetic circuit over a finite *ring* R, and accordingly, uses a generalization of Shamir's secret sharing scheme to any finite ring. This protocol has total communication complexity $O(t^2 \log t \cdot N_M(f_G) \cdot \ell(R))$ bits, where $N_M(f_G)$ is the number of multiplication operations in the circuit for f_G over R and $\ell(R) \geq \log |R|$ denotes the number of bits needed for representing elements of R. If we 'embed' group G in the ring $R = R(G)$, so that R inherits the multiplication operation of G, then $N_M(f_G) = n - 1$, and hence the protocol from [5] has total communication complexity $O(nt^2 \log t \cdot \ell(R(G)))$ bits, compared to $O(nt^2 \cdot \ell(G))$ bits for our (second) protocol (assuming $t < n/2.948$), where $\ell(G) \geq \log |G|$ is the representation length of elements of G. Hence, for $t < n/2.948$, the communication complexity of our protocol for f_G is smaller than the one from [5] by a factor $\Theta(\frac{\ell(R(G))}{\ell(G)} \cdot \log t)$ (for $n/2.948 < t < n/2$, the protocol of [5] is still asymptotically the most efficient known proven protocol). Note that, for any finite group G, we can always take $R(G)$ to be the *group algebra* (or group ring) of G over $GF(2)$, which can be viewed as a $|G|$-dimensional vector space over $GF(2)$ consisting of all linear combinations of the elements of G (the basis vectors) with coefficients from $GF(2)$ (the product operation of $R(G)$ is defined by the operation of G extended by linearity and associativity, and the addition operation of $R(G)$ is defined componentwise). However, for this generic choice of $R(G)$ we have $\ell(R(G)) = |G|$, so, assuming $\ell(G) = \log |G|$, our protocol reduces communication complexity by a factor $\Theta(\frac{|G|}{\log |G|} \cdot \log t)$, which is exponentially large in the representation length $\log |G|$. In the worst case, we may have $\ell(R(G)) = \Theta(\ell(G))$ and our protocol may only give a saving factor $O(\log t)$ over the protocol from [5], e.g. this is the case for $G = GL(k, 2)$ (the group of invertible $k \times k$ matrices over $GF(2)$). We remark that this $O(\log t)$ saving factor arises essentially from the fact that Shamir's secret sharing for $2t+1$ shares requires a ring of size greater than $2t + 1$, and hence, for a secret from $GF(2)$, the share length is greater than the secret length by a factor $\Theta(\log t)$ (whereas our approach does not use Shamir's sharing and hence does not suffer from this length expansion). On the other hand, for sharing a secret from $GF(q)$ for 'large' q ($q > 2t + 1$), Shamir's scheme is ideal, so for specific groups such as $G = GL(k, q)$ with $q > 2t + 1$, the communication cost of the protocols from [2,5] reduces to $O(nt^2 \cdot \ell(R(G)))$.

Organization. The paper is organized as follows. Section 2 contains definitions and results we use. In Section 3 we show that $t < n/2$ is necessary for secure computation of f_G. In Sections 4.2 and 4.3 we show how to construct a t-private protocol for f_G given a 't-Reliable' colouring of a planar graph. Then in Section 4.4, we present two constructions of such t-Reliable colourings. Finally, Section 4.5 summarizes some generalizations and extensions, and Section 5 concludes with some open problems. Some proofs are omitted from this version of the paper due to space limitations – they are available in the full version [6].

2   Preliminaries

We recall the definition of secure multi-party computation in the passive (semi-honest), computationally unbounded attack model, restricted to deterministic symmetric functionalities and perfect emulation [10]. Let $[n]$ denote the set $\{1, \ldots, n\}$.

Definition 1. *Let $f : (\{0,1\}^*)^n \to \{0,1\}^*$ denote an n-input, single-output function, and let Π be an n-party protocol for computing f. We denote the party input sequence by $\mathbf{x} = (x_1, \ldots, x_n)$, the joint protocol view of parties in subset $I \subseteq [n]$ by $\mathsf{VIEW}_I^\Pi(\mathbf{x})$, and the protocol output by $\mathsf{OUT}^\Pi(\mathbf{x})$. For $0 < t < n$, we say that Π is a t-private protocol for computing f if there exists a probabilistic polynomial-time algorithm S, such that, for every $I \subset [n]$ with $\#I \le t$ and every $\mathbf{x} \in (\{0,1\}^*)^n$, the random variables*

$$\langle \mathsf{S}(I, \mathbf{x}_I, f(\mathbf{x})), f(\mathbf{x}) \rangle \text{ and } \langle \mathsf{VIEW}_I^\Pi(\mathbf{x}), \mathsf{OUT}^\Pi(\mathbf{x}) \rangle$$

are identically distributed, where $\mathbf{x}_I$ denotes the projection of the n-ary sequence $\mathbf{x}$ on the coordinates in I.

To prove our result we will invoke a combinatorial characterization of 2-input functions for which a 1-private 2-party computation protocol exists, due to Kushilevitz [12]. To state this result, we need the following definitions.

Definition 2. *Let $M = C \times D$ be a matrix, where C is the set of rows and D is the set of columns. Define a binary relation $\sim$ on pairs of rows of M as follows: $x_1, x_2 \in C$ satisfy $x_1 \sim x_2$ if there exists $y \in D$ such that $M_{x_1,y} = M_{x_2,y}$. Let $\equiv$ denote the equivalence relation on the rows of M which is the transitive closure of $\sim$. Similarly, we define $\sim$ and $\equiv$ on the columns of M.*

Definition 3. *A matrix M is called* forbidden *if all its rows are equivalent, all its columns are equivalent, and not all entries of M are equal.*

Definition 4. *Let $f : \{0,1\}^n \times \{0,1\}^n \to \{0, \ldots, m-1\}$ be any 2-input function. A matrix M for f is a $2^n \times 2^n$ matrix with entries in $\{0, \ldots, m-1\}$, where each row x of f corresponds to a value for the first input to f, each column y corresponds to a value for the second input to f, and the entry $M_{x,y}$ contains the value $f(x,y)$.*

Theorem 1 (Kushilevitz [12]). *Let f be a 2-input function and let M be a matrix for f. Then a 1-private 2-party protocol for computing f exists if and only if M does not contain a forbidden submatrix.*

3   Honest Majority Is Necessary for n-Product in Non-abelian Groups

We show that an honest majority $t < n/2$ is necessary for secure computation of the n-product function in non-abelian groups.

Theorem 2. *Let $(G, \cdot)$ denote a finite non-abelian group and let $n \geq 4$. There does not exist a $\lceil \frac{n}{2} \rceil$-private protocol for computing $f_G(x_1, \ldots, x_n) = x_1 \cdot x_2 \cdots x_n$.*

Proof. The proof proceeds by contradiction; we show that if a $\lceil \frac{n}{2} \rceil$-private protocol Π exists for f_G for $n \geq 4$, then we can construct a 1-private 2-party protocol for a 2-input function f'_G whose matrix M' contains a forbidden submatrix, thus contradicting Theorem 1.

Lemma 1. *Suppose there exists a $\lceil \frac{n}{2} \rceil$-private n-party protocol Π for computing the n-input function $f_G : G^n \to G$ defined by $f_G(x_1, \ldots, x_n) = x_1 \cdots x_n$ for $n \geq 4$. Then we can construct a 1-private 2-party protocol Π' for computing the 2-input function $f'_G : G^2 \times G^2 \to G$ defined by $f'_G((x'_1, x'_3), (x'_2, x'_4)) = x'_1 \cdot x'_2 \cdot x'_3 \cdot x'_4$.*

Proof. Given party P'_1 with input (x'_1, x'_3) and party P'_2 with input (x'_2, x'_4), the protocol Π' runs as follows. First, if $n \geq 5$, we partition the set $\{5, \ldots, n\}$ into two disjoint subsets S'_1 and S'_2 such that the size of both S'_1 and S'_2 is at most $\lceil \frac{n}{2} \rceil - 2$ (namely, if n is even we take $\#S'_1 = \#S'_2 = n/2 - 2$, and if n is odd we take $\#S'_1 = (n-3)/2$ and $\#S'_2 = (n-5)/2$). Then $\Pi'(P'_1, P'_2)$ consists of running the n-party protocol $\Pi(P_1, \ldots, P_n)$ where:

- P'_1 plays the role of parties $(P_1, P_3, \{P_i\}_{i \in S'_1})$ in Π, and sets those parties inputs to be $x_1 = x'_1$, $x_3 = x'_3$, and $x_i = 1$ for all $i \in S'_1$, respectively.
- P'_2 plays the role of parties $(P_2, P_4, \{P_i\}_{i \in S'_2})$ in Π, and sets those parties inputs to be $x_2 = x'_2$, $x_4 = x'_4$ and $x_i = 1$ for all $i \in S'_2$, respectively.

The 1-privacy of protocol $\Pi'(P'_1, P'_2)$ for computing f_G follows from the $\lceil \frac{n}{2} \rceil$-privacy of protocol $\Pi(P_1, \ldots, P_n)$ for computing f_G because:

- $f_G(x'_1, x'_2, x'_3, x'_4, 1, \ldots, 1) = f'_G(x'_1, x'_2, x'_3, x'_4) = x'_1 \cdot x'_2 \cdot x'_3 \cdot x'_4$ for all $x'_1, x'_2, x'_3, x'_4 \in G$.
- For each (x'_1, x'_2, x'_3, x'_4), the view of P'_1 (resp. P'_2) in protocol $\Pi'(P'_1, P'_2)$ is identical to the view of a set of at most $\lceil \frac{n}{2} \rceil$ parties in protocol $\Pi(P_1, \ldots, P_n)$ whose inputs are known to P'_1 (resp. P'_2), with special settings of 1 for some inputs. Thus the same view simulator algorithm S of Π can be used to simulate the view in Π'.

This completes the proof. □

Lemma 2. *For any non-abelian group G, the matrix M for the 2-input function $f'_G : G^2 \times G^2 \to G$ defined by $f'_G((x'_1, x'_3), (x'_2, x'_4)) = x'_1 \cdot x'_2 \cdot x'_3 \cdot x'_4$ contains a 2×2 forbidden submatrix.*

Proof. Observe from Definitions 2 and 3 that any 2×2 matrix with 3 equal elements and a fourth distinct element is a forbidden matrix. Now recall that the rows of matrix M for f'_G are indexed by $(x'_1, x'_3) \in G^2$, the columns of M are indexed by $(x'_2, x'_4) \in G^2$, and the entry of M at row (x'_1, x'_3) and column (x'_2, x'_4) is $M_{(x'_1, x'_3),(x'_2, x'_4)} = x'_1 \cdot x'_2 \cdot x'_3 \cdot x'_4$. Also, since G is non-abelian, there exist a pair of elements a and b in G such that a and b do not commute and $a, b \neq 1$. Consider the 2×2 submatrix of M formed by the intersections of the 2 rows $(1,1)$ and (a, a^{-1}) and the 2 columns $(1,1)$ and (b, b^{-1}) (these row and column pairs are distinct because $a, b \neq 1$). We claim that this submatrix is forbidden. Indeed, three of the submatrix entries are equal because $M_{(1,1),(1,1)} = M_{(a,a^{-1}),(1,1)} = M_{(1,1),(b,b^{-1})} = 1$, and the remaining fourth entry is distinct because $M_{(a,a^{-1}),(b,b^{-1})} = a \cdot b \cdot a^{-1} \cdot b^{-1} = (a \cdot b) \cdot (b \cdot a)^{-1} \neq 1$ since a and b do not commute. This completes the proof. $\square$

Combining Lemma 1 and Lemma 2, we conclude that if a $\lceil \frac{n}{2} \rceil$-private protocol Π exists for f_G for $n \geq 4$, then we obtain a contradiction to Theorem 1. This completes the proof. $\square$

4   Constructions

4.1   Our Approach: Black Box Non-abelian Group Protocols

Our protocols will treat the group G as a black box in the sense that the only computations performed by players in our protocols will be one of the following three: Multiply (Given $x \in G$ and $y \in G$, compute $x \cdot y$), Inverse (Given $x \in G$, compute x^{-1}), and Random Sampling (Choose a uniformly random $x \in G$). It is easy to see that these three operations are sufficient for implementing a perfect k-of-k threshold secret sharing scheme. We use this k-of-k scheme as a fundamental building block in our protocols. The following proposition is easy to prove.

Proposition 1. *Fix $x \in G$ and integers k and $j \in [k]$, and suppose we create a k-of-k sharing $(s_x(1), s_x(2), \ldots, s_x(k))$ of x by picking the $k - 1$ shares $\{s_x(i)\}_{i \in [k] \setminus \{j\}}$ uniformly and independently at random from G, and computing $s_x(j)$ to be the unique element of G such that $x = s_x(1)s_x(2) \cdots s_x(k)$. Then the distribution of the shares $(s_x(1), s_x(2), \ldots, s_x(k))$ is independent of j.*

4.2   Construction of n-Product Protocol from a Shared 2-Product Subprotocol

We begin by reducing the problem of constructing a t-private protocol for the n-product function $f(x_1, \ldots, x_n) = x_1 \cdots x_n$ (where party P_i holds input x_i for $i = 1, \ldots, n$), to the problem of constructing a subprotocol for the *Shared 2-Product function* $f'(x, y) = x \cdot y$, where inputs x, y and output $z = x \cdot y$ are shared among the parties. We define for this subprotocol a so-called *strong t-privacy* definition, which will be needed later to prove the (standard) t-privacy of the full n-product protocol built from subprotocol Π_S. The definition of strong t-privacy requires the adversary's view simulator to simulate *all* output shares except one share not held by the adversary, in addition to simulating the internal subprotocol view of the adversary.

Definition 5 (Shared n-Party 2-Product Subprotocol). *A n-Party Shared 2-Product subprotocol Π_S with sharing parameter ℓ and share ownership functions $\mathcal{O}_x, \mathcal{O}_y, \mathcal{O}_z : [\ell] \to [n]$ has the following features:*

- *Input: For $j = 1, \ldots, \ell$, party $P_{\mathcal{O}_x(j)}$ holds jth share $s_x(j) \in G$ of x and party $P_{\mathcal{O}_y(j)}$ holds jth share $s_y(j) \in G$ of y, where $\mathbf{s}_x = (s_x(1), s_x(2), \ldots, s_x(\ell))$ and $\mathbf{s}_y = (s_y(1), s_y(2), \ldots, s_y(\ell))$ denote ℓ-of-ℓ sharing of $x \stackrel{\text{def}}{=} s_x(1) \cdot s_x(2) \cdots s_x(\ell)$ and $y \stackrel{\text{def}}{=} s_y(1) \cdot s_y(2) \cdots s_y(\ell)$, respectively.*
- *Output: For $j = 1, \ldots, \ell$, party $P_{\mathcal{O}_z(j)}$ holds jth share $s_z(j)$ of output product $z \stackrel{\text{def}}{=} s_z(1) \cdots s_z(\ell)$.*
- *Correctness: We say that that Π_S is correct if, for all protocol inputs $\mathbf{s}_x = (s_x(1), s_x(2), \ldots, s_x(\ell))$ and $\mathbf{s}_y = (s_y(1), s_y(2), \ldots, s_y(\ell))$, the output shares $\mathbf{s}_z = (s_z(1), s_z(2), \ldots, s_z(\ell))$ satisfy*

$$z = x \cdot y$$

where $x \stackrel{\text{def}}{=} s_x(1) \cdot s_x(2) \cdots s_x(\ell)$, $y \stackrel{\text{def}}{=} s_y(1) \cdot s_y(2) \cdots s_y(\ell)$ and $z \stackrel{\text{def}}{=} s_z(1) \cdots s_z(\ell)$.

- *Strong t-Privacy: We say that Π_S achieves **strong t-privacy** if there exists a probabilistic simulator algorithm S_{Π_S} such that for all $I \subset [n]$ with $\#I \leq t$, there exist $j^* \in [\ell]$ with $\mathcal{O}_x(j^*) \notin I$ and $\mathcal{O}_z(j^*) \notin I$, and $j_y^* \in [\ell]$ with $\mathcal{O}_y(j_y^*) \notin I$ such that for all protocol inputs $\mathbf{s}_x = (s_x(1), \ldots, s_x(\ell))$ and $\mathbf{s}_y = (s_y(1), \ldots, s_y(\ell))$, the random variables*

$$\langle \mathsf{S}_{\Pi_S}(I, \{\mathbf{s}_x(j)\}_{j \in [n] \setminus \{j^*\}}, \{\mathbf{s}_y(j)\}_{j \in [\ell] \setminus \{j_y^*\}}) \rangle \text{ and}$$

$$\langle \mathsf{VIEW}_I^{\Pi_S}(\mathbf{s}_x, \mathbf{s}_y), \{s_z(j)\}_{j \in [\ell] \setminus \{j^*\}} \rangle$$

*are identically distributed (over the random coins of Π_S). Here $\mathsf{VIEW}_I^{\Pi_S}(\mathbf{s}_x, \mathbf{s}_y)$ denotes the view of I in subprotocol Π_S run with input shares $\mathbf{s}_x, \mathbf{s}_y$, and $s_z(j)$ denotes the jth output share. If $j_y^* = j^*$ for all I, then we say Π_S achieves **symmetric strong t-privacy**.*

Remark 1. The share ownership functions $\mathcal{O}_x, \mathcal{O}_y, \mathcal{O}_z$ specify for each share index $j \in [\ell]$, the indices $\mathcal{O}_x(j), \mathcal{O}_y(j), \mathcal{O}_z(j)$ in $[n]$ of the party which holds the jth input shares $s_x(j)$ and $s_y(j)$ and jth output share $s_z(j)$, respectively.

Remark 2. The adversary view simulator S_{Π_S} for collusion I is given all input shares except the j^*th x-share $s_x(j^*)$ and j_y^*th y-share $s_y(j_y^*)$ (where $j^*, j_y^* \in [\ell]$, which depend on I, are indices of shares given to players *not* in I), and outputs all output shares except the j^*th share $s_z(j^*)$ of z. Because, for each I, the same value of index j^* is used for both x-input shares and output shares, this allows multiple simulator runs to be composed, using output shares of one subprotocol run as x-input shares in a following subprotocol run, as shown in the security proof of the following construction. If in addition, *symmetric* strong t-privacy is achieved, one can use output shares of one subprotocol run as either x-input or y-input shares for the following subprotocol run, allowing for more efficient protocols.

We now explain our construction of an n-Product Protocol $\Pi(T, \Pi_S)$ given a binary computation tree T for f_G with n leaf nodes corresponding to the n

protocol inputs (as illustrated in Fig. 1), and a Shared 2-Product subprotocol Π_S with sharing parameter ℓ and share ownership functions $\mathcal{O}_x, \mathcal{O}_y, \mathcal{O}_z$. The protocol Π begins with each party P_j computing an ℓ-of-ℓ sharing of its input x_j, and distributing out these shares to the n parties according to the share ownership functions $\mathcal{O}_x, \mathcal{O}_y$ of Π_S. Then protocol Π performs each of the internal node 2-product computations of the computation tree T on ℓ-of-ℓ sharings of the internal node's two children nodes by running the shared 2-product subprotocol Π_S, resulting in an ℓ-of-ℓ sharing of the internal node value. Eventually this recursive process gives an ℓ-of-ℓ sharing of the root node value $x_1 \cdots x_n$ of T, which is broadcast to all parties.

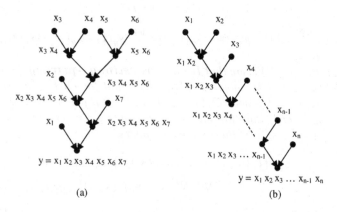

(a)   (b)

Fig. 1. (a) Example of a binary tree T with $n = 7$ leaves. (b) The *slanted linear tree* T_{slin} with n leaves.

The following Lemma establishes the t-privacy of protocol $\Pi(T, \Pi_S)$, assuming the correctness and strong t-privacy of subprotocol Π_S. Refer to [6] for a proof.

Lemma 3. *For any binary tree T with n leaves, if the n-party Shared 2-Product subprotocol Π_S satisfies correctness and symmetric strong t-privacy (see Definition 5), then protocol $\Pi(T, \Pi_S)$ is an n-party t-private protocol for computing n-Product function $f_G(x_1, \ldots, x_n) = x_1 \cdots x_n$. For the slanted linear binary tree T_{slin} shown in Fig 1(b), the above result holds even if Π_S satisfies (ordinary) strong t-privacy (i.e. symmetric strong t-privacy is not needed in this case).*

4.3 Construction of a t-Private n-Party Shared 2-Product Subprotocol from a t-Reliable n-Colouring of a Planar Graph

Next, we reduce the problem of constructing a t-Private n-Party Shared 2-Product Subprotocol Π_S to a combinatorial problem defined below of finding a 't-Reliable n-Colouring' of the nodes of a planar graph. We note that our notion of a 't-Reliable n-Colouring' is closely related to a similar notion defined in [7], and shown to be equivalent to the existence of private communication via a network graph in which each node is assigned one of n possible colours and the adversary controls all nodes with colours belonging to a t-colour subset I.

Consider a Planar Directed Acyclic Graph (PDAG) $\mathcal{G}$ having 2ℓ source (input) nodes drawn in a horizontal row at the top, ℓ sink (output) nodes drawn in a horizontal row at the bottom, and $\sigma_{\mathcal{G}}$ nodes overall. We use PDAG $\mathcal{G}$ to represent a blackbox protocol, where the input/output nodes are labelled by the protocol input/output group elements, and the internal graph nodes are labelled by intermediate protocol values. Each internal graph node is also assigned a *colour* specifying the player which computes the internal node value. The graph edges represent group elements sent from one player to another. The computation performed at each node is multiplication of the values on all incoming edges and resharing the product along the outgoing edges using the k-of-k secret sharing scheme in Proposition 1. All computations in the ith round of the 2-Product subprotocol correspond to the ith row (from the top) in the PDAG. Communications between nodes correspond to edges between consecutive rows.

Actually to construct a protocol for any non-abelian group our requirement on graph $\mathcal{G}$ is slightly stronger than planarity and can be precisely defined as follows.

Definition 6 (Admissible PDAG). *We call graph $\mathcal{G}$ an Admissible PDAG with share parameter ℓ and size parameter m if it has the following properties:*

- *Nodes of $\mathcal{G}$ are drawn on a square $m \times m$ grid of points (each node of $\mathcal{G}$ is located at a grid point but some grid points may not be occupied by nodes). Rows of the grid are indexed from top to bottom and columns from left to right by the integers $1, 2, \ldots, m$. A node of $\mathcal{G}$ at row i and column j is said to have index (i, j). $\mathcal{G}$ has 2ℓ source (input) nodes on top row 1, and ℓ sink (output) nodes on bottom row m.*
- *Incoming edges of a node on row i only come from nodes on row $i - 1$, and outgoing edges of a node on row i only go to nodes on row $i + 1$.*
- *For each row i and column j, let $\eta_1^{(i,j)} < \ldots < \eta_{q^{(i,j)}}^{(i,j)}$ denote the ordered column indices of the $q^{(i,j)} > 0$ nodes on level $i + 1$ which are connected to node (i, j) by an edge. Then, for each $j = 1, \ldots, m - 1$, we have*

$$\eta_{q^{(i,j)}}^{(i,j)} \leq \eta_1^{(i,j+1)}, \tag{1}$$

i.e. the rightmost node on level $i + 1$ connected to node (i, j) is to the left of (or equal to) the leftmost node on level $i + 1$ connected to node $(i, j + 1)$.

We call the left ℓ source nodes on row 1 (indexed $(1, 1), \ldots, (1, \ell)$) the '$x$-input' nodes and the last ℓ source nodes on row 1 (indexed $(1, \ell + 1), \ldots, (1, 2\ell)$) the '$y$-input' nodes. By ith x-input node, we mean the x-input node at position i from the left. We define the ith y-input and ith output node similarly.

Let $C : [m] \times [m] \to [n]$ be an n-Colouring function that associates to each node (i, j) of $\mathcal{G}$ a colour $C(i, j)$ chosen from a set of n possible colours $[n]$. We now define the notion of a t-Reliable n-Colouring.

Definition 7 (*t*-Reliable *n*-Colouring). *We say that $C : [m] \times [m] \to [n]$ is a t-Reliable n-Colouring for admissible PDAG $\mathcal{G}$ (with share parameter ℓ and size parameter m) if for each t-colour subset $I \subset [n]$, there exist $j^* \in [\ell]$ and $j_y^* \in [\ell]$ such that:*

- *There exists a path $PATH_x$ in $\mathcal{G}$ from the j^*th x-input node to the j^*th output node, such that none of the path node colours are in subset I (we call such a path I-avoiding), and*
- *There exists an I-avoiding path $PATH_y$ in $\mathcal{G}$ from the j_y^*th y-input node to the j^*th output node.*

If $j_y^* = j^*$ for all I, we say that C is a Symmetric t-Reliable n-Colouring.

Remark 3. The paths $PATH_x$ and $PATH_y$ in Definition 7 are free to move in *any* direction along each edge of directed graph $\mathcal{G}$, i.e. for this definition we regard $\mathcal{G}$ as an *undirected* graph (throughout the paper we assume that a path is *simple*, i.e. free of cycles; hence each node on the path is only visited once).

An example of an admissible PDAG with I-avoiding paths $PATH_x$ and $PATH_y$ is shown in Fig 2(a). Given an admissible PDAG $\mathcal{G}$ (with share parameter ℓ and size parameter m) and an associated t-Reliable n-Colouring $C : [m] \times [m] \to [n]$, we construct a t-Private n-Party Shared 2-Product Subprotocol $\Pi_S(\mathcal{G}, C)$.

Shared 2-Product Subprotocol $\Pi_S(\mathcal{G}, C)$

Input: We define the share ownership functions $\mathcal{O}_x, \mathcal{O}_y, \mathcal{O}_z$ of $\Pi_S(\mathcal{G}, C)$ according to the colours assigned by C to the input and output nodes of $\mathcal{G}$ (i.e. $\mathcal{O}_x(j) = C(1, j)$, $\mathcal{O}_y(j) = C(1, \ell + j)$, $\mathcal{O}_z(j) = C(m, j)$ for $j = 1, \ldots, \ell$). For $j = 1, \ldots, \ell$, party $P_{\mathcal{O}_x(j)}$ holds jth share $s_x(j) \in G$ of x and party $P_{\mathcal{O}_y(j)}$ holds jth share $s_y(j) \in G$ of y, where $\mathbf{s}_x = (s_x(1), s_x(2), \ldots, s_x(\ell))$ and $\mathbf{s}_y = (s_y(1), s_y(2), \ldots, s_y(\ell))$ denote ℓ-of-ℓ sharing of $x \stackrel{\text{def}}{=} s_x(1) \cdot s_x(2) \cdots s_x(\ell)$ and $y \stackrel{\text{def}}{=} s_y(1) \cdot s_y(2) \cdots s_y(\ell)$, respectively.

For each row $i = 1, \ldots, m$ and column $j = 1, \ldots, m$ of $\mathcal{G}$, party $P_{C(i,j)}$ does the following:

- $P_{C(i,j)}$ computes a label $v^{(i,j)}$ for node (i, j) of $\mathcal{G}$ as follows. If $i = 1$, $P_{C(i,j)}$ defines $v^{(i,j)} = s_x(j)$ for $j \leq \ell$ and $v^{(i,j)} = s_y(j)$ for $\ell + 1 \leq j \leq 2\ell$. If $i > 1$, $P_{C(i,j)}$ computes $v^{(i,j)}$ by multiplying the shares received from nodes at previous row $i - 1$ (labels of edges between a node on row $i - 1$ and node (i, j)), ordered from left to right according to the sender node column index.
- If $i = m$, $P_{C(m,j)}$ sets output share j to be the label $v^{(m,j)}$,
- else, if $i < m$, let $\eta_1^{(i,j)} < \ldots < \eta_{q^{(i,j)}}^{(i,j)}$ denote the ordered column indices of the nodes on level $i + 1$ which are connected to node (i, j) by an edge. $P_{C(i,j)}$ chooses $q^{(i,j)} - 1$ uniformly random elements from G and computes a $q^{(i,j)}$-of-$q^{(i,j)}$ secret sharing $s_1^{(i,j)}, \ldots, s_{q^{(i,j)}}^{(i,j)}$ of label $v^{(i,j)}$ such that:

$$v^{(i,j)} = s_1^{(i,j)} \cdots s_{q^{(i,j)}}^{(i,j)}.$$

– For $k = 1, \ldots, q^{(i,j)}$, $P_{C(i,j)}$ sends share $s_k^{(i,j)}$ to party $P_{C(i+1, \eta_k^{(i,j)})}$ and labels edge from node (i, j) to node $(i + 1, \eta_k^{(i,j)})$ by the share $s_k^{(i,j)}$.

Note that the correctness of Π_S follows from the fact that the product of node values at each row of PDAG $\mathcal{G}$ is preserved and hence equal to $x \cdot y$, thanks to condition (1) in Definition 6.

Lemma 4. *If $\mathcal{G}$ is an admissible PDAG and C is a t-Reliable n-Colouring for $\mathcal{G}$ then $\Pi_S(\mathcal{G}, C)$ achieves strong t-privacy. Moreover, if C is a Symmetric t-Reliable n-Colouring, then $\Pi_S(\mathcal{G}, C)$ achieves Symmetric strong t-privacy.*

Proof. (Sketch) The full proof of Lemma 4 can be found in [6]. Here we only explain the main idea by considering the case when the I-avoiding paths $PATH_x$ and $PATH_y$ only have downward edges (in [6] we extend the argument to paths with upward edges). Consider $PATH_x$ from the j^*th x-input node to the j^*th output node. At the first node $PATH_x(1)$ on the path, although the node value $v(1) = s_x(j^*)$ is not known to the view simulator S_{Π_S}, we may assume, by Proposition 1, that in the real subprotocol Π_S, when node $PATH_x(1)$ shares out its node label among its q outgoing edges, it sends new random elements (labels) r_i on each of the $q - 1$ outgoing edges *not* on $PATH_x$. Thus simulator S_{Π_S} can easily simulate all outgoing edge values of $PATH_x(1)$ which are not on $PATH_x$. The same argument shows that for all kth nodes $PATH_x(k)$ and $PATH_y(k)$ on $PATH_x$ and $PATH_y$ respectively, simulator S_{Π_S} can simulate all values on outgoing edges of $PATH_x(k)$ and $PATH_y(k)$ which are *not* on $PATH_x$ or $PATH_y$ by independent random elements. The values on edges along $PATH_x$ or $PATH_y$ depend on the inputs $s_x(j^*)$ and $s_y(j_y^*)$ which are not known to simulator S_{Π_S}, but since the paths $PATH_x$ and $PATH_y$ are I-avoiding, these values are not in the view of I and need not be simulated by S_{Π_S}. Since S_{Π_S} knows all inputs to Π_S it can compute all other edge values in the Π_S, including all outputs except the j^*th one (which is on $PATH_x$ and $PATH_y$), as required.   □

4.4 Constructions of t-Reliable n-Colourings of Planar Graphs

We now present two general constructions of t-Reliable n-Colourings of planar graphs which can be used to build t-Private n-Party protocols for the n-Product function in any finite group as explained in the previous sections. Our first deterministic construction achieves optimal collusion security $(t < n/2)$ but has exponential complexity $(\ell = \binom{n}{t})$. Our second probabilistic construction has a slightly suboptimal collusion security $(t < n/2.948)$ but has a very efficient linear complexity $(\ell = O(n))$.

The PDAG. The admissible PDAG $\mathcal{G}_{tri}(\ell', \ell)$ that we consider has sharing parameter ℓ and has $\ell' \times \ell$ nodes. It is shown in Fig. 2(b). The nodes of $\mathcal{G}_{tri}(\ell', \ell)$ are arranged in an $\ell' \times \ell$ node grid. Let (i, j) denote the node at row $i \in [\ell']$ (from the top) and column j (from the left). There are three types of edges in directed graph $\mathcal{G}_{tri}(\ell', \ell)$: (1) Horizontal edge: An edge connecting two adjacent nodes on the same row, directed from right to left (i.e. from node (i, j) to node $(i, j - 1)$, for $i \in [\ell']$, $j \in [\ell] \setminus \{1\}$), (2) Vertical edge: An edge connecting two adjacent nodes on the same column, directed from top to bottom (i.e. from node (i, j) to node $(i + 1, j)$, for $i \in [\ell'] \setminus \{\ell'\}$, $j \in [\ell]$), and (3) Diagonal edge: An edge connecting node (i, j) to node $(i + 1, j - 1)$, for $i \in [\ell'] \setminus \{\ell'\}$, $j \in [\ell] \setminus \{1\}$).

The ℓ nodes on the top row (row 1) of $\mathcal{G}_{tri}$ are the x-input nodes, indexed from left to right. The top ℓ nodes on the rightmost column of $\mathcal{G}_{tri}$ (column ℓ) are the y-input nodes, indexed from top to bottom.

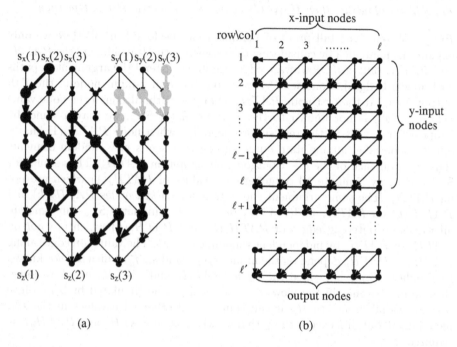

Fig. 2. (a) Example of an admissible PDAG $\mathcal{G}$ with sharing parameter $\ell = 3$ (node colours are not indicated). For a given collusion I, an example I-avoiding path $PATH_x$ is shown in heavy black, and an example I-avoiding path $PATH_y$ (until the meeting with $PATH_x$) is shown in heavy gray. In this example, we have $j^* = 2$ and $j_y^* = 3$. (b) The admissible PDAG $\mathcal{G}_{tri}(\ell', \ell)$.

Remark 4. The reader may notice that the above specification of $\mathcal{G}_{tri}$ does not formally satisfy the convention for drawing an admissible PDAG as defined in Def. 6, due to the horizontal edges and the fact that the y-input nodes are arranged along a column, rather than along the same row as the x-input nodes. However, it is easy to see that $\mathcal{G}_{tri}$ can also be drawn strictly according to Def. 6. Namely by rotating the drawing of $\mathcal{G}_{tri}$ in Fig. 2 by 45 degrees anticlockwise, the horizontal edges become diagonal edges, and x-inputs and y-inputs can be formally put on the same row by adding appropriate 'connecting' nodes of the same colour as the corresponding input nodes of $\mathcal{G}_{tri}$. These are only formal changes in drawing conventions, and there is no change in the protocol itself. In this section we use the drawing convention in Fig. 2 for clarity.

Remark 5. All diagonal edges in the definition of $\mathcal{G}_{tri}$ above are parallel (with a 'positive slope', when using the drawing convention in Fig 2). However, it is clear that the admissible PDAG requirements are still satisfied if we remove from $\mathcal{G}_{tri}$ some 'positive slope' diagonal edges and add some 'negative slope' diagonal edges (connecting a node (i, j) to node $(i + 1, j + 1)$, for some $i \in [\ell'] \setminus \{\ell'\}$, $j \in [\ell] \setminus \{\ell\}$), as long as planarity of $\mathcal{G}$ is preserved (no two diagonal edges intersect). We denote such 'generalised' PDAGs by $\mathcal{G}_{gtri}$.

First Construction C_{comb} $(t < n/2$ and $\ell = \binom{n}{t})$. We now present an explicit construction of a t-Reliable n-Colouring C_{comb} of the square graph $\mathcal{G}_{tri}(\ell, \ell)$. The construction applies for all $n \geq 2t + 1$ (i.e. $t \leq \lfloor \frac{n-1}{2} \rfloor$), and hence (by Section 3) the n-Product protocol constructed from it by the method of Sections 4.2 and 4.3 achieves $\lfloor \frac{n-1}{2} \rfloor$-privacy (which is optimal, as shown in Section 3). Unfortunately, the sharing parameter in this construction $\ell = \binom{n}{t}$, is exponential in t (and therefore the protocol communication cost is also exponential in t).

Colouring C_{comb} for graph $\mathcal{G}_{tri}(\ell, \ell)$ with $\ell = \binom{n}{t}$ and $n \geq 2t + 1$

1. Let $I_1, \ldots, I_\ell$ denote the sequence of all $\ell = \binom{n}{t}$ t-colour subsets of $[n]$ (in some ordering).
2. For each $(i, j) \in [\ell] \times [\ell]$, define the colour $C(i, j)$ of node (i, j) of $\mathcal{G}_{tri}(\ell, \ell)$ to be any colour in the set $S_{i,j} = [n] \setminus (I_i \bigcup I_j)$ (note that since $|I_i| = |I_j| = t$ and $n \geq 2t + 1$, the set $S_{i,j}$ contains at least $n - (|I_i| + |I_j|) \geq n - 2t \geq 1$ colours, so $S_{i,j}$ is never empty).

Lemma 5. *For $n \geq 2t + 1$, the colouring C_{comb} is a Symmetric t-Reliable n-Colouring for graph $\mathcal{G}_{tri}(\ell, \ell)$, with $\ell = \binom{n}{t}$.*

Proof. Given each t-colour subset $I \subseteq [n]$, let j^* denote the index of I in the sequence $I_1, \ldots, I_\ell$ of all t-colour subsets used to construct C_{comb}, i.e $I_{j^*} = I$. By construction of C_{comb}, none of the nodes of $\mathcal{G}_{tri}(\ell, \ell)$ along column j^* have colours in $I_{j^*} = I$. Hence one can take column j^* of $\mathcal{G}_{tri}(\ell, \ell)$ as $PATH_x$. Similarly, we also know that none of the nodes of $\mathcal{G}_{tri}(\ell, \ell)$ along row j^* have colours in $I_{j^*} = I$, so one can take $PATH_y$ to consist of all nodes on row j^* which are on columns $j \geq j^*$, followed by all nodes on column j^* which are on rows $i \geq j^*$. Thus C_{comb} is a Symmetric t-Reliable n-Colouring for graph $\mathcal{G}_{tri}(\ell, \ell)$, as required. $\square$

Remark 6. The colouring C_{comb} remains a Symmetric t-Reliable n-Colouring even if we remove all diagonal edges from $\mathcal{G}_{tri}(\ell, \ell)$ (since the paths $PATH_x$ and $PATH_y$ only contain vertical and horizontal edges).

Combining Lemma 5 (applied to a subset of $n' = 2t + 1 \leq n$ colours from $[n]$) with Lemmas 3 and 4, we have

Corollary 1. *For any $t < n/2$, there exists a black-box t-private protocol for f_G with communication complexity $O(n\binom{2t+1}{t}^2)$ group elements.*

Second Construction C_{rand} ($t < n/2.948$ **and** $\ell = O(n)$). It is natural to ask whether the exponentially large sharing parameter $\ell = \binom{n}{t}$ can be reduced. Our second construction C_{rand} shows that this is certainly the case when $t < n/2.948$, achieving a linear sharing parameter $\ell = O(n)$.

As a first step towards our second construction, we relax the properties required from C in Definition 7 to slightly simpler requirements for the square graph $\mathcal{G}_{tri}(\ell, \ell)$ (i.e. $\ell' = \ell$), as follows.

Definition 8 (Weakly t-Reliable n-Colouring). *We say that $C : [\ell] \times [\ell] \to [n]$ is a Weakly t-Reliable n-Colouring for graph $\mathcal{G}_{tri}(\ell, \ell)$ if for each t-colour subset $I \subset [n]$:*

- *There exists an I-avoiding path P_x in $\mathcal{G}$ from a node on the top row (row 1) to a node on the bottom row (row ℓ). We call such a path an I-avoiding top-bottom path.*
- *There exists an I-avoiding path P_y in $\mathcal{G}$ from a node on the rightmost column (column ℓ) to a node on the leftmost column (column 1). We call such a path an I-avoiding right-left path.*

Note that in the above definition of Weak t-Reliability, the index of the starting node of path P_x in the top row need not be the same as the index of the exit node of P_x in the bottom row (whereas in the definition of t-Reliability, $PATH_x$ must exit at the same position along the output row as the position in the top row where $PATH_x$ begins).

The following lemma shows that finding a Weakly t-Reliable n-Colouring for the square graph $\mathcal{G}_{tri}(\ell, \ell)$ is sufficient for constructing a (standard) t-Reliable n-Colouring for a rectangular graph $\mathcal{G}_{gtri}(2\ell - 1, \ell)$. The idea is to add $\ell - 1$ additional rows to $\mathcal{G}_{tri}(\ell, \ell)$ by appending a 'mirror image' (reflected about the last row) of itself, as shown in Fig. 3 (refer to [6] for the detailed proof).

Lemma 6. *Let $C : [\ell] \times [\ell] \to [n]$ be a Weakly t-Reliable n-Colouring (see Def. 8) for square admissible PDAG $\mathcal{G}_{tri}(\ell, \ell)$. Then we can construct a (standard) t-Reliable n-Colouring (see Def. 7) for a rectangular admissible PDAG $\mathcal{G}_{gtri}(2\ell - 1, \ell)$.*

For our second colouring construction, we use the 'probabilistic method' [1], namely we choose the colour of each node in the square graph $\mathcal{G}_{tri}(\ell, \ell)$ independently and uniformly at random from $[n]$. Although there is a finite error probability p that such a random n-Colouring will not be Weakly t-Reliable, we show that if $n/t > 2.948$ and we use a sufficiently large (but only *linear* in n) sharing parameter $\ell = O(n)$, then the error probability p can be made arbitrarily small. Moreover, p decreases exponentially fast with ℓ, so p can be easily made negligible.

Colouring C_{rand} for graph $\mathcal{G}_{tri}(\ell, \ell)$ with $\ell = O(n)$ and $n \geq 2.948t$

For each $(i, j) \in [\ell] \times [\ell]$, choose the colour $C(i, j)$ of node (i, j) of $\mathcal{G}_{tri}(\ell, \ell)$ independently and uniformly at random from $[n]$.

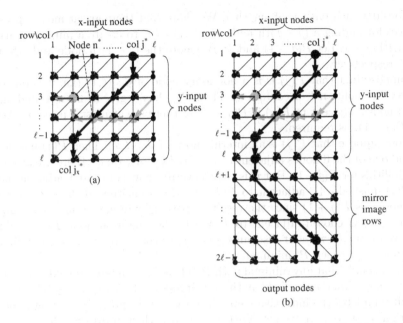

Fig. 3. (a) Example paths in square PDAG $\mathcal{G}_{tri}(\ell, \ell)$ for a given Weakly t-Reliable n-Colouring (P_x in heavy black, P_y in heavy gray). (b) Corresponding paths in rectangular PDAG $\mathcal{G}_{gtri}(2\ell - 1, \ell)$.

To analyse this construction, we will make use of the following counting Lemma. Here, for any right-left path in $\mathcal{G}_{tri}(\ell, \ell)$, we define its *length* as the number of nodes on the path. We say a path is *minimal* if removing any node from the path disconnects the path.

Lemma 7. *The number $N_P(k, \ell)$ of minimal right-left paths of length k in graph $\mathcal{G}_{tri}(\ell, \ell)$ is upper bounded as*

$$N_P(k, \ell) \le c(\mu) \cdot \ell \cdot \mu^k,$$

for some constants $\mu, c(\mu)$, with $\mu \le 2.948$. We call the minimal possible value for μ the connective constant *of $\mathcal{G}_{tri}(\ell, \ell)$.*

Proof. For a minimal right-left path, there are ℓ possible starting nodes on the rightmost column. We may assume without loss of generality that the first edge of the path is not a vertical edge. For the ith starting node on the rightmost column, there are at most 2 possibilities for the first path edge: a horizontal edge, or a diagonal edge. For $j \ge 1$, let $N_i(j)$ denote the number of minimal paths in $\mathcal{G}_{tri}(\ell, \ell)$ of length j starting at the ith node on the rightmost column. Note that the paths counted in $N_i(j)$ are not necessarily right-left paths, i.e. the last node in the path need not be on the leftmost column.

We use induction on j to show $N_i(j) \le 3^{j-1}$ for $j \ge 2$. We have already shown above the basis step $N_i(2) = 2 < 3$. For the induction step, suppose that $N_i(j) \le 3^{j-1}$ for some $j \ge 2$. We show that $N_i(j + 1) \le 3^j$.

Consider each path P of length j. We claim that there are at most 3 possible choices for adding a $(j+1)$th node $P(j+1)$ to P to create a minimal path P' of length $j+1$. Let $P(j-1)$ and $P(j)$ denote the $(j-1)$th node and jth node of P, respectively.

Suppose first that $P(j)$ is is a boundary node of $\mathcal{G}_{tri}(\ell, \ell)$ (i.e. it is on row 1 or row ℓ or column 1 or column ℓ). Then $P(j)$ has degree at most 4, and one of the 4 nodes adjacent to $P(j)$ is $P(j-1)$, so there are at most 3 possible choices for $P(j+1)$, as required.

Now suppose that $P(j)$ is an internal node of $\mathcal{G}_{tri}(\ell, \ell)$. Then $P(j)$ has degree 6, and one of the 6 nodes adjacent to $P(j)$ is $P(j-1)$. Hence there are at most 5 possibilities for $P(j+1)$. But it is easy to verify that 2 of those 5 adjacent nodes of $P(j)$ must also be adjacent to $P(j-1)$. Hence, neither of these 2 nodes can be chosen as $P(j+1)$ since the resulting path P' will not be minimal (indeed, if $P(j+1)$ is chosen adjacent to $P(j-1)$ then internal node $P(j)$ could be removed from P' without disconnecting it). So there are at most 3 possibilities for $P(j+1)$ to keep P' minimal.

We conclude that any minimal path P of length j can be extended in at most 3 ways to a minimal path P' of length $j+1$. It follows that $N_i(j+1) \leq 3N_i(j) \leq 3^j$, which completes the inductive step. Since there are ℓ possible starting nodes on the rightmost column, we get $N_P(k, \ell) \leq \ell \cdot 3^k$, which proves $\mu \leq 3$.

We now show how to improve the connective constant upper bound to $\mu \leq 2.948$. This improvement is based on the fact that the bound $\mu \leq 3$ only takes into account a '1 edge history' of the path to restrict the number of possible 'next' nodes by ruling out those which destroy the path minimality due to 3 node cycles. By taking into account m-edge history for larger $m > 1$, we can improve the bound by also ruling out m'-cycles for $m' > 3$. Here we examine the case of $m = 4$ edge history, ruling out $m' = 6$ node cycles, as well as $m' = 3$ node cycles (see [6] for some results with even larger m).

Consider the 6 node cycle C_6 in graph $\mathcal{G}_{tri}(\ell, \ell)$ shown in Fig. 4(a). For any minimal path P of length $j \geq 4$ whose last 4 edges match a sequence of 4 successive edges along C_6 (in either clockwise or anticlockwise sense, such as the 4 edges between nodes $P(j-4), P(j-3), P(j-2), P(j-1), P(j)$ in Fig. 4(a)), we have at most 2 possibilities (labelled n_1, n_2 in Fig. 4(a)) for choosing a $(j+1)$th node $P(j+1)$ to extend P to a minimal path P' of length $j+1$. This is because by minimality, only 3 possiblities are allowed for $P(j+1)$ to rule out 3-node cycles in P' (as shown above), and out of those 3 nodes, one (labelled n^* in Fig 4(a)) can be eliminated to rule out the 6-cycle C_6 from being contained in P'. This reduction from 3 to 2 possibilities for $P(j+1)$ when the last 4 edges of P match a sequence from C_6 will give us the improved upper bound on μ.

To analyse this improvement, let $S(j)$ denote the set of all minimal paths P in $\mathcal{G}_{tri}(\ell, \ell)$ of length j starting at the ith node on the rightmost column of $\mathcal{G}_{tri}(\ell, \ell)$. We partition $S(j)$ into 4 disjoint subsets $S_1(j), \ldots, S_4(j)$ according to the number of matches of the 4 last edges of P with a sequence of successive edges on C_6, namely:

- $S_4(j)$ denotes the subset of paths in $S(j)$ whose 4 last edges match a sequence of 4 successive edges along C_6 (in either clockwise or anticlockwise sense).
- For $k = 3, 2, 1$, $S_k(j)$ denotes the subset of paths in $S(j)$ which are not in $S_{k+1}(j)$, but whose k last edges match a sequence of k successive edges along C_6 (in either clockwise or anticlockwise sense).

For $j \geq 5$ and $k \in \{1, 2, 3, 4\}$, we say that a minimal path P of length j is in state k if $P \in S_k(j)$. We can now construct a finite state machine M whose state transition function specifies for each minimal path P of length j in state k, the possible 'next' state k' of a minimal path P' of length $j + 1$ formed by adding a $(j + 1)$th node to P. The state transition diagram of M is shown in Fig 4(b), where a label b on a transition from state k to k' indicates that there are b possibilities for the $(j + 1)$th node which lead to this state transition. For example, as shown in Fig 4(a), if P is in state 4, then there are 2 possibilities for node $P(j + 1)$: one (node labelled n_1) leads to a transition to state 1 (since no two successive edges in C_6 are in the same column), the other (node labelled n_2) leads to a transition to state 2 (since no three successive edges in C_6 are in the order 'horizontal, vertical, horizontal'). It is easy to verify that the same transition rule from state 4 holds for all paths P in state 4 (i.e. regardless of the particular sequence of 4 successive edges along C_6 which form the last 4 edges of P). The transition rules for the other three states are also easy to verify.

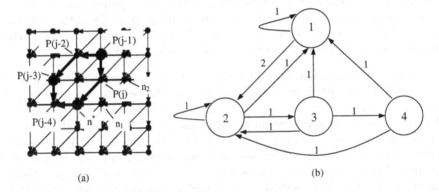

(a)   (b)

Fig. 4. (a) The 6 node cycle C_6 in $\mathcal{G}_{tri}(\ell, \ell)$ is shown in heavy black. (b) The state transition diagram of finite state machine M.

For $j \geq 5$ and $k \in \{1, 2, 3, 4\}$ let $N_k(j)$ denote the number of minimal paths (starting at ith node of the rightmost column of $\mathcal{G}_{tri}(\ell, \ell)$) of length j in state k. From the labelled state transition diagram of M in Fig 4(b), we immediately obtain the following recursive bound:

$$
\begin{pmatrix} N_1(j+1) \\ N_2(j+1) \\ N_3(j+1) \\ N_4(j+1) \end{pmatrix} \leq A_M \cdot \begin{pmatrix} N_1(j) \\ N_2(j) \\ N_3(j) \\ N_4(j) \end{pmatrix}, \text{ where } A_M = \begin{pmatrix} 1 & 1 & 1 & 1 \\ 2 & 1 & 1 & 1 \\ 0 & 1 & 0 & 0 \\ 0 & 0 & 1 & 0 \end{pmatrix}. \tag{2}
$$

It follows from (2) that the vector $\mathbf{N}(j) \stackrel{\text{def}}{=} [N_1(j) \; N_2(j) \; N_3(j) \; N_4(j)]^T$ satisfies

$$\mathbf{N}(j) \leq A_M^{j-5}\mathbf{N}(5) \tag{3}$$

for $j \geq 5$. The matrix A_M can be diagonalised into the form $A_M = Q \cdot D \cdot Q^{-1}$, where Q is a 4×4 invertible matrix having the eigenvectors of A_M as its columns, and D is a 4×4 diagonal matrix having the 4 eigenvalues $\lambda_1, \ldots, \lambda_4$ of A_M on the diagonal. Note that $A_M^{j-5} = Q \cdot D^{j-5} \cdot Q^{-1}$, and D^{j-5} is a diagonal matrix with diagonal elements λ_k^{j-5} for $k = 1, \ldots, 4$. Plugging into (3) and adding up the components of $\mathbf{N}(j)$, we get the following upper bound on the number $N_P(j) = N_1(j) + \cdots N_4(j)$ of minimal paths of length j, starting at the ith node in the rightmost column of $\mathcal{G}_{tri}(\ell, \ell)$:

$$N_P(j) \leq c_1\lambda_1^{j-5} + c_2\lambda_2^{j-5} + c_3\lambda_3^{j-5} + c_4\lambda_4^{j-5}, \tag{4}$$

where the constants $c_1, \ldots, c_4$ are determined from (3) by $\mathbf{N}(5)$ and the eigenvector matrix Q. It follows that $N_P(j) = O(\lambda^j)$, where $\lambda \stackrel{\text{def}}{=} \max_k |\lambda_k|$ is the largest eigenvalue magnitude of A_M. Numerical computation shows that $\lambda \leq 2.948$, and hence (considering the ℓ possible starting nodes on the rightmost column of $\mathcal{G}_{tri}(\ell, \ell)$), the claimed bound $N_P(k, \ell) \leq c(\mu) \cdot \ell \cdot \mu^k$ with $\mu = \lambda \leq 2.948$ follows, for some constant $c(\mu)$. □

Remark 7. Our terminology *connective constant* for μ comes from similar (although not identical) constants defined in combinatorial studies of the 'self avoiding walk' in a lattice [14,16]. However, the particular connective constant μ which arises in our work seems to not have been previously studied.

Remark 8. We have done some preliminary numerical eigenvalue computations using MATLAB with larger values of the 'edge history' parameter m on the path, extending our method for proving Lemma 7 (refer to [6] for more details). Using $m = 8$ we obtained the improved bound $\mu \leq 2.913$, although we are not yet certain about the accuracy of these MATLAB computations. We believe the efficient techniques from [14,16] can be useful to further improve our numerical computed upper bound on μ by using even larger values of the 'edge history' on the path. Also, our method of bounding μ does not take into account the restriction that the paths of length k are right-left paths, so further improvements might result by taking this restriction into account.

Now we are ready to prove the following result.

Theorem 3. *Let $\mu, c(\mu)$ denote the connective constants of $\mathcal{G}_{tri}(\ell, \ell)$ (see Lemma 7). For any real constant $R > \mu$, if $t \leq n/R$, there exists a Weakly t-Reliable n-Colouring for graph $\mathcal{G}_{tri}(\ell, \ell)$ for some $\ell = O(n)$. Moreover, for any constant $\delta > 0$, the probability p that the random n-Colouring C_{rand} is not Weakly t-Reliable is upper bounded by δ if we choose*

$$\ell \geq b \cdot \frac{\log\binom{n}{t}}{\log(R/\mu)},$$

for a constant b satisfying

$$b - \left(\frac{3}{\log R} \right) \log b \geq 1 + \frac{\log \left(2c(\mu)\delta^{-1} \left(\frac{\log R}{\log(R/\mu)} \right)^3 \right)}{\log R}. \tag{5}$$

Proof. Fix a t-colour subset I. We upper bound the probability $p(I)$, that if all ℓ^2 node colours of $\mathcal{G}_{tri}(\ell, \ell)$ are chosen uniformly and independently at random from $[n]$, the colouring C_{rand} is not Weakly t-Reliable, i.e. either an I-avoiding top-bottom path P_x doesn't exist, or an I-avoiding right-left path P_y doesn't exist.

Suppose that for a given colouring C, an I-avoiding top-bottom path P_x doesn't exist. This implies that the set $S(C)$ of graph nodes with colours in I must form a *top-bottom cutset*, which is defined as follows.

Definition 9 (Cutset/Minimal Cutset). *A set of nodes S in $\mathcal{G}_{tri}(\ell, \ell)$ is called a top-bottom cutset (resp. right-left cutset) if all top-bottom paths (resp. right-left paths) in $\mathcal{G}_{tri}(\ell, \ell)$ pass via a node in S. A cutset S is called minimal if removing any node from S destroys the cutset property.*

Note that the top-bottom cutset $S(C)$ must contain a minimal top-bottom cutset. The following intuitively obvious lemma shows that in order to count the minimal top-bottom cutsets of $\mathcal{G}_{tri}(\ell, \ell)$ it is enough to look at all minimal right-left paths in $\mathcal{G}_{tri}(\ell, \ell)$. Its formal proof can be found in [6].

Lemma 8 (Minimal Cutsets are Minimal Paths). *A set of nodes S in $\mathcal{G}_{tri}(\ell, \ell)$ is a minimal top-bottom cutset (resp. right-left cutset) if and only if it is a minimal right-left path (resp. top-bottom path).*

By Lemma 8, we conclude that if an I-avoiding top-bottom path doesn't exist for a colouring C then $S(C)$ contains a minimal right-left path $P_{c,x}$. Since $P_{c,x}$ is a subset of $S(C)$, its nodes only have colours in I. So, over the random choice of colouring C_{rand}, the probability that an I-avoiding top-bottom path doesn't exist is equal to the probability $p_x(I)$ that there exists a minimal right-left path $P_{c,x}$ whose node colours are all in t-collusion I.

Let $N_P(k, \ell)$ denote the total number of minimal right-left paths in $\mathcal{G}_{tri}(\ell, \ell)$ of length k. Since node colours are chosen independently and uniformly in $[n]$, each such path has probability $(t/n)^k$ to have all its node colours in I. It is clear that $\ell \leq k \leq \ell^2$. So, summing over all possible path lengths, we get the following upper bound: $p_x(I) \leq \sum_{k=\ell}^{\ell^2} N_P(k, \ell)(t/n)^k$. By symmetry, a similar argument gives the same upper bound on the probability $p_y(I)$ that a right-left I-avoiding path P_y does not exist. So we get the following upper bound on the probability $p(I)$ that either I-avoiding top-bottom path doesn't exist or an I-avoiding right-left path doesn't exist for each fixed t-subset I: $p(I) \leq 2 \sum_{k=\ell}^{\ell^2} N_P(k, \ell)(t/n)^k$. Finally, taking a union bound over all $\binom{n}{t}$ possible t-colour subsets I, we get an upper bound on the probability p that the colouring C_{rand} is not Weakly t-Reliable of the form $p \leq 2 \sum_{k=\ell}^{\ell^2} N_P(k, \ell)(t/n)^k \binom{n}{t}$. Using the bound on $N_P(k, \ell)$ from Lemma 7, we get

$$p \leq 2c(\mu)\ell^3(\mu t/n)^\ell \binom{n}{t}. \tag{6}$$

Since $n/t \geq R > \mu$, it is clear that this upper bound on p is less than 1 for sufficiently large ℓ. In fact, it suffices to take $\ell = O(\log(\binom{n}{t})/\log(n/(\mu t))) = O(n)$, as claimed. Now suppose we fix $\delta > 0$ and we want to find a lower bound on ℓ such that the error probability $p \leq \delta$. From (6) and using $n/t \geq R$ we see that $p \leq \delta$ is satisfied as long as

$$\ell \log(R/\mu) - 3\log(\ell) \geq \log(2c(\mu)N\delta^{-1}), \qquad (7)$$

where $N = \binom{n}{t}$. Take $\ell = b\log(N)/\log(R/\mu)$. Plugging this choice of ℓ into (7), and using the fact that $N \geq \binom{\lceil R \rceil}{1} \geq R$ for all $n \geq R$ (since $N = \binom{n}{n/R}$ increases monotonically with n), we conclude that (7) is satisfied if the constant b is sufficiently large such that (5) holds. This completes the proof. □

Combining Theorem 3 (applied with $n' = R \cdot t \leq n$ colours from $[n]$ for constant $R > \mu$) with Lemmas 3, 4, 6 and 7, we have

Corollary 2. *For any constant $R > 2.948$, if $t \leq n/R$, there exists a black box t-private protocol for f_G with communication complexity $O(nt^2)$ group elements. Moreover, for any $\delta > 0$, we can construct a probabilistic algorithm, with run-time polynomial in n and $\log(\delta^{-1})$, which outputs a protocol Π for f_G such that the communication complexity of Π is $O(nt^2 \log^2(\delta^{-1}))$ group elements and the probability that Π is not t-private is at most δ.*

Remark 9. Our computational experiments indicate that $t > n/2.948$ can be achieved with moderate values of ℓ – for example, for $n = 24$, $t = 11$ (i.e. $t \approx n/2.182$), we found a t-Reliable n-Colouring of $\mathcal{G}_{tri}(\ell, \ell)$ with $\ell = 350$, which is much smaller than $\binom{n}{t} \approx 2.5 \cdot 10^6$.

4.5 Generalisations and Other Results

General functions over G. Some applications may require n-party computation of more general functions over G (using only the group operation) instead of f_G. The most general such function is of the form $f'_G(x_1, \ldots, x_m) = x_1 \ldots, x_m$, where $m \geq n$ and each of the n parties holds one or more x_i's. Our reduction from Section 4.2 (and hence all our protocols) trivially extends to this most general case in the natural way.

General adversary structures. One may also consider more general adversary structures in place of the t-threshold structure. With the exception of our second construction in Section 4.4, all other results in the paper trivially generalise to the case of a Q^2 adversary structure $\mathcal{A}$, in which no pairwise union of collusions in $\mathcal{A}$ covers all n parties [11]. In particular, the generalisation of the first construction in Section 4.4 has communication complexity $O(n|\mathcal{A}|^2)$ group elements.

More efficient protocols for small t. For the cases $t \in \{1, 2\}$, we have managed to design explicit t-private black-box protocols for f_G with linear communication complexity ($O(n)$ group elements) and optimal collusion resistance. These protocols and their analysis can be found in [6]. We have also implemented a computer program for finding t-Reliable n-Colourings of a given graph, with which one can easily construct efficient protocols for small values of n, t (avoiding the error probability δ of Theorem 3).

5   Conclusions

We showed how to design black-box t-private protocols for computing the n-product function over any finite group by reducing the problem to a combinatorial graph colouring problem, using tools from communication security [7]. Our work raises some interesting combinatorial questions. For example, for our PDAG $\mathcal{G}_{tri}(\ell, \ell)$, what is the shape of the 'tradeoff' curve $R_{max}(\ell)$ relating the maximal achievable (using a suitable colouring) secure collusion resistance $R_{max} = t/n$ to the graph size ℓ? (we showed that $R_{max}(\ell) \geq 1/2.948$ for $\ell = O(t)$ and $R_{max}(\ell) = 1/2$ for $\ell \geq \binom{2t+1}{t}$). More generally, what is the largest collusion resistance achievable with an admissible PDAG of size polynomial in n, and what kind of PDAG achieves this optimum? There are also interesting cryptographic questions. First, can our black-box protocols be efficiently strengthened to yield black-box protocols secure against *active* adversaries? Second, can the communication complexity $O(nt^2)$ of our t-private protocols be reduced further? Third, does there exist an efficient (run-time polynomial in n) *deterministic* algorithm to generate a Weakly t-Reliable n-Colouring of $\mathcal{G}_{tri}(\ell, \ell)$ (or some other admissible PDAG) given n, t as input?

Acknowledgements. This research was supported by ARC research grants DP0451484, DP0558773, DP0663452 and DP0665035. Ron Steinfeld's research was supported in part by a Macquarie University Research Fellowship (MURF). Huaxiong Wang's research was supported in part by the Singapore Ministry of Education grant (T206B2204). Yvo Desmedt is grateful for the research visits to Macquarie University. The authors also thank Chris Charnes and Scott Contini for helpful discussions about this work.

References

1. Alon, N., Spencer, J.: The Probabilistic Method. Wiley-Interscience, New York (2000)
2. Bar-Ilan, J., Beaver, D.: Non-Cryptographic Fault-Tolerant Computing in a Constant Number of Rounds of Interaction. In: Symposium on Principles Of Distributed Computing (PODC), pp. 201–209. ACM Press, New York (1989)
3. Ben-Or, M., Goldwasser, S., Wigderson, A.: Completeness Theorems for Non-Cryptographic Fault-Tolerant Distributed Computation. In: Proc. 20-th STOC, pp. 1–10. ACM Press, New York (1988)
4. Chaum, D., Crépeau, C., Damgård, I.: Multiparty unconditionally secure protocols. In: Proceedings of the twentieth annual ACM Symp. Theory of Computing, STOC, May 2–4, 1988, pp. 11–19. ACM Press, New York (1988)

5. Cramer, R., Fehr, S., Ishai, Y., Kushilevitz, E.: Efficient Multi-Party Computation Over Rings. In: Biham, E. (ed.) EUROCRPYT 2003. LNCS, vol. 2656, pp. 596–613. Springer, Heidelberg (2003)

6. Desmedt, Y., Pieprzyk, J., Steinfeld, R., Wang, H.: On Secure Multi-Party Computation in Black-Box Groups. Full version of this paper(2007), Available at http://www.comp.mq.edu.au/~rons/

7. Desmedt, Y., Wang, Y., Burmester, M.: A Complete Characterization of Tolerable Adversary Structures for Secure Point-to-Point Transmissions. In: Deng, X., Du, D.-Z. (eds.) ISAAC 2005. LNCS, vol. 3827, pp. 277–287. Springer, Heidelberg (2005)

8. Diffie, W., Hellman, M.: New Directions in Cryptography. IEEE Trans. on Information Theory 22, 644–654 (1976)

9. ElGamal, T.: A Public Key Cryptosystem and a Signature Scheme Based on Discrete Logarithms. IEEE Tran. Info. Theory, IT 31(4), 469–472 (1985)

10. Goldreich, O.: Foundations of Cryptography, Volume II. Cambridge University Press, Cambridge (2004)

11. Hirt, M., Maurer, U.: Complete Characterization of Adversaries Tolerable in Secure Multi-Party Computation (Extended Abstract). In: Symposium on Principles Of Distributed Computing (PODC), pp. 25–34. ACM Press, New York (1997)

12. Kushilevitz, E.: Privacy and Communication Complexity. SIAM J. on Discrete Mathematics 5(2), 273–284 (1992)

13. Magliveras, S., Stinson, D., van Trung, T.: New approaches to Designing Public Key Cryptosystems using One-Way Functions and Trapdoors in Finite Groups. Journal of Cryptology 15, 285–297 (2002)

14. Noonan, J.: New Upper Bounds for the Connective Constants of Self-Avoiding Walks. Journal of Statistical Physics 91(5/6), 871–888 (1998)

15. Paeng, S., Ha, K., Kim, J., Chee, S., Park, C.: New Public Key Cryptosystem Using Finite Non Abelian Groups. In: Kilian, J. (ed.) CRYPTO 2001. LNCS, vol. 2139, pp. 470–485. Springer, Heidelberg (2001)

16. Pönitz, A., Tittmann, P.: Improved Upper Bounds for Self-Avoiding Walks in $\mathbb{Z}^d$. The Electronic Journal of Combinatorics 7 (2000)

17. Rivest, R.L., Shamir, A., Adleman, L.: A Method for Obtaining Digital Signatures and Public-Key Cryptosystems. Communications of the ACM 21(2), 120–128 (1978)

18. Shamir, A.: How To Share a Secret. Communications of the ACM 22, 612–613 (1979)

19. Shor, P.: Polynomial-Time Algorithms for Prime Factorization and Discrete Logarithms on a Quantum Computer. SIAM J. Comp. 26(5), 1484–1509 (1997)

A Note on Secure Computation of the Moore-Penrose Pseudoinverse and Its Application to Secure Linear Algebra

Ronald Cramer[1,2], Eike Kiltz[1,*], and Carles Padró[3,**]

[1] Cryptology and Information Security Research Theme
CWI Amsterdam
The Netherlands
{cramer,kiltz}@cwi.nl
[2] Mathematical Institute
Leiden University
The Netherlands
[3] Department of Applied Mathematics IV
Universitat Politècnica de Catalunya
Barcelona, Spain
cpadro@ma4.upc.edu

Abstract. This work deals with the communication complexity of secure multi-party protocols for linear algebra problems. In our model, complexity is measured in terms of the number of secure multiplications required and protocols terminate within a constant number of rounds of communication.

Previous work by Cramer and Damgård proposes secure protocols for solving systems $Ax = b$ of m linear equations in n variables over a finite field, with $m \leq n$. The complexity of those protocols is n^5.

We show a new upper bound of $m^4 + n^2 m$ secure multiplications for this problem, which is clearly asymptotically smaller. Our main point, however, is that the advantage can be substantial *in case m is much smaller than n*. Indeed, if $m = \sqrt{n}$, for example, the complexity goes down from n^5 to $n^{2.5}$.

Our secure protocols rely on some recent advances concerning the computation of the Moore-Penrose pseudo-inverse of matrices over fields of positive characteristic. These computations are based on the evaluation of a certain characteristic polynomial, in combination with variations on a well-known technique due to Mulmuley that helps to control the effects of non-zero characteristic. We also introduce a new method

* Supported by the research program Sentinels (http://www.sentinels.nl). Sentinels is being financed by Technology Foundation STW, the Netherlands Organization for Scientific Research (NWO), and the Dutch Ministry of Economic Affairs.
** Supported by the Spanish Ministry of Education and Science under projects TIC2003-00866 and TSI2006-02731. This work was done while this author was in a sabbatical stay at CWI, Amsterdam. This stay was funded by the *Secretaría de Estado de Educación y Universidades* of the Spanish Ministry of Education and Science.

A. Menezes (Ed.): CRYPTO 2007, LNCS 4622, pp. 613–629, 2007.

for secure polynomial evaluation that exploits properties of Chebychev polynomials, as well as a new secure protocol for computing the characteristic polynomial of a matrix based on Leverrier's lemma that exploits this new method.

1   Introduction

This paper deals with *secure multi-party computation* (MPC), that is, with the scenario in which n players want to compute an agreed function of their secret inputs in such a way that the correct result is obtained but no additional information about the inputs is released. These requirements should be achieved even in the presence of an *adversary* who is able to corrupt some players. The power of a *passive* adversary is limited to see all internal data of the corrupted adversaries, while an *active* one can control their behavior.

Multi-party computation protocols can be classified according to which model of communication is considered. In the *cryptographic* model, first considered in [21,11], the adversary can see all messages in the network and the security must rely on some computational assumption. Unconditional security can be achieved if the existence of a private channel between every pair of participants is assumed. This is the *information-theoretic* model introduced in [5,6]. It is well known that in both models any functionality can be securely evaluated — if evaluating the functionality is efficient, so is the secret multi-party protocol. However, generic solutions may need polynomial many rounds of communication between the participating players, whereas in practise one wants the round complexity to be as small as possible, preferably constant.

For conditionally secure multi-party protocols in the cryptographic model, every probabilistic polynomial-time functionality can be efficiently and privately evaluated in a constant number of communication rounds [22,3]. The situation is completely different for unconditionally secure multi-party protocols in the information-theoretic model. Up to now it is not known yet which class of functions can be efficiently computed in a constant number of rounds. Some progress in the direction of solving that question was made in [1,9,12,13,2] but, for instance it is not even known if all functions in basic classes like NC can be securely evaluated in constant rounds.

For specific functions of interest from linear algebra, Cramer and Damgård [7] proposed constant round multi-party computation protocols in the information-theoretic model. Among their considered functions are the determinant, the characteristic polynomial, the rank of a matrix, and solving a linear system of equations. The advantage with the approach from [7] is that all protocols could be tailor-made to the nature of the specific problem and, in contrast to the generic solutions, did not have to rely on circuit-based secure gate evaluation techniques.

1.1   Our Results

This work deals with the communication complexity of secure multi-party protocols for linear algebra problems. In our model, complexity is measured in terms of the number of secure multiplications required and protocols terminate within a constant number of rounds of communication.

Assuming a model in which constant round protocols for basic arithmetic operations are given as usual, previous work by Cramer and Damgård proposes secure protocols for solving systems $Ax = b$ of m linear equations in n variables over a finite field, with $m \leq n$. The complexity of those protocols is n^5. Since a solution in [7] could only be obtained for square matrices the general case of non-square matrices had to be reduced to solving linear systems for the case of an $n \times n$ matrix which is potentially huge compared to the original $m \times n$ matrix A. The protocol for the latter problem basically reduces to computing n times the characterstic polynomial which is shown in [7] to be securely computable in constant rounds and with (roughly) n^4 complexity (n^4 calls to the multiplication protocol). Therefore the overall complexity of the proposed protocol to solve the linear system $Ax = b$ is n^5.

We show a new upper bound of $m^4 + n^2 m$ secure multiplications for this problem, which is clearly asymptotically smaller. Our main point, however, is that the advantage can be substantial *in case m is much smaller than n*. Indeed, if $m = \sqrt{n}$, for example, the complexity goes down from n^5 to $n^{2.5}$.

As a concrete motivating application we consider the secure multi-party variant of the travelling salesman problem from combinatorial optimization. Given a number of t cities and the costs of travelling from any city to any other city, what is the cheapest round-trip route that visits each city exactly once and then returns to the starting city?[1] In a multi-party scenario the participating players may want to keep the travelling cost between two cities belonging to "their territory" secret such that only the concrete round-trip is revealed to everybody. It is well known [20, Vol 2, Chap. 58.4] that this problem can be reduced using integer linear programming to simultaneously solving two systems of linear equations, each of size $m \times n$, where $n = 2t \cdot 2^m \approx 2^m$ and $m \leq t^2$ is the number of edges in the graph representing the cost-matrix between the t cities. Hence, in this (admittingly extreme) example complexity of our protocol is $\approx (2^m)^2$, compared to the $(2^m)^5$ protocol from [7].

Our secure protocols rely on some recent advances concerning the computation of the Moore-Penrose pseudo-inverse of matrices over fields of positive characteristic. These computations are based on the evaluation of a certain characteristic polynomial, in combination with variations on a well-known technique due to Mulmuley that helps to control the effects of non-zero characteristic. We also introduce a new method for secure polynomial evaluation that exploits properties of Chebychev polynomials, as well as a new secure protocol for computing the characteristic polynomial of a matrix based on Leverrier's lemma that exploits this new method. These techniques may be of separate interest, and are central to our claimed improvements.

Below we give a more detailed overview of the techniques used. If A is an $n \times m$ matrix over a field $\mathbb{K}$, a *pseudoinverse* of A is any $m \times n$ matrix B such

---

[1] Since the travelling salesman problem is known to be NP-complete, for the purpose of this motivating example one may think of a small amount of cities t.

that $ABA = A$ and $BAB = B$. Note that in case A is a non-singular square matrix then $B = A^{-1}$. A linear system of equations $Ax = b$ can be easily solved if a pseudoinverse of A is given. First of all, the system has a solution if and only if $ABb = b$. In this case, $x_0 = Bb$ is a particular solution and, since the columns of the matrix $I_m - BA$ span $\ker A$, the general solution of the system is obtained.

Our secure MPC protocol to solve linear systems of equations applies the results and techniques from [10] about using of the Moore-Penrose pseudoinverse for solving linear systems of equations over arbitrary fields. Specifically, there is a polynomial that, evaluated on the Gram matrix $G = A^\top A$, (where $A^\top$ denotes the transpose of A) makes it possible to efficiently compute in MPC the Moore-Penrose pseudoinverse of A. The polynomial in turn is derived from the characteristic polynomial of $G = A^\top A$. Here our secure polynomial evaluation protocol based on Chebyshev polynomials can be used to perform the secure evaluation.

Nevertheless, the Moore-Penrose pseudoinverse of a matrix A exists if and only if the subspaces $\ker A$ and $\operatorname{Im} A$ have trivial intersection with their orthogonals, and unfortunately this may not be the case if the field has positive characteristic. This problem is solved by using some techniques from [10], based on results by Mulmuley [17]. Namely, there exists a random invertible diagonal matrix that, with high probability, transforms the matrix A into a matrix A' having the required properties on the subspaces $\ker A'$ and $\operatorname{Im} A'$.

Computing the Moore-Penrose pseudoinverse in particular involves secure evaluation of a public (or secret) polynomial in a secret field element (or a secret matrix). Motivated by this and maybe of independent interest, we present a constant round MPC protocol for the above task. If the field element (or the matrix) is guaranteed to be invertible this can be done using the well-known technique of unbounded fan-in multiplication [1]. In the general case of non-zero field elements a generic framework from [1] can be applied. However, the latter technique boosts the complexity of the resulting protocol from linear to quadratic in the degree d of the polynomial. One the other hand, if one admits some small probability of information leakage then the protocol can be made linear in d using certain randomization techniques.

We present an alternative protocol for the same task which is perfectly secure and has complexity linear in the degree d. The basic idea is explained in the following. Consider a matrix $M(x)$ whose entries are polynomials over a finite field $\mathbb{F}$ and such that $M(\alpha)$ is invertible for every $\alpha \in \mathbb{F}$. Specifically, we present a 2×2 matrix $M(x)$ such that in the top-left entry of $M(2x)M^{i-1}(x)$ we have the ith *Chebyshev polynomial* $T_i(x)$. Since the first $d+1$ Chebyshev polynomials $\{T_i(x)\}_{0 \le i \le d}$ form a basis of the polynomials of degree at most d, every polynomial of degree d is a linear combination the Chebyshev polynomials. Therefore we can securely compute, even if α may be zero, $F(\alpha)$ for every polynomial $F(x)$ with degree at most d by using the unbounded fan-in multiplication protocol to compute the needed powers of the matrix $M(\alpha)$.

1.2 Related Work

Nissim and Weinreb [19] also considered the problem of securely solving a set of linear equations in the *computational two-party model*, focusing on low (nearly optimal) communication complexity. Their protocol needs $O(n^{0.275})$ rounds of communication which was later improved to $O(\log n)$ [15].

2 Preliminaries

2.1 The Model

We assume that n parties are connected by perfectly secure channels in a synchronous network. Let $\mathbb{F}_p$ denote the finite field with p elements where p is a prime power. We will assume throughout that p is large enough because our protocols can only guarantee security with a probability $1 - O(n^2/p)$, where n is the maximum number of rows or columns in the matrices appearing in the linear systems of equations.

By $[a]$ we denote a secret sharing of $a \in \mathbb{F}_p$ over $\mathbb{F}_p$. We assume that the secret-sharing scheme allows to compute a sharing $[a+b]$ from $[a]$ and $[b]$ without communication, and that it allows to compute $[ab]$ from $a \in \mathbb{F}_p$ and $[b]$ without communication; we write

$$[a + b] \leftarrow [a] + [b] \text{ and } [ab] \leftarrow a[b]$$

for these operations. The secret-sharing scheme should of course also allow to take a sharing $[c]$ and reveal the value $c \in \mathbb{F}_p$ to all parties; We write $c \leftarrow$ REVEAL$([c])$.

We also assume that the secret sharing scheme allows to compute a sharing $[ab]$ from $[a]$ and $[b]$ with unconditional security. We denote the multiplication protocol by MULT, and write

$$[ab] \leftarrow \text{MULT}([a], [b]) .$$

We will express the protocols' round complexities as the number of sequential rounds of MULT invocations — and their communication complexities as the overall number of MULT invocations. I.e., if we first run a copies of MULT in parallel and then run b copies of MULT in parallel, then we say that we have round complexity 2 and communication complexity $a + b$. Note that standard linear (verifiable) secret-sharing schemes have efficient constant-rounds protocols for multiplication.

For a matrix $A \in \mathbb{F}_p^{n \times m} = (A_{ij})_{1 \leq i \leq n, 1 \leq j \leq m}$ we will use $[A] = ([A_{ij}])_{1 \leq i \leq n, 1 \leq j \leq m}$ for a sharing of a matrix. For multiplication of two matrices $A \in \mathbb{F}_p^{n \times k}$, $B \in \mathbb{F}_p^{k \times m}$ of matching dimensions we simply write $[C] \leftarrow$ MULT$([A], [B])$, where $C = AB \in \mathbb{F}_p^{n \times m}$. Matrix multiplication has to be understood componentwise and can be carried out in one round and nmk parallel invocations of the multiplication protocol.

For our protocols to be actively secure, the secret sharing scheme and the multiplication protocol should be actively secure. This in particular means that the adversary structure must be Q2. By the adversary structure we mean the set $\mathcal{A}$ of subsets $C \subset \{1, \ldots, n\}$ which the adversary might corrupt; It is Q2 if it holds for all $C \in \mathcal{A}$ that $\{1, \ldots, n\} \setminus C \notin \mathcal{A}$.

2.2   Some Known Techniques

The following known techniques will be of importance later on.

Random Elements. The parties can share a uniformly random, unknown field element. We write $[a] \leftarrow \text{RAN}()$. This is done by letting each party P_i deal a sharing $[a_i]$ of a uniformly random $a_i \in \mathbb{F}_p$. Then the parties compute the sharing $[a] = \sum_{i=1}^{n} [a_i]$. The communication complexity of this is given by n dealings, which we assume is upper bounded by the complexity of one invocation of the multiplication protocol.

If passive security is considered, this is trivially secure. If active security is considered and some party refuses to contribute with a dealing, the sum is just taken over the contributing parties. This means that the sum is at least taken over a_i for $i \in H$, where $H = \{1, \ldots, n\} \setminus C$ for some $C \in \mathcal{A}$. Since $\mathcal{A}$ is Q2 it follows that $H \notin \mathcal{A}$. So, at least one honest party will contribute to the sum, implying randomness and privacy of the sum.

Random Invertible Elements. Using [1] the parties can share a uniformly random, unknown, invertible field element along with a sharing of its inverse. We write $([a], [a^{-1}]) \leftarrow \text{RAN}^*()$ and it proceeds as follows: $[a] \leftarrow \text{RAN}()$ and $[b] \leftarrow \text{RAN}()$. $[c] = \text{MULT}([a], [b])$. $c \leftarrow \text{REVEAL}([c])$. If $c \notin \mathbb{F}_p^*$, then abort. Otherwise, proceed as follows: $[a^{-1}] \leftarrow c^{-1}[b]$. and output $([a], [a^{-1}])$.

The correctness is straightforward. As for privacy, if $c \in \mathbb{F}_p^*$, then (a, b) is a uniformly random element from $\mathbb{F}_p^* \times \mathbb{F}_p^*$ for which $ab = c$, and thus a is a uniformly random element in $\mathbb{F}_p^*$. If $c \notin \mathbb{F}_p^*$, then the algorithm aborts. This happens with probability less than $2/p$. The complexity is (at most) 2 rounds and 3 invocations of MULT.

Unbounded Fan-In Multiplication. Using the technique from [1] it is possible to do unbounded fan-in multiplication in constant rounds. For the special case where we compute all "prefix products" $\prod_{i=1}^{m} a_i$ ($m = 1, \ldots, \ell$), we write

$$([a_1], \ldots, [(a_1 a_2 \cdots a_\ell)]) \leftarrow \text{MULT}^*([a_1], \ldots, [a_\ell]) .$$

In the following, we only need the case where we have inputs $[a_1], \ldots, [a_\ell]$, where $a_i \in \mathbb{F}_p^*$. For $1 \leq i_0 \leq i_1 \leq \ell$, let $a_{i_0, i_1} = \prod_{i=i_0}^{i_1} a_i$. We are often only interested in computing $a_{1, \ell}$, but the method allows to compute any other a_{i_0, i_1} at the cost of one extra multiplication. For the complexity analysis, let A denote the number of a_{i_0, i_1}'s which we want to compute.

First run RAN^* $\ell + 1$ times in parallel, to generate $[b_0 \in_R \mathbb{F}_p^*], [b_1 \in_R \mathbb{F}_p^*], \ldots, [b_\ell \in_R \mathbb{F}_p^*]$, along with $[b_0^{-1}], [b_1^{-1}], \ldots, [b_\ell^{-1}]$, using 2 rounds and $3(\ell + 1)$ invocations of MULT. For simplicity we will use the estimate of 3ℓ invocations. Then for $i = 1, \ldots, \ell$ compute and reveal $[d_i] = \text{MULT}([b_{i-1}], [a_i], [b_i^{-1}])$, using 2 rounds and 2ℓ invocations of MULT. Now we have that $d_{i_0, i_1} = \prod_{i=i_0}^{i_1} d_i = b_{i_0-1}(\prod_{i=i_0}^{i_1} a_i) b_{i_1}^{-1} = b_{i_0-1} a_{i_0, i_1} b_{i_1}^{-1}$, so we can compute $[a_{i_0, i_1}] = d_{i_0, i_1} \text{MULT}([b_{i_0-1}^{-1}], [b_{i_1}])$, using 1 round and A invocations of MULT. The overall complexity is 5 rounds and $O(\ell+a)$ invocations of MULT.

The same concept generalizes to unbounded fan-in multiplication of matrices. Let shares $[M_i]$ of matrices $M_i \in \mathbb{F}_p^{m \times m}$ be given. Again we write

$$([M_1], \ldots, [(M_1 M_2 \cdots M_\ell)]) \leftarrow \text{MULT}^*([M_1], \ldots, [M_\ell]) .$$

for the special case where we compute all "prefix matrix products" $\prod_{i=1}^{k} M_i$ $(k = 1, \ldots, \ell)$. The above protocol generalizes to the matrix case, where a random invertible field element now translates to a random invertible matrix. Random invertible matrices are created using the same the method as generating a shared random invertible field element.

Equality. We define the equality function $\delta : \mathbb{F}_p \rightarrow \mathbb{F}_p$ as $\delta(x) = 0$ if $x = 0$ and $\delta(x) = 1$ otherwise. Given a shared value $[x]$, there exists a protocol [8,18] that computes, in a constant number of rounds and using $O(\log p)$ calls to the multiplication protocol MULT, shares $[\delta(x)]$. We write $[y] \leftarrow \text{EQ}([x])$.

3   Secure Polynomial Evaluation

In this section we are interested in the natural problem of secure polynomial evaluation: the players hold a public (shared) polynomial F of maximal public degree d and a shared field element x. The goal is to securely evaluate F in x, i.e. to compute shares $[F(x)]$.

Based on known techniques [1,4] the latter shares can be computed in constant rounds and quadratic complexity, i.e. the protocol makes $O(d^2)$ calls to the multiplication protocol.

Surprisingly, as we will show in this section, Chebyshev polynomials of the first and the second kind can be used as a mathematical tool to bring the complexity of the above problem down to linear. We will first consider the simpler case where the polynomial $F(X)$ is publicly known and later reduce the case of a shared polynomial to the latter one.

3.1   Known Solution

First we present a naïve protocol based on known techniques with linear complexity. Unfortunately, as we will see, the protocol leaks information for the interesting case when the polynomial gets evaluated at zero.

The protocol is given a shared value $[x]$, where $x \in \mathbb{F}_p^*$ and a public polynomial $F(X) = \sum_{i=0}^{d} a_i X^i$. The protocol's task is to compute shares $[F(x)]$. First, it computes $([x], [x^2], \ldots, [x^d]) \leftarrow \text{MULT}_p^*([x], \ldots, [x])$ and then the share $[F(x)]$

can be computed without interaction as $[F(x)] \leftarrow a_0 + a_1[x] + \sum_{i=2}^{d} a_i[x^i]$. The complexity is constant rounds and $6d = O(d)$ invocations of the multiplication protocol MULT. Privacy follows since we assumed $x \in \mathbb{F}_p^*$ and hence we can apply the protocol MULT* securely. On the other hand, if $x \notin \mathbb{F}_p^*$ then this fact will leak throughout the application of protocol MULT*.

As already done in [1], using a technique from [4], the general case (where the input may be equal to zero) can be reduced to unbounded fan-in multiplication of non-singular 3×3 matrices as we will sketch now. Later we will give an alternative protocol for the same task with improved running time. The main result from [4] states that every algebraic formula Φ of depth l can be expressed as the product of $O(4^l)$ non-singular 3×3 matrices over $\mathbb{F}_p$ (in the sense that the value $\Phi(x)$ can be read from the right top corner of the matrix product). Since any polynomial $F(X)$ of degree d can be expressed as an algebraic formula of depth $\log d$, $F(X)$ can be expressed as the product of $O(d^2)$ such non-singular 3×3 matrices. The appearing matrices are either one of five (known) constant matrices or are the identity matrix with x in the right upper corner. Using an efficient constant round protocol for multiplying non-singular constant size matrices we imply that there there exists a protocol that privately computes shares $[F(x)]$, where x may equal to zero. The protocol runs in a constant number of rounds and $O(d^2)$ invocations of MULT.

If we admit some small probability of information leakage we can get a $O(d)$ protocol for the same task as follows. First choose a random field element $[c] \leftarrow$ RAN() and compute the share $[x+c]$. Then compute $([x+c], [(x+c)^2], \ldots, [(x+c)^d]) \leftarrow \text{MULT}_p^*([x+c], \ldots, [x+c])$. This step is secure as long as $x+c \neq 0$ which happens with probability $1 - 1/p$ (over all coin tosses of the RAN protocol). Then open the share $[c]$ to obtain the field element c. Since the polynomials $(x+c)^i$ $(0 \leq i \leq d)$ form a basis for all polynomials of degree at most d we can compute $[F(x)]$ without interaction using $F(x) = \sum_{i=0}^{d} \lambda_i (x+c)^i$, where the coefficients λ_i can be computed by the players using only public information (including the value c). The protocol runs in a constant number of rounds and $O(d)$ invocations of MULT. However, it leaks information about x with probability $1/p$. In the remainder of this section we will develop a perfectly secure protocol in $O(d)$ invocations of MULT.

3.2   Chebyshev Polynomials

We use Chebyshev polynomials of the first kind which satisfy the linear recurrence

$$T_d(x) = 2xT_{d-1}(x) - T_{d-2}(x), \quad d \geq 2$$

with starting values $T_0(x) = 1$ and $T_1(x) = x$, and Chebyshev polynomials of the second kind

$$U_d(x) = 2xU_{d-1}(x) - U_{d-2}(x), \quad d \geq 2$$

with starting values $U_0(x) = 1$ and $U_1(x) = 2x$. For notational convenience we also set $T_d(x) = U_d(x) = 0$ for any $d < 0$. It is well known that the Chebyshev polynomials $T_i(x)$, $0 \leq i \leq d$ form a basis for all polynomials of degree at most

d. I.e., there exist coefficients $\lambda_i \in \mathbb{F}_p$ such that every polynomial F of degree at most d given in its monomial representation $F(x) = \sum_{i=0}^{d} a_i x^i$ can be expressed in the Chebyshev basis as

$$F(x) = \sum_{i=0}^{d} \lambda_i T_i(x) . \tag{1}$$

The coefficients λ_i only depend on the polynomial F, but not on x. (All λ_i's can be computed from the a_i's in $O(d^2 \log^2 p)$ bit operations using, for instance, the recursive formulas from [16].)

For $x \in \mathbb{F}_p$ define the 2×2 matrix $M(x)$ over $\mathbb{F}_p$ as

$$M(x) = \begin{pmatrix} x & -1 \\ 1 & 0 \end{pmatrix} ,$$

and note that since $\det(M(x)) = 1$, the matrix $M(x)$ is always non-singular.

Claim. The following identity holds for any integer $d \geq 1$:

$$M(x)M^{d-1}(2x) = \begin{pmatrix} T_d(x) & -T_{d-1}(x) \\ U_{d-1}(x) & -U_{d-2}(x) \end{pmatrix} . \tag{2}$$

We quickly prove the claim by induction over d. For $d = 1$ (2) is correct by definition. Now assume (2) holds for an integer $d \geq 1$. Then we have

$$M(x)M^d(2x) = M(x)M^{d-1}(2x) \cdot M(2x) = \begin{pmatrix} T_d(x) & -T_{d-1}(x) \\ U_{d-1}(x) & -U_{d-2}(x) \end{pmatrix} \cdot \begin{pmatrix} 2x & -1 \\ 1 & 0 \end{pmatrix} ,$$

$$= \begin{pmatrix} 2T_d(x) \cdot x - T_{d-1} & -T_d(x) \\ 2U_{d-1}(x) \cdot x - U_{d-2}(x) & -U_{d-1}(x) \end{pmatrix} .$$

This shows (2) for $d + 1$.

3.3 Perfectly Secure Polynomial Evaluation of a Shared Field Element

We now come to our improvement over the protocols from Section 3.1. We design an alternative protocol to evaluate a polynomial in a share with running time linear in the degree d (instead of quadratic). The protocol does not leak any information about the shared secret x. Using the results on the Chebychev polynomials from the last section a protocol to securely evaluate a given public polynomial $F \in \mathbb{F}_p[X]$ of degree d in a share $[x]$ is as follows: The players first locally create matrix-shares $[M(x)]$ and $[M(2x)]$ from the share $[x]$. Then they compute (component-wise and in parallel) matrix-shares $[M(x)M^{i-1}(2x)]$ for $1 \leq i \leq d$ by

$$([M(x)M(2x)], \ldots, [M(x)M^{d-1}(2x)]) \leftarrow \text{MULT}^*([M(x)], [M(2x)], \ldots, [M(2x)]).$$

Security is granted since $M(x)$ and $M(2x)$ are both non-singular. By Eq. (2), the share $[T_i(x)]$ can now be read in the upper left corner of the resulting matrices.

Once we are given shares of the Chebychev basis $\{T_i(x)\}_{1 \leq i \leq d}$ we can evaluate any given polynomial F of maximal degree d without interaction by computing $[F(x)] = \sum_{i=0}^{d} \lambda_i [T_i(x)]$. Here λ_i are the coefficients from (1) that are computed by each player in a precomputation phase. This leads to the following:

Proposition 1. *Let a set of ℓ public polynomials $F_i \in \mathbb{F}_p[X]$ be given, all of maximal degree d. There exists a multi-party protocol that, given shares $[x]$ (for any $x \in \mathbb{F}_p$, possibly $x = 0$), computes all shares $([F_1(x)], \dots, [F_\ell(x)])$. The protocol runs in constant rounds and $O(d)$ applications of the multiplication protocol.*

It is easy to see that the given techniques can be extended to evaluate a shared value x in a *shared* polynomial F, i.e. the shared F is given by shares of its coefficients $[a_i]$, $1 \leq i \leq d$. The protocol first computes shares $[x^i]$ for $1 \leq i \leq d$ with the methods from Proposition 1 (here the ith polynomial $F_i(X)$ is defined as $F_i(X) = X^i$). Then the polynomial F can be securely evaluated in x by first computing all shares $[a_i x^i]$ using d parallel applications of the multiplication protocol and finally summing the products all up.

Theorem 1. *Let a shared polynomial $[F(X)]$ of maximal degree d (i.e., shared field elements $[a_0], \dots, [a_d]$ such that $F(X) = \sum_{i=0}^{d} [a_i] X^i$) and a shared field element $[x]$ (for any $x \in \mathbb{F}_p$, possibly $x = 0$) be given. There exists a perfectly secure multi-party protocol that computes shares $[F(x)]$ in constant rounds and $O(d)$ applications of the multiplication protocol.*

3.4   Perfectly Secure Polynomial Evaluation of a Shared Matrix

In this section we generalize the results from the last sections to the case of evaluating a shared matrix in a known/shared polynomial. Let a share $[A]$ of a matrix $A \in \mathbb{F}_p^{m \times m}$ be given, together with a public polynomial $F(x)$ of degree d. We want to give a multi-party protocol that computed shares $[F(A)]$. With known techniques, similar to the finite field case from Section 3.1 this can be carried out using $O(d^2 m^3)$ applications of the multiplication protocol.

Analogously to Section 3.2, for an $m \times m$ matrix A we define the $2m \times 2m$ matrix $M(A)$ over $\mathbb{F}_p$ as

$$M(A) = \begin{pmatrix} A & -I_m \\ I_m & 0 \end{pmatrix},$$

where I_m is the $m \times m$ identity matrix. We note that since $\det(M(A)) = 1$, $M(A)$ is non-singular for each $A \in \mathbb{F}_p^{m \times m}$, including the special case of singular A. Then again the following identity is easy to show by induction over d:

$$M(A)M^{d-1}(2A) = \begin{pmatrix} T_d(A) & -T_{d-1}(A) \\ U_{d-1}(A) & -U_{d-2}(A) \end{pmatrix}.$$

Proposition 2. *Let a set of ℓ public polynomials $F_i \in \mathbb{F}_p[X]$ of maximal degree d and a shared $m \times m$ matrix $[A]$ be given. There exists a perfectly secure multi-party protocol that computes all shares $([F_1(A)], \dots, [F_l(A)])$ in constant rounds and $O(dm^3)$ applications of the multiplication protocol.*

Theorem 2. *Let a shared polynomial $[F(x)]$ of maximal degree d and a shared $m \times m$ matrix $[A]$ be given. There exists a perfectly secure multi-party protocol that computes shares $[F(x)]$ in constant rounds and $O(dm^3)$ applications of the multiplication protocol.*

4   Solving Linear Systems of Equations

In this section we provide the necessary mathematical framework for understanding our algorithm. In particular, we present here the probabilistic algorithm to solve linear systems of equations that will be implemented in Section 5 in a secure multi-party computation protocol. This algorithm is based on the methods presented in [10]. Specifically, we solve the linear system of equations $Ax = b$ by computing the Moore-Penrose pseudoinverse of the matrix A. Since we are dealing with finite fields, we have to use the results by Mulmuley [17] to avoid that certain subspaces have nontrivial intersection with their orthogonals.

4.1   Computing the Rank of a Matrix

Let $\mathbb{K}$ be a field. For every pair of vectors $u, v \in \mathbb{K}^n$, we notate $\langle u, v \rangle$ for the usual scalar product $\langle u, v \rangle = \sum_{i=1}^{n} u^i v^i$. If $V \subset K^n$ is a subspace, we notate $V^\perp = \{u \in \mathbb{K}^n : \langle u, v \rangle = 0 \text{ for every } v \in V\}$. Clearly, $\dim V^\perp = n - \dim V$. It is well known that $V^\perp \cap V = \{0\}$ if $\mathbb{K} = \mathbb{Q}$ or $\mathbb{K} = \mathbb{R}$. This does not hold in general if $\mathbb{K}$ has positive characteristic.

If A is an $n \times m$ matrix over the field $\mathbb{K}$, the *Gram matrix* of A is defined by $G(A) = A^\top A$, where $A^\top$ denotes the transpose of A. For every $i = 1, \dots, m$, we take the vector $u_i \in \mathbb{K}^n$ corresponding to the i-th column of A. Then, the entries of the Gram matrix are the scalar products of these vectors, that is, $G(A) = (\langle u_i, u_j \rangle)_{1 \leq i,j \leq m}$.

Consider the vector spaces $E = \mathbb{K}^m$ and $F = \mathbb{K}^n$ and let A be an $n \times m$ matrix over $\mathbb{K}$ representing a linear mapping $A \colon E \to F$. Then, the transpose matrix $A^\top$ corresponds to a linear mapping $A^\top \colon F \to E$ such that $\langle Ax, y \rangle = \langle x, A^\top y \rangle$ for every pair of vectors $x \in E$ and $y \in F$. Then, $\ker A^\top = (\operatorname{Im} A)^\perp$ and $\operatorname{Im} A^\top = (\ker A)^\perp$. The terminology we introduce in the following definition will simplify the presentation.

Definition 1. *A subspace $V \subset \mathbb{K}^n$ is said to be* suitable *if $V^\perp \cap V = \{0\}$. We say that a matrix A is* suitable *if $\operatorname{Im} A$ is a suitable subspace, that is, if $(\operatorname{Im} A)^\perp \cap \operatorname{Im} A = \{0\}$.*

Lemma 1. *Let A be an $n \times m$ matrix over $\mathbb{K}$ and let $G = A^\top A$ be its Gram matrix. Then A is a suitable matrix if and only if $\ker G = \ker A$.*

Proof. Observe that $\ker A \subset \ker G$. If A is suitable and $x \in \ker G$, then $Ax \in \operatorname{Im} A \cap \ker A^\top = \operatorname{Im} A \cap (\operatorname{Im} A)^\perp = \{0\}$. Conversely, if $\ker A = \ker G$ and $y = Ax \in \operatorname{Im} A \cap (\operatorname{Im} A)^\perp$, then $x \in \ker G = \ker A$, and hence $y = 0$.

Lemma 2. *Let A be an $n \times m$ matrix over $\mathbb{K}$ and suppose that A and $A^\top$ are both suitable matrices. Let $G = A^\top A$ and $H = AA^\top$ be the Gram matrices of A and $A^\top$, respectively. Then G and H are suitable matrices.*

Proof. Since G is a symmetric matrix, $\operatorname{Im} G = (\ker G)^\perp$. By applying Lemma 1, $\operatorname{Im} G = (\ker G)^\perp = (\ker A)^\perp = \operatorname{Im} A^\top$. Since $A^\top$ is suitable, $(\operatorname{Im} G)^\perp \cap \operatorname{Im} G = \{0\}$. Symmetrically, H is suitable as well.

Lemma 3. *Let G be a symmetric $m \times m$ matrix and assume G is suitable. Consider $P(X) = \det(XI_m - G) = X^m + a_1 X^{m-1} + \cdots + a_{m-1}X + a_m$, the characteristic polynomial of G. Then $\operatorname{rank} G = \max\{i : a_i \neq 0\}$.*

Proof. From Lemma 1, $\ker G^2 = \ker G$. If $r = \max\{i : a_i \neq 0\}$, the characteristic polynomial of G is of the form $P(X) = X^{m-r}Q(X)$ with $Q(0) \neq 0$. Then $\dim \ker G = \dim \ker G^{m-r} = m - r$, and hence $\operatorname{rank} G = r$.

From the previous lemmas the rank of the matrix A can be found by computing the characteristic polynomial of the Gram matrix $G(A) = A^\top A$. Nevertheless, we need that both A and $A^\top$ are suitable matrices. If we are dealing with a field with positive characteristic we cannot be sure that this is the case. We avoid this problem by applying a random transformation to the matrix A that, with high probability, produces a matrix with the same rank and verifying that property. This can be done by using Theorem 3, due to Mulmuley [17], and Propositions 3 and 4.

Theorem 3. *Consider the field $\mathbb{K}(x)$, a transcendental extension of the field $\mathbb{K}$, and the diagonal matrix $D_x = \operatorname{diag}(1, x, \ldots, x^{n-1})$, which defines a linear mapping $D_x \colon K(x)^n \to K(x)^n$. Then for every subspace $V \subset \mathbb{K}$, the subspace $V_x = D_x(V') \subset \mathbb{K}(x)^n$, where $V' \subset \mathbb{K}(x)^n$ is the natural extension of V, is suitable. As a consequence, for every $n \times m$ matrix A over the field $\mathbb{K}$, the matrix $D_x A$ (over the field $\mathbb{K}(x)$) is suitable.*

The proofs of the next two propositions will be given in the full version.

Proposition 3. *Let $\mathbb{K}$ be a finite field with $|\mathbb{K}| = p$. Consider the vector space $F = \mathbb{K}^n$ and a subspace $V \subset F$. For every $\alpha \in \mathbb{K}$, we consider the diagonal matrix $D_\alpha = \operatorname{diag}(1, \alpha, \alpha^2, \ldots, \alpha^{n-1})$. If an invertible element $\alpha \in \mathbb{K}^*$ is chosen uniformly at random, then the probability that the subspace $V_\alpha = D_\alpha(V) \subset F$ is suitable is at least $1 - 2n(n-1)/p$.*

Proposition 4. *Let $\mathbb{K}$ be a finite field with $|\mathbb{K}| = p$ and let A be an $n \times m$ matrix over the field $\mathbb{K}$. For every $\alpha \in \mathbb{K}$, take the diagonal matrices $D_{n,\alpha} = \operatorname{diag}(1, \alpha, \ldots, \alpha^{n-1})$ and $D_{m,\alpha} = \operatorname{diag}(1, \alpha, \ldots, \alpha^{m-1})$, and the matrix $A_\alpha = D_{n,\alpha} A D_{m,\alpha}$. Then the probability that both A_α and $A_\alpha^\top$ are suitable matrices if an invertible element $\alpha \in \mathbb{K}^*$ is chosen uniformly at random is at least $1 - (2/p)(n(n-1) + m(m-1))$.*

4.2 Moore-Penrose Pseudoinverse

Consider the vector spaces $E = \mathbb{K}^m$ and $F = \mathbb{K}^n$ and let A be an $n \times m$ matrix over $\mathbb{K}$ representing a linear mapping $A \colon E \to F$. A *pseudoinverse* of A is any $m \times n$ matrix $B \colon F \to E$ such that $ABA = A$ and $BAB = B$. Given two subspaces $V, W \subset E$, the notation $E = V \oplus W$ means that E is the *direct sum* of V and W, that is, $E = V + W$ and $V \cap W = \{0\}$.

There can exist many different pseudoinverses of a matrix. If B is a pseudoinverse of A, then $E = \operatorname{Im} B \oplus \ker A$ and $F = \operatorname{Im} A \oplus \ker B$. Moreover, for every pair of subspaces $V_1 \subset E$ and $V_2 \subset F$ such that $E = V_1 \oplus \ker A$ and $F = \operatorname{Im} A \oplus V_2$, there exists a unique pseudoinverse B of A such that $V_1 = \operatorname{Im} B$ and $V_2 = \ker B$. Finally, there is at most one pseudoinverse B of A such that AB and BA are symmetric matrices. This is the only pseudoinverse with $\operatorname{Im} B = (\ker A)^\perp$ and $\ker B = (\operatorname{Im} A)^\perp$. Of course, such a pseudoinverse exists if and only if $\ker A \subset E$ and $\operatorname{Im} A \subset F$ are suitable subspaces.

Definition 2. Let A be an $n \times m$ matrix corresponding to a linear mapping $A \colon E \to F$ such that $\ker A \cap (\ker A)^\perp = \{0\}$ and $\operatorname{Im} A \cap (\operatorname{Im} A)^\perp = \{0\}$, that is, A and $A^\top$ are suitable matrices. The *Moore-Penrose pseudoinverse* $A^\dagger$ of A is the unique pseudoinverse of A with $\operatorname{Im} A^\dagger = (\ker A)^\perp$ and $\ker A^\dagger = (\operatorname{Im} A)^\perp$. Actually, the Moore-Penrose pseudoinverse of A can be defined too as the unique $m \times n$ matrix $A^\dagger \colon F \to E$ such that $AA^\dagger A = A$, and $A^\dagger AA^\dagger = A^\dagger$, and $AA^\dagger$ and $A^\dagger A$ are symmetric matrices.

Observe that the Moore-Penrose pseudoinverse of A can be defined only if A and $A^\top$ are suitable matrices. Assume that we are in this situation. We consider $G = A^\top A$ and $H = AA^\top$, the Gram matrices of A and $A^\top$. From Lemma 2, G and H are suitable matrices with $\ker G = \ker A$ and $\ker H = \ker A^\top$.

We present next a useful expression for $A^\dagger$ in terms of the characteristic polynomial of H or the one of G. Let $f_0 \colon \operatorname{Im} A^\top \to \operatorname{Im} A$ be the linear mapping obtained from the restriction of $A \colon E \to F$ to $\operatorname{Im} A^\top$ and let $\pi \colon F \to \operatorname{Im} A$ be the orthogonal projection over $\operatorname{Im} A$. It is not difficult to check that f_0 is invertible and that $A^\dagger = f_0^{-1} \pi$. Consider $r = \operatorname{rank} A = \operatorname{rank} A^\top = \operatorname{rank} G = \operatorname{rank} H$. From Lemma 3, the characteristic polynomial of H is of the form $\det(X I_n - H) = X^n + a_1 X^{n-1} + \cdots + a_r X^{n-r}$ with $a_r \neq 0$. Moreover, the characteristic polynomial of G has the same coefficients as the one of H, that is, $\det(X I_m - G) = X^m + a_1 X^{m-1} + \cdots + a_r X^{m-r}$. Consider a vector $y \in F$ and take $z = H^r y + a_1 H^{r-1} y + \cdots + a_{r-1} H y + a_r y$. By applying Cayley-Hamilton and taking into account that $\ker H^2 = \ker H$, we get that $z \in \ker H$. Then, $y = a_r^{-1} z - a_r^{-1}(H^r y + a_1 H^{r-1} y + \cdots + a_{r-1} H y) = z_1 + z_2$ with $z_1 = a_r^{-1} z \in \ker H = \ker A^\top$ and $z_2 \in \operatorname{Im} H = \operatorname{Im} A$. Therefore, the orthogonal projection of $y \in F$ on $\operatorname{Im} A$ is $\pi(y) = -a_r^{-1}(H^r y + a_1 H^{r-1} y + \cdots + a_{r-1} H y)$. Now, taking into account that $f_0^{-1} AA^\top = A^\top$, we get that

$$A^\dagger = f_0^{-1} \pi = -a_r^{-1} f_0^{-1}(H^r + a_1 H^{r-1} + \cdots + a_{r-1} H) =$$

$$= -a_r^{-1} A^\top (H^{r-1} + a_1 H^{r-2} + \cdots + a_{r-1} I_n) = -a_r^{-1}(G^{r-1} + a_1 G^{r-2} + \cdots + a_{r-1} I_m) A^\top.$$

The Moore-Penrose pseudoinverse can be used to solve a linear system of equations of the form $Ax = b$, but we need that both A and $A^\top$ are suitable matrices. Nevertheless, by using Proposition 4, we can apply a random transformation to the matrix A to obtain a matrix A_α verifying this property with high probability.

4.3   The Algorithm

Given the theoretical results from the preceding sections, we extract the following probabilistic algorithm for solving linear systems of equations.

Algorithm Linsolve.

  The input is A, y, m, n, where $A \in \mathbb{F}_p^{n \times m}$, $y \in \mathbb{F}_p^n$, and $m \leq n$
  The output is x such that $Ax = y$ and a bit s indicating if the system is solvable.

  1. Pick random $\alpha \xleftarrow{R} \mathbb{F}_p$ and create the $n \times n$ matrix $D_{n,\alpha} = \mathrm{diag}(1, \alpha, \ldots, \alpha^{n-1})$ and the $m \times m$ matrix $D_{m,\alpha} = \mathrm{diag}(1, \alpha, \ldots, \alpha^{m-1})$.
  2. Compute $A_\alpha \leftarrow D_{n,\alpha} A D_{m,\alpha}$ and $y_\alpha \leftarrow D_{n,\alpha} y$.
  3. Compute $G \leftarrow A_\alpha^\top A_\alpha \in \mathbb{F}^{m \times m}$ //G is a symmetric $m \times m$ matrix
  4. Compute the coefficients $(a_1, \ldots, a_m)$ of the characteristic polynomial of G
  5. Compute the rank r of G
  6. Compute $A_\alpha^\dagger \leftarrow -a_r^{-1}(G^{r-1} + a_1 G^{r-2} + \cdots + a_{r-1} I_m) A_\alpha^\top$
  7. Check if $A_\alpha A_\alpha^\dagger y_\alpha = y_\alpha$. If not, the system has no solution, and the bit $s = 0$ is returned.
  8. If the system has a solution return $s = 1$ and $x \leftarrow D_{m,\alpha} A_\alpha^\dagger y_\alpha$.

Correctness of the algorithm is stated in the next lemma.

Lemma 4. *Let $A \in \mathbb{F}_p^{n \times m}$, $y \in \mathbb{F}_p^n$, and $m \leq n$. Suppose that $y \in \mathrm{Im}\, A$, that is, that the system has a solution. Let x be the output of the randomized algorithm Linsolve applied to A, y, m, n. Then, with probability at least $1 - (2/p)(n(n-1) + m(m-1))$, we have $Ax = y$.*

Proof. Clearly, $Ax = D_{n,\alpha}^{-1} A_\alpha D_{m,\alpha}^{-1} D_{m,\alpha} A_\alpha^\dagger y_\alpha = D_{n,\alpha}^{-1} A_\alpha A_\alpha^\dagger y_\alpha = D_{n,\alpha}^{-1} y_\alpha = y$.

Until now we assumed $m \leq n$. If $n \leq m$, we should adapt the algorithm by using $H = A_\alpha A_\alpha^\top$ instead of G to obtain $A_\alpha^\dagger$. Since the obtained solution depends on α, a random solution x_0 of the linear system of equations $Ax = y$ is obtained but, clearly, the probability distribution is not uniform on the set of all possible solutions. If we want the output of the algorithm to be uniformly distributed among all possible solutions of the system, we can take a random vector $z \in \mathbb{F}_p^m$ and compute $x_1 = x_0 + D_{m,\alpha}(I_m - A_\alpha^\dagger A_\alpha)z$. Finally, observe that by picking at random an $m \times m$ invertible matrix M and computing $D_{m,\alpha}(I_m - A_\alpha^\dagger A_\alpha)M$, we get a random element among all $m \times m$ matrices whose columns span $\ker A$.

5   The Secure Multi-party Protocols

Theorem 4. *Let shares $[A]$ of an $n \times m$ matrix and shares $[y]$ of an n-dimensional vector be given. There exists a multi-party protocol that, with probability at least $1 - O(n^2/p)$, securely computes shares $[x]$ of a solution to the system of linear equations $Ax = y$ and shares $[s]$ of a bit indication if the system is solvable. The protocol runs in constant rounds and uses $O(m^4 + m^2 n + m \log p)$ applications of the multiplication protocol.*

We remark that the above protocol can easily be extended to yield shares of a uniform solution of the system.

Theorem 5. *Assume $\mathbb{F}_p$ has characteristic at least m. Let shares $[A]$ of an $n \times m$ matrix be given. There exists a multi-party protocol that, with probability at least $1 - O(n^2/p)$, securely computes shares $([a_1], \ldots, [a_n])$ of the characteristic polynomial of A. The protocol runs in constant rounds and uses $O(m^4 + m^2 n)$ applications of the multiplication protocol.*

Proof of Theorem 4 (sketch). We show how to securely implement each step of the protocol LINSOLVE from Section 4.3 within the given complexity bounds. We remark that, as a by-result, we also get efficient constant-round protocols for securely computing the characteristic polynomial and the rank of a given shared matrix. Details of the protocol are given in the full version of the paper. Instead we give some intuition and mention the main techniques used.

For the first two steps the players jointly agree on a common public value α. Since α is public, for computing shares of the appearing matrices there is no further interaction needed. Computing shares of the characteristic polynomial in Step 4 is done with the protocol from Theorem 5. In Step 5, shares of the rank need to be computed that, by Lemma 3, can be derived from the characteristic polynomial. Here we have to use several sequential applications of the equality protocol EQ to finally compute the rank in unary representation. Step 6 computes shares of the Moore-Penrose pseudoinverse. Note that the formula to compute $A_\alpha^\dagger$ *explicitly* depends on the rank r of matrix G. Since we do not know r in the clear we need to develop a technique to obliviously evaluate the matrix G in the correct polynomial. A first approach is to evaluate $A_\alpha^\dagger$ for all the possible values of the rank $r \in \{1, \ldots, m\}$ and then sum the resulting matrices weighted with the respective bit indicating if the summation index equals the rank. Note that shares of the latter bits are known from the last step. However, the naive complexity of this approach is m^5. Using certain linearities in the coefficients of the sums of the above polynomials we develop an alternative approach to obtain the necessary complexity $O(m^4)$. Efficiency of this step heavily relies on our efficient polynomial evaluation protocols proposed in Section 3. The rest of the steps are more or less easy to implement. We mention that the complexity of the protocol is dominated by Steps 4 and 6 ($O(m^4)$), Step 5 ($O(m \cdot \log p)$ for in total $O(m)$ applications of EQ), and computing two products of an $m \times n$ with an $m \times m$ matrix ($O(m^2 n)$) in Steps 3 and 6. Security of the protocol follows by the security of the sub-protocols used.

Proof of Theorem 5 (sketch). We assume we are given shares of a symmetric square $m \times m$ matrix, if not apply the first three steps of the Linsolve protocol using $O(m^2 n)$ multiplications. Due to [7] there already exists a constant-round protocol for computing shares of the characteristic polynomial. We present an alternative and much simpler protocol based on Leverrier's Lemma (see Lemma 5) which basically says that the coefficients of the characteristic polynomial can be retrieved by inverting a certain non-singular lower-triangular matrix S, where each entry below the diagonal is the trace of the powers G^i of the matrix G. Leverrier's lemma is obtained by combining Newton's identities with the fact that these traces correspond to sums of powers of the characteristic roots.

Computing shares of all the m powers G^i of G can be done using the protocol from Proposition 2 in $O(m^4)$ applications of the multiplication protocol. All the traces of G^i can be locally computed by the players and assembled into the $m \times m$ matrix S. Finally the players compute the inverse of the non-singular matrix S using the protocol INV which enables them to compute the coefficients of the characteristic polynomial. The total complexity of the protocol is $O(m^4 + m^2 n)$ applications of the multiplication protocol and it runs in constant rounds. More details will be given in Appendix A. Security of the protocol follows by the security of the sub-protocols used.

Acknowledgments

The authors would like to thank an anonymous referee from CRYPTO 2006 who proposed the "small error protocol" for secure polynomial evaluation from Section 3.1.

References

1. Bar-Ilan, J., Beaver, D.: Non-cryptographic fault-tolerant computing in a constant number of rounds interaction. In: 8th ACM PODC, Edmonton, Alberta, Canada, August 14–16, 1989, pp. 201–209 (1989)
2. Beaver, D.: Minimal latency secure function evaluation. In: Preneel, B. (ed.) EUROCRYPT 2000. LNCS, vol. 1807, pp. 335–350. Springer, Heidelberg (2000)
3. Beaver, D., Micali, S., Rogaway, P.: The round complexity of secure protocols. In: 22nd ACM STOC, Baltimore, Maryland, USA, May 14–16, 1990, pp. 503–513. ACM Press, New York (1990)
4. Ben-Or, M., Cleve, R.: Computing algebraic formulas using a constant number of registers. SIAM J. Comput. 21(1), 54–58 (1992)
5. Ben-Or, M., Goldwasser, S., Wigderson, A.: Completeness theorems for noncryptographic fault-tolerant distributed computations. In: 20th ACM STOC, Chicago, Illinois, USA, May 2–4, 1988, pp. 1–10. ACM Press, New York (1988)
6. Chaum, D., Crépeau, C., Damgård, I.: Multiparty unconditionally secure protocols. In: 20th ACM STOC, Chicago, Illinois, USA, May 2–4, 1988, pp. 11–19. ACM Press, New York (1988)
7. Cramer, R., Damgård, I.: Secure distributed linear algebra in a constant number of rounds. In: Kilian, J. (ed.) CRYPTO 2001. LNCS, vol. 2139, pp. 119–136. Springer, Heidelberg (2001)

8. Damgård, I., Fitzi, M., Kiltz, E., Nielsen, J.B., Toft, T.: Unconditionally secure constant-rounds multi-party computation for equality, comparison, bits and exponentiation. In: Halevi, S., Rabin, T. (eds.) TCC 2006. LNCS, vol. 3876, pp. 285–304. Springer, Heidelberg (2006)
9. Feige, U., Kilian, J., Naor, M.: A minimal model for secure computation. In: 26th ACM STOC, Montréal, Québec, Canada, May 23–25, 1994, pp. 554–563. ACM Press, New York (1994)
10. Lombardi, H., Diaz-Toca, G.M., Gonzalez-Vega, L.: Generalizing cramer's rule: Solving uniformly linear systems of equations. SIAM J. Matrix Anal. Appl. 27, 621–637 (2005)
11. Goldreich, O., Micali, S., Wigderson, A.: How to play any mental game, or a completeness theorem for protocols with honest majority. In: 19th ACM STOC, May 25–27, 1987, pp. 218–229. ACM Press, New York (1987)
12. Ishai, Y., Kushilevitz, E.: Private simultaneous messages protocols with applications. In: Proc. 5th Israel Symposium on Theoretical Comp. Sc. ISTCS, pp. 174–183 (1997)
13. Ishai, Y., Kushilevitz, E.: Randomizing polynomials: A new paradigm for round-efficient secure computation. In: 41st FOCS, Las Vegas, Nevada, USA, November 12–14, 2000. IEEE Computer Society Press, Los Alamitos (2000)
14. Jájá, J.: An Introduction to Parallel Algorithms. Eddison-Wesley (1992)
15. Kiltz, E., Mohassel, P., Weinreb, E., Franklin, M.: Secure linear algebra using linearly recurrent sequences. In: Vadhan, S.P. (ed.) TCC 2007. LNCS, vol. 4392, pp. 291–310. Springer, Heidelberg (2007)
16. Krogh, F.T.: Efficient algorithms for polynomial interpolation and numerical differentiationi. Math. Comput. 24, 185–190 (1970)
17. Mulmuley, K.: A fast parallel algorithm to compute the rank of a matrix over an arbitrary field. Combinatorica 7, 101–104 (1987)
18. Nishide, T., Ohta, K.: Multiparty Computation for Interval, Equality, and Comparison without Bit-Decomposition Protocol. In: PKC 2007. LNCS, vol. 4450, pp. 343–360. Springer, Heidelberg (2007)
19. Nissim, K., Weinreb, E.: Communication efficient secure linear algebra. In: Halevi, S., Rabin, T. (eds.) TCC 2006. LNCS, vol. 3876. Springer, Heidelberg (2006)
20. Schrijver, A.: Combinatorial Optimization - Polyhedra and Efficiency. Springer, Heidelberg (2003)
21. Yao, A.: Protocols for secure computation. In: 23rd FOCS, Chicago, Illinois, November 3–5, 1982, pp. 160–164. IEEE Computer Society Press, Los Alamitos (1982)
22. Yao, A.: How to generate and exchange secrets. In: 27th FOCS, Toronto, Ontario, Canada, October 27–29, 1986, pp. 162–167. IEEE Computer Society Press, Los Alamitos (1986)

A   Protocol for the Characteristic Polynomial

We assume we are given shares of a symmetric square $m \times m$ matrix (possibly singular), if not apply the first three steps of the Linsolve protocol using $O(m^2 n)$ multiplications. We want to compute shares $([a_1], \ldots, [a_m])$ of the characteristic polynomial of G. With the techniques of Cramer and Damgård [7] this can be reduced to computing m times (in parallel) the determinant of a non-singular matrix and applying polynomial interpolation to reconstruct the coefficients.

Since securely computing the determinant can essentially be done by multiplying two shared $m \times m$ matrices, which can be carried out in constant rounds and using $O(m^3)$ applications of the multiplication protocol, the whole protocol runs in constant rounds and $O(m^4)$ applications of the multiplication protocol. We write

$$([a_1], \ldots, [a_m]) \leftarrow \text{CHARPOLY}([G]).$$

We now describe an alternative and more simple approach with roughly the same complexity based on Leverrier's Lemma [14, Chapter 8]. For this technique to work we will have to assume that the finite field's characteristic is at least m. Efficiency of this approach depends on the new secure polynomial evaluation technique from Section 3. We note that the use of Leverrier's Lemma in that context was first proposed by M. Rabin in [7]. Our algorithm retrieves the coefficients of the characteristic polynomial by inverting a certain lower-triangular matrix, where each entry below the diagonal is the trace of the powers G^i of the matrix G. The following lemma is obtained by combining Newton's identities with the fact that these traces correspond to sums of powers of the characteristic roots.

Lemma 5 (Leverrier's Lemma). *The coefficients $a_1, a_2, \ldots, a_m$ of the characteristic polynomial of a matrix G satisfy*

$$S \cdot \begin{pmatrix} a_1 \\ a_2 \\ a_3 \\ \vdots \\ a_{m-1} \\ a_m \end{pmatrix} = \begin{pmatrix} s_1 \\ s_2 \\ s_3 \\ \vdots \\ s_{m-1} \\ s_m \end{pmatrix}, \text{ where } S = \begin{bmatrix} 1 & 0 & 0 & \ldots & 0 & 0 \\ s_1 & 2 & 0 & \ldots & 0 & 0 \\ s_2 & s_1 & 3 & \ldots & 0 & 0 \\ \vdots & \vdots & \vdots & & \vdots & \vdots \\ s_{m-2} & s_{m-3} & s_{m-4} & \ldots & m-1 & 0 \\ s_{m-1} & s_{m-2} & s_{m-3} & \ldots & s_1 & m \end{bmatrix},$$

and $s_i = \text{tr}(G^i) = \sum_{j=1}^{m} G^i_{jj}$ for all $1 \le i \le m$.

Based on Leverrier's Lemma we can securely compute shares of the characteristic polynomial as follows: First the players compute shares of all the powers of G using the protocol from Proposition 2 and then they locally compute shares of the traces $[s_i]$. Then they apply the matrix inversion protocol to compute $[S^{-1}] \leftarrow \text{INV}([S])$, where S is the matrix from Lemma 5. Finally they calculate shares of the matrix-vector product $S^{-1} \cdot (s_1, s_2, \ldots, s_m)^\top$ to obtain shares of the characteristic polynomial. (Note that the matrix S is guaranteed to be non-singular since for it's determinant we have $\det(S) = \prod_{i=1}^{m} i$ which is non-zero by our assumption that $\mathbb{F}_p$ has characteristic at least m.) Using the protocol explained in Proposition 2 shares of all powers $G, G^2, \ldots, G^m$ can be computed in constant rounds and $O(m \cdot m^3) = O(m^4)$ applications of the multiplication protocol. The protocol is secure with probability at least $1 - O(m^2/p)$. This proves Theorem 5.

Author Index

Lecture Notes in Computer Science

For information about Vols. 1–4543

please contact your bookseller or Springer